RADIATIVE TRANSFER

RADIATIVE TRANSFER

AND INTERACTIONS WITH CONDUCTION AND CONVECTION

M. NECATI ÖZISIK

Professor, Mechanical and Aerospace Engineering
North Carolina State University

A Wiley-Interscience Publication

JOHN WILEY & SONS, New York . London . Sydney . Toronto

Library of Congress Cataloging in Publication Data:

Özişik, M Necati.
Radiative transfer and interactions with conduction and convection.

"A Wiley-Interscience publication."
Includes bibliographies.
1. Heat—Transmission. 2. Radiative transfer.
I. Title.

QC320.095 536'.2 72-12824
ISBN 0-471-65722-0

Printed in the United States of America

10 9 8 7 6 5 4 3 2 1

To Hakan

Preface

The rapid development of modern technology in recent years has significantly influenced the teaching of radiative heat transfer in engineering schools. Traditional courses on radiative heat transfer that dealt primarily with the nonparticipating media have been expanded to include the treatment of the absorbing, emitting, scattering media and the interaction of radiation with other modes of heat transfer. Radiative transfer in participating media has been studied extensively by astrophysicists in their investigations of stellar atmospheres and problems governed by the same transport equation have been investigated by physicists working in neutron transport theory. In engineering, the interest in this subject has increased very rapidly in the past decade. Although new methods have been developed and some of the mathematical techniques used in other disciplines for the treatment of the transport equation have been adopted for the analysis of radiative heat transfer problems, it would be beneficial to have a unified and systematic treatment of all these new developments readily available for graduate students, scientists, and researchers in engineering. In the field of engineering a book that presents a comprehensive, systematic, and unified treatment of fundamental concepts, basic theory, various methods of solution of radiative transfer problems in both nonparticipating and participating media, and the interactions of radiation with other modes of heat transfer is necessary. Therefore, this volume is intended to serve as a graduate level text on radiative transfer in engineering schools and as a reference book for scientists and engineers working in this field.

The first two chapters review the basic relations and fundamental concepts that are useful in the study of radiative transfer in nonparticipating and participating media and discuss the radiative properties of materials. The remaining twelve chapters can be separated into two distinct parts: radiative transfer in nonparticipating media (Chapters 3 through 7), and radiative transfer in participating media (Chapters 8 through 14). The mathematical treatment of radiative transfer in nonparticipating media is relatively simple and straightforward. When using this book as a text, the instructor can readily cover the material in Chapters 3 through 7 by focusing his attention on the general approach of the analysis and by providing some insight into the physical significance of the subject matter while the student concentrates on the details of the mathematical developments. However, the analysis of radiative heat transfer in participating media is more involved because of the mathematical difficulties associated with the solution of the governing equation; therefore, Chapters 8, 9, and 10 are devoted to the methods of solution of the equation of radiative transfer. In Chapter 8 formal solutions are obtained, by the classical approach, for such physical quantities as radiation intensity, incident radiation, and radiative flux, and in Chapter 9 various approximate methods of solution are presented. Chapter 10 introduces the normal-mode expansion (or singular eigenfunction expansion) technique basic to a more rigorous analysis of radiative transfer for absorbing, emitting, and isotropically scattering plane-parallel media. This powerful technique, developed by Case in 1960 for the exact solution of one-dimensional neutron transport problems, has been applied only recently in the field of radiative transfer. When presenting the material in Chapter 10 the instructor should place additional emphasis on analytical developments since some of the mathematical concepts discussed in this chapter are not commonly used in engineering applications. This chapter and its application in some sections of the following chapters can readily be omitted from the course without affecting the continuity of the presentation. The formal and approximate solutions of the equation of radiative transfer, given in Chapters 8 and 9, are used extensively in Chapters 11 through 14, which deal with its applications. Chapter 11 is devoted to the applications of radiative heat transfer in participating, plane-parallel media, while Chapters 12, 13, and 14 treat the interaction of radiation with conduction, boundary layer flow, and channel flow, respectively.

I gratefully acknowledge the assistance of several friends and colleagues during the preparation of this manuscript. I am particularly indebted to Dr. C. E. Siewert for his valuable suggestions on the exact solutions, to Drs. J. A. Clark, M. A. Heaslet, S. I. Pai, J. E. Sunderland, R. S. Thorsen, R. Viskanta, and J. C. Williams, III, for their very helpful comments, and to Y. Yener for his careful proofreading of the manuscript.

M. NECATI ÖZIŞIK

Raleigh, North Carolina
October 1972

Contents

RADIATIVE TRANSFER

CHAPTER 1

Basic Relations

In this chapter we review briefly the physics of thermal radiation and discuss some of the results obtained from the electromagnetic wave theory and the quantum theory that are important in the study of radiative heat transfer. According to Maxwell's classical electromagnetic theory radiant energy travels in the form of electromagnetic waves, and according to Planck's hypothesis it travels in the form of discrete photons. Both of these concepts have been utilized in the investigation of radiative heat transfer. For example, the results obtained from the electromagnetic wave theory have been used to predict the radiative properties of materials, such as reflectivity and emissivity; on the other hand, the results of the quantum theory have been utilized to determine the amount of radiative energy emitted by matter at any given frequency (or wavelength) because of its temperature. Here we present some of the implications of electromagnetic wave theory and of quantum mechanics that are of interest in engineering applications. The reader should refer to the texts by Planck [1]* and Milne [2] for detailed discussion of the physics and thermodynamics of radiation, and to the books by Clemmow [3], Jones [4], Stratton [5], and Van de Hulst [6] for complete treatments of the electromagnetic wave nature of radiation.

1-1 THE WAVE NATURE OF THERMAL RADIATION

Figure 1-1 shows the electromagnetic wave spectrum and a typical subdivision of the spectrum according to different applications. The *thermal*

* Bracketed numbers refer to references at the end of the chapter.

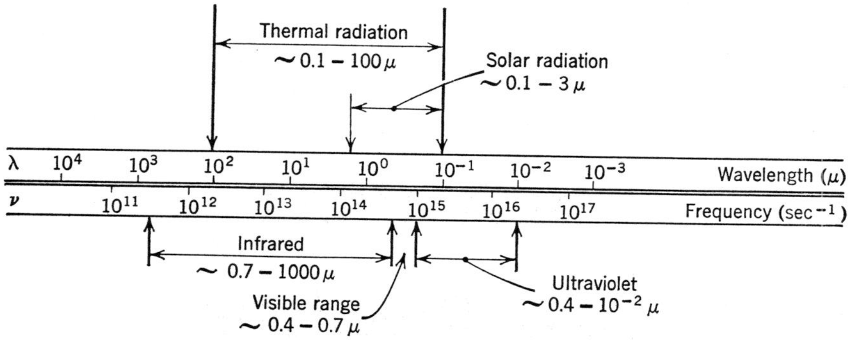

Fig. 1-1. Electromagnetic wave spectrum.

radiation refers to radiation emitted by bodies because of their own temperature; for practical purposes its range on the wave spectrum is considered to extend from 0.1 μ to 100 μ, where the major portion of the energy associated with thermal radiation lies, while the visible part ranges from 0.4 μ to 0.7 μ. Other types of radiation on the electromagnetic wave spectrum include X-rays, γ-rays, and cosmic rays, which have much shorter wavelengths than thermal radiation, and radio waves, which have much longer wavelengths. Different types of radiation are produced by different methods. For example, X-rays are produced by the bombardment of a metal with high-frequency electrons, and γ-rays by the fission of nuclei or by radioactive disintegration.

In 1865 Maxwell published his set of famous equations for the propagation of electromagnetic waves. When radiation is treated as an electromagnetic wave, its propagation can be described by the solution of Maxwell's equations. The reader should refer to the books on electromagnetic theory by Stratton [5] and Pugh and Pugh [7] for the derivation of these equations; we summarize below Maxwell's equations in their differential form for an isotropic, homogeneous medium:

$$\nabla \times \mathbf{H} = \varepsilon \frac{\partial \mathbf{E}}{\partial t} + \sigma \mathbf{E} \tag{1-1a}$$

$$\nabla \times \mathbf{E} = -\mu \frac{\partial \mathbf{H}}{\partial t} \tag{1-1b}$$

$$\nabla \cdot \mathbf{H} = 0 \tag{1-1c}$$

$$\nabla \cdot \mathbf{E} = 0 \tag{1-1d}$$

where **H** and **E** are the *magnetic field* and *electric field* vectors, and ε, μ, and σ are the electric inductive capacity, magnetic inductive capacity, and specific conductivity, respectively.

In analyzing the propagation of radiation as an electromagnetic wave it is customary to pay attention to *plane waves*, primarily because of the simplicity of solutions of Maxwell's equations for plane waves. The primary objective of the following analysis of the solution of Maxwell's equations is to demonstrate how the propagation of radiation can be represented as traveling plane waves and how the results of the analysis can be utilized in studying the reflection of radiation. We shall examine the propagation of plane waves in both a perfectly dielectric (i.e., nonconducting) medium and a conducting medium. Although a perfect dielectric does not exist in nature, the results obtained from such analysis are most useful in studying the properties of real dielectrics.

Dielectric Medium

Consider a plane wave propagating in the z direction in an isotropic, homogeneous, and perfectly nonconducting medium. We choose as the reference coordinates ol, or, oz rectangular axes, which form a right-handed system. Figure 1-2 shows schematically the electric and magnetic field vectors **E** and **H** in the coordinate system under consideration. For a plane wave the electric and magnetic field vectors are perpendicular to the direction of propagation oz and to each other, and all properties remain constant over any plane ol-or at any instant of time. Hence we have $\partial/\partial l = 0$, $\partial/\partial r = 0$, $E_z = 0$, $H_z = 0$, and for a dielectric medium $\sigma = 0$. When these restrictions

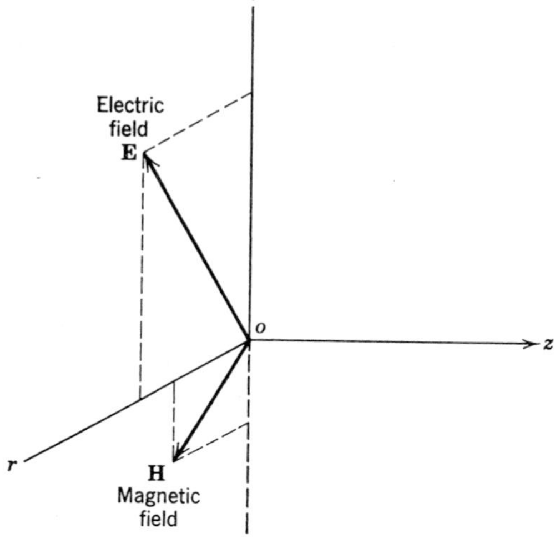

Fig. 1-2. Coordinates for propagation of a plane wave in the z direction.

are utilized, Maxwell's equations 1-1 simplify to[1]*

$$\frac{\partial^2 E_l}{\partial z^2} = \frac{1}{c^2}\frac{\partial^2 E_l}{\partial t^2} \tag{1-2a}$$

$$\frac{\partial^2 E_r}{\partial z^2} = \frac{1}{c^2}\frac{\partial^2 E_r}{\partial t^2} \tag{1-2b}$$

and similar relations can be obtained for the components H_l and H_r of the magnetic field vector. Here c is the speed of propagation of the plane wave in the medium and is related to μ and ε by

$$c^2 = \frac{1}{\mu\varepsilon} \tag{1-3a}$$

If the medium is a free space (i.e., a vacuum), the speed of propagation of the wave is equal to the speed of light in a vacuum, and Eq. 1-3a is written as

$$c_0^{\;2} = \frac{1}{\mu_0\varepsilon_0} \tag{1-3b}$$

where c_0 is the speed of light in a vacuum (i.e., $c_0 = 2.9979 \times 10^8$ m/sec), and ε_0 and μ_0 are the electric and magnetic inductive capacities for a vacuum.[2]

The solution of Eqs. 1-2 can be written in the form

$$E_l = a_l \exp\left\{i\left[\omega\left(t - \frac{z}{c}\right) + \gamma_l\right]\right\} \tag{1-4a}$$

$$E_r = a_r \exp\left\{i\left[\omega\left(t - \frac{z}{c}\right) + \gamma_r\right]\right\} \tag{1-4b}$$

where a_l and a_r are the amplitudes, γ_l and γ_r are the phase angles, ω is the circular frequency, and i is the unit imaginary number $\sqrt{-1}$.

It is convenient to define a *wave number*, k, as

$$k = \frac{2\pi}{\lambda} = \frac{\omega}{c} = \omega\sqrt{\mu\varepsilon} \tag{1-5}$$

and the *refractive index*, n, for a dielectric medium as

$$n = \frac{c_0}{c} = \frac{\lambda_0}{\lambda} = \frac{k}{k_0} = \sqrt{\frac{\varepsilon}{\varepsilon_0}\frac{\mu}{\mu_0}} = \sqrt{K_e K_m} \tag{1-6}$$

where λ is the wavelength, and the *dielectric constant*, K_e, and the *magnetic permeability*, K_m, are defined as

$$K_e = \frac{\varepsilon}{\varepsilon_0} \quad \text{and} \quad K_m = \frac{\mu}{\mu_0} \tag{1-7}$$

* Superior numbers refer to notes at the end of the chapter.

Here the quantities with a subscript 0 refer to a vacuum. Now, by utilizing the above definition of the refractive index n, the solutions given by Eqs. 1-4 can be written in the alternative form as

$$E_l = a_l \exp\{i[(\omega t - k_0 nz) + \gamma_l]\} \tag{1-8a}$$

$$E_r = a_r \exp\{i[(\omega t - k_0 nz) + \gamma_r]\} \tag{1-8b}$$

where k_0 is the wave number for a vacuum and is given by

$$k_0 = \frac{2\pi}{\lambda_0} = \frac{\omega}{c_0} = \omega\sqrt{\varepsilon_0 \mu_0} \tag{1-9}$$

For most dielectric materials the magnetic permeability K_m is so close to unity that the refractive index for such materials is obtained approximately from Eq. 1-6 as

$$n \simeq \sqrt{K_e} \tag{1-10}$$

Table 1-1 shows a comparison of the values of $\sqrt{K_e}$ for some gases measured

Table 1-1 Comparison of $\sqrt{K_e}$ with the Refractive Index for Gases[a]

Hydrogen	1.000132	1.000138–1.000142
Air	1.000295	1.000293
Carbon monoxide	1.000350	1.000335–1.000340
Carbon dioxide	1.000492	1.000448–1.000454
Nitrous oxide	1.000565	1.000516

[a] From E. M. Pugh and E. W. Pugh [7, p. 369].

at frequencies below 3×10^{10} Hz with the measured values of the refractive index n for visible light traveling through these gases. The agreement between the values of $\sqrt{K_e}$ and the measured values of n is very good for the gases considered, but the approximation of Eq. 1-10 is unsatisfactory for solids and liquids. An interesting feature of the data in Table 1-1 is that for the gases in question the refractive index is very close to unity, and this conclusion is applicable for all gases (i.e., $n \simeq 1$ for gases).

Conducting Medium

For a plane wave propagating in the z direction in an isotropic, homogeneous medium having a finite electrical conductivity σ, Maxwell's equations 1-1 can be simplified to[3]

$$\frac{\partial^2 E_l}{\partial z^2} = \mu\varepsilon \frac{\partial^2 E_l}{\partial t^2} + \mu\sigma \frac{\partial E_l}{\partial t} \tag{1-11a}$$

$$\frac{\partial^2 E_r}{\partial z^2} = \mu\varepsilon \frac{\partial^2 E_r}{\partial t^2} + \mu\sigma \frac{\partial E_r}{\partial t} \tag{1-11b}$$

and similar relations can be obtained for the components H_l and H_r of the magnetic field vector. Here we note that for $\sigma = 0$ (i.e., a dielectric medium) Eqs. 1-11 reduce to the expressions given by Eqs. 1-2 for a dielectric medium.

In seeking a solution to Eqs. 1-11, we shall assume, for simplicity in the analysis, that the electric field vector **E** lies along the *ol* axis; then the component E_r vanishes and only the component E_l must be considered.

The solution of Eq. 1-11a for E_l can be written in the form

$$E_l = a_l \exp\{i[(\omega t - k_0 nz)] - k_0 n' z\} \tag{1-12}$$

where

$$k_0 = \frac{2\pi}{\lambda_0} = \frac{\omega}{c_0} = \omega\sqrt{\mu_0\varepsilon_0}$$

and n and n' are real, positive quantities that are to be determined by constraining this solution to satisfy the differential equation. The term

$$\exp(-k_0 n' z)$$

in Eq. 1-12 implies that radiation propagating in a conducting medium undergoes attenuation with distance z.

Equation 1-12 can be written more compactly in the form

$$E_l = a_l \exp[i(\omega t - k_0 m z)] \tag{1-13a}$$

where we have defined the *complex refractive index* m for a conducting medium as[4]

$$m = n - in' \tag{1-13b}$$

The unknown coefficients n and n' can be determined by substituting the solution given by Eqs. 1-13 into the differential equation 1-11a and by requiring that it satisfy this differential equation. We obtain

$$-k_0^2(n - in')^2 = -\omega^2\mu\varepsilon + i\omega\mu\sigma \tag{1-14a}$$

or

$$n^2 - n'^2 - 2inn' = \frac{\omega^2}{k_0^2}\mu\varepsilon - i\frac{\omega}{k_0^2}\mu\sigma \tag{1-14b}$$

and, by utilizing the relations of Eqs. 1-7 and 1-9, we can write this result in the form

$$n^2 - n'^2 - 2inn' = K_m K_e\left(1 - i\frac{\sigma}{\omega\varepsilon}\right) \tag{1-15}$$

By equating the real and imaginary parts of Eq. 1-15 we obtain

$$n^2 - n'^2 = K_m K_e \tag{1-16a}$$

$$2nn' = K_m K_e \frac{\sigma}{\omega\varepsilon} \tag{1-16b}$$

When Eqs. 1-16 are solved simultaneously for n and n', we obtain

$$n^2 = \frac{K_e K_m}{2}\left[1 + \sqrt{1 + \left(\frac{\sigma}{\omega\varepsilon}\right)^2}\right] \tag{1-17a}$$

$$n'^2 = \frac{K_e K_m}{2}\left[-1 + \sqrt{1 + \left(\frac{\sigma}{\omega\varepsilon}\right)^2}\right] \tag{1-17b}$$

Here we have chosen the positive sign before the square roots because both n and n' have been assumed to be real and positive quantities.

Once the complex refractive index m for a conducting medium is known, the speed of propagation of a plane wave in that medium, c, is related to the speed of propagation in a vacuum, c_0, by

$$c = \frac{c_0}{|m|} = \frac{c_0}{\sqrt{n^2 + n'^2}} \tag{1-18}$$

For a dielectric medium n' is zero, and Eq. 1-18 reduces to Eq. 1-6.

The coefficients n and n' are not constants for any given material; they depend on both the wavelength and the temperature. For metals the electrical conductivity σ is generally very large; hence $\sigma/\omega\varepsilon \gg 1$ if ω is not very large. Then Eqs. 1-17 simplify to

$$n \cong n' \cong \left(\frac{K_e K_m}{2}\frac{\sigma}{\omega\varepsilon}\right)^{1/2} \quad \text{for} \quad \frac{\sigma}{\omega\varepsilon} \gg 1 \tag{1-19a}$$

and the complex refractive index becomes

$$m = n - in' \cong (1 - i)\left(\frac{K_e K_m}{2}\frac{\sigma}{\omega\varepsilon}\right)^{1/2} \quad \text{for} \quad \frac{\sigma}{\omega\varepsilon} \gg 1 \tag{1-19b}$$

1-2 THE ENERGY TRANSMITTED BY ELECTROMAGNETIC WAVES

Maxwell's equations predict the propagation of electromagnetic waves in a dielectric or a conducting medium. These electromagnetic waves should transmit energy; otherwise their detection would not be possible. The energy transported by the electromagnetic wave is given by the *Poynting vector* **S**, which is related to the electric field vector **E** and the magnetic field vector

H by the relation [5, p. 132; 7, p. 377]

$$\mathbf{S} = \mathbf{E} \times \mathbf{H} \tag{1-20a}$$

This quantity may be interpreted as the power of energy flow (i.e., the energy per unit time, crossing a unit area in the direction perpendicular to the plane of **E** and **H** vectors by the right-hand rule). Figure 1-3 shows the electric

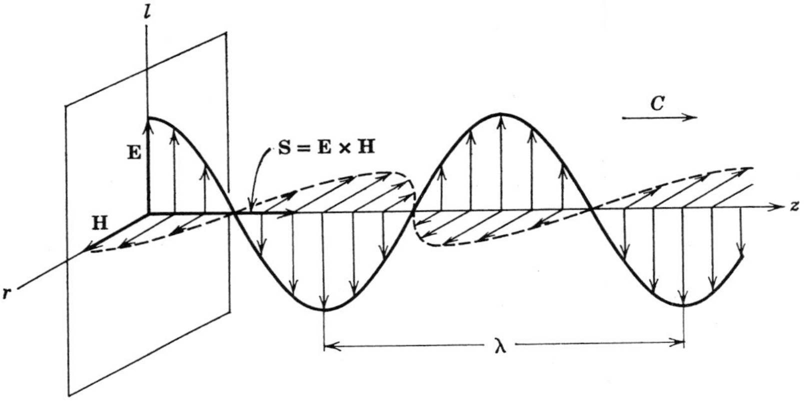

Fig. 1-3. The electric and magnetic field vectors and the direction of the corresponding Poynting vector.

and magnetic field vectors and the Poynting vector for a plane wave propagating in the positive z direction. For a plane wave, the components E_z and B_z being zero, the cross-product term on the right-hand side of Eq. 1-20a becomes

$$\mathbf{S} = \hat{\mathbf{k}}[E_l H_r - E_r H_l] \tag{1-20b}$$

where $\hat{\mathbf{k}}$ is the unit vector along the positive z axis. By utilizing the solution given by Eqs. 1-13 and the Maxwell's equations it can be shown that the components of the magnetic field vector, H_l and H_r, are proportional to E_r and E_l respectively; then Eq. 1-20b becomes[5]

$$\mathbf{S} = \hat{\mathbf{k}} \frac{m}{c_0 \mu} [E_l^2 + E_r^2] \tag{1-20c}$$

where m is the complex refractive index.

For a dielectric medium Eq. 1-20c simplifies to

$$\mathbf{S} = \hat{\mathbf{k}} \frac{n}{c_0 \mu} [E_l^2 + E_r^2] \tag{1-20d}$$

The instantaneous value of the energy flow associated with radiation traveling in a dielectric medium is obtained from Eq. 1-20d as[6]

$$|\mathbf{S}| = \frac{n}{c_0 \mu} [E_l E_l^* + E_r E_r^*] \tag{1-20e}$$

where the asterisk denotes the complex conjugate (i.e., the complex conjugate is obtained by replacing $i = \sqrt{-1}$ by $-i$).

An actual beam of radiation consists of many waves in rapid succession with amplitudes and phases subject to continual variations. For example, oscillations representing a beam of light may change irregularly millions of times a second; in such situations the average value of the energy, taken over a sufficiently short time, should be considered. The average power of radiation in a dielectric medium is given as

$$\overline{|\mathbf{S}|} = \frac{n}{c_0 \mu} [\overline{E_l E_l^*} + \overline{E_r E_r^*}] \equiv \frac{n}{c_0 \mu} (I_l + I_r) \tag{1-21}$$

Equation 1-21 implies that the energy transported by an electromagnetic plane wave is composed of two intensity components I_l and I_r, which are related to the electric field vibrations along the directions l and r, respectively.

1-3 THE STOKES PARAMETERS

The two intensity components I_l and I_r discussed above are not sufficient to describe the radiation field when the polarization effects are to be considered. Chandrasekhar [8, p. 40] formulated the equation of radiative transfer in the general form for polarized radiation by treating the radiation intensity as a four-component vector in terms of the four *Stokes parameters*. In other words, the four quantities I, Q, U, V, known as the *Stokes parameters*, or I_l, I_r, U, V, known as the *modified Stokes parameters*, give a complete description of the polarization properties of a stream of electromagnetic plane waves as far as measurable quantities are concerned. For example, the time-average intensity, the plane of polarization, the ellipticity, and the degree of polarization are the quantities of interest. Therefore the radiation intensity for a polarized radiation is, in general, a four-component vector:

$$\mathbf{I} = \begin{bmatrix} I \\ Q \\ U \\ V \end{bmatrix} \quad \text{or} \quad \mathbf{I} = \begin{bmatrix} I_l \\ I_r \\ U \\ V \end{bmatrix} \tag{1-22a}$$

These two sets are equivalent, because I and Q are related to I_l and I_r by

$$I = I_l + I_r \qquad \text{and} \qquad Q = I_l - I_r \tag{1-22b}$$

The reader should refer to the books by Van de Hulst [6, Chapter 5], Chandrasekhar [8, pp. 24–36], and Deirmendjian [9, Chapter 3] for a comprehensive treatment of the Stokes parameters; here we present only a brief account.

Consider a *completely polarized* stream of radiation whose electric field vector is represented by the two components E_l and E_r along the rectangular axes ol and or, respectively. The components E_l and E_r are given by Eqs. 1-8 as

$$E_l = a_l \exp\{i[(\omega t - k_0 nz) + \gamma_l]\} \tag{1-8a}$$

$$E_r = a_r \exp\{i[(\omega t - k_0 nz) + \gamma_r]\} \tag{1-8b}$$

where a_l and a_r are the maximum amplitudes along the axes ol and or, respectively, and γ_l and γ_r are the phase angles. Let the amplitudes a_l and a_r and the difference between the phase angles δ, that is, $\delta = \gamma_l - \gamma_r$, be constants.

The *Stokes parameters* are defined as

$$I = E_l E_l^* + E_r E_r^* = a_l^2 + a_r^2 = I_l + I_r \tag{1-23a}$$

$$Q = E_l E_l^* - E_r E_r^* = a_l^2 - a_r^2 = I_l - I_r \tag{1-23b}$$

$$U = \text{Re}\,[2E_l E_r^*] = 2a_l a_r \cos\delta \tag{1-23c}$$

$$V = \text{Im}\,[2E_l E_r^*] = 2a_l a_r \sin\delta \tag{1-23d}$$

where the asterisk denotes the complex conjugate. Alternatively, the *modified Stokes parameters* are defined as

$$I_l = E_l E_l^* \tag{1-23e}$$

$$I_r = E_r E_r^* \tag{1-23f}$$

$$U = \text{Re}\,[2E_l E_r^*] \tag{1-23g}$$

$$V = \text{Im}\,[2E_l E_r^*] \tag{1-23h}$$

For the completely polarized radiation considered here, it can be shown, by squaring the relations of Eqs. 1-23a through 1-23d and adding them, that the Stokes parameters satisfy among themselves the following equality:

$$I^2 = Q^2 + U^2 + V^2 \tag{1-23i}$$

The radiation is said to be completely polarized when the relation of Eq. 1-23i is valid among the four Stokes parameters.

In the above analysis we have considered a strictly monochromatic radiation in the sense that the amplitudes a_l and a_r and the difference in the phase angles δ remained constant. No actual radiation, however, is so perfectly monochromatic. The amplitudes and phases change irregularly millions of times a second; these variations are extremely swift in comparison with the duration of any measurement. Therefore it is possible to measure only the mean amplitudes and the mean phases. In that case the Stokes parameters correspond to the mean quantities and are defined as

$$I = \langle E_l E_l^* \rangle + \langle E_r E_r^* \rangle \tag{1-24a}$$

$$Q = \langle E_l E_l^* \rangle - \langle E_r E_r^* \rangle \tag{1-24b}$$

$$U = \mathrm{Re}\,[\langle 2E_l E_r^* \rangle] \tag{1-24c}$$

$$V = \mathrm{Im}\,[\langle 2E_l E_r^* \rangle] \tag{1-24d}$$

where the notation $\langle\ \rangle$ denotes the mean value with respect to time. It can be proved that, for any stream of radiation for which the Stokes parameters are defined as given by Eqs. 1-24, the Stokes parameters satisfy among themselves the following inequality [8, p. 32]:

$$I^2 \geq Q^2 + U^2 + V^2 \tag{1-24e}$$

We shall refer to this situation as *partial polarization*. The equality in Eq. 1-24e is valid only for completely polarized radiation.

The *natural* or *unpolarized radiation*, such as the Planckian radiation, is so random or chaotic in nature that it is not possible to measure, with the existing experimental techniques, any difference between the phase angles and the intensity components; the intensity in any direction in the transverse plane is the same. With these considerations in mind the natural or unpolarized radiation is specified in terms of the Stokes parameters as

$$Q = U = V = 0 \tag{1-25a}$$

or

$$[I, 0, 0, 0] \tag{1-25b}$$

Therefore the magnitude of the intensity I is sufficient to specify an unpolarized radiation.

A partially polarized radiation specified by the four Stokes parameters (I, Q, U, V) such that

$$I^2 \geq Q^2 + U^2 + V^2 \tag{1-26a}$$

may be considered to consist of a *completely polarized* radiation of intensity $(Q^2 + U^2 + V^2)^{1/2}$, specified by the Stokes parameters as

$$[(Q^2 + U^2 + V^2)^{1/2}, Q, U, V] \tag{1-26b}$$

and an unpolarized radiation of intensity $[I - (Q^2 + U^2 + V^2)^{1/2}]$, specified in terms of the Stokes parameters as

$$[I - (Q^2 + U^2 + V^2)^{1/2}, 0, 0, 0] \tag{1-26c}$$

The *degree of polarization* of a partially polarized radiation is defined as

$$0 \leq \frac{(Q^2 + U^2 + V^2)^{1/2}}{I} \leq 1 \tag{1-27}$$

which is the ratio of the intensity of the completely polarized radiation in the stream to the intensity of the stream as a whole.

We now discuss the characterization of *completely polarized* radiation. Consider the real parts of the components of the electric field vector, E_l and E_r of Eq. 1-8, written in the form

$$\text{Re}\,[E_l] \equiv A_l = a_l \cos \psi \tag{1-28a}$$

$$\text{Re}\,[E_r] \equiv A_r = a_r \cos (\psi - \delta) \tag{1-28b}$$

where we have defined

$$\psi \equiv \omega t - k_0 nz + \gamma_l \tag{1-28c}$$

$$\delta \equiv \gamma_l - \gamma_r \tag{1-28d}$$

Equations 1-28a and 1-28b are parametric representations of the two coupled harmonic oscillations along the axes *ol* and *or*. By eliminating ψ between Eqs. 1-28a and 1-28b we obtain[7]

$$\left(\frac{A_l}{a_l}\right)^2 + \left(\frac{A_r}{a_r}\right)^2 - 2\frac{A_l}{a_l}\frac{A_r}{a_r}\cos \delta = \sin^2 \delta \tag{1-29a}$$

which is the equation of an ellipse, whose major and minor axes do not necessarily coincide with the coordinate axes *ol* and *or*. In this case the electromagnetic wave is said to be *elliptically polarized*, because the end point of the electric field vector **E** will appear to trace an ellipse on the transverse plane, as shown in Fig. 1-4*a*. The direction of rotation depends on the sign of the difference in the phase angles δ.

Two special cases of Eq. 1-29a include a circle and a straight line. When the difference in the phase angles is $\pi/2$ (i.e., $\cos \delta = 0$) and the amplitudes are equal (i.e., $a_l = a_r \equiv a$), Eq. 1-29a simplifies to

$$\left(\frac{A_l}{a}\right)^2 + \left(\frac{A_r}{a}\right)^2 = 1 \tag{1-29b}$$

which is the equation of a circle of radius a. In this case the electromagnetic wave is said to be *circularly polarized* because the end point of the electric field vector **E** appears to trace a circle, as shown in Fig. 1-4*b*, to an observer facing the oncoming wave front.

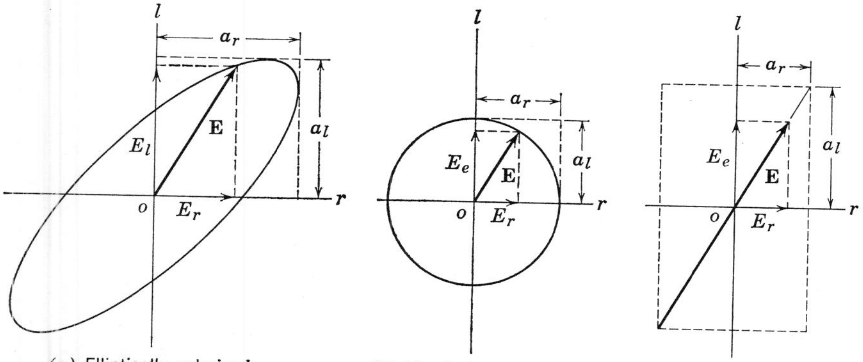

(a) Elliptically polarized (b) Circularly polarized (c) Linearly polarized

Fig. 1-4. The locus of the end points of the electric field vector **E** to an observer facing the oncoming wave front.

When the difference in the phase angles is zero or equal to π (i.e., $\sin \delta = 0$), Eq. 1-29a simplifies to

$$\frac{A_l}{a_l} \mp \frac{A_r}{a_r} = 0 \tag{1-29c}$$

which is the equation of a straight line. Then the end point of the electric field vector **E** appears to travel up and down the straight line shown in Fig. 1-4*c*, and the wave is said to be *linearly polarized.*

The general equation of an ellipse given by Eq. 1-29a takes a simple form if it is referred to the coordinate system $o\xi$ and $o\eta$, which lie along the major and minor axes, respectively, of the ellipse. Let χ be the angle between the axes $o\xi$ and ol, as shown in Fig. 1-5. We consider the transformation for the rotation of axes given in the form

$$A_\xi = A_l \cos \chi + A_r \sin \chi \tag{1-30a}$$

$$A_\eta = -A_l \sin \chi + A_r \cos \chi \tag{1-30b}$$

where the angle χ is to be determined.

We substitute A_l and A_r from Eqs. 1-28a and 1-28b into Eqs. 1-30 and obtain

$$A_\xi = B_1 \cos \psi + B_2 \sin \psi \tag{1-30c}$$

$$A_\eta = B_3 \cos \psi + B_4 \sin \psi \tag{1-30d}$$

where we have defined

$$B_1 \equiv a_l \cos \chi + a_r \cos \delta \sin \chi \tag{1-31a}$$

$$B_2 \equiv a_r \sin \delta \sin \chi \tag{1-31b}$$

$$B_3 \equiv -a_l \sin \chi + a_r \cos \delta \cos \chi \tag{1-31c}$$

$$B_4 \equiv a_r \sin \delta \cos \chi \tag{1-31d}$$

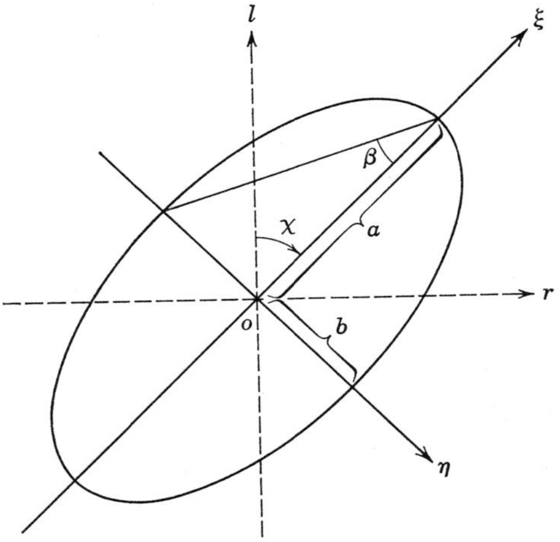

Fig. 1-5. The coordinates $o\xi$ and $o\eta$ along the major and minor axes of ellipse.

Eliminating ψ between Eqs. 1-30c and 1-30d and setting

$$B_1B_3 + B_2B_4 = 0 \tag{1-32}$$

we obtain the equation of the ellipse in the coordinate system $o\xi$ and $o\eta$ as[8]

$$\frac{A_\xi^{\,2}}{a^2} + \frac{A_\eta^{\,2}}{b^2} = 1 \tag{1-33a}$$

where we have defined

$$a^2 \equiv \frac{B^2}{B_3^{\,2} + B_4^{\,2}}, \quad b^2 \equiv \frac{B^2}{B_1^{\,2} + B_2^{\,2}}, \quad \text{and} \quad B^2 \equiv (B_1B_4 - B_2B_3)^2 \tag{1-33b}$$

The angle χ can be determined by substituting Eqs. 1-31 into the constrain of Eq. 1-32. We obtain

$$\tan 2\chi = \frac{2a_l a_r}{a_l^{\,2} - a_r^{\,2}} \cos \delta \tag{1-34}$$

We define the *ellipticity*, $\tan \beta$, as

$$\tan \beta = \frac{b}{a} \tag{1-35a}$$

Then

$$\sin 2\beta = \frac{2 \tan \beta}{1 + \tan^2 \beta} = \frac{2ab}{a^2 + b^2} \tag{1-35b}$$

$$\cos 2\beta = \frac{1 - \tan^2 \beta}{1 + \tan^2 \beta} = \frac{a^2 - b^2}{a^2 + b^2} \tag{1-35c}$$

It can be shown that for an ellipse the following relation holds:

$$a_l^2 + a_r^2 = a^2 + b^2 \tag{1-36a}$$

By utilizing this relation and Eqs. 1-33b, we obtain[9]

$$ab = \pm a_l a_r \sin\delta \tag{1-36b}$$

and from Eqs. 1-36a, 1-36b, and 1-35b we write

$$\frac{2a_l a_r \sin\delta}{a_l^2 + a_r^2} = \sin 2\beta \tag{1-36c}$$

We are now in a position to express the Stokes parameters (Eqs. 1-23) in terms of the angles β and χ. We find[10]

$$I = I_l + I_r \tag{1-37a}$$

$$Q = I_l - I_r = I\cos 2\beta \cos 2\chi \tag{1-37b}$$

$$U = (I_l - I_r)\tan 2\chi = I\cos 2\beta \sin 2\chi \tag{1-37c}$$

$$V = I\sin 2\beta \tag{1-37d}$$

1-4 THE INTENSITY OF RADIATION

The analysis of radiative transfer is complicated by the fact that the propagation of radiation at any point in a medium cannot be represented by a single vector as in the case of heat flow by conduction. To specify the radiation incident at a given point, it is necessary to know the radiation from all directions because radiation beams from all directions are independent of one another. Therefore a fundamental quantity frequently used in radiative transfer studies to describe the amount of radiation energy transmitted by the ray in any given direction per unit time is the *monochromatic* (or *spectral*) *radiation intensity*. To define this quantity we consider an elemental surface dA about the space coordinate $\mathbf{r}$ characterized by a normal unit direction vector $\hat{\mathbf{n}}$ as illustrated in Fig. 1-6. Let dE_ν denote the amount of radiative

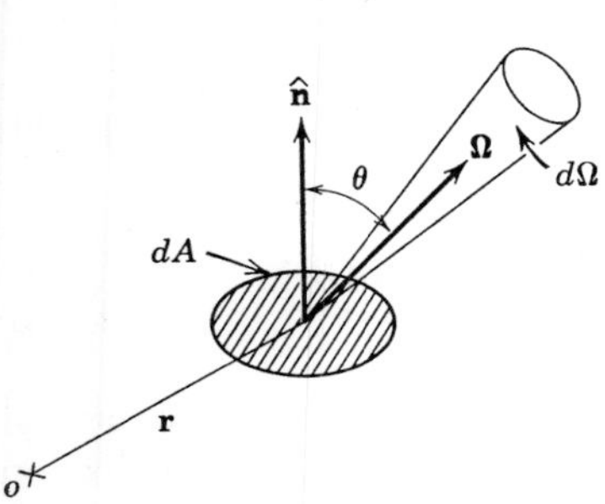

Fig. 1-6. Symbols for the definition of intensity.

energy in the frequency interval between ν and $\nu + d\nu$, confined to an element of solid angle $d\Omega$ around the direction of propagation $\hat{\mathbf{\Omega}}$ streaming through the elemental surface dA (i.e., transmitted through or emitted by and/or reflected from the surface) during the time interval between t and $t + dt$. Let θ be the polar angle between the unit direction vector $\hat{\mathbf{n}}$ and the direction of propagation $\hat{\mathbf{\Omega}}$. The monochromatic radiation intensity $I_\nu(\mathbf{r}, \hat{\mathbf{\Omega}}, t)$ is defined as

$$I_\nu(\mathbf{r}, \hat{\mathbf{\Omega}}, t) = \left[\frac{dE_\nu}{dA \cos\theta \, d\hat{\Omega} \, d\nu \, dt}\right]_{\lim dA, d\Omega, d\nu, dt \to 0} \tag{1-38}$$

In this relation $dA \cos\theta$ is the projection of the surface dA on a plane perpendicular to the direction $\hat{\mathbf{\Omega}}$; hence the intensity is defined on the basis of the projected area. According to Eq. 1-38 the *monochromatic intensity is the amount of radiative energy (in appropriate energy units) streaming through a unit area perpendicular to the direction of propagation* $\hat{\mathbf{\Omega}}$, *per unit solid angle around the direction* $\hat{\mathbf{\Omega}}$, *per unit frequency about the frequency* ν, *and per unit time about the time* t.

If the radiation intensity to or from a surface element is considered in the frequency range between ν_1 and ν_2, and through a solid angle between Ω_1 and Ω_2, then the quantity

$$\int_{\nu_1}^{\nu_2} \int_{\phi_1}^{\phi_2} \int_{\theta_1}^{\theta_2} I_\nu(\mathbf{r}, \theta, \phi, t) \cos\theta \sin\theta \, d\theta \, d\phi \, d\nu \tag{1-39a}$$

is the amount of radiative energy to or from the surface per unit area and per unit time in the frequency range ν_1 to ν_2, and through the solid angle between Ω_1 and Ω_2. An element of solid angle $d\Omega$ in the polar coordinates is represented by

$$d\Omega = \sin\theta \, d\theta \, d\phi \tag{1-39b}$$

where θ is the polar angle between the direction of the intensity and the normal to the surface, and ϕ is the azimuthal angle. Then relation 1-39a can be written as

$$\int_{\nu_1}^{\nu_2} \int_{\phi_1}^{\phi_2} \int_{\mu_2}^{\mu_1} I_\nu(\mathbf{r}, \mu, \phi, t)\mu \, d\mu \, d\phi \, d\nu \tag{1-39c}$$

where we have defined

$$\mu \equiv \cos\theta \tag{1-39d}$$

1-5 THE BLACK BODY AND BLACK-BODY RADIATION INTENSITY

The term *black body* is used to denote a body that possesses the property of allowing all incident radiation to enter the medium without surface reflection and without allowing it to leave the medium again. Therefore a black body must possess a surface that allows the incident radiation to enter without reflection. During the propagation of radiation in a medium each ray suffers a certain amount of weakening because of absorption; therefore a black body must have sufficient thickness, depending on its absorbing power, to ensure that rays will not leave the medium. A beam traveling in a medium is deviated from the original path and scattered in all directions because of the presence of small impurities and inhomogeneities. Although in the process of scattering of thermal radiation the energy is neither created nor destroyed, a black body must have no or negligibly small scattering properties to ensure that the radiation entering the medium will not be scattered out. These properties refer to radiation beams coming from all directions and for all wavelengths. Hence a black body absorbs all incident radiation from all directions and at all frequencies, without reflecting, transmitting, and scattering them out.

Emission of Radiation by a Black Body

We conclude from the foregoing discussion that a black body is a perfect absorber of radiation from all directions and at all frequencies. We now consider a black body inside an isothermal enclosure whose boundary absorbs and emits radiation, and we assume that after a period of time the black body and the enclosure reach thermal equilibrium and attain some uniform temperature. While in thermal equilibrium a body emits as much energy as it absorbs, and for a black body the emission of radiation must be maximum since it absorbs the maximum possible radiation from all directions and at all frequencies. Therefore the radiation emitted at any given temperature T is a maximum for a black body.

By considering a black body in thermal equilibrium inside an enclosure whose boundary emits and absorbs radiation only in a frequency interval $d\nu$ about ν, and by following a similar argument, it can be concluded that the radiation emitted by a black body at any given temperature T and frequency ν is a maximum. Furthermore the radiation emitted by a black body is isotropic.

The intensity of spectral (or monochromatic) radiation emitted by a black body at a temperature T into a vacuum was determined by Planck [1] and is given by

$$I_{\nu b,\mathrm{vac}}(T) = \frac{2h\nu^3}{c_0^2[\exp(h\nu/kT) - 1]} \tag{1-40}$$

where h and k are, respectively, the Planck and Boltzmann constants, c_0 is the speed of light in a vacuum, T is the absolute temperature, and ν is the frequency.

In many engineering applications wavelength is used more commonly than frequency to characterize the monochromatic intensity. We cannot transform Eq. 1-40 from frequency to wavelength by merely replacing ν by λ in this equation; but we can transform it by considering that the radiation energy emitted in the frequency interval $d\nu$ about ν should be equal to that in the wavelength interval $d\lambda_0$ about λ_0, that is,

$$I_\nu \, d\nu = -I_{\lambda_0} \, d\lambda_0 \tag{1-41}$$

Since the wavelength depends on the type of medium in which the radiation is traveling, we have used the subscript 0 to denote that the medium is a vacuum. The frequency, however, does not depend on the type of medium. The frequency and wavelength are related by

$$\nu = \frac{c_0}{\lambda_0} \tag{1-42a}$$

By differentiating we obtain

$$d\nu = -\frac{c_0}{\lambda_0^2} \, d\lambda_0 \qquad \text{and} \qquad d\lambda_0 = -\frac{c_0}{\nu^2} \, d\nu \tag{1-42b}$$

By utilizing Eqs. 1-41 and 1-42b we write

$$I_{\lambda_0 b,\text{vac}}(T) = -I_{\nu b,\text{vac}}(T) \frac{d\nu}{d\lambda_0} = I_{\nu b,\text{vac}}(T) \frac{\nu^2}{c_0} \tag{1-43a}$$

From Eqs. 1-40 and 1-43a we obtain Planck's black body radiation intensity function in terms of the wavelength:

$$I_{\lambda_0 b,\text{vac}}(T) = \frac{2hc_0^2}{\lambda_0^5[\exp(hc_0/\lambda_0 kT) - 1]} \tag{1-43b}$$

which represents the radiation intensity emitted by a black body into a pure vacuum. That is, it represents the radiative energy per unit projected area, per unit time, per unit solid angle, per unit wavelength about λ_0. If Btu units are used for the energy and microns for the wavelength, the intensity is given in Btu/hr ft^2 sr μ.

When radiant energy is emitted by a black body into a *medium other than a vacuum*, Eq. 1-40 should be replaced by

$$I_{\nu b}(T) = \frac{2h\nu^3}{c^2[\exp(h\nu/kT) - 1]} \tag{1-44a}$$

where c is the speed of propagation of radiation in the medium in question. For a *dielectric medium* Eq. 1-44a is written as

$$I_{\nu b}(T) = \frac{2h\nu^3 n^2}{c_0^2[\exp(h\nu/kT) - 1]} = n^2 I_{\nu b,\text{vac}}(T) \tag{1-44b}$$

since $c = c_0/n$. It is apparent from Eq. 1-44b that the intensity of radiation emitted by a black body at temperature T into a dielectric medium having a real refractive index n is n^2 times the intensity emitted by a black body at the same temperature into a vacuum.

Equation 1-44b can be expressed in terms of the wavelength by utilizing the relations

$$I_{\nu b}(T)\,d\nu = -I_{\lambda b}\,d\lambda \tag{1-45a}$$

and

$$\nu = \frac{c}{\lambda} = \frac{c_0}{n\lambda} \tag{1-45b}$$

Assuming that n is *independent of frequency*, Eq. 1-45b can be differentiated to yield

$$d\nu = -\frac{c_0}{n\lambda^2}\,d\lambda \tag{1-45c}$$

Then from Eqs. 1-45a and 1-45c we obtain

$$I_{\lambda b}(T) = -I_{\nu b}(T)\frac{d\nu}{d\lambda} = \frac{c_0}{n\lambda^2} I_{\nu b}(T) \tag{1-46}$$

Substituting Eq. 1-44b into 1-46, and replacing ν by λ according to Eq. 1-45b, we obtain

$$I_{\lambda b}(T) = \frac{2hc_0^{\,2}}{n^2\lambda^5[\exp(hc_0/n\lambda kT) - 1]} \tag{1-47}$$

where λ is the wavelength in the medium in question.

Total Black-Body Radiation Intensity

The intensity of radiation emitted by a black body over all frequencies (or wavelengths) is called the *total black-body radiation intensity* and is obtained by integrating the monochromatic black-body radiation intensity over the entire energy spectrum:

$$I_b(T) = \int_{\nu=0}^{\infty} I_{\nu b}(T)\,d\nu \tag{1-48a}$$

By substituting $I_{\nu b}(T)$ from Eq. 1-44b into 1-48a we obtain

$$I_b(T) = \frac{2h}{c_0^{\,2}} \int_{\nu=0}^{\infty} \frac{n^2\nu^3}{e^{h\nu/kT} - 1}\,d\nu \tag{1-48b}$$

If the refractive index n is assumed to be independent of frequency, Eq. 1-48b can be rearranged as

$$I_b(T) = \frac{2hn^2}{c_0^2}\left(\frac{kT}{h}\right)^4 \int_{\nu=0}^{\infty} \frac{(\nu h/kT)^3}{e^{\nu h/kT} - 1}\, d\left(\frac{\nu h}{kT}\right) \tag{1-48c}$$

or

$$I_b(T) = \frac{2k^4}{c_0^2 h^3}(n^2T^4)\int_{x=0}^{\infty} \frac{x^3}{e^x - 1}\, dx \tag{1-48d}$$

The integral in Eq. 1-48d can be evaluated by using the standard integral tables, and we obtain[11]

$$I_b(T) = n^2 \frac{\bar{\sigma}T^4}{\pi} \tag{1-48e}$$

where the Stefan-Boltzmann constant $\bar{\sigma}$ is defined as

$$\bar{\sigma} \equiv \frac{2\pi^5 k^4}{15 c_0^2 h^3} \tag{1-48f}$$

When Btu units are used, $I_b(T)$ is given in Btu/hr ft^2 sr.

1-6 THE BLACK-BODY EMISSIVE FLUX

In many engineering applications a physical quantity of interest is the *monochromatic* (or *spectral*) *black-body emissive flux* $q_{\lambda b}(T)$, defined as

$$q_{\lambda b}(T) = \int_{\phi=0}^{2\pi}\int_{\mu=0}^{1} I_{\lambda b}(T)\mu\, d\mu\, d\phi = \pi I_{\lambda b}(T) \tag{1-49a}$$

since $I_{\lambda b}(T)$ is independent of direction. Substituting the value of $I_{\lambda b}(T)$ from Eq. 1-47, we obtain

$$q_{\lambda b}(T) = \frac{c_1}{n^2\lambda^5[\exp(c_2/n\lambda T) - 1]} \tag{1-49b}$$

where we have defined

$$c_1 \equiv 2\pi h c_0^2 \qquad \text{and} \qquad c_2 \equiv \frac{hc_0}{k} \tag{1-49c}$$

Here we note that $q_{\lambda b}(T)$ represents the amount of radiative energy emitted by a black body at temperature T per unit area of its surface, per unit time, and per unit wavelength in all directions in the hemispherical space. If Btu units are used and wavelength is measured in microns, $q_{\lambda b}(T)$ is given in Btu/hr ft^2 μ.

The integration of $q_{\lambda b}(T)$ over all wavelengths from $\lambda = 0$ to infinity yields the *total black-body emissive flux* $q_b(T)$:

$$q_b(T) = \int_{\lambda=0}^{\infty} q_{\lambda b}(T)\, d\lambda = \pi\int_{\lambda=0}^{\infty} I_{\lambda b}(T)\, d\lambda = \pi I_b(T) = n^2\bar{\sigma}T^4 \tag{1-50}$$

Here we have used the relation given by Eq. 1-48e for $I_b(T)$.

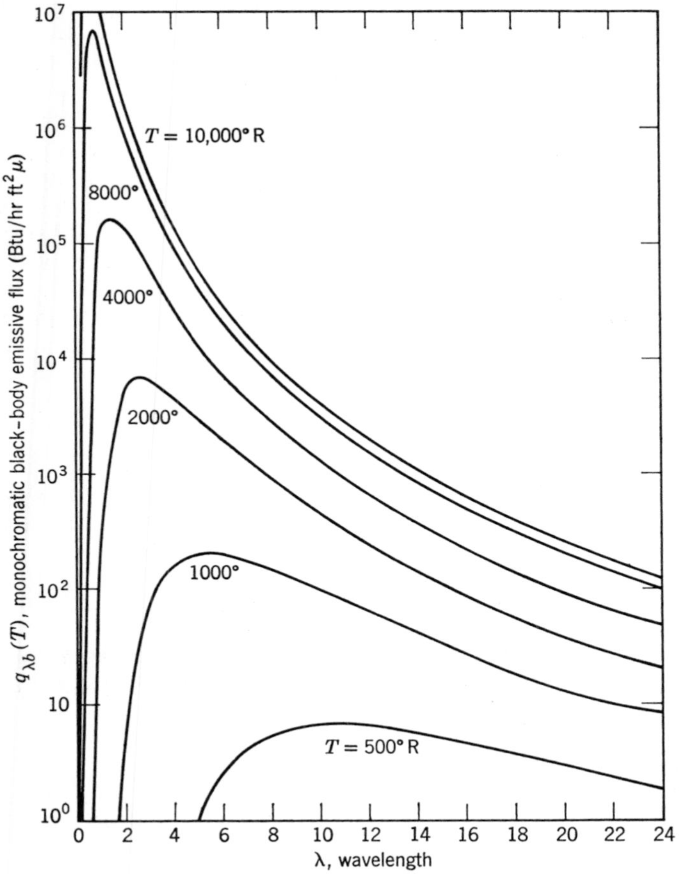

Fig. 1-7. Spectral distribution of monochromatic black-body emissive flux in a vacuum (i.e., $n = 1$).

Figure 1-7 shows the spectral distribution of monochromatic black-body emissive flux $q_{\lambda b}(T)$, evaluated from Eq. 1-49b for $n = 1$. It is evident from this figure that at any given wavelength the radiative energy emitted by a black body increases as the absolute temperature rises. Furthermore, each curve shows a peak, and these peaks tend to shift toward smaller wavelengths as the temperature rises. The locus of the peaks can be determined analytically by *Wien's displacement rule*, which is given as

$$(n\lambda T)_{q_{\lambda b,\max}} = 5215.6\ \mu\ {}^{\circ}\mathrm{R} \tag{1-51}$$

Finally we have summarized in Table 1-2 in different systems of units the

Table 1-2 Numerical Values of Radiation Constants[a]

Quantity	$(\text{cgs})_1$	$(\text{cgs})_2$	Btu
$q_{\lambda b}$	erg/sec cm^2 cm	W/cm^2 μ	Btu/hr ft^2 μ
q_b	erg/sec cm^2	W/cm^2	Btu/hr ft^2
λ	cm	μ (micron)	μ (micron)
c_1	3.704×10^{-5} erg cm^2/sec	37404 W μ^4/cm^2	1.1870×10^8 Btu μ^4/hr ft^2
c_2	1.4387 cm °K	14387 μ °K	25896 μ °R
$\bar{\sigma}$	5.6699×10^{-5} erg/sec cm^2 °K^4	5.6699×10^{-12} W/cm^2 °K^4	1.714×10^{-9} Btu/hr ft^2 °R^4
$(n\lambda T)_{\text{max}}$	0.28976 cm °K	2897.6 μ °K	5215.6 μ °R

[a] From N. W. Snyder [11].

numerical values of the constants c_1 and c_2 of Eq. 1-49b, the Stefan-Boltzmann constant $\bar{\sigma}$, Wien's displacement rule, and the corresponding units for $q_{\lambda b}$ and q_b.

The Fractional Functions

The fraction of the total black-body radiation intensity having wavelengths between 0 and λ is called the *fractional function of the first kind*, $f_{0-\lambda}(T)$, and is given as

$$f_{0-\lambda}(T) = \frac{\int_0^\lambda I_{\lambda' b}(T)\, d\lambda'}{I_b(T)} = \frac{\int_0^\lambda q_{\lambda' b}(T)\, d\lambda'}{q_b(T)} = \frac{q_{0-\lambda, b}(T)}{n^2 \bar{\sigma} T^4} \tag{1-52}$$

where we have defined

$$q_{0-\lambda, b}(T) \equiv \int_0^\lambda q_{\lambda' b}(T)\, d\lambda' \tag{1-53}$$

and assumed that the refractive index of the medium is independent of frequency. The fractional function of the first kind is given in Table 1-3.

The *fractional function of the second kind*, $f^*_{0-\lambda}(T)$, has been defined by Czerny and Walther [13] as

$$f^*_{0-\lambda}(T) = \int_0^\lambda \left.\frac{\partial I_{\lambda b}(T)}{\partial I_b(T)}\right|_\lambda d\lambda = \int_0^\lambda \left.\frac{\partial q_{\lambda b}(T)}{\partial q_b(T)}\right|_\lambda d\lambda \tag{1-54a}$$

This function can be related to the fractional function of the first kind, $f_{0-\lambda}(T)$, by[12]

$$f^*_{0-\lambda}(T) = f_{0-\lambda}(T) + \frac{\lambda T}{4} \frac{df_{0-\lambda}(T)}{d(\lambda T)} \tag{1-54b}$$

Table 1-3 Fractional Function of the First Kind[a]

$n\lambda T^b$	$f_{0-\lambda} = \dfrac{q_{0-\lambda,b}}{n^2\bar{\sigma}T^4}$	$n\lambda T^b$	$f_{0-\lambda} = \dfrac{q_{0-\lambda,b}}{n^2\bar{\sigma}T^4}$	$n\lambda T^b$	$f_{0-\lambda} = \dfrac{q_{0-\lambda,b}}{n^2\bar{\sigma}T^4}$
1000	0	7200	0.4809	13400	0.8317
1200	0	7400	0.5007	13600	0.8370
1400	0	7600	0.5199	13800	0.8421
1600	0.0001	7800	0.5381	14000	0.8470
1800	0.0003	8000	0.5558	14200	0.8517
2000	0.0009	8200	0.5727	14400	0.8563
2200	0.0025	8400	0.5890	14600	0.8606
2400	0.0053	8600	0.6045	14800	0.8648
2600	0.0098	8800	0.6195	15000	0.8688
2800	0.0164	9000	0.6337	16000	0.8868
3000	0.0254	9200	0.6474	17000	0.9017
3200	0.0368	9400	0.6606	18000	0.9142
3400	0.0506	9600	0.6731	19000	0.9247
3600	0.0667	9800	0.6851	20000	0.9335
3800	0.0850	10000	0.6966	21000	0.9411
4000	0.1051	10200	0.7076	22000	0.9475
4200	0.1267	10400	0.7181	23000	0.9531
4400	0.1496	10600	0.7282	24000	0.9589
4600	0.1734	10800	0.7378	25000	0.9621
4800	0.1979	11000	0.7474	26000	0.9657
5000	0.2229	11200	0.7559	27000	0.9689
5200	0.2481	11400	0.7643	28000	0.9718
5400	0.2733	11600	0.7724	29000	0.9742
5600	0.2983	11800	0.7802	30000	0.9765
5800	0.3230	12000	0.7876	40000	0.9881
6000	0.3474	12200	0.7947	50000	0.9941
6200	0.3712	12400	0.8015	60000	0.9963
6400	0.3945	12600	0.8081	70000	0.9981
6600	0.4171	12800	0.8144	80000	0.9987
6800	0.4391	13000	0.8204	90000	0.9990
7000	0.4604	13200	0.8262	100000	0.9992
				∞	1.0000

[a] From R. V. Dunkle [12].

[b] In this table λT is based on wavelength in microns and temperature in degrees Rankine.

A comprehensive tabulation of the fractional function of the second kind has been given by Czerny and Walther [13]. Table 1-4 gives the numerical values of $f^*_{0-\lambda}(T)$ as a function of the dimensionless parameter $\lambda T/c_2$, where $c_2 =$ 1.438 cm °K, λ is in centimeters, and T is in degrees Kelvin.

The utility of fractional functions has been illustrated with examples by Czerny and Walther [13] and Edwards [14]. Consider, for instance, the mean value of absorptivity α (this will be discussed in Section 1-8), defined as

$$\alpha = \frac{\int_0^\infty \alpha_\lambda I_{\lambda b}(T)\, d\lambda}{\int_0^\infty I_{\lambda b}(T)\, d\lambda} = \frac{\int_0^\infty \alpha_\lambda I_{\lambda b}(T)\, d\lambda}{I_b(T)} \tag{1-55a}$$

Table 1-4 Fractional Function of the Second Kind[a]

$\frac{\lambda T}{c_2}$[b]	$f^*_{0-\lambda}(T)$	$\frac{\lambda T}{c_2}$[b]	$f^*_{0-\lambda}(T)$	$\frac{\lambda T}{c_2}$[b]	$f^*_{0-\lambda}(T)$
0.0	0.0	0.26	0.61572	0.80	0.97679
0.05	0.0156×10^{-3}	0.27	0.64248	0.85	0.98048
0.06	0.2211×10^{-3}	0.28	0.66722	0.90	0.98344
0.07	1.350×10^{-3}	0.29	0.69008	0.95	0.98583
0.08	4.938×10^{-3}	0.30	0.71117	1.00	0.98778
0.09	1.294×10^{-2}	0.31	0.73063	1.05	0.98940
0.10	2.702×10^{-2}	0.32	0.74859	1.10	0.99074
0.11	4.802×10^{-2}	0.33	0.76514	1.15	0.99187
0.12	7.584×10^{-2}	0.34	0.78041	1.20	0.99283
0.13	0.10967	0.35	0.79451	1.25	0.99364
0.14	0.14824	0.36	0.80752	1.30	0.99433
0.15	0.19016	0.37	0.81954	1.40	0.99544
0.16	0.23405	0.38	0.83066	1.50	0.99628
0.17	0.27873	0.39	0.84093	1.60	0.99693
0.18	0.32320	0.40	0.85045	1.70	0.99743
0.19	0.36670	0.45	0.88863	1.80	0.99783
0.20	0.40869	0.50	0.91526	1.90	0.99815
0.21	0.44877	0.55	0.93925	2.00	0.99842
0.22	0.48671	0.60	0.94809	2.10	0.99863
0.23	0.52238	0.65	0.95837	2.20	0.99881
0.24	0.55576	0.70	0.96614	3.00	0.99953
0.25	0.58685	0.75	0.97212	∞	1.00000

[a] From M. Czerny and A. Walther [13].

[b] In this table $c_2 = 1.438$ cm °K, λ is in centimeters, and T is in degrees Kelvin.

Suppose that the entire wavelength spectrum is divided into M wavelength bands such that over each band the value of α_λ can be approximated as constant. Then the integral in Eq. 1-55a can be represented as a summation, and we obtain

$$\alpha = \frac{\sum_{m=1}^{M} \alpha_m \int_{\lambda_{m-1}}^{\lambda_m} I_{\lambda b}(T)\, d\lambda}{I_b(T)} = \sum_{m=1}^{M} \alpha_m \left[\frac{\int_0^{\lambda_m} I_{\lambda b}(T)\, d\lambda}{I_b(T)} - \frac{\int_0^{\lambda_{m-1}} I_{\lambda b}(T)\, d\lambda}{I_b(T)} \right]$$
$$= \sum_{m=1}^{M} \alpha_m [f_{0-\lambda_m}(T) - f_{0-\lambda_{m-1}}(T)] \qquad \text{(1-55b)}$$

where α_m is the mean value of α_λ in the wavelength interval from λ_{m-1} to λ_m. Thus the integration has been replaced by a summation involving the fractional functions of the first kind, which can be obtained from Table 1-3.

To illustrate the utility of the fractional function of the second kind we consider the evaluation of the *Rosseland mean extinction coefficient* β_R (this quantity will be discussed in Chapter 9), defined as

$$\frac{1}{\beta_R} = \int_0^\infty \frac{1}{\beta_\lambda} \frac{\partial I_{\lambda b}(T)}{\partial I_b(T)}\, d\lambda \qquad \text{(1-56a)}$$

where β_λ is the spectral extinction coefficient. We now assume that the entire wavelength spectrum is divided into M bands and that over each band the value of β_λ can be approximated as uniform. Then Eq. 1-56a can be written in the form

$$\frac{1}{\beta_R} = \sum_{m=1}^{M} \frac{1}{\beta_m} \int_{\lambda_{m-1}}^{\lambda_m} \frac{\partial I_{\lambda b}(T)}{\partial I_b(T)}\, d\lambda = \sum_{m=1}^{M} \frac{1}{\beta_m} \left[\int_0^{\lambda_m} \frac{\partial I_{\lambda b}(T)}{\partial I_b(T)}\, d\lambda - \int_0^{\lambda_{m-1}} \frac{\partial I_{\lambda b}(T)}{\partial I_b(T)}\, d\lambda \right]$$
$$= \sum_{m=1}^{M} \frac{1}{\beta_m} [f^*_{0-\lambda_m}(T) - f^*_{0-\lambda_{m-1}}(T)] \qquad \text{(1-56b)}$$

where β_m is the mean value of β_λ in the wavelength interval from λ_{m-1} to λ_m. Thus the integration has been replaced by a summation involving the fractional functions of the second kind.

1-7 RADIATION TO AND FROM A VOLUME ELEMENT

Consider a beam of radiation of intensity $I_\nu(\mathbf{r}, \hat{\mathbf{\Omega}})$, propagating in a participating medium at a given direction. The energy of the radiation will be attenuated because of the absorption of radiation by the matter, will be weakened because a certain fraction of it is deviated from its original path by scattering in all directions, and will be augmented as a result of emission of radiation by the matter. The absorption, scattering, and emission properties of the matter influence the energy of a beam of radiation traveling through the matter, and a comprehensive discussion of the characterization of absorption, scattering, and emission of radiation by matter can be found in the publications by Chandrasekhar [8], Kourganoff [15], and Viskanta [16, 17]. In this section we present a brief account of the interaction of radiation with a volume element.

Absorption of Radiation

We will consider a beam of monochromatic radiation of intensity $I_\nu(\mathbf{r}, \hat{\mathbf{\Omega}}')$, confined to an element of solid angle $d\Omega'$ that is incident normally on the surface dA of a slab of thickness ds as illustrated in Fig. 1-8. As the incident

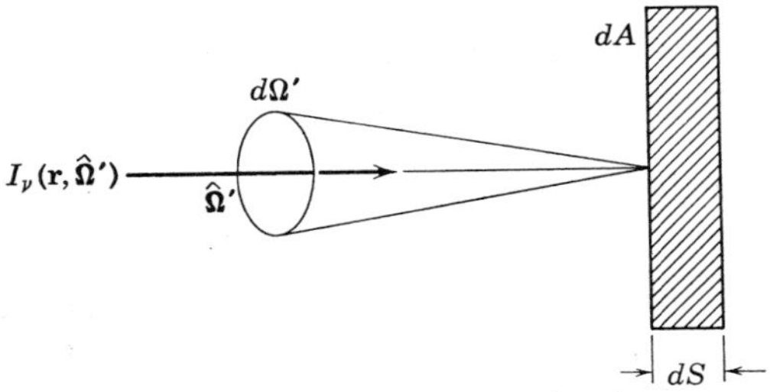

Fig. 1-8. Symbols for absorption of radiation.

radiation propagates through the material, part of it will be absorbed by the material. Let $\kappa_\nu(\mathbf{r})$ denote the *spectral volumetric absorption coefficient*,[13] which represents the fraction of the incident radiation that is absorbed by the matter per unit length along the path of the beam and has a unit of $(\text{length})^{-1}$. Then the quantity

$$\kappa_\nu(\mathbf{r}) I_\nu(\mathbf{r}, \hat{\mathbf{\Omega}}')\, d\Omega' \tag{1-57}$$

is the absorption of the incident radiation, $I_\nu(\mathbf{r}, \hat{\mathbf{\Omega}}')\, d\Omega'$, by the matter per unit time, per unit volume, per unit frequency. If the radiation is incident on the volume element from all directions in the spherical space, we integrate Eq. 1-57 over all solid angles and obtain

$$\kappa_\nu(\mathbf{r}) \int_{\phi=0}^{2\pi} \int_{\mu'=-1}^{1} I_\nu(\mathbf{r}, \mu', \phi')\, d\mu'\, d\phi' \tag{1-58}$$

This relation represents absorption by the matter per unit time, per unit volume, per unit frequency of radiation incident on the volume element from all directions in the spherical space (i.e., Btu/hr ft^3 frequency).

Emission of Radiation

In the problems of radiative transfer in participating media the assumption of *local thermodynamic equilibrium* (LTE) is almost generally introduced in order to simplify the characterization of the emission of radiation by a volume element. It means that any small volume element in the medium is assumed to be in local thermodynamic equilibrium so that each point can be characterized by a local temperature $T(\mathbf{r})$. This is valid when the atomic collision in the matter is so predominant that it establishes a local thermodynamic equilibrium at each point $\mathbf{r}$ in the medium. In such situations the emission of radiative energy from a volume element can be characterized in terms of the Planck function. If $J_\nu^e(\mathbf{r})$ denotes radiation emitted by the matter per unit time, per unit volume, per unit solid angle, and per unit frequency, then the emission of radiation by the matter is related to Planck's black-body radiation intensity by

$$J_\nu^e(\mathbf{r}) = \kappa_\nu(\mathbf{r}) I_{\nu b}[T(\mathbf{r})] \tag{1-59}$$

where $I_{\nu b}(T)$ is given by Eq. 1-44b, and the Kirchhoff law[14] is assumed to be valid.

If LTE cannot be assumed, then the emission of radiative energy becomes a function of the energy states in the gas.

Scattering of Radiation

A beam of radiation traveling through a medium will be scattered in all directions if inhomogeneities, such as extremely small particles, are present in the matter. For example, dust particles or water droplets in the atmosphere scatter light traveling through such a medium. Very small bubbles inside a semitransparent plastic material scatter thermal radiation traveling through the medium. Even the smallest elements of space may exhibit discontinuities due to their atomic structure and hence scatter radiation; we see blue sky because of the scattering of the solar rays by the air molecules, and a rainbow is due to scattering by water droplets. In fact no substance is homogeneous in the absolute sense except an absolute vacuum. A medium may be regarded as optically homogeneous, however, provided that the linear dimension of the inhomogeneities is sufficiently small compared with the wavelength of the radiation. A *coherent* and an *incoherent* scattering should also be distinguished. The scattering is called *coherent* when the scattered radiation has the same frequency as the incident radiation, and *incoherent* when the frequency of the scattered radiation is different from that of the incident radiation. In the following analysis we are concerned only with coherent scattering.

Consider a beam of monochromatic radiation of intensity $I_\nu(\mathbf{r}, \hat{\mathbf{\Omega}}')$, confined to an element of solid angle $d\Omega'$ about the direction $\hat{\mathbf{\Omega}}'$ and incident normally on the surface dA of an elemental slab of thickness ds, as shown in Fig. 1-9. As the incident radiation travels through the medium, part of it will be scattered by the material. Let $\sigma_\nu(\mathbf{r})$ denote the *spectral volumetric scattering coefficient*,[15] which represents the fraction of the incident radiation that is scattered by the matter in all directions per unit length along the path of the beam and has a unit of $(\text{length})^{-1}$. Then the quantity

$$\sigma_\nu(\mathbf{r}) I_\nu(\mathbf{r}, \hat{\mathbf{\Omega}}')\, d\Omega' \tag{1-60}$$

is the scattering of the incident radiation, $I_\nu(\mathbf{r}, \hat{\mathbf{\Omega}}')\, d\Omega'$, by the matter in all directions in the spherical space per unit time, per unit volume, per unit frequency. However, Eq. 1-60 does not provide any information as to the directional distribution of scattered radiation. The directional distribution can be described by a *phase function*, $p_\nu(\hat{\mathbf{\Omega}}' \to \hat{\mathbf{\Omega}})$, introduced by Hopf [18] and normalized so that

$$\frac{1}{4\pi} \int_{\Omega=4\pi} p_\nu(\hat{\mathbf{\Omega}}' \to \hat{\mathbf{\Omega}})\, d\Omega = 1 \tag{1-61a}$$

or

$$\frac{1}{4\pi} \int_{\phi=0}^{2\pi} \int_{\mu=-1}^{1} p_\nu(\mu', \phi' \to \mu, \phi)\, d\mu\, d\phi = 1 \tag{1-61b}$$

Here we note that the quantity

$$\frac{1}{4\pi}\, p_\nu(\hat{\mathbf{\Omega}}' \to \hat{\mathbf{\Omega}})\, d\Omega \tag{1-62}$$

represents the probability that the incident radiation at $\hat{\mathbf{\Omega}}'$ will be scattered into an element of solid angle $d\Omega$ about the direction $\hat{\mathbf{\Omega}}$. Then

$$[\sigma_\nu(\mathbf{r}) I_\nu(\mathbf{r}, \hat{\mathbf{\Omega}}')\, d\Omega'] \frac{1}{4\pi}\, p_\nu(\hat{\mathbf{\Omega}}' \to \hat{\mathbf{\Omega}})\, d\Omega \tag{1-63}$$

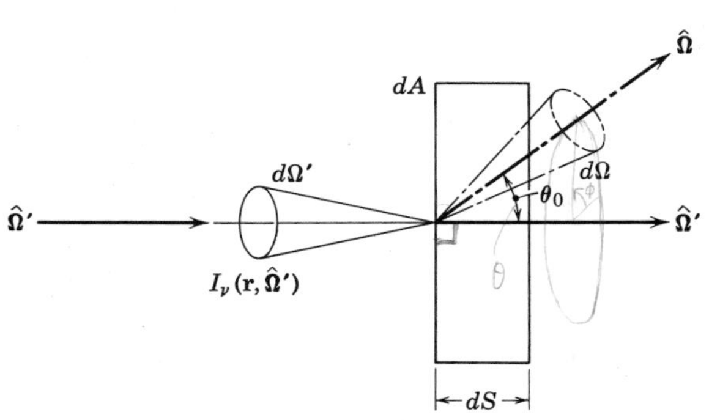

Fig. 1-9. Symbols for scattering of radiation.

is the scattering of the incident radiation, $I_\nu(\mathbf{r}, \hat{\mathbf{\Omega}}')\, d\Omega'$, by the matter *per unit time, per unit volume, per unit frequency* into an element of solid angle $d\Omega$ about the direction $\hat{\mathbf{\Omega}}$.

When radiation is incident on the volume element from all directions in the spherical space, the integration of Eq. 1-63 over all solid angles of incidence yields

$$\frac{1}{4\pi}\sigma_\nu(\mathbf{r})\, d\Omega \int_{\Omega'=4\pi} I_\nu(\mathbf{r}, \hat{\mathbf{\Omega}}')p_\nu(\hat{\mathbf{\Omega}}' \to \hat{\mathbf{\Omega}})\, d\Omega' \tag{1-64}$$

which represents the scattering of radiation incident on the volume element from all directions in the spherical space into a solid angle $d\Omega$ about the direction $\hat{\mathbf{\Omega}}$ per unit time, per unit volume, per unit frequency.

When the scattering particles in the medium are composed of homogeneous and isotropic material with perfect spherical symmetry and the medium has no preferential direction for scattering, then the phase function depends only on the angle θ_0 between the directions $\hat{\mathbf{\Omega}}'$ and $\hat{\mathbf{\Omega}}$. It can be shown from solid geometry that the angle θ_0 between the incident and scattered rays is given by the relation

$$\cos\theta_0 = \cos\theta\cos\theta' + \sin\theta\sin\theta'\cos(\phi - \phi') \tag{1-65a}$$

or

$$\mu_0 = \mu\mu' + \sqrt{1-\mu^2}\sqrt{1-\mu'^2}\cos(\phi - \phi') \tag{1-65b}$$

where θ, ϕ and θ', ϕ' are the polar coordinates characterizing the directions $\hat{\mathbf{\Omega}}$ and $\hat{\mathbf{\Omega}}'$, and μ, μ', μ_0 represent $\cos\theta$, $\cos\theta'$, $\cos\theta_0$, respectively.

When the phase function depends on the angle θ_0 only, Eq. 1-64 can be written as

$$\frac{1}{4\pi}\sigma_\nu(\mathbf{r})\, d\Omega \int_{\phi'=0}^{2\pi}\int_{\mu'=-1}^{1} I_\nu(\mathbf{r}, \mu', \phi')p_\nu(\mu_0)\, d\mu'\, d\phi' \tag{1-66}$$

where μ_0 has been defined by Eq. 1-65b.

The simplest phase function is for isotropic scattering and is given as

$$p_\nu = 1 \tag{1-67}$$

The Radiation Leaving a Volume Element

The radiation leaving a volume element per unit time, per unit volume, per unit frequency, and per unit solid angle about any given direction $\hat{\mathbf{\Omega}}$ is composed of an emitted and a scattered component and is given in the form

$$J_\nu^e(\mathbf{r}) + \frac{1}{4\pi}\sigma_\nu(\mathbf{r})\int_{\Omega'=4\pi} I_\nu(\mathbf{r}, \hat{\mathbf{\Omega}}')p_\nu(\hat{\mathbf{\Omega}}' \to \hat{\mathbf{\Omega}})\, d\Omega' \tag{1-68}$$

When the Kirchhoff law is valid and the medium has no preferential direction for scattering, this relation becomes

$$\kappa_\nu(\mathbf{r})I_{\nu b}[T(\mathbf{r})] + \frac{1}{4\pi}\,\sigma_\nu(\mathbf{r})\int_{\phi'=0}^{2\pi}\int_{\mu'=-1}^{1} I_\nu(\mathbf{r}, \mu', \phi')p_\nu(\mu_0)\,d\mu'\,d\phi' \quad (1\text{-}69)$$

Here the first term is the radiation emitted by the material due to its temperature, and the second term is the radiation incident on the volume element from all directions in the spherical space that it scattered in the direction $\hat{\mathbf{\Omega}}$.

The Radiation Energy Density

A volume element receiving radiation from all directions in the spherical space contains certain amounts of radiative energy at any instant t. The amount of radiative energy contained within a volume element per unit volume and per unit frequency is called the *spectral radiation energy density* and is denoted by the symbol $u_\nu(\mathbf{r})$. When radiation of intensity $I_\nu(\mathbf{r}, \hat{\mathbf{\Omega}}')$ is incident on a volume element from all directions, the spectral radiation energy density is given by

$$u_\nu(\mathbf{r}) = \frac{1}{c}\int_{\phi'=0}^{2\pi}\int_{\mu'=-1}^{1} I_\nu(\mathbf{r}, \mu', \phi')\,d\mu'\,d\phi' \quad (1\text{-}70)$$

where c is the speed of propagation of radiation in the medium. For a dielectric medium $c = c_0/n$, and Eq. 1-70 becomes

$$u_\nu(\mathbf{r}) = \frac{n}{c_0}\int_{\phi'=0}^{2\pi}\int_{\mu'=-1}^{1} I_\nu(\mathbf{r}, \mu', \phi')\,d\mu'\,d\phi' \quad (1\text{-}71)$$

If we assume that the radiation intensity is independent of direction, the integration can be performed and Eq. 1-71 becomes

$$u_\nu(\mathbf{r}) = \frac{4\pi n}{c_0}\,I_\nu(\mathbf{r}) \quad (1\text{-}72)$$

The *total radiation energy density* $u(\mathbf{r})$ is obtained by integrating $u_\nu(\mathbf{r})$ over all frequencies:

$$u(\mathbf{r}) = \int_{\nu=0}^{\infty} u_\nu(\mathbf{r})\,d\nu \quad (1\text{-}73)$$

The Net Radiative Flux

The *spectral radiative flux vector* $\mathbf{q}_\nu(\mathbf{r})$ is defined as the integration of $\hat{\mathbf{\Omega}}I_\nu(\mathbf{r}, \hat{\mathbf{\Omega}})$ over the entire spherical space and is given as

$$\mathbf{q}_\nu(\mathbf{r}) = \int_{\Omega=4\pi} I_\nu(\mathbf{r}, \hat{\mathbf{\Omega}})\hat{\mathbf{\Omega}}\,d\Omega \quad (1\text{-}74)$$

The component of this vector in any given direction $\hat{\mathbf{n}}$ is the *spectral net radiative flux* in the direction $\hat{\mathbf{n}}$ and is given as

$$q_{\nu n}(\mathbf{r}) = \hat{\mathbf{n}} \cdot \mathbf{q}_\nu(\mathbf{r}) = \int_{\Omega=4\pi} I_\nu(\mathbf{r}, \hat{\mathbf{\Omega}})\hat{\mathbf{\Omega}} \cdot \hat{\mathbf{n}}\, d\Omega \tag{1-75}$$

Let θ be the angle between the directions $\hat{\mathbf{\Omega}}$ and $\hat{\mathbf{n}}$, as illustrated in Fig. 1-10. Then

$$\hat{\mathbf{\Omega}} \cdot \hat{\mathbf{n}} = \cos\theta \equiv \mu \tag{1-76}$$

and Eq. 1-75 becomes

$$q_{\nu n}(\mathbf{r}) = \int_{\phi=0}^{2\pi} \int_{\mu=-1}^{1} I_\nu(\mathbf{r}, \mu, \phi)\mu\, d\mu\, d\phi \tag{1-77}$$

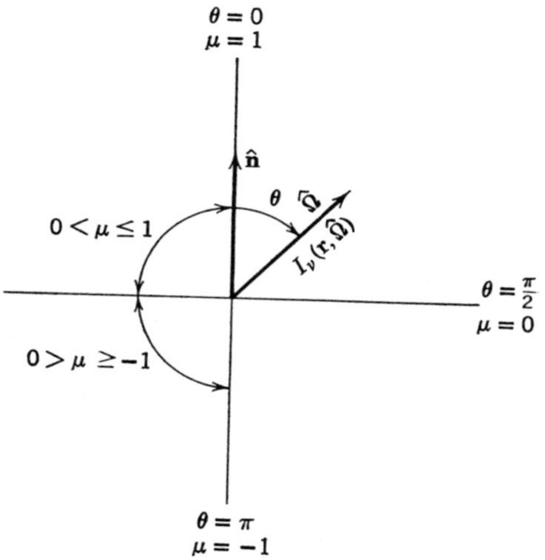

Fig. 1-10. Coordinates for net radiative heat flux.

Here $q_{\nu n}(\mathbf{r})$, as defined above, represents the net flow of monochromatic radiative energy across a surface normal to the direction $\hat{\mathbf{n}}$, per unit time, per unit area, per unit frequency, due to radiation incident from all directions in the spherical space. When the net flow of radiative energy is in the positive $\hat{\mathbf{n}}$ direction, $q_{\nu n}(\mathbf{r})$ is considered to be a positive quantity.

Equation 1-77 can be written as the difference between two spectral *half-range radiative fluxes* in the form

$$q_{\nu n}(\mathbf{r}) = q_\nu^+(\mathbf{r}) - q_\nu^-(\mathbf{r}) \tag{1-78}$$

where we have defined the spectral half-range radiative fluxes as

$$q_\nu^+(\mathbf{r}) \equiv \int_{\phi=0}^{2\pi} \int_{\mu=0}^{1} I_\nu(\mathbf{r}, \mu, \phi)\mu \, d\mu \, d\phi \tag{1-79a}$$

$$q_\nu^-(\mathbf{r}) \equiv -\int_{\phi=0}^{2\pi} \int_{\mu=-1}^{0} I_\nu(\mathbf{r}, \mu, \phi)\mu \, d\mu \, d\phi \tag{1-79b}$$

Here the half-range flux q_ν^+ represents the part of the net radiative flux due to radiation from the direction $\hat{\mathbf{\Omega}}$ such that $\hat{\mathbf{n}} \cdot \hat{\mathbf{\Omega}} > 0$ (i.e., $0 < \mu \leq 1$), and q_ν^- represents the part due to radiation from the direction $\hat{\mathbf{\Omega}}$ such that $\hat{\mathbf{n}} \cdot \hat{\mathbf{\Omega}} < 0$ (i.e., $-1 \leq \mu < 0$).

When the intensity of radiation $I_\nu(\mathbf{r}, \hat{\mathbf{\Omega}})$ is independent of the azimuthal angle ϕ, Eq. 1-77 becomes

$$q_{\nu n}(\mathbf{r}) = 2\pi \int_{\mu=-1}^{1} I_\nu(\mathbf{r}, \mu)\mu \, d\mu \tag{1-80}$$

When the intensity of radiation is independent of direction, Eq. 1-77 (or Eq. 1-80) has a value of zero.

The Radiation Pressure

The radiative energy contained in a volume element is expected to exert pressure on the walls of the container because radiation has an energy density. The *net radiation pressure* at any point $\mathbf{r}$ in the medium is defined as the net rate of transfer of momentum normally occurring across a unit area of an arbitrary surface chosen to pass through $\mathbf{r}$, and it can be determined as described below.

Consider a beam of monochromatic radiation $I_\nu(\mathbf{r}, \hat{\mathbf{\Omega}}) \, d\Omega$ incident on an elemental surface dA within the volume element at an angle θ with the normal to the surface. The radiative energy incident at the surface per unit area, per unit time, per unit frequency is given as

$$I_\nu(\mathbf{r}, \hat{\mathbf{\Omega}}) \cos\theta \, d\Omega \tag{1-81}$$

The net rate of transfer of momentum is obtained by dividing this expression by the speed of propagation c, and the normal component across dA is obtained by multiplying it by $\cos\theta$:

$$\frac{1}{c} I_\nu(\mathbf{r}, \hat{\mathbf{\Omega}}) \cos^2\theta \, d\Omega \tag{1-82}$$

If the radiation is incident from all directions, we integrate Eq. 1-82 over all solid angles in the spherical space and obtain the *spectral radiation pressure* as

$$p_\nu(\mathbf{r}) = \frac{1}{c}\int_{\phi=0}^{2\pi}\int_{\mu=-1}^{1} I_\nu(\mathbf{r}, \mu, \phi)\mu^2\, d\mu\, d\phi \tag{1-83}$$

When the radiation intensity is independent of direction, Eq. 1-83 simplifies to

$$p_\nu(\mathbf{r}) = \frac{4\pi}{3c} I_\nu(\mathbf{r}) = \frac{4\pi n}{3c_0} I_\nu(\mathbf{r}) = \tfrac{1}{3}u_\nu(\mathbf{r}) \tag{1-84}$$

where the spectral radiation energy density $u_\nu(\mathbf{r})$ has been defined by Eq. 1-72.

The total net radiation pressure $P(\mathbf{r})$ is obtained by integrating $P_\nu(\mathbf{r})$ over all frequencies:

$$P(\mathbf{r}) = \int_{\nu=0}^{\infty} P_\nu(\mathbf{r})\, d\nu \tag{1-85}$$

The Total Quantities

The total energy quantities are obtained by integrating the spectral energy quantities over the entire frequency spectrum from $\nu = 0$ to infinity (or over the entire wavelength spectrum from $\lambda = 0$ to infinity). If Z_ν denotes a spectral energy quantity, the total energy quantity Z is given by

$$Z = \int_{\nu=0}^{\infty} Z_\nu\, d\nu \tag{1-86}$$

where Z_ν = the spectral radiation intensity, radiation energy density, radiation pressure, radiation flux, or radiation leaving a volume element.

The Average (Total) Radiative Properties

The spectral radiative properties, such as κ_ν and σ_ν, are valid only for the frequency indicated. In many engineering applications the *average values* of these properties (also referred to as the *total properties*) are needed over the entire range of the spectrum from $\nu = 0$ to infinity. The average values of σ_ν and κ_ν (when κ_ν is used for the absorption of radiation) over the entire frequency spectrum are given by

$$Z = \frac{\int_{\nu=0}^{\infty} Z_\nu G_\nu\, d\nu}{G} \tag{1-87}$$

where we have defined

$$Z_\nu \equiv \sigma_\nu \quad \text{or} \quad \kappa_\nu \text{ (for absorption)} \tag{1-88a}$$

and the weight function G_ν is taken as

$$G_\nu = \int_{4\pi} I_\nu \, d\Omega = \int_{\phi=0}^{2\pi} \int_{\mu=-1}^{1} I_\nu \, d\mu \, d\phi \tag{1-88b}$$

and

$$G = \int_{\nu=0}^{\infty} G_\nu \, d\nu \tag{1-88c}$$

Here G_ν is called the *spectral incident radiation*, and G is the *incident radiation*.[16]

When the Kirchhoff law is valid and κ_ν is used for the *emission of radiation*, that is, $J_\nu^{\,e} = \kappa_\nu I_{\nu b}(T)$, the average value of κ_ν over the entire frequency spectrum should be defined as

$$\kappa_e = \frac{\int_{\nu=0}^{\infty} \kappa_\nu \left[\int_{4\pi} I_{\nu b}(T) \, d\Omega \right] d\nu}{\int_{\nu=0}^{\infty} \left[\int_{4\pi} I_{\nu b}(T) \, d\Omega \right] d\nu} = \frac{\int_{\nu=0}^{\infty} \kappa_\nu I_{\nu b}(T) \, d\nu}{\int_{\nu=0}^{\infty} I_{\nu b}(T) \, d\nu} = \frac{\int_{\nu=0}^{\infty} \kappa_\nu I_{\nu b}(T) \, d\nu}{I_b(T)} \tag{1-89}$$

Here κ_e denotes the average value for the emission of radiation. The mean coefficient as defined by Eq. 1-89 is sometimes referred to the *Planck mean absorption coefficient;* it should be used only for the emission of radiation.

1-8 RADIATION TO AND FROM A SURFACE ELEMENT

The emission and absorption of radiation by a body constitute a bulk process; strictly speaking, the surface of a body alone never emits and absorbs radiation. It allows part of the radiation coming from the interior to pass through it, or the radiation incident on the surface penetrates into the medium, where it is attenuated. In situations where a large proportion of the incident radiation is attenuated within a very short distance from the surface, we speak, for the sake of convenience in the analysis, of radiation as being absorbed at the surface. Similarly we speak of radiation as being emitted by the surface. With this limitation to the concept of emission and absorption of radiation by a surface, we proceed now to discuss various definitions concerning radiation to and from a surface element. The basic difference between the radiation to and from a surface element and that to and from a volume element is that the former involves radiation in the hemispherical space (i.e., $\Omega = 2\pi$); the latter, radiation in the spherical space (i.e., $\Omega = 4\pi$). In this section we present definitions for the absorption, emission, and reflection of radiation by a surface element.

Reflection of Radiation

The reflectivity[17] of a surface depends not only on the temperature and the properties of a surface but also on the directions of the incident and reflected radiation. Therefore several different definitions have been used in the literature to describe the reflection of radiation from a surface element. A discussion of various concepts can be found in the publications by Dunkle [19], Bevans and Edwards [20], Birkebak and Eckert [21], Torrance and Sparrow [22], and McNicholas [23]. Some of these concepts that characterize the reflection of radiation from a surface will now be presented here.

The Reflection Distribution Function

Consider a beam of monochromatic radiation, $I_\nu(\mathbf{r}, \hat{\mathbf{\Omega}}')\,d\Omega'$, incident on a surface element dA. Let θ' be the angle between the incident ray and the normal to the surface, as illustrated in Fig. 1-11. The amount of radiative

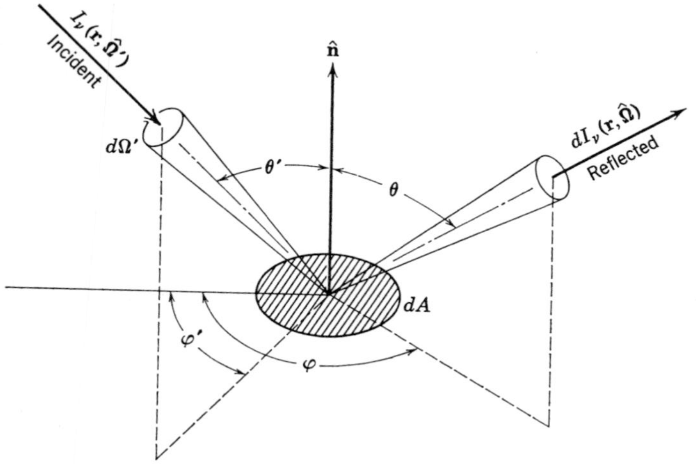

Fig. 1-11. Symbols for the definition of reflection distribution function, $f_\nu(\mathbf{r}, \hat{\mathbf{\Omega}}', \hat{\mathbf{\Omega}})$.

energy incident on the surface element per unit time, per unit area, per unit frequency is

$$I_\nu(\mathbf{r}, \hat{\mathbf{\Omega}}') \cos\theta'\, d\Omega' \tag{1-90}$$

Part of this radiation will be reflected by the surface in all directions in the hemispherical space. Let $dI_\nu(\mathbf{r}, \hat{\mathbf{\Omega}})$ be the intensity of radiation reflected in the direction $\hat{\mathbf{\Omega}}$. The intensity of the reflected radiation is related to the incident radiative energy by the *spectral reflection distribution function*,[18]

$f_\nu(\mathbf{r}, \hat{\mathbf{\Omega}}', \hat{\mathbf{\Omega}})$, defined by McNicholas [23] as

$$f_\nu(\mathbf{r}, \hat{\mathbf{\Omega}}', \hat{\mathbf{\Omega}}) = \frac{dI_\nu(\mathbf{r}, \hat{\mathbf{\Omega}})}{I_\nu(\mathbf{r}, \hat{\mathbf{\Omega}}') \cos\theta' \, d\Omega'} \tag{1-91}$$

This quantity can be larger or smaller than unity, depending on the magnitude of $dI_\nu(\mathbf{r}, \hat{\mathbf{\Omega}})$. For a specularly reflecting surface, for example, all of the incident energy contained within the solid angle $d\Omega'$ is reflected into a solid angle $d\Omega = d\Omega'$ about the direction $\theta = \theta'$ and $\phi = \phi' \pm \pi$. In that case dI_ν is not an infinitesimal quantity, but is of the same order of magnitude as I_ν. Then f_ν becomes large as $d\Omega'$ is chosen very small.

By utilizing the general reciprocity theorem first stated by Helmholtz [26] it can be shown that the reflection distribution function is symmetrical with respect to the directions of incidence and reflection [23]:

$$f_\nu(\mathbf{r}, \hat{\mathbf{\Omega}}', \hat{\mathbf{\Omega}}) = f_\nu(\mathbf{r}, \hat{\mathbf{\Omega}}, \hat{\mathbf{\Omega}}') \tag{1-92a}$$

or

$$f_\nu(\mathbf{r}; \theta', \phi'; \theta, \phi) = f_\nu(\mathbf{r}; \theta, \phi; \theta', \phi') \tag{1-92b}$$

The reflection properties of a surface are completely characterized if the reflection distribution function is specified in all directions in the hemispherical space. However, to obtain such information by experimental means is very laborious; hence the reflection distribution function is not a practical quantity for engineering purposes. The definitions that characterize the averaged reflection properties of a surface are given below.

The Directional-Hemispherical Reflectivity

Consider a beam of monochromatic radiation, $I_\nu(\mathbf{r}, \hat{\mathbf{\Omega}}') \, d\Omega'$, incident on a surface element dA. The amount of radiative energy incident on the surface per unit area, per unit time, per unit frequency is

$$I_\nu(\mathbf{r}, \hat{\mathbf{\Omega}}') \cos\theta' \, d\Omega' \tag{1-93}$$

where θ' is the angle between the incident ray and the normal to the surface. If $dI_\nu(\mathbf{r}, \hat{\mathbf{\Omega}})$ is the intensity of radiation reflected in the direction $\hat{\mathbf{\Omega}}$, the radiative energy that is reflected back into the entire hemispherical space is given by

$$\int_{\Omega=2\pi} dI_\nu(\mathbf{r}, \hat{\mathbf{\Omega}}) \cos\theta \, d\Omega \tag{1-94}$$

where θ is the angle between the direction of reflection $\hat{\mathbf{\Omega}}$ and the normal $\hat{\mathbf{n}}$ to the surface. Then the spectral directional-hemispherical reflectivity,[19] $\rho_\nu(\mathbf{r}, \hat{\mathbf{\Omega}}' \to 2\pi)$, is defined as

$$\rho_\nu(\mathbf{r}, \hat{\mathbf{\Omega}}' \to 2\pi) = \frac{\int_{\Omega=2\pi} dI_\nu(\mathbf{r}, \hat{\mathbf{\Omega}}) \cos\theta \, d\Omega}{I_\nu(\mathbf{r}, \hat{\mathbf{\Omega}}') \cos\theta' \, d\Omega'} \tag{1-95}$$

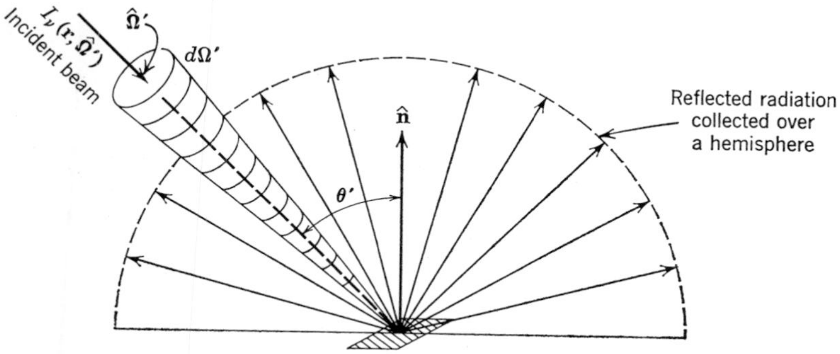

Fig. 1-12. Symbols for the definition of directional-hemispherical reflectivity, $\rho_\nu(\mathbf{r}, \hat{\mathbf{\Omega}}' \to 2\pi)$.

Figure 1-12 illustrates the concept for the definition of $\rho_\nu(\mathbf{r}, \hat{\mathbf{\Omega}}' \to 2\pi)$; that is, the surface is irradiated by a beam of radiation, $I_\nu(\mathbf{r}, \hat{\mathbf{\Omega}}')\, d\Omega'$, in any given direction $\hat{\mathbf{\Omega}}'$, and the reflected radiation is collected over the entire hemispherical space. The definition of Eq. 1-95 is the ratio of the collected to the incident radiative energy.

By utilizing the definition of the reflection distribution function of Eq. 1-91, we can relate $\rho_\nu(\mathbf{r},\hat{\mathbf{\Omega}}' \to 2\pi)$ to $f_\nu(\mathbf{r}, \hat{\mathbf{\Omega}}',\hat{\mathbf{\Omega}})$ as

$$\rho_\nu(\mathbf{r}, \hat{\mathbf{\Omega}}' \to 2\pi) = \int_{\Omega=2\pi} \frac{dI_\nu(\mathbf{r}, \hat{\mathbf{\Omega}})}{I_\nu(\mathbf{r}, \hat{\mathbf{\Omega}}') \cos \theta' \, d\Omega'} \cos \theta \, d\Omega \tag{1-96a}$$

$$= \int_{\Omega=2\pi} f_\nu(\mathbf{r}, \hat{\mathbf{\Omega}}', \hat{\mathbf{\Omega}}) \cos \theta \, d\Omega \tag{1-96b}$$

$$= \int_{\phi=0}^{2\pi} \int_{\mu=0}^{1} f_\nu(\mathbf{r}; \mu', \phi'; \mu, \phi)\mu \, d\mu \, d\phi \tag{1-96c}$$

The Hemispherical-Directional Reflectivity

We now consider the reverse of the situation described above. The radiation intensity $I_\nu(\mathbf{r}, \hat{\mathbf{\Omega}}')$ is incident on the surface element from all directions in the hemispherical space, and the reflected intensity $I_\nu(\mathbf{r}, \hat{\mathbf{\Omega}})$ is measured in any given direction $\hat{\mathbf{\Omega}}$. Figure 1-13 illustrates the concept for the definition of hemispherical-directional reflectivity. The intensity of radiation $I_\nu(\mathbf{r},\hat{\mathbf{\Omega}})$ reflected in any given direction $\hat{\mathbf{\Omega}}$ is given by

$$I_\nu(\mathbf{r}, \hat{\mathbf{\Omega}}) = \int_{\Omega'=2\pi} f_\nu(\mathbf{r}, \hat{\mathbf{\Omega}}', \hat{\mathbf{\Omega}}) I_\nu(\mathbf{r}, \hat{\mathbf{\Omega}}') \cos \theta' \, d\Omega' \tag{1-97}$$

which is obtained by integrating Eq. 1-91 over the hemispherical space.

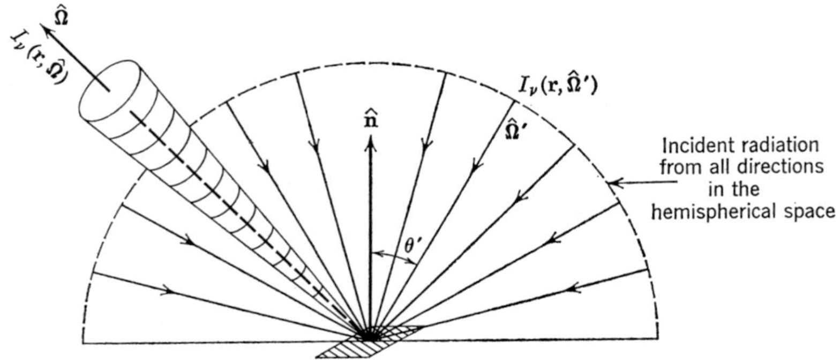

Fig. 1-13. Symbols for the definition of hemispherical-directional reflectivity, $\rho_\nu(\mathbf{r}, 2\pi \rightarrow \hat{\mathbf{\Omega}})$.

Equation 1-97 represents the radiative energy reflected from the surface per unit time, area, steradian, and frequency.

The radiative energy incident on the surface per unit time, area, and frequency from all directions in the hemispherical space is

$$\int_{\Omega'=2\pi} I_\nu(\mathbf{r}, \hat{\mathbf{\Omega}}') \cos\theta' \, d\Omega' \tag{1-98}$$

or the quantity

$$\frac{1}{\pi}\int_{\Omega'=2\pi} I_\nu(\mathbf{r}, \hat{\mathbf{\Omega}}') \cos\theta' \, d\Omega' \tag{1-99}$$

represents the radiative energy incident on the surface per unit time, area, steradian, and frequency. Then the spectral hemispherical-directional reflectivity, $\rho_\nu(\mathbf{r}, 2\pi \rightarrow \hat{\mathbf{\Omega}})$, is defined as

$$\rho_\nu(\mathbf{r}, 2\pi \rightarrow \hat{\mathbf{\Omega}}) = \frac{\displaystyle\int_{\Omega'=2\pi} f_\nu(\mathbf{r}, \hat{\mathbf{\Omega}}', \hat{\mathbf{\Omega}}) I_\nu(\mathbf{r}, \hat{\mathbf{\Omega}}') \cos\theta' \, d\Omega'}{\displaystyle(1/\pi)\int_{\Omega'=2\pi} I_\nu(\mathbf{r}, \hat{\mathbf{\Omega}}\,) \cos\theta' \, d\Omega'} \tag{1-100}$$

When the intensity of the incident radiation is independent of direction, Eq. 1-100 simplifies to

$$\begin{aligned}\rho_\nu(\mathbf{r}, 2\pi \rightarrow \hat{\mathbf{\Omega}}) &= \int_{\Omega'=2\pi} f_\nu(\mathbf{r}, \hat{\mathbf{\Omega}}', \hat{\mathbf{\Omega}}) \cos\theta' \, d\Omega' \\ &= \int_{\phi'=0}^{2\pi}\int_{\mu'=0}^{1} f_\nu(\mathbf{r}; \mu', \phi'; \mu, \phi)\mu' \, d\mu' \, d\phi' \end{aligned} \tag{1-101}$$

By comparing Eqs. 1-96 and 1-101 we note that

$$\rho_\nu(\mathbf{r}, 2\pi \rightarrow \hat{\mathbf{\Omega}}) = \rho_\nu(\mathbf{r}, \hat{\mathbf{\Omega}}' \rightarrow 2\pi) \tag{1-102}$$

which is valid for $\phi = \phi'$, $\theta = \theta'$, and incident radiation that is independent of direction.

The Hemispherical Reflectivity

We now consider a situation in which radiation is incident on the surface from all directions in the hemispherical space and the reflected energy is collected over the entire hemispherical space. Then the radiative energy incident on the surface per unit time, area, and frequency is

$$\int_{\Omega'=2\pi} I_\nu(\mathbf{r}, \hat{\mathbf{\Omega}}') \cos\theta' \, d\Omega' \tag{1-103}$$

The radiative energy that is reflected in all directions in the hemispherical space per unit time, area, and frequency is

$$\int_{\Omega'=2\pi} \rho_\nu(\mathbf{r}, \hat{\mathbf{\Omega}}' \to 2\pi) I_\nu(\mathbf{r}, \hat{\mathbf{\Omega}}') \cos\theta' \, d\Omega' \tag{1-104}$$

The spectral hemispherical reflectivity $\rho_\nu(\mathbf{r})$ is defined as

$$\rho_\nu(\mathbf{r}) = \frac{\displaystyle\int_{\Omega'=2\pi} \rho_\nu(\mathbf{r}, \hat{\mathbf{\Omega}}' \to 2\pi) I_\nu(\mathbf{r}, \hat{\mathbf{\Omega}}') \cos\theta' \, d\Omega'}{\displaystyle\int_{\Omega'=2\pi} I_\nu(\mathbf{r}, \hat{\mathbf{\Omega}}') \cos\theta' \, d\Omega'} \tag{1-105}$$

When the incident radiation is independent of direction, Eq. 1-105 simplifies to

$$\rho_\nu(\mathbf{r}) = \frac{1}{\pi} \int_{\Omega'=2\pi} \rho_\nu(\mathbf{r}, \hat{\mathbf{\Omega}}' \to 2\pi) \cos\theta' \, d\Omega' \tag{1-106a}$$

$$= \frac{1}{\pi} \int_{\phi'=0}^{2\pi} \int_{\mu'=0}^{1} \rho_\nu(\mathbf{r}; \mu', \phi'; 2\pi) \mu' \, d\mu' \, d\phi' \tag{1-106b}$$

Substituting for $\rho_\nu(\mathbf{r}, \hat{\mathbf{\Omega}}' \to 2\pi)$ from Eq. 1-96b, we obtain

$$\rho_\nu(\mathbf{r}) = \frac{1}{\pi} \int_{\Omega'=2\pi} \left[\int_{\Omega=2\pi} f_\nu(\mathbf{r}, \hat{\mathbf{\Omega}}', \hat{\mathbf{\Omega}}) \cos\theta \, d\Omega \right] \cos\theta' \, d\Omega' \tag{1-107}$$

When the reflection distribution function f_ν is independent of direction, Eq. 1-107 simplifies to

$$\rho_\nu(\mathbf{r}) = \pi f_\nu(\mathbf{r}) \tag{1-108}$$

Diffuse and Specular Reflections

A surface is called a *diffuse reflector* if the intensity of the reflected radiation is constant for all angles of reflection and is independent of the direction of the incident radiation. A surface is a *specular reflector* if the incident and the reflected rays lie symmetrically with respect to the normal at the point of incidence and the reflected beam is contained within the solid angle $d\Omega$ equal to the solid angle of incidence $d\Omega'$ (i.e., $d\Omega = d\Omega'$). The assumption of diffuse and specular reflection has been frequently used in the analysis of radiative heat transfer problems because it introduces significant simplification, but a real surface is neither a perfectly diffuse nor a perfectly specular reflector.

Absorption of Radiation

We present now some of the definitions that describe the absorption of radiation by surfaces.

Directional Absorptivity

Consider a beam of monochromatic radiation, $I_\nu(\mathbf{r},\hat{\mathbf{\Omega}}')\,d\Omega'$, incident on a surface element dA, as illustrated in Fig. 1-14. The amount of radiative energy

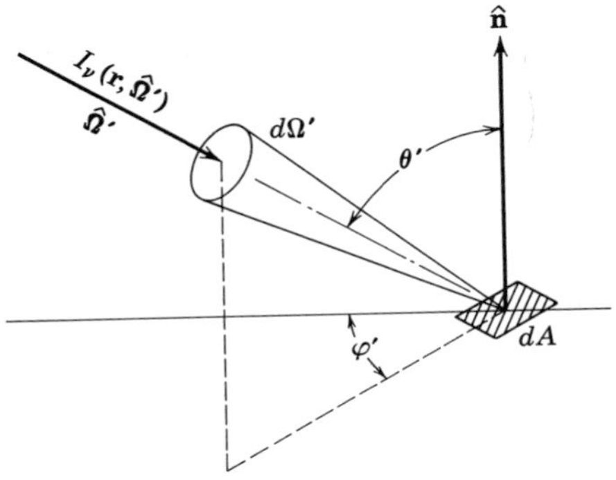

Fig. 1-14. Symbols for the definition of directional absorptivity, $\alpha_\nu(\mathbf{r},\hat{\mathbf{\Omega}}')$.

incident on the surface per unit time, area, frequency is

$$I_\nu(\mathbf{r},\hat{\mathbf{\Omega}}')\cos\theta'\,d\Omega'$$

where θ' is the angle between the direction of incidence and the normal to the surface, $\hat{\mathbf{n}}$. Let dq_ν be the amount of radiative energy absorbed by the surface per unit time, area, and frequency. Then the *spectral directional absorptivity* $\alpha_\nu(\mathbf{r},\hat{\mathbf{\Omega}}')$, is defined as

$$\alpha_\nu(\mathbf{r},\hat{\mathbf{\Omega}}') = \frac{dq_\nu}{I_\nu(\mathbf{r},\hat{\mathbf{\Omega}}')\cos\theta'\,d\Omega'} \tag{1-109}$$

If we assume that the surface is *opaque*, that is, the surface absorbs and reflects radiation but does not transmit it, we write

$$\begin{pmatrix}\text{Radiative}\\ \text{energy absorbed}\end{pmatrix} = \begin{pmatrix}\text{Radiative}\\ \text{energy incident}\end{pmatrix} - \begin{pmatrix}\text{Radiative}\\ \text{energy reflected}\end{pmatrix} \quad (1\text{-}110)$$

or

$$dq_\nu = I_\nu(\mathbf{r}, \hat{\mathbf{\Omega}}') \cos\theta'\, d\Omega' - \rho_\nu(\mathbf{r}, \hat{\mathbf{\Omega}}' \to 2\pi) I_\nu(\mathbf{r}, \hat{\mathbf{\Omega}}') \cos\theta'\, d\Omega' \quad (1\text{-}111)$$

Substituting Eq. 1-111 into Eq. 1-109, we obtain

$$\alpha_\nu(\mathbf{r}, \hat{\mathbf{\Omega}}') = 1 - \rho_\nu(\mathbf{r}, \hat{\mathbf{\Omega}}' \to 2\pi) \quad (1\text{-}112)$$

Hemispherical Absorptivity

The radiative energy incident on a surface per unit time, area, and frequency from all directions in the hemispherical space is given by

$$\int_{\Omega'=2\pi} I_\nu(\mathbf{r}, \hat{\mathbf{\Omega}}') \cos\theta'\, d\Omega' \quad (1\text{-}113)$$

The amount absorbed by the surface per unit time, area, and frequency is given by

$$\int_{\Omega'=2\pi} \alpha_\nu(\mathbf{r}, \hat{\mathbf{\Omega}}') I_\nu(\mathbf{r}, \hat{\mathbf{\Omega}}') \cos\theta'\, d\Omega' \quad (1\text{-}114)$$

which is obtained by integrating de_ν in Eq. 1-109 over the entire hemispherical space. Then the *spectral hemispherical absorptivity*, $\alpha_\nu(\mathbf{r})$, is defined as

$$\alpha_\nu(\mathbf{r}) = \frac{\displaystyle\int_{\Omega'=2\pi} \alpha_\nu(\mathbf{r}, \hat{\mathbf{\Omega}}') I_\nu(\mathbf{r}, \hat{\mathbf{\Omega}}') \cos\theta'\, d\Omega'}{\displaystyle\int_{\Omega'=2\pi} I_\nu(\mathbf{r}, \hat{\mathbf{\Omega}}') \cos\theta'\, d\Omega'} \quad (1\text{-}115)$$

For an *opaque* surface we substitute $\alpha_\nu(\mathbf{r}, \hat{\mathbf{\Omega}}')$ from Eq. 1-112 into Eq. 1-115 and obtain

$$\alpha_\nu(\mathbf{r}) = 1 - \frac{\displaystyle\int_{\Omega'=2\pi} \rho_\nu(\mathbf{r}, \hat{\mathbf{\Omega}}' \to 2\pi) I_\nu(\mathbf{r}, \hat{\mathbf{\Omega}}') \cos\theta'\, d\Omega'}{\displaystyle\int_{\Omega'=2\pi} I_\nu(\mathbf{r}, \hat{\mathbf{\Omega}}') \cos\theta'\, d\Omega'} \quad (1\text{-}116)$$

By Eq. 1-105 this relation is simplified to

$$\alpha_\nu(\mathbf{r}) = 1 - \rho_\nu(\mathbf{r}) \quad (1\text{-}117)$$

Emission of Radiation

The intensity of radiation emitted by a real surface at any frequency ν and temperature T is always less than that emitted by a black surface at the same frequency and temperature. The ratio of the energy flux emitted by a surface to that from a black body at the same temperature is called the *emissivity* of a surface; however, the emissivity may have different values, depending on the type of measurement made. We present now various definitions of the emissivity of a surface.

Spectral Directional Emissivity

If $I_\nu(\mathbf{r},\hat{\mathbf{\Omega}})$ is the intensity of monochromatic radiation emitted in the direction $\hat{\mathbf{\Omega}}$ at a frequency ν by a surface at temperature T, the *spectral directional emissivity* of the surface, $\varepsilon_\nu(\mathbf{r}, \hat{\mathbf{\Omega}})$, is defined as

$$\varepsilon_\nu(\mathbf{r}, \hat{\mathbf{\Omega}}) = \frac{I_\nu(\mathbf{r}, \hat{\mathbf{\Omega}})}{I_{\nu b}(T)} \tag{1-118}$$

where $I_{\nu b}(T)$ is the black-body radiation intensity at frequency ν and temperature T.

When the Kirchhoff law is applicable, the emission and absorption characteristics of a surface are considered the same, and we write

$$\varepsilon_\nu(\mathbf{r}, \hat{\mathbf{\Omega}}) = \alpha_\nu(\mathbf{r}, \hat{\mathbf{\Omega}}) \tag{1-119}$$

If we assume also that the surface is opaque, we have

$$\varepsilon_\nu(\mathbf{r}, \hat{\mathbf{\Omega}}) = \alpha_\nu(\mathbf{r},\hat{\mathbf{\Omega}}) = 1 - \rho_\nu(\mathbf{r}, \hat{\mathbf{\Omega}} \to 2\pi) \tag{1-120}$$

To obtain this relation we utilized Eq. 1-112.

Spectral Hemispherical Emissivity

The radiative energy emitted at frequency ν by a real surface at temperature T per unit time, area, and frequency in all directions in the hemispherical space is given as

$$\int_{\Omega=2\pi} I_\nu(\mathbf{r}, \hat{\mathbf{\Omega}}) \cos\theta \, d\Omega \tag{1-121}$$

If the surface were a black surface, this emission would be

$$\int_{\Omega=2\pi} I_{\nu b}(T) \cos\theta \, d\Omega \tag{1-122}$$

Then the *spectral hemispherical emissivity*, $\varepsilon_\nu(\mathbf{r})$, of the real surface is

$$\varepsilon_\nu(\mathbf{r}) = \frac{\int_{2\pi} I_\nu(\mathbf{r}, \hat{\mathbf{\Omega}}) \cos\theta \, d\Omega}{\int_{2\pi} I_{\nu b}(T) \cos\theta \, d\Omega} \tag{1-123a}$$

$$= \frac{1}{\pi} \int_{2\pi} \frac{I_\nu(\mathbf{r}, \hat{\mathbf{\Omega}})}{I_{\nu b}(T)} \cos\theta \, d\Omega \tag{1-123b}$$

since $I_{\nu b}(T)$ is independent of direction. By utilizing relation 1-118, we obtain

$$\varepsilon_\nu(\mathbf{r}) = \frac{1}{\pi}\int_{2\pi} \varepsilon_\nu(\mathbf{r}, \hat{\mathbf{\Omega}}) \cos\theta \, d\Omega = \frac{1}{\pi}\int_{\phi=0}^{2\pi}\int_{\mu=0}^{1} \varepsilon_\nu(\mathbf{r}, \mu, \phi)\mu \, d\mu \, d\phi \quad (1\text{-}124)$$

When the Kirchhoff law is valid, the emissivity is taken as equal to the absorptivity:

$$\varepsilon_\nu(\mathbf{r}) = \alpha_\nu(\mathbf{r}) \quad (1\text{-}125)$$

If we assume also that the surface is opaque, the emissivity is related to the reflectivity by

$$\varepsilon_\nu(\mathbf{r}) = \alpha_\nu(\mathbf{r}) = 1 - \rho_\nu(\mathbf{r}) \quad (1\text{-}126)$$

Radiation Leaving a Surface Element

The intensity of monochromatic radiation leaving an opaque surface element in any given direction $\hat{\mathbf{\Omega}}$ is composed of an emitted and a reflected component, that is,

$$\underset{\text{leaving}}{I_\nu(\mathbf{r}, \hat{\mathbf{\Omega}})} = \underset{\text{emitted}}{I_\nu(\mathbf{r}, \hat{\mathbf{\Omega}})} + \underset{\text{reflected}}{I_\nu(\mathbf{r}, \hat{\mathbf{\Omega}})} \quad (1\text{-}127)$$

This concept is illustrated in Fig. 1-15.

For a surface element irradiated from all directions in the hemispherical space, the intensity of the reflected radiation in any direction $\hat{\mathbf{\Omega}}$ can be obtained from Eq. 1-97. If the surface is at temperature T and has a spectral emissivity $\varepsilon_\nu(\mathbf{r}, \hat{\mathbf{\Omega}})$, the intensity of the emitted radiation is given by Eq. 1-118.

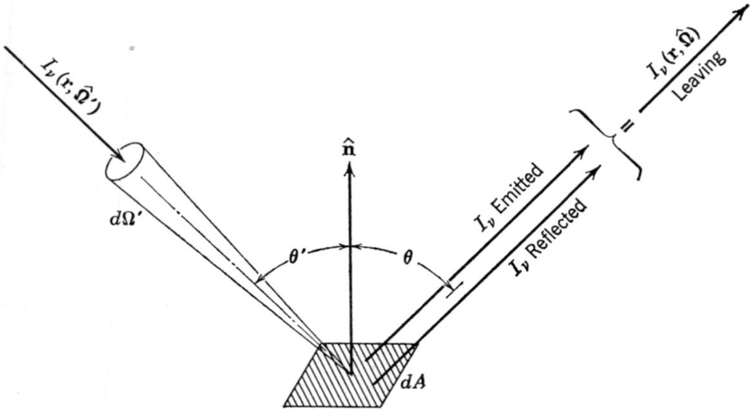

Fig. 1-15. Radiation leaving a surface element.

Then Eq. 1-127 becomes

$$\underset{\text{leaving}}{I_\nu(\mathbf{r}, \hat{\mathbf{\Omega}})} = \varepsilon_\nu(\mathbf{r}, \hat{\mathbf{\Omega}}) I_{\nu b}(T) + \int_{\Omega'=2\pi} f_\nu(\mathbf{r}, \hat{\mathbf{\Omega}}', \hat{\mathbf{\Omega}}) I_\nu(\mathbf{r}, \hat{\mathbf{\Omega}}') \cos\theta' \, d\Omega' \tag{1-128a}$$

or

$$\underset{\text{leaving}}{I_\nu(\mathbf{r}, \hat{\mathbf{\Omega}})} = \varepsilon_\nu(\mathbf{r}, \hat{\mathbf{\Omega}}) I_{\nu b}(T) + \int_{\phi'=0}^{2\pi} \int_{\mu'=0}^{1} f_\nu(\mathbf{r}; \mu', \phi'; \mu, \phi) I_\nu(\mathbf{r}, \mu', \phi') \mu' \, d\mu' \, d\phi' \tag{1-128b}$$

Let $R_\nu(\mathbf{r})$ denote the spectral radiative energy flux leaving a surface element normally in the direction $\hat{\mathbf{n}}$ per unit time, area, and frequency into the hemispherical space. Then $R_\nu(\mathbf{r})$ is given by the relation

$$R_\nu(\mathbf{r}) = \int_{\Omega=2\pi} \underset{\text{leaving}}{I_\nu(\mathbf{r}, \hat{\mathbf{\Omega}})} \cos\theta \, d\Omega = \int_{\phi=0}^{2\pi} \int_{\mu=0}^{1} \underset{\text{leaving}}{I_\nu(\mathbf{r}, \mu, \phi)} \mu \, d\mu \, d\phi \tag{1-129}$$

where $\mu = \cos\theta$, and θ is the angle between the direction $\hat{\mathbf{\Omega}}$ and the normal $\hat{\mathbf{n}}$ to the surface. By substituting Eq. 1-128 into Eq. 1-129 we obtain

$$\begin{aligned} R_\nu(\mathbf{r}) = {} & I_{\nu b}(T) \int_{\phi=0}^{2\pi} \int_{\mu=0}^{1} \varepsilon_\nu(\mathbf{r}, \mu, \phi) \mu \, d\mu \, d\phi \\ & + \int_{\phi=0}^{2\pi} \int_{\mu=0}^{1} \left[\int_{\phi'=0}^{2\pi} \int_{\mu'=0}^{1} f_\nu(\mathbf{r}; \mu', \phi'; \mu, \phi) I_\nu(\mathbf{r}, \mu', \phi') \mu' \, d\mu' \, d\phi' \right] \\ & \times \mu \, d\mu \, d\phi \end{aligned} \tag{1-130}$$

For a surface that is a diffuse emitter and diffuse reflector, both ε_ν and f_ν are independent of direction, integrations over μ and ϕ can be performed, and Eq. 1-130 simplifies to

$$R_\nu(\mathbf{r}) = \varepsilon_\nu(\mathbf{r}) \pi I_{\nu b}(T) + \rho_\nu(\mathbf{r}) \int_{\phi'=0}^{2\pi} \int_{\mu'=0}^{1} I_\nu(\mathbf{r}, \mu', \phi') \mu' \, d\mu' \, d\phi' \tag{1-131}$$

Here $\varepsilon_\nu(\mathbf{r})$ and $\rho_\nu(\mathbf{r})$ are the spectral hemispherical emissivity and spectral hemispherical reflectivity, respectively. We utilized Eq. 1-108 to relate f_ν to ρ_ν.

In engineering literature the radiative energy leaving a surface element normally per unit time, area, and frequency is called the *spectral radiosity* when the considered surface is a diffuse emitter and diffuse reflector. Therefore $R_\nu(\mathbf{r})$, as defined by Eq. 1-131 for the special case of diffuse emission and diffuse reflection, is the *spectral radiosity*.

The Net Radiative Heat Flux

Consider an opaque surface element dA characterized by a normal unit direction vector $\hat{\mathbf{n}}$ as illustrated in Fig. 1-16. The *spectral net radiative flux*, $q_{\nu n}(\mathbf{r})$, is defined as the net flow of spectral radiative energy normal to the

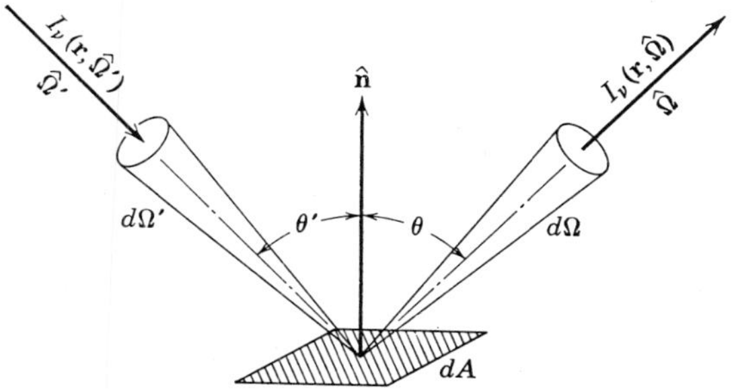

Fig. 1-16. Symbols for the definition of net radiative heat flux at a surface.

surface in the direction $\hat{\mathbf{n}}$ per unit time, area, and frequency into the hemispherical space. In an analogous manner to that given by Eq. 1-78, $q_{\nu n}(\mathbf{r})$ can be expressed as the difference between the two half-range fluxes in the form

$$q_{\nu n}(\mathbf{r}) = q_\nu^+(\mathbf{r}) - q_\nu^-(\mathbf{r}) \tag{1-132}$$

where we have defined the half-range fluxes as

$$q_\nu^+(\mathbf{r}) = \int_{\phi=0}^{2\pi} \int_{\mu=0}^{1} \underset{\text{leaving}}{I_\nu(\mathbf{r}, \mu, \phi)}\mu \, d\mu \, d\phi \tag{1-133a}$$

$$q_\nu^-(\mathbf{r}) = \int_{\phi'=0}^{2\pi} \int_{\mu'=0}^{1} \underset{\text{incident}}{I_\nu(\mathbf{r}, \mu', \phi')}\mu' \, d\mu' \, d\phi' \tag{1-133b}$$

Here we note that the half-range flux $q_\nu^+(\mathbf{r})$ is the same as the spectral radiative flux leaving a surface normally, $R_\nu(\mathbf{r})$, defined by Eq. 1-129. The half-range flux $q_\nu^-(\mathbf{r})$ represents the spectral radiative flux incident on the surface normally from all directions in the hemispherical space. Therefore the spectral net radiative flux $q_{\nu n}$ is the difference between the spectral radiative fluxes leaving and incident on the surface normally. When $q_{\nu n}$ is a positive quantity, the net flow of radiative energy is in the positive $\hat{\mathbf{n}}$ direction.

For an opaque surface, an alternative expression for the spectral net radiative flux $q_{\nu n}$ can be written as the difference between the spectral radiative energy emitted and that absorbed by the surface, that is,

$$q_{\nu n}(\mathbf{r}, t) = \begin{pmatrix} \text{Radiative energy} \\ \text{emitted per unit time,} \\ \text{area, frequency} \end{pmatrix} - \begin{pmatrix} \text{Radiative energy} \\ \text{absorbed per unit time,} \\ \text{area, frequency} \end{pmatrix} \tag{1-134a}$$

or

$$q_{\nu n}(\mathbf{r}) = I_{\nu b}(T) \int_{\phi=0}^{2\pi} \int_{\mu=0}^{1} \varepsilon_\nu(\mathbf{r}, \mu, \phi)\mu \, d\mu \, d\phi - \int_{\phi'=0}^{2\pi} \int_{\mu'=0}^{1} \alpha_\nu(\mathbf{r}, \mu', \phi') I_\nu(\mathbf{r}, \mu', \phi')\mu' \, d\mu' \, d\phi' \tag{1-134b}$$

By utilizing the relation of Eq. 1-124 we write

$$q_{\nu n}(\mathbf{r}) = \varepsilon_\nu(\mathbf{r})\pi I_{\nu b}(T) - \int_{\phi'=0}^{2\pi}\int_{\mu'=0}^{1} \alpha_\nu(\mathbf{r}, \mu', \phi') I_\nu(\mathbf{r}, \mu', \phi')\mu' \, d\mu' \, d\phi' \quad (1\text{-}135)$$

If we assume that the absorptivity is independent of direction, Eq. 1-135 simplifies to

$$q_{\nu n}(\mathbf{r}) = \varepsilon_\nu(\mathbf{r})\pi I_{\nu b}(T) - \alpha_\nu(\mathbf{r}) \int_{\phi'=0}^{2\pi}\int_{\mu'=0}^{1} I_\nu(\mathbf{r}, \mu', \phi')\mu' \, d\mu' \, d\phi' \quad (1\text{-}136)$$

When the Kirchhoff law is applicable, we set $\varepsilon_\nu(\mathbf{r}) = \alpha_\nu(\mathbf{r})$.

The Total Quantities

The total energy quantities are obtained by integrating the spectral energy quantities over the entire frequency spectrum from $\nu = 0$ to infinity. If Z_ν denotes a spectral energy quantity, the total quantity Z is given by

$$Z = \int_{\nu=0}^{\infty} Z_\nu \, d\nu \quad (1\text{-}137)$$

where $Z_\nu \equiv$ the spectral radiation intensity, radiative flux, radiosity, or radiation leaving a surface element.

The Average (Total) Radiative Properties

The average radiative property of a surface over the frequency spectrum from $\nu = 0$ to infinity is needed in most engineering applications. Since the spectral radiative property depends on the frequency of radiation, the averages are weighted with respect to the appropriate weight factor. For example, the spectral reflectivity and absorptivity depend on the frequency of the incident radiation; hence the appropriate weight factor in this case is the incident radiation. When absorptivity describes the emission of radiation, it depends on the frequency of the emitted radiation; then the black-body radiation intensity is taken as the weight factor. We summarize now the average radiative properties (or the so-called total radiative properties) of surfaces for the reflection, absorption, and emission of radiation.

Average Reflection Properties

The average value of the spectral reflection distribution function over the entire spectrum from $\nu = 0$ to infinity is defined as

$$f(\mathbf{r}, \hat{\mathbf{\Omega}}', \hat{\mathbf{\Omega}}) = \frac{\int_{\nu=0}^{\infty} f_\nu(\mathbf{r}, \hat{\mathbf{\Omega}}', \hat{\mathbf{\Omega}}) I_\nu(\mathbf{r}, \hat{\mathbf{\Omega}}') \, d\nu}{\int_{\nu=0}^{\infty} I_\nu(\mathbf{r}, \hat{\mathbf{\Omega}}') \, d\nu} \quad (1\text{-}138)$$

where the intensity of the incident radiation, $I_\nu(\mathbf{r}, \hat{\mathbf{\Omega}}')$, is taken as the weight factor.

The average directional-hemispherical reflectivity is given as

$$\rho(\mathbf{r}, \hat{\mathbf{\Omega}}' \to 2\pi) = \frac{\displaystyle\int_{\nu=0}^{\infty} \rho_\nu(\mathbf{r}, \hat{\mathbf{\Omega}}' \to 2\pi) I_\nu(\mathbf{r}, \hat{\mathbf{\Omega}}')\, d\nu}{\displaystyle\int_{\nu=0}^{\infty} I_\nu(\mathbf{r}, \hat{\mathbf{\Omega}}')\, d\nu} \tag{1-139}$$

and the average hemispherical-directional reflectivity as

$$\rho(\mathbf{r}, 2\pi \to \hat{\mathbf{\Omega}}) = \frac{\displaystyle\int_{\nu=0}^{\infty} \rho_\nu(\mathbf{r}, 2\pi \to \hat{\mathbf{\Omega}}) \left[\int_{2\pi} I_\nu(\mathbf{r}, \hat{\mathbf{\Omega}}') \cos\theta'\, d\Omega'\right] d\nu}{\displaystyle\int_{\nu=0}^{\infty} \left[\int_{2\pi} I_\nu(\mathbf{r}, \hat{\mathbf{\Omega}}') \cos\theta'\, d\Omega'\right] d\nu} \tag{1-140a}$$

For uniform incident radiation Eq. 1-140a simplifies to

$$\rho(\mathbf{r}, 2\pi \to \hat{\mathbf{\Omega}}) = \frac{\displaystyle\int_{\nu=0}^{\infty} \rho_\nu(\mathbf{r}, 2\pi \to \hat{\mathbf{\Omega}}) I_\nu(\mathbf{r})\, d\nu}{\displaystyle\int_{\nu=0}^{\infty} I_\nu(\mathbf{r})\, d\nu} \tag{1-140b}$$

The average hemispherical reflectivity is defined as

$$\rho(\mathbf{r}) = \frac{\displaystyle\int_{\nu=0}^{\infty} \rho_\nu(\mathbf{r}) \left[\int_{2\pi} I_\nu(\mathbf{r}, \hat{\mathbf{\Omega}}') \cos\theta'\, d\Omega'\right] d\nu}{\displaystyle\int_{\nu=0}^{\infty} \left[\int_{2\pi} I_\nu(\mathbf{r}, \hat{\mathbf{\Omega}}') \cos\theta'\, d\Omega'\right] d\nu} \tag{1-141}$$

Average Absorption Properties

The average directional absorptivity is given as

$$\alpha(\mathbf{r}, \hat{\mathbf{\Omega}}') = \frac{\displaystyle\int_{\nu=0}^{\infty} \alpha_\nu(\mathbf{r}, \hat{\mathbf{\Omega}}') I_\nu(\mathbf{r}, \hat{\mathbf{\Omega}}')\, d\nu}{\displaystyle\int_{\nu=0}^{\infty} I_\nu(\mathbf{r}, \hat{\mathbf{\Omega}}')\, d\nu} \tag{1-142}$$

and the average hemispherical absorptivity as

$$\alpha(\mathbf{r}) = \frac{\displaystyle\int_{\nu=0}^{\infty} \alpha_\nu(\mathbf{r}) \left[\int_{2\pi} I_\nu(\mathbf{r}, \hat{\mathbf{\Omega}}') \cos\theta'\, d\Omega'\right] d\nu}{\displaystyle\int_{\nu=0}^{\infty} \left[\int_{2\pi} I_\nu(\mathbf{r}, \hat{\mathbf{\Omega}}') \cos\theta'\, d\Omega'\right] d\nu} \tag{1-143}$$

(c) Average Emission Properties

The average directional emissivity is defined as

$$\varepsilon(\mathbf{r}, \hat{\mathbf{\Omega}}) = \frac{\int_{\nu=0}^{\infty} \varepsilon_\nu(\mathbf{r}, \hat{\mathbf{\Omega}}) I_{\nu b}(T)\, d\nu}{\int_{\nu=0}^{\infty} I_{\nu b}(T)\, d\nu} \tag{1-144a}$$

$$= \frac{\int_{\nu=0}^{\infty} \varepsilon_\nu(\mathbf{r}, \hat{\mathbf{\Omega}}) I_{\nu b}(T)\, d\nu}{I_b(T)} \tag{1-144b}$$

and the average hemispherical emissivity as

$$\varepsilon(\mathbf{r}) = \frac{\int_{\nu=0}^{\infty} \varepsilon_\nu(\mathbf{r}) \left[\int_{2\pi} I_{\nu b}(T) \cos\theta\, d\Omega \right] d\nu}{\int_{\nu=0}^{\infty} \left[\int_{2\pi} I_{\nu b}(T) \cos\theta\, d\Omega \right] d\nu} \tag{1-145a}$$

$$= \frac{\int_{\nu=0}^{\infty} \varepsilon_\nu(\mathbf{r}) I_{\nu b}(T)\, d\nu}{\int_{\nu=0}^{\infty} I_{\nu b}(T)\, d\nu} \tag{1-145b}$$

$$= \frac{\int_{\nu=0}^{\infty} \varepsilon_\nu(\mathbf{r}) I_{\nu b}(T)\, d\nu}{I_b(T)} \tag{1-145c}$$

When the Kirchhoff law is applicable, the absorptivity is used in place of the emissivity. When the absorptivity describes the emission of radiation, the average directional absorptivity and hemispherical absorptivity should be defined as in Eqs. 1-144 and 1-145, respectively, not as in Eqs. 1-142 and 1-143.

REFERENCES

1. M. Planck, *The Theory of Heat Radiation*, Dover Publications, New York, 1959.
2. E. A. Milne, "Thermodynamics of Stars," in *Handbuch der Astrophysik*, edited by G. Eberhard et al., Vol. 3, Part I, pp. 65–255, Springer-Verlag, Berlin, 1930.
3. P. C. Clemmow, *The Plane Wave Spectrum Representation of Electromagnetic Fields*, Pergamon Press, New York, 1966.
4. D. S. Jones, *The Theory of Electromagnetism*, Macmillan Co., New York, 1964.
5. J. A. Stratton, *Electromagnetic Theory*, McGraw-Hill Book Co., New York, 1941.

6. H. C. Van de Hulst, *Light Scattering by Small Particles*, John Wiley and Sons, New York, 1957.

7. E. M. Pugh and E. W. Pugh, *Principles of Electricity and Magnetism*, Addison-Wesley Publishing Co., Reading, Mass., 1960.

8. S. Chandrasekhar, *Radiative Transfer*, Oxford University Press, London, 1950; also Dover Publications, New York, 1960.

9. D. Deirmendjian, *Electromagnetic Scattering on Spherical Polydispersions*, American Elsevier Publishing Co., New York, 1969.

10. H. B. Dwight, *Tables of Integrals and Other Mathematical Data*, Macmillan Co., New York, 1961.

11. N. W. Snyder, "A Review of Thermal Radiation Constants," *Trans. ASME*, **76**, 537–540, 1954.

12. R. V. Dunkle, "Thermal Radiation Tables and Applications," *Trans. ASME*, **76**, 549–552, 1954.

13. M. Czerny and A. Walther, *Tables of Fractional Functions for the Planck Radiation Law*, Springer-Verlag, Berlin, 1961.

14. D. K. Edwards, "Radiative Characteristics of Materials," *J. Heat Transfer*, **91C**, 1–15, 1969.

15. V. Kourganoff, *Basic Methods in Transfer Problems*, Dover Publications, New York, 1963.

16. R. Viskanta, *Heat Transfer in Thermal Radiation Absorbing and Scattering Media*, ANL-6170, Argonne National Laboratory, Argonne, Ill., 1960.

17. R. Viskanta, "Radiation Transfer and Interaction of Convection with Radiation Heat Transfer," in *Advances in Heat Transfer*, edited by T. F. Irvine and J. P. Hartnett, Academic Press, New York, 1966.

18. E. Hopf, *Mathematical Problems of Radiative Equilibrium*, Cambridge University Press, London, 1934.

19. R. V. Dunkle, "Thermal Radiation Characteristics of Surfaces," in *Theory and Fundamental Research in Heat Transfer*, edited by J. A. Clark, pp. 1–31, Pergamon Press, New York, 1963.

20. J. T. Bevans and D. K. Edwards, "Radiation Exchange in Enclosures with Directional Wall Properties," *J. Heat Transfer*, **87C**, 388–396, 1965.

21. R. C. Birkebak and E. R. G. Eckert, "Effects of Roughness of Metal Surfaces on Angular Distribution of Monochromatic Reflected Radiation," *J. Heat Transfer*, **87C**, 85–94, 1965.

22. K. E. Torrance and E. M. Sparrow, Discussion in reference 21, pp. 93–94.

23. H. J. McNicholas, "Absolute Methods of Reflectometry," *J. Res. Natl. Bur. Std.*, **1**, 29–72, 1928.

24. E. M. Sparrow and R. D. Cess, *Radiation Transfer*, Brooks/Cole Publishing Co., Belmont, Calif., 1966.

25. R. Siegel and J. R. Howell, *Thermal Radiation Heat Transfer*, Vol. 1, NASA SP-164, U.S. Government Printing Office, Washington, D.C., 1968.

26. H. L. Helmholtz, *Physiological Optics*, 3rd ed., 1909; translated by J. P. C. Southall and published in *J. Opt. Soc. Am.*, **1**, 231, 1924.

NOTES

[1] By setting $\partial/\partial l = 0$, $\partial/\partial r = 0$, $H_z = 0$, $E_z = 0$, and $\sigma = 0$, Maxwell's equations 1-1a and 1-1b simplify respectively to

$$-\frac{\partial H_r}{\partial z} = \varepsilon \frac{\partial E_l}{\partial t} \quad (1a) \qquad -\frac{\partial E_r}{\partial z} = -\mu \frac{\partial H_l}{\partial t} \quad (2a)$$

$$\frac{\partial H_l}{\partial z} = \varepsilon \frac{\partial E_r}{\partial t} \quad (1b) \qquad \frac{\partial E_l}{\partial z} = -\mu \frac{\partial H_r}{\partial t} \quad (2b)$$

When Eqs. 1 are differentiated with respect to t, Eqs. 2 are differentiated with respect to z, and H_l and H_r are eliminated from the resulting equations, we obtain

$$\frac{\partial^2 E_r}{\partial z^2} = \mu\varepsilon \frac{\partial^2 E_r}{\partial t^2} \equiv \frac{1}{c^2} \frac{\partial^2 E_l}{\partial t^2} \quad (3a)$$

$$\frac{\partial^2 E_l}{\partial z^2} = \mu\varepsilon \frac{\partial^2 E_l}{\partial t^2} \equiv \frac{1}{c^2} \frac{\partial^2 E_l}{\partial t^2} \quad (3b)$$

where we have defined $c^2 \equiv 1/\mu\varepsilon$.

[2] If the mks system of units is used, the vacuum magnetic inductive capacity is given exactly by

$$\mu_0 = 4\pi \times 10^{-7} \text{ henry/meter}$$

Then the value of ε_0 for a vacuum can be evaluated from Eq. 1-3b since the speed of light in a vacuum is known.

[3] For a plane wave propagating in the direction z in a conducting medium, we set $\partial/\partial l = 0$, $\partial/\partial r = 0$, $H_z = 0$, and $E_z = 0$. Then Maxwell's equations 1-1a and 1-1b simplify respectively to

$$-\frac{\partial H_r}{\partial z} = \varepsilon \frac{\partial E_l}{\partial t} + \sigma E_l \quad (1a) \qquad -\frac{\partial E_r}{\partial z} = -\mu \frac{\partial H_l}{\partial t} \quad (2a)$$

$$\frac{\partial H_l}{\partial z} = \varepsilon \frac{\partial E_r}{\partial t} + \sigma E_r \quad (1b) \qquad \frac{\partial E_l}{\partial z} = -\mu \frac{\partial H_r}{\partial t} \quad (2b)$$

When H_l and H_r are eliminated we obtain

$$\frac{\partial^2 E_l}{\partial z^2} = \mu\varepsilon \frac{\partial^2 E_l}{\partial t^2} + \mu\sigma \frac{\partial E_l}{\partial t} \quad (3a)$$

$$\frac{\partial^2 E_r}{\partial z^2} = \mu\varepsilon \frac{\partial^2 E_r}{\partial t^2} + \mu\sigma \frac{\partial E_r}{\partial t} \quad (3b)$$

[4] In the present analysis we have defined the complex refractive index in the form $m = n - in'$. An alternative form, $m = n - in\kappa = n(1 - i\kappa)$, has also been used in the literature. Therefore, when the numerical values are determined for these coefficients from tabulated data in the literature, care must be exercised in selecting the definition used in the tables.

Consider Eqs. 2a and 2b of not 3:

$$\frac{\partial E_r}{\partial z} = \mu \frac{\partial H_l}{\partial t} \tag{1a}$$

$$\frac{\partial E_l}{\partial z} = -\mu \frac{\partial H_r}{\partial t} \tag{1b}$$

The solutions for E_l and E_r can be given in the form

$$E_l = a_l \exp \{i[(\omega t - k_0 mz) + \gamma_l]\} \tag{2a}$$

$$E_r = a_r \exp \{i[(\omega t - k_0 mz) + \gamma_r]\} \tag{2b}$$

and, substituting Eqs. 2 into Eqs. 1, we obtain

$$\mu \frac{\partial H_l}{\partial t} = -ik_0 m a_r \exp \{i[(\omega t - k_0 mz) + \gamma_r]\} \tag{3a}$$

$$\mu \frac{\partial H_r}{\partial t} = ik_0 m a_l \exp \{i[(\omega t - k_0 mz) + \gamma_l]\} \tag{3b}$$

Integrating Eqs. 3 with respect to t and omitting the constants of integration, we obtain

$$H_l = -\frac{k_0 m}{\omega\mu} E_r \tag{4a}$$

$$H_r = \frac{k_0 m}{\omega\mu} E_l \tag{4b}$$

Substituting Eqs. 4 into Eq. 1-20b, we obtain

$$\mathbf{S} = \hat{\mathbf{k}} \frac{k_0 m}{\omega\mu} [E_l^2 + E_r^2] = \hat{\mathbf{k}} \frac{m}{\mu c_0} [E_l^2 + E_r^2] \tag{5}$$

which is the result given by Eq. 1-20c.

[6] The square of the magnitude of a complex quantity Z is obtained by multiplying it by its complex conjugate Z^*, that is,

$$|Z|^2 = ZZ^*$$

[7] We write Eq. 1-28b in the form

$$\frac{A_r}{a_r} = \cos\psi \cos\delta + \sin\psi \sin\delta \tag{1}$$

By substituting in this equation $\cos\psi = A_l/a_l$ from Eq. 1-28a, hence $\sin\psi = [1 - (A_l/a_l)^2]^{1/2}$, we obtain

$$\left[1 - \left(\frac{A_l}{a_l}\right)^2\right]^{1/2} \sin\delta = \frac{A_r}{a_r} - \frac{A_l}{a_l}\cos\delta \tag{2}$$

When this equation is squared, the result in Eq. 1-29a is obtained.

[8] We write Eqs. 1-30c and 1-30d, respectively, as

$$\cos\psi = \frac{A_\xi - B_2 \sin\psi}{B_1} \quad \text{and} \quad \sin\psi = \frac{A_\eta - B_3 \cos\psi}{B_4} \tag{1}$$

A simultaneous solution of Eqs. 1 for sin ψ and cos ψ yields

$$\sin\psi = \frac{A_\eta B_1 - A_\xi B_3}{B_1 B_4 - B_2 B_3} \qquad \text{and} \qquad \cos\psi = \frac{A_\xi B_4 - A_\eta B_2}{B_1 B_4 - B_2 B_3} \tag{2}$$

By squaring Eqs. 1 and adding the resulting expressions, we obtain

$$\frac{(A_\xi - B_2 \sin\psi)^2}{B_1^2} + \frac{(A_\eta - B_3 \cos\psi)^2}{B_4^2} = 1 \tag{3}$$

Substituting Eqs. 2 into Eq. 3, we obtain after simplifying

$$A_\xi^2 \frac{B_3^2 + B_4^2}{B^2} + A_\eta^2 \frac{B_1^2 + B_2^2}{B^2} - 2A_\xi A_\eta \frac{B_1 B_3 + B_2 B_4}{B^2} = 1 \tag{4a}$$

where we have defined

$$B^2 \equiv (B_1 B_4 - B_2 B_3)^2 \tag{4b}$$

Now, if we set

$$B_1 B_3 + B_2 B_4 = 0 \tag{5}$$

Eq. 4 simplifies to

$$\frac{A_\xi^2}{a^2} + \frac{A_\eta^2}{b^2} = 1 \tag{6a}$$

where we have defined

$$a^2 \equiv \frac{B^2}{B_3^2 + B_4^2} \qquad \text{and} \qquad b^2 \equiv \frac{B^2}{B_1^2 + B_2^2} \tag{6b}$$

Equations 6 are the same as Eqs. 1-33.

[9] From Eqs. 1-33b we write

$$\frac{1}{a^2} + \frac{1}{b^2} = \frac{B_1^2 + B_2^2 + B_3^2 + B_4^2}{B^2} \tag{1}$$

By using the definitions of Eq. 1-31 we obtain

$$B_1^2 + B_2^2 + B_3^2 + B_4^2 = a_l^2 + a_r^2 \tag{2a}$$

and

$$B^2 = (B_1 B_4 - B_2 B_3)^2 = (a_l a_r \sin\delta)^2 \tag{2b}$$

Substituting Eqs. 2 into Eq. 1, we find

$$\frac{a^2 + b^2}{a^2 b^2} = \frac{a_l^2 + a_r^2}{(a_l a_r \sin\delta)^2} \tag{3}$$

Since we have $a^2 + b^2 = a_l^2 + a_r^2$, then from Eq. 3 we obtain

$$ab = \pm a_l a_r \sin\delta \tag{4}$$

[10] From Eqs. 1-23b, 1-23c, and 1-34 we obtain

$$\frac{U}{Q} = \frac{2a_l a_r \cos\delta}{a_l^2 - a_r^2} = \tan 2\chi \tag{1}$$

From Eqs. 1-23a, 1-23d, and 1-36c we find

$$\frac{V}{I} = \frac{2a_l a_r \sin\delta}{a_l^2 + a_r^2} = \sin 2\beta \qquad \text{(i.e., Eq. 1-37d)} \tag{2}$$

Substituting Eqs. 1 and 2 into Eq. 1-23i, we obtain

$$I^2 = Q^2 + Q^2 \tan^2 2\chi + I^2 \sin^2 2\beta$$

or

$$Q^2 = (I \cos 2\beta \cos 2\chi)^2$$

or

$$Q = I \cos 2\beta \cos 2\chi \qquad \text{(i.e., Eq. 1-37b)} \tag{3}$$

From Eqs. 1 and 3 we have

$$U = I \cos 2\beta \sin 2\chi \qquad \text{(i.e., Eq. 1-37c)} \tag{4}$$

[11] From the integral tables by Dwight [10, p. 231] we have

$$\int_{x=0}^{\infty} \frac{x^3}{e^x - 1}\, dx = \frac{\pi^4}{15}$$

[12] Consider the $f_{0-\lambda}(T)$ function as defined by Eq. 1-52:

$$\int_0^{\lambda} q'_{\lambda b}(T)\, d\lambda' = f_{0-\lambda}(T) q_b(T) \tag{1}$$

Differentiating this relation with respect to e_b for λ constant, we obtain

$$\int_0^{\lambda} \left[\frac{\partial q_{\lambda b}(T)}{\partial q_b(T)}\right]_{\lambda} d\lambda = f_{0-\lambda}(T) + q_b(T) \left.\frac{\partial f_{0-\lambda}(T)}{\partial q_b(T)}\right|_{\lambda} \tag{2}$$

The left-hand side of Eq. 2 is $f^*_{0-\lambda}(T)$. Then

$$f^*_{0-\lambda}(T) = f_{0-\lambda}(T) + q_b(T) \left.\frac{\partial f_{0-\lambda}(T)}{\partial q_b(T)}\right|_{\lambda} \tag{3}$$

Since

$$q_b(T) = \bar{\sigma} T^4 \tag{4a}$$

$$\partial q_b(T) = 4\bar{\sigma} T^3\, \partial T \tag{4b}$$

$$\frac{\partial q_b(T)}{q_b(T)} = 4 \frac{\partial T}{T} \tag{4c}$$

Equation 3 becomes

$$f^*_{\lambda-0}(T) = f_{0-\lambda}(T) + \tfrac{1}{4} T \left.\frac{\partial f_{0-\lambda}(T)}{\partial T}\right|_{\lambda} = f_{0-\lambda}(T) + \tfrac{1}{4}\lambda T \left.\frac{\partial f_{0-\lambda}(T)}{\partial(\lambda T)}\right|_{\lambda} \tag{5}$$

[13] The volumetric absorption coefficient $\kappa_\nu(\mathbf{r})$ is related to the mass absorption coefficient $\kappa_{\nu m}(\mathbf{r})$ used in references 8 and 15 by the relation

$$\kappa_\nu(\mathbf{r}) = \rho \kappa_{\nu m}(\mathbf{r})$$

where ρ is the density of the material.

[14] The Kirchhoff law is concerned with the emission and absorption of radiation in matter in thermodynamic equilibrium. It states that the emission of radiation $J_\nu{}^e(\mathbf{r})$ in any direction in the matter in thermodynamic equilibrium is related to the spectral absorption coefficient $\kappa_\nu(\mathbf{r})$ and the black-body radiation intensity $I_{\nu b}[T]$ at the temperature $T(\mathbf{r})$ of the matter as follows:

$$J_\nu{}^e(\mathbf{r}) = \kappa_\nu(\mathbf{r}) I_{\nu b}[T(\mathbf{r})]$$

[15] The volumetric scattering coefficient $\sigma_\nu(\mathbf{r})$ is related to the mass scattering coefficient $\sigma_{\nu m}(\mathbf{r})$ used in references 8 and 15 by the expression

$$\sigma_\nu(\mathbf{r}) = \rho\sigma_{\nu m}(\mathbf{r})$$

where ρ is the density of the material.

[16] The terminology *incident radiation* is used in the references on engineering applications. This quantity divided by 4π is referred to as *mean intensity* in the books on astrophysics.

[17] Some authors use the suffix "-ivity" (i.e., reflectivity, absorptivity, emissivity) to characterize the radiation properties of *ideal surfaces* (i.e., those optically smooth and perfectly uncontaminated) and the suffix "-ance" (i.e., reflectance, absorptance, emittance) to characterize the radiation properties of *real surfaces*. We believe that there is no need for a change in nomenclature when the surface condition is changed and prefer to use the suffix "-ivity" for all cases.

[18] The *reflection distribution function* as defined by Eq. 1-91 is called *biangular reflectance* in references 19, 22, and 24, and *bidirectional reflectivity* in reference 25. We prefer the nomenclature reflection distribution function because f_ν, as defined by Eq. 1-91, can be larger than unity, whereas reflectivity, defined in the usual sense, is understood to be equal to or less than unity.

[19] The *directional-hemispherical reflectivity* as defined above is called *angular-hemispherical reflectance* in references 21, 22, and 24; *angular reflectance* in reference 19; and *directional reflectance* in reference 20.

CHAPTER 2

Radiative Properties of Materials

When an electromagnetic plane wave is incident upon the interface between two media, it will be reflected and refracted. If the interface is an *ideal surface* (i.e., a surface that is optically smooth and perfectly clean), the reflectivity and the absorptivity of the interface can be predicted from the electromagnetic theory. However, for *real surfaces* encountered in engineering applications there is large deviation between the measured and the predicted values of reflectivity and absorptivity.

When an electromagnetic wave travels through a medium containing inhomogeneities, say a gas with suspended particles, the wave will be attenuated by both the gas and the particles in suspension. Several models are available in the literature to determine the absorption of radiation by gases. The absorption and scattering properties of suspended particles having simple geometries, such as a sphere, can be predicted from the electromagnetic theory.

In this chapter we discuss the prediction from the electromagnetic theory of the radiative properties of ideal surfaces, present results on the absorption and scattering characteristics of spherical particles, and describe various models for the absorption and emission of radiation by gases.

2-1 LAWS OF REFLECTION AND REFRACTION

Consider that a beam of an electromagnetic plane wave traveling in medium 1 in the direction $\hat{\mathbf{\Omega}}_1$ is incident upon the interface between media 1 and 2 at

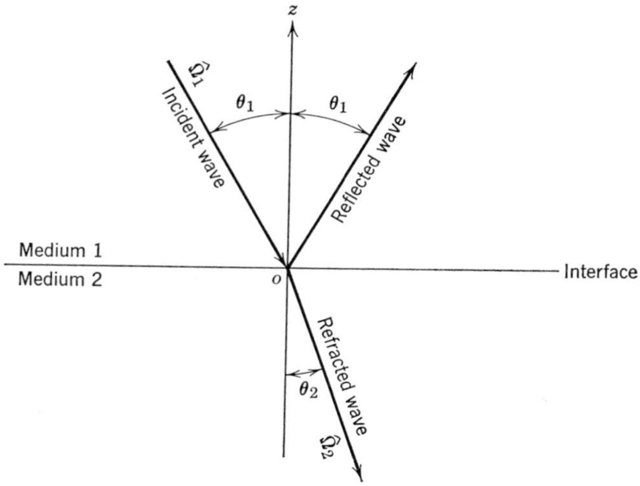

Fig. 2-1. Reflection and refraction of incident plane wave at the interface between two media when the interface is an ideal surface.

an *angle of incidence* θ_1 (i.e., the acute angle between the direction of propagation $\hat{\mathbf{\Omega}}_1$ and the normal to the interface). Part of this radiation will be reflected, and the rest will propagate in medium 2 in the direction $\hat{\mathbf{\Omega}}_2$ at an *angle of refraction* θ_2 (i.e., the acute angle between the direction $\hat{\mathbf{\Omega}}_2$ and the normal to the interface). Figure 2-1 illustrates the angles of incidence θ_1 and refraction θ_2. If the interface is an ideal surface, the laws of reflection and refraction can be determined by utilizing the results obtained from Maxwell's equations. The derivations of these laws can be found in standard textbooks on optics and electromagnetic theory, for example, those by Drude [1] and Jones [2].

The first of these laws is the *law of reflection*, which states that the incident and the reflected beams lie symmetrical with respect to the normal at the point of incidence; that is, the reflection is *specular*.

The second one is the *law of refraction*, which relates the angle of refraction θ_2 to the angle of incidence θ_1. When both media 1 and 2 are dielectric, having refractive indices n_1 and n_2, respectively, angles θ_1 and θ_2 are related by

$$\frac{\sin \theta_2}{\sin \theta_1} = \frac{n_1}{n_2} \tag{2-1}$$

This relation is called *Snell's law of refraction*.

When the two media are electric conductors having complex refractive indices m_1 and m_2, Snell's law of refraction takes the form

$$\frac{\sin\theta_2}{\sin\theta_1} = \frac{m_1}{m_2} = \frac{n_1 - in_1'}{n_2 - in_2'} \tag{2-2}$$

Here we note that $\sin\theta_2$ is a complex quantity since m_1 and m_2 are complex quantities. This complex ratio characterizes a change in both amplitude and phase of the wave propagating in medium 2.

2-2 PREDICTION OF RADIATIVE PROPERTIES OF IDEAL SURFACES

Consider that a beam of an electromagnetic plane wave propagating in medium 1 in the direction $\hat{\boldsymbol{\Omega}}_1$ is incident upon the interface between media 1 and 2 as illustrated in Fig. 2-2. Let the incident and the reflected electric field vectors each be resolved into two components of polarization, one perpendicular to the *plane of incidence* (i.e., the plane that includes the direction of incidence and the normal to the interface at the point of incidence) and the other parallel to the plane of incidence. The parallel and perpendicular components of the incident beam will be denoted by $E_{i,\parallel}$ and $E_{i,\perp}$, and those of the reflected beam by $E_{r,\parallel}$ and $E_{r,\perp}$, respectively. The components of the reflected beam are related to the components of the incident beam by

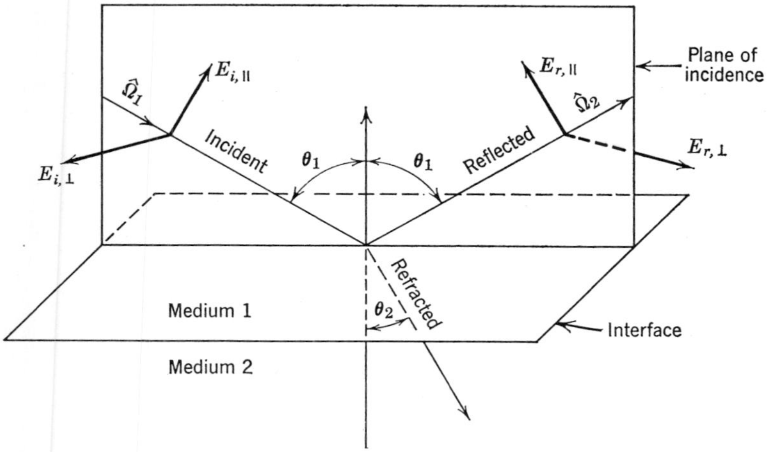

Fig. 2-2. Two components of polarization of the incident and reflected rays.

Fresnel's reflection equations [1, pp. 282–283; 2, pp. 319–320] and are given as

$$\frac{E_{r,\perp}}{E_{i,\perp}} = -\frac{\sin(\theta_1 - \theta_2)}{\sin(\theta_1 + \theta_2)} \tag{2-3a}$$

$$\frac{E_{r,\parallel}}{E_{i,\parallel}} = \frac{\tan(\theta_1 - \theta_2)}{\tan(\theta_1 + \theta_2)} \tag{2-3b}$$

where θ_1 and θ_2 are the angles of incidence and refraction, respectively.

By utilizing the relations given by Eqs. 2-2, Fresnel's relations can be written in the alternative forms as

$$\frac{E_{r,\perp}}{E_{i,\perp}} = -\frac{(\cos\theta_2/\cos\theta_1) - (m_1/m_2)}{(\cos\theta_2/\cos\theta_1) + (m_1/m_2)} \tag{2-4a}$$

$$\frac{E_{r,\parallel}}{E_{i,\parallel}} = \frac{(\cos\theta_1/\cos\theta_2) - (m_1/m_2)}{(\cos\theta_1/\cos\theta_2) + (m_1/m_2)} \tag{2-4b}$$

where

$$m_1 \equiv n_1 - in_1' \text{ and } m_2 \equiv n_2 - in_2'. \tag{2-4c}$$

Recalling that the energy transported by the electromagnetic plane wave is proportional to the square of the components of the electric field vector (see Eq. 1-20c), we define the spectral directional reflectivities for the perpendicular and parallel components of the incident radiation as

$$\bar{\rho}_{\nu,\perp} = \left(\frac{E_{r,\perp}}{E_{i,\perp}}\right)^2 \tag{2-5a}$$

$$\bar{\rho}_{\nu,\parallel} = \left(\frac{E_{r,\parallel}}{E_{i,\parallel}}\right)^2 \tag{2-5b}$$

The bar denotes that the quantity is a complex number. Substituting Eqs. 2-3 into Eqs. 2-5, we obtain

$$\bar{\rho}_{\nu,\perp} = \left[\frac{\sin(\theta_1 - \theta_2)}{\sin(\theta_1 + \theta_2)}\right]^2 \tag{2-6a}$$

$$\bar{\rho}_{\nu,\parallel} = \left[\frac{\tan(\theta_1 - \theta_2)}{\tan(\theta_1 + \theta_2)}\right]^2 \tag{2-6b}$$

Alternatively, by substituting Eq. 2-4 into Eq. 2-5, we obtain

$$\bar{\rho}_{\nu,\perp} = \left[\frac{(\cos\theta_2/\cos\theta_1) - (m_1/m_2)}{(\cos\theta_2/\cos\theta_1) + (m_1/m_2)}\right]^2 \tag{2-7a}$$

$$\bar{\rho}_{\nu,\parallel} = \left[\frac{(\cos\theta_1/\cos\theta_2) - (m_1/m_2)}{(\cos\theta_1/\cos\theta_2) + (m_1/m_2)}\right]^2 \tag{2-7b}$$

where m_1 and m_2 were defined by Eq. 2-4c.

For unpolarized radiation the parallel and perpendicular components of the incident radiation are of equal intensity. Then the reflectivity for unpolarized radiation is taken as the average of $\bar{\rho}_{\nu,\perp}$ and $\bar{\rho}_{\nu,\parallel}$, that is,

$$\bar{\rho}_\nu = \tfrac{1}{2}(\bar{\rho}_{\nu,\perp} + \bar{\rho}_{\nu,\parallel}) = \frac{1}{2}\left[\frac{\sin^2(\theta_1 - \theta_2)}{\sin^2(\theta_1 + \theta_2)} + \frac{\tan^2(\theta_1 - \theta_2)}{\tan^2(\theta_1 + \theta_2)}\right] \tag{2-8}$$

The foregoing relations for reflectivities are complex quantities unless both media are dielectric. When they are complex, their absolute magnitudes describe the reflectivity.

We now examine the application of the above relations to predict the reflectivity when (a) both media are dielectric, and (b) medium 1 is dielectric and medium 2 is a conductor.

Reflectivity When Both Media Are Dielectric

Consider two dielectric media 1 and 2 characterized by the real refractive indices n_1 and n_2, respectively. The angle of refraction θ_2 is given by (see Eq. 2-1)

$$\sin\theta_2 = \frac{n_1}{n_2}\sin\theta_1 \equiv \frac{1}{n}\sin\theta_1 \tag{2-9a}$$

where we have defined

$$n \equiv \frac{n_2}{n_1} \tag{2-9b}$$

Here n is called the refractive index of medium 2 relative to that of medium 1, or simply the *relative refractive index.*

Substituting Eq. 2-9a into Eqs. 2-6 and eliminating θ_2, we obtain the spectral reflectivity components for the perpendicular and parallel polarization, respectively, as

$$\rho^s_{\nu,\perp}(\theta_1) = \left(\frac{\sqrt{n^2 - \sin^2\theta_1} - \cos\theta_1}{\sqrt{n^2 - \sin^2\theta_1} + \cos\theta_1}\right)^2 \tag{2-10a}$$

$$\rho^s_{\nu,\parallel}(\theta_1) = \rho^s_{\nu,\perp}(\theta_1)\left(\frac{\sqrt{n^2 - \sin^2\theta_1} - \sin\theta_1\tan\theta_1}{\sqrt{n^2 - \sin^2\theta_1} + \sin\theta_1\tan\theta_1}\right)^2 \tag{2-10b}$$

$$= \left(\frac{n^2\cos\theta_1 - \sqrt{n^2 - \sin^2\theta_1}}{n^2\cos\theta_1 + \sqrt{n^2 - \sin^2\theta_1}}\right)^2 \tag{2-10c}$$

where θ_1 is the angle of incidence, and the superscript s denotes specular reflectivity.

The reflectivities of Eq. 2-10 are real quantities since refractive indices for both media are real and are applicable at the interface for radiation traveling from medium 1 into medium 2.

At normal incidence ($\theta_1 = 0$), Eqs. 2-10 simplify to

$$\rho_{\nu,\perp}^{s}(0) = \rho_{\nu,\parallel}^{s}(0) = \left(\frac{n-1}{n+1}\right)^2 \tag{2-11}$$

and at $\theta_1 = \pi/2$ reduce to

$$\rho_{\nu,\perp}^{s}\left(\frac{\pi}{2}\right) = \rho_{\nu,\parallel}^{s}\left(\frac{\pi}{2}\right) = 1 \tag{2-12}$$

In most engineering applications the reflectivity of a material in air is of interest. In such cases medium 1 is chosen as air, and we set $n_1 = 1$.

Generally the *thermal radiation is unpolarized.* The reflectivity for unpolarized radiation, $\rho_\nu^{s}(\theta_1)$, is obtained by taking the arithmetic average of the reflectivity components, $\rho_{\nu,\perp}^{s}(\theta_1)$ and $\rho_{\nu,\parallel}^{s}(\theta_1)$. We obtain

$$\rho_\nu^{s}(\theta_1) = \tfrac{1}{2}[\rho_{\nu,\perp}^{s}(\theta_1) + \rho_{\nu,\parallel}^{s}(\theta_1)] \tag{2-13}$$

For normal incidence Eq. 2-13 simplifies to

$$\rho_\nu^{s}(0) = \left(\frac{n-1}{n+1}\right)^2 \tag{2-14}$$

For unpolarized radiation of uniform intensity incident upon a surface from all directions in the hemispherical space, the spectral hemispherical reflectivity ρ_ν is evaluated from (see Eq. 1-106b)

$$\rho_\nu = \frac{1}{\pi}\int_{\phi=0}^{2\pi}\int_{\mu=0}^{1} \rho_\nu^{s}(\mu)\mu\, d\mu\, d\phi = 2\int_{\mu=0}^{1} \rho_\nu^{s}(\mu)\mu\, d\mu \tag{2-15}$$

where $\mu \equiv \cos\theta_1$, and $\rho_\nu^{s}(\mu)$ is given by Eqs. 2-10 and 2-13. The integration in Eq. 2-15 was evaluated by Walsh [3], and the resulting spectral hemispherical reflectivity of an ideal surface for dielectric media was expressed by Dunkle [4] in the form

$$\rho_\nu = \frac{1}{2} + \frac{(n-1)(3n+1)}{6(n+1)^2} - \frac{2n^3(n^2+2n-1)}{(n^2+1)(n^4-1)} + \frac{8n^4(n^4+1)}{(n^2+1)(n^4-1)^2}\ln(n) + \frac{n^2(n^2-1)^2}{(n^2+1)^3}\ln\left(\frac{n-1}{n+1}\right) \tag{2-16}$$

Figure 2-3 shows the reflectivity components $\rho_{\nu,\perp}^{s}(\theta_1)$ and $\rho_{\nu,\parallel}^{s}(\theta_1)$ of Eqs. 2-10 and their arithmetic mean value $\rho_\nu^{s}(\theta_1)$ of Eq. 2-13 as a function of the incidence angle for a dielectric medium having a relative refractive index $n_2/n_1 \equiv n = 1.5$.

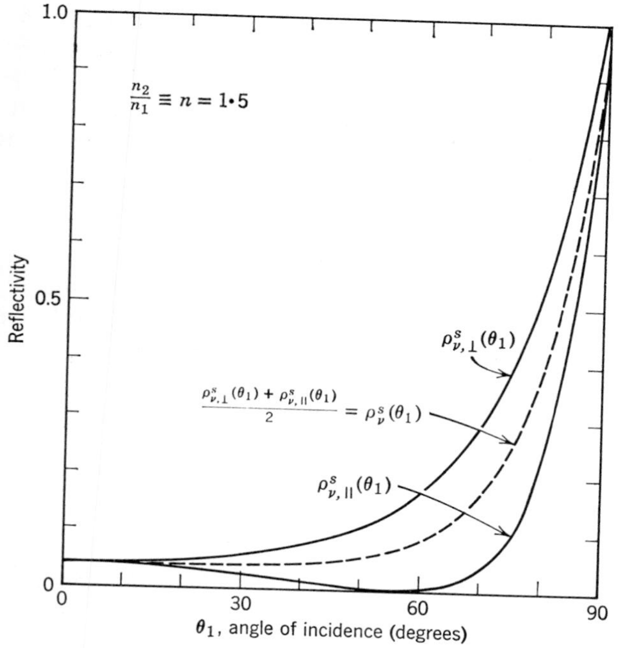

Fig. 2-3. Reflectivities $\rho^s_{\nu,\perp}(\theta_1)$ and $\rho_{\nu,\parallel}(\theta_1)$ calculated from electromagnetic theory. Both media are dielectric.

Reflectivity When Medium 1 Is Dielectric, Medium 2 Conductor

Consider an interface between two media, 1 and 2, which are characterized by a real refractive index n_1 and a complex refractive index $m_2 = n_2 - in_2'$, respectively. For an electromagnetic plane wave propagating from medium 1 into medium 2, Snell's law becomes (see Eq. 2-2)

$$\frac{\sin\theta_2}{\sin\theta_1} = \frac{n_1}{n_2 - in_2'} \tag{2-17a}$$

or

$$\sin\theta_2 = \frac{1}{m}\sin\theta_1 \tag{2-17b}$$

where we have defined the *relative complex refractive index* of medium 2 as

$$m \equiv \frac{n_2}{n_1} - i\,\frac{n_2'}{n_1} \tag{2-17c}$$

Here we note that the relation for $\sin\theta_2$ given by Eq. 2-17b is of exactly the same form as that given by Eq. 2-9a, except that in the former m is a complex quantity. When Eq. 2-17b is substituted into Eqs. 2-6 and θ_2 is eliminated, the resulting expressions for the reflectivity components will be of the same form as those given by Eqs. 2-10, but they will be complex quantities since m is complex. The absolute magnitude of these complex quantities, which represent the reflectivities, can be determined readily by introducing the identity defined by König [5]:

$$a - ib \equiv (m^2 - \sin^2\theta_1)^{1/2} \tag{2-18}$$

where a and b are real numbers and are yet to be determined. In this case the relations for the complex reflectivity components $\bar{\rho}_{\nu,\perp}(\theta_1)$ and $\bar{\rho}_{\nu,\parallel}(\theta_1)$ can be obtained from Eqs. 2-10a and 2-10b, respectively, by replacing in these equations n by m, and then by introducing the identity of Eq. 2-18. We obtain

$$\bar{\rho}_{\nu,\perp}(\theta_1) = \left[\frac{(a - ib) - \cos\theta_1}{(a - ib) + \cos\theta_1}\right]^2 \tag{2-19a}$$

$$\bar{\rho}_{\nu,\parallel}(\theta_1) = \bar{\rho}_{\nu,\perp}(\theta_1)\left[\frac{(a - ib) - \sin\theta_1\tan\theta_1}{(a - ib) + \sin\theta_1\tan\theta_1}\right]^2 \tag{2-19b}$$

Here the bar denotes the complex quantities.

The absolute values of the complex expressions in Eqs. 2-19 yield the reflectivity components for the perpendicular and parallel polarization, respectively, as[1]

$$\rho^{s}_{\nu,\perp}(\theta_1) = \frac{(a - \cos\theta_1)^2 + b^2}{(a + \cos\theta_1)^2 + b^2} \tag{2-20a}$$

$$\rho^{s}_{\nu,\parallel}(\theta_1) = \rho^{s}_{\nu,\perp}(\theta_1)\,\frac{(a - \sin\theta_1\tan\theta_1)^2 + b^2}{(a + \sin\theta_1\tan\theta_1)^2 + b^2} \tag{2-20b}$$

The parameters a and b can be determined by utilizing the identity of Eq. 2-18. By squaring both sides of Eq. 2-18 and then equating the real and imaginary parts we obtain

$$a^2 - b^2 = n^2 - n'^2 - \sin^2\theta_1 \tag{2-21a}$$

$$ab = nn' \tag{2-21b}$$

where we have defined

$$n \equiv \frac{n_2}{n_1} \quad \text{and} \quad n' \equiv \frac{n_2'}{n_1} \tag{2-22}$$

A simultaneous solution of Eqs. 2-21 yields

$$a^2 = \tfrac{1}{2}\left[(n^2 - n'^2 - \sin^2\theta_1) + \sqrt{(n^2 - n'^2 - \sin^2\theta_1)^2 + 4n^2n'^2}\right] \tag{2-23a}$$

$$b^2 = \tfrac{1}{2}\left[-(n^2 - n'^2 - \sin^2\theta_1) + \sqrt{(n^2 - n'^2 - \sin^2\theta_1)^2 + 4n^2n'^2}\right] \tag{2-23b}$$

where θ_1 is the angle of incidence. When the reflectivity of a material in air is of interest, we choose medium 1 as air and set $n_1 = 1$.

When medium 2 is dielectric, we set $n' = 0$; then Eqs. 2-23 simplify to

$$a^2 = n^2 - \sin^2 \theta_1 \tag{2-24a}$$

$$b^2 = 0 \tag{2-24b}$$

Substituting a and b from Eq. 2-24 into Eqs. 2-20, we obtain the reflectivities of Eqs. 2-10 for dielectric media, as expected.

The reflectivities $\rho^s_{v,\perp}(\theta_1)$ and $\rho^s_{v,\parallel}(\theta_1)$ of Eqs. 2-20 have been tabulated by Holl [6] for all combinations of $n = 0.1$ to 4.0 (with 0.1) and $n' = 0.1$ to 6.0 (with 0.1), and for angles of incidence $\theta_1 = 0°$ to 85° (with 5°). Figure 2-4 shows the reflectivities $\rho^s_{v,\perp}(\theta_1)$ and $\rho^s_{v,\parallel}(\theta_1)$ for $n = 4$ with n' varied

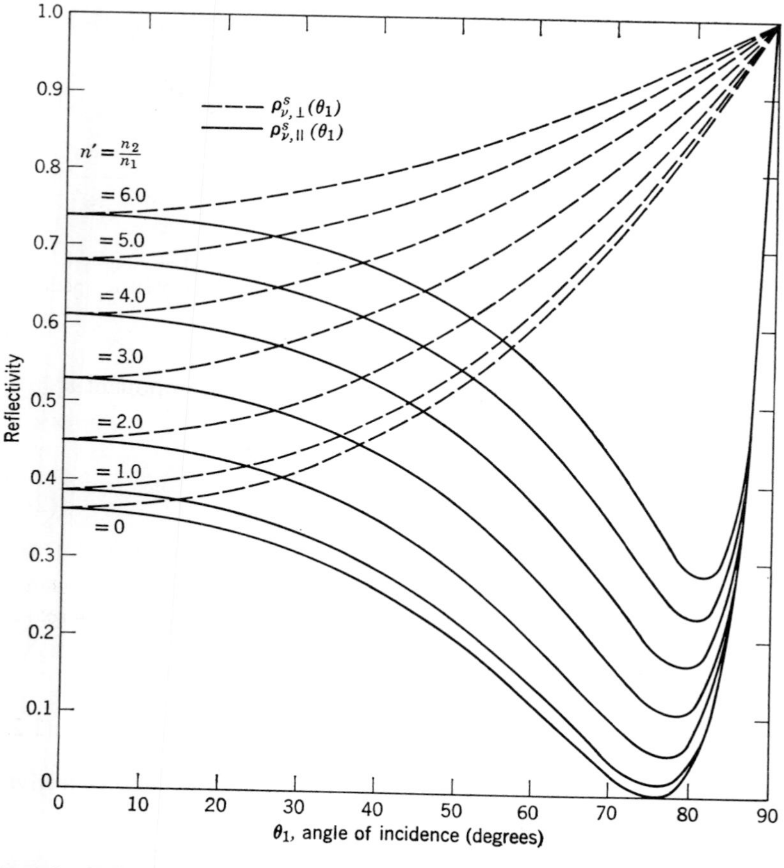

Fig. 2-4. Reflectivities $\rho^s_{v,\perp}(\theta_1)$ and $\rho^s_{v,\parallel}(\theta_1)$ calculated from electromagnetic theory for $m = 4 - in'$ when medium 1 is dielectric, medium 2 is conductor. (From H. B. Holl [7]).

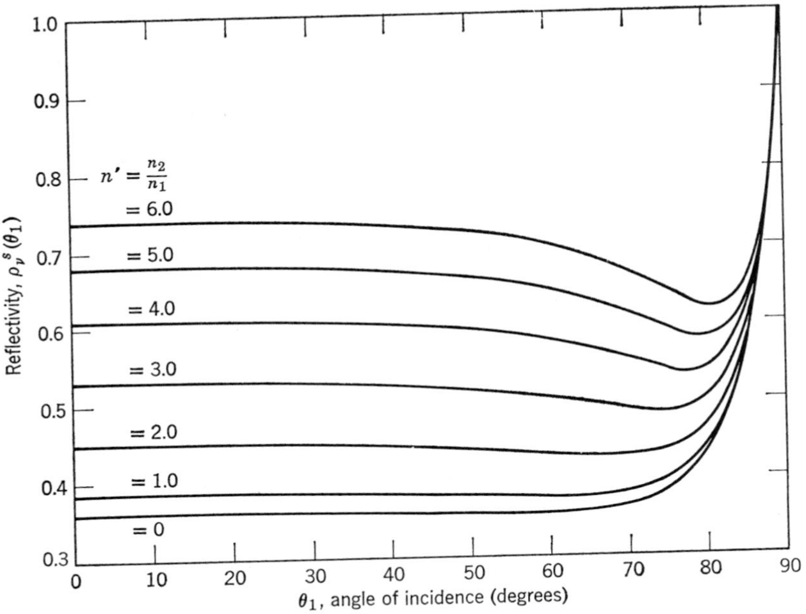

Fig. 2-5. Reflectivity for unpolarized radiation, $\rho_\nu(\theta_1) = \frac{1}{2}[\rho^s_{\nu,\perp}(\theta_1) + \rho^s_{\nu,\parallel}(\theta_1)]$, calculated from electromagnetic theory for $m = 4 - in'$ when medium 1 is dielectric, medium 2 is conductor. (From H. B. Holl [7].)

from 0 to 6. We note from this figure that the reflectivity component $\rho^s_{\nu,\parallel}(\theta_1)$ displays a dip at large angles of incidence.

For unpolarized radiation the reflectivity is given by

$$\rho_\nu{}^s(\theta_1) = \tfrac{1}{2}[\rho^s_{\nu,\perp}(\theta_1) + \rho^s_{\nu,\parallel}(\theta_1)] \tag{2-25}$$

Figure 2-5 shows the reflectivity $\rho_\nu{}^s(\theta_1)$ based on the data of Fig. 2-4. It is apparent from this figure that the reflectivity displays a dip at large angles of incidence, and that the dip becomes steeper as the value of n increases; for dielectric media (i.e., $n' = 0$) the dip is barely visible.

For normal incidence (i.e., $\theta_1 = 0$), Eqs. 2-23 simplify to

$$a^2 = n^2 \qquad \text{and} \qquad b^2 = n'^2 \tag{2-26}$$

When these values of a and b are substituted into Eqs. 2-20, the reflectivity components for normal incidence become

$$\rho_{\nu,\perp}(0) = \rho_{\nu,\parallel}(0) = \frac{(n-1)^2 + n'^2}{(n+1)^2 + n'^2} \tag{2-27}$$

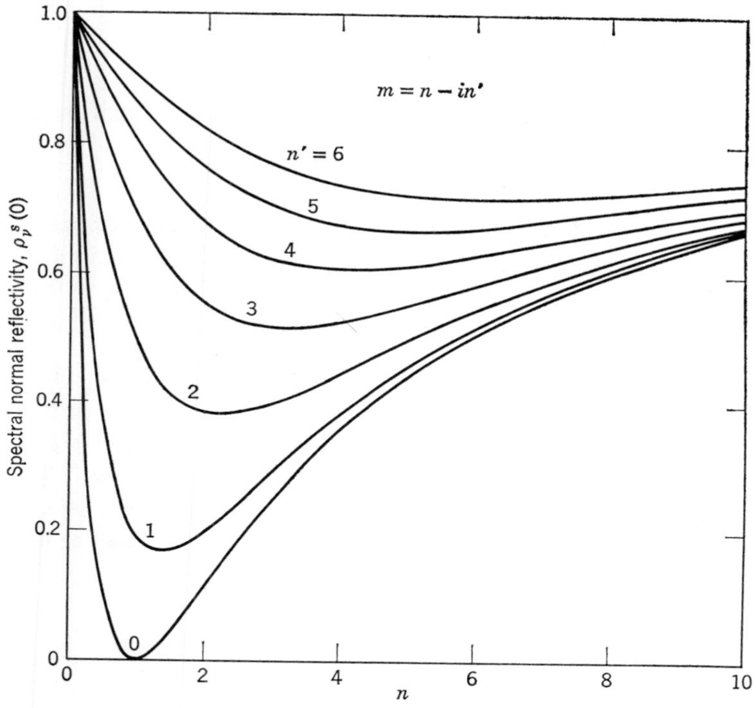

Fig. 2-6. Spectral normal reflectivity predicted from electromagnetic theory.

For a dielectric medium $n' = 0$; then Eqs. 2-27 simplifies to the relation given by Eq. 2-14, as expected.

Figure 2-6 shows a plot of Eq. 2-27 as a function of n for several different values of $n' = 0$ to 6. The reflectivity given in this figure also represents the spectral reflectivity of an ideal surface for unpolarized radiation at normal incidence, since $\rho_\nu^s(0)$ is equal to the arithmetic mean of $\rho^s_{\nu,\perp}(0)$and $\rho^s_{\nu,\parallel}(0)$ as given by Eq. 2-25.

Metals in Air

For the reflectivity of metals in air the foregoing relations can be simplified because for metals n_2 and n_2' are quite large at wavelengths greater than about 0.5 μ, and $n_1 \simeq 1$. We write Snell's law (i.e., Eq. 2-2) as

$$\sin \theta_2 = \frac{\sin \theta_1}{(n_2/n_1) - i(n_2'/n_1)} \cong \frac{\sin \theta_1}{n_2 - in_2'} \tag{2-28}$$

since $n_1 \simeq 1$ and $n_1' = 0$ for air. The absolute value of this complex quantity is given by

$$|\sin^2 \theta_2| = \frac{\sin^2 \theta_1}{n^2 + n_2'^2} \leq \frac{1}{n_2^2 + n_2'^2} \tag{2-29}$$

When n_2 and n_2' are large, as in the case of metals, $\cos \theta_2$ can be approximated by

$$\cos \theta_2 = \sqrt{1 - \sin^2 \theta_2} \cong 1 \tag{2-30}$$

with an error of less than 5 per cent for $n_2^2 + n_2'^2 \geq 10$. Then, for metals in air, the complex reflectivity components given by Eqs. 2-7 are simplified by setting in these equations $\cos \theta_2 \cong 1$, $m_1 = n_1 \simeq 1$, and $m_2 = n_2 - in_2'$. We obtain

$$\bar{\rho}_{\nu,\perp}(\theta_1) = \left[\frac{(n_2 - in_2') - \cos \theta_1}{(n_2 - in_2') + \cos \theta_1}\right]^2 \tag{2-31a}$$

$$\bar{\rho}_{\nu,\parallel}(\theta_1) = \left[\frac{(n_2 - in_2') - (1/\cos \theta_1)}{(n_2 - in_2') + (1/\cos \theta_1)}\right]^2 \tag{2-31b}$$

The absolute values of these complex expressions yield the reflectivity components for perpendicular and parallel polarization for metals in air.

$$\rho^s_{\nu,\perp}(\theta_1) = \frac{(n_2 - \cos \theta_1)^2 + n_2'^2}{(n_2 + \cos \theta_1)^2 + n_2'^2} \tag{2-32a}$$

$$\rho^s_{\nu,\parallel}(\theta_1) = \frac{[n_2 - (1/\cos \theta_1)]^2 + n_2'^2}{[n_2 + (1/\cos \theta_1)]^2 + n_2'^2} \tag{2-32b}$$

which are valid when n_2 and n_2' are large.

Emissivity and Absorptivity

When Kirchhoff's law is valid, the emissivity and absorptivity of a material are taken to be equal, and this result is applicable to each component of polarization of directional emissivity and absorptivity. We write

$$\varepsilon_{\nu,\perp}(\theta_1) = \alpha_{\nu,\perp}(\theta_1) \tag{2-33a}$$

$$\varepsilon_{\nu,\parallel}(\theta_1) = \alpha_{\nu,\parallel}(\theta_1) \tag{2-33b}$$

If we assume also that the material is *opaque*, the absorptivity and reflectivity are related by

$$\varepsilon_{\nu,\perp}(\theta_1) = \alpha_{\nu,\perp}(\theta_1) = 1 - \rho^s_{\nu,\perp}(\theta_1) \tag{2-34a}$$

$$\varepsilon_{\nu,\parallel}(\theta_1) = \alpha_{\nu,\parallel}(\theta_1) = 1 - \rho^s_{\nu,\parallel}(\theta_1) \tag{2-34b}$$

In the case of unpolarized radiation we have

$$\varepsilon_\nu(\theta_1) = \alpha_\nu(\theta_1) = 1 - \rho_\nu^s(\theta_1) = 1 - \tfrac{1}{2}[\rho^s_{\nu,\perp}(\theta_1) + \rho^s_{\nu,\parallel}(\theta_1)] \tag{2-35}$$

The above relations can be utilized to determine the emissivity from the reflectivity data. We consider now the determination of emissivity for the interface of two dielectric media and for metals in air.

Two Dielectric Media

The spectral normal reflectivity at the interface of two dielectric media was given by Eq. 2-14. When this relation is utilized, the spectral normal emissivity $\varepsilon_\nu(0)$ is given by

$$\varepsilon_\nu(0) = 1 - \rho_\nu{}^s(0) = \frac{4n}{(n+1)^2} \tag{2-36}$$

The spectral hemispherical emissivity ε_ν can be determined by utilizing Eq. 2-16 for the spectral hemispherical reflectivity. We obtain

$$\varepsilon_\nu = 1 - \rho_\nu = \frac{4n+2}{3(n+1)^2} + \frac{2n^3(n^2+2n-1)}{(n^2+1)(n^4-1)} - \frac{8n^4(n^4+1)}{(n^2+1)(n^4-1)^2}\ln(n) - \frac{n^2(n^2-1)^2}{(n^2+1)^3}\ln\left(\frac{n-1}{n+1}\right) \tag{2-37}$$

The ratio of hemispherical emissivity to normal emissivity is obtained by dividing Eq. 2-37 by Eq. 2-36:

$$\frac{\varepsilon_\nu}{\varepsilon_\nu(0)} = \frac{1}{3} + \frac{1}{6n} + \frac{n^2(n+1)(n^2+2n-1)}{2(n^2+1)^2(n-1)} - \frac{2n^3(n^4+1)}{(n^2+1)^3(n-1)^2}\ln(n) - \frac{n(n+1)^2(n^2-1)^2}{4(n^2+1)^3}\ln\left(\frac{n-1}{n+1}\right) \tag{2-38}$$

where $n \equiv n_2/n_1$.

Care must be exercised in the numerical evaluation of Eqs. 2-16, 2-37, and 2-38 because a small number is determined from the algebraic sum of many large numbers. To obtain accurate results many significant digits should be included in the calculations.

Metals in Air

The emissivity components for perpendicular and parallel polarization for metals in air can be determined from the reflectivity components given by Eqs. 2-32. We obtain

$$\varepsilon_{\nu,\perp}(\mu) = 1 - \rho^s_{\nu,\perp}(\mu) = \frac{4n\mu}{(n+\mu)^2 + n'^2} = \frac{4n\mu}{\mu^2 + 2n\mu + n^2 + n'^2} \tag{2-39a}$$

$$\varepsilon_{\nu,\parallel}(\mu) = 1 - \rho^s_{\nu,\parallel}(\mu) = \frac{4n/\mu}{(n+1/\mu)^2 + n'^2} = \frac{4n\mu}{(n^2+n'^2)\mu^2 + 2n\mu + 1} \tag{2-39b}$$

where we have defined $\mu \equiv \cos\theta_1$, and for simplicity have removed the subscript 2 from both n and n'.

The spectral hemispherical emissivity for perpendicular and parallel polarization can be evaluated from (see Eq. 1-124):

$$\varepsilon_{\nu,j} = 2\int_{\mu=0}^{1} \varepsilon_{\nu,j}(\mu)\mu\, d\mu, \qquad j \equiv \perp \quad \text{or} \quad \parallel \tag{2-40}$$

The integrals in Eq. 2-40 were evaluated by Dunkle [8], and the results are given as[2]

$$\varepsilon_{\nu,\perp} = 8n - 8n^2 \ln\left(\frac{1 + 2n + n^2 + n'^2}{n^2 + n'^2}\right) + \frac{8n(n^2 - n'^2)}{n'}\tan^{-1}\left(\frac{n'}{n + n^2 + n'^2}\right) \tag{2-41a}$$

$$\varepsilon_{\nu,\parallel} = \frac{8n}{n^2 + n'^2} - \frac{8n^2}{(n^2 + n'^2)^2}\ln(1 + 2n + n^2 + n'^2) + \frac{8n(n^2 - n'^2)}{n'(n^2 + n'^2)^2}\tan^{-1}\left(\frac{n'}{1 + n}\right) \tag{2-41b}$$

For unpolarized radiation the spectral directional emissivity is obtained from

$$\varepsilon_\nu(\mu) = \tfrac{1}{2}[\varepsilon_{\nu,\perp}(\mu) + \varepsilon_{\nu,\parallel}(\mu)] \tag{2-42}$$

where $\varepsilon_{\nu,\perp}(\mu)$ and $\varepsilon_{\nu,\parallel}(\mu)$ are given by Eqs. 2-39, and the spectral hemispherical emissivity is obtained from

$$\varepsilon_\nu = \tfrac{1}{2}(\varepsilon_{\nu,\perp} + \varepsilon_{\nu,\parallel}) \tag{2-43}$$

where $\varepsilon_{\nu,\perp}$ and $\varepsilon_{\nu,\parallel}$ are given by Eqs. 2-41.

The spectral normal emissivity is determined from Eqs. 2-42 and 2-39 as

$$\varepsilon_\nu(\mu = 1) = \frac{4n}{1 + 2n + n^2 + n'^2} \tag{2-44}$$

since $\mu = \cos\theta_1 = 1$ for $\theta_1 = 0$. Equation 2-44 is related to the data in Fig. 2-6 by $\varepsilon_\nu(\mu = 1) = \varepsilon_\nu(\theta_1 = 0) = 1 - \rho_\nu(0)$.

The spectral hemispherical emissivity is obtained from Eqs. 2-43 and 2-41 as

$$\begin{aligned}\varepsilon_\nu = 4n\Bigg[1 &- n\ln\left(\frac{1 + 2n + n^2 + n'^2}{n^2 + n'^2}\right) \\ &+ \frac{n^2 - n'^2}{n'}\tan^{-1}\left(\frac{n'}{n + n^2 + n'^2}\right) + \frac{1}{n^2 + n'^2} \\ &- \frac{n}{(n^2 + n'^2)^2}\ln(1 + 2n + n^2 + n'^2) + \frac{n^2 - n'^2}{n'(n^2 + n'^2)^2}\tan^{-1}\left(\frac{n}{1 + n}\right)\Bigg]\end{aligned} \tag{2-45}$$

The above relations for emissivity are strictly applicable for opaque materials. For opaque materials the radiative properties of the interface are determined by an extremely thin layer at the interface since radiation is attenuated within a very short distance in the medium. Metals, for example, are opaque to thermal radiation since the attenuation coefficient n_2' in that frequency range is sufficiently large.

A material is called *transparent* or *semitransparent* to radiation if the attenuation coefficient n_2' is zero or extremely small so that radiation penetrates for great depths within the material. For example, glass is transparent to thermal radiation over a certain portion of the spectrum.

The foregoing relations for reflectivity and emissivity are applicable only to ideal surfaces, that is, surfaces that are optically smooth, free of roughness, oxidation, and contamination, and that have been prepared by a process such as chemical deposition or electrodeposition which does not damage the material. They cannot be used to predict the reflectivity and emissivity of real surfaces encountered in engineering applications.

2-3 PREDICTION OF RADIATIVE PROPERTIES OF ROUGH SURFACES

The surfaces encountered in engineering applications deviate from ideality as a result of roughness, oxidization, and contamination; hence the radiative properties of these real surfaces differ greatly from those predicted by the electromagnetic theory. In this section we discuss briefly the effects of roughness on the radiative properties of opaque surfaces.

In regard to problems of radiative heat transfer the roughness of real surfaces can be divided into two categories: (1) small surface irregularities, such that the incident radiation cannot undergo more than a single reflection, and (2) deep cavities in which the incident radiation undergoes multiple reflections.

Effects of Small Surface Irregularities

Figure 2-7 illustrates schematically the reflection of normally incident radiation from a rough surface having small irregularities. The incident beam is reflected partly specularly and partly diffusely, because the facets of the roughness behave like small mirrors pointed in various directions. The treatment of the problem of reflection from such surfaces involves geometrical optics. Davies [9] used a statistical approach for the reflection of electromagnetic waves from a rough, conducting surface and derived analytical expressions for the reflectivity as a function of the root-mean-square (rms) roughness.[3]

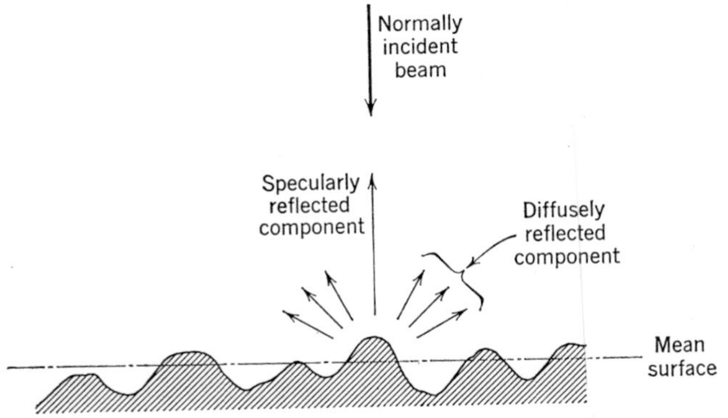

Fig. 2-7. Reflection of normally incident radiation at a surface with small irregularities.

In the analysis Davies assumed that the material was perfectly conducting, the distribution of heights of surface irregularities was Gaussian about the mean, and no multiple reflections occurred. The resulting expression for the relative *specular reflectivity* of a rough surface, for both the slightly rough (i.e., $h/\lambda \ll 1$) and the very rough case (i.e., $h/\lambda \gg 1$), is given in the form

$$\frac{\rho^s(\theta_i)}{\rho_0(\theta_i)} = \exp\left[-\left(4\pi \frac{h}{\lambda} \cos \theta_i\right)^2\right] \tag{2-46}$$

where $\rho^s(\theta_i)$ = specular reflectivity of rough surface,
$\rho_0(\theta_i)$ = reflectivity of an ideal surface of the same material,
h/λ = ratio of rms height of roughness to wavelength of radiation,
θ_i = angle of incidence (or reflection).

For normal incidence ($\theta_i = 0$), Eq. 2-46 simplifies to

$$\frac{\rho^s(0)}{\rho_0(0)} = \exp\left[-\left(4\pi \frac{h}{\lambda}\right)^2\right] \tag{2-47}$$

These results are applicable for materials in which reflection occurs very near the surface, such as reflection from metallic surfaces; they are not applicable to electrical nonconductors, in which reflection occurs not only from the surface but also from the interior of the material.

The results given above do not include the contribution due to diffuse reflection in the direction of specular reflection. For diffuse reflection, Davies' relation for the relative *directional-hemispherical reflectivity* can be expressed in the form

$$\frac{\rho^d(\theta_i \to 2\pi)}{\rho_0(\theta_i)} = \frac{1}{\cos \theta_i} \int_{\phi=0}^{2\pi} \int_{\theta=0}^{\pi/2} \frac{\rho(\theta_i \to \theta, \phi)}{\rho_0(\theta_i)} \sin \theta \, d\theta \, d\phi \tag{2-48a}$$

where the function under the integral is given for a slightly rough surface as

$$\frac{\rho(\theta_i \to \theta, \phi)}{\rho_0(\theta_i)} = \pi^3 \left(\frac{a}{\lambda}\right)^2 \left(\frac{h}{\lambda}\right)^2 (\cos\theta + \cos\theta_i)^4$$
$$\times \exp\left\{-\left(\frac{\pi a}{\lambda}\right)^2 [(\sin\theta\cos\phi - \sin\theta_i)^2 + \sin^2\theta\sin^2\phi]\right\} \quad \text{for } \frac{h}{\lambda} \ll 1 \tag{2-48b}$$

and for a very rough surface as

$$\frac{\rho(\theta_i \to \theta, \phi)}{\rho_0(\theta_i)} = \frac{1}{32\pi^2}\left(\frac{a}{h}\right)^2 (\cos\theta + \cos\theta_i)^2$$
$$\times \exp\left\{-\frac{1}{2}\left(\frac{a}{h}\right)^2 \left[\frac{(\sin\theta\cos\phi - \sin\theta_i)^2 + \sin^2\theta\sin^2\phi}{(\cos\theta + \cos\theta_i)^2}\right]\right\} \quad \text{for } \frac{h}{\lambda} > 1 \tag{2-48c}$$

where we have defined the following:

$\rho^d(\theta_i \to 2\pi)$ = directional-hemispherical reflectivity,
$\rho_0(\theta_i)$ = reflectivity of an ideal surface of the same material,
θ_i = angle of incidence,
h = rms roughness height,
a = autocovariance length.

The autocovariance length a is related to the rms roughness height h and the rms slope of the surface profile m by the relation [10]

$$a = \frac{\sqrt{2}\, h}{m} \tag{2-48d}$$

For normal incidence ($\theta_i = 0$) Eqs. 2-48 simplify to

$$\frac{\rho^d(0 \to 2\pi)}{\rho_0(0)} = 2\pi \int_{\theta=0}^{\pi/2} \frac{\rho(0 \to \theta)}{\rho_0(0)} \sin\theta \, d\theta \tag{2-49a}$$

where the function under the integral is given as

$$\frac{\rho(0 \to \theta)}{\rho_0(0)} = \pi^3 \left(\frac{a}{\lambda}\right)^2 \left(\frac{h}{\lambda}\right)^2 (1 + \cos\theta)^4 \exp\left[-\left(\frac{\pi a}{\lambda}\sin\theta\right)^2\right], \quad \text{for } \frac{h}{\lambda} \ll 1 \tag{2-49b}$$

and

$$\frac{\rho(0 \to \theta)}{\rho_0(0)} = \frac{1}{32\pi^2}\left(\frac{a}{h}\right)^2 (1 + \cos\theta)^2 \exp\left[-\frac{1}{2}\left(\frac{a}{h}\right)^2 \left(\frac{\sin\theta}{1 + \cos\theta}\right)^2\right], \quad \text{for } \frac{h}{\lambda} > 1 \tag{2-49c}$$

Porteus [11] cautions about using Eq. 2-48c (also 2-49c) at wavelengths much shorter than the surface roughness height. For wavelengths that are very short compared with the average dimension of the facets, each facet of the roughness will behave as an independent plane reflector of infinite extent. Then the reflectivity will approach that of an ideal surface.

The effects of roughness on the reflectivity of surfaces have been investigated experimentally by Bennett [12], Birkebak et al. [13], Torrance and Sparrow [14, 15], Birkebak and Eckert [16], and Safwat and Parmer [17], and the validity of the foregoing analysis has been examined.

Bennett [12] measured the specular component of reflectivity at normal incidence (i.e., $\theta_i = 0$), using aluminized ground-glass samples for roughness $h/\lambda < 1$. Figure 2-8 shows a comparison of the experimental and predicted (from Eq. 2-47) values of the relative specular reflectivity, $\rho^s(0)/\rho_0(0)$, as a function of h/λ. The measured and the theoretical results agree reasonably well over the range of roughnesses considered.

Birkebak et al. [13] determined experimentally the diffuse component of reflectivity for metallic surfaces with controlled uniform roughness. The test surfaces were prepared by depositing films of pure aluminum, gold, platinum, and nickel on ground-glass and nickel samples. The ratios of the hemispherical-directional reflectivity[4] for the roughened surface to the hemispherical-directional reflectivity for an ideal surface (i.e., $h = 0.003\ \mu$) of the same material were examined over the range of roughness parameter h/λ both less than and greater than unity, and for several different angles of incidence. It was pointed out by Birkebak et al. [13] that Davies' equations

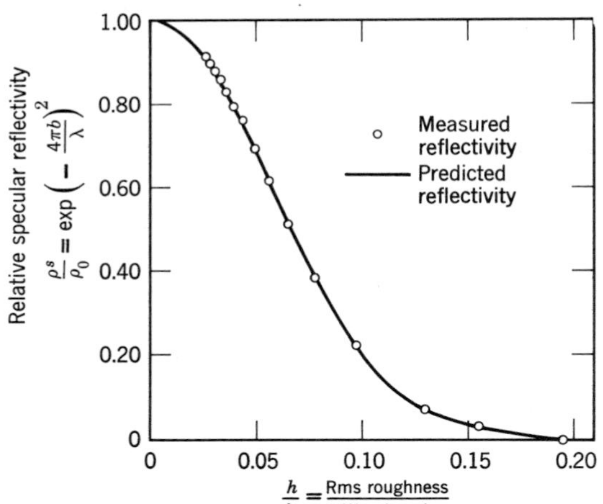

Fig. 2-8. Comparison of the predicted and the measured relative specular component of reflectivity for reflection from aluminized ground glass. (From H. E. Bennett [12]).

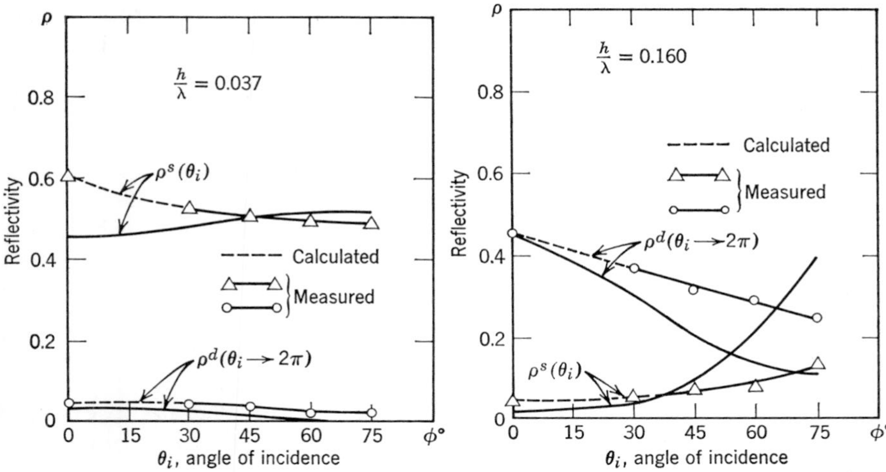

Fig. 2-9. Comparison of measured and predicted values of specular and directional-hemispherical (diffuse) components of reflectivity for roughened steel samples. (From H. H. Safwat and J. F. Parmer [17].)

2-48a and 2-48c could be conservatively used for angles of incidence up to 20°.

Figure 2-9 shows a comparison of measured and predicted (from Eqs. 2-46 and 2-48) values of the specular and diffuse components of reflectivity for roughened steel samples reported by Safwat and Parmer [17]. In these experiments roughnesses ranged from 0.05 μ to 0.3 μ, h/λ from 0.037 to 0.160, and angles of incidence from 0 to 75°. Agreement between theory and experiment appears to be better at smaller angles of incidence.

Torrance and Sparrow [14] determined experimentally the specular and diffuse components of reflectivity for electrical nonconductors (i.e., fused polycrystalline magnesium oxide) having roughnesses ranging from 0.16 μ to 5.8 μ, over wavelengths from 0.5 μ to 12 μ at two angles of incidence, $\theta_i =$ 10° and 45°. Since no theory is available for the reflection of radiation from the rough surface of an electric nonconductor, they tried Davies' theory to correlate the experiments. Although this theory is strictly applicable to electrical conductors, the correlation of the specular reflectivity component appeared, surprisingly, to be good for the range of rms roughnesses $0 < h/\lambda < 0.05$; deviation occurred for larger values of h/λ, however, and no correlation was possible for the diffuse component of reflectivity.

It appears that further work is needed in the form of both theoretical and experimental investigations of the effects of surface roughness on reflectivity.

Porteus [11] points out the inadequacy of such parameters as the rms roughness and the rms slope, especially, when the surface roughness distribution is non-Gaussian. The reader should refer to the book by Beckmann and Spizzichino [18, Chapter 3] for further theoretical developments concerning the reflection of radiation from the rough surface of a conducting material. A discussion of the effects of surface imperfections on the radiative properties of surfaces can be found in the *Symposium on Thermal Radiation in Solids* [19].

Real surfaces deviate from ideality not only because of roughness but also as a result of contamination, oxidation, and the like. Presently no adequate theory is available to account for the effects of deviation from ideality; therefore experimental measurements are the only means of determining the radiative properties of real surfaces. A vast number of experimental data on the reflectivity, emissivity, and absorptivity of real surfaces have been reported in the literature; for example, extensive compilations of experimental data can be found in references [21–25]. We discuss these experimental results at the end of this chapter.

Effects of Deep Cavities

When the surface irregularities are in the form of deep cavities, radiation incident upon the surface undergoes multiple reflections. Since each additional reflection results in additional absorption of the incident radiation, the reflectivity of a cavity is less than that of a plane surface of identical material stretched over the opening of the cavity. The determination of the absorption and emission characteristics of a cavity involves the solution of the integral equation for radiative exchange over the entire surface of the cavity. The treatment of such problems will be examined in Chapter 5.

2-4 SCATTERING AND ABSORPTION OF RADIATION BY SPHERICAL PARTICLES

A beam of radiation traveling through a medium containing inhomogeneities, such as a gas with a cloud of particles dispersed throughout, is absorbed and scattered by the inhomogeneities, as well as being absorbed by the medium itself. The investigation of the scattering of electromagnetic waves has been an interdisciplinary activity. Astrophysicists are generally concerned with the scattering of radiation by interplanetary dust; chemists and biochemists, with light scattering by colloidal suspensions; physicists and electrical engineers, with the scattering of microwaves and radio waves; and so forth. An important problem in scattering by particulate matter is how to relate the properties of particles (i.e., size, shape, refractive index) to the angular distribution of the scattered radiation and to the absorption of radiation by the particulate matter. Therefore numerous theoretical and experimental investigations have been performed to study the scattering of electromagnetic waves.

Rayleigh obtained a simple solution for the scattering of radiation by spheres whose dimensions are very small compared with the wavelength of the radiation. This work was followed by a more general theory formulated by Mie [26] for the absorption and scattering of radiation by small, homogeneous particles having a simple geometry such as a sphere or a circular cylinder. The Mie theory, which is based on the solution of Maxwell's equations, is concerned with a very idealized situation, namely, a simple sphere composed of homogeneous, isotropic material embedded in a homogeneous, isotropic, dielectric, infinite medium irradiated by plane waves propagating in a specified direction. For a dielectric sphere no absorption is involved; for a conducting sphere the incident radiation is partly absorbed, partly scattered, and partly transmitted. The derivation of the Mie solution and the mathematical and physical considerations for the analysis can be found in detail, in addition to the original reference, in the books by Van de Hulst [27], Stratton [28], and Kerker [29]. The resulting solutions for the amplitude of the scattered wave are in the form of complicated series that involve the Riccati-Bessel functions and Riccati-Hankel functions of increasing order. A discussion of the adaptation of the Mie solution to machine computation can be found in the book by Deirmendjian [30]. The results from the Mie solution have proved to be most useful in predicting the absorption and scattering coefficients and the phase function for spherical particles suspended in a dielectric medium, provided that the particles are located sufficiently far from each other. Several experiments have also been performed to determine the minimum spacing between the spheres to ensure independent scattering. It appears that, to eliminate interference, the center-to-center spacing between the spheres should not be less than about 3 diameters. In most practical problems, however, the particles are separated by much larger distances. The limitations to the Mie theory should also be recognized. It considers a highly idealized situation, namely, a single spherical particle that acts as an independent point scatterer in an infinite medium, whereas the scatterers encountered in most engineering applications have arbitrary geometries. No adequate theory is yet available to predict the scattering of radiation from particles of random shape and orientation, variable properties, and composite structure. Therefore the experimental approach is the only means to determine scattering for such situations.

We discuss now the results from the Mie theory, since it is the only general theory presently available and the results are useful for many idealized situations.

The Parameters in the Mie Solution

In general, a single spherical particle placed in a beam of a plane electromagnetic wave both scatters and absorbs a certain amount of this energy. The ratio of the rate of energy scattered by the sphere to the incident energy flow rate per unit area is called the *scattering cross section* at the frequency considered and is denoted by the symbol C_s. Similarly one can define the *absorption cross section*, C_a, and the *extinction cross section*, C_e. By definition the sum of the absorption and scattering cross sections is equal to the extinction cross section; therefore we write

$$C_a + C_s = C_e \tag{2-50a}$$

The cross section has the dimension of an area.

The ratio of the cross section to the geometric cross section is called the *efficiency factor* and is denoted by the symbol Q_i, $i = a, s$, or e (i.e., absorption, scattering, or extinction, respectively). Thus we write

$$Q_a = \frac{C_a}{\pi r^2} = \text{efficiency factor for absorption} \tag{2-50b}$$

$$Q_s = \frac{C_s}{\pi r^2} = \text{efficiency factor for scattering} \tag{2-50c}$$

$$Q_e = \frac{C_e}{\pi r^2} = \text{efficiency factor for extinction} \tag{2-50d}$$

where r is the radius of the sphere. The efficiency factors satisfy among themselves the following relation:

$$Q_a + Q_s = Q_e \tag{2-50e}$$

The analytical expressions for the efficiency factor for a sphere are obtainable, in addition to the original derivation by Mie [26], from a comprehensive treatment of this subject in the books by Van de Hulst [27] and Stratton [28]; a summary of these equations is given by Deirmendjian [30] for the purpose of adapting them to computational form. In order to give the reader some idea of the results of the Mie solution and of the parameters involved in the problem, we present below the expressions for the efficiency factors for scattering and extinction [27, pp. 127–128]:

$$Q_s = \frac{2}{x^2}\sum_{n=1}^{\infty}(2n+1)(|a_n|^2 + |b_n|^2) \tag{2-51a}$$

$$Q_e = \frac{2}{x^2}\sum_{n=1}^{\infty}(2n+1)\,\text{Re}\,\{a_n + b_n\} \tag{2-51b}$$

where Re indicates the real part of the summation. If the particle does not absorb any of the incident radiation (i.e., the index of refraction is real and the particle is pure scatterer), Eqs. 2-51a and 2-51b lead to identical results. If the particle absorbs, then the index of refraction is complex and the efficiency factor for absorption, Q_a, is obtained from the definition of Q_e as

$$Q_a = Q_e - Q_s \tag{2-51c}$$

In Eqs. 2-51 the terms a_n and b_n are called the *Mie coefficients;* these are complicated functions of the Riccati-Bessel functions and are given in the from [27, p. 123]

$$a_n = \frac{\psi_n(x)[\psi_n'(y)/\psi_n(y)] - m\psi_n'(x)}{\xi_n(x)[\psi_n'(y)/\psi_n(y)] - m\xi_n'(x)} \tag{2-52a}$$

$$b_n = \frac{m\psi_n(x)[\psi_n'(y)/\psi_n(y)] - \psi_n'(x)}{m\xi_n(x)[\psi_n'(y)/\psi_n(y)] - \xi_n'(x)} \tag{2-52b}$$

where primes denote differentiation with respect to the argument concerned. The Riccati-Bessel functions $\psi_n(z)$ and $\xi_n(z)$ are related to the Bessel function of half-integral order as

$$\psi_n(z) = \left(\frac{\pi z}{2}\right)^{1/2} J_{n+1/2}^{(z)} \tag{2-52c}$$

$$\xi_n(z) = \left(\frac{\pi z}{2}\right)^{1/2} J_{n+1/2}^{(z)} + (-1)^n i J_{-n-1/2}^{(z)}, \quad i = \sqrt{-1} \tag{2-52d}$$

here $z \equiv x$ or y, and the arguments x and y are defined as

$$x = \frac{\pi D}{\lambda} \quad \text{and} \quad y = mx \tag{2-52e}$$

where D is the diameter of the sphere, λ is the wavelength of the incident radiation in the surrounding medium, and $m = n - in'$ is the complex refractive index of the sphere relative to the surrounding medium. Whenever the index of refraction m is complex, the function $\psi_n'(y)/\psi_n(y)$ involves Bessel functions of complex argument.

The angular distribution of the scattered radiation, that is, the phase function, is also available from the Mie solution. Since the sphere is a symmetrical particle, the scattering is independent of the azimuthal angle ϕ but is a function of the scattering angle θ, which is the angle between the forward direction of the incident ray and the forward direction of the scattered ray.

It is apparent from the foregoing discussion of the Mie solution for the scattering of radiation by a spherical particle that the solution involves three basic parameters: (1) the index of refraction of the sphere, $m = n - in'$, relative to the surrounding medium; (2) a dimensionless size parameter x, defined as $x = \pi D/\lambda$; and (3) the scattering angle θ. The numerical calculation of the Mie coefficients, however, is complicated because of the absence of tabulations of Bessel functions of complex arguments.

When a beam of radiation is propagating through a medium containing N spherical particles per unit volume of the same composition and uniform size, each of radius R, then the absorption and scattering cross sections C_a and C_s (or absorption and scattering efficiency factors Q_a and Q_s) may be related to the volumetric spectral absorption and scattering coefficients κ_λ (Eq. 1-57) and σ_λ (Eq. 1-60) by

$$\kappa_\lambda = C_a N = \pi R^2 Q_a N \tag{2-53a}$$

$$\sigma_\lambda = C_s N = \pi R^2 Q_s N \tag{2-53b}$$

When the medium contains a cloud of spherical particles of the same composition but of different sizes, the volumetric spectral absorption and scattering coefficients may be computed from

$$\kappa_\lambda = \int_0^\infty C_a N(r)\, dr = \int_0^\infty \pi r^2 Q_a N(r)\, dr \tag{2-53c}$$

$$\sigma_\lambda = \int_0^\infty C_s N(r)\, dr = \int_0^\infty \pi r^2 Q_s N(r)\, dr \tag{2-53d}$$

where $N(r)\, dr$ is the number of particles having radii between r and $r + dr$ per unit volume. Here the cross sections or the efficiency factors depend on radius because of the argument $x = 2\pi r/\lambda$. Sometimes it is desirable to change the integration variable r to x through the relation $x = 2\pi r/\lambda$. If the size distribution of particles is broken up into a class of increments of radius r_j, $j = 1, 2, \ldots$, then the above integrals may be represented by a summation.

The Mie Scattering Region

The size parameter x appearing in the Mie solution can have values from 0 to infinity, and the refractive index m can have values from 1 to infinity for a sphere in a vacuum. The refractive index m can be smaller than unity if the medium surrounding the sphere is not a vacuum (i.e., the refractive index for an air bubble in water is less than unity). Although the Mie solution is applicable over the entire m-x domain, it has been found that numerical computations of the phase function and efficiency factors become very difficult for arbitrary m and x. For example, the convergence of the series defining the Mie coefficients becomes very slow as the relative size of the sphere increases with respect to the wavelength of the incident radiation.

Another difficulty is the irregularity of the values of the coefficients a_m and b_m, which makes interpolation rather inaccurate. Fortunately the Mie calculations need not be performed over the entire m-x domain; over certain regions the limiting values of the Mie solution can be determined by simplified methods of computation. The reader may refer to the book by Van de Hulst [27, pp. 131–134] for a discussion of these limiting cases. For example, for large values of the size parameter x (i.e., a large sphere compared with the wavelength) the convergence of the exact Mie solution becomes very difficult; however, the principles of geometric optics are applicable to determine the phase function and the efficiency factor for such cases, and the resulting expressions are very simple.

For very small values of x the exact Mie formulation is simplified by introducing a power series expansion of the spherical Bessel functions into the Mie coefficients a_m and b_m. Van de Hulst [27, p. 270] and Penndorf [32a] developed power series expansions for these coefficients. In such expansions the leading term represents the Rayleigh scattering law. The power series expansion of the Mie solution for small x is given in the form [27, p. 270]

$$Q_e = -\text{Im}\left\{4x\,\frac{m^2-1}{m^2+2} + \frac{4}{15}\,x^2\left(\frac{m^2-1}{m^2+2}\right)^2 \frac{m^4+27m^2+38}{2m^2+3} + \cdots\right\}$$
$$+ \text{Re}\left\{\frac{8}{3}\,x^4\left(\frac{m^2-1}{m^2+2}\right)^2 + \cdots\right\} \quad (2\text{-}54)$$

The first term characterizes the efficiency factor for absorption; the second term, the efficiency factor for scattering. The result is valid of $x \ll 1$ as well as $mx \ll 1$.

Therefore the exact Mie solution is expected to approach the results obtained from the principles of geometric optics for large values of x and to those obtained by the Rayleigh scattering law for small values of x. It is of interest to know the range of applicability of these two limiting cases, since the numerical computation of the Mie solution is very laborious. To investigate this matter Penndorf [32b] computed the scattering in the forward scattering area (i.e., $\theta = 0$) from the Mie formulation for spheres having real refractive indices n of 1.05 to 2 over a wide range of size parameters and compared the results with those obtained from geometric optics and the Rayleigh scattering law. The phase function obtained from the Mie solution is found to differ considerably from the constant value of 1.5 obtained from the Rayleigh phase function for forward scattering [i.e., $p(\theta) = \frac{3}{4}(1 + \cos^2\theta)$ for $\theta = 0$]. For the size parameter $x = 0.5$, the phase function from the Mie solution is about 10 per cent greater than that obtained from the Rayleigh phase function. Hence the Rayleigh region for phase function does not extend beyond about $x = 0.5$. A comparison of the scattering coefficients shows that for small values of x the Rayleigh scattering coefficient is smaller than that obtained from the Mie solution; however, there is a specific value of x, which depends on the refractive index, where crossover occurs, and beyond this crossover the Rayleigh scattering coefficient is always larger than that obtained from the Mie solution. For a size parameter x greater than about 20 to 30, depending on the refractive index, the phase function predicted from the laws of geometric optics equals or falls below 25 per cent of that predicted from the Mie solution. The intermediate range of size parameters for which neither the law of Rayleigh scattering nor the principles of geometric optics are applicable is generally referred to as the *Mie scattering region;* here lie most cases of practical interest.

A complete determination of the phase function from the Mie solution requires that the calculations be carried out for a large number of scattering angles. To alleviate this difficulty Chu and Churchill [33] expressed the phase function for unpolarized radiation in a series of Legendre polynomials in the form

$$p(\cos \theta_0) = 1 + \sum_{j=1}^{\infty} A_j P_j(\cos \theta_0) \tag{2-55}$$

where θ_0 = the scattering angle,
$P_j(\cos \theta_0)$ = Legendre polynomials of order j and argument $\cos \theta_0$,
A_j = the expansion coefficients,

and the coefficients A_j can be calculated from a_n and b_n of Eqs. 2-52 by means of exact expressions derived from the Mie equations. The coefficients A_j are a function of the size parameter x and the relative refractive index m of the sphere only. The advantage of expressing the phase function $p(\cos \theta_0)$ as in Eq. 2-55 lies in the fact that angular distribution is related to the familiar Legendre polynomials $P_j(\cos \theta_0)$.

The reader should refer to the publications by Kerker [29] and Rowell and Stein [34] for detailed treatment of recent theoretical developments on scattering.

Results from the Mie Theory for Spherical Particles

The first step in determining the phase function for spherical particles from the Mie theory consists of evaluating the coefficients a_n and b_n from Eqs. 2-52 by using appropriate Riccati-Bessel functions. The phase function and the scattering and absorption coefficients (or efficiency factors) are then readily calculated. These computations are very complicated for particles with a complex refractive index because in that case the Riccati-Bessel functions have complex arguments; they are also very laborious for large particles because of convergence difficulties. Early computations, therefore, were limited to very specific situations. With the advent of high-speed digital computers more comprehensive tabulations of phase functions have been published. Here we present a brief survey of the literature and discuss some of the results obtained on the efficiency factors for absorption and scattering and the phase function for scattering by spherical particles.

An extensive tabulation of the Mie solution for spheres with real refractive indices, including some cases with complex refractive indices, has been published by Lowan [35]. In these tables, the angular distribution of the scattered radiation is given as a function of the particle-size parameter and the refractive index. The efficiency factors for scattering are also included in these tables.

Chu, Clark, and Churchill [36] evaluated from the Mie solution the coefficients A_j in Eq. 2-55 for nonabsorbing (i.e., dielectric) spheres for the range of size parameters from $x = 1$ to 18 and real refractive indices from $n = 0.9$ to 2.0 and for $n = \infty$. The numerical values of these coefficients for a limited number of cases are presented in Table 2-1 as a function of the size parameter x and the real refractive index n of the sphere relative to the surrounding medium. The phase function for a conducting sphere having a complex refractive index $m = n - in'$ varies insignificantly from that for a dielectric sphere, that is, $n' = 0$, when the value of n' is very small. Therefore the tables prepared by Chu et al. [36] for dielectric spheres may be used for spheres with complex refractive indices if the value of n' is less than about 0.001.

Deirmendjian, Clasen, and Viezee [37] computed the Mie solution for spheres with complex refractive indices; Plass [38, 39] and Gryvnak and Burch [40] evaluated the absorption and scattering cross section for aluminum oxide and magnesium oxide spheres. Plass [41] and Kattawar and Plass [42] determined the absorption and scattering cross sections for spheres over a wide range of complex refractive indices. Stull and Plass [43] evaluated these cross sections for spherical carbon particles, and Herman [44] for water spheres. The reader may refer to Penndorf [32c] and Kerker [29] for comprehensive bibliographies of references for phase functions for spheres having real and complex refractive indices.

The absorption and scattering coefficients are of interest in many applications. Figures 2-10 and 2-11 show the efficiency factors for absorption Q_a and for scattering Q_s as a function of the size parameter $x = \pi D/\lambda$ for spheres having a complex refractive index for $n = 1.01$ and several different values of n'. The efficiency factors for absorption Q_a for $n' = 1$ and 10 reach maximum values and approach their limiting values for large $\pi D/\lambda$ from above, whereas for $n' < 0.1$ they gradually approach their limiting values for large $\pi D/\lambda$ from below.

Table 2-1 Coefficient A_j in Phase Function $p(\cos\theta_0) = 1 + \sum_{j=1} A_j P_j(\cos\theta_0)$ for Scattering of Radiation by Dielectric Spheres[a]

Real Refractive Index		$x = \pi D/\lambda =$					
n	j	1	2	3	4	5	10
0.9	1	0.48520	1.73113	2.39672	2.58641	2.72099	2.89817
	2	0.53927	1.29297	2.63876	3.38188	3.83141	4.55427
	3	0.14000	0.67360	1.93958	3.30004	4.19870	5.88635
	4	0.00004	0.23940	1.08155	2.53827	3.92475	6.85728
	5	0.00000	0.05651	0.47225	1.58466	3.14307	7.47188
	6		0.00786	0.15850	0.81473	2.15398	7.74557
	7		0.00000	0.04049	0.34363	1.26356	7.69956
	8			0.00748	0.11784	0.63122	7.38616
	9			0.00102	0.03225	0.26621	6.83159
	10			0.00000	0.00733	0.09391	6.08745
1.05	1	0.50204	1.86816	2.37480	2.62473	2.72473	
	2	0.57171	1.39597	2.74822	3.46657	3.89046	
	3	0.10745	0.69384	2.07069	3.48400	4.36782	
	4	0.00005	0.24136	1.13263	2.82054	4.18566	
	5	0.00000	0.05077	0.47765	1.79228	3.53192	
	6		0.01428	0.15457	0.90537	2.54688	
	7		0.00000	0.03743	0.36828	1.51696	
	8			0.00630	0.11971	0.74438	
	9			0.00001	0.03159	0.30149	
	10			0.00000	0.00640	0.10088	
1.20	1	0.52716	1.98398	2.35789	2.56819	2.68018	2.78197
	2	0.55397	1.50823	2.76628	3.44748	3.80783	4.25856
	3	0.12224	0.70075	2.20142	3.47905	4.29631	5.38683
	4	0.01621	0.23489	1.24514	2.96720	4.23484	6.19015
	5	0.00000	0.05133	0.51215	2.03079	3.69187	6.74492
	6		0.00760	0.16096	1.08340	2.87668	7.06711
	7		0.00048	0.03778	0.43822	1.88439	7.20999
	8		0.00000	0.00667	0.13982	0.99757	7.20063
	9			0.00081	0.03508	0.40817	7.03629
	10			0.00000	0.00698	0.13472	6.76587
1.40	1	0.57024	1.97663	2.31152	2.41512	2.41817	1.71249
	2	0.56134	1.55151	2.68810	3.18581	3.36029	2.28988
	3	0.11297	0.64590	2.30714	3.32459	3.71488	1.74098
	4	0.01002	0.21298	1.41315	2.94263	3.72531	2.12773
	5	0.00000	0.04466	0.55647	2.27513	3.52403	1.88003

Table 2-1 (continued)

Real Refractive Index n	j	$x = \pi D/\lambda =$ 1	2	3	4	5	10
	6		0.00623	0.18332	1.34790	2.95734	2.09325
	7		0.00052	0.04206	0.54935	2.28926	2.18715
	8		0.00000	0.00713	0.21007	1.38851	2.19825
	9			0.00093	0.05301	0.62456	2.26705
	10			0.00008	0.01031	0.30062	2.04777
1.60	1	0.62895	1.75119	2.02621	1.93460	1.71743	2.24483
	2	0.56832	1.35594	2.44677	2.70129	2.32943	3.33876
	3	0.10652	0.57923	2.03180	2.51662	1.92162	3.49424
	4	0.00905	0.21611	1.30053	2.63819	2.44117	4.13703
	5	0.00000	0.04368	0.65860	2.29733	2.23808	3.86252
	6		0.00591	0.31736	1.85084	2.71408	4.42809
	7		0.00056	0.06657	1.25977	2.63798	4.11586
	8		0.00000	0.01076	0.76693	2.39188	4.66127
	9			0.00134	0.11094	1.67654	4.38576
	10			0.00013	0.02125	1.05571	4.90345
2.00	1	0.82864	1.51676	1.17222	0.97103	1.81854	
	2	0.58827	1.44500	1.62448	1.82581	2.68487	
	3	0.09472	1.16386	1.26038	0.16239	2.57600	
	4	0.00805	0.58621	1.51925	0.77877	3.19198	
	5	0.00042	0.06755	0.89405	−0.39807	2.40581	
	6	0.00000	0.00796	0.92815	0.91924	2.37422	
	7		0.00069	−0.01436	0.07775	1.10021	
	8		0.00005	0.03124	1.58783	0.80641	
	9		0.00000	0.00427	−0.01633	−0.12512	
	10			0.00041	0.64361	−0.03497	
∞	1	−0.56524	0.84664	1.17355	1.30275	1.36775	1.46512
	2	0.29783	0.03635	1.15420	1.61283	1.85395	2.25639
	3	0.08571	−0.04477	0.80626	1.56313	2.03007	2.88513
	4	0.01003	0.33367	0.01043	1.26056	1.97993	3.36955
	5	0.00063	0.13727	0.19199	0.68088	1.63259	3.71899
	6	−0.00000	0.02852	0.38893	0.10855	1.28883	3.93916
	7		0.00353	0.18424	0.33797	0.60845	4.04894
	8		0.00027	0.05003	0.43523	0.21785	4.03870
	9		−0.00000	0.00877	0.22315	0.43845	3.92468
	10		−0.00000	0.00113	0.07110	0.47312	3.72343

[a] From C. M. Chu, G. C. Clark, and S. W. Churchill [36].

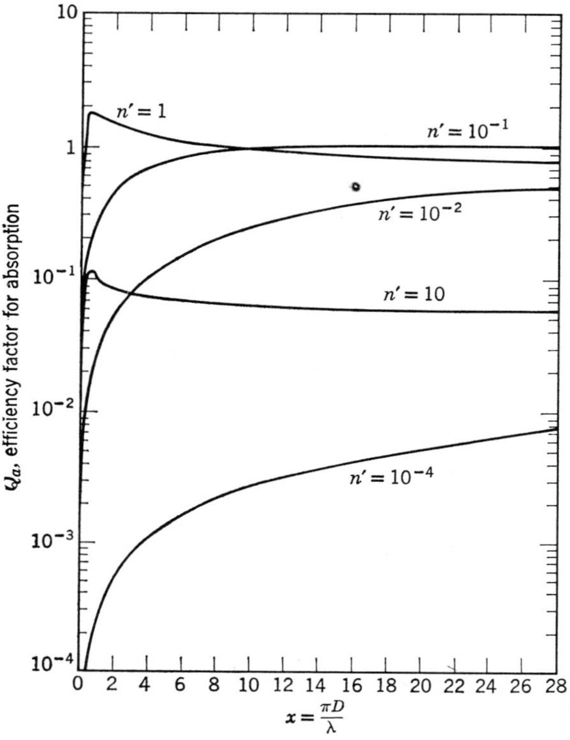

Fig. 2-10. Effects of size parameter x and n' on efficiency factor for absorption for $m = 1.01 - in'$. (From G. N. Plass [41]).

Figures 2-12 and 2-13 show the efficiency factors for absorption and scattering for $\pi D/\lambda = 1$ as a function of n' for several different values of n. It is apparent from Fig. 2-12 that the efficiency factor for absorption Q_a is rather insensitive to the values of n. The efficiency factor for scattering Q_s in Fig. 2-13 appears to be independent of n' until the latter exceeds about 10^{-2}. For small values of n', the efficiency factor for scattering is orders of magnitude higher than the efficiency factor for absorption.

2-5 REFRACTIVE INDICES OF MATERIALS

The refractive index of a material affects its absorption, scattering, and reflection properties; hence a precise determination of the refractive index is essential for the evaluation of the radiative properties of a material by

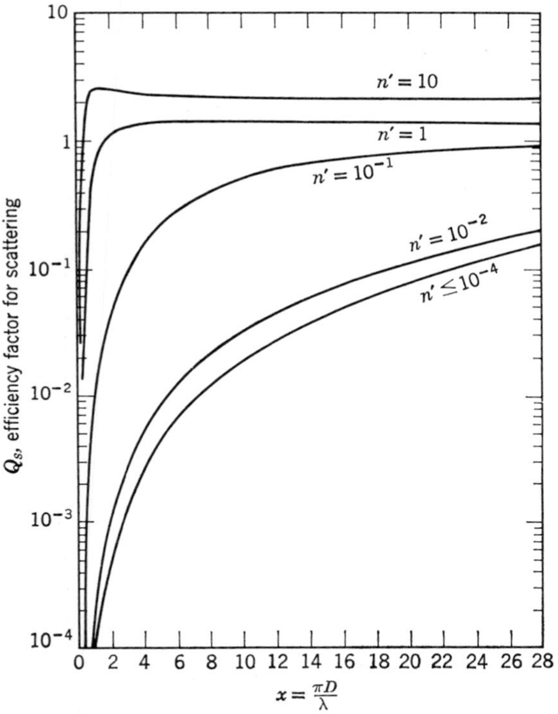

Fig. 2-11. Effects of size parameter x and n' on efficiency factor for scattering for $m = 1.01 - in'$. (From G. N. Plass [41]).

theoretical considerations. A considerable number of data are reported in the literature on the refractive indices of conducting and nonconducting materials; the *American Institute of Physics Handbook* [45] and the *International Critical Tables* [46] give comprehensive tabulations. Almost all of the experimental data on refractive indices reported in the literature are relative to air, because most optical systems have air as the surrounding medium. A close scrutiny of the experimental data on the refractive indices for metals and partially conducting materials reveals that in some cases there is disagreement among the values reported for the same material by different investigators. These differences arise from the extreme sensitivity of the optical constants to the purity of the sample, the method of preparation, and the experimental setup. The optical constants vary with the chemical composition of the material and the wavelength of the incident radiation.

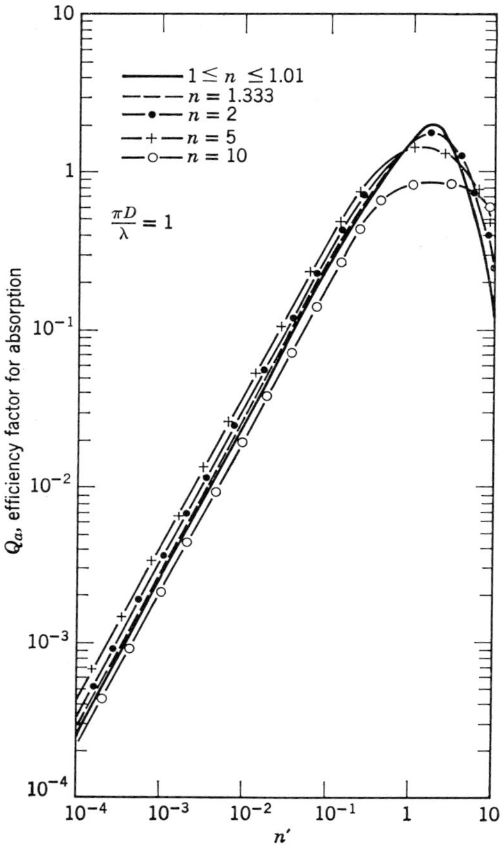

Fig. 2-12. Effects of n and n' on efficiency factor for absorption for $x = \pi D/\lambda = 1$. (From G. W. Kattawar and G. N. Plass [42]).

Figure 2-14 shows the refractive indices of transparent materials such as crystals and glasses as a function of the wavelength. Table 2-2 gives the complex refractive index $m = n - in'$ for a number of absorbing materials at ordinary temperatures. To illustrate some of the inconsistencies in the measured values of refractive indices for conducting materials, we have included in this table three different sets of refractive index data for iron obtained from three independent sources. The results differ almost by a factor of 2. Deirmendjian [30, p. 85] pointed out that recently reported experimental data for liquid water around $\lambda = 3.0\ \mu$ showed an absorption some 5 times the old value. Therefore the accuracy of theoretical predictions of the absorption, scattering, and reflection properties of materials is limited

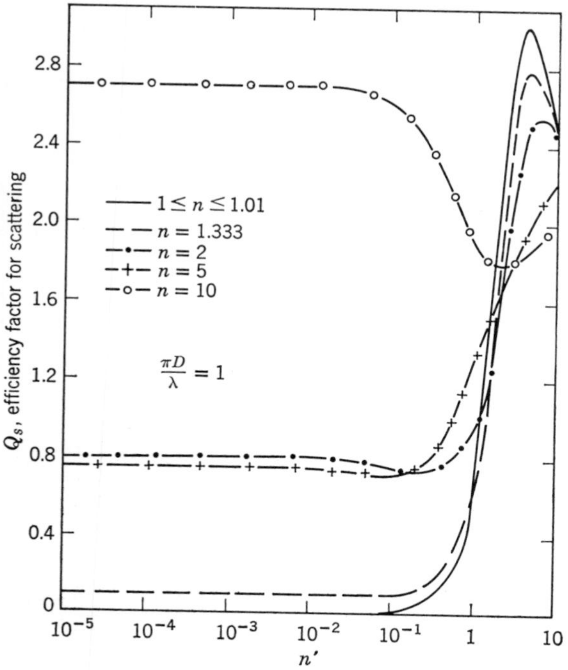

Fig. 2-13. Effects of n and n' on efficiency factor for scattering for $x = \pi D/\lambda = 1$. (From G. W. Kattawar and G. N. Plass [42]).

in most cases by the degree of accuracy of the available data on refractive index.

For an absorbing medium having a complex refractive index $m = n - in'$, the complex part n' is related to the spectral absorption coefficient κ_λ by the relation [27, p. 267]

$$\kappa_\lambda = \frac{4\pi n'}{\lambda} \tag{2-56}$$

which has dimensions in reciprocal centimeters (cm^{-1}).

2-6 ABSORPTION OF RADIATION BY GASES

The absorption (or emission) of radiation by gases does not take place continuously over the entire spectrum; rather, it occurs over a large number

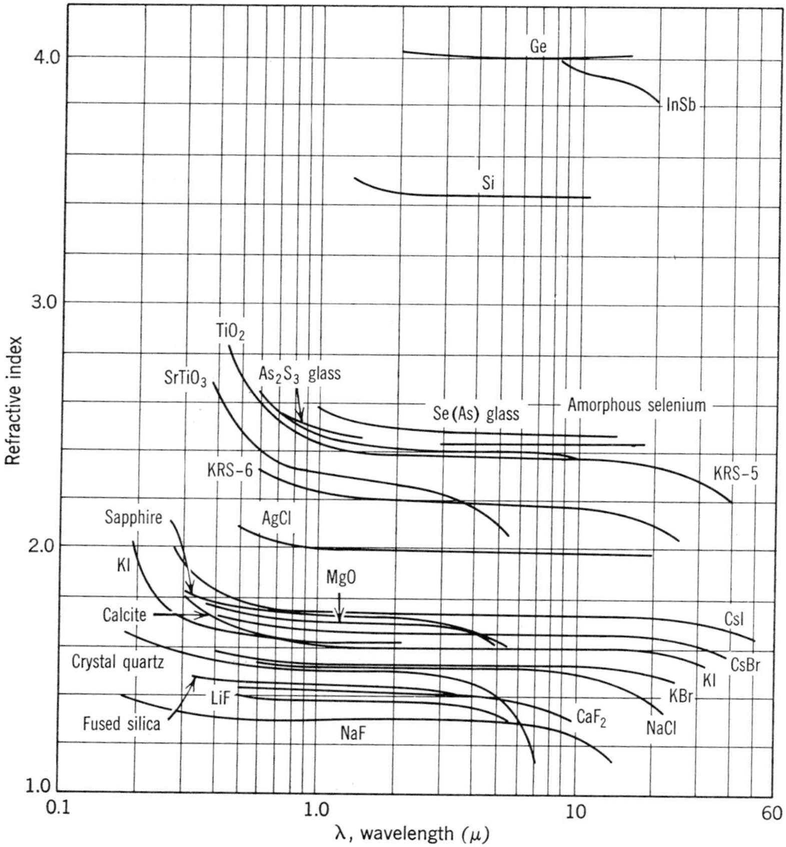

Fig. 2-14. Refractive indices of various transparent materials such as crystals and glasses as a function of wavelength. (From reference 45).

of relatively narrow strips of intense absorption (or emission). Figure 2-15 shows the absorption spectrum for water vapor in the far-infrared region (i.e., $\lambda = 18\ \mu$ to $75\ \mu$) from the measurements by Randall et al. [47]. The spectrum consists of a large number of peaks. Figure 2-16 shows the absorption spectrum for carbon dioxide from the data by Edwards [48]. Here the spectrum is composed of four absorption bands positioned approximately at wavelengths 15, 4.3, 2.7, and 1.9 μ.

In this section we discuss briefly the absorption and emission characteristics of gases and describe various models for the representation of gas absorption. The reader should refer to the books by Penner [49] and Herzberg [50] for comprehensive treatments of this subject.

Table 2-2 Complex Indices of Refraction, $m = n - in'$, of Absorbing Materials at Ordinary Temperature

Substance	λ, μ	n	n'
Aluminum[a] (evap.)	0.22	0.14	2.35
	0.4	0.40	4.45
	0.9	1.96	7.7
	2.0	2.3	16.5
	6.0	10.8	42.6
	10.0	26.0	67.3
Copper[a] (evap.)	0.5	2.42	0.88
	1.0	6.27	0.197
	5.0	27.45	2.92
	10.25	60.6	11.0
Germanium[a] (evap.)	0.4	2.3	2.8
	1.0	5.1	0.45
	2.0	4.35	0.03
	5.0	4.3	
	10.0	4.3	
Gold[a] (evap.)	0.2	1.24	0.92
	0.5	0.84	1.84
	1.0	0.179	6.04
	1.95	1.3	10.7
	6.65	12.9	35.5
	9.9	25.2	55.9
Iron[b]	0.441	2.66	3.84
	0.559	3.46	3.88
	0.668	3.57	4.03
Iron[a] (bulk)	0.589	2.36	3.20
Iron[c]	0.441	1.28	1.37
	0.589	1.51	1.63
	0.668	1.70	1.84
Mercury[a] (liquid)	0.4	0.73	3.01
	0.6	1.39	4.39
	0.8	2.14	5.33
Nickel[a] (evap.)	1.12	2.63	4.28
	2.0	3.74	8.80
	4.4	4.35	10.59
	6.75	5.86	15.2
	10.5	8.86	22.5
Silver[a] (evap.)	0.3	1.2	0.8
	0.6	0.060	3.75
	1.0	0.129	6.83
	4.0	1.89	28.7
	10.0	10.69	69.0
Silver[a] (bulk)	0.316	1.13	0.43
	0.500	0.17	2.94
	0.589	0.18	3.94
Silver[a] (chemically deposited)	0.75	0.17	5.16
	1.00	0.24	6.96
	1.50	0.45	10.7
	4.37	4.34	32.6
Sodium[a] (vacuum deposited)	0.254	0.026	0.621
	0.365	0.042	1.44
Tungsten[a] (bulk)	0.579	2.76	2.71
	0.589	3.46	3.25
Water[d] (liquid)	0.45	1.34	0.0
	0.70	1.33	0.0
	1.61	1.315	0.0
	2.25	1.290	0.0
	3.90	1.353	0.0059
	5.30	1.315	0.0143
	8.15	1.29	0.0472
	10.00	1.212	0.0601
Zinc[a]	0.257	0.554	0.612
	0.361	0.720	2.610
	0.468	1.049	3.485
	0.668	2.618	5.083

[a] From *American Institute of Physics Handbook* [45].
[b] From D. Deirmendjian, R. Clasen, and W. Viezee [37] (based on data from Yolken and Kruger).
[c] From H. C. Van de Hulst [27].
[d] From D. Deirmendjian, R. Clasen, and W.Viezee. [37] (based on data from M. Centeno)

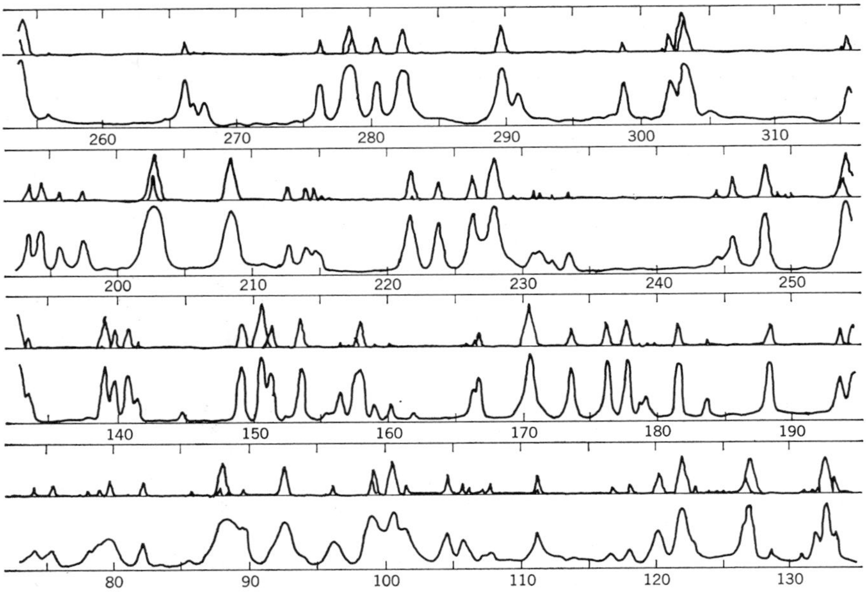

Fig. 2-15. Absorption spectra for water vapor. (From H. M. Randall et al. [47]).

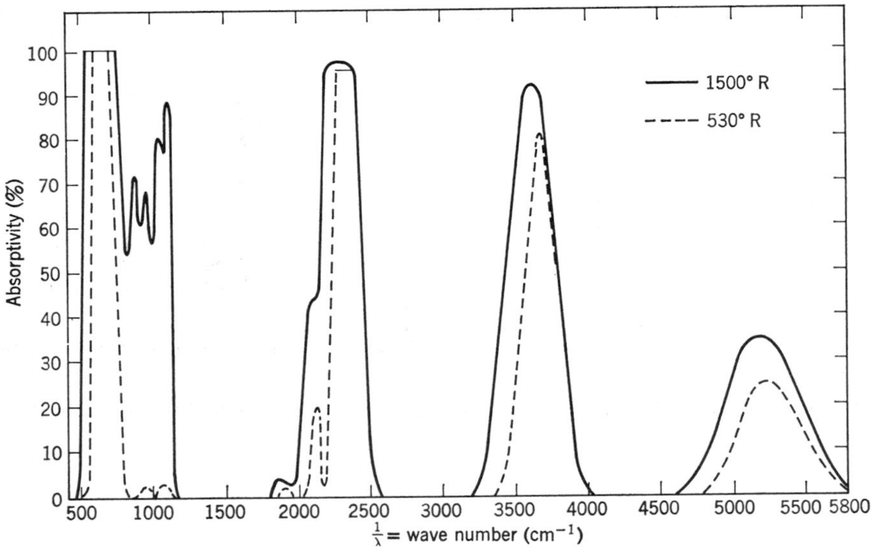

Fig. 2-16. Absorption spectra for carbon dioxide. (From D. K. Edwards [48]).

The absorption (or emission) of radiation by gases is caused by changes in the energy levels of the molecules, which involve *electronic*, *vibrational*, and *rotational* changes. In general, the transition between electronic levels gives rise to spectral lines in the visible and shorter wavelengths (i.e., ultraviolet); the transition between vibrational levels, to spectral lines in the infrared region; and the transition between rotational levels, to lines in the far-infrared region. When the frequency of the energy has the proper value, the changes in the vibrational and rotational levels are coupled, and the transition occurs simultaneously. Since vibrational energies are larger than rotational energies, the resulting spectrum consists of closely spaced spectral lines within a narrow wavelength; this is called the *vibration rotation band.* Therefore a proper description of the absorption characteristic of a gas as a function of the wavelength is very complicated. Consider, for example, a beam of monochromatic radiation of intensity I_ν traveling in a gas layer in the direction $\hat{\mathbf{\Omega}}$. If the scattering of radiation by the gas molecules is negligible and no scattering particles exist in the gas, the radiation intensity will be attenuated as a result of absorption and augmented as a result of emission of radiation by the gas molecules. Let dI_ν be the net increase in the intensity of radiation in traveling an elemental distance ds in the direction $\hat{\mathbf{\Omega}}$.

By assuming the Kirchhoff law is valid the following energy balance equation can be written:

$$dI_\nu = \kappa_\nu I_{\nu b}(T)\, ds - \kappa_\nu I_\nu\, ds \tag{2-57}$$

Here the first term on the right hand side represents the increase in the intensity due to emission by the matter, and the second term represents the attenuation of intensity due to absorption by the matter in traveling a distance ds in the direction $\hat{\mathbf{\Omega}}$. Equation 2-57 can be rewritten in the form[5]

$$\frac{dI_\nu}{ds} + \kappa_\nu I_\nu = \kappa_\nu I_{\nu b}(T) \tag{2-58}$$

where s is the distance measured in the direction of propagation $\hat{\mathbf{\Omega}}$, $I_{\nu b}(T)$ is the Planck function, and κ_ν is the spectral absorption coefficient. Let the condition at $s = 0$ be given as

$$I_\nu = I_{\nu 0} \quad \text{at } s = 0 \tag{2-59}$$

The solution of Eq. 2-58 subject to the condition of Eq. 2-59 yields

$$I_\nu = I_{\nu 0} e^{-\int_0^s \kappa_\nu\, ds'} + \int_0^s \kappa_\nu I_{\nu b}(T) e^{-\int_{s'}^s \kappa_\nu\, ds''}\, ds' \tag{2-60}$$

If the pressure, temperature, and composition of the gas are assumed to be constant, κ_ν and $I_{\nu b}(T)$ are considered independent of position, and Eq. 2-60 simplifies to

$$I_\nu = I_{\nu 0} e^{-\kappa_\nu s} + I_{\nu b}(T)(1 - e^{-\kappa_\nu s}) \tag{2-61}$$

By referring to the solution given by Eq. 2-61, the *spectral transmissivity* Γ_ν of a layer of uniformly distributed absorbing gas of thickness s, measured in the direction of propagation of radiation, is defined as

$$\Gamma_\nu \equiv e^{-\kappa_\nu s} \tag{2-62a}$$

and the *spectral absorptivity* α_ν as

$$\alpha_\nu \equiv 1 - e^{-\kappa_\nu s} = 1 - \Gamma_\nu \tag{2-62b}$$

When the Kirchhoff law is applicable, the spectral absorptivity α_ν and the spectral emissivity ε_ν are equal, and Eq. 2-62b characterizes also the spectral emissivity of a gas layer of thickness s.

We integrate Eq. 2-61 over an absorption band in a finite frequency interval $\Delta\nu$ and obtain

$$I \equiv \int_{\Delta\nu} I_\nu \, d\nu = \int_{\Delta\nu} I_{\nu 0} e^{-\kappa_\nu s} \, d\nu + \int_{\Delta\nu} I_{\nu b}(T)(1 - e^{-\kappa_\nu s}) \, d\nu \tag{2-63}$$

If the frequency interval $\Delta\nu$ is large enough to contain many spectral lines but small enough to replace $I_{\nu 0}$ and $I_{\nu b}(T)$ by their average values $\bar{I}_{\nu 0}$ and $\bar{I}_{\nu b}(T)$, respectively, over the frequency interval $\Delta\nu$, then $\bar{I}_{\nu 0}$ and $\bar{I}_{\nu b}(T)$ can be taken outside the integral sign, and Eq. 2-63 becomes

$$I = \bar{I}_{\nu 0} \int_{\Delta\nu} e^{-\kappa_\nu s} \, d\nu + \bar{I}_{\nu b}(T) \int_{\Delta\nu} (1 - e^{-\kappa_\nu s}) \, d\nu \tag{2-64a}$$

By utilizing the above definitions of spectral transmissivity and absorptivity of a gas layer, Eq. 2-64a can be written as

$$I = \bar{I}_{\nu 0} \int_{\Delta\nu} \Gamma_\nu \, d\nu + \bar{I}_{\nu b}(T) \int_{\Delta\nu} \alpha_\nu \, d\nu \tag{2-64b}$$

The evaluation of the integrals in Eqs. 2-64 is very involved because for gases κ_ν is a complex function of frequency. The absorption of radiation by a single isolated line offers the simplest situation for evaluating these integrals. The absorption by a vibration-rotation band, however, is very difficult to analyze. Therefore several different models have been developed to represent the variation of κ_ν with frequency. We consider now some of these models to characterize gas absorption by a single isolated line and by a vibration-rotation band.

Single-Isolated-Line Models

Among a number of factors that influence the shape of a spectral line, *collision broadening* and *Doppler broadening* are the two important ones in the infrared region. Collision broadening results from the disturbing effects of neighboring gas molecules and is sometimes referred to as pressure broadening. Doppler broadening results from the thermal motion of the radiating gas molecules. Doppler effects control the shape of a line at high temperatures and/or at low pressures. Conversely, collision effects are important at low temperatures and/or at high pressures. When collision effects are dominant (i.e., Doppler effects are negligible), the shape of a single line is described approximately by the Lorentz [51] formula, given in the form [49, p. 42]

$$\kappa_\nu = \frac{K}{\pi}\left[\frac{a_c}{(\nu - \nu_0)^2 + a_c^2}\right] \tag{2-65}$$

where ν_0 is the center of the line, a_c is the *collision half-width* as illustrated in Fig. 2-17, and K is the *integrated intensity* of the line, defined as

$$K = \int_{-\infty}^{\infty} \kappa_\nu \, d\nu \tag{2-66}$$

The maximum value of κ_ν in Eq. 2-65 is obtained by setting $\nu = \nu_0$ in that equation:

$$\kappa_{\nu,\max} = \frac{K}{\pi a_c} \tag{2-67}$$

For $\nu - \nu_0 = \pm a_c$, the value of κ_ν in Eq. 2-65 becomes

$$\kappa_\nu = \frac{K}{2\pi a_c} \quad \text{for } \nu - \nu_0 = \pm a_c \tag{2-68}$$

which is equal to half the value of $\kappa_{\nu,\max}$.

When Doppler effects are dominant (i.e., collision broadening is negligible), the shape of the line is described by the relation

$$\kappa_\nu = \frac{K}{a_d\sqrt{\pi}} \exp\left[-\left(\frac{\nu - \nu_0}{a_d}\right)^2\right] \tag{2-69a}$$

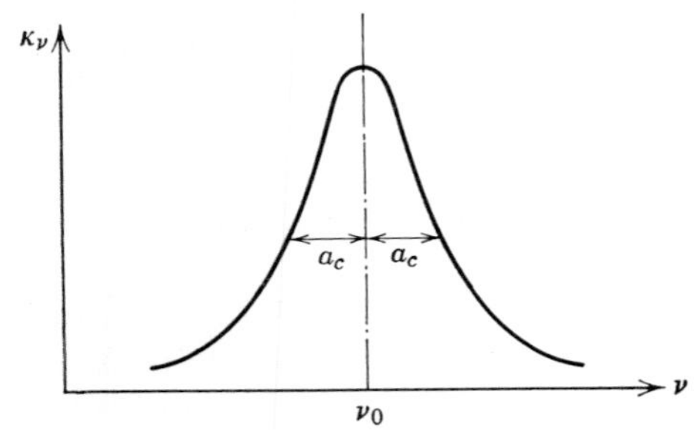

Fig. 2-17. A simple spectral line.

where a_d is the *Doppler half-width*, defined as

$$a_d = \frac{\nu}{c}\left(\frac{2kT}{m}\right)^{1/2} \tag{2-69b}$$

Here k is the Boltzmann constant, T the absolute temperature, and m the mass of the particle. Equation 2-69b shows that the shape of the spectral line is Gaussian for Doppler broadening.

The integrated emission of a single spectral line for a gas layer of thickness y is given by

$$E_l = \int_{\nu=0}^{\infty} I_{\nu b}(T)(1 - e^{-\kappa_\nu y})\, d\nu \tag{2-70a}$$

The Planck function $I_{\nu b}(T)$ in this equation can be approximated by its mean value at the frequency ν_0 of the line center, since the line emission occurs in a narrow region about ν_0. Then Eq. 2-70a becomes

$$\frac{E_l}{I_{\nu_0 b}(T)} = \int_{\nu=0}^{\infty} (1 - e^{-\kappa_\nu y})\, d\nu \equiv W_l \tag{2-70b}$$

where W_l, called the *equivalent width of a single line*, is a dimensional quantity; its units depend on the units of the integration variable (i.e., whether it is frequency, wave number, etc.).

The equivalent width of a single line for a collision-broadened line can be determined from Eq. 2-70b by substituting in that equation κ_ν from Eq. 2-65 and performing the integration. Similarly, W_l for a Doppler-broadened line can be determined by substituting in Eq. 2-70b the value of κ_ν from Eq. 2-69 and performing the integration. The details of these calculations and the resulting expressions for W_l can be found in the book by Penner [49, pp. 38–45].

Band Models

The absorption spectrum of a vibration-rotation band is made up of a number of spectral lines of varying spacing and intensity concentrated over a narrow wavelength band. Each of these lines contributes to the absorption at any frequency ν. Then the spectral absorption coefficient κ_ν at the frequency ν is taken as the sum of these individual contributions in the form

$$\kappa_\nu = \sum_i \kappa_i[\nu - \nu_i, K_i, a_i] \tag{2-71}$$

where $\kappa_i[\nu - \nu_i, K_i, a_i]$ is the contribution of line i centered at frequency ν_i, of half-width a_i, and of integrated intensity K_i.

Various models have been developed to determine the absorption coefficient for a vibration-rotation band. One of the earliest models is due to Schack [52]. The principal band models used in engineering applications are the *Elsasser model* [53], *the Mayer-Goody statistical model* [54, 55], and the *random superposition of Elsasser bands* [56].

The integrated emission of a vibration-rotation band over a frequency range from $\nu = \nu_1$ to $\nu = \nu_2$ for a gas layer of thickness y is given by

$$E_b = \int_{\nu_1}^{\nu_2} I_{\nu b}(T)(1 - e^{-\kappa_\nu y})\, d\nu \tag{2-72a}$$

where κ_ν is the spectral absorption coefficient for the band. The Planck function $I_{\nu b}(T)$ in this equation, if approximated by its average value $\bar{I}_{\nu b}(T)$ over the frequency interval ν_1 to ν_2, can be taken out of the integral. Then Eq. 2-72a becomes

$$\frac{E_b}{\bar{I}_{\nu b}(T)} = \int_{\nu_1}^{\nu_2} (1 - e^{-\kappa_\nu y})\, d\nu \equiv W_B \tag{2-72b}$$

where W_B is called the *equivalent width of a band* over the frequency interval $\Delta = \nu_2 - \nu_1$; this is analogous to the equivalent width of a single line, W_l, defined by Eq. 2-70b.

To evaluate the integral in Eq. 2-72b a knowledge of the spectral absorption coefficient of a band is needed. We present now a brief discussion of various band models to determine the mean absorptivity and mean transmissivity of a vibration-rotation band for a gas layer of thickness y.

(a) Elsasser Model

This model, developed by Elsasser [53], assumes that a vibration-rotation band consists of an infinite number of equal and equidistant spectral lines having a periodical pattern as illustrated in Fig. 2-18. Each line is assumed to have a shape described by the Lorentz formula, Eq. 2-65. The spectral absorption coefficient κ_ν at the frequency ν is obtained by summing up the

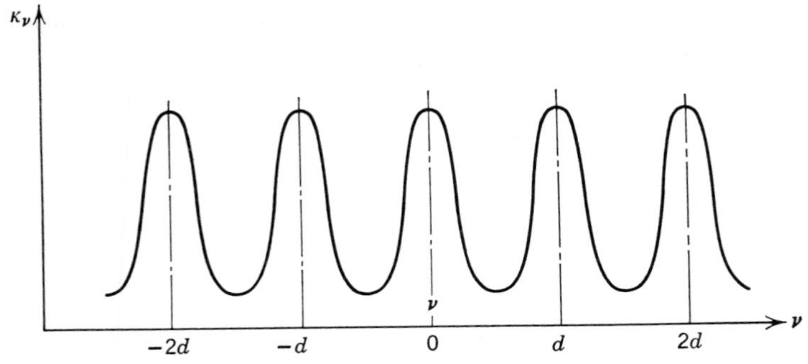

Fig. 2-18. Periodic pattern for equal and equidistant spectral lines.

contribution of each line to the absorption at ν (see Eq. 2-72):

$$\kappa_\nu = \sum_{n=-\infty}^{\infty} \frac{K}{\pi} \frac{a}{(\nu - nd)^2 + a^2} \tag{2-73}$$

where d is the frequency interval between the centers of the neighboring lines, a is the half-width of the line, and K is the integrated intensity of a single line.

The series in Eq. 2-73 can be expressed in terms of trigonometric functions as [53; 57, Chapter 7-4]

$$\kappa_\nu = \frac{K}{d} \frac{\sinh \beta}{\cosh \beta - \cos \xi} \tag{2-74a}$$

where

$$\beta \equiv \frac{2\pi a}{d} \qquad \text{and} \qquad \xi \equiv \frac{2\pi \nu}{d} \tag{2-74b}$$

The mean transmissivity over a frequency interval d for a gas layer of thickness y is defined as

$$\Gamma = \frac{1}{d} \int_{-d/2}^{d/2} e^{-\kappa_\nu y} \, d\nu \tag{2-75}$$

where κ_ν is given by Eq. 2-74. Equation 2-75 is applicable to any multiple of d because the lines are assumed to have a periodical pattern; therefore it represents the entire vibration-rotation band based on the Elsasser model. Equation 2-75 can be written in the form

$$\Gamma = \frac{1}{2\pi} \int_{-\pi}^{\pi} e^{-\kappa_\nu y} \, d\xi \tag{2-76a}$$

where we have defined

$$\xi = \frac{2\pi \nu}{d} \tag{2-76b}$$

Introducing κ_ν from Eq. 2-74a into Eq. 2-76a, we obtain the mean transmissivity of a vibration-rotation band as

$$\Gamma = \frac{1}{2\pi} \int_{-\pi}^{\pi} \exp\left(-\frac{Z\beta \sinh \beta}{\cosh \beta - \cos \xi}\right) d\xi \tag{2-77}$$

where we have defined

$$Z \equiv \frac{Ky}{2\pi a}, \qquad \beta = \frac{2\pi a}{d} \tag{2-78}$$

The absorptivity α for the band is evaluated from

$$\alpha = 1 - \Gamma \tag{2-79}$$

Although the integral in Eq. 2-77 cannot be evaluated analytically, simple approximate expressions can be obtained for the limiting cases. For $a \ll d$ we have $\beta \ll 1$; then, by setting $\cosh \beta \simeq 1$ and $\sinh \beta \simeq \beta$, the transmissivity of Eq. 2-77 simplifies to [53][6]

$$\Gamma = 1 - \operatorname{erf}\,[(\tfrac{1}{2}\beta^2 Z)^{1/2}] = 1 - \operatorname{erf}\left[\left(\frac{\pi a K y}{d^2}\right)^{1/2}\right] \tag{2-80}$$

and the absorptivity becomes

$$\alpha = \operatorname{erf}\,[(\tfrac{1}{2}\beta^2 Z)^{1/2}] \tag{2-81}$$

For example, at low pressures a is small compared with d and hence $\beta \ll 1$.

The transmissivity and absorptivity given by Eqs. 2-80 and 2-81, respectively, are correct to within 10 per cent when $Z > 1.25$ and $\beta < 0.3$ [56].

The Elsasser model described above can be used for a vibration-rotation band when the spacing between the lines is approximately uniform and the integrated line intensity is a slowly varying function of the frequency. The two parameters K and d can be determined from spectroscopic measurements.

(b) The Mayer-Goody Statistical Model

Mayer [54] and Goody [55] worked out independently a statistical model for the treatment of a vibration-rotation band when the absorption spectrum appears as a number of irregularly spaced lines having a random distribution of integrated line intensity. The model assumes that there is no correlation between line positions and line intensities. The derivation of this model is given, in addition to the original references, in the book by Penner [49, pp. 317–320]. Here we present some of the results obtained from this model.

The mean transmissivity of a band having an infinite number of lines with average spacing d is given as

$$\Gamma = \exp\left\{-\frac{1}{d}\int_{-\infty}^{\infty}\int_{0}^{\infty} p(\nu)[1 - \exp(-\nu s y)]\, d\nu\, d\nu'\right\} \tag{2-82}$$

where $p(\nu)\, d\nu$ is the probability that a line has an intensity in the frequency range ν to $\nu + d$, s is a line-shape parameter,[7] and y is the thickness of the layer. Here the probability function is so normalized that

$$\int_{0}^{\infty} p(\nu)\, d\nu = 1$$

To evaluate Γ from Eq. 2-82 the distribution function $p(\nu)$ and the line-shape parameter s must be known. For an exponential distribution and lines of Lorentz shape Eq. 2-82 simplifies to

$$\Gamma = \exp\left[-\frac{Ky}{d(1 + Ky/\pi a)^{1/2}}\right] \tag{2-83}$$

where K is the integrated line intensity, and a the line half-width.

For $Ky/\pi a \gg 1$, Eq. 2-83 simplifies to

$$\Gamma = \exp\left[-\left(\frac{\pi a K y}{d^2}\right)^{1/2}\right] = \exp\,(-\tfrac{1}{2}\beta^2 Z)^{1/2} \tag{2-84}$$

where β and Z were defined by Eqs. 2-78.

The mean absorptivity (or emissivity) is evaluated from

$$\alpha = 1 - \Gamma \tag{2-85}$$

(c) Random Superposition of Elsasser Bands

The Elsasser model assumes a situation in which the spectral lines are arranged at regular intervals with a periodic pattern, and the Mayer-Goody statistical model assumes that the lines are completely random. In an actual situation, however, the arrangement of lines may be neither completely regular nor completely random. For example, a vibration-rotation band may be composed of several bands superposed with band centers arbitrarily positioned. Plass [56] extended the statistical model to include random superposition of any number of Elsasser bands, each of which may have different integrated line intensity, line half-width, and line spacing. Consider a vibration-rotation band consisting of random superposition of N Elsasser bands. If $W_{B,i}$ is the equivalent width of an Elsasser band over an intensity distribution Δ_i for band i (see Eq. 2-72b), then the absorptivity of a gas layer for the vibration-rotation band obtained from the random-superposition-of-Elsasser-bands model is given as [56]

$$\alpha = 1 - \prod_{i=1}^{N}\left(1 - \frac{W_{B,i}}{\Delta_i}\right) \tag{2-86}$$

Comparison of Band Models

Figure 2-19 shows the absorptivity of a layer of gas as a function of the parameter[8] $\beta Z = Ky/d$ for $\beta = 2\pi a/d = 0.1$, determined from the random-superposition-of-Elsasser-bands model (i.e., $N = 1, 2, 5$). In one of the cases with $N = 2$ the integrated line intensities for the two Elsasser bands are taken as equal (i.e., $K_1 = K_2$), and in the other the integrated intensity for one of the Elsasser bands is taken as 10 times the other (i.e., $K_1 = 10K_2$). In the case with $N = 5$ the integrated intensities are taken as equal for all the bands (i.e., $K_1 = K_2 = \cdots = K_5$). Included in Fig. 2-19 are the absorptivities determined from the Elsasser model and the statistical model. It is apparent from this figure that the random-superposition-of-Elsasser-bands model provides an absorptivity which lies between the values predicted by the

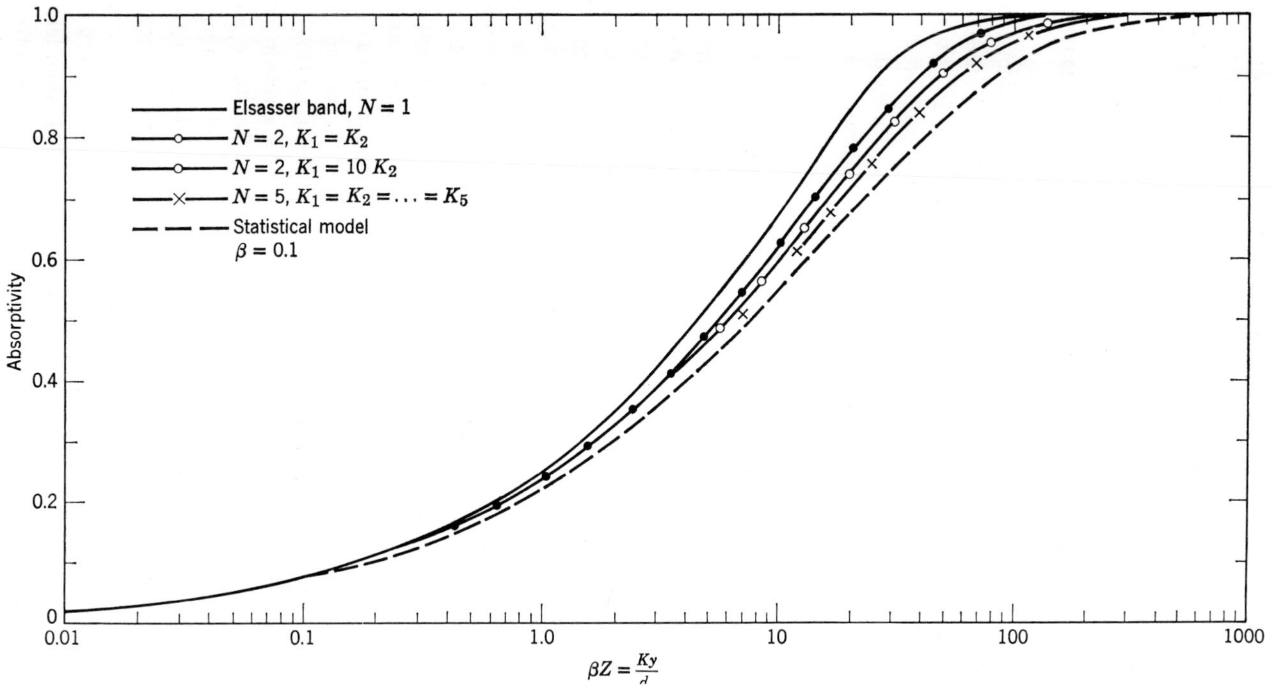

Fig. 2-19. Comparison of absorptivity evaluated from Elsasser model, statistical model, and random superposition of Elsasser bands model. (From G. N. Plass [56]).

Elsasser and the statistical models. Its limiting case for $N = 1$ corresponds to the Elsasser model, and as the number of bands becomes large the absorptivity obtained from the random-superposition-of-Elsasser-bands model approaches that from the statistical model.

2-7 EXPERIMENTAL DATA ON RADIATIVE PROPERTIES

In this section we discuss the experimental data on the reflectivity and emissivity of real surfaces, and the results of measurements of the absorption, emission, and scattering of radiation by matter. However, it is not possible to reproduce in the limited space available here the vast number of experimental data given in the literature. Only pertinent publications will be cited, and some of the results will be presented in order to illustrate the effects of various parameters on radiative properties.

Reflectivity and Emissivity of Real Surfaces

The definitions of reflectivity, emissivity, and absorptivity have already been introduced. Recalling the results from the electromagnetic theory on the propagation of plane waves, we note that the penetration of incident radiation into matter depends strongly on the absorption characteristics of the material. In the case of metals, the thermal radiation incident on the surface will not travel more than a few hundred angstroms before it is completely absorbed, because metals are strong absorbers. Therefore, for metals, the surface condition strongly influences the reflectivity and emissivity of the material. In the case of dielectric materials, however, it has been shown by Richmond [58] that radiative properties are less sensitive to surface conditions. Real surfaces deviate from ideality because of roughness, oxidization, and contamination. In the case of metals, therefore, it is most important to describe the surface conditions when reporting experimental data on reflectivity, emissivity, or absorptivity. Unfortunately, no standardized method is yet available to describe the actual condition of a surface. As a result most of the data reported in the literature on radiative properties of metallic surfaces suffer from such shortcomings; hence care must be exercised in interpreting experimental data when surface conditions are not adequately described.

The literature contains vast numbers of data on radiative properties of surfaces. For example, Gubareff, Janssen, and Torborg [20] give comprehensive compilations of emissivity and reflectivity from nearly 320 references. Hottel [21], Svet [22], Wood, Deem, and Lucks [23], and Fishenden (data compiled by Singham) [24] have published tabulations of radiative properties of surfaces. The most comprehensive tabulation of such properties has been given recently by Touloukian and DeWitt [25] as a function of wavelength of radiation together with information describing surface conditions.

In order to illustrate the effects of various parameters on radiative properties of materials we present here a limited number of data on the reflectivity and emissivity of metallic surfaces.

Figure 2-20 shows the influence of crystal structure on spectral reflectivity at normal incidence for an electropolished germanium single crystal and an evaporated germanium film (amorphous). The difference in reflectivity for the two cases apparently results from a difference in the structure of the germanium samples.

Figure 2-21 shows the specular reflectivity at normal incidence for optically smooth, flat samples of copper cut from the same high-purity ingot. One of these samples was prepared by electropolishing, and the other by conventional mechanical polishing. Although both samples are smooth and optically flat, the reflectivity of the mechanically polished sample appears to be lower than that of the electropolished sample. Bennett [59] attributed this difference to the surface damage resulting from the mechanical polishing process, since during electropolishing only a minimum amount of damage is introduced.

Figure 2-22 illustrates the effects of oxidation on total hemispherical emissivity. We note that oxidation increases emissivity at all wavelengths.

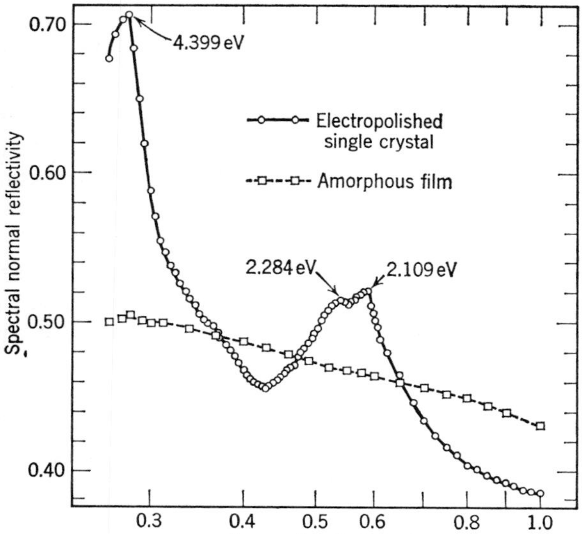

Fig. 2-20. Influence of crystal structure on spectral reflectivity at normal incidence of germanium sample. (From H. E. Bennett [59]).

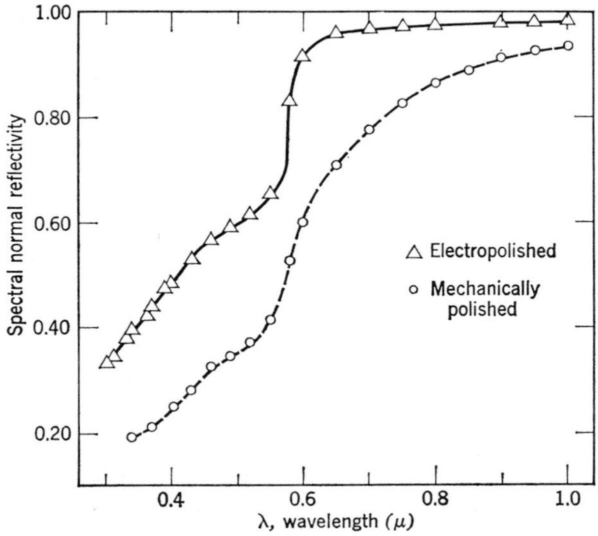

Fig. 2-21. Spectral normal reflectivity of electropolished and mechanically polished copper. (From H. E. Bennett [59]).

Figure 2-23 shows that not only oxidation, but also the degree of oxidation, is an important factor that influences emissivity.

Absorption, Emission, and Scattering Data

The absorption and emission of radiation by gases such as CO, CO_2, water vapor, and NH_3 are very important in heat transfer from flames in combustion chambers and furnaces. The radiation from high-temperature air is significant in nuclear detonations, high-speed flights, re-entry vehicles, and missiles. The transmission of infrared radiation through the earth's atmosphere is of interest in solving the problems encountered in astrophysics and meteorology. As a result, a considerable amount of theoretical and experimental work has been performed to predict the absorption, emission, and transmission of radiation through gases. We have already discussed the theoretical work in this area. The reader can find in references 60, 61, and 62 an extensive survey of spectral absorption coefficients for gases, determined either theoretically or experimentally. We present now some of the representative data on the absorption, emission, and scattering of radiation by matter, discuss the results, and cite pertinent literature.

Carbon Monoxide

Figure 2-24 shows the spectral absorption coefficient for the first overtone of carbon monoxide as a function of the wave number at room temperature.

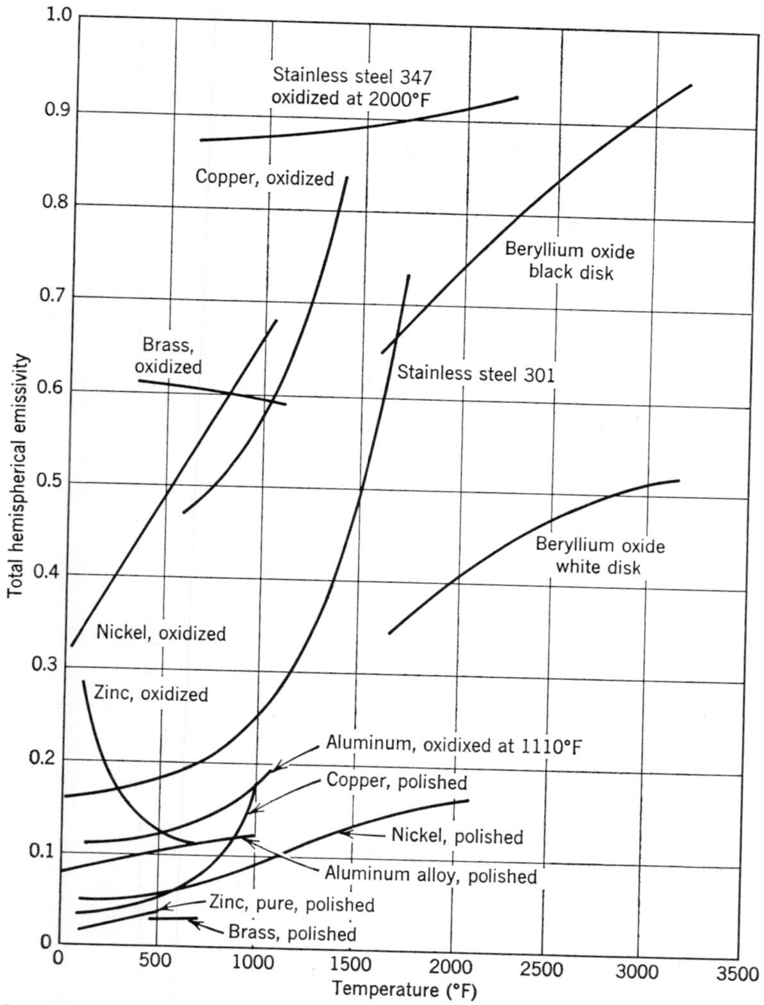

Fig. 2-22. Effects of temperature and oxidation on total hemispherical emissivity of metals. (Based on data from G. G. Gubareff, J. E. Janssen, and R. H. Torborg [20]).

Davies [63] reported spectral absorption measurements at high temperatures at three different wavelengths, Breeze and Ferriso [64] determined total band absorption at high temperatures, and Abu-Romia and Tien [65] measured absorption at temperatures from 300° to 1500°K.

In many engineering applications Planck mean and Rosseland mean absorption coefficients are needed. Figures 2-25*a* and 2-25*b* show the Planck mean and Rosseland mean absorption coefficients (calculated according to

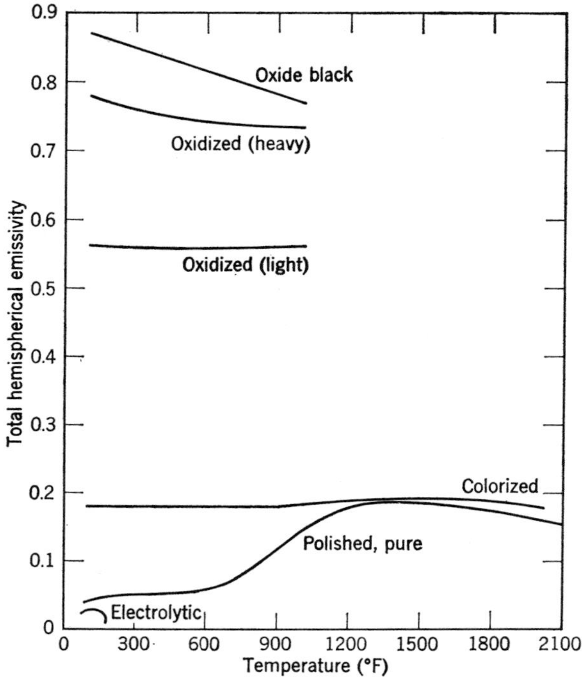

Fig. 2-23. Effects of various degrees of oxidation on the total hemispherical emissivity of copper. (From G. G. Gubareff, J. E. Janssen, and R. H. Torborg [**20**]).

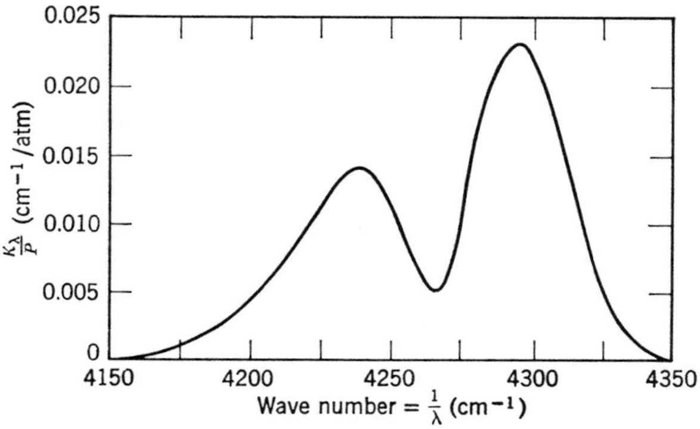

Fig. 2-24. Spectral absorption coefficient of carbon monoxide at room temperature. (From S. S. Penner [49]).

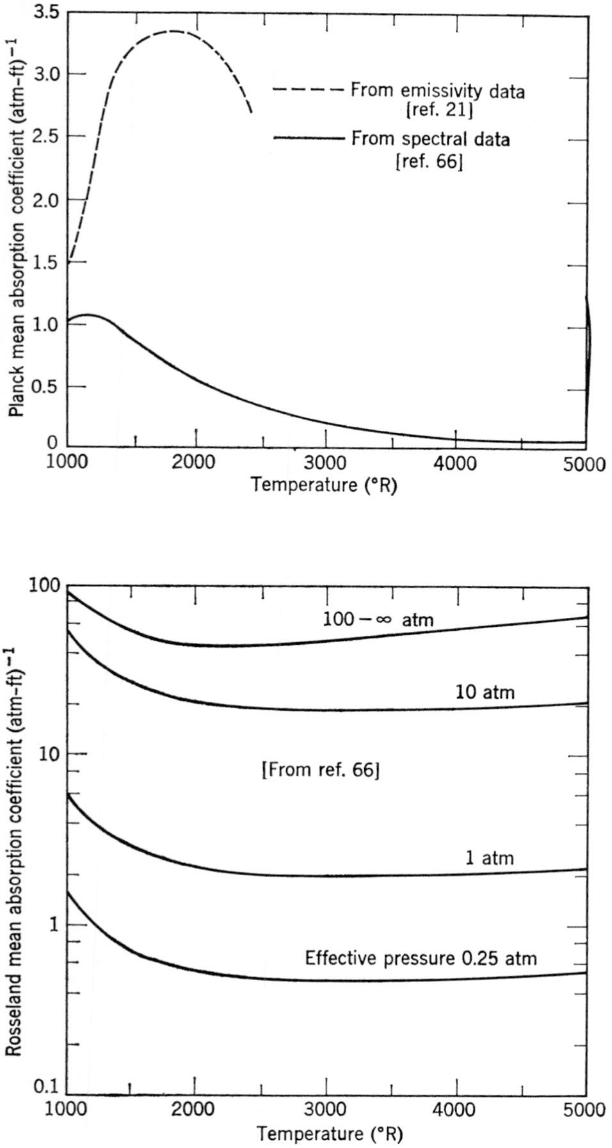

Fig. 2-25. (*a*) Planck mean absorption coefficient of carbon monoxide. (*b*) Rosseland mean absorption coefficient of carbon monoxide.

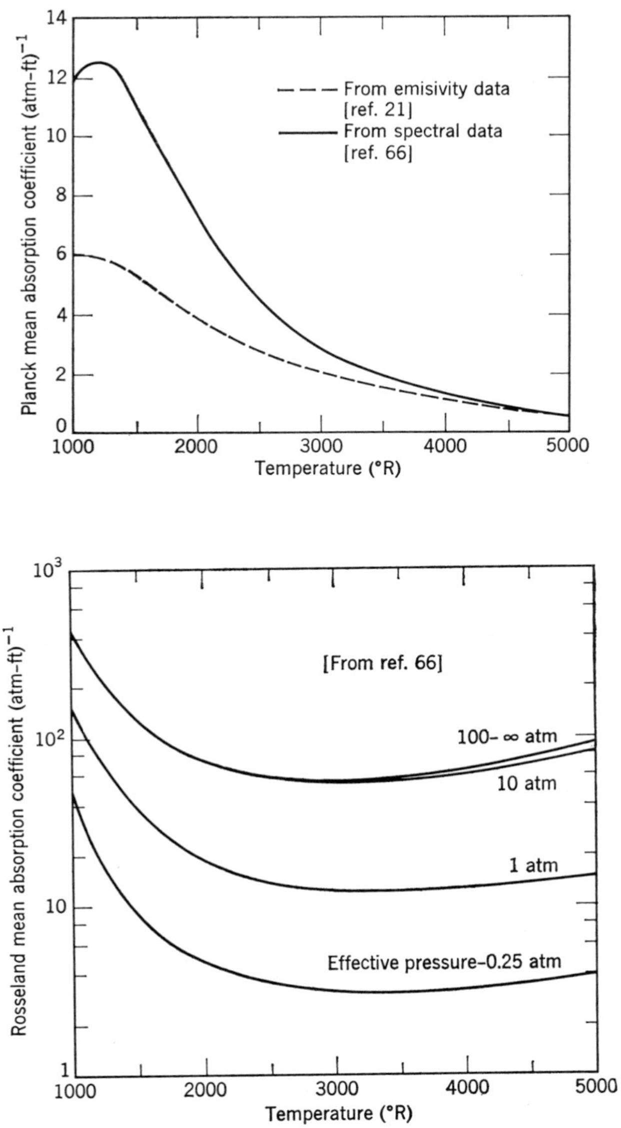

Fig. 2-26. (*a*) Planck mean absortion coefficient of carbon dioxide (*b*) Rosseland mean absorption coefficient of carbon dioxide.

Eqs. 1-89 and 9-21) for CO in the infrared region. As can be seen from this figure, there is a large deviation between the Planck mean values calculated from the spectral data given in reference 66 and those from the emissivity data of reference 21. The data by Abu-Romia and Tien [66] are recommended.

Carbon Dioxide

Figure 2-16 shows the absorption spectrum for carbon dioxide. As already mentioned, the spectrum is composed of bands positioned approximately at wavelengths 15, 4.3, 2.7, and 1.9 μ. Plass [67] reported the spectral emissivity of CO_2 as a function of temperature in the range 1800 to 2500 cm^{-1}; Edwards [68] presented experimental data and empirical correlations for absorption by infrared bands of CO_2 at elevated pressures and temperatures. Figures 2-26*a* and 2-26*b* show the Planck mean and Rosseland mean absorption coefficient for CO_2 for infrared radiation. Of the two Planck mean absorption coefficients shown in Fig. 2-26*a* the data from reference 66 are recommended.

Water Vapor

Water vapor contributes to the emission and absorption of radiation in industrial furnaces, rocket exhausts, combustion chambers, and the atmosphere of the earth. The infrared absorption or emission of water vapor in the 1 to 3 μ region has been measured by Howard, Burch, and Williams [69] at low temperatures, while the strong absorption or emission in the 2.7 μ region has been studied by several investigators [70–72]. Edwards et al. [73] presented measurements of total absorptance in the 1.38, 1.87, 2.7, and 6.3 μ regions at temperatures from 300°K to 1100°K. Figure 2-27 shows the spectral absorption coefficient for water vapor at 1000°K for the 2.7 μ region,

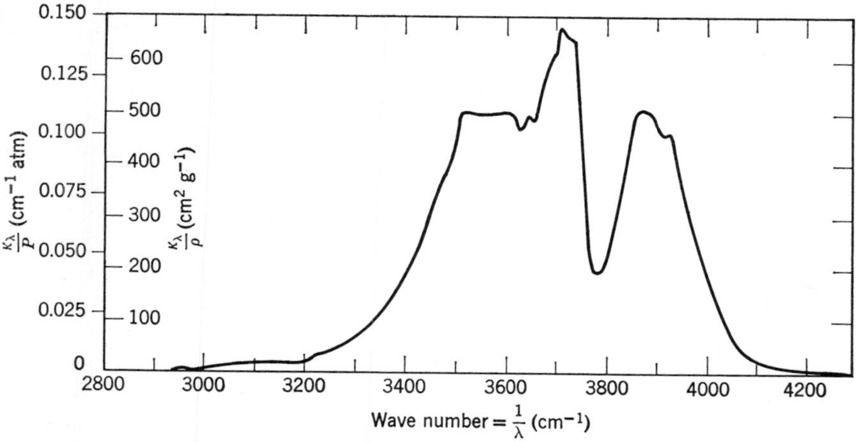

Fig. 2-27. Spectral absorption coefficient for water vapor at 1000°K for 2.7 μ region. (From R. Goldstein [58]).

obtained from measurements by Goldstein [74]. Figures 2-28*a* and 2-28*b* show the Planck mean and Rosseland mean absorption coefficients for infrared radiation.

Air

Interest in the properties of high-temperature air has increased in recent years because of heat transfer problems encountered in radiation from nuclear detonations, the re-entry of missiles and satellites, and so forth. As a result a considerable amount of work has been performed to determine the absorption and emission of radiation by air [75–77]. We present in Table 2-3 the Planck mean absorption coefficient (see Eq. 1-89) for air in the temperature range from 1000°K to 12000°K, of densities ρ/ρ_0 from 10 to 10^{-6} (i.e., $\rho_0 = 1.293 \times 10^{-3}$ g/cm³), calculated by Armstrong et al. [78]. The wavelength range covered in these calculations is 1167 Å to 19837 Å. The calculations of the Planck mean absorption coefficient by Kivel and Bailey [79] differ from the results shown in Table 2-3 by a factor of 10 or 100 at 3000°F and are in substantial agreement above 6000°F.

Liquid Water

The absorption spectrum of liquid water has been the subject of numerous investigations. The absorption in the spectral region from 2.00 to 2.65 μ was measured by Collins [80], from 2.5 to 7.5 μ by Fox and Martin [81], from 0.76 to 2.50 μ by Curcio and Petty [82], from 42 to 2000 μ by Stanevich and Yaroslavskii [83], from 0.58 to 0.79 μ by Sullivan [84], from 2200 to 3000 cm^{-1} and from 3700 to 7600 cm^{-1} by Goldstein and Penner [85], and from 30 to 330 μ by Draegert et al. [86]. Figure 2-29 shows the spectral absorption coefficients in the near-infrared spectrum at temperatures of 27, 89, 159, and 209°C, obtained from the measurements by Goldstein and Penner [85]. The regions of strong absorption are localized between 4600 and 5900 cm^{-1} and between 5900 and 7800 cm^{-1}. The spectral absorption coefficients of Curcio and Petty [82] are about 15 per cent lower than the data in Fig. 2-29, while the data of Collins [80] are about 10 per cent lower.

Experiments performed on the penetration of light into natural waters [87–90] have shown that there is also scattering in natural waters as a result of the presence of organisms, gas bubbles, and detritus. This is especially true of waters having a high degree of biological activity and of coastal waters, where detritus is stirred up by wave action.

Glass

Figure 2-30 shows the spectral absorption coefficient for window glass, obtained from measurements by Neuroth [91]. It is apparent from this figure that the glass is transparent to visible radiation but is almost opaque to long-wavelength infrared radiation.

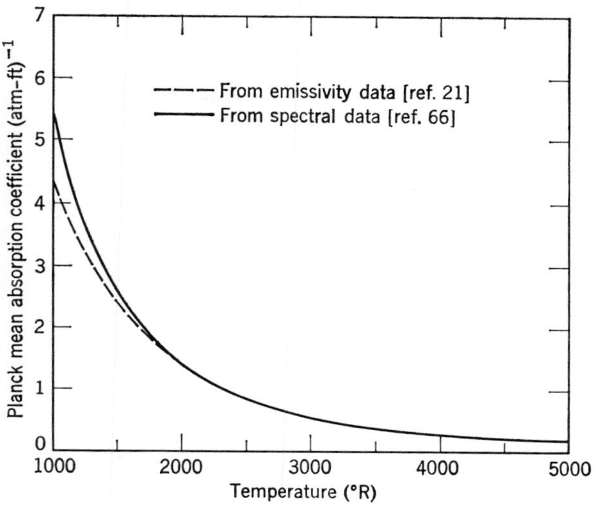

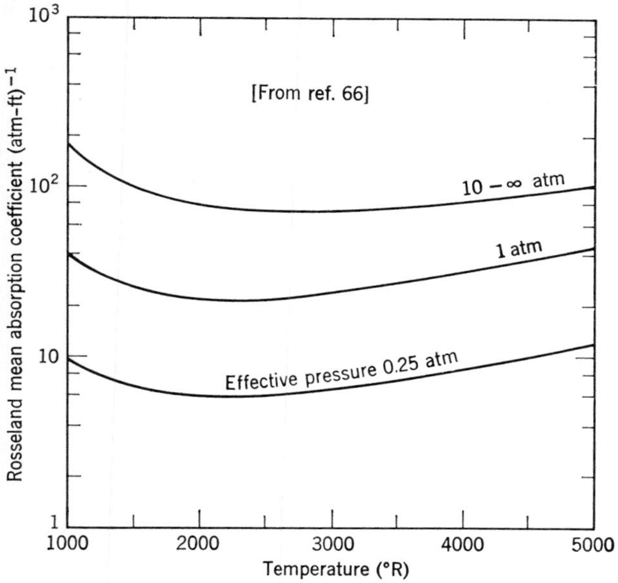

Fig. 2-28. (*a*) Planck mean absorption coefficient of water vapor. (*b*) Rosseland mean absorption coefficient of water vapor.

Table 2-3 Planck Mean Absorption Coefficient κ_m (cm^{-1}) for Various Temperatures and Densities,[a] ρ/ρ_0 ($\rho_0 = 1.293 \times 10^{-3}$ g/cm^3)
(The numbers in parantheses denote multiplication by the corresponding power of 10)

T (°K)	Density, ρ/ρ_0: 10^1	10^0	10^{-1}	10^{-2}	10^{-3}	10^{-4}	10^{-5}
3000	1.82 (−4)	1.78 (−5)	1.65 (−6)	1.30 (−7)	6.24 (−9)	1.62 (−10)	3.00 (−12)
4000	1.46 (−2)	1.14 (−3)	5.65 (−5)	1.47 (−6)	3.30 (−8)	9.37 (−10)	3.68 (−11)
6000	2.91 (−1)	9.44 (−3)	2.76 (−4)	1.14 (−5)	9.07 (−7)	6.70 (−8)	2.53 (−9)
8000	5.9 (−1)	2.15 (−2)	1.12 (−3)	6.04 (−5)	2.06 (−6)	6.35 (−8)	1.91 (−9)
12000	2.60	1.31 (−1)	8.28 (−3)	6.27 (−4)	4.58 (−5)	2.05 (−6)	3.80 (−8)
18000	6.51	5.77			2.11 (−4)		

[a] From B. H. Armstrong et al. [78].

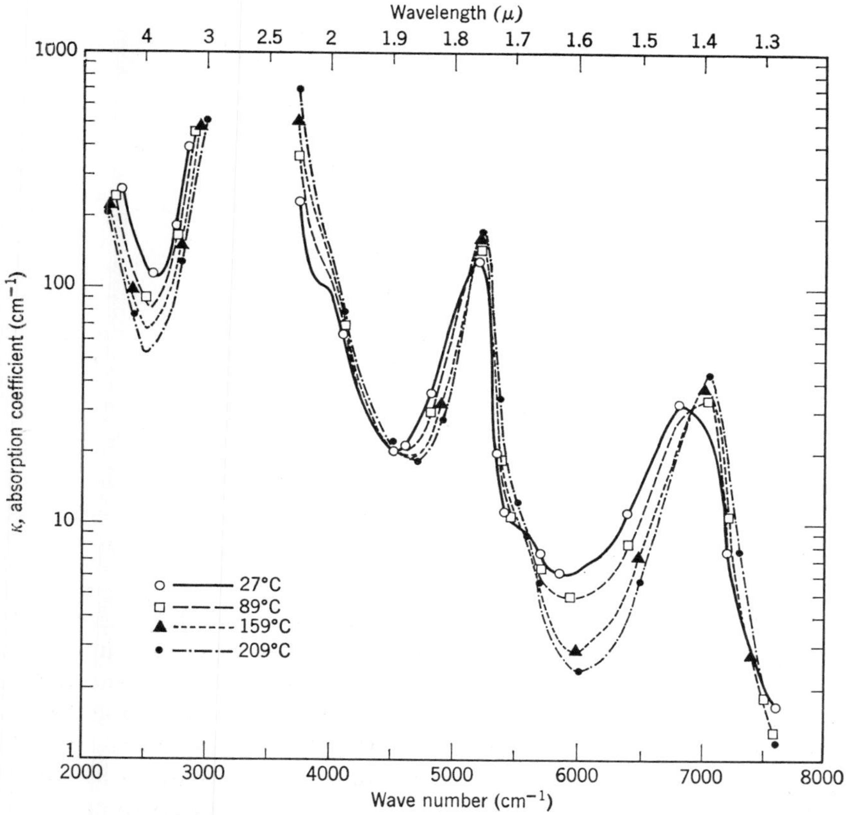

Fig. 2-29. Spectral absorption coefficient of liquid water at temperatures of 27, 89, 159, and 209°C. (From R. Goldstein and S. S. Penner [85]).

Absorption and Scattering of Radiation by Particles

For particles having a simple geometry such as a sphere the absorption and scattering coefficients and the phase function for scattering can be predicted from the Mie theory. Several experiments performed with spherical particles to investigate the validity of the Mie theory have been reported in the literature. Here we discuss some of the results from these investigations in order to illustrate the extent of agreement between the theory and experiments. Lewis and Lothian [92] measured the scattering coefficient for spherical particles (barium sulphate and *Lycoperdon pyrifarme*) of radii 4.6 and 1.8 μ in the wavelength range 0.4 to 2.0 μ; their results were in approximate agreement with the theoretical predictions. Sinclair's [93] work on scattering from a

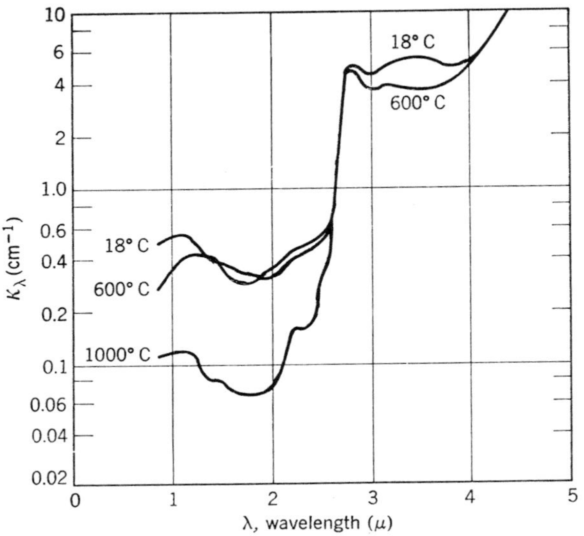

Fig. 2-30. Spectral absorption coefficient of window glass. (From N. Neuroth [91].)

series of monodisperse stearic acid fogs (refractive index 1.43) agreed with the Mie theory within experimental error of about 20 to 30 per cent over the range of size parameters from $x = 2.4$ to 12. Hodkins [94] extended these observations to higher values of the size parameter ($x \simeq 120$) and found good agreement with the Mie theory. Hepplestone and Lewis [95, 96] reported measurements with latex spheres in water for size parameters in the range 25 to 28 and for polystyrene spheres for size parameters from 1 to 20; they found good agreement between the shapes of the experimental curves and those obtained from the theory. For the spherical particles considered in all these investigations, the predictions from the Mie theory appear to be in reasonably good agreement with the results of experiments.

In many important engineering applications, however, particles are highly irregular in shape. For example, particles seeded in gas for the thermal radiation shielding of rockets, in advanced reactor concepts, and aerosols encountered in atmospheric pollution are not spherical. In such situations an experimental approach is the only way to determine the absorption and scattering properties of a cloud of particles in suspension in a gas. Several experiments have been reported in the literature on the prediction of the radiative properties of a cloud of particles of irregular shape. Lanzo and Ragsdale [97] measured the absorption of thermal radiation by submicron-sized refractory particles suspended in an air stream as a function of size and concentration. The air stream seeded by carbon particles absorbed more radiant energy from the electric arc than an unseeded stream. Burkig [98] studied the absorption of radiation by submicron-sized particles of carbon, iron, and tantalum carbide seeded in helium and hydrogen, and Love [99] determined the phase function and the extinction coefficient for micron-sized aluminum oxide particles for wavelengths in the infrared region between 4 and 6 μ. Williams [100, 101] reported measured values of extinction coefficients, and the phase functions for submicron-sized particles of tungsten, silicon, carbon, tungsten carbide, and silicon carbide. His results also demonstrated the highly forward nature of scattering by such clouds of particles.

A close scrutiny of these experimental results reveals that the Mie theory cannot be directly applied to predict the absorption and scattering properties of particles having irregular shape. Furthermore, when scattering is important, anisotropy is involved in the scattering of radiation by particles of irregular shape.

Rayleigh Scattering by Matter

Rayleigh scattering takes place when the size of the scattering particles is very small compared with the wavelength of radiation, that is, $x \equiv \pi D/\lambda \ll 1$. For example, the scattering of thermal radiation by atoms and molecules of a gas is in the range of Rayleigh scattering because the diameter of a molecule is orders of magnitude smaller than the wavelength of thermal radiation. Rayleigh obtained explicit relations for the scattering cross section and the phase function for the scattering of radiation by small particles. The main features of Rayleigh's derivation can be found, in addition to the original paper, in reference 29. The scattering cross section C_s for Rayleigh scattering is given [29, p. 37] by

$$C_s = \frac{128\pi^5 R^6}{3\lambda^4}\left(\frac{n^2 - 1}{n^2 + 2}\right)^2 = \frac{24\pi^3 V^2}{\lambda^4}\left(\frac{n^2 - 1}{n^2 + 2}\right)^2 \tag{2-87}$$

where n is the relative refractive index, λ is the wavelength in the medium, R is the particle radius, and V is the particle volume.

In the case of gases, if N is the number of gas molecules per cubic centimeter, then the spectral volumetric scattering coefficient σ_λ for Rayleigh scattering is obtained from Eqs. 2-53b and 2-87 as

$$\sigma_\lambda = C_s N = \frac{24\pi^3 V^2}{\lambda^4}\left(\frac{n^2 - 1}{n^2 + 2}\right)^2 \frac{1}{N} \cong \frac{8\pi^3}{3}\frac{(n^2 - 1)^2}{\lambda^4 N} \tag{2-88}$$

since $n^2 + 2 \simeq 3$ for a gas. The phase function for Rayleigh scattering is given [31, p. 6] as

$$p(\theta) = \frac{3}{4}(1 + \cos^2 \theta) \tag{2-89}$$

For most engineering applications the Rayleigh scattering of thermal radiation by gas molecules or atoms is negligible because the scattering coefficient σ_λ is very small. For example, for CO_2 at 32°F at 1 atm pressure for $\lambda = 2\mu = 2 \times 10^{-4}$ cm (i.e., infrared region) and $n = 1.00045$ calculations from Eq. 2-88 give $\sigma_\lambda \simeq 1.5 \times 10^{-9}\ \text{cm}^{-1}$, which is quite small for most practical purposes.

In a perfectly crystaline solid all molecules are fixed in a completely ordered array; hence the material behaves as a perfectly homogeneous medium, and there is no scattering. However, deviations from ideality due to imperfections in the crystal lattice (i.e., unoccupied, displaced, or interchanged lattice points) may give rise to Raleigh scattering, although the scattering of thermal radiation due to such causes is negligibly weak.

In the case of liquids the situation is intermediate between that for a gas and that for a solid. The liquids may exhibit Rayleigh scattering due to so-called density fluctuations resulting from the thermal motion within the body. The reader may refer to reference 29 [Chapter 9] for a detailed account of scattering due to density fluctuations; however, such scattering effects are negligible for thermal radiation purposes.

Radiative Properties of Semitransparent Materials

Radiation incident upon the surface of an opaque material never penetrates into greater depths; similarly radiation originating in the interior of an opaque body never reaches the surface. Therefore, for opaque materials, the absorption, emission, and reflection of radiation are surface phenomena. For a semitransparent material, however, the absorption and emission of radiation are bulk rather than surface phenomena. Consider, for example, a sheet of glass at a prescribed temperature. The rate of emission of radiation at its surface depends on the thickness of the layer, the distribution of temperature within the body, and the radiative properties of the material, such as the absorption coefficient, the scattering coefficient (if scattering particles exist), and the index of refraction.

Determination of the emission and reflection characteristics of semitransparent materials requires the solution of the integrodifferential equation of radiative transfer within the considered medium, subject to appropriate boundary conditions. The mathematical formulation and the solutions to problems of this type will be presented in Chapters 8 through 11.

REFERENCES

1. P. Drude, *The Theory of Optics*, Dover Publications, New York, 1959.
2. D. S. Jones, *The Theory of Electromagnetism*, Macmillan Co., New York, 1964.
3. John T. W. Walsh, Appendix of A. K. Taylor and C. J. W. Frievenson, "The Transmission Factor of Commercial Window Glasses," *Dept. Sci. Ind. Res., Illum. Res. Tech. Paper No.* 2, 1926.
4. R. V. Dunkle, "Thermal Radiation Characteristics of Surfaces," in *Theory and Fundamental Research in Heat Transfer*, edited by J. A. Clark, pp. 1–31, Pergamon Press, New York, 1963.
5. W. König, *Handbuch der Physik*, Vol. 20, pp. 190–192, Springer, Berlin, 1928.
6. Herbert B. Holl, The Reflection of Electromagnetic Radiation (Based on Classical Electrodynamics), *Army Missile Command Rept.* No. RF-TR-63-4, Mar. 15, 1963. Vol. II, Appendix: Tables of Radiation Reflection Functions.
7. Herbert B. Holl, "Numerical Solutions of Fresnel Equations in the Optical Region," in *Symposium on Thermal Radiation of Solids*, edited by S. Katzoff, NASA SP-55, pp. 45–61, U.S. Government Printing Office, Washington, D.C., 1965.
8. R. V. Dunkle, "Emissivity and Inter-reflection Relationships for Infinite Parallel Specular Surfaces," in *Symposium on Thermal Radiation of Solids*, edited by S. Katzoff, NASA SP-55, pp. 39–44, U.S. Government Printing Office, Washington, D.C., 1965.
9. H. Davies, "Reflection of Electromagnetic Waves from Rough Surfaces," *Proc. Inst. Elec. Engrs. (London)*, **101**, 209–214, 1954.
10. H. E. Bennett and J. O. Porteus, "Relation Between Surface Roughness and Specular Reflectance at Normal Incidence," *J. Opt. Soc. Am.*, **51**, 123–129, 1961.
11. J. O. Porteus, "Relation Between the Height Distribution of a Rough Surface and the Reflectance at Normal Incidence," *J. Opt. Soc. Am.*, **53**, 1394–1402, 1963.
12. H. E. Bennett, "Specular Reflection of Aluminized Ground Glass and the Height Distribution of Surface Irregularities," *J. Opt. Soc. Am.*, **53**, 1389–1394, 1963.
13. R. C. Birkebak, J. P. Dawson, B. A. McCullough, and B. E. Wood, "Hemispherical Reflectance of Metal Surfaces as a Function of Wavelength and Surface Roughness," *Intern. J. Heat Mass Transfer*, **10**, 1225–1232, 1967.
14. K. E. Torrance and E. M. Sparrow, "Biangular Reflectance of an Electric Nonconductor as a Function of Wavelength and Surface Roughness," *J. Heat Transfer*, **87C**, 283–292, 1965.
15. K. E. Torrance and E. M. Sparrow, "Off-Specular Peaks in the Directional Distribution of Reflected Thermal Radiation," *J. Heat Transfer*, **88C**, 223–230, 1966.
16. R. C. Birkebak and E. R. G. Eckert, "Effects of Roughness of Metal Surfaces on Angular Distribution of Monochromatic Reflected Radiation," *J. Heat Transfer*, **87C**, 85–94, 1965.
17. H. H. Safwat and J. F. Parmer, "Effect of Surface Roughness on Specular and Diffuse Reflectance Components of Carbon Steel in Visible Wavelength Range," *ASME Paper* No. 69-WA/HT-42, 1969.
18. Petr Beckmann and André Spizzichino, *The Scattering of Electromagnetic Waves from Rough Surfaces*, Macmillan Co., New York, 1960.

19. *Symposium on Thermal Radiation of Solids*, edited by S. Katzoff, NASA SP-55, U.S. Government Printing Office, Washington, D.C., 1965.

20. G. G. Gubareff, J. E. Janssen, and R. H. Torborg, *Thermal Radiation Properties Survey*, Honeywell Research Center, Minneapolis-Honeywell Regulator Company, Minneapolis, Minn., 1960.

21. H. Hottel, "Radiant Heat Transmission," in *Heat Transmission*, edited by W. H. McAdams, 3rd ed., McGraw-Hill Book Co., New York, 1954.

22. Darii Yakovlevich Svet, *Thermal Radiation*, Metals, Semiconductors, Ceramics, Partly Transparent Bodies and Films, Consultants Bureau, 1965.

23. W. D. Wood, H. W. Deem, and C. F. Lucks, *Thermal Radiative Properties*, **3**, Plenum Press, New York, 1964.

24. J. R. Singham, "Tables of Emissivity of Surfaces," *Intern. J. Heat Mass Transfer*, **5**, 67–76, 1962.

25. Y. S. Touloukian and D. P. DeWitt, *Thermal Radiative Properties*, Vol. 7; *Metallic Elements and Alloys*, 1F1/Plenum, New York, 1970.

26. G. A. Mie, "Beitrage zur Optic truber Medien Speziell Kolloidale Metollosungen," *Ann. Physik*, **25**, 377–445, 1908.

27. H. C. Van de Hulst, *Light Scattering by Small Particles*, John Wiley and Sons, New York, 1957.

28. J. A. Stratton, *Electromagnetic Theory*, McGraw-Hill Book Co., New York, 1941.

29. M. Kerker, *The Scattering of Light*, Academic Press, London, 1969.

30. D. Deirmendjian, *Electromagnetic Scattering on Spherical Polydispersions*, American Elsevier Publishing Co., New York, 1969.

31. S. Chandrasekhar, *Radiative Transfer*, Oxford University Press, London, 1950; also Dover Publications, New York, 1960.

32a. R. Penndorf, "Scattering and Extinction Coefficients for Small Absorbing and Non-absorbing Aerosols," *J. Opt. Soc. Am.*, **52**, 896–905, 1962.

32b. R. Penndorf, "Mie Scattering in the Forward Area," *Infrared Phys.*, **2**, 85–102, 1962.

32c. R. Penndorf, in *Electromagnetic Scattering*, edited by M. Kerker, Pergamon Press, Oxford, 1963.

33. Chiao-Min Chu and S. W. Churchill, "Representation of the Angular Distribution of Radiation Scattered by a Spherical Particle," *J. Opt. Soc. Am.*, **45**, 958–962, 1955.

34. R. L. Rowell and R. S. Stein, eds., *Electromagnetic Scattering*, Gordon and Breach Science Publishers, New York, 1967.

35. A. N. Lowan, *Tables of Scattering Functions for Spherical Particles*, National Bureau of Standards, Applied Mathematics Series No. 4, U.S. Government Printing Office, Washington, D.C., 1948.

36. C. M. Chu, G. C. Clark, and S. W. Churchill, *Tables of Angular Distribution Coefficients for Light Scattering by Spheres*, University of Michigan Press, Ann Arbor, Mich., 1957.

37. D. Deirmendjian, R. Clasen, and W. Viezee, "Mie Scattering with Complex Index of Refraction," *J. Opt. Soc. Am.*, **51**, 620–633, 1961.

38. G. N. Plass, "Mie Scattering and Absorption Cross Sections for Aluminum Oxide and Magnesium Oxide," *Appl. Opt.*, **3**, 867–782, 1964.

39. G. N. Plass, "Temperature Dependence of the Mie Scattering and Absorption Cross Sections for Aluminum Oxide," *Appl. Opt.*, **4**, 1616–1619, 1965.

40. D. A. Gryvnak and D. E. Burch, "Optical and Infrared Properties of Al_2O_3 at Elevated Temperatures," *J. Opt. Soc. Am.*, **55,** 625–629, 1969.

41. G. N. Plass, "Mie Scattering and Absorption Cross Sections for Absorbing Particles," *Appl. Opt.*, **5,** 279–285, 1966.

42. G. W. Kattawar and G. N. Plass, "Electromagnetic Scattering from Absorbing Spheres," *Appl. Opt.*, **6,** 1377–1382, 1967.

43. V. R. Stull and G. N. Plass, "Emissivity of Dispersed Carbon Particles," *J. Opt. Soc. Am.*, **50,** 121–129, 1960.

44. B. M. Herman, "Infra-red Absorption, Scattering, and Total Attenuation Cross-Sections for Water Spheres," *Quart. J. Roy. Meteorol. Soc.*, **88,** 143–150, 1962.

45. *American Institute of Physics Handbook*, 2nd ed., Section 6, pp. 11–131, McGraw-Hill Book Co., New York, 1963.

46. *International Critical Tables*, Vol. V, pp. 248–252, McGraw-Hill Book Co., New York, 1929.

47. H. M. Randall, D. M. Dennison, N. Ginsburg, and L. R. Weber, "The Far Infrared Spectrum of Water Vapor," *Phys. Rev.*, **52,** 160–174, 1937.

48. D. K. Edwards, "Radiation Interchange in a Nongray Enclosure Containing an Isothermal Carbon Dioxide-Nitrogen Gas Mixture," *J. Heat Transfer*, **84C,** 1–11, 1962.

49. S. S. Penner, *Quantitative Molecular Spectroscopy and Gas Emissivities*, Addison-Wesley Publishing Co., Reading, Mass., 1959.

50. G. Herzberg, *Molecular Spectra and Molecular Structure*, Vols. I and II, D. Van Nostrand Co., Princeton, N.J., 1950 and 1954.

51. H. A. Lorentz, "The Absorption and Emission Lines of Gaseous Bodies," *Acad. Sci. (Amsterdam)*, **8,** 591–611, 1906.

52. A. Schack, *Industrial Heat Transfer* (English translation), pp. 182–200, John Wiley and Sons, New York, 1933.

53. W. M. Elsasser, "Heat Transfer by Infrared Radiation in the Atmosphere," *Harvard Meteorological Studies*, No. 6, Harvard University, Blue Hill Meteorological Observatory, Milton, Mass., 1942.

54. H. Mayer, *Methods of Opacity Calculations*, LA-647, Los Alamos, N.M., Oct. 3 1947.

55. R. M. Goody, "A Statistical Model for Water-Vapor Absorption," *Quart. J. Roy. Meteorol. Soc.*, **78,** 165–169, 1952.

56. G. N. Plass, "Models for Spectral Band Absorption," *J. Opt. Soc. Am.*, **48,** 690–703, 1958.

57. E. T. Whittaker and G. N. Watson, *A Course of Modern Analysis*, 4th ed., Cambridge University Press, Cambridge, 1927.

58. J. C. Richmond, "Effect of Surface Roughness on Emittance of Nonmetals," *J. Opt. Soc. Am.*, **56,** 253–254, 1966.

59. H. E. Bennett, "Influence of Surface Roughness, Surface Damage, and Oxide Films on Emittance," in reference 19, pp. 145–152.

60. R. M. Goody, *Atmospheric Radiation*, Clarendon Press, Oxford, 1964.

61. W. Malkmus and A. Thomson, "Infrared Emissivity of Diatomic Gases for the Anharmonic Vibrating-Rotator Model," *J. Quant. Spectry. Radiative Transfer*, **2,** 17–39, 1962.

62. J. A. Jamieson, R. M. McFee, G. N. Plass, R. H. Grube, and R. G. Richards, *Infrared Physics and Engineering*, McGraw-Hill Book Co., New York, 1963.

63. W. O. Davies, "Infrared Absorption by Carbon Monoxide at High Temperature," *J. Chem. Phys.*, **36**, 292–297, 1962.

64. J. C. Breeze and C. C. Ferriso, General Dynamics/Corvair, Report No. GD/C-DBE 65-007, May 1965.

65. M. M. Abu-Romia and C. L. Tien, "Measurement and Correlations of Infrared Radiation of Carbon Monoxide at Elevated Temperatures," *J. Quant. Spectry. Radiative Transfer*, **6**, 143–167, 1966.

66. M. M. Abu-Romia and C. L. Tien, "Appropriate Mean Absorption Coefficients for Infrared Radiation of Gases," *J. Heat Transfer*, **89C**, 321–327, 1967.

67. G. N. Plass, "Spectral Emissivity of Carbon Dioxide from 1800–2500 cm^{-1}," *J. Opt. Soc. Am.*, **49**, 821–828, 1959.

68. D. K. Edwards, "Absorption by Infrared Bands of Carbon Dioxide at Elevated Pressures and Temperatures," *J. Opt. Soc. Am.*, **50**, 617–626, 1960.

69. J. N. Howard, D. E. Burch, and D. Williams, "Infrared Transmission of Synthetic Atmospheres, III: Absorption by Water Vapor," *J. Opt. Soc. Am.*, **46**, 242–245, 1956.

70. R. Goldstein, "Preliminary Absolute Intensity Measurements for 1.38, 1.87 and 2.7 μ Bands of Water Vapor between 125 and 200°C," *J. Quant. Spectry. Radiative Transfer*, **3**, 91–93, 1963.

71. J. H. Jaffe and U. S. Benedict, "The Strength of the ν_3-Vibration of H_2O," *J. Quant. Spectry. Radiative Transfer*, **3**, 87–88, 1963.

72. C. C. Ferriso and C. B. Ludwig, "Spectral Emissivities and Integrated Intensities of the 2.7 μ H_2O Band between 530° and 2200°K," *J. Quant. Spectry. Radiative Transfer*, **4**, 215–227, 1964.

73. D. K. Edwards, B. J. Flornes, L. K. Glassen, and W. Sun," Correlation of Absorption by Water at Temperatures from 300°K to 1100°K," *Appl. Opt.*, **4**, 715–721, 1965.

74. R. Goldstein, "Measurements of Infrared Absorption by Water Vapor at Temperatures to 1000°K," *J. Quant. Spectry. Radiative Transfer*, **4**, 343–352, 1964.

75. B. H. Armstrong, D. H. Holland, and R. E. Meyerott, "Absorption Coefficients of Air from 22,000° to 220,000°," Air Force Special Weapons Center, Report No. TR 58-36, Kirkland Air Force Base, Albuquerque, N.M., 1958.

76. B. H. Armstrong, "Mean Absorption Coefficients of Air, Nitrogen, and Oxygen from 22,000° to 220,000°," Lockheed Missiles and Space Division, Report No. LMSD 49759, Palo Alto, Calif., 1959.

77. R. E. Meyerott, J. Sokoloff, and R. W. Nicholls, "Absorption Coefficients of Air," Air Force Cambridge Research Center, Geophysics Research Paper No. 68, Bedford, Mass., 1960.

78. B. H. Armstrong et al., "Radiative Properties of High Temperature Air," *J. Quant. Spectry Radiative Transfer*, **1**, 143–162, 1961.

79. B. Kivel and K. Bailey, "Tables of Radiation from High Temperature Air," *AVCO-Everett Res. Lab. Res. Rept.* No. 21, 1957.

80. J. R. Collins, "A New Infra-Red Absorption Band of Liquid Water at 2.52 μ," *Phys. Rev.*, **55**, 470–472, 1939.

81. J. J. Fox and A. E. Martin, "Investigation of Infra-Red Spectra (2.5–7.5 μ) Absorption of Water," *Proc. Roy. Soc. London*, **A174,** 234–262, 1960.

82. J. A. Curcio and C. C. Petty, "The Near Infrared Absorption Spectrum of Liquid Water," *J. Opt. Soc. Am.*, **41,** 302–304, 1951.

83. A. E. Stanevich and N. G. Yaroslavskii, "Absorption of Liquid Water in the Long-Wavelength Part of the Infrared Spectrum (42–2000 μ)," *Opt. i Spektroskopiia*, **10,** No. 4, 278–279, 1961.

84. S. A. Sullivan, "Experimental Study of the Absorption in Distilled Water, Artificial Sea Water, and Heavy Water in the Visible Region of the Spectrum," *J. Opt. Soc. Am.*, **53,** 962–968, 1963.

85. R. Goldstein and S. S. Penner, "The Near-Infrared Absorption of Liquid Water at Temperatures between 27 and 209°C," *J. Quant. Spectry. Radiative Transfer*, **4,** 441–451, 1964.

86. D. A. Draegert, N. W. B. Stone, B. Curnutte, and D. Williams, "Far-Infrared Spectrum of Liquid Water," *J. Opt. Soc. Am.*, **56,** 64–69, 1966.

87. J. E. Tyler, "Monochromatic Measurement of the Volume Scattering of Natural Waters," *J. Opt. Soc. Am.*, **47,** 745–747, 1957.

88. J. E. Tyler and W. H. Richardson, "Nephelometer for the Measurement of Volume Scattering Function in Situ," *J. Opt. Soc. Am.*, **48,** 354–357, 1958.

89. J. E. Tyler, "Scattering Properties of Distilled and Natural Waters," *Limnol. Oceanog.*, **6,** 451–456, 1961.

90. A. F. Spilhaus, "Observations of Light Scattering in Sea Water," *Limnol. Oceanog.*, **13,** 418–422, 1968.

91. N. Neuroth, "Das Einfluss der Temperatur auf die Spektrale Absorption von Glasern in Ultraroter, *I*," *Glastech Ber.*, **25,** 242–249, 1952.

92. P. C. Lewis and G. F. Lothian, "Photoextinction Measurements on Spherical Particles," *Brit. J. Appl. Phys.*, Suppl. No. 3, pp. 71–74, 1954.

93. D. Sinclair, "Light Scattering by Spherical Particles," *J. Opt. Soc. Am.*, **37,** 475–480, 1947.

94. J. R. Hodkins, "Some Observations on Light Extinction by Spherical Particles," *Brit. J. Appl. Phys.*, **14,** 931–932, 1963.

95. G. W. Hepplestone and P. C. Lewis, "Light Transmission Measurements in Suspensions," *Brit. J. Appl. Phys.*, **18,** 1321–1325, 1967.

96. G. W. Hepplestone and P. C. Lewis, "Experimental Observations on the Angular Distribution of Scattered Radiation from Suspensions Containing Particles of Size Comparable with Wavelength," *Brit. J. Appl. Phys.*, (*J. Phys.* D), **1,** Ser. 2, 199–206, 1968.

97. C. D. Lanzo and R. G. Ragsdale, "Experimental Determination of Spectral and Total Transmissivities of Clouds of Small Particles," *NASA Tech. Note* TN D-1405, September 1962.

98. V. C. Burkig, "Theoretical Absorption in Seeded Gas," Douglas Aircraft Co., Report No. DAC-59985, NASW-1310, Los Angeles, Calif., January 1967.

99. T. J. Love, "An Experimental Investigation of Infra-Red Scattering by Clouds of Particles," *Aerospace Res. Lab. Rept.* ARL-64-109, June 1964.

100. J. R. Williams, "Thermal Radiation Transport in Particle-Seeded Gases," *Trans. Am. Nucl. Soc.*, **12,** 811–812, December 1969.

101. J. R. Williams, "Radiant Heat Absorption by Particle Seeded Gases," in *Proceedings of the Sixth Southeastern Seminar on Thermal Sciences*, edited by J. K. Ferrel, M. N. Özişik, and J. E. Sunderland, Raleigh, N.C., April 1970.

NOTES

[1] The complex expressions of Eq. 2-19 can be written formally as

$$\bar{\rho} = \left(\frac{A - ib}{B + ib}\right)^2 = \left[\frac{(AB + b^2) - ib(B - A)}{B^2 + b^2}\right]^2 \tag{1}$$

The complex conjugate of this relation is

$$\bar{\rho}^* = \left[\frac{(AB + b^2) + ib(B - A)}{B^2 + b^2}\right]^2 \tag{2}$$

Then the absolute value of $\bar{\rho}$ is given as

$$|\bar{\rho}| = [\bar{\rho}\bar{\rho}^*]^{1/2} = \frac{(AB + b^2)^2 + b^2(B - A)^2}{(B^2 + b^2)^2} = \frac{A^2 + b^2}{B^2 + b^2} \tag{3}$$

[2] Dunkle [8] defined the complex refractive index in the form $m = n(1 - ik)$, whereas in the present analysis it was defined as $m = n - in'$. These two definitions are related by $nk = n'$.

[3] The root-mean-square roughness is defined as the root-mean-square deviation of the surface from the mean level.

[4] We recall that when the surface is irradiated uniformly the following reciprocity relation will hold (see Eq. 1-102):

$$\rho(\mathbf{r}; 2\pi \rightarrow \theta, \phi) = \rho(\mathbf{r}; \theta_i, \phi_i \rightarrow 2\pi)$$

for $\theta = \theta_i$ and $\phi = \phi_i$.

[5] For an absorbing, emitting, and scattering medium the equation of radiative transfer 2-58 takes the form

$$\frac{dI_\nu}{ds} + (\kappa_\nu + \sigma_\nu)I_\nu = \kappa_\nu I_{\nu b}(T) + \frac{\sigma_\nu}{4\pi}\int_{\phi=0}^{2\pi}\int_{\mu=-1}^{1} p(\mu, \mu')I_\nu(s, \mu')\, d\mu'\, d\phi'$$

where $p(\mu, \mu')$ is the phase function, and σ_ν the spectral scattering coefficient. The equation of radiative transfer for the scattering medium will be discussed in Chapter 8.

[6] The error function integral is defined as

$$\text{erf } x = \frac{2}{\sqrt{\pi}}\int_0^x e^{-\xi^2}\, d\xi$$

[7] For a line described by the Lorentz formula of Eq. 2-65, for example, the line-shape parameter is given as

$$s = \frac{a/\pi}{(\nu - \nu_0)^2 + a^2}$$

where a is the line half-width.

[8] Various symbols have been defined previously: a is the half-width of a line, d the spacing between the lines, K the integrated intensity of a line, and y the thickness of the gas layer.

CHAPTER 3

View Factors

In problems of radiative heat exchange among surfaces separated by a nonparticipating medium (i.e., a medium that does not absorb, emit, or scatter radiation) the geometric orientation of the surfaces with respect to each other affects the radiative interchange. For convenience in the analysis of radiative heat exchange in such situations, the so-called *view factor* is defined to characterize the effects of geometric arrangement on radiative heat exchange between two surfaces. The terms *shape factor* and *angle factor* have also been used in the literature. We shall use the term *diffuse view factor* when the surfaces are diffuse reflectors and diffuse emitters, and *specular view factor* when the surfaces are diffuse emitters and specular reflectors. In this chapter we present the definitions and the methods of determination of these view factors.

3-1 DEFINITION OF DIFFUSE VIEW FACTORS

Consider two diffusely emitting, diffusely reflecting surfaces A_1 and A_2 maintained at uniform temperatures T_1 and T_2, respectively. Let dA_1 and dA_2 be two elemental surfaces on A_1 and A_2, respectively, and $\mathbf{r}_{12}$ (i.e., $\mathbf{r}_{12} = -\mathbf{r}_{21}$) be the vector joining dA_1 and dA_2. The unit vectors $\hat{\mathbf{n}}_1$ and $\hat{\mathbf{n}}_2$ drawn normal to dA_1 and dA_2 make angles θ_1 and θ_2, respectively, with the line joining these two elemental surfaces. Figure 3-1 shows the geometry and coordinates here considered.

The relations will now be derived to define the *diffuse view factor* between (a) two elemental surfaces dA_1 and dA_2, (b) an elemental surface dA_1 and a finite surface A_2, and (c) two finite surfaces A_1 and A_2.

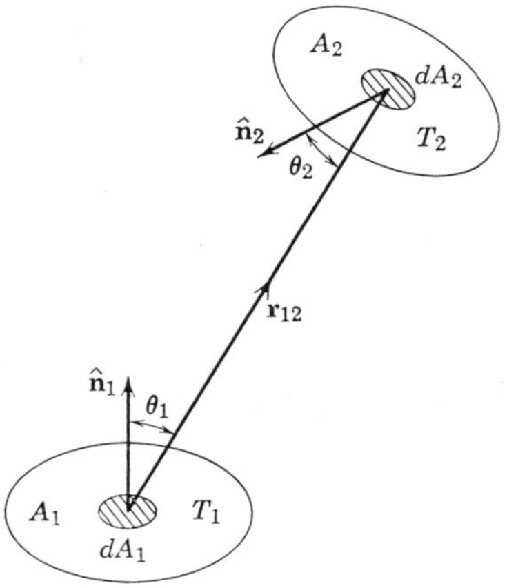

Fig. 3-1. Coordinates for the definition of diffuse view factor.

(a) Diffuse View Factor Between Two Elemental Surfaces

The diffuse view factor from an elemental surface dA_1 to an elemental surface dA_2 will be designated by the symbol $dF_{dA_1-dA_2}$. It represents the ratio of the radiative energy leaving surface element dA_1 that strikes surface dA_2 directly to the radiative energy leaving surface element dA_1 in the entire hemispherical space.

Let I_1 be the radiation intensity (spectral or total) leaving surface element dA_1, and $d\Omega_{12}$ be the solid angle under which an observer at dA_1 sees dA_2. The amount of radiation energy dq_1 leaving surface dA_1 that strikes dA_2 directly per unit time is immediately obtained from the definition of intensity given by Eq. 1-38. We find

$$dq_1 = dA_1 I_1 \cos\theta_1 \, d\Omega_{12} \tag{3-1}$$

where $d\Omega_{12}$ is given as

$$d\Omega_{12} = \frac{dA_2 \cos\theta_2}{r^2} \tag{3-2}$$

Here r is the length of the vector $\mathbf{r}_{12}$ joining dA_1 and dA_2. By combining Eqs. 3-1 and 3-2 we obtain

$$dq_1 = dA_1 I_1 \frac{\cos\theta_1 \cos\theta_2 \, dA_2}{r^2} \tag{3-3}$$

The radiative energy leaving dA_1 in all directions in the hemispherical space is given by

$$q_1 = dA_1 \int_{\phi=0}^{2\pi} \int_{\theta_1=0}^{\pi/2} I_1 \cos\theta_1 \sin\theta_1 \, d\theta_1 \, d\phi = \pi I_1 \, dA_1 \tag{3-4}$$

since the intensity I_1 is independent of direction for a diffusely emitting, diffusely reflecting surface.

The relation defining $dF_{dA_1-dA_2}$ is obtained from Eqs. 3-3 and 3-4 as

$$dF_{dA_1-dA_2} = \frac{dq_1}{q_1} = \frac{\cos\theta_1 \cos\theta_2 \, dA_2}{\pi r^2} \tag{3-5}$$

Conversely, $dF_{dA_2-dA_1}$ represents the fraction of radiative energy leaving surface element dA_2 in all directions in the hemispherical space that strikes surface dA_1 directly. The relation defining $dF_{dA_2-dA_1}$ is immediately obtained from Eq. 3-5 by interchanging the subscripts 1 and 2, since both surfaces are diffuse emitters and diffuse reflectors:

$$dF_{dA_2-dA_1} = \frac{\cos\theta_1 \cos\theta_2 \, dA_1}{\pi r^2} \tag{3-6}$$

The reciprocity relation between the diffuse view factors $dF_{dA_1-dA_2}$ and $dF_{dA_2-dA_1}$ is obtained from Eqs. 3-5 and 3-6 as

$$dA_1 \, dF_{dA_1-dA_2} = dA_2 \, dF_{dA_2-dA_1} \tag{3-7}$$

(b) Diffuse View Factor Between Surfaces dA_1 and A_2

The symbol $F_{dA_1-A_2}$ will be used to designate the diffuse view factor from an elemental surface dA_1 to a finite surface A_2 shown in Fig. 3-1. It represents the fraction of the radiative energy leaving elemental surface dA_2 in all directions in the hemispherical space that strikes surface A_2 directly. Therefore $F_{dA_1-A_2}$ can be determined by integrating elemental view factor $dF_{dA_1-dA_2}$ over surface A_1. We obtain

$$F_{dA_1-A_2} = \int_{A_2} dF_{dA_1-dA_2} = \int_{A_2} \frac{\cos\theta_1 \cos\theta_2}{\pi r^2} \, dA_2 \tag{3-8}$$

Conversely, $F_{A_2-dA_1}$ is the fraction of radiative energy leaving surface A_2 in all directions in the hemispherical space that strikes surface dA_1 directly. If I_2 is the intensity of radiation leaving surface A_2, the view factor $F_{A_2-dA_1}$ is written according to this definition as

$$F_{A_2-dA_1} = \frac{\displaystyle\int_{A_2} \left(\frac{I_2 \cos\theta_1 \cos\theta_2 \, dA_1}{r^2} \right) dA_2}{\displaystyle\int_{A_2} \left(\int_{\phi=0}^{2\pi} \int_{\theta_2=0}^{\pi/2} I_2 \cos\theta_2 \sin\theta_2 \, d\theta_2 \, d\phi \right) dA_2} \tag{3-9}$$

When I_2 is independent of direction and position over A_2, Eq. 3-9 simplifies to

$$F_{A_2-dA_1} = \frac{dA_1}{A_2} \int_{A_2} \frac{\cos\theta_1 \cos\theta_2}{\pi r^2} \, dA_2 \tag{3-10}$$

The following reciprocity relation is obtained from Eqs. 3-8 and 3-10:

$$dA_1\, F_{dA_1-A_2} = A_2 F_{A_2-dA_1} \tag{3-11}$$

(c) Diffuse View Factor Between Two Finite Surfaces A_1 and A_2

The diffuse view factor $F_{A_1-A_2}$ from a finite surface A_1 to a finite surface A_2 shown in Fig. 3-1 is defined as

$$F_{A_1-A_2} = \frac{\begin{bmatrix}\text{Radiative energy leaving surface} \\ A_1 \text{ that strikes } A_2 \text{ directly}\end{bmatrix}}{\begin{bmatrix}\text{Radiative energy leaving} \\ \text{surface } A_1 \text{ in all directions} \\ \text{in the hemispherical space}\end{bmatrix}} \tag{3-12}$$

When the intensity I_1 is independent of direction and position over A_1, Eq. 3-12 yields

$$F_{A_1-A_2} = \frac{1}{A_1} \int_{A_1} \int_{A_2} \frac{\cos\theta_1 \cos\theta_2}{\pi r^2}\, dA_2\, dA_1 \tag{3-13}$$

The diffuse view factor from surface A_2 to A_1, when the radiation intensity I_2 is independent of direction and position over A_2, is immediately obtained from Eq. 3-13 by interchanging the subscripts 1 and 2:

$$F_{A_2-A_1} = \frac{1}{A_2} \int_{A_1} \int_{A_2} \frac{\cos\theta_1 \cos\theta_2}{\pi r^2}\, dA_2\, dA_1 \tag{3-14}$$

The following reciprocity relation is obtained from Eqs. 3-13 and 3-14:

$$A_1 F_{A_1-A_2} = A_2 F_{A_2-A_1} \tag{3-15}$$

Properties of Diffuse View Factors for an Enclosure

In the foregoing analysis we focused our attention on diffuse view factors between two surfaces. We now consider the properties of diffuse view factors between the surfaces of an enclosure consisting of N zones. We assume that each zone is an isothermal, diffuse reflector and a diffuse emitter, that it has a finite surface area A_i, $i = 1, 2, \ldots, N$, and that the intensity of radiation leaving each surface is independent of position over that zone. The reciprocity relation for the view factor between any two surfaces A_i and A_j of the enclosure is given as

$$A_i F_{A_i-A_j} = A_j F_{A_j-A_i} \tag{3-16a}$$

This is written more compactly in the form

$$A_i F_{i-j} = A_j F_{j-i} \tag{3-16b}$$

The view factors for an enclosure obey the following *summation relationship:*

$$\sum_{k=1}^{N} F_{i-k} = 1 \tag{3-17}$$

The physical significance of this relation is obvious from the definition of the view factor. If surface A_i is flat or convex, none of the radiation leaving this surface will strike itself directly. If surface A_i is concave, part of the radiation leaving this surface will strike itself directly. Hence we have

$$F_{ii} = 0 \quad \text{(plane or convex surface)} \tag{3-18a}$$

$$F_{ii} \neq 0 \quad \text{(concave surface)} \tag{3-18b}$$

Methods of Determination of Diffuse View Factors

Determination of the diffuse view factor between two elemental surfaces as defined by Eq. 3-5 poses no problem. However, evaluation of the diffuse view factor between an elemental and a finite surface requires an area integral, and that between two finite surfaces requires two area integrals. Such integrals are difficult to perform analytically except for very simple geometries. Hamilton and Morgan [1] evaluated diffuse view factors for simple configurations involving rectangles, triangles, and cylinders and presented the results in the form of comprehensive tables and charts. Leuenberger and Pearson [2], Kreith [3], and Sparrow and Cess [4] compiled view factors for various simple geometries. A systematic tabulation of references for view factors has been given by Howell and Siegel [5]. Other compilations of view factors have been published by Hottel [6], Mackey et al. [7], and Feingold and Gupta [8]. Various analytical and experimental methods for the determination of diffuse view factors have been discussed by Jacob [9]. Recently a FORTRAN program was prepared to calculate diffuse view factors for cylindrical pins [10]. In the following sections we present some of the analytical techniques available for the determination of diffuse view factors.

3-2 DETERMINATION OF DIFFUSE VIEW FACTOR BY DIRECT INTEGRATION

In this section we illustrate the direct integration technique to evaluate the diffuse view factor between simple geometries with an example considered by Hamilton and Morgan [1]. The diffuse view factor $F_{dA_1-A_2}$ will be determined between an elemental surface dA_1 and a finite surface A_2, inclined at an angle ϕ ($0 < \phi < 180°$) and positioned as shown in Fig. 3-2. Let dA_2 be an elemental area positioned at (x, y) on the surface of A_2, and $\mathbf{r}_{12}$ be the vector joining the elemental surfaces dA_1 and dA_2. The unit vectors

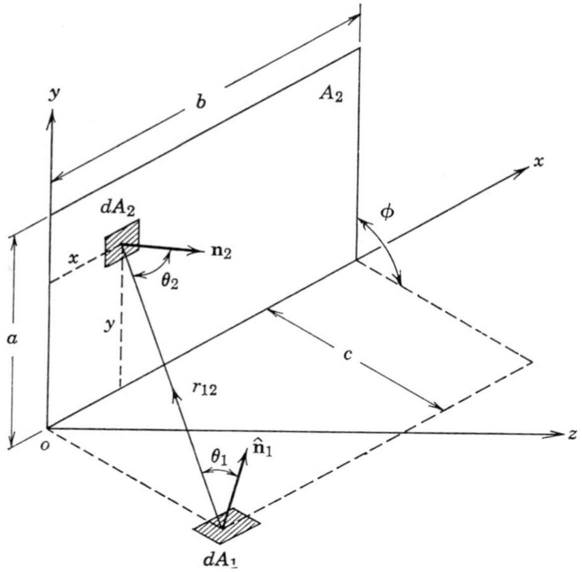

Fig. 3-2. Coordinates for determination of view factor between surfaces dA_1 and A_2.

$\hat{\mathbf{n}}_1$ and $\hat{\mathbf{n}}_2$ drawn normal to surfaces dA_1 and dA_2 make angles θ_1 and θ_2, respectively, with the line joining these two elemental surfaces. The diffuse view factor from dA_1 to A_2 is given by (see Eq. 3-8)

$$F_{dA_1-A_2} = \int_{A_2} \frac{\cos\theta_1 \cos\theta_2}{\pi r^2}\, dA_2 \tag{3-19}$$

where r is the length of the vector $\mathbf{r}_{12}$.

By referring to the x, y, z coordinate system shown in Fig. 3-2, the coordinates of elemental surfaces dA_1 and dA_2 are taken as

$$dA_1\colon x_1 = 0,\ y_1 = c\cos\phi, \quad \text{and} \quad z_1 = c\sin\phi \tag{3-20a}$$

$$dA_2\colon x_2 = x,\ y_2 = y, \quad \text{and} \quad z_2 = 0 \tag{3-20b}$$

The unit normal vectors $\hat{\mathbf{n}}_1$ and $\hat{\mathbf{n}}_2$ are defined as

$$\hat{\mathbf{n}}_1 = \hat{\mathbf{i}} l_1 + \hat{\mathbf{j}} m_1 + \hat{\mathbf{k}} n_1 \tag{3-21a}$$

$$\hat{\mathbf{n}}_2 = \hat{\mathbf{i}} l_2 + \hat{\mathbf{j}} m_2 + \hat{\mathbf{k}} n_2 \tag{3-21b}$$

where $\hat{\mathbf{i}}$, $\hat{\mathbf{j}}$, and $\hat{\mathbf{k}}$ are the unit vectors along the ox, oy, and oz axes, respectively. The direction cosines are given as

$$l_1 = 0,\ m_1 = \sin\phi, \quad \text{and} \quad n_1 = -\cos\phi \tag{3-22a}$$

$$l_2 = 0,\ m_2 = 0, \quad \text{and} \quad n_2 = 1 \tag{3-22b}$$

and the vector $\mathbf{r}_{12}$ as

$$\mathbf{r}_{12} = \hat{\mathbf{i}}(x_2 - x_1) + \hat{\mathbf{j}}(y_2 - y_1) + \hat{\mathbf{k}}(z_2 - z_1) \tag{3-23a}$$

$$\mathbf{r}_{21} = -\mathbf{r}_{12} \tag{3-23b}$$

Then the quantities r^2, $\cos\theta_1$, $\cos\theta_2$, and dA_2 in Eq. 3-19 are expressed in terms of the x, y, z variables as

$$\begin{aligned} r^2 = |\mathbf{r}_{12}|^2 &= (x_1 - x_2)^2 + (y_1 - y_2)^2 + (z_1 - z_2)^2 \\ &= (0 - x)^2 + (c\cos\phi - y)^2 + (c\sin\phi - 0)^2 \\ &= x^2 + y^2 + c^2 - 2cy\cos\phi \end{aligned} \tag{3-24}$$

$$\begin{aligned} \cos\theta_1 &= \frac{\hat{\mathbf{n}}_1 \cdot \mathbf{r}_{12}}{|\mathbf{r}_{12}|} = \frac{(x_2 - x_1)l_1 + (y_2 - y_1)m_1 + (z_2 - z_1)n_1}{r} \\ &= \frac{(x - 0)0 + (y - c\cos\phi)\sin\phi - (0 - c\sin\phi)\cos\phi}{r} \\ &= \frac{y\sin\phi}{r} \end{aligned} \tag{3-25}$$

$$\begin{aligned} \cos\theta_2 &= \frac{\hat{\mathbf{n}}_2 \cdot \mathbf{r}_{21}}{|\mathbf{r}_{21}|} = \frac{(x_1 - x_2)l_2 + (y_1 - y_2)m_2 + (z_1 - z_2)n_2}{r} \\ &= \frac{(0 - x)0 + (c\cos\phi - y)0 + (c\sin\phi - 0)1}{r} \\ &= \frac{c\sin\phi}{r} \end{aligned} \tag{3-26}$$

$$dA_2 = dx\,dy \tag{3-27}$$

The range of integration in Eq. 3-19 will be $0 \le x \le b$ and $0 \le y \le a$.

Substituting Eqs. 3-24 through 3-27 into Eq. 3-19, we obtain

$$\begin{aligned} F_{dA_1-A_2} &= \int_{x=0}^{b}\int_{y=0}^{a} \frac{y\sin\phi}{r}\,\frac{c\sin\phi}{r}\,\frac{1}{\pi r^2}\,dx\,dy \\ &= \frac{c\sin^2\phi}{\pi}\int_{x=0}^{b}\int_{y=0}^{a} \frac{y}{(x^2 + y^2 + c^2 - 2cy\cos\phi)^2}\,dx\,dy \end{aligned} \tag{3-28a}$$

The integrations in Eq. 3-28a have been performed by Hamilton and Morgan [1], and the resulting expression for the view factor is given as

$$\begin{aligned} F_{dA_1-A_2} &= \frac{1}{2\pi}\left\{\tan^{-1}M + \left(\frac{H\cos\phi - 1}{\sqrt{1 + H^2 - 2H\cos\phi}}\right)\tan^{-1}\left(\frac{M}{\sqrt{1 + H^2 - 2H\cos\phi}}\right)\right. \\ &\quad \left. + \left(\frac{M\cos\phi}{\sqrt{M^2 + \sin^2\phi}}\right)\left[\tan^{-1}\left(\frac{H - \cos\phi}{\sqrt{M^2 + \sin^2\phi}}\right) + \tan^{-1}\left(\frac{\cos\phi}{\sqrt{M^2 + \sin^2\phi}}\right)\right]\right\} \end{aligned} \tag{3-28b}$$

where

$$H \equiv \frac{a}{c}, \qquad M \equiv \frac{b}{c}$$

3-3 DETERMINATION OF DIFFUSE VIEW FACTOR BY CONTOUR INTEGRATION

The evaluation of the diffuse view factor by direct integration requires that a double or a quadruple integral be performed. This is extremely difficult except for very simple geometries. However, the surface integral can be reduced to a contour integral by application of the Stokes theorem. This approach is the basis of the contour integration technique for the determination of diffuse view factors. The technique was originally applied by Moon [11, pp. 312–315] and later by Moon and Spencer [12] for diffuse view factors in illuminating engineering. Sparrow [13] and Sparrow and Cess [4] utilized this technique in diffuse view factor calculations for heat transfer problems.

The Stokes theorem states that the circulation of a vector **V** around the boundary S of a closed surface A is equal to the flux of the curl of the vector **V** over the surface A; it is given as

$$\int_{\text{surface } A} \hat{\mathbf{n}} \cdot (\nabla \times \mathbf{V})\, dA = \oint_{\substack{\text{contour}\\ \text{of } A}} \mathbf{V} \cdot d\mathbf{S} \tag{3-29}$$

The convention for the direction of circulation around the contour follows the direction of rotation of a right-handed screw moving in the same direction as the unit normal vector $\hat{\mathbf{n}}$ drawn to the surface as illustrated in Fig. 3-3.

We write vectors **V** and $\hat{\mathbf{n}}$ in the forms

$$\mathbf{V} = \hat{\mathbf{i}}V_x + \hat{\mathbf{j}}V_y + \hat{\mathbf{k}}V_z \tag{3-30a}$$

$$\hat{\mathbf{n}} = \hat{\mathbf{i}}l + \hat{\mathbf{j}}m + \hat{\mathbf{k}}n \tag{3-30b}$$

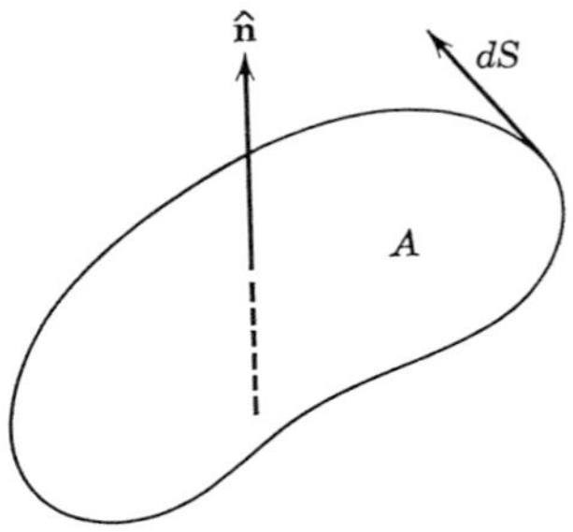

Fig. 3-3. Convention for the direction of circulation in the Stokes theorem.

where the components V_x, V_y, V_z are twice differentiable functions of x, y, z, and l, m, n are the direction cosines. Then the Stokes theorem given by Eq. 3-29 is written as

$$\iint_{\text{surface } A} \left[\left(\frac{\partial V_z}{\partial y} - \frac{\partial V_y}{\partial z}\right) l + \left(\frac{\partial V_x}{\partial z} - \frac{\partial V_z}{\partial x}\right) m + \left(\frac{\partial V_y}{\partial x} - \frac{\partial V_x}{\partial y}\right) n\right] dA$$

$$= \oint_{\substack{\text{contour}\\ \text{of } A}} (V_x\, dx + V_y\, dy + V_z\, dz) \qquad (3\text{-}31)$$

For proof of the Stokes theorem the reader should refer to any standard textbook in mathematics. The contour integration technique is now applied to determine the diffuse view factor between (a) an elemental and a finite surface, and (b) two finite surfaces.

(a) Diffuse View Factor Between Surfaces dA_1 and A_2

Consider an elemental surface dA_1 and a finite surface A_2 positioned as shown in Fig. 3-4. The diffuse view factor from dA_1 to A_2 is defined as (see Eq. 3-8)

$$F_{dA_1-A_2} = \oint_{A_2} \frac{\cos\theta_1 \cos\theta_2}{\pi r^2}\, dA_2 \qquad (3\text{-}32)$$

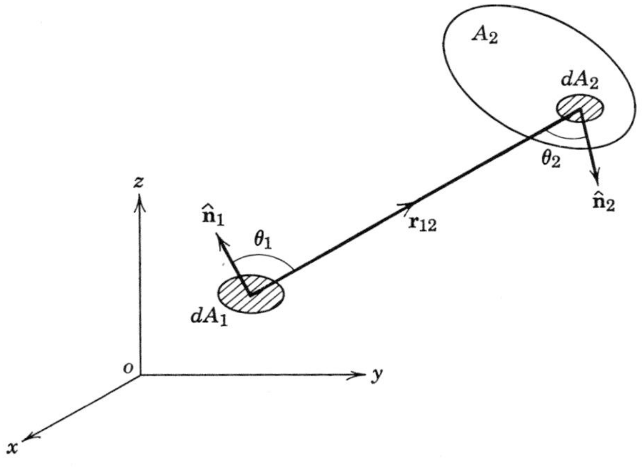

Fig. 3-4. Application of the Stokes theorem to determine diffuse view factor $F_{dA_1-dA_2}$.

and various quantities have already been defined. The cosines of angles θ_1 and θ_2 can be written as

$$\cos\theta_1 = \frac{\hat{\mathbf{n}}_1 \cdot \mathbf{r}_{12}}{r} \tag{3-33a}$$

$$\cos\theta_2 = \frac{\hat{\mathbf{n}}_2 \cdot \mathbf{r}_{21}}{r} = -\frac{\hat{\mathbf{n}}_2 \cdot \mathbf{r}_{12}}{r} \tag{3-33b}$$

where

$$r = |\mathbf{r}_{12}|.$$

Substituting Eqs. 3-33 into 3-32, we obtain

$$F_{dA_1-A_2} = -\frac{1}{\pi}\int_{A_2}\left(\frac{\hat{\mathbf{n}}_1 \cdot \mathbf{r}_{12}}{r^2}\right)\left(\frac{\hat{\mathbf{n}}_2 \cdot \mathbf{r}_{12}}{r^2}\right) dA_2 \tag{3-34}$$

which can be rearranged as

$$F_{dA_1-A_2} = -\frac{1}{\pi}\int_{A_2}\hat{\mathbf{n}}_2 \cdot \left[\frac{\mathbf{r}_{12}}{r^2}\left(\frac{\hat{\mathbf{n}}_1 \cdot \mathbf{r}_{12}}{r^2}\right)\right] dA_2 \tag{3-35}$$

We now consider the following identity [11, p. 313]:[1]

$$-\frac{\mathbf{r}_{12}}{r^2}\left(\frac{\hat{\mathbf{n}}_1 \cdot \mathbf{r}_{12}}{r^2}\right) = \tfrac{1}{2}\nabla \times \left(\frac{\mathbf{r}_{12} \times \hat{\mathbf{n}}_1}{r^2}\right) \tag{3-36}$$

Then Eq. 3-35 can be written in the form

$$F_{dA_1-A_2} = \frac{1}{2\pi}\int_{A_2}\hat{\mathbf{n}}_2 \cdot \left[\nabla \times \left(\frac{\mathbf{r}_{12} \times \hat{\mathbf{n}}_1}{r^2}\right)\right] dA_2 \tag{3-37}$$

The surface integral in this equation can be transformed into a contour integral by the Stokes theorem, Eq. 3-29. We find

$$F_{dA_1-A_2} = \frac{1}{2\pi}\oint_{\substack{\text{contour}\\ \text{of } A_2}}\left(\frac{\mathbf{r}_{12} \times \hat{\mathbf{n}}_1}{r^2}\right) \cdot d\mathbf{s} \tag{3-38}$$

In the x, y, z, rectangular coordinate system we have

$$\mathbf{r}_{12} = (x_2 - x_1)\hat{\mathbf{i}} + (y_2 - y_1)\hat{\mathbf{j}} + (z_2 - z_1)\hat{\mathbf{k}} \tag{3-39a}$$

$$\hat{\mathbf{n}}_1 = l_1\hat{\mathbf{i}} + m_1\hat{\mathbf{j}} + n_1\hat{\mathbf{k}} \tag{3-39b}$$

$$d\mathbf{s} = dx_2\,\hat{\mathbf{i}} + dy_2\,\hat{\mathbf{j}} + dz_2\,\hat{\mathbf{k}} \tag{3-39c}$$

Substituting Eqs. 3-39 into Eq. 3-38, we obtain

$$F_{dA_1-A_2} = \frac{l_1}{\pi} \oint_{\substack{\text{contour}\\ \text{of } A_2}} \frac{(z_2 - z_1)\, dy_2 - (y_2 - y_1)\, dz_2}{r^2}$$

$$+ \frac{m_1}{2\pi} \oint_{\substack{\text{contour}\\ \text{of } A_2}} \frac{(x_2 - x_1)\, dz_2 - (z_2 - z_1)\, dx_2}{r^2}$$

$$+ \frac{n_1}{2\pi} \oint_{\substack{\text{contour}\\ \text{of } A_2}} \frac{(y_2 - y_1)\, dx_2 - (x_2 - x_1)\, dy_2}{r^2} \qquad (3\text{-}40)$$

where $r^2 = (x_2 - x_1)^2 + (y_2 - y_1)^2 + (z_2 - z_1)^2$, and l_1, m_1, and n_1 are the direction cosines. The contour integrations in Eq. 3-40 should be performed around A_2 by the convention described previously.

If the coordinate axes are so oriented that the unit normal vector $\hat{\mathbf{n}}_1$ to surface element dA_1 lies along one of the coordinate axes, the direction cosines of $\hat{\mathbf{n}}_1$ with respect to the other two axes become zero; hence two of the integrals in Eq. 3-40 vanish. If, in addition, any one of the boundaries of surface A_2 is parallel to the coordinate axis, the integration is further simplified.

(b) The Diffuse View Factor Between A_1 and A_2

We now consider the diffuse view factor $F_{A_1-A_2}$ from a surface A_1 to a surface A_2 defined by[2]

$$A_1 F_{A_1-A_2} = \int_{A_1} F_{dA_1-A_2}\, dA_1 \qquad (3\text{-}41)$$

We substitute $F_{dA_1-A_2}$ from Eq. 3-40 into Eq. 3-41, rearrange the integrals, and obtain

$$A_1 F_{A_1-A_2} = \frac{1}{2\pi} \oint_{\substack{\text{contour}\\ \text{of } A_2}} \left[\int_{A_1} \frac{(y_2 - y_1)n_1 - (z_2 - z_1)m_1}{r^2}\, dA_1 \right] dx_2$$

$$+ \frac{1}{2\pi} \oint_{\substack{\text{contour}\\ \text{of } A_2}} \left[\int_{A_1} \frac{(z_2 - z_1)l_1 - (x_2 - x_1)n_1}{r^2}\, dA_1 \right] dy_2$$

$$+ \frac{1}{2\pi} \oint_{\substack{\text{contour}\\ \text{of } A_2}} \left[\int_{A_1} \frac{(x_2 - x_1)m_1 - (y_2 - y_1)l_1}{r^2}\, dA_1 \right] dz_2 \qquad (3\text{-}42)$$

The surface integrals in Eq. 3-42 will now be changed into contour integrals. The first surface integral on the right-hand side can be written as

$$\int_{A_1} \frac{(y_2 - y_1)n_1 - (z_2 - z_1)m_1}{r^2}\, dA_1 = \int_{A_1} \hat{\mathbf{n}}_1 \cdot (\nabla \times \mathbf{V}_1)\, dA_1 \tag{3-43}$$

where we have defined

$$\hat{\mathbf{n}}_1 \equiv \hat{\mathbf{i}} l_1 + \hat{\mathbf{j}} m_1 + \hat{\mathbf{k}} n_1, \qquad \nabla \equiv \hat{\mathbf{i}} \frac{\partial}{\partial x_1} + \hat{\mathbf{j}} \frac{\partial}{\partial y_1} + \hat{\mathbf{k}} \frac{\partial}{\partial z_1}$$

$$\mathbf{V}_1 \equiv \hat{\mathbf{i}} \ln r$$

By applying the Stokes theorem we obtain for the surface integral on the right-hand side of Eq. 3-43

$$\int_{A_1} \frac{(y_2 - y_1)n_1 - (z_2 - z_1)m_1}{r^2}\, dA_1 = \oint_{\substack{\text{contour}\\ \text{of } A_1}} \mathbf{V}_1 \cdot d\mathbf{s}_1 = \oint_{\substack{\text{contour}\\ \text{of } A_1}} \ln r\, dx_1 \tag{3-44}$$

since $d\mathbf{s}_1 = \hat{\mathbf{i}}\, dx_1 + \hat{\mathbf{j}}\, dy_1 + \hat{\mathbf{k}}\, dz_1$. In a similar manner it can be shown that

$$\int_{A_1} \frac{(z_2 - z_1)l_1 - (x_2 - x_1)n_1}{r^2}\, dA_1 = \oint_{\substack{\text{contour}\\ \text{of } A_1}} \mathbf{V}_2 \cdot d\mathbf{s}_1 = \oint_{\substack{\text{contour}\\ \text{of } A_1}} \ln r\, dy_1 \tag{3-45}$$

$$\int_{A_1} \frac{(x_2 - x_1)m_1 - (y_2 - y_1)l_1}{r^2}\, dA_1 = \oint_{\substack{\text{contour}\\ \text{of } A_1}} \mathbf{V}_3 \cdot d\mathbf{s}_1 = \oint_{\substack{\text{contour}\\ \text{of } A_1}} \ln r\, dz_1 \tag{3-46}$$

where we have defined

$$\mathbf{V}_2 = \hat{\mathbf{j}} \ln r \qquad \text{and} \qquad \mathbf{V}_3 = \hat{\mathbf{k}} \ln r$$

Substituting Eqs. 3-44 through 3-46 into Eq. 3-42, we obtain

$$\begin{aligned} A_1 F_{A_1-A_2} &= \frac{1}{2\pi} \oint_{\substack{\text{contour}\\ \text{of } A_2}} \left(\oint_{\substack{\text{contour}\\ \text{of } A_1}} \ln r\, dx_1 \right) dx_2 + \frac{1}{2\pi} \oint_{\substack{\text{contour}\\ \text{of } A_2}} \left(\oint_{\substack{\text{contour}\\ \text{of } A_1}} \ln r\, dy_1 \right) dy_2 \\ &\quad + \frac{1}{2\pi} \oint_{\substack{\text{contour}\\ \text{of } A_2}} \left(\oint_{\substack{\text{contour}\\ \text{of } A_1}} \ln r\, dz_1 \right) dz_2 \end{aligned} \tag{3-47a}$$

which can be rearranged as

$$A_1 F_{A_1-A_2} = \frac{1}{2\pi} \oint_{\substack{\text{contour}\\ \text{of } A_1}} \oint_{\substack{\text{contour}\\ \text{of } A_2}} (\ln r\, dx_2\, dx_1 + \ln r\, dy_2\, dy_1 + \ln r\, dz_2\, dz_1) \tag{3-47b}$$

where

$$r = \sqrt{(x_2 - x_1)^2 + (y_2 - y_1)^2 + (z_2 - z_1)^2}$$

To illustrate the application of the contour integration technique we consider in the following examples the determination of $F_{dA_1-A_2}$ and $F_{A_1-A_2}$ for simple geometries.

Example 1. Consider an elemental surface dA_1 and a finite rectangular surface A_2, which are parallel to each other and positioned as shown in Fig. 3-5. Surface dA_1 is parallel to the xy plane and positioned on the oy axis at a distance d from the origin. Surface A_2 has one corner at the oz axis, and its sides a and b are parallel to the ox and oy axes, respectively.

The coordinates of dA_1 are $x_1 = 0$, $y_1 = d$, and $z_1 = 0$. The direction cosines of the unit normal vector $\hat{\mathbf{n}}_1$ to the surface dA_1 are $l_1 = 0$, $m_1 = 0$, and $n_1 = 1$. By substituting these coordinates and the direction cosines into Eq. 3-40 we obtain

$$F_{dA_1-A_2} = \frac{1}{2\pi} \oint_{\substack{\text{contour} \\ \text{of } A_2}} \frac{(y_2 - d)\, dx_2 - x_2\, dy_2}{x_2^2 + (y_2 - d)^2 + c^2} \tag{3-48}$$

where x_2, y_2, c are the coordinates of any point on the surface of A_2. To perform this integration we divide the contour of A_2 into four segments, I, II, III, IV, and follow an integration path in the direction shown in Fig. 3-5. The limits of integration on each of these segments are given as:

I: $x_2 = 0$, $dx_2 = 0$; then the integral vanishes identically.
II: $y_2 = b$, $dy_2 = 0$, and x_2 varies from 0 to a.
III: $x_2 = a$, $dx_2 = 0$, and y_2 varies from b to 0.
IV: $y_2 = 0$, $dy_2 = 0$, and x_2 varies from a to 0.

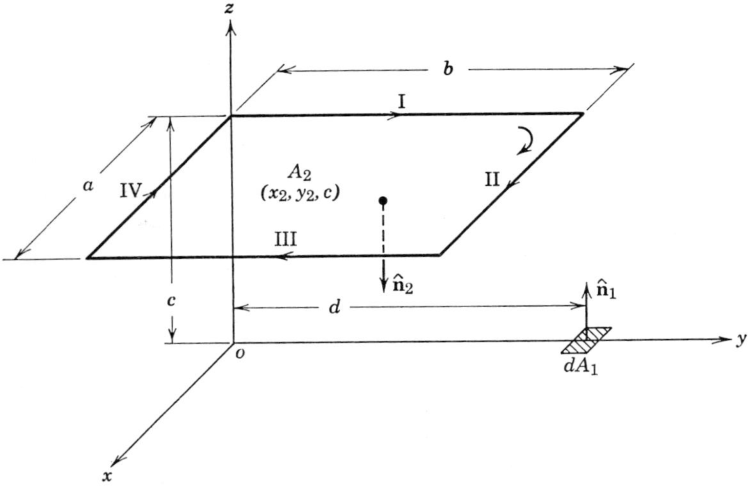

Fig. 3-5. Evaluation of diffuse view factor $F_{dA_1-A_2}$ by contour integration.

With the limits of integration as given above, Eq. 3-48 becomes

$$F_{dA_1-A_2} = \frac{1}{2\pi}\left[0 + \int_{x_2=0}^{a} \frac{b-d}{x_2^2+(b-d)^2+c^2}\,dx_2 - \int_{y_2=b}^{0} \frac{a}{a^2+(y_2-d)^2+c^2}\,dy_2 - \int_{x_2=a}^{0} \frac{d}{x_2^2+d^2+c^2}\,dx_2\right] \tag{3-49}$$

The evaluation of these integrals is straightforward.

Example 2. The contour integration technique will now be applied to determine the diffuse view factor $F_{A_1-A_2}$ between two parallel rectangular finite surfaces A_1 and A_2, separated by a distance c as illustrated in Fig. 3-6.

Equation 3-47b will be utilized to evaluate this view factor. For the geometry considered, $z_1 = 0$ and $z_2 = c$ for surfaces A_1 and A_2, respectively; then $dz_1 = dz_2 = 0$ and Eq. 3-47b simplifies to

$$2\pi A_1 F_{A_1-A_2} = \oint_{\substack{\text{contour}\\ \text{of } A_1}} \oint_{\substack{\text{contour}\\ \text{of } A_2}} (\ln r\,dx_2\,dx_1 + \ln r\,dy_2\,dy_1) \tag{3-50}$$

Each of the contours of surfaces A_1 and A_2 will be separated into four segments as shown in Fig. 3-6, and the integrations in Eq. 3-50 will be performed successively around the contours of A_2 and A_1. Following the path around A_2, we note that $dx_2 = 0$ for segments I and III, and $dy_2 = 0$

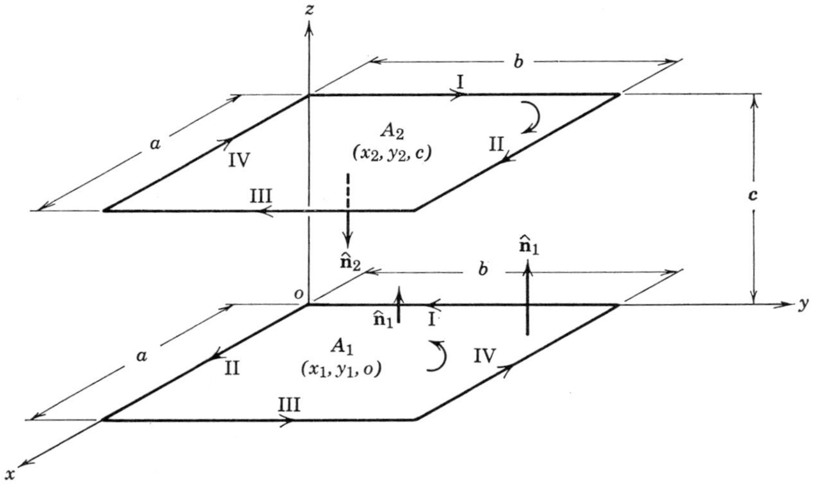

Fig. 3-6. Evaluation of diffuse view factor $F_{A_1-A_2}$ by contour integration.

for segments II and IV. Then Eq. 3-50 becomes

$$2\pi ab F_{A_1-A_2} = \oint_{\substack{\text{contour}\\ \text{of } A_1}} \left\{ \int_{y_2=0}^{b} \ln\,[x_1^{\,2} + (y_2 - y_1)^2 + c^2]^{1/2}\, dy_2 \right\} dy_1$$
$$+ \oint_{\substack{\text{contour}\\ \text{of } A_1}} \left\{ \int_{x_2=0}^{a} \ln\,[(x_2 - x_1)^2 + (b - y_1)^2 + c^2]^{1/2}\, dx_2 \right\} dx_1$$
$$+ \oint_{\substack{\text{contour}\\ \text{of } A_1}} \left\{ \int_{y_2=b}^{0} \ln\,[(a - x_1)^2 + (y_2 - y_1)^2 + c^2]^{1/2}\, dy_2 \right\} dy_1$$
$$+ \oint_{\substack{\text{contour}\\ \text{of } A_1}} \left\{ \int_{x_2=a}^{0} \ln\,[(x_2 - x_1)^2 + y_1^{\,2} + c^2]^{1/2}\, dx_2 \right\} dx_1 \qquad (3\text{-}51)$$

In a similar manner we follow the path of integration around the contour of A_1, and Eq. 3-51 becomes

$$2\pi ab F_{A_1-A_2} = \int_{y_1=b}^{0} \int_{y_2=0}^{b} \ln\,[(y_2 - y_1)^2 + c^2]^{1/2}\, dy_2\, dy_1$$
$$+ \int_{y_1=b}^{0} \int_{y_2=b}^{0} \ln\,[a^2 + (y_2 - y_1)^2 + c^2]^{1/2}\, dy_2\, dy_1$$
$$+ \text{Integrals for segments II, III, and IV} \qquad (3\text{-}52)$$

The resulting integrals can be performed using standard integral tables.

3-4 DIFFUSE VIEW FACTOR BETWEEN AN ELEMENTAL SURFACE AND AN INFINITELY LONG STRIP

Consider an elemental surface dA_1 lying on the xy plane at the origin and an infinitely long strip A_2 whose generating lines are parallel to the x axis as illustrated in Fig. 3-7. Let ab be the contour of the intersection of strip A_2 with the yz plane, and ϕ_a and ϕ_b be the angles between the oz axis and the lines oa and ob, respectively. The view factor $dF_{dA_1-\text{strip } dA_2}$ between the elemental surface dA_1 and the infinitely long elemental strip dA_2 is given by the following simple relation:

$$dF_{dA_1-\text{strip } dA_2} = \tfrac{1}{2} \cos\phi\, d\phi = \tfrac{1}{2} d(\sin\phi) \qquad (3\text{-}53\text{a})$$

and the view factor $F_{dA_1-\text{strip } A_2}$ between the elemental surface dA_1 and the infinitely long strip A_2 is obtained by integrating Eq. 3-53a as

$$F_{dA_1-\text{strip } A_2} = \int_{\phi_a}^{\phi_b} \tfrac{1}{2} \cos\phi\, d\phi = \tfrac{1}{2}[\sin\phi_b - \sin\phi_a] \qquad (3\text{-}53\text{b})$$

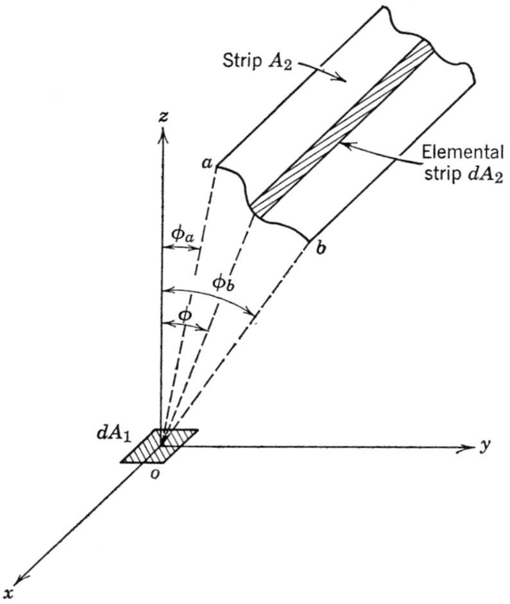

Fig. 3-7. Diffuse view factor between an elemental surface dA_1 and an infinitely long strip A_2.

originally derived by Hottel; its derivation is given in detail by Jacob [9, pp. 19–21]. This relation is applicable also to the case in which the elemental surface dA_1 is an element strip on the xy plane parallel to the ox axis.

3-5 THE DIFFUSE VIEW FACTOR ALGEBRA

The diffuse view factor for a complex geometry can be expressed in terms of the known view factors for simpler geometries by utilizing the principle of superposition and the reciprocity relations for the view factors. Several examples of this approach, which is known as *diffuse view factor algebra*, have been given by Hamilton and Morgan [1] and Mackey et al. [7]. We present now the application of diffuse view factor algebra with simple examples.

(a) Diffuse View Factor Between Surfaces dA_1 and A_2'

Consider an elemental surface dA_1 and a finite rectangular surface A_2', which are parallel to each other and positioned as shown in Fig. 3-8. We shall utilize diffuse view factor algebra to determine the diffuse view factor $F_{dA_1-A_2'}$ between surfaces dA_1 and A_2' in terms of the known diffuse view factors $F_{dA_1-A_2}$ between surfaces dA_1 and A_2, positioned as shown in Fig. 3-9.

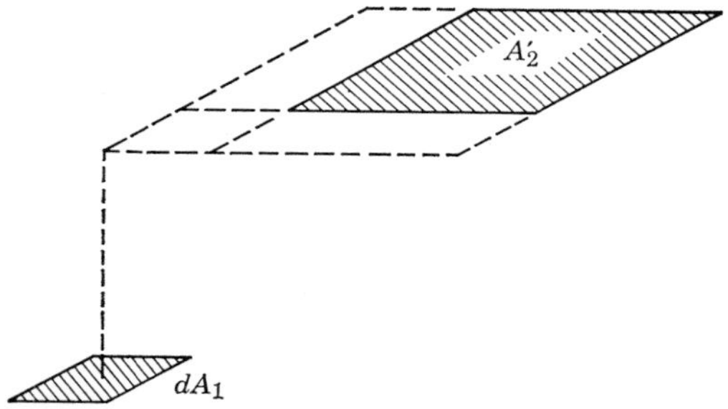

Fig. 3-8. Arrangement of surfaces dA_1 and A_2'.

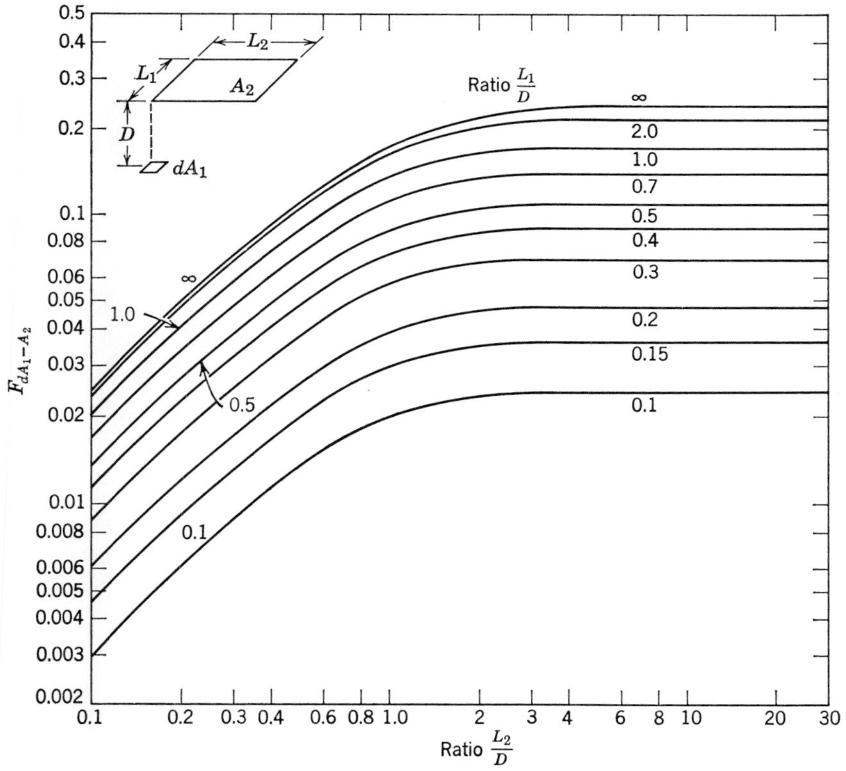

Fig. 3-9. Diffuse view factor $F_{dA_1-A_2}$. (From C. O. Mackey et al. [7]).

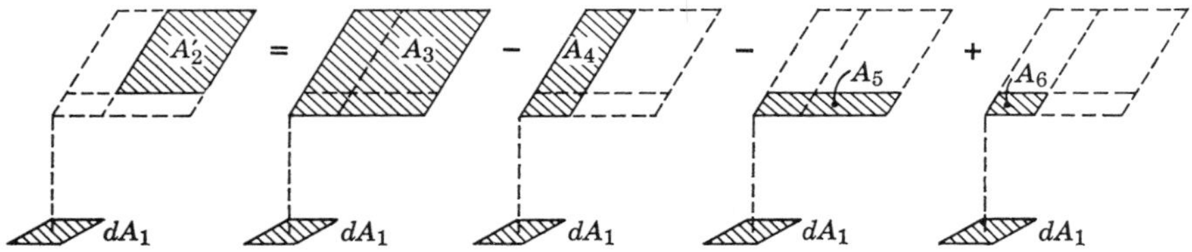

Fig. 3-10. Determination of the diffuse view factor $F_{dA_1-A'_2}$ by superposition.

The first step in the analysis is to express A_2' as an algebraic sum of subareas A_i ($i = 3, 4, 5, 6$) as shown in Fig. 3-10, so that each diffuse view factor $F_{dA_1-A_i}$ ($i = 3, 4, 5, 6$) can be determined from the chart in Fig. 3-9. The conservation of radiant energy from dA_1 to A_2' implies that $F_{dA_1-A_2'}$ can be obtained by the superposition of view factors $F_{dA_1-A_i}$ as

$$F_{dA_1-A_2'} = F_{dA_1-A_3} - F_{dA_1-A_4} - F_{dA_1-A_5} + F_{dA_1-A_6} \tag{3-54}$$

since

$$A_2' = A_3 - A_4 - A_5 + A_6 \tag{3-55}$$

The view factors on the right-hand side of Eq. 3-54 are obtainable from the chart in Fig. 3-9.

(b) Diffuse View Factor Between Surfaces A_1 and $A_{2'}$

Consider two finite rectangular surfaces A_1 and $A_{2'}$, positioned as shown in Fig. 3-11. The diffuse view factor $F_{A_1-A_{2'}}$ between surfaces A_1 and $A_{2'}$ will be determined by the superposition of view factors $F_{A_1-A_2}$ between two rectangular surfaces A_1 and A_2, positioned as shown in Fig. 3-12.

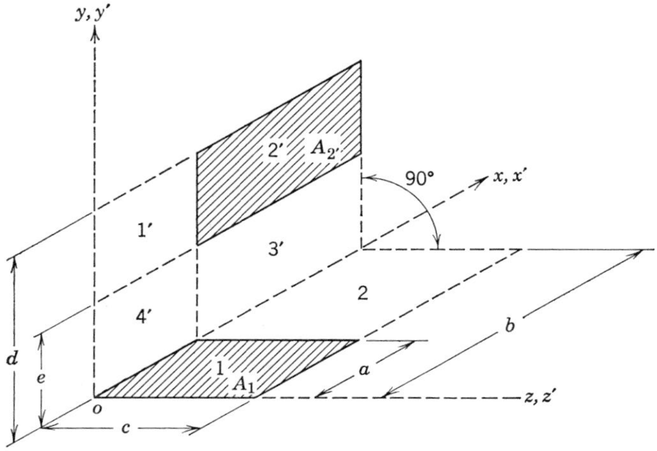

Fig. 3-11. Arrangement of surfaces for diffuse view factor $F_{A_1-A_{2'}}$.

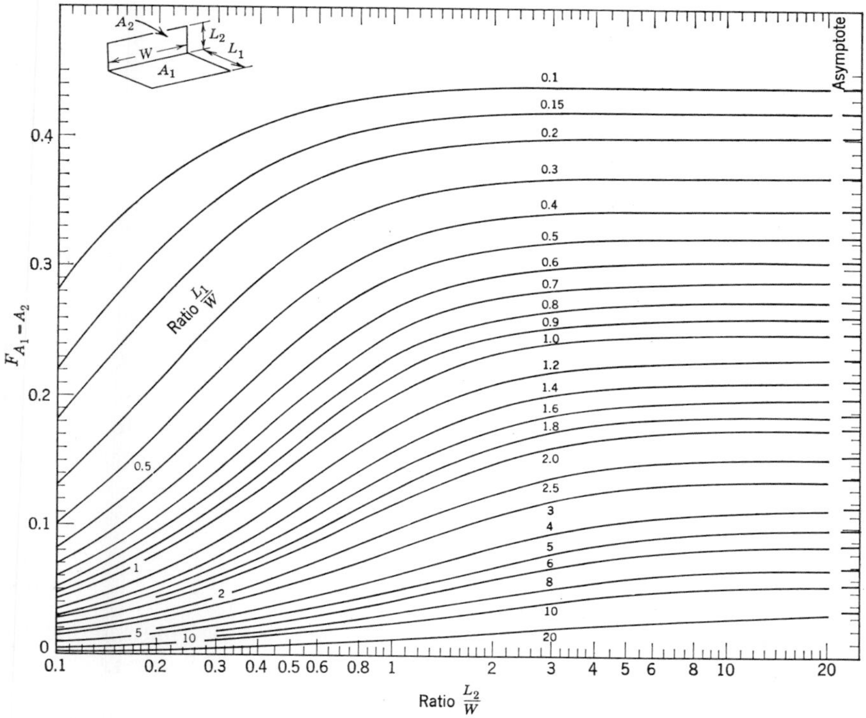

Fig. 3-12. Diffuse view factor $F_{A_1-A_2}$. (From C. O. Mackey et al. [7]).

For convenience in the analysis we define

$$G_{\alpha-\beta} \equiv A_\alpha F_{\alpha-\beta} \tag{3-56}$$

Then the reciprocity relation for the diffuse view factors between surfaces A_α and A_β can be written as

$$G_{\alpha-\beta} = G_{\beta-\alpha} \tag{3-57}$$

The laws of arithmetic that are applicable to view factors for a composite system can be written compactly by utilizing the notation given by Eqs. 3-56 and 3-57. For example, if surfaces A_1 and A_2 are subdivided as

$$A_1 = A_i + A_j \tag{3-58}$$

$$A_2 = A_k + A_l \tag{3-59}$$

then the diffuse view factors between the surfaces of this composite system obey the following laws of arithmetic:

$$G_{1-2} \equiv G_{ij-kl} = G_{ij-k} + G_{ij-l} \tag{3-60}$$

$$= G_{i-kl} + G_{j-kl} \tag{3-61}$$

$$= G_{i-k} + G_{i-l} + G_{j-k} + G_{j-l} \tag{3-62}$$

where we have defined

$$G_{ij-kl} \equiv (A_i + A_j)F_{(A_i+A_j)-(A_k+A_l)} \tag{3-63}$$

$$G_{i-kl} \equiv A_i F_{A_i-(A_k+A_l)} \tag{3-64}$$

etc.

The above relations will be used in the following example.

Consider the subdivided geometry shown in Fig. 3-11. We expand view factor $G_{12-1'2'3'4'}$ according to the above laws of arithmetic as

$$\begin{aligned} G_{12-1'2'3'4'} &= G_{12-1'2'} + G_{12-3'4'} \\ &= (G_{1-1'} + G_{1-2'} + G_{2-1'} + G_{2-2'}) + G_{12-3'4'} \end{aligned} \tag{3-65}$$

In Eq. 3-65, $G_{1-2'}$ is the diffuse view factor sought. View factors $G_{12-1'2'3'4'}$ and $G_{12-3'4'}$ are of the form that can be evaluated directly from the chart given in Fig. 3-12. View factors $G_{1-1'}$, $G_{2-1'}$, and $G_{2-2'}$, however, cannot be determined from Fig. 3-12. We rearrange view factors $G_{1-1'}$ and $G_{2-2'}$ in the forms

$$G_{1-1'} = G_{1-1'4'} - G_{1-4} \tag{3-66a}$$

$$G_{2-2'} = G_{2-2'3'} - G_{2-3'} \tag{3-66b}$$

Then the terms on the right-hand sides of Eqs. 3-66 can be obtained directly from Fig. 3-12; hence view factors $G_{1-1'}$ and $G_{2-2'}$ are also known.

It can be shown that view factor $G_{2-1'}$ is equivalent to[3]

$$G_{2-1'} = G_{1-2'} \tag{3-67}$$

Substituting Eqs. 3-66 and 3-67 into Eq. 3-65, we obtain

$$2G_{1-2'} = G_{12-1'2'3'4'} + G_{1-4'} + G_{2-3'} - G_{1-1'4'} - G_{2-2'3'} - G_{12-3'4'} \tag{3-68}$$

Hence $G_{1-2'}$ can be evaluated from Eq. 3-68, since all the terms on the right-hand side of this equation are obtainable from Eq. 3-12.

(c) Diffuse View Factor for Long Enclosures

Hottel [6, pp. 65–68] determined the diffuse view factor between the surfaces of a long enclosure by utilizing view factor algebra. Consider an enclosure, as shown in Fig. 3-13, consisting of three surfaces very long in the direction

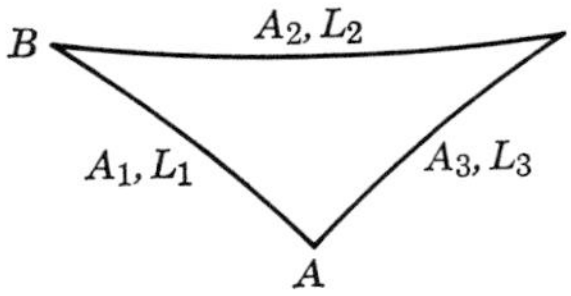

Fig. 3-13. Diffuse view factor between the surfaces of a long enclosure.

perpendicular to the plane of the figure. The summation rule for diffuse view factors between the surfaces of this enclosure can be written as

$$\sum_{j=1}^{3} F_{i-j} = 1, \quad i = 1, 2, 3 \tag{3-69a}$$

with

$$F_{jj} = 0 \tag{3-69b}$$

where we have assumed that the surfaces of the enclosure are flat or convex. The reciprocity relation is given as

$$A_i F_{i-j} = A_j F_{j-i}, \quad i, j = 1, 2, 3 \tag{3-70}$$

Suppose that we wish to determine the diffuse view factor F_{1-2}. By solving Eqs. 3-69 and 3-70 we obtain

$$A_1 F_{1-2} = \frac{A_1 + A_2 - A_3}{2} \tag{3-71a}$$

which can be written as

$$L_1 F_{1-2} = \frac{L_1 + L_2 - L_3}{2} \tag{3-71b}$$

where L_1, L_2, and L_3 are the lengths of the arcs AB, BC, and CA, respectively.

The relation given by Eq. 3-71b will now be utilized to determine the diffuse view factor between the surfaces of an enclosure, as shown in Fig. 3-14, consisting of four surfaces very long in the direction perpendicular to the plane of the figure. In the following analysis the surfaces of the enclosure in Fig. 3-14 can be flat, convex, or concave (i.e., the restriction $F_{jj} = 0$ is removed). We assume that imaginary strings (shown by dotted lines in Fig. 3-14) are tightly stretched among the corners A, B, C, and D of the enclosure. Let L_i, $i = 1, 2, 3, 4, 5, 6$, denote the lengths of the strings joining the corners A–B, B–C, C–D, D–A, D–B, and A–C, respectively. Suppose that we

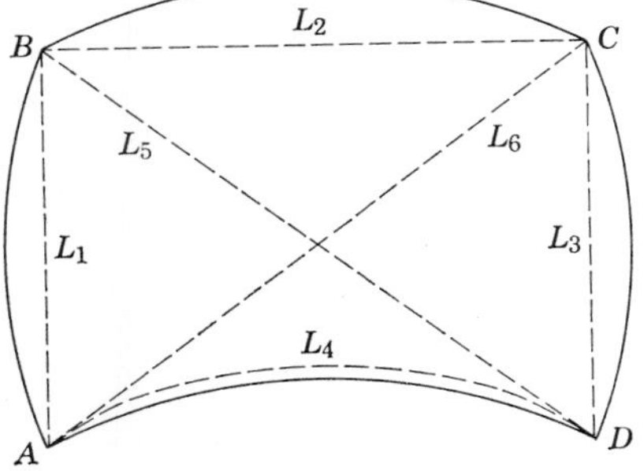

Fig. 3-14. Diffuse view factor between the surfaces of a long enclosure.

wish to determine the diffuse view factor $F_{AB\text{-}CD}$ from surface AB to surface CD. We consider the imaginary enclosures ABC and ABD formed by the imaginary strings. By the application of Eq. 3-71b we determine L_1F_{1-2} and L_1F_{1-4} for the imaginary enclosures ABC and ABD, respectively. The summation rule for the diffuse view factors can be written as $F_{1-2} + F_{1-3} + F_{1-4} = 1$. Substitution of F_{1-2} and F_{1-4} evaluated according to Eq. 3-71b into this summation relation yields F_{1-3} as

$$L_1F_{1-3} = \frac{(L_5 + L_6) - (L_2 + L_4)}{2} \tag{3-72}$$

It can readily be shown that the left-hand side of Eq. 3-72 is equivalent to $AB\ F_{AB\text{-}CD}$, where AB and CD characterize the curved surfaces. We note that in Eq. 3-72 the term $(L_5 + L_6)$ is the sum of the lengths of the crossed strings and $(L_2 + L_4)$ is the sum of the lengths of the uncrossed strings.

3-6 DETERMINATION OF DIFFUSE VIEW FACTOR BY DIFFERENTIATION

The diffuse view factor between two elemental surfaces can be determined by a short-cut method based on differentiation of the diffuse view factor between two finite surfaces. Since no general rule can be set up for this method, its application will be illustrated by representative examples.

(a) Diffuse View Factor Between an Elemental Circular Band and an Elemental Circular Ring

In problems of radiative exchange inside a cylindrical cavity the diffuse view factor is needed between an *elemental circular band* and an *elemental circular ring*. Figure 3-15 shows a cylindrical cavity of radius a, an elemental circular

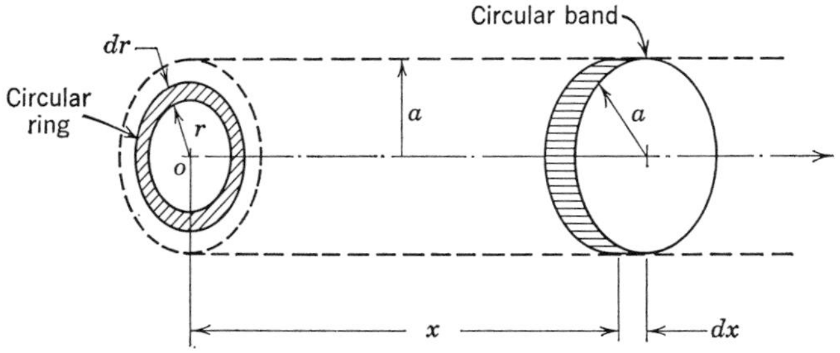

Fig. 3-15. Diffuse view factor between an elemental circular band and an elemental circular ring.

band of width dx, and an elemental circular ring of width dr. For convenience in the subsequent analysis we shall use the following notation to characterize various diffuse view factors:

$dF_{dr-dx,x}$ = elemental diffuse view factor from a *ring* (r, dr) to a *band* (a, dx) at a distance x apart.

$F_{dr-a,x}$ = diffuse view factor from a *ring* (r, dr) to a *disk* of radius a at a distance x apart.

$F_{a-dr,x}$ = diffuse view factor from a disk of radius a to a *ring* (r, dr) at a distance x apart.

$F_{a-r,x}$ = diffuse view factor from a disk of radius a to a disk of radius r, which are parallel and concentric and are at a distance x apart.

Here we have used *band* (a, dx) to denote an elemental circular band of radius a and width dx, positioned at x on the cylindrical surface, and *ring* (r, dr) to denote an elemental ring of radius r and width dr, positioned at the base of the cylindrical cavity as shown in Fig. 3-15.

The elemental diffuse view factor $dF_{dr-dx,x}$ will now be determined. It has been shown by Sparrow, Albers, and Eckert [14] that this diffuse view factor can be obtained by differentiating the diffuse view factor $F_{a-r,x}$ according to the relation

$$dF_{dr-dx,x} = -\frac{a^2}{2r}\frac{\partial}{\partial x}\left(\frac{\partial}{\partial r}F_{a-r,x}\right)dx \tag{3-73}$$

where $F_{a-r,x}$ is given by Jacob [9, p. 14] as

$$F_{a-r,x} = \frac{a^2 + r^2 + x^2 - \sqrt{(a^2 + r^2 + x^2)^2 - 4a^2r^2}}{2a^2} \tag{3-74}$$

Equation 3-73 will now be proved. By the law of conservation of energy we write

$$dF_{dr-dx,x} = F_{dr-a,x} - F_{dr-a,(x+dx)} = -\frac{\partial}{\partial x}(F_{dr-a,x})\,dx \tag{3-75}$$

The physical significance of Eq. 3-75 is that the radiative energy leaving the *ring* (r, dr) that strikes the *band* (a, dx), that is, $dF_{dr-dx,x}$, is equal to the radiative energy leaving the *ring* (r, dr) that strikes the disk of radius a at x, that is, $F_{dr-a,x}$, minus the radiative energy leaving the *ring* (r, dr) that strikes the disk of radius a at $x + dx$, that is, $F_{dr-a,(x+dx)}$.

From the reciprocity relation we have

$$(2\pi r\,dr)F_{dr-a,x} = (\pi a^2)F_{a-dr,x} \tag{3-76a}$$

or

$$F_{dr-a,x} = \frac{a^2}{2r\,dr}F_{a-dr,x} \tag{3-76b}$$

By the law of conservation of energy $F_{a-dr,x}$ can be written as

$$F_{a-dr,x} = F_{a-(r+dr),x} - F_{a-r,x} = \frac{\partial}{\partial r}(F_{a-r,x})\,dr \tag{3-77}$$

where $F_{a-(r+dr),x}$ is the view factor from the disk of radius a to the disk of radius $r + dr$ at a distance x apart, and so forth.

By substituting Eqs. 3-76b and 3-77 into Eq. 3-75 we obtain

$$dF_{dr-dx,x} = -\frac{a^2}{2r}\frac{\partial}{\partial x}\left(\frac{\partial}{\partial r}F_{a-r,x}\right)dx \tag{3-78}$$

which proves the validity of Eq. 3-73.

Substituting Eq. 3-74 into 3-78 and performing the differentiation, we obtain

$$dF_{dr-dx,x} = -2xa^2\frac{x^2 + r^2 - a^2}{[(x^2 + r^2 + a^2)^2 - 4r^2a^2]^{3/2}}\,dx \tag{3-79}$$

The elemental view factor $dF_{dx-dr,x}$ from the band (a, dx) to the ring (r, dr) at a distance x apart can now be evaluated by utilizing the reciprocity relation, that is,

$$(2\pi r\,dr)\,dF_{dr-dx,x} = (2\pi a\,dx)\,dF_{dx-dr,x}$$

or

$$dF_{dx-dr,x} = \frac{r\,dr}{a\,dx}\,dF_{dr-dx,x} \tag{3-80}$$

(b) Diffuse View Factor Between Two Elemental Coaxial Circular Bands

Consider two elemental circular bands (a, dx_1) at x_1 and (a, dx_2) at x_2 at a distance h apart on the surface of a circular cylinder of radius a as shown in Fig. 3-16. We introduce the following notation to characterize various

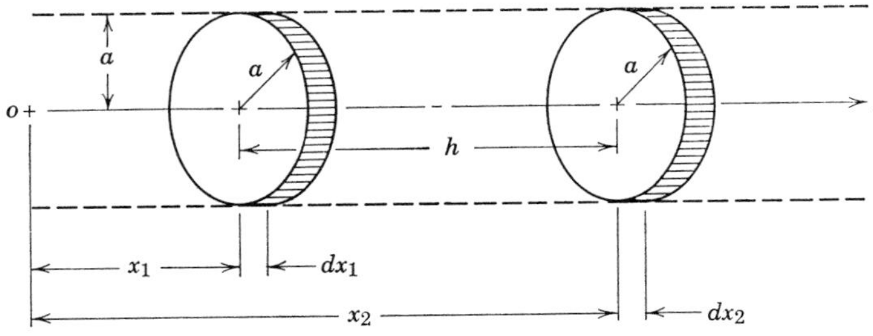

Fig. 3-16. Diffuse view factor between two elemental coaxial circular bands.

diffuse view factors:

$dF_{dx_1-dx_2,h}$ = elemental diffuse view factor from a *band* (a, dx_1) to a *band* (a, dx_2) at a distance h apart.

$dF_{dx_2-dx_1,h}$ = elemental diffuse view factor from a *band* (a, dx_2) to a *band* (a, dx_1) at a distance h apart.

We shall now evaluate the elemental diffuse view factor $dF_{dx_2-dx_1,h}$. By the conservation of energy we write

$$dF_{dx_2-dx_1,h} = F_{dx_2-a,\text{disk at }(x_1+dx_1)} - F_{dx_2-a,\text{disk at }x_1} = \frac{\partial}{\partial x_1}(F_{dx_2-a,h})\,dx_1 \tag{3-81}$$

where

$F_{dx_2-a,h}$ = diffuse view factor from a *band* (a, dx_2) to a disk of radius a at a distance h apart.

From the reciprocity relation we have

$$(2\pi a\, dx_2)F_{dx_2-a,h} = (\pi a^2)F_{a-dx_2,h} \tag{3-82a}$$

or

$$F_{dx_2-a,h} = \frac{a}{2dx_2}F_{a-dx_2,h} \tag{3-82b}$$

where

$F_{a-dx_2,h}$ = diffuse view factor from a disk of radius a at x_1 to a *band* (a, dx_2) at x at a distance $x_2 - x_1 = h$ apart.

By the law of conservation of energy we write

$$F_{a-dx_2,h} = F_{a-a,\text{disk at }x_2} - F_{a-a,\text{disk at }(x_2+dx_2)} = -\frac{\partial}{\partial x_2}(F_{a-a,h})\,dx_2 \tag{3-83}$$

Substituting Eqs. 3-82b and 3-83 into 3-81, we obtain

$$dF_{dx_2-dx_1,h} = -\frac{a}{2}\frac{\partial^2}{\partial x_1\,\partial x_2}(F_{a-a,h})\,dx_1 \tag{3-84}$$

The elemental view factor $dF_{dx_1-dx_2,h}$ is obtained by utilizing the reciprocity relation, that is,

$$(2\pi a\, dx_1)\,dF_{dx_1-dx_2,h} = (2\pi a\, dx_2)\,dF_{dx_2-dx_1,h} \tag{3-85a}$$

or

$$dF_{dx_1-dx_2,h} = \frac{dx_2}{dx_1}\,dF_{dx_2-dx_1,h} \tag{3-85b}$$

Substituting Eq. 3-84 into 3-85b yields

$$dF_{dx_1-dx_2,h} = -\frac{a}{2}\frac{\partial^2}{\partial x_1\,\partial x_2}(F_{a-a,h})\,dx_2 \tag{3-86}$$

The diffuse view factor $F_{a-a,h}$ is obtained from Eq. 3-74 by replacing in that equation r by a and x by h:

$$F_{a-a,h} = \frac{2a^2 + h^2 - \sqrt{(2a^2 + h^2)^2 - 4a^4}}{2a^2} \tag{3-87}$$

By substituting Eq. 3-87 into Eq. 3-86 and performing the differentiation, we obtain

$$dF_{dx_1-dx_2,(x_2-x_1)} = \frac{1}{2a}\left[1 - |x_2 - x_1|\,\frac{(x_2 - x_1)^2 + 6a^2}{[4a^2 + (x_2 - x_1)^2]^{3/2}}\right] dx_2 \tag{3-88a}$$

Here we have used the absolute value $|x_2 - x_1|$ because the view factor depends on the absolute value of the distance of separation, $h = |x_2 - x_1|$, between the bands. Equation 3-88a can be written in terms of the dimensionless variables as

$$dF_{d\xi_1-d\xi_2,(\xi_2-\xi_1)} = \left\{1 - |\xi_2 - \xi_1|\,\frac{(\xi_2 - \xi_1)^2 + \frac{3}{2}}{[1 + (\xi_2 - \xi_1)^2]^{3/2}}\right\} d\xi_2 \tag{3-88b}$$

where $\xi_i = x_i/2a$, $i = 1$ or 2.

3-7 THE SPECULAR VIEW FACTORS

In this section we examine *specular view factors*,[4] which are applicable to diffusely emitting and purely specularly reflecting surfaces. The relations defining the specular view factor between surfaces can be derived by utilizing the so-called *mirror-image method*, which is well known in the fields of optics and illuminating engineering. Eckert and Sparrow [15] were the first to apply this method to determine the specular view factors for radiative heat transfer applications.

The concept behind the mirror-image method is best envisioned by examining the interreflections of radiation between two finite surfaces A_1 and A_2, positioned as shown in Fig. 3-17. For simplicity we assume that surface A_1 is a purely diffuse reflector and surface A_2 a purely specular reflector. Both surfaces are considered to be purely diffuse emitters. Let us consider the radiative energy leaving diffusely an elemental area dA_1 on A_1 and reaching surface A_1 after one specular reflection at surface A_2. The path followed by the radiation during this exchange is schematically illustrated

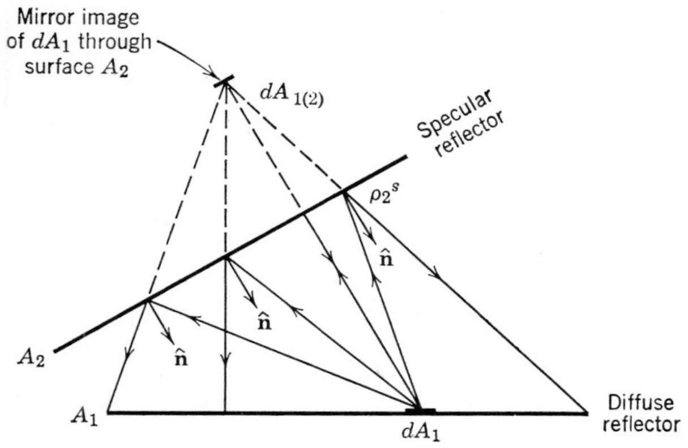

Fig. 3-17. Surface A_1 receives radiation from dA_1 after one specular reflection by A_2.

in Fig. 3-17. A close examination of this path reveals that the radiative energy leaving dA_1 and reaching A_1 after one specular reflection at A_2 *appears to originate from a diffuse radiation source* $dA_{1(2)}$, which is the mirror image of dA_1 through the mirror (i.e., the specularly reflecting surface) A_2. This concept is the basis of the *mirror-image method.*

If surface A_2 is a perfect specular reflector, that is, $\rho_2{}^s = 1$, then the fraction of radiative energy leaving dA_1 in all directions in the hemispherical space that strikes A_1 after one specular reflection at A_2 is given as

$$F_{dA_{1(2)}-A_1} \tag{3-89}$$

where $F_{dA_{1(2)}-A_1}$ denotes the diffuse view factor from the mirror image $dA_{1(2)}$ to surface A_1. If surface A_2 is not a perfect specular reflector, that is, $\rho_2{}^s < 1$, the radiative energy will be attenuated by a factor $\rho_2{}^s$ when reflected by surface A_2. Then the fraction of radiative energy leaving dA_1 that is received by A_1 after one specular reflection at A_2 becomes[5]

$$\rho_2{}^s F_{dA_{1(2)}-A_1} \tag{3-90}$$

Figure 3-18 shows a different arrangement of surfaces A_1 and A_2, in which A_1 is a diffuse reflector and A_2 a specular reflector. In this case the arrangement of the surfaces is such that radiation leaving dA_1 strikes only a portion of surface A_1 after one specular reflection at A_2. Then the fraction of radiative energy leaving dA_1 that reaches A_1 after one specular reflection at A_2 is given by

$$\rho_2{}^s \, {}^*F_{dA_{1(2)}-A_1} \tag{3-91}$$

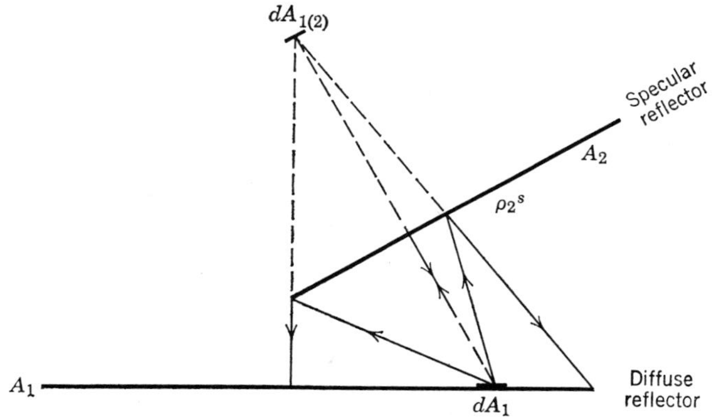

Fig. 3-18. Portion of A_1 receives radiation from dA_1 after one specular reflection by A_2.

where the asterisk denotes that the diffuse view factor corresponds only to the portion of the surface that is visible from $dA_{1(2)}$ through A_2. The term $^*F_{dA_{1(2)}-A_1}$ is called a *partial diffuse view factor* from $dA_{1(2)}$ to a portion of surface A_1.

Two Specularly Reflecting Surfaces

When two diffusely emitting surfaces A_1 and A_2 are both specular reflectors, there will be multiple specular reflections between the surfaces and successive images will be formed. Figure 3-19 illustrates the formation of successive

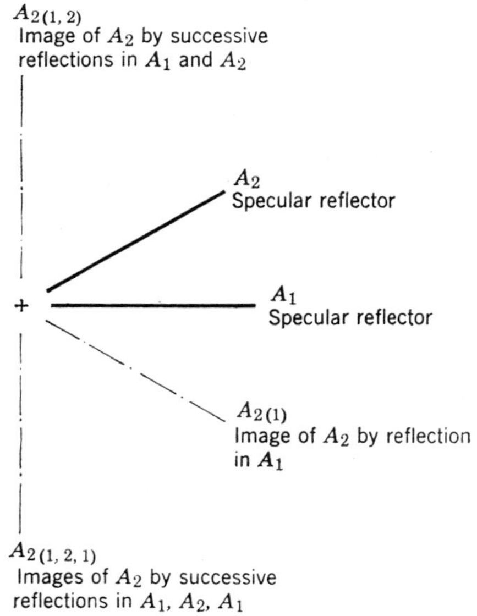

Fig. 3-19. Images of A_2 formed by successive specular reflections in surfaces A_1 and A_2.

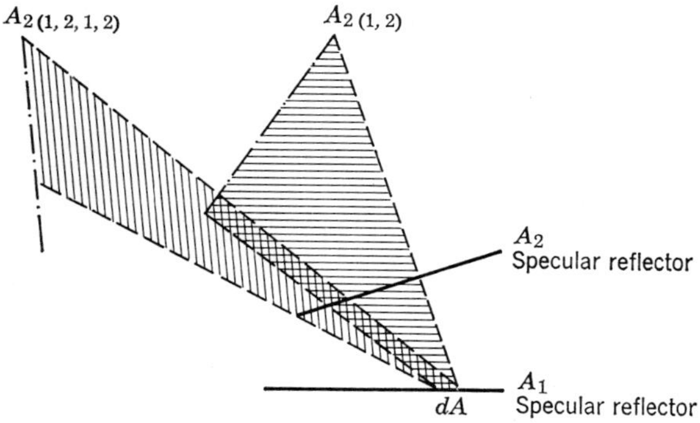

Fig. 3-20. Surface dA_1 receives radiation from A_2 directly and by multiple specular reflections in A_1 and A_2.

images of A_2 by specular reflections in surfaces A_1 and A_2. Image $A_{2(1)}$ refers to the image of A_2 formed by specular reflection in A_1, and $A_{2(1,2)}$ denotes the image of A_2 formed by successive specular reflections in A_1 and A_2. Similarly, $A_{2(1,2,1)}$ corresponds to the image of A_2 formed by successive specular reflections in A_1, A_2, and A_1.

A different arrangement is shown in Fig. 3-20 for surfaces A_1 and A_2, which are diffuse emitters and purely specular reflectors. We shall determine the fraction of radiative energy leaving surface A_2 diffusely that reaches an elemental surface dA_1 on A_1 both directly and by all possible intervening specular reflections in A_1 and A_2. It is composed of the following components.

(1) The Direct Component: This is the fraction of radiative energy that leaves surface A_2 diffusely and strikes dA_1 directly. It is given by

$$F_{A_2-dA_1} \tag{3-92a}$$

which is the diffuse view factor from A_2 to dA_1.

(2) Two Reflections: The fraction of radiative energy leaving A_2 diffusely that reaches dA_1 after two successive specular reflections in A_1 and A_2 is given by

$$\rho_1{}^s \rho_2{}^s \, F_{A_{2(1,2)}-dA_1} \tag{3-92b}$$

where $\rho_1{}^s$ and $\rho_2{}^s$ are the specular reflectivities, and $F_{A_{2(1,2)}-dA_1}$ is the diffuse view factor from image $A_{2(1,2)}$ to surface dA_1.

(3) Four Reflections: The fraction of radiative energy leaving A_2 diffusely that reaches dA_1 after four successive specular reflections in surfaces A_1, A_2, A_1, and A_2 is

$$(\rho_1{}^s)^2(\rho_2{}^s)^2 \; {}^*F_{A_{2(1,2,1,2)}-dA_1} \tag{3-92c}$$

where the asterisk denotes that a partial view factor is involved; that is, looking through the surface A_2 (i.e., the last surface in the sequence of reflections) only a portion of image $A_{2(1,2,1,2)}$ will view dA_1.

The procedure is repeated until the image of A_2 cannot view dA_1 looking through the last surface in the reflectance sequence.

Adding all the above components, we obtain

$$F^s_{A_2-dA_1} = F_{A_2-dA_1} + \rho_1{}^s\rho_2{}^s\, F_{A_{2(1,2)}-dA_1} + (\rho_1{}^s)^2(\rho_2{}^s)^2\, {}^*F_{A_{2(1,2,1,2)}-dA_1} + \cdots \tag{3-93}$$

where $F^s_{A_2-dA_1}$ is called the *specular view factor* between surfaces A_2 and dA_1. It represents the fraction of radiative energy leaving surface A_2 diffusely that reaches surface dA_2 directly and by all possible specular interreflections. Here we note that, for diffusely reflecting surfaces $\rho_1{}^s = \rho_2{}^s = 0$, all the terms on the right-hand side of Eq. 3-93 vanish except the first term; then the specular view factor reduces to the diffuse view factor.

The General Form of Specular View Factor

In the foregoing example the specular view factor was given for a specific geometry that involved only two specularly reflecting surfaces. For an enclosure containing several specularly reflecting, diffusely emitting surfaces, the general expression for the specular view factor between a surface i and a surface j may be written in the form

$$F^s_{i-j} = F_{i-j} + \rho_1{}^s\, F_{i(1)-j} + \rho_1{}^s\rho_2{}^s\, F_{i(1,2)-j} + \rho_1{}^s\rho_2{}^s\rho_3{}^s\, F_{i(1,2,3)-j} + \cdots \tag{3-94a}$$

or in the form

$$F^s_{i-j} = F_{i-j} + \sum \rho^s_m\, \rho^s_n\, \rho^s_p\, \cdots F_{i(m,n,p,\ldots)-j} \tag{3-94b}$$

The specular view factors obey the reciprocity theorem in a way similar to that for the diffuse view factors, that is,

$$A_i F^s_{A_i-A_j} = A_j F^s_{A_j-A_i} \tag{3-95a}$$

$$dA_i F^s_{dA_i-A_j} = A_j F^s_{A_j-dA_i} \tag{3-95b}$$

$$dA_i\, dF^s_{dA_i-dA_j} = dA_j\, dF^s_{dA_j-dA_i} \tag{3-95c}$$

The validity of these relations is readily proved by utilizing the reciprocity relations for the diffuse view factors, that is,

$$A_i F_{i-j} = A_j F_{j-i} \tag{3-96a}$$

$$A_i F_{i(m,n,p,\ldots)-j} = A_j F_{j-i(m,n,p,\ldots)} \tag{3-96b}$$

In the foregoing analysis we considered surfaces that are plane, diffusely emitting, and purely specularly reflecting. When surfaces are curved and have arbitrary shapes, no general procedure is yet available to determine the specular view factor. Lin and Sparrow [16] described a method for determining the specular view factor for curved surfaces having axisymmetric configuration. Polgar and Howell [17] applied the method to radiative exchange in specularly reflecting conical cavities, but the evaluation of specular view factors was rather involved. To illustrate the basic approach we consider a cylindrical geometry as shown in Fig. 3-21 and determine the elemental specular view factor $dF_{dA-dA'}$ from a circular *band* (a, dx) positioned at x of area dA to a circular *band* (a, dx') positioned at x' of area dA' on the surface of the cylinder. By definition the specular view factor $dF_{dA-dA'}$ represents the fraction of the radiative energy leaving surface dA that reaches surface dA' both directly and by all possible specular interreflections; for the geometry considered it can be written in the form

$$dF^s_{dA-dA'} = \begin{pmatrix} \text{Fraction of radiation} \\ \text{leaving } dA \text{ that reaches} \\ dA' \text{ directly} \end{pmatrix} + \sum_{n=1}^{\infty} \begin{pmatrix} \text{Fraction of radiation} \\ \text{leaving } dA \text{ that reaches} \\ dA' \text{with } n \text{ intervening} \\ \text{specular reflections} \end{pmatrix} \tag{3-97}$$

The first term on the right-hand side of Eq. 3-97 is the elemental diffuse view factor from the band of surface dA to the band of surface dA' and is given as

$$dF_{dA-dA'} \tag{3-98}$$

The terms of the series in Eq. 3-97 can be evaluated with the following consideration. Let dA_n^* denote the surface area of a circular band positioned somewhere between bands dA and dA', so that the diffuse radiation leaving band dA and reaching band dA' with n intervening specular reflections makes the first of these reflections at band dA_n^*. Then the fraction of radiative

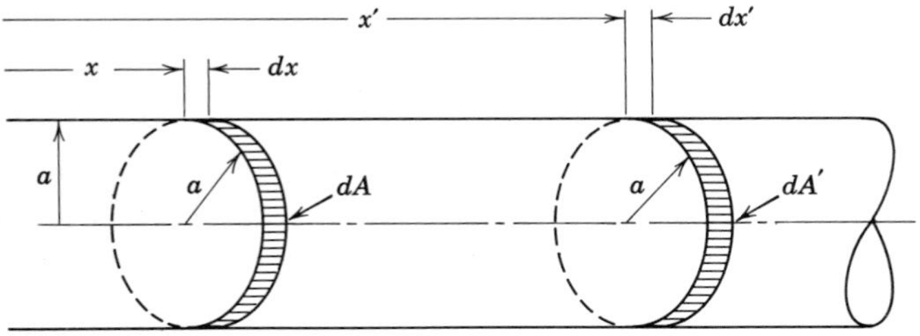

Fig. 3-21. The specular view factor between two elemental bands dA and dA' for axisymmetric configuration.

energy leaving dA that reaches dA' with n intervening specular reflections is given as

$$(\rho^s)^n \, dF_{dA-dA_n^*} \tag{3-99}$$

where $dF_{dA-dA_n^*}$ is the elemental diffuse view factor from band dA to band dA_n^*.

Substituting Eqs. 3-98 and 3-99 into Eq. 3-97, we obtain

$$dF^s_{dA-dA'} = dF_{dA-dA'} + \sum_{n=1}^{\infty} (\rho^s)^n \, dF_{dA-dA_n^*} \tag{3-100}$$

The determination of the diffuse view factor $dF_{dA-dA_n^*}$ will be illustrated with an example in Chapter 5.

Several applications of the image method for radiative heat transfer among specularly reflecting surfaces are given by Sparrow and Lin [18], Sarofim and Hottel [19], and Sparrow, Eckert, and Jonsson [20].

REFERENCES

1. D. C. Hamilton and W. R. Morgan, "Radiant Interchange Configuration Factors," *NACA TN* 2836, U.S. Government Printing Office, Washington, D.C., 1952.
2. H. Leuenberger and R. A. Pearson, "Compilation of Radiant Shape Factors for Cylindrical Assemblies," *ASME Paper* No. 56-A-144, 1956.
3. F. Kreith, *Radiation Heat Transfer for Spacecraft and Solar Power Design*, International Textbook Co., Scranton, Penn., 1962.
4. E. M. Sparrow and R. D. Cess, *Radiation Heat Transfer*, Brooks/Cole Publishing Co., Belmont, Calif., 1966.
5. J. R. Howell and R. Siegel, *Thermal Radiation Heat Transfer*, Vol. II, NASA SP-164, U.S. Government Printing Office, Washington, D.C., 1969.
6. H. C. Hottel, "Radiant Heat Transmission," in *Heat Transmission*, by W. H. McAdams, McGraw-Hill Book Co., New York, 1954.
7. C. O. Mackey, L. T. Wright, R. E. Clark, and N. R. Gay, Radiant Heating and Cooling, Part I, *Cornell Univ. Eng. Expt. Stat. Bull.* No. 22, 1943.
8. A. Feingold and K. G. Gupta, "New Analytical Approach to the Evaluation of Configuration Factors in Radiation from Spheres and Infinitely Long Cylinders," *J. Heat Transfer*, **92C**, 69–76, 1970.
9. M. Jacob, *Heat Transfer*, Vol. 2, John Wiley and Sons, New York, 1957.
10. G. L. Singer, "Viewpin: a FORTRAN Program to Calculate View Factors for Cylindrical Pins," Aerojet Nuclear Co., Report No. ANCR-1054, Idaho Falls, Idaho, March 1972.
11. Parry Moon, *Scientific Basis of Illuminating Engineering*, McGraw-Hill Book Co., New York, 1936; also Dover Publications, New York, 1961.
12. Parry Moon and D. Eberle Spencer, "Optical Transmittance of Louver Systems," *J. Franklin Inst.*, **273**, 1–24, 1962.

13. E. M. Sparrow, "A New and Simpler Formulation for Radiative Angle Factors," *J. Heat Transfer*, **85C**, 81–88, 1963.

14. E. M. Sparrow, L. U. Albers, and E. R. G. Eckert, "Thermal Radiation Characteristics of Cylindrical Enclosures," *J. Heat Transfer*, **84C**, 73–81, 1962.

15. E. R. G. Eckert and E. M. Sparrow, "Radiative Heat Exchange Between Surfaces with Specular Reflection," *Intern. J. Heat Mass Transfer*, **3**, 42–54, 1961.

16. S. H. Lin and E. M. Sparrow, "Radiant Interchange Among Curved Specularly Reflecting Surfaces—Application to Cylindrical and Conical Cavities," *J. Heat Transfer*, **87C**, 299–307, 1965.

17. L. G. Polgar and J. R. Howell, "Directional Thermal-Radiative Properties of Conical Cavities," *NASA Tech. Note* TN D-2904, June 1965.

18. E. M. Sparrow and S. L. Lin, "Radiation Heat Transfer at a Surface Having Both Specular and Diffuse Reflectance Components, *Intern. J. Heat Mass Transfer*, **8**, 769–779, 1965.

19. A. F. Sarofim and H. C. Hottel, "Radiative Exchange among Non-Lambert Surfaces," *J. Heat Transfer*, **88C**, 37–44, 1966.

20. E. M. Sparrow, E. R. G. Eckert, and V. K. Jonsson, "An Enclosure Theory for Radiative Exchange Between Specularly and Diffusely Reflecting Surfaces," *J. Heat Transfer*, **84C**, 294–300, 1962.

NOTES

[1] The validity of the identity of Eq. 3-36 can be proved by direct expansion. Let

$$\hat{\mathbf{n}}_1 = \hat{\mathbf{i}} l_1 + \hat{\mathbf{j}} m_1 + \hat{\mathbf{k}} n_1 \tag{1}$$

$$\mathbf{r} = \hat{\mathbf{i}}(x_2 - x_1) + \hat{\mathbf{j}}(y_2 - y_1) + \hat{\mathbf{k}}(z_2 - z_1) \equiv \hat{\mathbf{i}}x + \hat{\mathbf{j}}y + \hat{\mathbf{k}}z \tag{2}$$

$$r = (x^2 + y^2 + z^2)^{1/2} \tag{3}$$

Then the term $[(\mathbf{r}_{12} \times \hat{\mathbf{n}}_1)/r^2]$ in Eq. 3-36 becomes

$$\frac{\mathbf{r}_{12} \times \hat{\mathbf{n}}_1}{r^2} = \frac{1}{r^2}\,[(yn_1 - zm_1)\hat{\mathbf{i}} + (zl_1 - xn_1)\hat{\mathbf{j}} + (xm_1 - yl_1)\hat{\mathbf{k}}] \tag{4}$$

To evaluate the right-hand side of Eq. 3-36, the vector $\hat{\mathbf{n}}$ is treated as a constant in differentiation with respect to x, y, z since it is an arbitrary quantity. We obtain

$$\tfrac{1}{2}\nabla \times \left(\frac{\mathbf{r}_{12} \times \hat{\mathbf{n}}_1}{r^2}\right) = -\frac{1}{r^4}\,[(x^2 l_1 + yxm_1 + zxn_1)\hat{\mathbf{i}} + (xyl_1 + y^2 m_1 + zyn_1)\hat{\mathbf{j}} + (xzl_1 + yzm_1 + z^2 n_1)\hat{\mathbf{k}}] \tag{5}$$

Now we expand the left-hand side of Eq. 3-36 and obtain

$$-\frac{1}{r^4}\mathbf{r}_{12}(\hat{\mathbf{n}}_1 \cdot \mathbf{r}_{12}) = -\frac{1}{r^4}\,[(x^2 l_1 + yxm_1 + zxn_1)\hat{\mathbf{i}} + (xyl_1 + y^2 m_1 + zyn_1)\hat{\mathbf{j}} + (xzl_1 + yzm_1 + z^2 n_1)\hat{\mathbf{k}}] \tag{6}$$

The equality of the results in Eqs. 5 and 6 proves the validity of Eq. 3-36.

[2] From Eq. 3-13 we have

$$A_1 F_{A_1-A_2} = \int_{A_1}\left(\int_{A_2} \frac{\cos\theta_1 \cos\theta_2}{\pi r^2}\,dA_2\right) dA_1 \tag{1}$$

The term inside the parentheses, by Eq. 3-8, is equivalent to $F_{dA_1-A_2}$. Then we obtain

$$A_1F_{A_1-A_2} = \int_{A_1} F_{dA_1-A_2}\, dA_1 \tag{2}$$

[3] The validity of Eq. 3-67 can be shown by utilizing the definition of a diffuse view factor between two finite surfaces given by Eq. 3-14. If x', y', z' and x, y, z are the coordinates for the primed and unprimed surfaces, respectively, $G_{2-1'}$ and $G_{1-2'}$ are written as

$$\begin{aligned} G_{2-1'} &= \int_{A_2}\int_{A_1} \frac{\cos\theta_2 \cos\theta_{1'}}{\pi r^2}\, dA_2\, dA_{1'} \\ &= \frac{1}{\pi}\int_{x=a}^{b}\int_{z=0}^{c}\int_{x'=0}^{a}\int_{y'=e}^{d} \frac{y'z}{[(x-x')^2+y'^2+z^2]^2}\, dx'\, dy'\, dx\, dy \\ G_{1-2'} &= \int_{A_1}\int_{A_{2'}} \frac{\cos\theta_1 \cos\theta_{2'}}{\pi r^2}\, dA_{2'}\, dA_1 \\ &= \frac{1}{\pi}\int_{x=0}^{a}\int_{z=0}^{c}\int_{x'=a}^{b}\int_{y'=e}^{d} \frac{y'z}{[(x-x')^2+y'^2+z^2]^2}\, dx'\, dy'\, dx\, dy \end{aligned}$$

The right-hand side of these two equations are identical because of the symmetry of the integrand. This proves the equivalence of $G_{2-1'}$ and $G_{1-2'}$.

[4] The term *exchange factor* is used by Sparrow and Cess [4], and *specular configuration factor* by Howell and Siegel [5].

[5] If surface A_2 were a diffuse reflector, the fraction of radiative energy leaving dA_1 that was received by A_1 after one diffuse reflection at A_2 would have been

$$F_{dA_1-A_2}\rho_2^d F_{A_2}$$

CHAPTER 4

Radiative Exchange in an Enclosure—Simplified Zone Analysis

In this chapter we examine radiative exchange among the surfaces of an enclosure that contains a nonparticipating medium (i.e., a medium that does not absorb, emit, or scatter radiation and hence has no effect on the radiation traveling through it). A vacuum is a perfectly nonparticipating medium, and air at low or moderate temperatures is considered a nonparticipating medium. Here the term *enclosure* is used to designate a region completely surrounded by a set of surfaces that are characterized by their radiative properties and temperatures (or heat fluxes), so that a full account can be made of the incoming and outgoing radiation at any of these surfaces. For example, an opening in an enclosure is considered as an *imaginary surface*, and the amount of radiative energy streaming into the enclosure through the opening characterizes the emissive power of the imaginary surface.

In the general problem of radiative exchange in an enclosure the radiative properties of surfaces may vary with direction, frequency, and position, and the temperature may vary from point to point over the surface; however, the solution of such a problem is very complicated. The analysis of radiative, heat exchange in an enclosure can be simplified significantly by dividing the entire surface of the enclosure into a finite number of zones and by assuming that the following conditions are satisfied at the surface of each zone:

1. The radiative properties are uniform and independent of direction.
2. Either a uniform temperature or a uniform heat flux is prescribed over the surface of each zone.
3. The surfaces are diffuse emitters and diffuse reflectors.
4. The radiosity (i.e., the radiant energy leaving the surface) is uniform over the surface of each zone.
5. The surfaces are opaque, that is, $\rho = 1 - \alpha$.

Several investigators have utilized similar simplifying assumptions to develop methods for the solution of the radiative heat transfer problem in an enclosure. Hottel [1] introduced the so-called script $\mathscr{F}$ or total view factor method, which is based on the radiosity concept, Eckert and Drake [2] applied the radiosity approach, Gebhart [3a, 3b] used the concept of an absorption factor, Oppenheim [4] developed the method of electrical network analogy, Sparrow [5] and Sparrow and Cess [6] introduced an alternative formulation, and Clark and Korybalski [7] presented an approach, based on the concept of radiosity, similar to that of Hottel [1]. However, a close scrutiny of all these methods reveals that there is no significant difference among them, since all are based on the same simplifying assumptions discussed above, and that for a given physical system each will provide the same answer. The principal difference between these several approaches lies in the method of formulation of the problem, and the advantage of one method over any other may be accessed by its computational merits. The method described in references 5 and 6 appears to be more straightforward and offers some computational advantages. A general formulation is given by Bevans and Edwards [8].

In this chapter we first present the basic equations of radiative heat exchange for a general enclosure. The equations are developed in the initial stages of the analysis from the concept of radiation intensity, not from the concept of radiosity as has been done in most of the formulations available in the literature. The equations obtained in this manner are in the general form, and by introducing into these equations the assumptions stated above the simplified formulations in terms of radiosity are readily obtained. The advantage of presenting such general relations is that they help the reader to envision more clearly the role of radiation intensity in relation to the concept of radiosity, which will be used in the treatment of radiative transfer in nonparticipating media (Chapters 4 to 7); to the concept of radiation intensity will be used in the formulation of radiative transfer in participating media (Chapters 8 to 14).

The rest of the chapter is devoted to the formulation and treatment of radiative heat transfer problems in enclosures by the application of simplified analysis.

4-1 EQUATIONS OF RADIATIVE HEAT EXCHANGE FOR A GENERAL ENCLOSURE

Consider an enclosure filled by a nonparticipating medium. For generality we assume that the radiative properties of surfaces are dependent on the direction and frequency of radiation and on position, and that the surface temperature varies from point to point. The equations for radiative exchange within the enclosure can be derived by writing an energy balance equation for the incident, reflected, and emitted radiation for an elemental area dA on the surface of the enclosure.

Figure 4-1 shows an elemental surface dA, characterized by a unit normal vector $\hat{\mathbf{n}}$ on the inside surface of an enclosure. Let $T(\mathbf{r})$ be the temperature, $\varepsilon_\nu(\mathbf{r}, \hat{\mathbf{\Omega}})$ the spectral directional emissivity, and $f_\nu(\mathbf{r}, \hat{\mathbf{\Omega}}', \hat{\mathbf{\Omega}})$ the spectral reflection distribution function at dA. The spectral radiation intensities incident and leaving this surface element will be denoted by $I_\nu^i(\mathbf{r}, \hat{\mathbf{\Omega}}')$ and $I_\nu(\mathbf{r}, \hat{\mathbf{\Omega}})$, respectively.

The spectral radiation intensity $I_\nu(\mathbf{r}, \hat{\mathbf{\Omega}})$ leaving the surface element in any direction $\hat{\mathbf{\Omega}}$ is equal to the sum of the emitted and reflected intensities and can be written in the form

$$I_\nu(\mathbf{r}, \hat{\mathbf{\Omega}}) = \varepsilon_\nu(\mathbf{r}, \hat{\mathbf{\Omega}}) I_{\nu b}[T(\mathbf{r})] + \int_{\Omega'=2\pi} f_\nu(\mathbf{r}, \hat{\mathbf{\Omega}}', \hat{\mathbf{\Omega}}) I_\nu^i(\mathbf{r}, \hat{\mathbf{\Omega}}') \cos\theta' \, d\Omega' \quad (4\text{-}1)$$

where $I_{\nu b}[T]$ is the Planck function at temperature $T(\mathbf{r})$, and θ' is the acute angle between the incident ray and the normal to the surface. The first term on the right-hand side of Eq. 4-1 is the directional emission from the surface due to its temperature, and the second term is the radiative energy incident on the surface from all directions in the enclosure that is reflected in the direction $\hat{\mathbf{\Omega}}$.

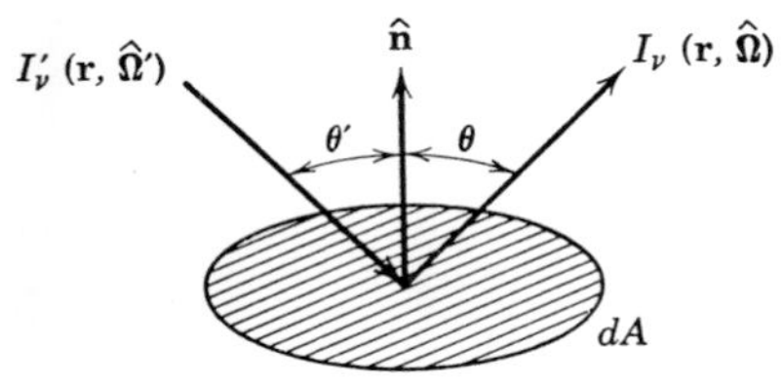

Fig. 4-1. Radiation incident and leaving an elemental surface dA.

The *net radiative heat flux* $q(\mathbf{r})$ at the surface element is equal to the difference between the radiative heat fluxes leaving and incident at the surface, and is given as

$$q(\mathbf{r}) = \int_{\nu=0}^{\infty}\left[\int_{\Omega=2\pi} I_\nu(\mathbf{r}, \hat{\mathbf{\Omega}}) \cos\theta \, d\Omega\right] d\nu - \int_{\nu=0}^{\infty}\left[\int_{\Omega'=2\pi} I'_\nu(\mathbf{r}, \hat{\mathbf{\Omega}}') \cos\theta' \, d\Omega'\right] d\nu \tag{4-2}$$

where the first and the second terms on the right-hand side are, respectively, the radiative heat flux leaving and incident normally on the surface. When the flow of net radiative heat flux is in the positive $\hat{\mathbf{n}}$ direction, $q(\mathbf{r})$ is considered positive.

The angular distribution of radiation intensity $I_\nu(\mathbf{r}, \hat{\mathbf{\Omega}})$ over the surface of the enclosure, in principle, can be determined from the solution of Eq. 4-1 if the temperature and radiative properties are prescribed over the entire surface of the enclosure. Then the distribution of radiation intensity is known, the net radiative heat flux can be evaluated from Eq. 4-2. However, Eq. 4-1 is an integral equation and its solution over the entire surface of an enclosure is extremely difficult. Furthermore, very few data are available for the reflection distribution function $f_\nu(\mathbf{r}, \hat{\mathbf{\Omega}}', \hat{\mathbf{\Omega}})$ for real surfaces to justify the solution of such a complex problem. Therefore simplified versions of these equations are used in practice. Simplification of these equations will now be considered.

4-2 SIMPLIFIED ZONE ANALYSIS FOR ENCLOSURES WITH DIFFUSELY REFLECTING SURFACES

Consider an enclosure whose surfaces are divided into N zones so that the radiative properties, the temperature, and the intensity of radiation leaving the surfaces are uniform and independent of direction over the surface of each zone. These assumptions imply also that the surfaces are diffuse emitters and diffuse reflectors. Let dA_i and dA_j be two elemental surfaces at zones A_i and A_j, respectively, and r_{ij} be the length of the straight line joining dA_i and dA_j as shown in Fig. 4-2. We write Eqs. 4-1 and 4-2 for the elemental surface dA_i at zone A_i. When these assumptions are applied, Eq. 4-1 simplifies to

$$I_{i,\nu} = \varepsilon_{i,\nu} I_{\nu b}(T_i) + f_{i,\nu}\sum_{j=1}^{N} I_{j,\nu}\int_{A_j} \frac{\cos\theta_i' \cos\theta_j}{r_{ij}^2}\, dA_j \tag{4-3}$$

and Eq. 4-2 to

$$q_i = \int_{\nu=0}^{\infty} q_{i,\nu}\, d\nu \tag{4-4a}$$

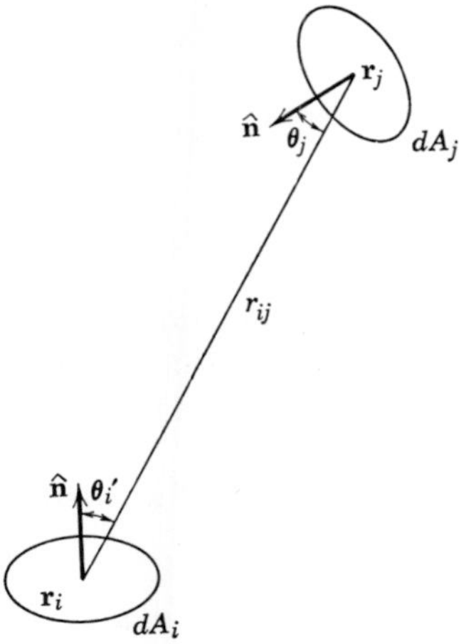

Fig. 4-2. Coordinates for radiative exchange in an enclosure.

where

$$q_{i,\nu} = \pi I_{i,\nu} - \sum_{j=1}^{N} I_{j,\nu} \int_{A_j} \frac{\cos\theta_i' \cos\theta_j}{r_{ij}^2}\, dA_j, \quad i = 1, 2, \ldots, N \tag{4-4b}$$

since

$$d\Omega_j' = \frac{\cos\theta_j\, dA_j}{r_{ij}^2} \tag{4-5}$$

$$\int_{\Omega=2\pi} \cos\theta\, d\Omega = \pi \tag{4-6}$$

In the above equations $f_{i,\nu}$ and $I_{i,\nu}$ are both independent of direction. Then $f_{i,\nu}$ is related to the spectral hemispherical reflectivity by (see Eq. 1-108)

$$\pi f_{i,\nu} = \rho_{i,\nu} \tag{4-7}$$

and $I_{i,\nu}$ is related to the spectral radiosity $R_{i,\nu}$ by (see: Eq. 1-129)

$$\pi I_{i,\nu} = R_{i,\nu} \tag{4-8}$$

Substituting Eqs. 4-7 and 4-8 into Eqs. 4-3 and 4-4, integrating the resulting expressions over surface A_i of zone i, and noting that F_{i-j}, by definition, is given as (see Eq. 3-14)

$$\frac{1}{A_i} \int_{A_i} \int_{A_j} \frac{\cos\theta_i' \cos\theta_j}{\pi r_{ij}^2}\, dA_j\, dA_i = F_{A_i-A_j} \equiv F_{i-j} \tag{4-9}$$

we find that Eq. 4-3 simplifies to

$$R_{i,\nu} = \varepsilon_{i,\nu} \pi I_{\nu b}(T_i) + \rho_{i,\nu} \sum_{j=1}^{N} R_{j,\nu} F_{i-j} \tag{4-10}$$

and Eq. 4-4 to

$$q_i = \int_{\nu=0}^{\infty} q_{i,\nu}\, d\nu \tag{4-11a}$$

where

$$q_{i,\nu} = R_{i,\nu} - \sum_{j=1}^{N} R_{j,\nu} F_{i-j}, \quad i = 1, 2, \ldots, N \tag{4-11b}$$

Equations 4-10 and 4-11 provide the mathematical formulation of the problem of radiative heat exchange in an enclosure for simplified zone analysis. Equation 4-11b for $q_{i,\nu}$ can be written in alternative forms as follows.

By eliminating the summation term from Eq. 4-11b by means of Eq. 4-10, we obtain

$$q_{i,\nu} = \frac{\varepsilon_{i,\nu}\pi I_{\nu b}(T_i) - (1 - \rho_{i,\nu})R_{i,\nu}}{\rho_{i,\nu}} \quad \text{for } \rho_{i,\nu} \neq 0 \tag{4-12a}$$

or, by eliminating $R_{i,\nu}$ between Eqs. 4-12b and 4-12c, we obtain

$$q_{i,\nu} = \varepsilon_{i,\nu}\pi I_{\nu b}(T_i) - (1 - \rho_{i,\nu}) \sum_{j=1}^{N} R_{j,\nu} F_{i-j} \tag{4-12b}$$

When frequency-dependent analysis is to be performed, the solution of Eqs. 4-10 and 4-11 in their present forms is too laborious for practical purposes. To simplify the analysis, the entire energy spectrum is divided into a finite number of bands over entire frequency (or wavelength) and the radiative properties are assumed to be uniform over each band. This approach, called the *band approximation*, will be applied to Eqs. 4-10 and 4-11.

The Band Approximation

We assume that the entire energy spectrum is divided into a finite number of frequency bands $\Delta\nu_k$, $k = 1, 2, \ldots, K$, so that the radiative properties are uniform over each band. The application of the band approximation to Eqs. 4-10 and 4-11 yields respectively

$$R_{i,k} = \varepsilon_{i,k}\pi I_{b,k}(T_i) + \rho_{i,k} \sum_{j=1}^{N} R_{j,k} F_{i-j} \tag{4-13}$$

$$q_i = \sum_{k=1}^{K} q_{i,k} \tag{4-14a}$$

where

$$q_{i,k} = R_{i,k} - \sum_{j=1}^{N} R_{j,k} F_{i-j} \tag{4-14b}$$

and the alternative forms, Eqs. 4-12a and 4-12b, respectively, become

$$q_{i,k} = \frac{\varepsilon_{i,k}\pi I_{b,k}(T_i) - (1 - \rho_{i,k})R_{i,k}}{\rho_{i,k}} \quad \text{for } \rho_{i,k} \neq 0 \tag{4-14c}$$

$$q_{i,k} = \varepsilon_{i,k}\pi I_{b,k}(T_i) - (1 - \rho_{i,k})\sum_{j=1}^{N} R_{j,k}F_{i-j} \tag{4-14d}$$

where $i = 1, 2, 3, \ldots, N$, $k = 1, 2, 3, \ldots, K$, we have defined

$$R_{i,k} \equiv \int_{\Delta\nu_k} R_{i,\nu}\, d\nu \tag{4-15a}$$

$$I_{b,k}(T_i) \equiv \int_{\Delta\nu_k} I_{\nu b}(T_i)\, d\nu \tag{4-15b}$$

$$q_{i,k} \equiv \int_{\Delta\nu_k} q_{i,\nu}\, d\nu \tag{4-15c}$$

and $\varepsilon_{i,k}$, $\rho_{i,k}$ are the uniform values of the spectral emissivity and reflectivity over the frequency interval $\Delta\nu_k$.

When the temperature and the radiative properties are prescribed for each zone, Eqs. 4-13 provide N simultaneous algebraic equations for the N unknown radiosities R_i, $i = 1, 2, \ldots, N$, for each frequency band $\Delta\nu_k$, $k = 1, 2, \ldots, K$. Once these radiosities have been determined, the net radiative heat flux at each zone is determined from Eq. 4-14.

The Gray-Body Approximation

When the radiative properties are independent of frequency, the integration of Eqs. 4-10 and 4-11 over the entire frequency yields respectively

$$R_i = \varepsilon_i \bar{\sigma} T_i^4 + \rho_i \sum_{j=1}^{N} R_j F_{i-j} \tag{4-16}$$

$$q_i = R_i - \sum_{j=1}^{N} R_j F_{i-j} \tag{4-17a}$$

and Eq. 4–17a can be written in alternative forms as

$$q_i = \frac{\varepsilon_i \bar{\sigma} T_i^4 - (1 - \rho_i)R_i}{\rho_i}, \quad \rho_i \neq 0 \tag{4-17b}$$

or

$$q_i = \varepsilon_i \bar{\sigma} T_i^4 - (1 - \rho_i)\sum_{j=1}^{N} R_j F_{i-j}, \quad i = 1, 2, \ldots, N \tag{4-17c}$$

since

$$\pi I_b(T_i) = \bar{\sigma} T_i^4 \tag{4-18}$$

General Application

The problem of radiative heat exchange in an enclosure is concerned with the determination of net radiative heat flux at a zone with prescribed temperature, or conversely with the determination of temperature at a zone with prescribed heat flux. We examine below general applications of the simplified analysis to several representative situations and discuss the methods of solution of the resulting equations.

(a) Gray Enclosure, Prescribed Temperature at Surfaces of All Zones

When the temperature is prescribed at the surfaces of all zones and the radiative properties are independent of frequency, Eq. 4-16 provides N simultaneous algebraic equations for the N unknown radiosities R_i, $i = 1, 2, \ldots, N$. Once these radiosities are determined from the solution of these equations, the net radiative heat fluxes at the zones are determined from Eq. 4-17a or 4-17b or 4-17c.

Equations 4-16 can be written in the form

$$\frac{R_i}{\varepsilon_i} - \frac{\rho_i}{\varepsilon_i}\sum_{j=1}^{N} R_j F_{i-j} = \bar{\sigma} T_i^4 \tag{4-19a}$$

or

$$\sum_{j=1}^{N}\left(\frac{\delta_{ij} - \rho_i F_{i-j}}{\varepsilon_i}\right) R_j = \bar{\sigma} T_i^4, \quad i = 1, 2, \ldots, N \tag{4-19b}$$

where

$$\delta_{ij} = \begin{cases} 1 & \text{for } i = j \\ 0 & \text{otherwise} \end{cases}$$

It is convenient to write Eqs. 4-19b in the matrix form

$$\mathbf{MR} = \mathbf{T} \tag{4-20a}$$

where we have defined

$$\mathbf{M} \equiv \begin{bmatrix} m_{11} & m_{12} & m_{13} & \cdots & m_{1N} \\ m_{21} & m_{22} & m_{23} & \cdots & m_{2N} \\ \cdot & & & & \\ \cdot & & & & \\ \cdot & & & & \\ m_{N1} & m_{N2} & m_{N3} & \cdots & m_{NN} \end{bmatrix} \tag{4-20b}$$

$$m_{ij} \equiv \frac{\delta_{ij} - \rho_i F_{i-j}}{\varepsilon_i} \tag{4-20c}$$

$$\mathbf{R} \equiv \begin{bmatrix} R_1 \\ R_2 \\ \cdot \\ \cdot \\ \cdot \\ R_N \end{bmatrix} \quad \text{and} \quad \mathbf{T} \equiv \begin{bmatrix} \bar{\sigma}T_1^4 \\ \bar{\sigma}T_2^4 \\ \cdot \\ \cdot \\ \cdot \\ \bar{\sigma}T_N^4 \end{bmatrix} \tag{4-20d}$$

The solution of Eqs. 4-20a yields the radiosities in the form

$$\mathbf{R} = \mathbf{M}^{-1}\mathbf{T} \tag{4-21}$$

where $\mathbf{M}^{-1}$ is the inverse of the matrix $\mathbf{M}$. When the matrix $\mathbf{M}$ is given, its inverse $\mathbf{M}^{-1}$ can be determined readily by a digital computer, using the standard matrix inversion subroutines. Let m'_{ij} be the elements of the inverse matrix $\mathbf{M}^{-1}$, that is,

$$\mathbf{M}^{-1} \equiv \begin{bmatrix} m'_{11} & m'_{12} & \cdots & m'_{1N} \\ m'_{21} & m'_{22} & \cdots & m'_{2N} \\ \cdot & & & \\ \cdot & & & \\ \cdot & & & \\ m'_{N1} & m'_{N2} & \cdots & m'_{NN} \end{bmatrix} \tag{4-22}$$

where m'_{ij} are considered known quantities. Then the solution for the radiosities, by Eq. 4-21, is written in the form

$$R_i = \sum_{j=1}^{N} m'_{ij}\bar{\sigma}T_j^4, \quad i = 1, 2, \ldots, N \tag{4-23}$$

Knowing the radiosities, we can evaluate the net radiative heat flux at the zones by any one of Eqs. 4-17.

(b) Gray Enclosure, Prescribed Surface Temperature at Some Zones and Prescribed Heat Flux at Others

Consider an enclosure in which the surface temperatures T_i are prescribed for zones $i = 1, 2, \ldots, r$, and the net heat fluxes q_i are prescribed at the surface for zones $i = r + 1, r + 2, \ldots, N$, respectively. The net radiative heat fluxes will be determined for the zones with prescribed temperatures, and the temperatures will be determined for the zones with prescribed heat fluxes.

From Eqs. 4-16 and 4-17a we write

$$R_i = \varepsilon_i\bar{\sigma}T_i^4 + \rho_i\sum_{j=1}^{N} R_jF_{i-j}, \quad i = 1, 2, \ldots, r \tag{4-24}$$

$$R_i = q_i + \sum_{j=1}^{N} R_jF_{i-j}, \quad i = r + 1, r + 2, \ldots, N \tag{4-25}$$

Equations 4-24 and 25 are N simultaneous algebraic equations for the N unknown radiosities R_i, $i = 1, 2, \ldots, N$. Once the radiosities are determined from the solution of this system of equations, the net radiative heat fluxes q_i at the surfaces with prescribed temperatures are evaluated from Eqs. 4-17; the temperatures at the surfaces with prescribed heat fluxes are determined from Eq. 4-16 or 4-17b or 4-17c.

When the net radiative heat flux is zero, say, at zone j, we set $q_j = 0$. Such a surface is called a *reradiating* or an *adiabatic* surface because it does not participate in the net radiative heat exchange. An adiabatic surface behaves as a perfectly reflecting surface (i.e., $\rho_j = 1$ or $\varepsilon_j = 0$) since the energy absorbed is equal to the energy emitted.

(c) The Band Approximation: Prescribed Surface Temperature at All Zones

When the radiative properties vary significantly with frequency, the gray assumption is invalid. Here we consider a nongray problem in which the frequency dependence of radiative properties is represented by the band approximation and surface temperatures are prescribed for all zones. We assume that the entire energy spectrum is divided into K frequency bands, each of width $\Delta\nu_k$, $k = 1, 2, \ldots, K$. Then Eqs. 4-13 provide N simultaneous algebraic equations for the N unknown radiosities $R_{i,k}$ $(i = 1, 2, \ldots, N)$ for each frequency band $\Delta\nu_k$ $(k = 1, 2, \ldots, K)$. Once these radiosities have been determined for all frequency bands, the net radiative heat fluxes are determined from Eqs. 4-14.

We write Eqs. 4-13 in the form

$$\sum_{j=1}^{N}\left[\frac{\delta_{ij} - \rho_{i,k}F_{i-j}}{\varepsilon_{i,k}}\right]R_{j,k} = \pi I_{b,k}(T_i), \quad i = 1, 2, \ldots, N;\ k = 1, 2, \ldots, K \tag{4-26}$$

which can be written in the matrix form as

$$\mathbf{M}(k)\mathbf{R}(k) = \mathbf{I}(k), \quad k = 1, 2, \ldots, K \tag{4-27a}$$

where we have defined

$$\mathbf{M}(k) \equiv \begin{bmatrix} m_{11}(k) & m_{12}(k) & \cdots & m_{1N}(k) \\ \cdot & & & \\ \cdot & & & \\ \cdot & & & \\ m_{N1}(k) & m_{N2}(k) & \cdots & m_{NN}(k) \end{bmatrix} \tag{4-27b}$$

$$m_{ij}(k) \equiv \frac{\delta_{ij} - \rho_{i,k}F_{i-j}}{\varepsilon_{i,k}} \tag{4-27c}$$

$$\mathbf{R}(k) \equiv \begin{bmatrix} R_{1k} \\ R_{2k} \\ \cdot \\ \cdot \\ \cdot \\ R_{Nk} \end{bmatrix} \quad \text{and} \quad \mathbf{I}(k) \equiv \begin{bmatrix} \pi I_{b,k}(T_1) \\ \pi I_{b,k}(T_2) \\ \cdot \\ \cdot \\ \cdot \\ \pi I_{b,k}(T_N) \end{bmatrix} \tag{4-27d}$$

The solution of Eqs. 4-27a can be written as

$$\mathbf{R}(k) = \mathbf{M}^{-1}(k)\mathbf{I}(k), \quad k = 1, 2, \ldots, K \tag{4-28}$$

where the inverse matrix $\mathbf{M}^{-1}(k)$ can be evaluated with a digital computer, using a matrix inversion subroutine. Then the solution for the radiosity $R_{i,k}$ is given by

$$R_{i,k} = \sum_{j=1}^{N} m'_{ij}(k)\pi I_{b,k}(T_j), \quad k = 1, 2, \ldots, K; \; i = 1, 2, \ldots, N \tag{4-29}$$

where $m'_{ij}(k)$ are the elements of the inverse matrix $\mathbf{M}^{-1}(k)$.

By substituting the radiosities given by Eqs. 4-29 into Eqs. 4-14a and 4-14c we obtain the net radiative heat flux at the surfaces of zones as

$$q_i = \sum_{k=1}^{K} \frac{\varepsilon_{i,k}\pi I_{b,k}(T_i) - (1 - \rho_{i,k})\sum_{j=1}^{N} m'_{ij}(k)\pi I_{b,k}(T_j)}{\rho_{i,k}}, \quad \rho_{i,k} \neq 0 \tag{4-30a}$$

which can be rearranged, upon setting $\rho_{i,k} = 1 - \varepsilon_{i,k}$, as

$$q_i = \sum_{k=1}^{K} \sum_{j=1}^{N} [\delta_{ij} - m'_{ij}(k)] \frac{\varepsilon_{i,k}}{1 - \varepsilon_{i,k}} \pi I_{b,k}(T_j) \tag{4-30b}$$

Thus the net radiative heat flux q_i at the surfaces of zones $i = 1, 2, \ldots, N$ can be evaluated from Eq. 4-30.

(d) The Band Approximation: Prescribed Heat Flux at Some Zones, Prescribed Temperature at Others

The solution of the problem of radiative heat exchange for a nongray enclosure is rather involved when temperatures are prescribed for the surfaces of some zones and heat fluxes for the others. Here we consider a simple situation in which the temperatures T_i are prescribed for the surfaces $i = 1, 2, \ldots, N - 1$, and the net heat flux q_N is prescribed for surface N in an N-zone enclosure: We assume that the entire energy spectrum is divided into K bands.

First we determine the temperature T_N for zone N, for which heat flux q_N is prescribed. When Eq. 4-30b is written for $i = N$ and rearranged, we obtain

$$q_N - \sum_{k=1}^{K} \sum_{j=1}^{N-1} C_{i,j}(k) I_{b,k}(T_j) = \sum_{k=1}^{K} C_{N,N}(k) I_{b,k}(T_N) \tag{4-31a}$$

where

$$C_{i,j}(k) \equiv \pi[\delta_{ij} - m'_{ij}(k)] \frac{\varepsilon_{i,k}}{1 - \varepsilon_{i,k}} \tag{4-31b}$$

The left-hand side of Eq. 4-31a is known; the right-hand side involves the unknown temperature T_N through the function $I_{b,k}(T_N)$, which is defined by Eq. 4-15b, that is,

$$I_{b,k}(T_N) \equiv \int_{\Delta\nu_k} I_{\nu b}(T_N)\, d\nu \tag{4-32}$$

The function $I_{b,k}(T)$ can be evaluated for a given temperature T over a frequency band $\Delta\nu_k$, since the Planck function $I_{\nu b}(T)$ is known. Then T_N can be determined from Eq. 4-31 by trial and error.

Once the temperature T_N of zone N is determined, temperatures are known for all the zones in the enclosure. Then the procedure described in case (*c*) is applicable to determine the heat flux at each zone.

4-3 SIMPLIFIED ZONE ANALYSIS FOR GRAY ENCLOSURES WITH SURFACES HAVING BOTH DIFFUSE AND SPECULAR REFLECTIVITY COMPONENTS

Real surfaces possess both diffuse and specular reflectivity components, but the computational problems become enormous if the actual reflectivity of a real surface is used in the analysis of radiative heat exchange. Therefore there is need for a reflectivity model that is simple enough to avoid computational difficulties and yet closely approximates the physical situation. Seban [9] suggested that the hemispherical reflectivity can be separated into a diffuse component ρ^d and a specular component ρ^s in the form

$$\rho = \rho^d + \rho^s \tag{4-33}$$

This reflectivity concept is better envisioned by referring to Fig. 4-3. This figure, given by Sarofim and Hottel [10a] and based on data from Munch [10b], shows the reflectivity of an oxidized brass sample for reflection in the plane of incidence. On this diagram the reflected radiation is subdivided into a specular component (the shaded area) and a diffuse component (the clear area).

Experimental values of ρ^s and ρ^d determined in a similar manner have been reported by Birkebak et al. [11]. However, measurements by Torrance

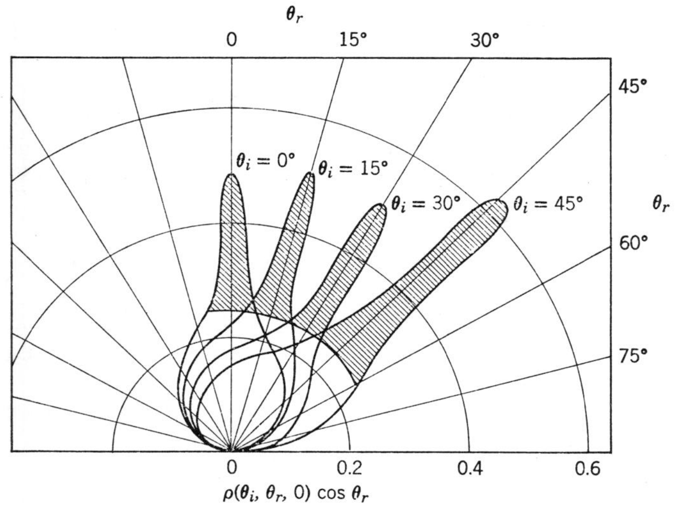

Fig. 4-3. Subdivision of radiation reflected by an oxidized brass sample into a specular component (shaded area) and a diffuse component (clear area). Values taken from reference 10 for reflection in plane of incidence. (From A. F. Sarofim and H. C. Hottel [10a]).

and Sparrow [12] show that representation of reflectivity by a simple summation, as given by Eq. 4-33, is, under certain conditions, inadequate to describe the reflectivity of real surfaces. Therefore care must be exercised when reflectivity is represented by a simple sum of diffuse and specular reflectivity components.

Consider an enclosure whose surfaces are opaque, gray, diffuse emitters and have reflectivity that is represented as the sum of a diffuse and a specular reflectivity component. The enclosure is divided into N zones such that the temperature and the radiative properties are uniform over the surface of each zone. The reflectivity for zone i is given as

$$\rho_i = \rho_i^{\,d} + \rho_i^{\,s} \tag{4-34}$$

The radiosity R_i for zone i, by definition, is equal to the sum of the radiant energy diffusely emitted and the radiant energy incident on the surface that is diffusely reflected. It is given as

$$R_i = \varepsilon_i \bar{\sigma} T_i^4 + \rho_i^{\,d} \sum_{j=1}^{N} R_j F_{i-j}^s \tag{4-35}$$

The first term on the right-hand side is the diffusely emitted energy, and the second term the diffusely reflected energy per unit time and per unit area. Equation 4-35 is analogous to Eq. 4-16 except that the specular view factor is used in the former because the incident radiant energy contains contributions from both specular and diffuse reflections.

The net radiative heat flux q_i at zone i can be represented as the difference between the radiative energies emitted and absorbed by the surface per unit time and per unit area. The emitted and absorbed radiative energies per unit time and per unit area are given respectively by $\varepsilon_i \bar{\sigma} T_i^4$ and $\alpha_i \sum_{j=1}^{N} R_j F_{i-j}^s$. Then the net radiative heat flux becomes

$$q_i = \varepsilon_i \bar{\sigma} T_i^4 - (1 - \rho_i) \sum_{j=1}^{N} R_j F_{i-j}^s \tag{4-36a}$$

where we have replaced α_i by $(1 - \rho_i)$ for opaque surfaces. Equation 4-36a reduces to Eq. 4-17c when all surfaces in the enclosure are diffuse reflectors.

Alternative expressions can be obtained for q_i as follows.

By eliminating the summation term from Eq. 4-36a by means of Eq. 4-35 and replacing $(1 + \rho_i^d - \rho_i)$ by $(1 - \rho_i^s)$ we obtain

$$q_i = \frac{1}{\rho_i^d} [\varepsilon_i (1 - \rho_i^s) \bar{\sigma} T_i^4 - (1 - \rho_i) R_i], \quad \rho_i^d \neq 0 \tag{4-36b}$$

which reduces to Eq. 4-17b for $\rho_i^s = 0$.

Or, by eliminating $\bar{\sigma} T_i^4$ from Eq. 4-36a by means of Eq. 4-35, we obtain

$$q_i = R_i - (1 - \rho_i^s) \sum_{j=1}^{N} R_j F_{i-j}^s \tag{4-36c}$$

which simplifies to Eq. 4-17a when all the zones are diffuse reflectors.

Equations 4-35 and 4-36 give the complete mathematical formulation of the simplified zone analysis for radiative exchange in a gray enclosure with surfaces having diffuse and specular reflectivity components.

An Enclosure with Prescribed Surface Temperature at Some Zones and Prescribed Heat Flux at Others

Consider an N-zone enclosure in which surface temperatures T_i are prescribed for the zones $i = 1, 2, \ldots, r$, and net heat fluxes q_i are prescribed for the zones $i = r + 1, r + 2, \ldots, N$, respectively. Temperatures are required for the zones with prescribed heat fluxes, and heat fluxes are required for the zones with prescribed temperatures.

In this case we write from Eq. 4-35

$$R_i = \varepsilon_i \bar{\sigma} T_i^4 + \rho_i^d \sum_{j=1}^{N} R_j F_{i-j}^s, \quad i = 1, 2, \ldots, r \tag{4-37a}$$

and from Eq. 4-36c

$$R_i = q_i + (1 - \rho_i^s) \sum_{j=1}^{N} R_j F_{i-j}^s, \quad i = r + 1, r + 2, \ldots, N \tag{4-37b}$$

Equations 4-37 are N simultaneous algebraic equations for the N unknown radiosities R_i. Once the R_i have been determined from the solution of Eqs. 4-37, the heat fluxes at the zones with prescribed temperatures are determined from any one of Eqs. 4-36. The temperatures of the zones with prescribed heat fluxes are determined from Eq. 4-35 or 4-36a or 4-36b, whichever is convenient.

If a zone with prescribed temperature, say zone 1, is a purely specular reflector, we set $\rho_1^d = 0$ for that zone, and Eq. 4-37a for $i = 1$ simplifies to $R_1 = \varepsilon_1 \bar{\sigma} T_1^4$. Thus the radiosity for a purely specularly reflecting zone with prescribed temperature is immediately available. The number of equations to be solved simultaneously is reduced in Eqs. 4-37a accordingly if more zones are purely specular reflectors.

Various methods have been used in the literature to solve the radiative heat transfer problem discussed here. The reader may refer to the papers by Sarofim and Hottel [10a], Holman [13], Sparrow and Lin [14], and Bobco [15].

4-4 APPLICATIONS OF SIMPLIFIED ZONE ANALYSIS

The applications of simplified zone analysis will be illustrated in this section with specific examples involving simple geometries.

(a) Two Parallel Plates

Consider an enclosure composed of two parallel, infinite, opaque plates as shown in Fig. 4-4. Surfaces 1 and 2 are kept at uniform temperatures T_1 and T_2 such that $T_1 > T_2$, and have spectral hemispherical emissivities $\varepsilon_{1\lambda}$ and $\varepsilon_{2\lambda}$ and spectral diffuse hemispherical reflectivities $\rho_{1\lambda}$ and $\rho_{2\lambda}$, respectively. The medium between the plates is nonparticipating.

The problem is to determine the net radiative heat flux at the plates (or the radiative heat exchange between the surfaces) since the temperatures are prescribed at both zones. The equations for the spectral radiosities are obtained from Eq. 4-10 by setting $i = 1$ or 2:

$$R_{1\lambda} = \varepsilon_{1\lambda} \pi I_{\lambda b}(T_1) + \rho_{1\lambda}(R_{1\lambda} F_{1-1} + R_{2\lambda} F_{1-2}) \tag{4-38a}$$

$$R_{2\lambda} = \varepsilon_{2\lambda} \pi I_{\lambda b}(T_2) + \rho_{2\lambda}(R_{1\lambda} F_{2-1} + R_{2\lambda} F_{2-2}) \tag{4-38b}$$

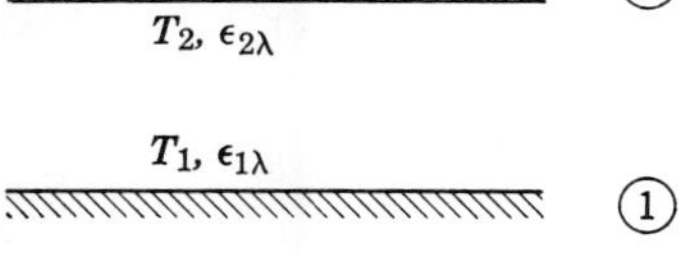

Fig. 4-4. An enclosure composed of two infinite parallel plates.

Here we have used wavelength instead of frequency to characterize the spectral variations. The diffuse view factors between two parallel infinite plates are given as

$$F_{1-1} = F_{2-2} = 0 \qquad \text{and} \qquad F_{1-2} = F_{2-1} = 1 \tag{4-39}$$

Then Eqs. 4-38 become

$$R_{1\lambda} - (1 - \varepsilon_{1\lambda})R_{2\lambda} = \varepsilon_{1\lambda}\pi I_{\lambda b}(T_1) \tag{4-40a}$$

$$-(1 - \varepsilon_{2\lambda})R_{1\lambda} + R_{2\lambda} = \varepsilon_{2\lambda}\pi I_{\lambda b}(T_2) \tag{4-40b}$$

Here we have assumed that the Kirchhoff law is valid and hence have replaced $\rho_{i\lambda}$ by $(1 - \varepsilon_{i\lambda})$, $i = 1$ or 2. Simultaneous solution of Eqs. 4-40 yields

$$R_{1\lambda} = \frac{\varepsilon_{2\lambda}(1 - \varepsilon_{1\lambda})\pi I_{\lambda b}(T_2) + \varepsilon_{1\lambda}\pi I_{\lambda b}(T_1)}{1 - (1 - \varepsilon_{1\lambda})(1 - \varepsilon_{2\lambda})} \tag{4-41a}$$

$$R_{2\lambda} = \frac{\varepsilon_{1\lambda}(1 - \varepsilon_{2\lambda})\pi I_{\lambda b}(T_1) + \varepsilon_{2\lambda}\pi I_{\lambda b}(T_2)}{1 - (1 - \varepsilon_{1\lambda})(1 - \varepsilon_{2\lambda})} \tag{4-41b}$$

The spectral radiative heat flux at the surfaces can be evaluated by Eq. 4-11b:

$$q_{1\lambda} = R_{1\lambda} - \sum_{j=1}^{2} R_{j\lambda}F_{1-j} = R_{1\lambda} - R_{2\lambda} \tag{4-42a}$$

$$q_{2\lambda} = R_{2\lambda} - \sum_{j=1}^{2} R_{j\lambda}F_{2-j} = R_{2\lambda} - R_{1\lambda} \tag{4-42b}$$

Equations 4-42 imply that for the geometry considered the spectral heat fluxes at two surfaces are related by

$$q_{1\lambda} = -q_{2\lambda} \tag{4-43}$$

To evaluate the spectral heat flux, say, $q_{1\lambda}$, we substitute the radiosities from Eqs. 4-41 into 4-42a and obtain

$$q_{1\lambda} = \frac{\pi I_{\lambda b}(T_1) - \pi I_{\lambda b}(T_2)}{(1/\varepsilon_{1\lambda}) + (1/\varepsilon_{2\lambda}) - 1} \tag{4-44}$$

The net radiative heat flux at surface 1 is determined by integrating Eq. 4-44 over all wavelengths

$$q_1 = \int_{\lambda=0}^{\infty} \frac{\pi I_{\lambda b}(T_1) - \pi I_{\lambda b}(T_2)}{(1/\varepsilon_{1\lambda}) + (1/\varepsilon_{2\lambda}) - 1}\, d\lambda \tag{4-45}$$

where the Planck black-body radiation intensity is given by (see Eq. 1-49 and Table 1-2)

$$\pi I_{\lambda b}(T) = \frac{37404}{\lambda^5[\exp(14387/\lambda T) - 1]}\ \text{W}/\mu\text{cm}^2 \tag{4-46}$$

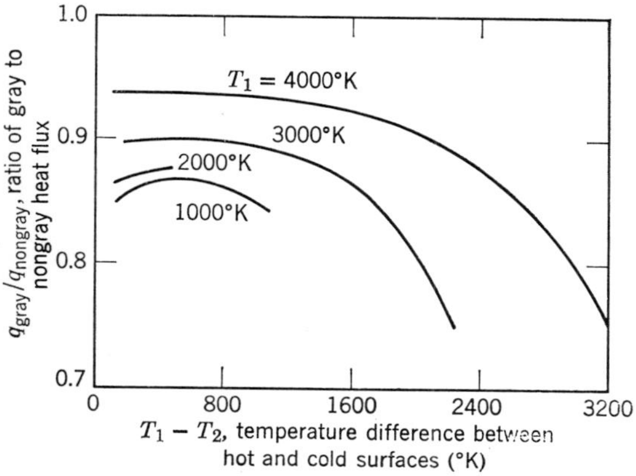

Fig. 4-5. Comparison of gray and nongray analysis for net radiative heat flux between two parallel plates. (From J. R. Branstetter [16].)

Branstetter [16] evaluated the integral in Eq. 4-45 for tungsten plates by by summing up the net spectral heat fluxes over about 400 increments of wavelength for several different temperatures. This result was then compared with the net heat flux evaluated by the gray analysis, that is,[1]

$$q_1 = \frac{\bar{\sigma} T_1^4 - \bar{\sigma} T_2^4}{(1/\varepsilon_1) + (1/\varepsilon_2) - 1} \tag{4-47}$$

where the emissivity of hot surface ε_1 was evaluated at T_1, and the emissivity of cold surface ε_2 was evaluated at the geometrical mean temperature $\sqrt{T_1 T_2}$, according to the empirical method recommended for metals.

Figure 4-5 shows a comparison of the gray and nongray radiative heat fluxes for tungsten plates evaluated by Branstetter [16]. It appears that the radiative heat flux determined by the gray analysis is approximately 8 to 25 per cent too low for the specific case considered.

(b) Two Concentric Spheres (or Very Long Coaxial Cylinders)

Consider an enclosure composed of two concentric spheres or two very long coaxial cylinders as shown in Fig. 4-6. The surfaces are opaque, diffuse emitters and diffuse reflectors. Let A_1 and A_2 be the surface areas, $\varepsilon_{1\lambda}$ and $\varepsilon_{2\lambda}$ be the spectral emissivities, and T_1 and T_2 be the temperatures of the inner and outer surfaces, respectively. We shall determine the net radiative heat fluxes at the surfaces.

Since the enclosure considered has two zones and the surface temperatures are prescribed, the equations for radiosities are of exactly the same form as

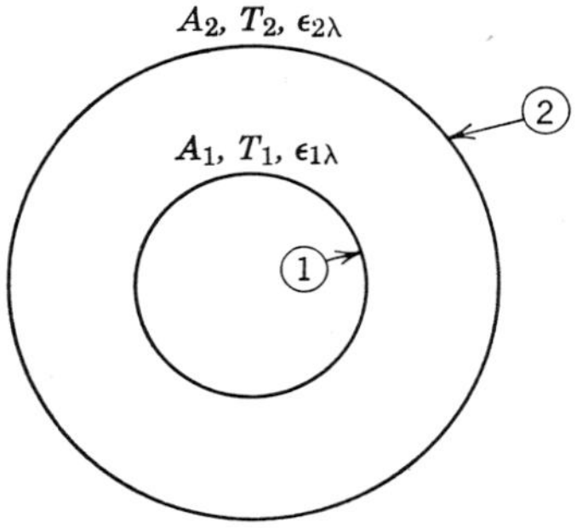

Fig. 4-6. Radiation exchange between two concentric spheres or two very long, coaxial cylinders.

Eqs. 4-38 for parallel plates. We write

$$R_{1\lambda} = \varepsilon_{1\lambda}\pi I_{\lambda b}(T_1) + \rho_{1\lambda}(R_{1\lambda}F_{1-1} + R_{2\lambda}F_{1-2}) \tag{4-48a}$$

$$R_{2\lambda} = \varepsilon_{2\lambda}\pi I_{\lambda b}(T_2) + \rho_{2\lambda}(R_{1\lambda}F_{2-1} + R_{2\lambda}F_{2-2}) \tag{4-48b}$$

The diffuse view factors for two concentric spheres (or for two very long coaxial cylinders) are given as[2]

$$F_{1-2} = 1, \qquad F_{1-1} = 0, \qquad F_{2-1} = \frac{A_1}{A_2}, \qquad \text{and} \qquad F_{2-2} = 1 - \frac{A_1}{A_2} \tag{4-49}$$

Substituting Eqs. 4-49 into Eqs. 4-48, we obtain

$$R_{1\lambda} - \rho_{2\lambda}R_{2\lambda} = \varepsilon_{1\lambda}\pi I_{\lambda b}(T_1) \tag{4-50a}$$

$$-\rho_{2\lambda}\frac{A_1}{A_2}R_{1\lambda} + \left[1 - \rho_{2\lambda}\left(1 - \frac{A_1}{A_2}\right)\right]R_{2\lambda} = \varepsilon_{2\lambda}\pi I_{\lambda b}(T_2) \tag{4-50b}$$

The solution of Eqs. 4-50 yields the spectral radiosities; knowing these, we can evaluate the spectral radiative heat fluxes from (see Eq. 4-11b)

$$q_{1\lambda} = R_{1\lambda} - R_{2\lambda} \tag{4-51a}$$

$$q_{2\lambda} = -\frac{A_1}{A_2}(R_{1\lambda} - R_{2\lambda}) \tag{4-51b}$$

and determine the net radiative heat fluxes from

$$q_1 = \int_{\lambda=0}^{\infty} q_{1\lambda}\, d\lambda \qquad \text{and} \qquad q_2 = \int_{\lambda=0}^{\infty} q_{2\lambda}\, d\lambda \tag{4-52}$$

The total net heat transfer rates at the surfaces are given by

$$Q_1 = A_1 q_1 \qquad \text{and} \qquad Q_2 = A_2 q_2 \tag{4-53}$$

From Eqs. 4-51 and 4-53 we note that the total net heat transfer rates and the net radiative heat fluxes at the inner and outer surfaces are related by

$$Q_1 = -Q_2 \qquad \text{and} \qquad A_1 q_1 = -A_2 q_2 \tag{4-54}$$

This result is to be expected from the consideration of conservation of energy.

Assuming that the Kirchhoff law is valid, we can replace ρ_λ by $1 - \varepsilon_\lambda$; then the spectral radiative heat flux at the inner surface is given by

$$q_{1\lambda} = \frac{\pi I_{\lambda b}(T_1) - \pi I_{\lambda b}(T_2)}{(1/\varepsilon_{1\lambda}) + (A_1/A_2)[(1/\varepsilon_{2\lambda}) - 1]} \tag{4-55}$$

and the net radiative heat flux by

$$q_1 = \int_{\lambda=0}^{\infty} \frac{\pi I_{\lambda b}(T_1) - \pi I_{\lambda b}(T_2)}{(1/\varepsilon_{1\lambda}) + (A_1/A_2)[(1/\varepsilon_{2\lambda}) - 1]}\, d\lambda \tag{4-56}$$

For the gray-body approximation Eq. 4-56 simplifies to[3]

$$q_1 = \frac{\bar{\sigma}T_1^4 - \bar{\sigma}T_2^4}{(1/\varepsilon_1) + (A_1/A_2)[(1/\varepsilon_2) - 1]} \tag{4-57}$$

In the foregoing relations the ratio A_1/A_2 can be replaced by r_1/r_2 for coaxial cylinders and by $(r_1/r_2)^2$ for concentric spheres, where r_1 and r_2 are the inner and outer radii, respectively.

Chupp and Viskanta [17] compared the net radiative heat fluxes obtained from the gray and nongray analysis for type 303 stainless steel concentric spheres (or very long coaxial cylinders). The inner surface was taken as a gray body at 2000°K and the outer surface as nongray at 300°K, with prescribed spectral emissivity. The net radiative heat flux from the nongray analysis was found to be larger than that from the gray analysis.

(c) A Rectangular Enclosure with Two Specularly Reflecting Surfaces

Consider a rectangular enclosure as shown in Fig. 4-7, which is very long in the direction perpendicular to the plane of the figure.

The surfaces are opaque, gray, and diffuse emitters. The two adjacent surfaces A_1 and A_2 are purely specular reflectors having reflectivities ρ_1^s and ρ_2^s, while the remaining surfaces A_2 and A_4 are purely diffuse reflectors having reflectivities ρ_3^d and ρ_4^d. The surfaces are kept at temperatures T_1, T_2, T_3, and T_4 and have emissivities ε_1, ε_2, ε_3, and ε_4. The determination of net radiative heat flux at the surfaces of such an enclosure was examined by Sparrow, Eckert, and Jonsson [18].

If the radiosities at the surfaces of the enclosure are known, the net radiative heat flux at diffusely reflecting surfaces 3 and 4 can be evaluated from (see Eq. 4-36b for $\rho_i^s = 0$ and $\varepsilon_i = 1 - \rho_i$)

$$q_i = \frac{\varepsilon_i}{\rho_i^d}(\bar{\sigma}T_i^4 - R_i), \quad i = 3, 4 \text{ and } \rho_i \neq 0 \tag{4-58}$$

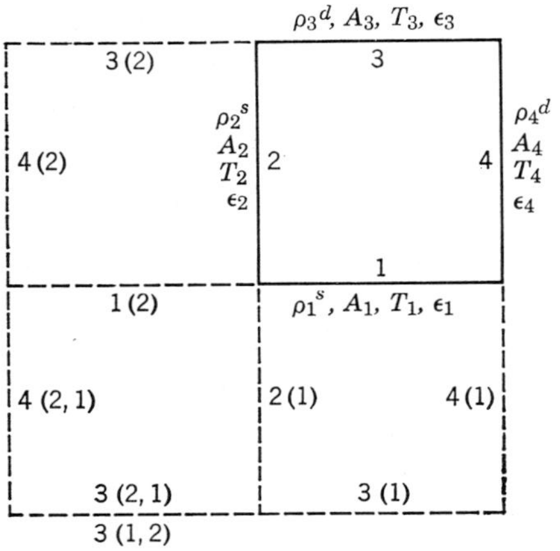

Fig. 4-7. Rectangular enclosure with two adjacent specularly reflecting and two diffusely reflecting surfaces.

and the net radiative heat flux at specularly reflecting surfaces 1 and 2 can be evaluated from (see Eq. 4-36a for $\varepsilon_i = 1 - \rho_i$)

$$q_i = \varepsilon_i\left(\bar{\sigma}T_i^4 - \sum_{j=1}^{4} R_j F_{i-j}^s\right), \quad i = 1, 2 \tag{4-59}$$

The equations for radiosities are obtained from Eqs. 4-35 as

$$R_i = \varepsilon_i\bar{\sigma}T_i^4 + \rho_i^d \sum_{j=1}^{4} R_j F_{i-j}^s, \quad i = 1, 2, 3, 4 \tag{4-60}$$

since temperatures are prescribed at all surfaces. Equations 4-60 can be simplified further since $\rho_1{}^d = \rho_2{}^d = 0$ for the purely specularly reflecting surfaces 1 and 2. Then Eqs. 4-60 for $i = 1$ and 2 simplify to

$$R_1 = \varepsilon_1\bar{\sigma}T_1{}^4 \tag{4-61a}$$

$$R_2 = \varepsilon_2\bar{\sigma}T_2{}^4 \tag{4-61b}$$

Thus the radiosities for surfaces 1 and 2 are available. The radiosities R_3 and R_4 for the diffusely reflecting surfaces are obtained from the solutions of Eqs. 4-60 for $i = 3$ and 4, that is,

$$R_3 = \varepsilon_3\bar{\sigma}T_3{}^4 + \rho_3{}^d \sum_{j=1}^{4} R_j F_{3-j}^s \tag{4-62}$$

$$R_4 = \varepsilon_4\bar{\sigma}T_4{}^4 + \rho_4{}^d \sum_{j=1}^{4} R_j F_{4-j}^s \tag{4-63}$$

These equations involve specular view factors which can be determined as described below.

F^s_{3-1}: This represents the fraction of radiative energy leaving surface A_3 that reaches surface A_1 both directly and by all possible specular interreflections. It is given by

$$F^s_{3-1} = F_{3-1} + \rho_2{}^s F_{3(2)-1} \tag{4-64}$$

Here $F_{3(2)-1}$ is the diffuse view factor from image surface $A_{3(2)}$ to surface A_1. We note that by symmetry $F_{3(2)-1}$ is equal to $F_{3-1(2)}$. No additional terms are needed in Eq. 4-64, because further specular reflections from surface A_3 do not reach A_1.

F^s_{3-2}: The fraction of radiative energy leaving A_3 that reaches A_2 both directly and by all possible specular interreflections is given by

$$F^s_{3-2} = F_{3-2} + \rho_1{}^s F_{3(1)-2} \tag{4-65}$$

We note that by symmetry considerations $F_{3(1)-2}$ is equal to $F_{3-2(1)}$.

F^s_{3-3}: The fraction of radiative energy leaving A_3 that returns to A_3 by all possible specular reflections is given by

$$F^s_{3-3} = \rho_1{}^s F_{3(1)-3} + \rho_1{}^s \rho_2{}^s F^*_{3(1,2)-3} + \rho_2{}^s \rho_1{}^s F^*_{3(2,1)-3} \tag{4-66}$$

The first term on the right-hand side is the fraction of radiative energy leaving A_3 that returns to A_3 after one specular reflection at A_1. The second term is the fraction of radiative energy leaving A_3 that returns to A_3 after two consecutive specular reflections at A_1 and A_2. The third term can be interpreted in a similar manner. Here the asterisk on F denotes a partial view of A_3 when looking through the last surface in the sequence of specular reflections.

F^s_{3-4}: The fraction of radiative energy leaving A_3 that reaches A_4 both directly and by all possible specular interreflections is given by

$$F^s_{3-4} = F_{3-4} + \rho_1{}^s F_{3(1)-4} + \rho_2{}^s F_{3(2)-4} + \rho_2{}^s \rho_1{}^s F_{3(2,1)-4} \tag{4-67}$$

Here $F_{3(1)-4}$, $F_{3(2)-4}$, and $F_{3(2,1)-4}$ are the diffuse view factors from images $A_{3(1)}$, $A_{3(2)}$, and $A_{3(2,1)}$, respectively, to surface A_4.

The specular view factors F^s_{4-j}, in Eq. 4-63, are evaluated in a similar manner.

Equations 4-62 and 4-63 can now be solved since the specular view factors appearing in these equations can be evaluated as described above. Once the radiosities R_3 and R_4 have been determined, the net radiative heat fluxes can be evaluated from Eqs. 4-58 and 4-59.

REFERENCES

1. H. C. Hottel, "Radiant Heat Transmission," in *Heat Transmission*, by W. H. McAdams, 3rd ed., pp. 72–79, McGraw-Hill Book Co., New York, 1954.

2. E. R. Eckert and R. M. Drake, *Heat and Mass Transfer*, pp. 407–411, McGraw-Hill Book Co., New York, 1959.

3a. B. Gebhart, "A New Method for Calculating Radiant Exchanges," *Heating, Piping, Air Conditioning*, **30**, 131–135, July 1958.

3b. B. Gebhart, "Surface Temperature Calculations in Radiant Surroundings of Arbitrary Complexity for Gray, Diffuse Radiation," *Intern. J. Heat Mass Transfer*, **3**, 341–346, 1961.

4. A. K. Oppenheim, "Radiation Analysis by the Network Method," *Trans. ASME* **78**, 725–735, 1956.

5. E. M. Sparrow, "Radiation Heat Transfer Between Surfaces," in *Advances in Heat Transfer*, edited by James P. Hartnett and T. F. Irvine, Jr., pp. 407–411, Academic Press, New York, 1965.

6. E. M. Sparrow and R. D. Cess, *Radiation Heat Transfer*, Brooks/Cole Publishing Co., Belmont, Calif., 1965.

7. J. A. Clark and E. Korybalski, "Radiation Heat Transfer in an Enclosure Having Surfaces Which Are Adiabatic or of Known Temperature," *First National Heat and Mass Transfer Conference, Madras, India*, December 1971.

8. J. T. Bevans and D. K. Edwards, "Radiation Exchange in an Enclosure with Directional Wall Properties," *J. Heat Transfer*, **87C**, 388–396, 1965.

9. R. A. Seban, Discussion of paper by E. M. Sparrow, E. R. G. Eckert, and R. V. Jonsson, "An Enclosure Theory for Radiative Exchange Between Specularly and Diffusely Reflecting Surfaces," *Trans. ASME*, **84C**, 294–300, 1962.

10a. A. F. Sarofim and H. C. Hottel, "Radiation Exchange among non-Lambert Surfaces," *J. Heat Transfer*, **88C**, 37–44, 1964.

10b. B. Munch, "Die Richtungsverteilung bei der Reflexion von Wärmestrachlung und ihr Einfluss auf die Wärmeübertragung, "Mitteilungen aus den Institut für Thermodynamik und Verbrenmungsmotor-enbau an der Eidgenössischen Technischen Hochschule in Zürich, No. 16, 1955.

11. R. C. Birkebak, E. M. Sparrow, E. R. G. Eckert, and J. W. Ramsey, "Effect of Surface Roughness on the Total Hemispherical and Specular Reflectance of Metallic Surfaces," *J. Heat Transfer*, **86C**, 193–199, 1964.

12. K. E. Torrance and E. M. Sparrow, "Off-Specular Peaks in the Directional Distribution of Reflected Thermal Radiation," *J. Heat Transfer*, **88C**, 223–230, 1966.

13. J. P. Holman, "Radiation Networks for Specular-Diffuse Reflecting and Transmitting Surfaces," *ASME Paper* No. 66-WA/HT-9, 1966.

14. E. M. Sparrow and S. L. Lin, "Radiation Heat Transfer at a Surface Having Both Specular and Diffuse Reflectance Components," *Intern. J. Heat Mass Transfer*, **8**, 769–779, 1965.

15. R. P. Bobco, "Radiation Transfer in Semigray Enclosures with Specularly and Diffusely Reflecting Surfaces," *J. Heat Transfer*, **86C**, 123–130, 1964.

16. J. R. Branstetter, "Radiant Heat Transfer Between Nongray Parallel Plates of Tungsten," *NASA Tech. Note* TN-D-1088, 1961.

17. R. E. Chupp and R. Viskanta, "Radiant Heat Transfer Between Concentric Spheres and Coaxial Cylinders," *J. Heat Transfer*, **88C**, 326–327, 1966.

18. E. M. Sparrow, E. R. G. Eckert, and V. K. Jonsson, "An Enclosure Theory for Radiative Exchange Between Specularly and Diffusely Reflecting Surfaces," *J. Heat Transfer*, **84C**, 294–300, 1962.

NOTES

[1] Equation 4-47 is applicable also for two parallel, infinite, gray plates with *specularly reflecting* surfaces. The reason is that all the radiative energy leaving surface 1 will reach surface 2 (similarly from surface 2 to surface 1) whether the surfaces are specular or diffuse reflectors.

[2] Let A_1 and A_2 be the inner and outer surface areas of two diffusely reflecting concentric spheres. Since all the radiative energy leaving A_1 strikes A_2, we have

$$F_{1-2} = 1 \tag{1}$$

From the reciprocity relation

$$A_2 F_{2-1} = A_1 F_{1-2} \qquad \text{or} \qquad F_{2-1} = \frac{A_1}{A_2} \tag{2}$$

Noting that

$$F_{2-1} + F_{2-2} = 1 \tag{3}$$

we have

$$F_{2-2} = 1 - F_{2-1} = 1 - \frac{A_1}{A_2} \tag{4}$$

[3] Equation 4-57 is applicable also when the outer surface is a *diffuse* reflector and the inner surface a *diffuse* or *specular* reflector. However, when both surfaces are specular reflectors, the radiative energy is reflected between the surfaces indefinitely as if the surfaces were infinite parallel plates; for this special case the net radiative heat flux is the same as that for two parallel plates, given by Eq. 4-47.

CHAPTER 5

Radiative Exchange in an Enclosure—Generalized Zone Analysis

The utility of the simplified zone analysis described in Chapter 4 is severely restricted because of the assumption of uniform radiosity over the surface of each zone. The radiosity will vary over the surface of each zone when the spacing between the zones is small in comparison to their dimensions. In such situations the heat transfer results may be in error unless the effects of the variation of radiosity over the surface of each zone are included in the analysis. There are many engineering applications in which an accurate calculation of radiative heat transfer in an enclosure is important. Some of these include radiative heat transfer associated with space-vehicle environmental control; heat rejection from space-vehicle power plants; emission from surfaces that may not be smooth because of the presence of depressions, holes, or pits; and the design of black bodies. Therefore in the present analysis the assumption of uniform radiosity and of uniform temperature (or heat flux) over the surface of each zone is removed, the mathematical formulation of such problems is presented, methods of solution are discussed, and applications are illustrated with specific example.

5-1 GENERALIZED ZONE ANALYSIS FOR ENCLOSURES WITH DIFFUSELY REFLECTING SURFACES

Consider an enclosure whose surfaces are diffuse emitters and diffuse reflectors and are divided into N zones such that the radiative properties are uniform over the surface of each zone. Let $\mathbf{r}_i$ be the space coordinates for the zone A_i, $i = 1, 2, \ldots, N$. Under these assumptions the equations for the radiation intensity and the net radiative heat flux at zone A_i, given by Eqs. 4-1 and 4-2, respectively, simplify to

$$I_{i,\nu}(\mathbf{r}_i) = \varepsilon_{i,\nu} I_{\nu b}[T_i(\mathbf{r}_i)] + f_{i,\nu} \sum_{j=1}^{N} \int_{A_j} I_{j,\nu}(\mathbf{r}_j) \frac{\cos\theta_i' \cos\theta_j}{r_{ij}^2} dA_j \tag{5-1}$$

and

$$q_i(\mathbf{r}_i) = \int_{\nu=0}^{\infty} q_{i,\nu}(\mathbf{r}_i)\, d\nu \tag{5-2a}$$

where

$$q_{i,\nu}(\mathbf{r}_i) = \pi I_{i,\nu}(\mathbf{r}_i) - \sum_{j=1}^{N} \int_{A_j} I_{j,\nu}(\mathbf{r}_j) \frac{\cos\theta_i' \cos\theta_j}{r_{ij}^2} dA_j, \quad i = 1, 2, \ldots, N \tag{5-2b}$$

since

$$d\Omega_j' = \frac{\cos\theta_j \, dA_j}{r_{ij}^2}$$

Here r_{ij} is the length of the straight line joining the elemental surfaces dA_i and dA_j. For diffuse reflection the spectral reflection distribution function $f_{i,\nu}$ and the spectral radiation intensity $I_{i,\nu}(\mathbf{r}_i)$ leaving the surface are independent of direction; then $f_{i,\nu}$ and $I_{i,\nu}(\mathbf{r}_i)$ are related to the spectral hemispherical reflectivity and spectral radiosity, respectively, by

$$\pi f_{i,\nu} = \rho_{i,\nu} \qquad \text{(see Eq. 1-108)} \tag{5-3a}$$

$$\pi I_{i,\nu}(\mathbf{r}_i) = R_{i,\nu}(\mathbf{r}_i) \qquad \text{(see Eq. 1-129)} \tag{5-3b}$$

By the definition of diffuse view factor we have

$$dF_{dA_i - dA_j} = \frac{\cos\theta_i' \cos\theta_j}{\pi r_{ij}^2} dA_j \tag{5-3c}$$

Substitution of Eqs. 5-3 into Eqs. 5-1 and 5-2 yields respectively

$$R_{i,\nu}(\mathbf{r}_i) = \varepsilon_{i,\nu} \pi I_{\nu b}[T_i(\mathbf{r}_i)] + \rho_{i,\nu} \sum_{j=1}^{N} \int_{A_j} R_{j,\nu}(\mathbf{r}_j)\, dF_{dA_i - dA_j} \tag{5-4}$$

and

$$q_i(\mathbf{r}_i) = \int_{\nu=0}^{\infty} q_{i,\nu}(\mathbf{r}_i)\, d\nu \tag{5-5a}$$

where

$$q_{i,\nu}(\mathbf{r}_i) = R_{i,\nu}(\mathbf{r}_i) - \sum_{j=1}^{N} \int_{A_j} R_{j,\nu}(\mathbf{r}_j)\, dF_{dA_i-dA_j} \tag{5-5b}$$

Alternative expressions can be obtained for $q_{i,\nu}(\mathbf{r}_i)$ as follows. By eliminating the summation term from Eq. 5-5b by means of Eq. 5-4, we obtain

$$q_{i,\nu}(\mathbf{r}_i) = \frac{1}{\rho_{i,\nu}} \{\varepsilon_{i,\nu}\pi I_{\nu b}[T_i(\mathbf{r}_i)] - (1 - \rho_{i,\nu})R_{i,\nu}(\mathbf{r}_i)\} \qquad \rho_{i,\nu} \neq 0 \tag{5-5c}$$

or, by eliminating $R_{i,\nu}(\mathbf{r}_i)$ between Eqs. 5-5b and 5-4, we obtain

$$q_{i,\nu}(\mathbf{r}_i) = \varepsilon_{i,\nu}\pi I_{\nu b}[T_i(\mathbf{r}_i)] - (1 - \rho_{i,\nu})\sum_{j=1}^{N} \int_{A_j} R_{j,\nu}(\mathbf{r}_j)\, dF_{dA_i-dA_j} \qquad i = 1, 2, \ldots, N \tag{5-5d}$$

Equations 5-4 and 5-5 give the complete mathematical formulation of radiative exchange in an enclosure for the *generalized zone analysis*. These equations will simplify to those given by Eqs. 4-10 and 4-11 of the simplified zone analysis if the radiosities and temperatures are assumed to be uniform over the surface of each zone.

The Band Approximation

The foregoing equations for the generalized zone analysis can be simplified further by the application of the band approximation. We assume that the entire energy spectrum is divided into a finite number of frequency bands $\Delta\nu_k$, $k = 1, 2, \ldots, K$, such that the radiative properties of the surfaces are considered uniform over each band. Then the integration of Eqs. 5-4 and 5-5 over each frequency band $\Delta\nu_k$ yields respectively

$$R_{i,k}(\mathbf{r}_i) = \varepsilon_{i,k}\pi I_{b,k}[T_i(\mathbf{r}_i)] + \rho_{i,k}\sum_{j=1}^{N} \int_{A_j} R_{j,k}(\mathbf{r}_j)\, dF_{dA_i-dA_j} \tag{5-6}$$

and

$$q_i(\mathbf{r}_i) = \sum_{k=1}^{K} q_{i,k}(\mathbf{r}_i) \tag{5-7a}$$

where

$$q_{i,k}(\mathbf{r}_i) = R_{i,k}(\mathbf{r}_i) - \sum_{j=1}^{N} \int_{A} R_{j,k}(\mathbf{r}_j)\, dF_{dA_i-dA_j} \tag{5-7b}$$

or

$$q_{i,k}(\mathbf{r}_i) = \frac{1}{\rho_{i,k}} \{\varepsilon_{i,k}\pi I_{b,k}[T_i(\mathbf{r}_i)] - (1 - \rho_{i,k})R_{i,k}(\mathbf{r}_i)\} \quad \rho_{i,k} \neq 0 \tag{5-7c}$$

or

$$q_{i,k}(\mathbf{r}_i) = \varepsilon_{i,k}\pi I_{b,k}[T_i(\mathbf{r}_i)] - (1 - \rho_{i,k})\sum_{j=1}^{N} \int_{A_j} R_{j,k}(\mathbf{r}_j)\, dF_{dA_i-dA_j}$$

$$i = 1, 2, \ldots, N;\ k = 1\ 2, \ldots, K \tag{5-7d}$$

Here we have defined

$$I_{b,k}[T(\mathbf{r}_i)] \equiv \int_{\Delta\nu_k} I_{b\nu}[T(\mathbf{r}_i)]\, d\nu \tag{5-8a}$$

$$q_{i,k}(\mathbf{r}_i) \equiv \int_{\Delta\nu_k} q_{i,\nu}(\mathbf{r}_i)\, d\nu \tag{5-8b}$$

$$R_{i,k}(\mathbf{r}_i) \equiv \int_{\Delta\nu_k} R_{i,\nu}(\mathbf{r}_i)\, d\nu \tag{5-8c}$$

and $\varepsilon_{i,k}$ and $\rho_{i,k}$ are the mean values of the spectral emissivity and spectral reflectivity, respectively, over the frequency interval $\Delta\nu_k$. Here we note that the equations for each frequency band are independent of those for other bands.

The Gray-Body Approximation

If it is assumed that the radiative properties of the surfaces of the enclosure are independent of frequency, integration of Eqs. 5-4 and 5-5 over the entire frequency range yields respectively

$$R_i(\mathbf{r}_i) = \varepsilon_i \bar{\sigma} T_i^4(\mathbf{r}_i) + \rho_i \sum_{j=1}^{N} \int_{A_j} R_j(\mathbf{r}_j)\, dF_{dA_i - dA_j} \tag{5-9}$$

and

$$q_i(\mathbf{r}_i) = R_i(\mathbf{r}_i) - \sum_{j=1}^{N} \int_{A_j} R_j(\mathbf{r}_j)\, dF_{dA_i - dA_j} \tag{5-10a}$$

$$q_i(\mathbf{r}_i) = \frac{1}{\rho_i} [\varepsilon_i \bar{\sigma} T_i^4(\mathbf{r}_i) - (1 - \rho_i) R_i(\mathbf{r}_i)], \quad \rho_i \neq 0 \tag{5-10b}$$

or

$$q_i(\mathbf{r}_i) = \varepsilon_i \bar{\sigma} T_i^4(\mathbf{r}_i) - (1 - \rho_i) \sum_{j=1}^{N} \int_{A_j} R_j(\mathbf{r}_j)\, dF_{dA_i - dA_j} \quad i = 1, 2, \ldots, N \tag{5-10c}$$

since

$$\int_{\nu=0}^{\infty} R_{i,\nu}\, d\nu \equiv R_i$$

$$\pi \int_{\nu=0}^{\infty} I_{\nu b}(T)\, d\nu = \pi I_b(T) = \bar{\sigma} T^4$$

Suppose that the temperatures are prescribed at the surfaces of all zones of the enclosure; then Eqs. 5-9 provide N simultaneous integral equations for the N unknown radiosity functions $R_i(\mathbf{r}_i)$, $i = 1, 2, \ldots, N$. Once the radiosities have been determined, the distribution of the net radiative heat flux over the surface of each zone is evaluated from Eqs. 5-10.

Suppose that temperatures $T_i(\mathbf{r}_i)$ are prescribed for the surfaces of zones $i = 1, 2, \ldots, r$ and heat fluxes $q_i(\mathbf{r}_i)$ are prescribed for zones $i = r + 1$,

$r + 2, \ldots, N$ in an N-zone enclosure. In this problem the temperature distribution is required for the zones with prescribed heat fluxes, and the heat flux distribution is required for the zones with prescribed temperatures. The equations for radiosities are obtained from Eq. 5-9 for zones with prescribed surface temperature and from Eqs 5-10a for zones with prescribed heat fluxes. We have

$$R_i(\mathbf{r}_i) = \varepsilon_i \bar{\sigma} T_i^4(\mathbf{r}_i) + \rho_i \sum_{j=1}^{N} \int_{A_j} R_j(\mathbf{r}_j)\, dF_{dA_i - dA_j}, \quad i = 1, 2, \ldots, r \tag{5-11}$$

$$R_i(\mathbf{r}_i) = q_i(\mathbf{r}_i) + \sum_{j=1}^{N} \int_{A_j} R_j(\mathbf{r}_j)\, dF_{dA_i - dA_j}, \quad i = r + 1, r + 2, \ldots, N \tag{5-12}$$

Equations 5-11 and 5-12 are N simultaneous integral equations for the N unknown radiosity functions $R_i(\mathbf{r}_i)$, $i = 1, 2, \ldots, N$. Once the radiosities are determined, the net heat fluxes for the zones with prescribed temperatures (i.e., $i = 1, 2, \ldots, r$) can be evaluated from Eqs. 5-12 (or any one of Eqs. 5-10); and the temperatures for the zones with prescribed heat fluxes (i.e., $i = r + 1, r + 2, \ldots, N$) can be evaluated from Eq. 5-11 (or Eqs 5-10b or 5-10c).

5-2 GENERALIZED ZONE ANALYSIS FOR A GRAY ENCLOSURE WITH SURFACES HAVING BOTH DIFFUSE AND SPECULAR REFLECTIVITY COMPONENTS

Consider an N-zone enclosure in which the radiative properties over the surface of each zone are uniform, and the surfaces are gray, diffuse emitters and have reflectivities that can be represented as the sum of a diffuse and a specular reflectivity component in the form

$$\rho_i = \rho_i^s + \rho_i^d \tag{5-13}$$

Since the surfaces have both diffuse and specular reflectivity components, the equations for radiosities are obtained from Eq. 5-9, by replacing in that equations ρ_i by ρ_i^d and the diffuse view factor by the specular view factor:

$$R_i(\mathbf{r}_i) = \varepsilon_i \bar{\sigma} T_i^4(\mathbf{r}_i) + \rho_i^d \sum_{j=1}^{N} \int_{A_j} R_j(\mathbf{r}_j)\, dF^s_{dA_i - dA_j} \tag{5-14}$$

Here the first and second terms on the right-hand side represent respectively the diffusely emitted and diffusely reflected radiative energies per unit time and per unit area at position $\mathbf{r}_i$ on zone A_i. If the temperatures and radiosities are assumed to be uniform over the surface of each zone, Eq. 5-14 reduces to Eq. 4-35 of the simplified zone analysis.

The net radiative heat flux $q_i(\mathbf{r}_i)$ at any position $\mathbf{r}_i$ on zone A_i can be represented as the difference between the emitted and absorbed radiative energies per unit time and per unit area and given as

$$q_i(\mathbf{r}_i) = \varepsilon_i \bar{\sigma} T_i^4(\mathbf{r}_i) - (1 - \rho_i) \sum_{j=1}^{N} \int_{A_j} R_j(\mathbf{r}_j)\, dF^s_{dA_i - dA_j} \tag{5-15a}$$

This equation is analogous to Eq. 5-10c except that we have used here the specular view factor. Alternative forms of Eq. 5-15a are obtained as follows. By eliminating the summation term from Eq. 5-15a by means of Eq. 5-14, we obtain

$$q_i(\mathbf{r}_i) = \frac{1}{\rho_i^d} [\varepsilon_i(1 - \rho_i^s)\bar{\sigma} T_i^4(\mathbf{r}_i) - (1 - \rho_i) R_i(\mathbf{r}_i)], \quad \rho_i^d \neq 0 \tag{5-15b}$$

which reduces to Eq. 5-10b for $\rho_i^s = 0$.

Or, by eliminating $\bar{\sigma} T_i^4(\mathbf{r}_i)$ between Eqs. 5-15a and 5-15b, we obtain

$$q_i(\mathbf{r}_i) = R_i(\mathbf{r}_i) - (1 - \rho_i^s) \sum_{j=1}^{N} \int_{A_j} R_j(\mathbf{r}_j)\, dF^s_{dA_i - dA_j} \tag{5-15c}$$

which simplifies to Eq. 5-10a for $\rho_i^s = 0$.

Equations 5-14 and 5-15 constitute the complete mathematical formulation of the problem of radiative exchange for the generalized zone analysis in an N-zone, gray enclosure having surfaces with reflectivities represented as the sum of a diffuse and a specular component.

5-3 METHOD OF SOLUTION OF FREDHOLM TYPE INTEGRAL EQUATIONS

The generalized zone analysis described in Section 5-2 resulted in a set of coupled integral equations for the radiosity functions. In this section we present a brief account of the methods of solution of the *Fredholm type integral equations* that are encountered in such formulations. For comprehensive treatment of integral equations the reader should refer to the books by Courant and Hilbert [1], Hildebrand [2], and Lovitt [3].

Consider a linear integral equation of the form

$$\phi(x) = f(x) + \lambda \int_a^b K(x, \eta)\phi(\eta)\, d\eta \tag{5-16}$$

where $f(x)$ and $K(x, \eta)$ are given functions, λ, a, and b are constants, and the function $\phi(x)$ is to be determined. Equation 5-16 is a linear integral equation known as the *Fredholm integral equation of the second kind.* The function $K(x, \eta)$ is called the *kernel* of the integral equation; the equation is said to be homogeneous when the free term $f(x)$ is equal to zero.

(a) Method of Successive Approximations

The Fredholm integral equation 5-16 with $K(x, \eta)$ and $f(x)$ real continuous functions in the interval $a \leq x \leq b$, $a \leq \eta \leq b$, and a, b, λ real constants can be solved by the method of successive approximations (or by an iterative method) described in the books by Hildebrand [2, pp. 421–424] and Lovitt [3, p. 15]. The method consists of replacing the function $\phi(\eta)$ under the integral sign on the right-hand side of Eq. 5-16 by an initial guess $\phi_0(\theta)$ in order to obtain the first approximation $\phi_1(x)$, given by

$$\phi_1(x) = f(x) + \lambda \int_a^b K(x, \eta)\phi_0(\eta)\, d\eta \tag{5-17}$$

The first approximation $\phi_1(\eta)$ is introduced on the right-hand side of Eq. 5-16 to obtain the second approximation $\phi_2(x)$:

$$\phi_2(x) = f(x) + \lambda \int_a^b K(x, \eta)\phi_1(\eta)\, d\eta \tag{5-18}$$

The second approximation is used to determine the third approximation, and continuing in this manner we obtain

$$\phi_3(x) = f(x) + \lambda \int_a^b K(x, \eta)\phi_2(\eta)\, d\eta \tag{5-19}$$

$$\vdots$$

$$\phi_n(x) = f(x) + \lambda \int_a^b K(x, \eta)\phi_{n-1}(\eta)\, d\eta \tag{5-20}$$

When these equations are combined, the nth approximation $\phi_n(x)$ is given as

$$\phi_n(x) = f(x) + \lambda L f(x) + \lambda^2 L^2 f(x) + \lambda^3 L^3 f(x) + \cdots + \lambda^{n-1} L^{n-1} f(x) + \lambda^n L^n \phi_0(x) \tag{5-21}$$

Here we have defined the operator L as

$$Lf(x) \equiv \int_a^b K(x, \eta) f(\eta)\, d\eta \tag{5-22}$$

$$L^2 f(x) \equiv \int_a^b K(x, \eta) \int_a^b K(\eta, \eta_1) f(\eta_1)\, d\eta_1\, d\eta \tag{5-23}$$

$$L^3 f(x) \equiv \int_a^b K(x, \eta) \int_a^b K(\eta, \eta_1) \int_a^b K(\eta_1, \eta_2) f(\eta_2)\, d\eta_2\, d\eta_1\, d\eta \tag{5-24}$$

$$\cdots\cdots$$

It has been shown by Hildebrand [2, pp. 421–424] and Lovitt [3, pp. 11–16] that as $n \to \infty$ the solution given by Eq. 5-21 will converge to the desired solution of the integral equation 5-16, that is,

$$\lim_{n \to \infty} \phi_n(x) = \phi(x) \tag{5-25}$$

if the following condition is satisfied

$$|\lambda| < \frac{1}{(b - a)M} \tag{5-26}$$

where M is the maximum value of the kernel $K(x, \eta)$ [i.e., $|K(x, \eta)|$], in the interval $a \leq x \leq b$, $a \leq \eta \leq b$. The convergence of the series will be rapid if $\lambda \ll 1$.

In practice only a few stages of the above integrations can be carried out analytically since the higher-order analysis becomes extremely involved, but to perform these integrations numerically with a high-speed computer would pose no problem.

(b) Reduction to Sets of Algebraic Equations

The integral equation 5-16 can be approximated by a set of linear algebraic equations by representing the integral term on the right-hand side by a summation. The Simpson rule, the trapezoidal rule, or the Gaussian quadrature formula can be used to approximate the integral by a summation, and the resulting simultaneous, algebraic linear equations can be solved by standard computer subroutines. (The application of Gaussian quadrature will be discussed in Chapter 11.)

(c) A Variational Method for the Solution of Integral Equation

It is shown by Courant and Hilbert [1, p. 205] that a function $\phi(x)$ that gives an extremum (i.e., maximum or minimum) in the following variational expression:

$$I \equiv \lambda \int_a^b \int_a^b K(x, y)\phi(x)\phi(\eta)\, dx\, d\eta + 2\int_a^b \phi(x) f(x)\, dx - \int_a^b [\phi(x)]^2\, dx \tag{5-27}$$

will also lead to the requirement that $\phi(x)$ satisfy the integral equation 5-16. Although it is difficult to find the exact function $\phi(x)$ that will give an extremum in the variational expression 5-27, an approximate solution can be obtained by utilizing the *Ritz method* described by Hildebrand [2, p. 187].

Let the desired solution $\phi(x)$ be represented approximately as a linear combination of n suitably chosen functions $\Psi_i(x)$ in the form

$$\phi(x) = \sum_{k=1}^{n} c_k \Psi_k(x) \tag{5-28}$$

The choice of functions $\Psi_k(x)$ may be arbitrary, but the selection should be made with physical insight into the nature of the problem. The constants c_k are to be determined from the requirement that the function given by Eq. 5-28, when substituted into the variational expression Eq. 5-27, should yield an extremum in that expression. Let

$$I \equiv I(c_1, c_2, \ldots, c_n) \tag{5-29}$$

denote the resulting expression when Eq. 5-28 is substituted into Eq. 5-27. The coefficients $c_1, c_2, \ldots, c_n$ that will lead to an extremum for $I(c_1, c_2, \ldots, c_n)$ can be determined by differentiating Eq. 5-29 with respect to each coefficient c_k and setting each of the resulting expressions equal to zero for $k = 1, 2, \ldots, n$, that is,

$$\frac{\partial I}{\partial c_k} = 0, \quad k = 1, 2, \ldots, n \tag{5-30}$$

Equations 5-30 yield n simultaneous algebraic equations for the determination of n unknown coefficients c_k.

In practice, the above procedure is applied by starting with only a few terms in the expression given by Eq. 5-28; the number of terms is increased until convergence to a desired accuracy is achieved.

The variational method has been applied by Sparrow [4], Usiskin and Siegel [5], and Perlmutter and Siegel [6] to solve the problem of radiative exchange in an enclosure.

(d) A Generalized Variational Method

The formulation of radiative exchange for an enclosure, using the generalized zone analysis, involves simultaneous integral equations of the form (see Eq. 5-11)

$$\phi_i(\mathbf{r}_i) = f_i(\mathbf{r}_i) + \lambda_i \sum_{j=1}^{N} \int_{A_j} \phi_j(\mathbf{r}_j) K_{ij}\, dA_j, \quad i = 1, 2, \ldots, N \tag{5-31}$$

where the functions $\phi_i(\mathbf{r}_i)$, which characterize the radiosities, are to be determined. Here $f_i(\mathbf{r}_i)$ is a prescribed function (i.e., the distribution of temperature or heat flux at zone i), λ_i is a constant, and the kernel K_{ij} is related to the elemental diffuse view factor by

$$dF_{dA_i - dA_j} \equiv K_{ij}\, dA_j \tag{5-32}$$

hence K_{ij} is known for a specified geometry.

Sparrow and Haji-Sheikh [7] showed that the solution of the integral equations 5-31 is equivalent to finding the extremum for the following variational expression:

$$I = \sum_{j=1}^{N} \int_{A_j} \int_{A_j} \phi_j^2 K_{jj} \, dA_j \, dA_j + 2 \sum_{j=2}^{N} \left(\sum_{i=1}^{j-1} \int_{Ai} \int_{Aj} \phi_j \phi_i K_{ij} \, dA_j \, dA_i \right)$$
$$- \sum_{j=1}^{N} \frac{1}{\lambda_j} \int_{A_j} \phi_j^2 \, dA_j + 2 \sum_{j=1}^{N} \int_{Aj} f_j \phi_j \, dA_j \qquad (5\text{-}33)$$

If the functions $\phi_1, \phi_2, \ldots, \phi_N$ are determined in such a way that they lead to an extremum in Eq. 5-33, then these functions are also the solutions of the integral equations 5-31. It is very difficult to find the exact functions that will satisfy these requirements; however, the Ritz method can be applied to obtain approximate solutions. It will be assumed that each of the functions $\phi_i(\mathbf{r}_i)$ can be approximated by a linear combination of M suitably chosen functions $\Psi_{im}(\mathbf{r}_i)$ in the form

$$\phi_i(\mathbf{r}_i) = \sum_{m=1}^{M} c_{im} \Psi_{im}(\mathbf{r}_i), \quad i = 1, 2, \ldots, N \qquad (5\text{-}34)$$

where the constants c_{im}, $i = 1, 2, \ldots, N$ and $m = 1, 2, \ldots, M$, are to be determined from the requirement that when Eq. 5-34 is substituted into Eq. 5-33 it should yield an extremum in I. Let

$$I \equiv I(c_{11}, c_{12}, \ldots, c_{im}, \ldots) \qquad (5\text{-}35)$$

denote the resulting expression when Eq. 5-34 is substituted into Eq. 5-33 and the indicated operations are performed. To determine the coefficients c_{im} that will lead to an extremum in I, we differentiate Eq. 5-35 with respect to each coefficient c_{im} and set the resulting expressions equal to zero:

$$\frac{\partial I}{\partial c_{im}} = 0, \quad i = 1, 2, \ldots, N; \; m = 1, 2, \ldots, M \qquad (5\text{-}36)$$

This system yields $M \times N$ simultaneous algebraic equations for the determination of $M \times N$ unknown coefficients c_{im}. Although the accuracy of the assumed solution can be improved by increasing the number of terms in the summation in Eq. 5-34, the difficulty of obtaining an accurate solution to a large number of simultaneous equations should be recognized.

(e) Approximation of the Kernel

The integral equation 5-16 can be transformed into a second-order ordinary differential equation if the kernel $K(x, \eta)$ is approximated by an exponential function in the form

$$K(x, \eta) \cong c e^{-\beta |x - \eta|} \qquad (5\text{-}37)$$

To illustrate the application, we substitute Eq. 5-37 into Eq. 5-16 and obtain

$$\phi(x) = f(x) + \lambda c\left[\int_a^x e^{-\beta(x-\eta)}\phi(\eta)\,d\eta + \int_x^b e^{-\beta(\eta-x)}\phi(\eta)\,d\eta\right] \tag{5-38}$$

We differentiate Eq. 5-38 with respect to x twice, eliminate the integral term from the resulting expression by means of Eq. 5-38, and obtain

$$\frac{d^2\phi(x)}{dx^2} - \beta(2-\beta)\phi(x) = \frac{d^2f(x)}{dx^2} - \beta^2 f(x) \tag{5-39}$$

This ordinary differential equation for $\phi(x)$ can be solved numerically or analytically, depending on the complexity of function $f(x)$. The boundary conditions required to solve Eq. 5-39 are determined from the original integral equation 5-16 by evaluating it at the boundary points.

The kernel $K(x, \eta)$ can be approximated more accurately if it is represented as a sum of two exponentials in the form

$$K(x, \eta) \simeq c_1 e^{-\beta_1|x-\eta|} + c_2 e^{-\beta_2|x-\eta|} \tag{5-40}$$

When the kernel is represented as in Eq. 5-40, the integral equation 5-16 will transform into a fourth-order ordinary differential equation for the function $\phi(x)$.

The exponential kernel approximation has been applied by Buckley [8], Sparrow and Jonsson [9], and Perlmutter and Siegel [6] to the problems of radiative heat exchange inside a cavity.

5-4 RADIATIVE EXCHANGE BETWEEN TWO PARALLEL PLATES

The generalized zone analysis has been applied to the problem of radiative exchange between two parallel plates by Sparrow [4], Sparrow, Gregg, Szel, and Manos [10], and Sparrow and Haji-Shiekh [7]. In this section we present the formulation of the problem of radiative exchange for a parallel plate system and discuss the results.

Consider two parallel plates as shown in Fig. 5-1, each of length L, separated by a distance h, and infinitely long in the direction perpendicular to the plane of the figure. The plates are opaque, gray, diffuse emitters and diffuse reflectors, have the same emissivity ε, and are kept at a uniform, identical temperature T. For simplicity in the analysis it will be assumed that the environment that the plates see through the gap between them is at zero temperature. The determination of the distribution of net radiative heat flux over the surfaces of the plates is of interest in this problem.

We choose the mid-points of the plates, O_1 and O_2, as origins for the x_1 and x_2 coordinates, respectively, since the problem involves symmetry about

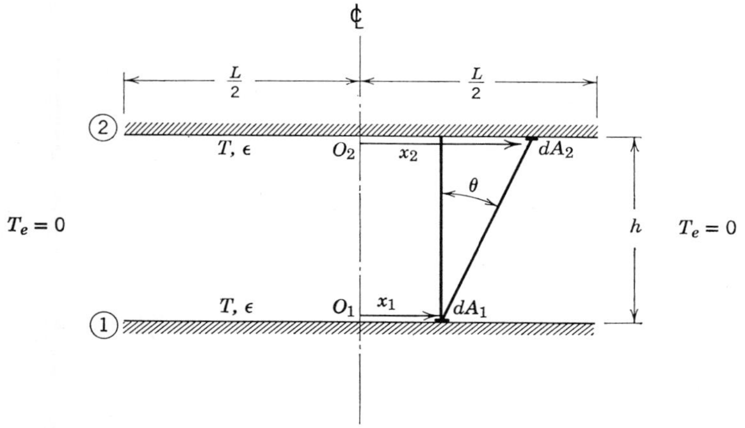

Fig. 5-1. Radiation interchange between parallel plates each at a uniform and identical temperature T.

the mid-points of the plates. The integral equations for the radiosities $R_1(x_1)$ and $R_2(x_2)$ for plates 1 and 2 are immediately obtained from Eq. 5-9 as

$$R_1(x_1) = \varepsilon\bar{\sigma}T^4 + (1 - \varepsilon)\int_{-L/2}^{L/2} R_2(x_2)\, dF_{dx_1-dx_2} \tag{5-41}$$

$$R_2(x_2) = \varepsilon\bar{\sigma}T^4 + (1 - \varepsilon)\int_{-L/2}^{L/2} R_1(x_1)\, dF_{dx_2-dx_1} \tag{5-42}$$

Here we have replaced the view factors $dF_{dA_1-dA_2}$ and $dF_{dA_2-dA_1}$ by $dF_{dx_1-dx_2}$ and $dF_{dx_2-dx_1}$, respectively, and have substituted $(1 - \varepsilon)$ for ρ.

Because of symmetry we have $R_1(x_1) = R_2(x_2)$ for $x_1 = x_2$. Then only one of Eqs. 5-41 and 5-42 should be considered, and the equation to be solved becomes

$$R(x_1) = \varepsilon\bar{\sigma}T^4 + (1 - \varepsilon)\int_{-L/2}^{L/2} R(x_2)\, dF_{dx_1-dx_2} \tag{5-43}$$

Here we have removed the subscripts from the radiosities. The diffuse view factor between two elongated, parallel strips dx_1 and dx_2 is given by (see Eqs. 3-53a)

$$dF_{dx_1-dx_2} = \tfrac{1}{2}d(\sin\theta) \tag{5-44}$$

where

$$\sin\theta = \frac{x_2 - x_1}{[(x_2 - x_1)^2 + h^2]^{1/2}} \tag{5-45}$$

Then the diffuse view factor becomes

$$dF_{dx_1-dx_2} = \frac{1}{2}\frac{h^2}{[(x_2 - x_1)^2 + h^2]^{3/2}}\, dx_2 \tag{5-46}$$

Substituting Eq. 5-46 into 5-43, we obtain the following equation for the radiosity:

$$R(x_1) = \varepsilon\bar{\sigma}T^4 + \tfrac{1}{2}(1-\varepsilon)\int_{-L/2}^{L/2} \frac{h^2}{[(x_2 - x_1)^2 + h^2]^{3/2}} R(x_2)\, dx_2 \qquad (5\text{-}47)$$

which can be expressed in the dimensionless form as

$$\phi(x) = 1 + \lambda \int_{-1/2}^{1/2} K(x, \eta)\phi(\eta)\, d\eta \qquad (5\text{-}48)$$

where we have defined

$$x \equiv \frac{x_1}{L}, \qquad \eta \equiv \frac{x_2}{L}, \qquad \gamma \equiv \frac{h}{L} \qquad (5\text{-}49a)$$

$$\phi \equiv \frac{R}{\varepsilon\bar{\sigma}T^4}, \qquad \lambda \equiv (1-\varepsilon)\frac{\gamma^2}{2} \qquad (5\text{-}49b)$$

$$K(x, \eta) \equiv \frac{1}{[(x-\eta)^2 + \gamma^2]^{3/2}} \qquad (5\text{-}49c)$$

and we note that the kernel $K(x, \eta)$ is a symmetrical one.

Once the integral equation 5-48 is solved and the radiosity function $\phi(x)$ determined, the distribution of the net radiative heat flux over the surface of the plate is evaluated from (see Eq. 5-10b)

$$q(x) = \frac{\varepsilon}{1-\varepsilon}[\bar{\sigma}T^4 - R(x)], \quad \varepsilon \neq 1 \qquad (5\text{-}50a)$$

or

$$\frac{q(x)}{\varepsilon\bar{\sigma}T^4} = \frac{1}{1-\varepsilon}[1 - \varepsilon\phi(x)], \quad \varepsilon \neq 1 \qquad (5\text{-}50b)$$

The total amount of heat Q leaving the plate per unit time and per unit width into the external environment is obtained from

$$Q = \int_{-L/2}^{L/2} q(x)\, dx = \frac{\varepsilon}{1-\varepsilon}\left[\bar{\sigma}T^4 L - \int_{-L/2}^{L/2} R(x)\, dx\right] \qquad (5\text{-}51a)$$

or

$$\frac{Q/L}{\varepsilon\bar{\sigma}T^4} = \frac{1}{1-\varepsilon}\left[1 - \varepsilon\int_{-L/2}^{L/2} \phi(x)\, dx\right], \quad \varepsilon \neq 1 \qquad (5\text{-}51b)$$

Solution by Variational Method

The variational expression appropriate to Eq. 5-48 is given in the form (see Eq. 5-27)

$$I = \lambda \int_{-1/2}^{1/2}\int_{-1/2}^{1/2} K(x, \eta)\phi(x)\phi(\eta)\, dx\, d\eta + 2\int_{-1/2}^{1/2} \phi(x)\, dx - \int_{-1/2}^{1/2} [\phi(x)]^2\, dx \quad (5\text{-}52)$$

where λ and $K(x, \eta)$ have already been defined. The function $\phi(x)$ will now be represented as a polynomial in x, but the odd powers of x will be excluded by symmetry considerations. For a two-term polynomial representation of $\phi(x)$ we have

$$\phi(x) = c_1 + c_2 x^2 \tag{5-53}$$

where the constants c_1 and c_2 are to be determined.

Substituting Eq. 5-53 into Eq. 5-52 and performing the indicated operations, we obtain

$$I = (1 - \varepsilon)(c_1^2 a_1 + c_1 c_2 a_2 + c_2^2 a_3) - c_1^2 - \tfrac{1}{6} c_1 c_2 - \tfrac{1}{80} c_2^2 + 2c_1 + \frac{c_2}{6} \tag{5-54}$$

where a_1, a_2, a_3 are known constants that depend on γ and are given in reference 4. Differentiating Eq. 5-53 with respect to c_1 and c_2 and then equating the resulting expressions to zero yields respectively

$$2c_1[a_1(1 - \varepsilon) - 1] + c_2[a_2(1 - \varepsilon) - \tfrac{1}{6}] = -2 \tag{5-55}$$

$$c_1[a_2(1 - \varepsilon) - \tfrac{1}{6}] + 2c_2[a_3(1 - \varepsilon) - \tfrac{1}{80}] = -\tfrac{1}{6} \tag{5-56}$$

Simultaneous solution of Eqs. 5-55 and 5-56 gives the values of the coefficients c_1 and c_2. Knowing c_1 and c_2, we can evaluate the distribution of the net radiative heat flux and the total amount of heat leaving the plane per unit time and per unit width from Eqs. 5-50b and 5-51b, respectively, that is,

$$\frac{q(x)}{\varepsilon \bar{\sigma} T^4} = \frac{1}{1 - \varepsilon}[1 - \varepsilon(c_1 + c_2 x^2)] \tag{5-57}$$

$$\frac{Q/L}{\varepsilon \bar{\sigma} T^4} = \frac{1}{1 - \varepsilon}\left[1 - \varepsilon\left(c_1 + \frac{c_2}{12}\right)\right] \tag{5-58}$$

Accuracy of Variational Solution

The accuracy of the variational solution can be improved by using higher-order terms in representing the function $\phi(x)$. However, the degree of accuracy cannot be established unless the results are compared with the exact solution. Table 5-1 shows a comparison of the values of function $\phi(x)$ obtained by the variational method, using a second- and a fourth-degree polynomial representation, and by numerical integration (i.e., exact solution). The variational solution with a fourth-degree polynomial representation appears to be accurate enough for most engineering applications.

Table 5-1 Comparison of Function $\phi(x)$ Determined by the Variational Method and Numerical Integration[a]

$x = x_1/L$	0	0.1	0.2	0.3	0.4	0.5
		$\gamma = 1,\ \rho = 0.9$				
Second degree	1.642	1.637	1.620	1.592	1.554	1.504
Exact	1.644	1.638	1.620	1.590	1.552	1.508
		$\gamma = 0.1,\ \rho = 0.9$				
Second degree	7.39	7.22	6.73	5.90	4.74	3.25
Fourth degree	7.21	7.10	6.76	6.07	4.88	2.95
Exact	7.22	7.11	6.75	6.07	4.90	2.97

[a] From E. M. Sparrow [4].

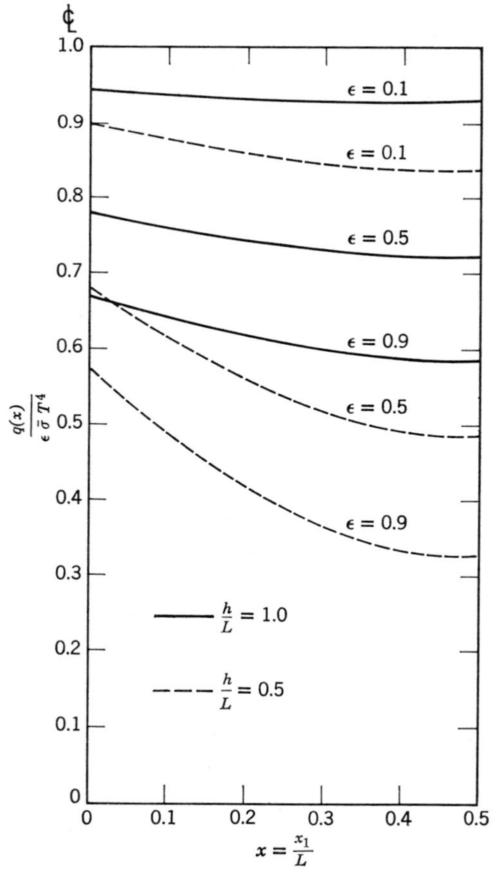

Fig. 5-2. Effects of spacing and emissivity on local heat transfer rate for parallel plate systems. (From E. M. Sparrow et al. [10].)

Table 5-2 Comparison of Total Heat Transfer Rate $(Q/L)/\bar{\sigma}T^4$ for Parallel Plate Geometry, Obtained from Generalized and Simplified Zone Analysis[a]

$\frac{L}{h}$	$\varepsilon = 0.1$		$\varepsilon = 0.5$		$\varepsilon = 0.9$	
	Generalized Analysis	Simplified Analysis	Generalized Analysis	Simplified Analysis	Generalized Analysis	Simplified Analysis
1.0	0.09338	0.09340	0.3692	0.3604	0.5500	0.5500
0.5	0.08576	0.08607	0.2747	0.2764	0.3658	0.3664
0.1	0.0442	0.05122	0.07964	0.08677	0.09269	0.09402
0.05	0.0252	0.03388	0.04128	0.04649	0.04751	0.04848

[a] From E. M. Sparrow et al. [10].

Figure 5-2 shows the effects of plate spacing and of emissivity on the local heat flux, $q(x)/\varepsilon\bar{\sigma}T^4$. As the spacing between the plates decreases, there is a large variation in the distribution of local heat flux.

Table 5-2 shows a comparison of the total heat transfer rate $(Q/L)/\bar{\sigma}T^4$, as obtained from the generalized zone analysis and the simplified zone analysis. These results show that the prediction of the total heat transfer rate by the simplified zone analysis is remarkably good for $h/L > 0.5$. However, as the spacing between the plates becomes smaller, the simplified zone analysis grossly overestimates the heat transfer rate.

5-5 RADIATIVE EXCHANGE BETWEEN TWO PARALLEL, COAXIAL, CIRCULAR DISKS

Consider two parallel, coaxial circular disks separated by a distance h, each having the same radius a as shown in Fig. 5-3. The surfaces are opaque, gray, diffuse emitters and diffuse reflectors and have the same emissivity ε. The lower and upper disks are kept at uniform but different temperatures T_1 and T_2, respectively, and the external environment is assumed to be at zero temperature (i.e., a negligible amount of radiative energy passes inward from the external environment to the plates). The distribution of net radiative heat flux over the surfaces of the disks is of interest in the problem under consideration. Walsh [11] and Sparrow and Gregg [12] investigated the radiative transfer for similar problems. Here we present the formulation of the governing equations and discuss some of the heat transfer results.

Let the disk centers O_1 and O_2 be the origins for the r_1 and r_2 coordinates, respectively. The integral equations for the radiosities $R_1(r_1)$ and $R_2(r_2)$ for

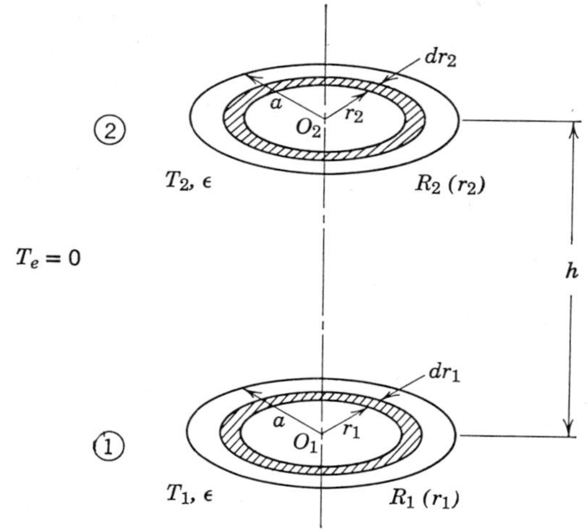

Fig. 5-3. Radiation interchange between two circular disks at uniform but different temperatures T_1 and T_2.

the lower and upper disks, respectively, are obtained from Eq. 5-9 as

$$R_1(r_1) = \varepsilon\bar{\sigma}T_1^4 + (1 - \varepsilon)\int_{r_2=0}^{a} R_2(r_2)\, dF_{dr_1-dr_2} \tag{5-59}$$

$$R_2(r_2) = \varepsilon\bar{\sigma}T_2^4 + (1 - \varepsilon)\int_{r_1=0}^{a} R_1(r_1)\, dF_{dr_2-dr_1} \tag{5-60}$$

where $dF_{dr_1-dr_2}$ is the diffuse view factor from the elemental ring (dr_1, r_1) on disk 1 to the elemental ring (dr_2, r_2) on disk 2. The elemental diffuse view factor $dF_{dr_1-dr_2}$ can be determined by the method described in Section 3-6 and is given as

$$dF_{dr_1-dr_2} = \frac{2h^4 + 2h^2r_1^2 + 2h^2r_2^2}{[(h^2 + r_1^2 + r_2^2)^2 - 4r_1^2r_2^2]}\, r_2\, dr_2 \tag{5-61}$$

and $dF_{dr_2-dr_1}$ is obtained from the reciprocity relation as

$$dF_{dr_2-dr_1} = \frac{r_1\, dr_1}{r_2\, dr_2}\, dF_{dr_1-dr_2} \tag{5-62}$$

Equations 5-59 and 5-60 are two coupled integral equations for the radiosity functions $R_1(r_1)$ and $R_2(r_2)$ and must be solved simultaneously.

Subdivision into Simpler Problems

Since Eqs. 5-59 and 5-60 are linear in T_i^4, $i = 1$ or 2, they can be subdivided into simpler problems and the complete solution obtained by superposition.

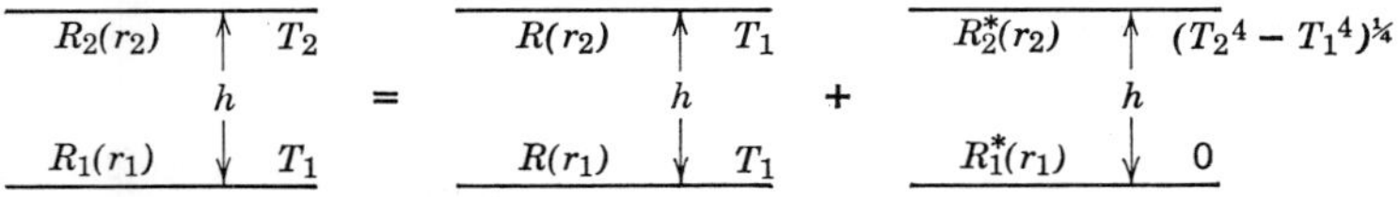

Fig. 5-4. Subdivision of the problem for two parallel circular disks into two simpler problems.

Figure 5-4 illustrates the subdivision of the problem considered into two simpler problems by assuming that $T_2 > T_1$. One of these problems is the radiative exchange between two circular disks that have the same geometry and radiative properties as in Fig. 5-3, but are kept at the same uniform temperature T_1, with the external environment at zero temperature. In this case the distribution of radiosity, $R(r_1)$ or $R(r_2)$, at the surfaces of the disks satisfies the following integral equation:

$$R(r_1) = \varepsilon\bar{\sigma}T_1^4 + (1 - \varepsilon)\int_0^a R(r_2)\, dF_{dr_1-dr_2} \tag{5-63}$$

since by symmetry the radiosities at both disks are equal for $r_1 = r_2$.

The second problem is the radiative exchange between two disks having the same geometry and radiative properties as in Fig. 5-3, but disc 1 is at zero temperature, disc 2 at temperature $(T_2^4 - T_1^4)^{1/4}$, and the external environment at zero temperature. In this case the radiosities $R_1^*(r_1)$ and $R_2^*(r_2)$ satisfy the following two coupled integral equations:

$$R_1^*(r_1) = (1 - \varepsilon)\int_0^a R_2^*(r_2)\, dF_{dr_1-dr_2} \tag{5-64a}$$

$$R_2^*(r_2) = \varepsilon\bar{\sigma}(T_2^4 - T_1^4) + (1 - \varepsilon)\int_0^a R_1^*(r_1)\, dF_{dr_2-dr_1} \tag{5-64b}$$

Then the radiosities $R_1(r_1)$ and $R_2(r_2)$ for the original problem, that is, Eqs. 5-59 and 5-60, are obtained by the superposition of radiosities for the above simpler problems as

$$R_1(r_1) = R(r_1) + R_1^*(r_1) \tag{5-65a}$$

$$R_2(r_2) = R(r_2) + R_2^*(r_2) \tag{5-65b}$$

Equations 5-63 and 5-64 can be written in the dimensionless form as, respectively,

$$\phi(\eta_1) = 1 + \lambda\int_{\eta_2=0}^1 K(\eta_1, \eta_2)\phi(\eta_2)\, d\eta_2 \tag{5-66}$$

and

$$\phi_1^*(\eta_1) = \lambda\int_{\eta_2=0}^1 K(\eta_1, \eta_2)\phi_2^*(\eta_2)\, d\eta_2 \tag{5-67a}$$

$$\phi_2^*(\eta_2) = 1 + \lambda\int_{\eta_1=0}^1 K(\eta_2, \eta_1)\phi_1^*(\eta_1)\, d\eta_1 \tag{5-67b}$$

where we have defined

$$\phi(\eta_i) = \frac{R(\eta_i)}{\varepsilon\bar{\sigma}T_1^4}, \qquad \phi_i^*(\eta_i) = \frac{R_i^*(\eta_i)}{\varepsilon\bar{\sigma}[T_2^4 - T_1^4]}, \quad i = 1 \text{ or } 2 \tag{5-68a}$$

$$\lambda = 2(1-\varepsilon)\gamma^2, \qquad \gamma = \frac{h}{a}, \qquad \eta_i = \frac{r_i}{a}, \quad i = 1 \text{ or } 2 \tag{5-68b}$$

$$K(\eta_1, \eta_2) = \frac{(\gamma^2 + \eta_1^2 + \eta_2^2)\eta_2}{[(\gamma^2 + \eta_1^2 + \eta_2^2)^2 - 4\eta_1^2\eta_2^2]^{3/2}} \tag{5-68c}$$

and $K(\eta_2, \eta_1)$ is obtainable from Eq. 5-68c by interchanging subscripts 1 and 2.

Once the dimensionless radiosities are determined from the solution of Eqs. 5-66 and 5-67, the radiosities for the original problem (i.e., Eqs. 5-59 and 5-60) are given by

$$R_1 = R + R_1^* = \varepsilon\bar{\sigma}[T_1^4\phi + (T_2^4 - T_1^4)\phi_1^*] \tag{5-69a}$$

$$R_2 = R + R_2^* = \varepsilon\bar{\sigma}[T_1^4\phi + (T_2^4 - T_1^4)\phi_2^*] \tag{5-69b}$$

The net radiative heat flux for the simpler problem described by Eq. 5-66 is given by (see Eq. 5-10b)

$$\frac{q(\eta_1)}{\varepsilon\bar{\sigma}T_1^4} = \frac{1}{1-\varepsilon}[1 - \varepsilon\phi(\eta_1)] \tag{5-70}$$

which is applicable for both disks.

The net radiative heat fluxes for the simpler problem described by Eqs. 5-67 are given as

$$\frac{q_1^*(\eta_1)}{\varepsilon\bar{\sigma}T_1^4} = -\frac{\varepsilon}{1-\varepsilon}\phi_1^*(\eta_1) \qquad \text{(for disk 1)} \tag{5-71}$$

$$\frac{q_2^*(\eta_2)}{\varepsilon\bar{\sigma}T^{*4}} = \frac{1}{1-\varepsilon}[1 - \varepsilon\phi_2^*(\eta_2)] \quad \text{(for disk 2)} \tag{5-72}$$

where

$$T^{*4} \equiv T_2^4 - T_1^4 \tag{5-73}$$

Results

The simpler problems described by Eqs. 5-66 and 5-67 have been solved numerically by Sparrow and Gregg [12]. Figure 5-5 shows the distribution of net radiative heat fluxes over the surfaces of disks 1 and 2 for the simpler problem described by Eqs. 5-67 for several different values of spacing and emissivity.

It is apparent from these figures that for an aspect ratio (i.e., h/a) greater than about 5 the distribution of the net radiative heat flux over the surfaces of disks is almost uniform. As the aspect ratio becomes smaller, the distribution of the net radiative heat flux becomes increasingly nonuniform.

5-6 RADIATIVE EXCHANGE INSIDE A CYLINDRICAL ENCLOSURE

The problem of radiative exchange inside cylindrical enclosures has been studied by various investigators. Buckley [8] appears to have been the first to solve the problem of radiative transfer inside a long cylinder open at one end, with lateral surfaces kept at uniform temperature. He used the exponential kernel approximation. Sparrow and Albers [13] solved a similar problem numerically by the method of successive approximations. Usiskin and Siegel [5] and Sparrow, Albers, and Eckert [14] considered a finite cylinder with

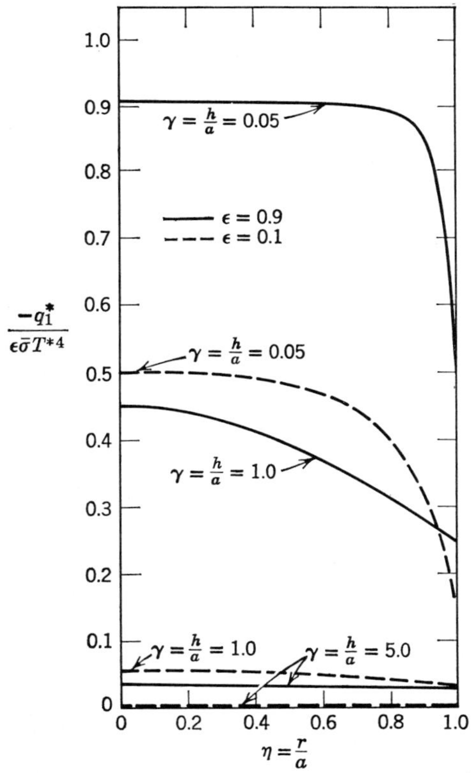

Fig. 5-5(a)

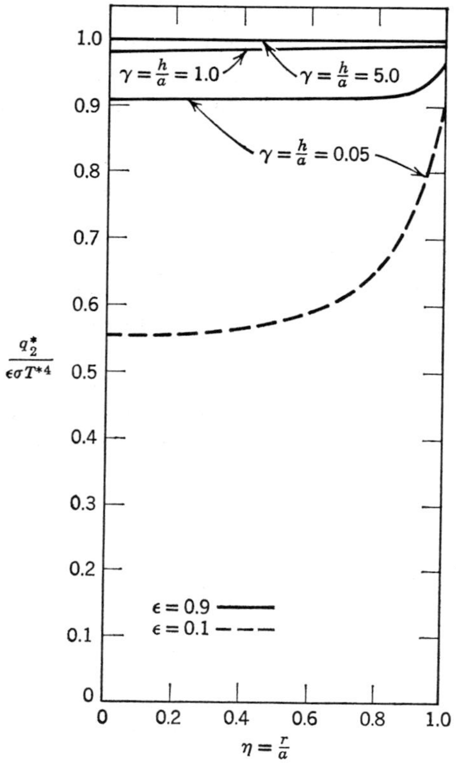

Fig. 5-5(b)

Fig. 5-5. Local net radiative heat flux at disks 1 and 2 for the simpler problem when disk 1 is at zero temperature, disk 2 at temperature $T^* = (T_2^4 - T_1^4)^{1/4}$. (From E. M. Sparrow and J. L. Gregg [12].)

uniform surface heat flux and uniform surface temperature, respectively. Perlmutter and Siegel [6] investigated the effects of specular reflection for a finite cylinder open at both ends, with uniformly applied heat flux at the walls.

In this section we consider representative problems for radiative exchange inside a cylindrical enclosure with diffusely reflecting and diffusely emitting walls for both prescribed surface temperature and prescribed surface heat flux.

(a) Cylindrical Cavity with Uniform Surface Temperature

Consider a cylindrical cavity of radius a and length L, closed at one end and open at the other end to an environment at zero temperature, as shown

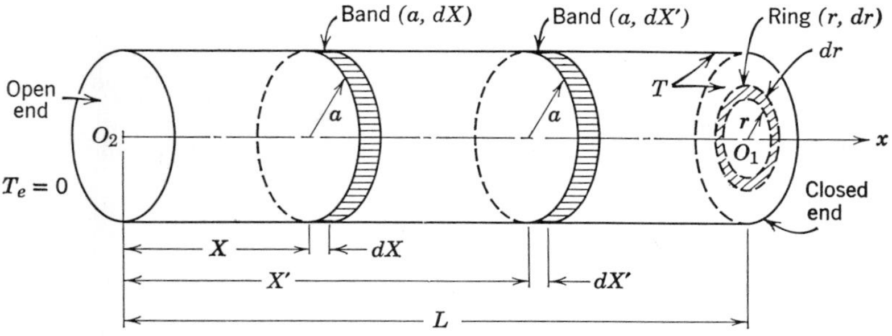

Fig. 5-6. Radiative interchange inside a cylindrical cavity with uniform surface temperature.

in Fig. 5-6. The inside surfaces are assumed to be opaque, gray, diffuse reflectors and diffuse emitters, to have a uniform emissivity ε, and to be kept at uniform temperature T throughout. In the present analysis we shall be concerned with the determination of the total heat transfer from the open end of the cylindrical hole.

We choose center O_1 of the closed end as the origin of the radial coordinate r for the closed end, and center O_2 of the open end as the origin of the axial coordinate x for the cylindrical surface. The problem possesses cylindrical symmetry since the temperatures and radiative properties are uniform over the surface of each zone.

Let $R_1(r)$ and $R_2(x)$ be the radiosities for the closed end and the cylindrical surface, respectively. The equations for radiosities are immediately written according to Eq. 5-9 as

$$R_1(r) = \varepsilon\bar{\sigma}T^4 + (1-\varepsilon)\int_{x'=0}^{L} R_2(x')\, dF_{dr-dx',(L-x')} \tag{5-74}$$

$$R_2(x) = \varepsilon\bar{\sigma}T^4 + (1-\varepsilon)\int_{r'=0}^{a} R_1(r')\, dF_{dx-dr',(L-x)} + (1-\varepsilon)\int_{x'=0}^{L} R_2(x')\, dF_{dx-dx',(x'-x)} \tag{5-75}$$

where $dF_{dr-dx',(L-x')}$ = elemental diffuse view factor from the ring (r, dr) to the band (a, dx') at a distance $L - x'$,

$dF_{dx-dr,(L-x)}$ = elemental diffuse view factor from the band (a, dx) to the ring (r, dr) at a distance $L - x$,

$dF_{dx-dx',(x'-x)}$ = elemental diffuse view factor from the band (a, dx) to the band (a, dx') at a distance $x' - x$.

The diffuse view factors in Eqs. 5-74 and 5-75 can be obtained readily from the relations given in Chapter 3. First, $dF_{dr-dx',(L-x')}$ is given by (see

Eq. 3-79 for $x = L - x'$)

$$dF_{dr-dx',(L-x')} = 2(L - x')a^2 \frac{(L - x')^2 + r^2 - a^2}{\{[(L - x')^2 + r^2 + a^2]^2 - 4r^2a^2\}^{3/2}} dx' \quad (5\text{-}76)$$

Then, $dF_{dx-dr',(L-x)}$ can be determined by the reciprocity relation

$$dF_{dx-dr,(L-x)} = \frac{r}{a}\frac{dr}{dx} dF_{dr-dx,(L-x)} \quad (5\text{-}77)$$

Substituting Eq. 5-76 into 5-77, we obtain

$$dF_{dx-dr,(L-x)} = 2(L - x)ar \frac{(L - x)^2 + r^2 - a^2}{\{[(L - x)^2 + r^2 + a^2]^2 - 4r^2a^2\}^{3/2}} dr \quad (5\text{-}78)$$

Finally, $dF_{dx-dx',(x'-x)}$ is given by (see Eq. 3-88a)

$$dF_{dx-dx',(x'-x)} = \frac{1}{2a}\left[1 - |x' - x| \frac{(x' - x)^2 + 6a^2}{\{(x' - x)^2 + 4a^2\}^{3/2}}\right] dx' \quad (5\text{-}79)$$

Here the absolute value $|x' - x|$ is used because the view factor depends only on the magnitude of separation between the bands.

Equations 5-74 and 5-75 with the view factors as given above provide two simultaneous integral equations for the two radiosity functions $R_1(r)$ and $R_2(x)$. Once these equations are solved and the radiosities are determined, the local net radiative heat fluxes $q_1(r)$ and $q_2(x)$ at the closed end of the cylinder and at the cylindrical surface, respectively, are readily evaluated according to Eq. 5-10b.

The amount of radiative energy streaming out of the open end of the cylinder, Q, is determined by integrating the local radiative heat fluxes over the entire surface of the cavity, that is,

$$Q = \int_{r=0}^{a} 2\pi r q_1(r)\, dr + \int_{x=0}^{L} 2\pi a q_2(x)\, dx \quad (5\text{-}80)$$

In many engineering applications the *apparent hemispherical emissivity* ε_a of the cavity, defined as the ratio of the actual radiative energy streaming out of the open end of the cylinder to the radiative energy that would have been emitted by a black surface at temperature T, having the same area as the opening, is of interest. This quantity is obtainable from Eq. 5-80 as

$$\varepsilon_a \equiv \frac{Q}{\pi a^2 \bar{\sigma} T^4} \quad (5\text{-}81)$$

Sparrow, Albers, and Eckert [14] solved the integral equations 5-74 and 5-75 numerically by an iterative scheme and determined the distribution of radiative heat flux inside the cylinder and the apparent hemispherical emissivity of the cavity. Table 5-3 shows the apparent hemispherical emissivity

Table 5-3 Apparent Hemispherical Emissivity of a Cylindrical Cavity[a]

$\frac{L}{2a}$	Apparent Hemispherical Emissivity ε_a		
	$\varepsilon = 0.9$	$\varepsilon = 0.75$	$\varepsilon = 0.5$
0.25	0.9434	0.8491	0.6569
0.5	0.9618	0.8948	0.7424
1.0	0.9720	0.9229	0.8084
2.0	0.9746	0.9308	0.8331
4.0	0.9749	0.9317	0.8367

[a] From E. M. Sparrow, L. U. Albers, and E. R. G. Eckert [14].

of the cavity for several different values of the actual emissivity of the surface, ε, and the depth of the hole, $L/2a$. It is apparent from this table that as the cavity becomes deeper the apparent emissivity ε_a approaches a constant value. For example, for $\varepsilon = 0.9$, there is little change in ε_a for $L/2a > 2$.

(b) Cylindrical Hole with Uniform Surface Heat Flux

Consider a finite cylinder of length L and radius a with uniformly applied heat flux at the cylindrical surface. The ends at $x = 0$ and $x = L$ are open to external environments that are at temperatures T_1 and T_2, respectively, as illustrated in Fig. 5-7. The cylindrical surface is an opaque, gray, diffuse reflector and a diffuse emitter and has uniform emissivity ε. The determination of the temperature distribution over the cylindrical surface is of interest.

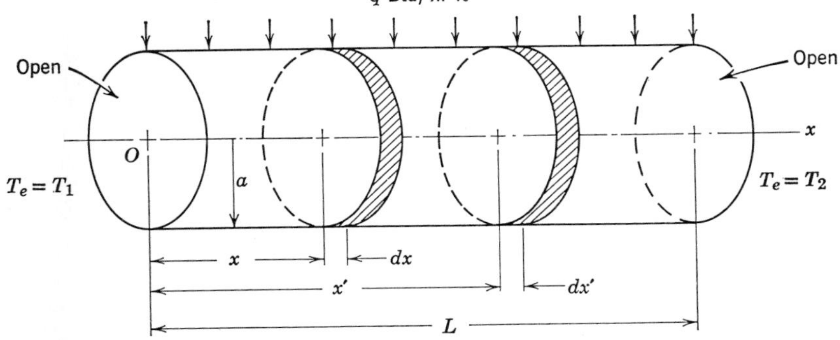

Fig. 5-7. Radiative interchange inside a cylinder with uniform heat flux.

For the problem in question the cylindrical surface is the only zone for which the distribution of radiosity is unknown. Since the heat flux q is prescribed over this surface, the equation for radiosity $R(x)$ is immediately obtained from Eq. 5-10a as

$$R(x) = q + \bar{\sigma} T_1^4 F_{dx-a,x} + \bar{\sigma} T_2^4 F_{dx-a,(L-x)} + \int_{x'=0}^{L} R(x')\, dF_{dx-dx',(x'-x)} \qquad (5\text{-}82)$$

where $F_{dx-a,x}$ = diffuse view factor from the band (a, dx) at x to the opening at $x = 0$,

$F_{dx-a,(L-x)}$ = diffuse view factor from the band (a, dx) at x to the opening at $x = L$,

$dF_{dx-dx',(x'-x)}$ = elemental diffuse view factor from the band (a, dx) to the band (a, dx') at a distance $x' - x$ apart.

The diffuse view factor $F_{dx-a,x}$ can be determined from Eq. 3-83 by applying the reciprocity relation. We obtain

$$F_{dx-a,x} = -\frac{a}{2}\frac{\partial}{\partial x}(F_{a-a,x}) \qquad (5\text{-}83a)$$

where $F_{a-a,x}$ is given by (see Eq. 3-87)

$$F_{a-a,x} = \frac{2a^2 + x^2 - \sqrt{(2a^2 + x^2)^2 - 4a^4}}{2a^2} \qquad (5\text{-}83b)$$

Performing the differentiation, we obtain

$$F_{dx-a,x} = \frac{\frac{1}{2} + (x/2a)^2}{\sqrt{1 + (x/2a)^2}} - \frac{x}{2a} \qquad (5\text{-}84a)$$

Similarly we write

$$F_{dx-a,(L-x)} = \frac{\frac{1}{2} + [(L - x)/2a]^2}{\sqrt{1 + [(L - x)/2a]^2}} - \frac{L - x}{2a} \qquad (5\text{-}84b)$$

and from Eq. 5-79 we have

$$dF_{dx-dx',(x'-x)} = \frac{1}{2a}\left[1 - |x' - x|\frac{(x' - x)^2 + 6a^2}{[(x' - x)^2 + 4a^2]^{3/2}}\right] dx' \qquad (5\text{-}85)$$

The solution of the integral equation 5-82 with view factors as defined above gives the distribution of radiosity $R(x)$ over the cylindrical surface. Once the radiosity is known, the distribution of temperature over the cylindrical surface is evaluated from Eq. 5-10b. Usiskin and Siegel [5] solved Eq. 5-82 by using an exponential kernel approximation, the variational method, and a numerical integration scheme. Table 5-4 shows the results of their calculation of dimensionless wall radiosity $R(x)/q$ for specified q at

Table 5-4 Distribution[a] of $R(x)/q$

L	$\frac{x}{L}$	Numerical	Exponential Kernel	Variational
1	0	2.06	2.00	2.06
	$\frac{1}{8}$	2.31	2.22	2.31
	$\frac{1}{4}$	2.49	2.38	2.49
	$\frac{3}{8}$	2.60	2.47	2.60
	$\frac{1}{2}$	2.64	2.50	2.63
4	0	4.95	5.00	4.98
	$\frac{1}{8}$	8.61	8.50	8.60
	$\frac{1}{4}$	11.2	11.0	11.2
	$\frac{3}{8}$	12.6	12.5	12.7
	$\frac{1}{2}$	13.2	13.0	13.2

[a] From C. M. Usiskin and R. Siegel [5].

walls and zero temperature at the ends of the enclosure. The results with the variational method appears to agree better with the numerical solution than with the exponential kernel approximation.

5-7 EFFECTS OF SPECULAR REFLECTION ON RADIATIVE HEAT TRANSFER

In the preceding sections consideration was given to radiative heat transfer in enclosures with purely diffuse reflecting surfaces. However, actual surfaces have both specular and diffuse reflectivity components; therefore enclosures with purely specularly reflecting surfaces represent another extreme situation.

Radiative exchange inside a cylindrical enclosure with specularly reflecting surfaces has been investigated by Perlmutter and Siegel [6], Sparrow and Jonsson [9], and Krishnan [15, 16].[1] To illustrate the effects of specular reflection on radiative transfer we present here the problem for a cylindrical enclosure studied by Perlmutter and Siegel [6].

Consider a finite cylinder of radius a and length L, with uniformly applied heat flux q at the cylindrical surface. The ends at $x = 0$ and $x = L$ are open to external environments at temperatures T_1 and T_2, respectively as illustrated in Fig. 5-8. The cylindrical surface is an opaque, gray, diffuse emitter and a purely specular reflector.

The distribution of temperature over the cylindrical surface is of interest.

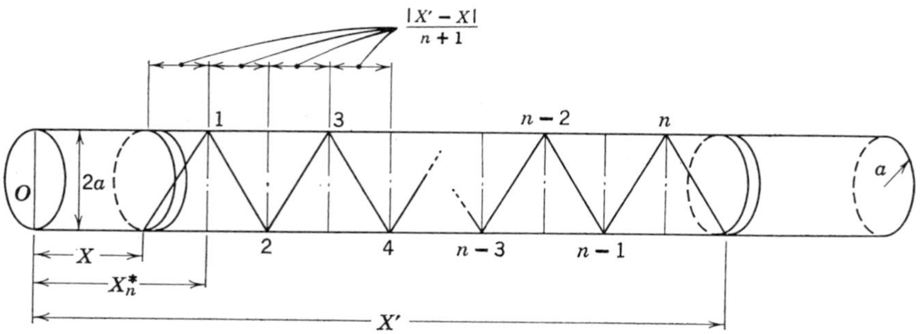

Fig. 5-8. Radiant energy leaving position x and arriving at position x' by n intervening specular reflections.

The integral equation for radiosity $R(x)$ is obtained from Eq. 5-15c as

$$R(x) = q + (1 - \rho^s)\left[\bar{\sigma}T_1^{\,4}F^s_{dx-a,x} + \bar{\sigma}T_2^{\,4}F^s_{dx-a,(L-x)} + \int_{x'=0}^{L} R(x')\,dF^s_{dx-dx',(x'-x)}\right] \tag{5-86}$$

where $F^s_{dx-a,x}$ = specular view factor from the band (a, dx) at x to the opening at $x = 0$ at a distance x,

$F^s_{dx-a,(L-x)}$ = specular view factor from the band (a, dx) at x to the opening at $x = L$ at a distance $L - x$,

$dF^s_{dx-dx',(x'-x)}$ = elemental specular view factor from the band (a, dx) to the band (a, dx') at a distance $x' - x$ apart.

The distribution of temperature $T(x)$ at the cylindrical surface is related to the radiosity $R(x)$ by Eq. 5-15b as[2]

$$\bar{\sigma}T^4(x) = \frac{R(x)}{\varepsilon} \tag{5-87}$$

By eliminating $R(x)$ from Eq. 5-86 by means of Eq. 5-87 we obtain

$$\varepsilon\bar{\sigma}T^4(x) = q + \varepsilon\left[\bar{\sigma}T_1^{\,4}F^s_{dx-a,x} + \bar{\sigma}T_2^{\,4}F^s_{dx-a,(L-x)} + \varepsilon\int_{x'=0}^{L} \bar{\sigma}T^4(x')\,dF^s_{dx-dx',(x'-x)}\right] \tag{5-88}$$

where we have replaced $1 - \rho^s$ by ε. Equation 5-88 is an integral equation for the determination of the temperature distribution $T^4(x)$. Various specular view factors entering this equation can be evaluated as described below.

(a) The Specular View Factor $dF^s_{dx-dx',(x'-x)}$: The elemental specular view factor between the bands (a, dx) and (a, dx') at a distance $x' - x$ apart can be obtained from Eq. 3-100 as

$$dF^s_{dx-dx',(x'-x)} = dF_{dx-dx',(x'-x)} + \sum_{n=1}^{\infty} (\rho^s)^n \, dF_{dx-dx^*_n,(x^*_n-x)} \tag{5-89}$$

where $dF_{dx-dx',(x'-x)}$ = elemental diffuse view factor from the band (a, dx) to the band (a, dx') at a distance $x' - x$ apart,

$dF_{dx-dx^*_n,(x^*_n-x)}$ = elemental diffuse view factor from the band (a, dx) to the band (a, dx^*_n) at a distance $x^*_n - x$ apart,

x^*_n = axial position at the cylindrical surface such that the rays leaving the band (a, dx) and reaching the band (a, dx') with n intervening specular reflections make the first of these reflections at the band (a, dx^*_n) positioned at x^*_n.

The location of x^*_n can be determined by the following consideration. For a beam that leaves x and arrives at x' with n intervening specular reflections, the spacing between specular reflections is $|x' - x|/(n + 1)$. Then the location of x^*_n is at a distance $|x' - x|/(n + 1)$ away from x, as shown in Fig. 5-8.

The elemental diffuse view factors entering Eq. 5-89 can now be determined by utilizing Eq. 5-85 as follows:

$$dF_{dx-dx',(x'-x)} = \left[1 - \left| \frac{x' - x}{2a} \right| \frac{\left(\dfrac{x' - x}{2a} \right)^2 + \dfrac{3}{2}}{\left[\left(\dfrac{x' - x}{2a} \right)^2 + 1 \right]^{3/2}} \right] d\left(\frac{x'}{2a} \right) \tag{5-90}$$

$$dF_{dx-dx^*_n,(x^*_n-x)} = \left[1 - \left| \frac{x' - x}{2(n+1)a} \right| \frac{\left(\dfrac{x' - x}{2(n+1)a} \right)^2 + \dfrac{3}{2}}{\left[\left(\dfrac{x' - x}{2(n+1)a} \right)^2 + 1 \right]^{3/2}} \right] d\left(\frac{x'}{2(n+1)a} \right) \tag{5-91}$$

(b) The Specular View Factor $F^s_{dx-a,x}$: The specular view factor from the band (a, dx) to the opening at $x = 0$ can be evaluated by summing the contributions from all reflections and can be given in the form

$$F^s_{dx-a,x} = F_{dx-a,x} + \sum_{n=1}^{\infty} (\rho^s)^n [F_{dx-a,x/(n+1)} - F_{dx-a,x/n}] \tag{5-92}$$

To prove the validity of this relation we consider the radiative energy from the band (a, dx) that leaves the tube through the opening at $x = 0$ as illustrated in Fig. 5-9. It is composed of the following components:

1. The fraction of the radiative energy from the band (a, dx) that leaves the opening directly (i.e., no reflections) is (see Fig. 5-9a)

$$F_{dx-a,x} \tag{5-93}$$

which is the diffuse view factor from the band (a, dx) to the opening at $x = 0$ at a distance x apart.

2. The fraction of the radiative energy from the band (a, dx) that leaves the opening at $x = 0$ after one specular reflection in the cylinder is

$$\rho^s[F_{dx-a,x/2} - F_{dx-a,x}] \tag{5-94}$$

The terms in the bracket represent the fraction of the radiative energy from the band (a, dx) that strikes the tube surface at the shaded region between x and $x/2$ illustrated in Fig. 5-9b.

3. The fraction of the radiative energy from the band (a, dx) that leaves the opening after two specular reflections in the cylinder is

$$(\rho^s)^2[F_{dx-a,x/3} - F_{dx-a,x/2}] \tag{5-95}$$

The terms in the bracket represent the fraction of the radiative energy from the band (a, dx) that strikes the tube surface at the shaded region between $x/2$ and $x/3$, as shown in Fig. 5-9c.

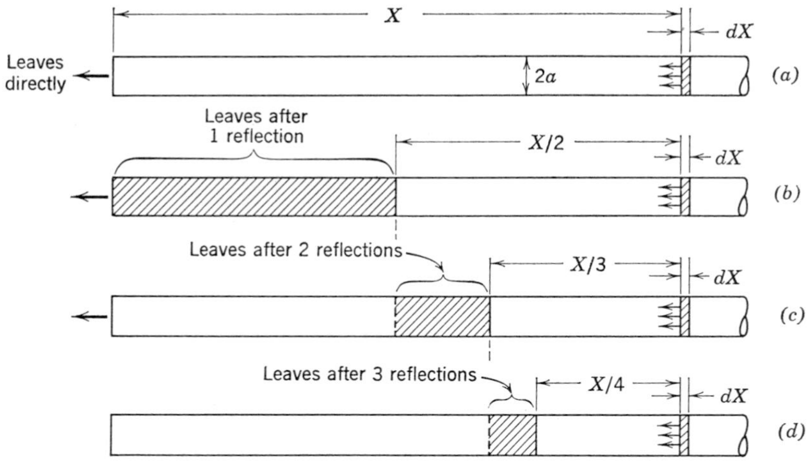

Fig. 5-9. Evaluation of specular view factor between the band dX and the opening.

4. Proceeding as above, we can give the fraction of the radiative energy from the band (a, dx) that leaves the opening after n specular reflections as

$$(\rho^s)^n [F_{dx-a,x/n+1} - F_{dx-a,x/n}] \tag{5-96}$$

The terms in the bracket represent the fraction of the radiative energy from the band (a, dx) that strikes the cylinder surface at the region between $x/(n + 1)$ and x/n.

Adding the components given by Eqs. 5-93 through 5-96, we obtain the specular view factor given by Eq. 5-92. The diffuse view factor between a band (a, dx) and a disk of radius a at a distance z apart is given as (see Eq. 5-84a)

$$F_{dx-a,z} = \frac{\frac{1}{2} + (z/2a)^2}{\sqrt{1 + (z/2a)^2}} - \frac{z}{2a} \tag{5-97}$$

(c) The Specular View Factor $F^s_{dx-a,(L-x)}$: The specular view factor from the band (a, dx) and the opening at $x = L$ at a distance $L - x$ apart is immediately obtained from Eq. 5-92 by replacing x in that equation by $L - x$. We obtain

$$F^s_{dx-a,(L-x)} = F_{dx-a,(L-x)} + \sum_{n=1}^{\infty} (\rho^s)^n [F_{dx-a,(L-x)/(n+1)} - F_{dx-a,(L-x)/n}] \tag{5-98}$$

Subdivision into Simpler Problems

The problem of radiative exchange described by Eq. 5-88 can be subdivided into simple problems and the complete solution obtained by superposition since Eq. 5-88 is linear in $\bar{\sigma}T^4$.

One of these simple problems is for radiative exchange in a finite cylinder with uniform wall heat flux q, and the ends open to an environment at zero temperature. In this case the temperature at the cylindrical surface $T^*(x)$ satisfies the following integral equation:

$$\varepsilon\bar{\sigma}T^{*4}(x) = q + \varepsilon^2 \int_{x'=0}^{L} \bar{\sigma}T^{*4}(x')\, dF^s_{dx-dx',(x'-x)} \tag{5-99}$$

The second of these simple problems is for radiative exchange in a finite cylinder with zero wall heat flux (i.e., $q = 0$) and with the environment at $x = 0$ kept at temperature $(T_1^4 - T_2^4)^{1/4}$, $T_1 > T_2$, and the environment at $x = L$ at zero temperature. The temperature at the cylindrical surface, $T^{**}(x)$, for this problem satisfies the following integral equation:

$$\varepsilon\bar{\sigma}T^{**4}(x) = \varepsilon[\bar{\sigma}T_1^4 - \bar{\sigma}T_2^4]F^s_{dx-a,x} + \varepsilon^2 \int_{x'=0}^{L} \bar{\sigma}T^{**4}(x')\, dF^s_{dx-dx',(x'-x)} \quad \text{for } T_1 > T_2 \tag{5-100}$$

The third of these simple problems is for a finite cylinder with zero wall heat flux (i.e., $q = 0$) and with the ends at $x = 0$ and $x = L$ open to an environment kept at temperature T_2. In this case the cylinder surface must be at the same temperature as the environment. Hence the temperature of the cylindrical surface is equal to T_2.

The solution of the problem described by Eq. 5-88 is obtained by the superposition of the temperatures for the three simple problems described above, that is,

$$T^4(x) = T^{*4}(x) + T^{**4}(x) + T_2^4 \tag{5-101}$$

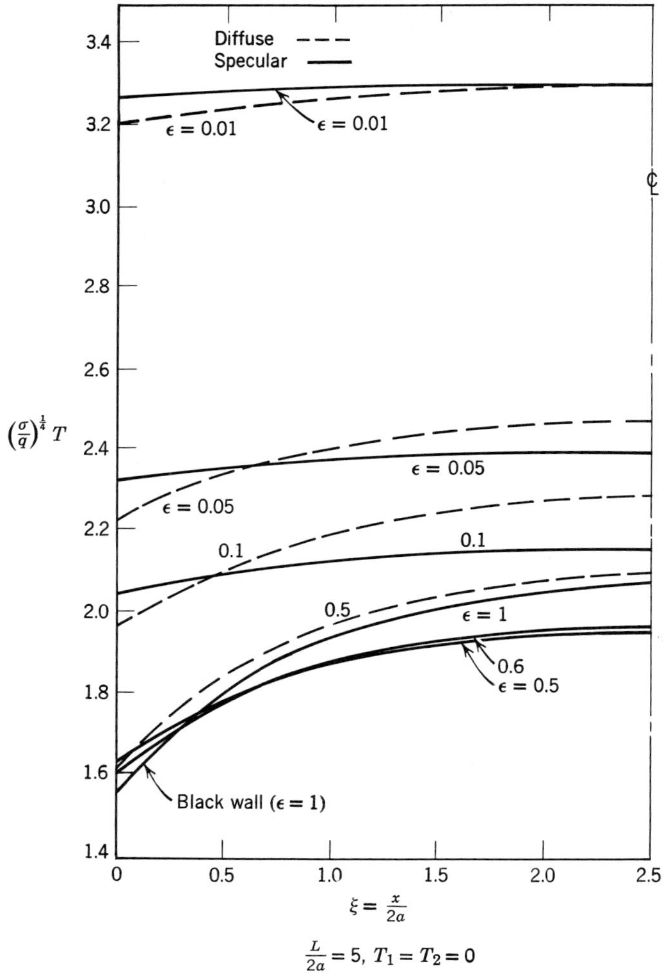

Fig. 5-10a

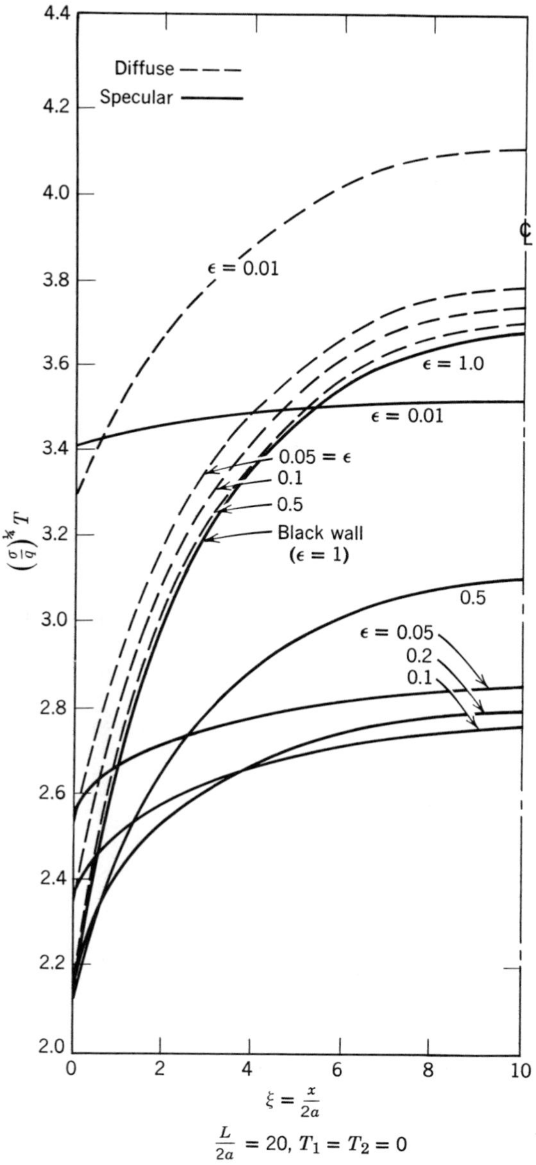

Fig. 5-10b. Effects of specular reflection on wall temperature for a cylindrical enclosure with uniform wall heat flux. (From M. Perlmutter and R. Siegel [6].)

The validity of this result can be proved by the superposition of these three simple problems.

Results

Perlmutter and Siegel [6] solved numerically the integral equations characterizing the simple problems. Figure 5-10 shows a comparison of the temperature distribution with purely specularly reflecting and purely diffuse reflecting walls for a cylinder with uniformly applied wall heat flux q, when the external environments are kept at zero temperature (i.e., $T_1 = T_2 = 0$). For a purely diffusely reflecting surface the wall temperature increases monotonically as the emissivity decreases from that of a black wall ($\varepsilon = 1$) to a highly reflecting case ($\varepsilon = 0.001$). However, for a purely specularly reflecting surface the wall temperature goes through a minimum for emissivities in the range 0.1 to 0.5. The reason for the minimum may be explained as follows. The outer surface of the tube being insulated, all the heat supplied through the walls must be dissipated by radiation through the two end openings. The temperature at the tube center will be minimized when the optimum dissipation rate is achieved between this part of the tube and the end openings.

REFERENCES

1. R. Courant and D. Hilbert, *Methods of Mathematical Physics*, 1st English ed., Vol. 1, Interscience Publishers, New York, 1953.
2. F. B. Hildebrand, *Methods of Applied Mathematics*, Prentice-Hall, Englewood Cliffs, N.J., 1952.
3. W. V. Lovitt, *Linear Integral Equations*, McGraw-Hill Book Co., New York, 1924.
4. E. M. Sparrow, "Application of Variational Methods to Radiation Heat Transfer Calculations," *J. Heat Transfer*, **82C**, 375–380, 1960.
5. C. M. Usiskin and R. Siegel, "Thermal Radiation from a Cylindrical Enclosure with Specified Heat Flux," *J. Heat Transfer*, **82C**, 369–374, 1960.
6. M. Perlmutter and R. Siegel, "Effects of Specularly-Reflecting Gray Surface on Thermal Radiation through a Tube and from Its Heated Wall," *J. Heat Transfer*, **85C**, 55–62, 1963.
7. E. M. Sparrow and A. Haji-Sheikh, "A Generalized Variational Method for Calculating Radiant Interchange Between Surfaces," *J. Heat Transfer*, **87C**, 103–109, 1965.
8. H. Buckley, "On the Radiation from Inside of a Circular Cylinder," *Phil. Mag.*, **4**, 753–762, 1927.
9. E. M. Sparrow and V. K. Jonsson, "Thermal Radiation Absorption in Rectangular-Groove Cavities," *J. Appl. Mech.*, **30E**, 237–244, 1963.
10. E. M. Sparrow, J. L. Gregg, J. V. Szel, and P. Manos, "Analysis, Results, and Interpretation for Radiation Between Simply Arranged Gray Surfaces," *J. Heat Transfer*, **83C**, 207–214, 1961.
11. J. W. T. Walsh, "Radiation from a Perfectly Diffusing Circular Disc," *Phys. Soc. (London)*, **32**, 59–71, 1919–1920.

12. E. M. Sparrow and J. L. Gregg, "Radiant Interchange Between Circular Discs Having Arbitrarily Different Temperatures," *J. Heat Transfer*, **83C,** 494–502, 1961.

13. E. M. Sparrow and L. U. Albers, "Apparent Emissivity and Heat Transfer in a Long Cylindrical Hole," *J. Heat Transfer*, **82C,** 253–255, 1960.

14. E. M. Sparrow, L. U. Albers, and E. R. G. Eckert, "Thermal Radiation Characteristics of Cylindrical Enclosures," *J. Heat Transfer*, **84C,** 73–81, 1962.

15. K. S. Krishnan, "Effects of Specular Reflections on the Radiation Flux from a Heated Tube," *Nature*, **187,** 135, 1960.

16. K. S. Krishnan, "Effects of Specular Reflections on the Radiation Flux from a Heated Tube," *Nature*, **188,** 652–653, 1960.

NOTES

[1] An error in Equation 7 of Krishnan's [15] work was subsequently corrected by Krishnan [16].

[2] From Eq. 5-15b we write

$$\bar{\sigma}T^4(x) = \frac{\rho^d q + (1-\rho)R(x)}{\varepsilon(1-\rho^s)} \tag{1}$$

For the problem considered we have $\rho^d = 0$ and $\rho = \rho^s$. Then this relation simplifies to

$$\bar{\sigma}T^4(x) = \frac{R(x)}{\varepsilon} \tag{2}$$

CHAPTER 6

Radiation and Conduction in Nonparticipating Media

Heat transfer by radiation has been an important means of disposing of waste heat in outer-space applications. For example, in space vehicles the waste heat from the power plant, electronic equipment, and various component is carried by a coolant fluid to space radiators, where it is transported by conduction to the fin surface and then by thermal radiation into the atmosphere-free space. Since radiators are probably among the heaviest components in the cooling system of a space vehicle, it is desirable to choose fin geometries that augment heat transfer by radiation and to determine accurately the heat transfer characteristics of the radiator in order to minimize the weight. Figure 6-1 shows typical radiators for space-vehicle applications; the reader should refer to the publications by Callinan and Berggren [1] and Fraas and Özışık [2, pp. 215–222] for discussion of the wide range of design problems associated with radiators for space vehicles. The basic mechanism of heat transfer in a space radiator is conduction combined with radiation in a nonparticipating medium, and the heat transfer characteristics of simple, one-dimensional, radiating fins have been studied extensively [3–14]. For fin geometries such as those shown in Figs. 6-1c and 6-1*d* the radiative heat exchange between the fin surface and the fin base is negligible; hence the analysis is relatively simple. However, for fin geometries such as those shown in Figs. 6-1*a*, 6-1*b*, and 6-1*e* the radiative heat exchange between the fin surface and the fin base needs to be included in the analysis thus making

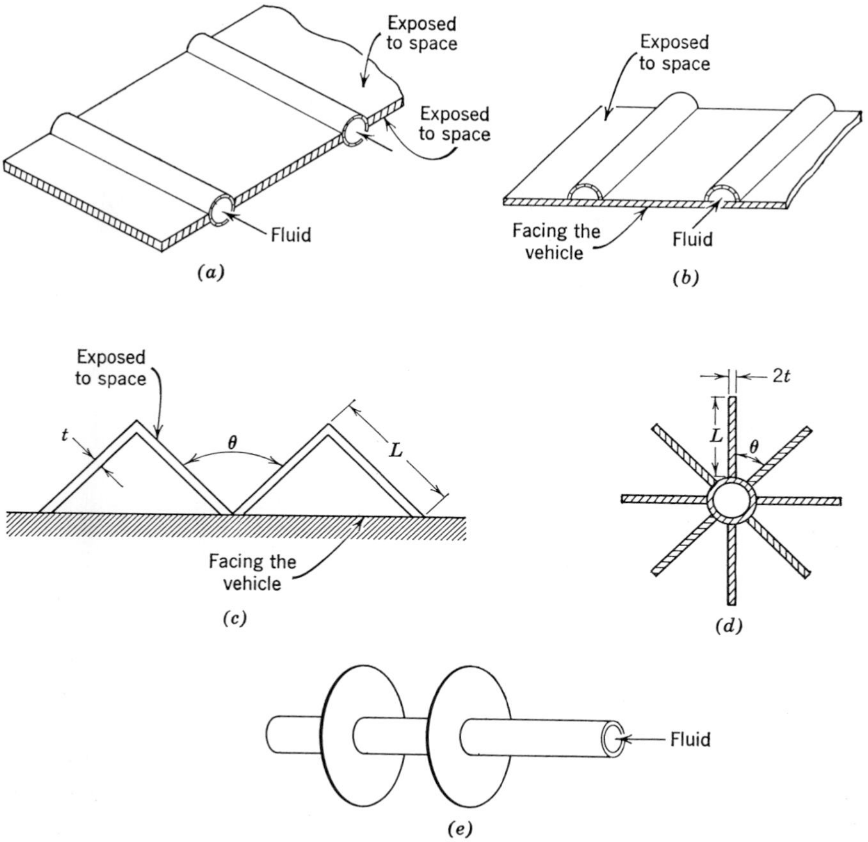

Fig. 6-1. Typical configurations for radiators for space vehicles.

the treatment of the problem more difficult computationally. The optimization of fin weight is also important in engineering applications; this problem has been considered by some of the investigators who have studied the heat transfer characteristics of the radiating extended surfaces.

In this chapter we demonstrate with specific examples the formulation of the heat transfer problem for radiating fins, and discuss the methods of solution and the heat transfer results.

6-1 RADIATING LONGITUDINAL PLATE FINS

Consideration is now given to the analysis of heat transfer by simultaneous radiation and conduction from a longitudinal plate fin geometry, shown in

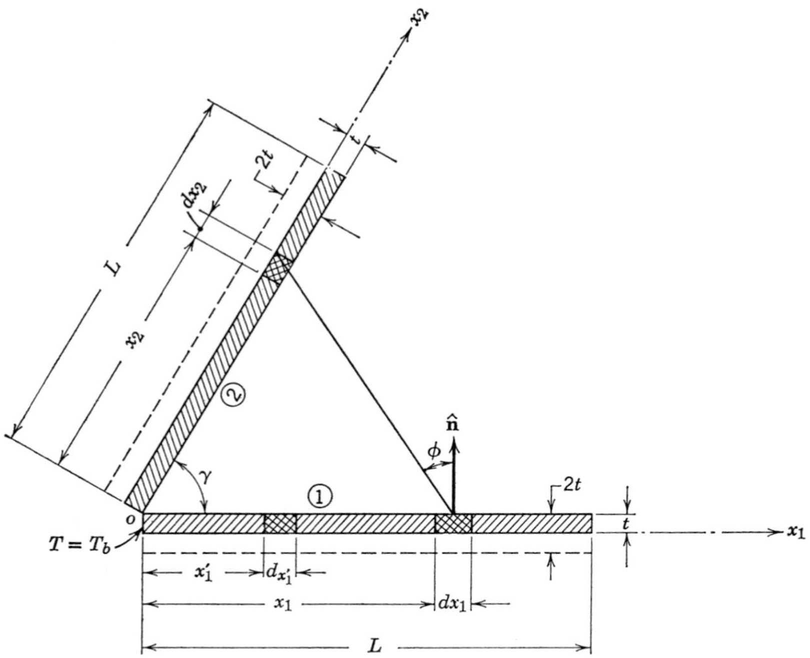

Fig. 6-2. Longitudinal plate fin.

Fig. 6-2. This arrangement characterizes the fin configuration of Fig. 6-1*d* for a fin plate thickness $2t$. To formulate the problem the following assumptions are made.

1. The dimension of the plate normal to the plane of the figure is sufficiently long so that there is no temperature variation in that direction.
2. $L \gg t$ so that the temperature in the plate is a function of the axial coordinate only, that is, $T_1(x_1)$ and $T_2(x_2)$.
3. The radiative energy incident on the fin surface from the external environment is negligible.
4. The temperature at the fin base is uniform throughout, that is, $T_1(0) = T_2(0) = T_b$.
5. Heat loss from the fin tips is negligible, that is, $dT_1(L)/dx_1 = 0$ and $dT_2(L)/dx_2 = 0$.
6. The surfaces are opaque, gray, diffuse emitters and have uniform emissivity ε. The thermal conductivity k is uniform through the plates.
7. Kirchhoff's law is applicable.
8. The surfaces are diffuse reflectors.

The steady-state energy balance equation for a differential volume element of the fin can be stated as

$$\text{(Net conductive heat gains)} + \text{(Net radiative heat gains)} = 0 \tag{6-1a}$$

which can be written more compactly as

$$dQ^c + dQ^r = 0 \tag{6-1b}$$

Let w be the width of the plates normal to the plane of Fig. 6-2. The conductive net gain in plate 1 for a volume element $tw\,dx$ is given as

$$dQ^c = ktw\,dx_1 \frac{d^2T_1}{dx_1^{\,2}} \tag{6-2a}$$

and the radiative heat gain as

$$dQ^r = -dx_1\,wq_1{}^r(x_1) \tag{6-2b}$$

since $w \gg t$.

The minus sign is used in Eq. 6-2b because $q_1{}^r(x_1)$ represents the net radiative heat flux leaving the fin surface into the space. Substituting Eqs. 6-2 into Eq. 6-1b, we obtain the energy equation as

$$\frac{d^2T_1(x_1)}{dx_1^{\,2}} = \frac{1}{kt}\,q_1{}^r(x_1) \tag{6-3}$$

The net radiative heat flux $q_1{}^r(x_1)$ can be determined from the generalized zone analysis (Eq. 5-10a) as[1]

$$q_1{}^r(x_1) = R_1(x_1) - \int_{x_2=0}^{L} R_2(x_2)\,dF_{dx_1-dx_2} \tag{6-4}$$

Substitution of Eq. 6-4 into Eq. 6-3 yields the following integrodifferential equation for the temperature distribution $T_1(x_1)$ in plate 1:

$$\frac{d^2T_1(x_1)}{dx_1^{\,2}} = \frac{1}{kt}\left[R_1(x_1) - \int_{x_2=0}^{L} R_2(x_2)\,dF_{dx_1-dx_2}\right] \quad \text{in } 0 \le x_1 \le L \tag{6-5}$$

with boundary conditions obtained from assumptions 4 and 5 above as

$$T_1(x_1) = T_b \quad \text{at } x_1 = 0 \tag{6-6a}$$

$$\frac{dT_1(x_1)}{dx_1} = 0 \quad \text{at } x_1 = L \tag{6-6b}$$

The equation for radiosity $R_1(x_1)$ is obtained from Eq. 5-9 as

$$R_1(x_1) = \varepsilon\bar{\sigma}T_1^{\,4}(x_1) + (1-\varepsilon)\int_{x_2=0}^{L} R_2(x_2)\,dF_{dx_1-dx_2} \tag{6-7}$$

where ρ is replaced by $1 - \varepsilon$ by assumption 7.

A set of relations similar to Eqs. 6-5 through 6-7 can be written for the temperature distribution $T_2(x_2)$ and the radiosity $R_2(x_2)$ at plate 2. However, these additional relations for plate 2 are not needed because the problem possesses symmetry, that is, $R_1(x_1) = R_2(x_2)$ and $T_1(x_1) = T_2(x_2)$ for $x_1 = x_2$. Then subscripts 1 and 2 can be removed from the radiosity and temperature functions in Eqs. 6-5 through 6-7, and the resulting equations can be written in the dimensionless form as

$$\frac{d^2\theta(\xi_1)}{d\xi_1^{\,2}} = \frac{1}{N_c}\left[\beta(\xi_1) - \int_{\xi_2=0}^{1} \beta(\xi_2)\, dF_{d\xi_1-d\xi_2}\right] \quad \text{in } 0 \le \xi_1 \le 1 \tag{6-8}$$

$$\theta(\xi_1) = 1 \quad \text{at } \xi_1 = 0 \tag{6-9}$$

$$\frac{d\theta(\xi_1)}{d\xi_1} = 0 \quad \text{at } \xi_1 = 1 \tag{6-10}$$

and

$$\beta(\xi_1) = \varepsilon\theta^4(\xi_1) + (1-\varepsilon)\int_{\xi_2=0}^{1} \beta(\xi_2)\, dF_{d\xi_1-d\xi_2} \tag{6-11}$$

where the dimensionless quantities have been defined as

$$\theta \equiv \frac{T}{T_b}, \qquad \beta \equiv \frac{R}{\bar{\sigma}T_b^{\,4}}, \qquad N_c \equiv \frac{kt}{L^2\bar{\sigma}T_b^{\,3}}, \qquad \xi_1 \equiv \frac{x_1}{L}, \qquad \text{and} \qquad \xi_2 \equiv \frac{x_2}{L} \tag{6-12}$$

The quantity N_c is called the conduction-to-radiation parameter;[2] it is a measure of the importance of conduction as compared with radiation.

The large values of N_c correspond to the conduction-dominant case, and small values to the radiation-dominant case. When $N_c \to \infty$, Eqs. 6-8 simplify to the case of pure conduction.

The elemental diffuse view factor $dF_{d\xi_1-d\xi_2}$ represents the view factor from a strip $d\xi_1$ on plate 1 to the strip $d\xi_2$ on plate 2 and can be determined by means of the relation given by Eq. 3-53; that is,

$$dF_{d\xi_1-d\xi_2} = \frac{1}{2}\, d(\sin\phi) \tag{6-13}$$

where ϕ is the angle between the normal to the strip $d\xi_1$ and the straight line joining the strips $d\xi_1$ and $d\xi_2$ as shown in Fig. 6-2, and is given by

$$\begin{aligned}\sin\phi &= \frac{x_1 - x_2\cos\gamma}{[(x_1 - x_2\cos\gamma)^2 + (x_2\sin\gamma)^2]^{1/2}} \\ &= \frac{x_1 - x_2\cos\gamma}{(x_1^{\,2} - 2x_1x_2\cos\gamma + x_2^{\,2})^{1/2}}\end{aligned} \tag{6-14}$$

Then

$$dF_{dx_1-dx_2} = \frac{1}{2} \frac{x_1 x_2 \sin^2 \gamma}{(x_1^2 - 2x_1 x_2 \cos \gamma + x_2^2)^{3/2}} dx_2 \tag{6-15a}$$

or

$$dF_{d\xi_1-d\xi_2} = \frac{1}{2} \frac{\xi_1 \xi_2 \sin^2 \gamma}{(\xi_1^2 - 2\xi_1 \xi_2 \cos \gamma + \xi_2^2)^{3/2}} d\xi_2 \tag{6-15b}$$

Once Eqs. 6-8 through 6-11 have been solved and the dimensionless radiosity function $\beta(\xi_1)$ has been determined, the distribution of the net radiative heat flux on the surface of the fin is given by Eq. 6-4, can be written in the dimensionless form as

$$\frac{q^r(\xi_1)}{\bar{\sigma} T_b^4} = \beta(\xi_1) - \int_{\xi_2=0}^{1} \beta(\xi_2)\, dF_{d\xi_1-d\xi_2} \tag{6-16}$$

The net rate of heat dissipation by radiation Q^r, from the surface of one of the fin plates per unit width normal to the plane of Fig. 6-2, is given by

$$Q^r = \int_{x_1=0}^{L} q^r(x_1)\, dx_1$$

or

$$\frac{Q^r}{\bar{\sigma} T_b^4} = L \int_{\xi_1=0}^{1} \left[\beta(\xi_1) - \int_{\xi_2=0}^{1} \beta(\xi_2)\, dF_{d\xi_1-d\xi_2} \right] d\xi_1 \tag{6-17b}$$

Now we consider an ideal situation in which the fin surfaces are black ($\varepsilon = 1$) and are maintained at a uniform temperature T_b everywhere. The rate of ideal heat dissipation by radiation from the surface of one of the fin plates per unit width normal to the plane of Fig. 6-2 is given as

$$Q^r_{\text{ideal}} = \bar{\sigma} T_b^4 \left[L \sin \frac{\gamma}{2} \right] \tag{6-18}$$

Then the *radiative effectiveness* η is defined as

$$\eta \equiv \frac{Q^r}{Q^r_{\text{ideal}}} = \frac{1}{\sin(\gamma/2)} \int_{\xi_1=0}^{1} \left[\beta(\xi_1) - \int_{\xi_2=0}^{1} \beta(\xi_2)\, dF_{d\xi_1-d\xi_2} \right] d\xi_1 \tag{6-19}$$

Discussion of Results

Equations 6-8 and 6-11 are two coupled integrodifferential equations which must be solved simultaneously for the unknowns, $\theta(\xi_1)$ and $\beta(\xi_1)$. It is highly unlikely that analytical solutions can be found to such a system, but the equations can be solved numerically by an iterative scheme, using a high-speed digital computer for prescribed values of the parameters ε, γ, and N_c.

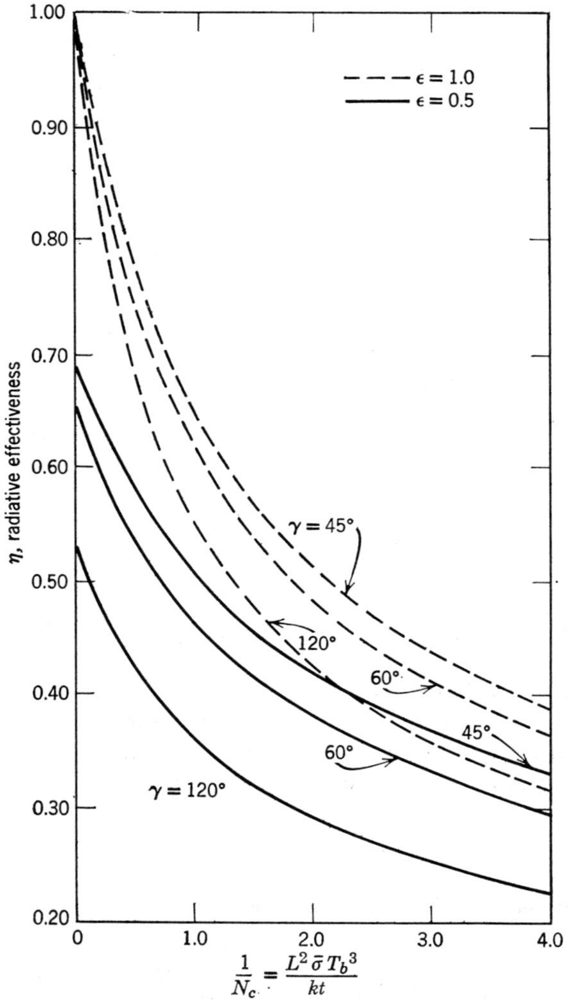

Fig. 6-3. Radiative effectiveness for longitudinal plate fins. (From E. M. Sparrow, E. R. G. Eckert, and T. F. Irvine [7].)

Figure 6-3 shows the radiative effectiveness of the fin calculated by Sparrow, Eckert, and Irvine [7], plotted as a function of the conduction-to-radiation parameter N_c for $\varepsilon = 1.0$ and 0.5, and for several different values of the opening angle γ. The curves for $\varepsilon = 1.0$ converge to the greatest possible heat loss value as $N_c \to \infty$ (i.e., as thermal conductivity becomes very high). However, for $\varepsilon = 0.5$ this ideal heat transfer is not reached as $N_c \to \infty$ because the fin surfaces are nonblack. Fin effectiveness decreases as N_c

decreases (i.e., radiation effects increase). For any given N_c the fin effectiveness is always greater for smaller opening angles.

Knowing the radiative effectiveness, we can determine the net radiative heat transfer Q^r from one surface of the fin per unit width normal to the plane of the figure from Eqs. 6-18 and 6-19 as

$$Q^r = \eta Q^r_{\text{ideal}} = \eta \bar{\sigma} T_b^{\,4} L \sin \frac{\gamma}{2} \tag{6-20}$$

The addition of fins to a radiator surface increases not only the heat transfer area but also the radiator weight. Therefore it is important to determine the conditions that will yield maximum heat dissipation for given values of ε and γ and a fixed profile, that is, $A = Lt =$ constant.

When the values of ε and γ are fixed, the radiative effectiveness is a function of the conduction-to-radiation parameter N_c only, that is, $\eta \equiv \eta(N_c)$. Then Eq. 6-20 can be written as

$$Q^r = \bar{\sigma} T_b^{\,4} \frac{A}{t} \eta(N_c) \sin \frac{\gamma}{2} \tag{6-21}$$

since $A = Lt$, and N_c can be related to t by

$$N_c = \frac{kt^3}{A^2 \bar{\sigma} T_b^{\,3}} \tag{6-22}$$

In Eq. 6-21 the fin thickness t is the only variable. To maximize Q^r we differentiate Eq. 6-21 with respect to t and equate the resulting expression to zero:

$$\frac{\partial Q^r}{\partial t} = \bar{\sigma} T_b^{\,4} A \sin \frac{\gamma}{2} \frac{\partial}{\partial t}\left[\frac{\eta(N_c)}{t}\right] = 0 \tag{6-23a}$$

or

$$-\frac{1}{t^2}\eta(N_c) + \frac{1}{t}\frac{d\eta(N_c)}{dN_c}\frac{\partial N_c}{\partial t} = 0 \tag{6-23b}$$

Differentiation of Eq. 6-22 with respect to t yields

$$\frac{\partial N_c}{\partial t} = \frac{3kt^2}{A^2 \bar{\sigma} T_b^{\,3}} = \frac{3N_c}{t} \tag{6-24}$$

Eliminating $\partial N_c/\partial t$ between Eqs. 6-23b and 6-24, we obtain

$$(N_c)_{\text{opt.}} = \frac{1}{3}\frac{\eta}{(\partial \eta/\partial N_c)} \tag{6-25}$$

where $(N_c)_{\text{opt.}}$ is the value of the conduction-to-radiation parameter that will give maximum heat transfer Q^r for given values of ε and γ.

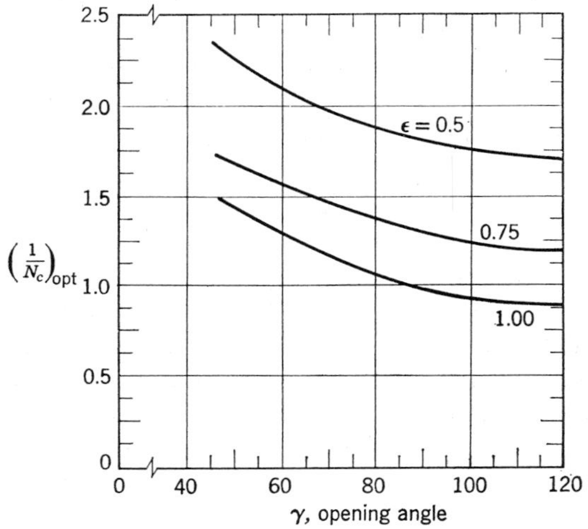

Fig. 6-4. Optimum values of N_c corresponding to maximum heat transfer rate from fin surface. (From E. M. Sparrow, E. R. G. Eckert, and T. F. Irvine [7].)

Figure 6-4 shows the values of $(N_c)_{\text{opt.}}$ as a function of the opening angle γ for three different values of emissivity. The optimum value of $1/N_c$ is seen to decrease with increasing opening and increasing surface emissivity. This means that, for a given thickness, optimum performance is realized for shorter fin heights as the opening angle or the emissivity increases.

6-2 EFFECTS OF SPECULAR REFLECTION ON RADIATIVE HEAT TRANSFER

To illustrate the effects of specular reflection on radiative heat transfer we consider here a problem similar to that described in Section 6-1 but involving purely specularly reflecting surfaces. Assumptions 1 through 7 of Section 6-1 remain unchanged, but assumption 8 is replaced by the assumption that the surfaces are purely specular reflectors. The configuration of Fig. 6-2 is applicable for the problem.

The energy equation for plate 1 is given as

$$\frac{d^2T_1(x_1)}{dx_1^{\,2}} = \frac{1}{kt}\, q_1^{\,r}(x_1) \tag{6-26}$$

The relation for the net radiative heat flux $q_1^r(x_1)$ at plate 1 for specular reflection is obtained from Eq. 5-15c as

$$q_1^r(x_1) = R_1(x_1) - (1 - \rho^s)\int_{x_2=0}^{L} R_2(x_2)\, dF^s_{dx_1-dx_2} - (1 - \rho^s)\int_{x_1'=0}^{L} R_1(x_1')\, dF^s_{dx_1-dx_1'} \quad (6\text{-}27)$$

For diffusely reflecting surfaces $\rho^s = 0$; then Eq. 6-27 simplifies to Eq. 6-4 since the specular view factors reduce to diffuse view factors and $dF_{dx_1-dx_1'}$ vanishes for a flat plate.

Substituting Eq. 6-27 into Eq. 6-26 and replacing $1 - \rho^s$ by ε, we obtain

$$\frac{d^2T_1(x_1)}{dx_1^{\,2}} = \frac{1}{kt}\left[R_1(x_1) - \varepsilon\int_{x_2=0}^{L} R_2(x_2)\, dF^s_{dx_1-dx_2} - \varepsilon\int_{x_1'=0}^{L} R_1(x_1')\, dF^s_{dx_1-dx_1'}\right] \quad \text{in } 0 \le x \le L \quad (6\text{-}28a)$$

with the boundary conditions

$$T_1(x_1) = T_b \quad \text{at } x_1 = 0 \quad (6\text{-}28b)$$

$$\frac{dT_1(x_1)}{dx_1} = 0 \quad \text{at } x_1 = L \quad (6\text{-}28c)$$

The relation for the radiosity $R_1(x_1)$ is obtained from Eq. 5-14 as

$$R_1(x_1) = \varepsilon\bar{\sigma}T_1^{\,4}(x_1) \quad (6\text{-}29)$$

since $\rho^d = 0$ for purely specularly reflecting surfaces. Relations similar to Eqs. 6-28 and 6-29 can be written for plate 2, but they are not needed in the present problem by symmetry considerations, that is, $R_1(x_1) = R_2(x_2)$ and $T_1(x_1) = T_2(x_2)$ for $x_1 = x_2$. Then subscripts 1 and 2 can be removed from the temperature and radiosity functions in Eqs. 6-28 and 6-29, and the resulting equations for temperature can be written in the dimensionless form as

$$\frac{d^2\theta(\xi_1)}{d\xi_1^{\,2}} = \frac{\varepsilon}{N_c}\left[\theta^4(\xi_1) - \varepsilon\int_{\xi_2=0}^{1} \theta^4(\xi_2)\, dF^s_{d\xi_1-d\xi_2} - \varepsilon\int_{\xi_1'=0}^{1} \theta^4(\xi_1')\, dF^s_{d\xi_1-d\xi_1'}\right] \quad \text{in } 0 \le \xi_1 \le 1 \quad (6\text{-}30)$$

with the boundary conditions

$$\theta(\xi_1) = 1 \quad \text{at } \xi_1 = 0 \quad (6\text{-}31a)$$

$$\frac{d\theta(\xi_1)}{d\xi_1} = 0 \quad \text{at } \xi_1 = 1 \quad (6\text{-}31b)$$

where we have defined

$$\theta = \frac{T}{T_b}, \qquad N_c = \frac{kt}{L^2\bar{\sigma}T_b^{\,3}}, \qquad \xi_1 = \frac{x_1}{L}, \qquad \xi_2 = \frac{x_2}{L}, \qquad \xi_1 = \frac{x_1'}{L} \quad (6\text{-}32)$$

The specular view factors in Eq. 6-30 can be determined by the method of image formation as discussed in Chapter 3. We describe now the evaluation of these specular view factors for certain values of the opening angle γ.

$90° \leq \gamma < 180°$: The specular view factor $dF^s_{d\xi_1-d\xi_2}$ is composed of the directly transported component only, because all specular reflections will leave through the opening without reaching strip $d\xi_2$. Hence

$$dF^s_{d\xi_1-d\xi_2} = dF_{d\xi_1-d\xi_2,\gamma} \quad \text{for } 90° \leq \gamma < 180° \tag{6-33}$$

where $dF_{d\xi_1-d\xi_2,\gamma}$ is the diffuse view factor from strip $d\xi_1$ to strip $d\xi_2$ for an opening angle γ.

The specular view factor $dF^s_{d\xi_1-d\xi_1'}$ is also zero because none of the radiation leaving strip $d\xi_1$ will reach back to plate 1 after a specular reflection.

$60° \leq \gamma < 90°$: The specular view factor $dF^s_{d\xi_1-d\xi_2}$ is equal to the directly transported component only, since all the specularly reflected components will leave through the opening without reaching strip $d\xi_2$. Hence

$$dF^s_{d\xi_1-d\xi_2} = dF_{d\xi_1-d\xi_2,\gamma} \quad \text{for } 60° \leq \gamma < 90° \tag{6-34}$$

The specular view factor $dF^s_{d\xi_1-d\xi_1'}$ is evaluated according to the rules of image formation and is given as

$$dF^s_{d\xi_1-d\xi'} = \rho^s \, dF_{d\xi_1(2)-d\xi_1,2\gamma} \quad \text{for } 60° \leq \gamma < 90° \tag{6-35}$$

where $d\xi_1(2)$ is the mirror image of $d\xi_1$ through plate 2; hence $dF_{d\xi_1(2)-d\xi_1',2\gamma}$ is the diffuse view factor from image strip $d\xi_1(2)$ to strip $d\xi_1'$ on plate 1 for an opening angle of 2γ. No other terms are included in Eq. 6-35 because further specular reflections leave through the opening without reaching plate 1.

$45° \leq \gamma < 60°$: The components of $dF^s_{d\xi_1-d\xi_2}$ include the radiative energy directly transported from strip $d\xi_1$ to $d\xi_2$, and the radiative energy from $d\xi_1$ that reaches $d\xi_2$ after consecutive specular reflections at plates 2 and 1. Hence

$$dF^s_{d\xi_1-d\xi_2} = dF_{d\xi_1-d\xi_2,\gamma} + (\rho^s)^2 \, dF_{d\xi_1(2-1)-d\xi_2,3\gamma} \quad \text{for } 45° \leq \gamma < 60° \tag{6-36}$$

where $d\xi_1(2-1)$ is the image of $d\xi_1$ after consecutive specular reflections at plates 2 and 1; hence $dF_{d\xi_1(2-1)-d\xi_2,3\gamma}$ is the diffuse view factor from image strip $d\xi_1(2-1)$ to strip $d\xi_2$ for an opening angle 3γ.

The view factor $dF^s_{d\xi_1-d\xi_1'}$ is given as

$$dF^s_{d\xi_1-d\xi_1} = \rho^s \, dF_{d\xi_1(2)-d\xi_1,2\gamma} \quad \text{for } 45° \leq \gamma < 60° \tag{6-37}$$

The specular view factors for smaller opening angles can be evaluated in a similar manner.

The diffuse view factor from strip $d\xi_1$ to strip $d\xi_2$ for an opening angle of $n\gamma$ is obtained from Eq. 6-15b as

$$dF_{d\xi_1-d\xi_2,n\gamma} = \frac{1}{2} \frac{\xi_1\xi_2 \sin^2 n\gamma}{(\xi_1^2 + \xi_2^2 - \xi_1\xi_2 \cos n\gamma)^{3/2}} \, d\xi_2, \quad n = 1, 2, 3, \ldots \tag{6-38}$$

Substituting the above specular view factors into Eqs. 6-30, we can write the resulting equations in the forms

$$\frac{d^2\theta(\xi_1)}{d\xi_1^{\,2}} = \frac{\varepsilon}{N_c}\left[\theta^4(\xi_1) - \varepsilon\int_{\xi_2=0}^{1}\theta^4(\xi_2)G_\gamma(\xi_1, \xi_2)\,d\xi_2\right] \quad \text{in } 0 \le \xi_1 \le 1 \qquad \text{(6-39a)}$$

$$\theta(\xi_1) = 1 \quad \text{at } \xi_1 = 0 \qquad \text{(6-39b)}$$

$$\frac{d\theta(\xi_1)}{d\xi_1} = 0 \quad \text{at } \xi_1 = 1 \qquad \text{(6-39c)}$$

where

$$G_\gamma(\xi_1, \xi_2) \equiv \begin{cases} f_\gamma(\xi_1, \xi_2) & \dfrac{180^\circ}{2} \le \gamma < 180^\circ \quad \text{(6-40a)} \\ f_\gamma(\xi_1, \xi_2) + \rho^s f_{2\gamma}(\xi_1, \xi_2) & \dfrac{180^\circ}{3} \le \gamma < \dfrac{180^\circ}{2} \quad \text{(6-40b)} \\ f_\gamma(\xi_1, \xi_2) + \rho_s f_{2\gamma}(\xi_1, \xi_2) & \dfrac{180^\circ}{4} \le \gamma < \dfrac{180^\circ}{3} \quad \text{(6-40c)} \\ \qquad + (\rho^s)^2 f_{3\gamma}(\xi_1, \xi_2) & \\ \vdots & \\ \sum_{n=1}^{N}(\rho_s)^{n-1} f_{n\gamma}(\xi_1, \xi_2) & \dfrac{180^\circ}{N+1} \le \gamma < \dfrac{180^\circ}{N} \quad \text{(6-40d)} \end{cases}$$

$$f_{n\gamma}(\xi_1, \xi_2) = \frac{1}{2}\,\frac{\xi_1\xi_2 \sin^2 n\gamma}{(\xi_1^{\,2} + \xi_2^{\,2} - 2\xi_1\xi_2 \cos n\gamma)^{3/2}} \qquad \text{(6-40e)}$$

and

$$\rho^s = 1 - \varepsilon \qquad \text{(6-40f)}$$

Once the temperature distribution has been determined from the solution of Eqs. 6-39, the distribution of net radiative heat flux over the fin surface can be evaluated from

$$\frac{q^r(\xi_1)}{\varepsilon\bar{\sigma}T_b^{\,4}} = \theta^4(\xi_1) - \varepsilon\int_{\xi_2=0}^{1}\theta^4(\xi_2)\,dF^s_{d\xi_1-d\xi_2} - \varepsilon\int_{\xi_1'=0}^{1}\theta^4(\xi_1')\,dF^s_{d\xi_1-d\xi_1'} \qquad \text{(6-41a)}$$

which is obtained by combining Eqs. 6-27 and 6-29. By utilizing the $G_\gamma(\xi_1, \xi_2)$ function as defined above, Eq. 6-41a can be written as

$$\frac{q^r(\xi_1)}{\varepsilon\bar{\sigma}T_b^{\,4}} = \theta^4(\xi_1) - \varepsilon\int_{\xi_2=0}^{1}\theta^4(\xi_2)G_\gamma(\xi_1, \xi_2)\,d\xi_2 \qquad \text{(6-41b)}$$

The total heat transfer from the surface of one plate per unit width can be evaluated from

$$Q^r = L\int_{\xi_1=0}^{1} q^r(\xi_1)\,d\xi_1 \qquad \text{(6-42)}$$

The radiative effectiveness η of the fin is determined from (see Eq. 6-20)

$$\eta = \frac{Q^r}{Q^r_{\text{ideal}}} = \frac{Q^r}{\bar{\sigma} T_b^4 L \sin(\gamma/2)} \tag{6-43}$$

A Numerical Solution Technique

The integrodifferential equation 6-39a can be transformed into an integral equation by integrating it twice and utilizing the boundary conditions 6-39b and 6-39c. The first integration of Eq. 6-39a from $\xi_1' = 1$ to ξ_1 and the application of the boundary condition 6-39c gives

$$\frac{d\theta(\xi_1)}{d\xi_1} = \frac{\varepsilon}{N_c}\left[\int_{\xi_1'=1}^{\xi_1} \theta^4(\xi_1')\,d\xi_1' - \varepsilon\int_{\xi_1'=1}^{\xi_1}\int_{\xi_2=0}^{1} \theta^4(\xi_2)G_\gamma(\xi_1', \xi_2)\,d\xi_2\,d\xi_1'\right] \tag{6-44}$$

and the integration from $\xi_1' = 0$ to ξ_1 and the application of the boundary condition 6-39b yields

$$\theta(\xi_1) = 1 + \frac{\varepsilon}{N_c}\left[\int_{\xi_1'=0}^{\xi_1}\int_{\xi_1''=1}^{\xi_1'} \theta^4(\xi_1'')\,d\xi_1''\,d\xi_1' - \varepsilon\int_{\xi_1'=0}^{\xi_1}\int_{\xi_1''=1}^{\xi_1'}\int_{\xi_2=0}^{1} \theta^4(\xi_2)G_\gamma(\xi_1'', \xi_2)\,d\xi_2\,d\xi_1''\,d\xi_1'\right] \tag{6-45}$$

Here the double- and the third-order integrals on the right-hand side of Eq. 6-45 are transformed into first-order integrals; then Eq. 6-45 simplifies to

$$\theta(\xi_1) = 1 + \frac{\varepsilon}{N_c}\left[\int_{\xi_1'=0}^{\xi_1} (\xi_1 - \xi_1')\theta^4(\xi_1')\,d\xi_1' - \int_{\xi_2=0}^{1} K_\gamma(\xi_1, \xi_2)\theta^4(\xi_2)\,d\xi_2\right] \tag{6-46}$$

where

$$K_\gamma(\xi_1, \xi_2) \equiv \begin{cases} \xi_1 + \varepsilon J_\gamma(\xi_1, \xi_2) & \dfrac{180^\circ}{2} \le \gamma < 180^\circ \\ \xi_1 + \varepsilon[J_\gamma(\xi_1, \xi_2) + \rho^s J_{2\gamma}(\xi_1, \xi_2)] & \dfrac{180^\circ}{3} \le \gamma < \dfrac{180^\circ}{2} \\ \xi_1 + \varepsilon[J_\gamma(\xi_1, \xi_2) + \rho^s J_{2\gamma}(\xi_1, \xi_2) + (\rho^s)^2 J_{3\gamma}(\xi_1, \xi_2)] & \dfrac{180^\circ}{4} \le \gamma < \dfrac{180^\circ}{3} \\ \cdot \\ \cdot \\ \cdot \end{cases} \tag{6-47}$$

$$J_{n\gamma}(\xi_1, \xi_2) \equiv \int_{\xi_1'=0}^{\xi_1}\int_{\xi_1''=1}^{\xi_1'} f_{n\gamma}(\xi_1'', \xi_2)\,d\xi_1''\,d\xi_1' \tag{6-48}$$

and the integral in Eq. 6-48 can be evaluated analytically.

Hering [8] solved integral equation 6-48 by an iterative scheme and determined the local net radiative heat flux, the total heat transfer from the fin surface, and the radiative effectiveness of the fin. In the process of numerical calculation of the net radiative heat flux from Eq. 6-41b difficulty may arise as the corner is approached because the kernel $G_\gamma(\xi_1, \xi_2)$ becomes indeterminant for $\xi_1 \to 0$, $\xi_2 \to 0$. This difficulty may be bypassed by taking the limiting value of the diffuse view factor by the physical argument employed in reference 7 or by the numerical techniques used in reference 14. Then, for $\xi_1 \to 0$, Eq. 6-41b simplifies to

$$\frac{q^r(0)}{\varepsilon\bar{\sigma}T_b^{\,4}} = 1 - \varepsilon L_\gamma(\varepsilon) \tag{6-49}$$

where

$$L_\gamma(\varepsilon) \equiv \begin{cases} \frac{1}{2}(1 + \cos\gamma) & \frac{180^\circ}{2} \le \gamma < 180^\circ \\ \frac{1}{2}(1 + \cos\gamma) + \rho^s(1 + \cos 2\gamma) & \frac{180^\circ}{3} \le \gamma < \frac{180^\circ}{2} \\ \frac{1}{2}(1 + \cos\gamma) + \rho^s(1 + \cos 2\gamma) + (\rho^s)^2(1 + \cos 3\gamma) & \frac{180^\circ}{4} \le \gamma < \frac{180^\circ}{3} \\ \cdot \\ \cdot \\ \cdot \end{cases} \tag{6-50}$$

The results of these calculations, presented in Fig. 6-5, show that the radiative effectiveness of specularly reflecting fins, when plotted as a function of the conduction-to-radiation parameter N_c, follows the same general trend as that for diffusely reflecting fins. However, the effectiveness of specularly reflecting fins is greater than that of diffusely reflecting fins and the increase is more pronounced at smaller values of the opening angle and at lower emissivities. The curves have their maximum value for pure conduction (i.e., $N_c \to \infty$); the effectiveness is reduced as N_c decreases (i.e., the radiation effect increases).

An Approximate Solution Technique

The Kármán-Pohlhausen technique described by Schlichting [15, pp. 192–201] for the solution of boundary layer flow problems was used by Tien [16] to solve Eq. 6-39 approximately in linearized form. To linearize Eqs. 6-39 a new dimensionless temperature function $\psi(\xi)$ is defined as

$$\theta(\xi) = 1 - \psi(\xi) \tag{6-51}$$

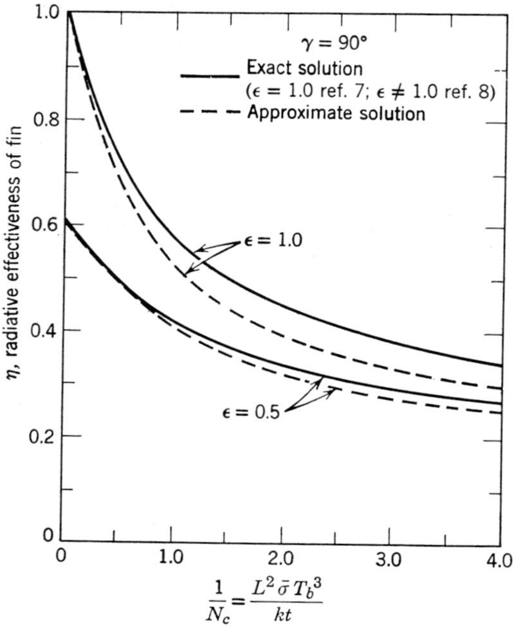

Fig. 6-5. A comparison of exact and approximate solutions. (From C. L. Tien [16]).

For $\psi(\zeta) \ll 1$ we write

$$\theta^4(\xi) = [1 - \psi(\xi)]^4 \cong 1 - 4\psi(\xi) \tag{6-52}$$

where we have replaced ξ_1 by ξ for simplicity.

Then the linearized forms of Eqs. 6-39 become

$$-\frac{d^2\psi(\xi)}{d\xi^2} = \frac{\varepsilon}{N_c}\left\{1 - 4\psi(\xi) - \varepsilon\int_{\xi'=0}^{1}[1 - \psi(\xi')]G_\gamma(\xi, \xi')\,d\xi'\right\} \tag{6-53a}$$

$$\psi(\xi) = 0 \quad \text{at } \xi = 0 \tag{6-53b}$$

$$\frac{d\psi(\xi)}{d\xi} = 0 \quad \text{at } \xi = 1 \tag{6-53c}$$

To solve Eqs. 6-53 by the Kármán-Pohlhausen technique a suitable profile is chosen to describe the temperature distribution in the plate. Assuming a third-degree polynomial representation, we write

$$\psi(\xi) = a_0 + a_1\xi + a_2\xi^2 + a_3\xi^3 \tag{6-54}$$

Four conditions are needed to evaluate the four unknown coefficients, a_0, a_1, a_2, and a_3. Two of these conditions are taken as the boundary conditions

6-53b and 6-53c; a third condition is obtained by evaluating the integro-differential equation 6-53a at $\xi = 0$, giving (see Eq. 6-49)

$$-\left.\frac{d^2\psi(\xi)}{d\xi^2}\right|_{\xi=0} = \frac{\varepsilon}{N_c}[1 - \varepsilon L_\gamma(\varepsilon)] \tag{6-55}$$

where $L_\gamma(\varepsilon)$ was defined by Eq. 6-50.

A fourth condition is determined by the gross energy-balance equation obtained by integrating Eq. 6-53a along the entire length of the plate (i.e., from $\xi = 0$ to 1). This yields

$$\left.\frac{d\psi(\xi)}{d\xi}\right|_{\xi=0} = \frac{\varepsilon}{N_c}\int_{\xi=0}^{1}[1 - 4\psi(\xi)]\,d\xi - \frac{\varepsilon^2}{N_c}\int_{\xi'=0}^{1}[1 - 4\psi(\xi')]\left[\int_{\xi=0}^{1} G_\gamma(\xi, \xi')\,d\xi\right]d\xi' \tag{6-56}$$

The four conditions given by Eqs. 6-53b, 6-53c, 6-55, and 6-56 are sufficient to determine the four unknown coefficients in Eq. 6-54.

The total heat transfer from the surface of one plate per unit width is given by

$$Q^r = L\int_{\xi=0}^{1} q^r(\xi)\,d\xi \tag{6-57}$$

From Eqs. 6-26 and 6-51 we have

$$q^r(\xi) = N_c\bar{\sigma}T_b^4\frac{d^2\theta(\xi)}{d\xi^2} = -N_c\bar{\sigma}T_b^4\frac{d^2\psi(\xi)}{d\xi^2} \tag{6-58}$$

Substituting Eq. 6-58 into Eq. 6-57 and utilizing the boundary condition 6-53c yields

$$Q^r = LN_c\bar{\sigma}T_b^4\left.\frac{d\psi(\xi)}{d\xi}\right|_{\xi=0} \tag{6-59a}$$

By introducing Eq.6-54 into Eq. 6-59a we obtain

$$Q^r = (LN_c\bar{\sigma}T_b^4)a_1 \tag{6-59b}$$

Then the radiative effectiveness of the fin is given by

$$\eta = \frac{Q^r}{Q^r_{\text{ideal}}} = \frac{LN_c\bar{\sigma}T_b^4a_1}{\bar{\sigma}T_b^4L\sin(\gamma/2)} = \frac{a_1N_c}{\sin(\gamma/2)} \tag{6-60}$$

Thus, once the coefficient a_1 is known, the total heat transfer and the radiative fin effectiveness can be readily calculated from the relations given above.

Figure 6-5 shows a comparison of the fin effectiveness values obtained by the approximate integral approach and the exact numerical solution. The agreement between the exact and approximate solutions is good only for large values of N_c(i.e., when radiation effects are small).

6-3 RADIATING PLATE FINS WITH INTERACTION BETWEEN THE FIN BASE AND FIN SURFACE

When the area exposed by the fin base is sufficiently large in comparison to the fin surfaces, the radiation from the fin base becomes important and should be included in the analysis. Figure 6-6 shows a rectangular fin plate geometry in which the fin base has appreciable exposed surface. To formulate the problem of radiative exchange for the fin geometry shown in this figure we make the same assumptions as those given in Section 6-1 for the fin surfaces and the external environment; in addition the fin base is assumed to be an opaque, gray, diffuse emitter and a diffuse reflector and to have the same emissivity as the fin surfaces.

Consideration will be given to only half the thickness of the fin plate because of symmetry. Let o_1x_1, o_2x_2, and o_3x_3 be the coordinates for plates 1 and 2 and the fin base, respectively, as illustrated in Fig. 6-6. No heat conduction is involved for the fin base, which is at uniform temperature.

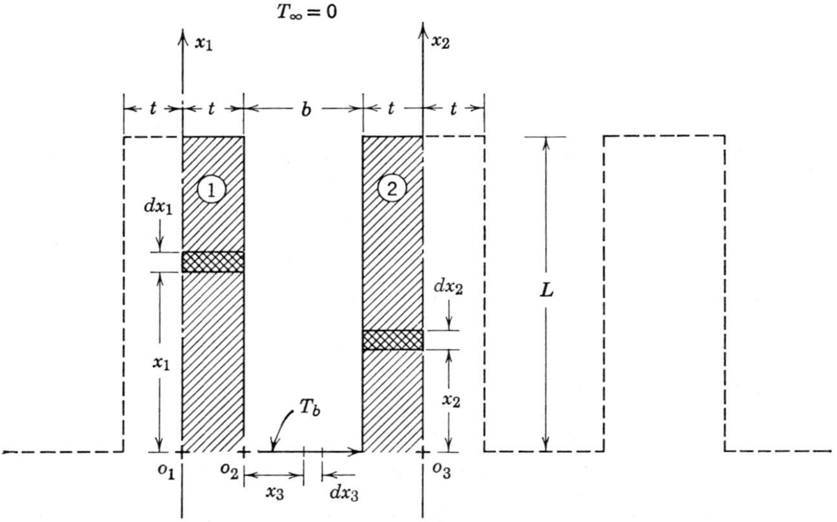

Fig. 6-6. Rectangular plate fin geometry.

The energy equation for plate 1 is given as (see Eq. 6-3)

$$\frac{d^2T_1(x_1)}{dx_1^{\,2}} = \frac{1}{kt}\, q^r(x_1) \quad \text{in } 0 \leq x_1 \leq L \tag{6-61a}$$

with the boundary conditions

$$T_1(x_1) = T_b \quad \text{at } x_1 = 0 \tag{6-61b}$$

$$\frac{dT_1(x_1)}{dx_1} = 0 \quad \text{at } x_1 = L \tag{6-61c}$$

The net radiative heat flux $q_1{}^r(x_1)$ at the surface of plate 1 is obtained from Eq. 5-10a as

$$q_1{}^r(x_1) = R_1(x_1) - \int_{x_2=0}^{L} R_2(x_2)\, dF_{dx_1-dx_2} - \int_{x_3=0}^{b} R_3(x_3)\, dF_{dx_1-dx_3} \tag{6-62}$$

where the second and third terms on the right-hand side represent radiation from fin plate 2 and the fin base, respectively.

The relations for the radiosities $R_1(x_1)$ and $R_3(x_3)$ are obtained from Eq. 5-9 as

$$R_1(x_1) = \varepsilon\bar{\sigma}T_1^{\,4}(x_1) + (1-\varepsilon)\int_{x_2=0}^{L} R_2(x_2)\, dF_{dx_1-dx_2} + (1-\varepsilon)\int_{x_3=0}^{b} R_3(x_3)\, dF_{dx_1-dx_3} \tag{6-63}$$

$$R_3(x_3) = \varepsilon\bar{\sigma}T_b^{\,4} + (1-\varepsilon)\int_{x_1=0}^{L} R_1(x_1)\, dF_{dx_3-dx_1} + (1-\varepsilon)\int_{x_2=0}^{L} R_2(x_2)\, dF_{dx_3-dx_2} \tag{6-64}$$

Similar equations can be written for the temperature $T_2(x_2)$ and radiosity $R_2(x_2)$ of plate 2, but they are not needed because of symmetry, that is, $T_1(x_1) = T_2(x_2)$ and $R_1(x_1) = R_2(x_2)$ for $x_1 = x_2$. Therefore subscripts 1 and 2 can be removed from temperature $T_1(x_1)$ and radiosities $R_1(x_1)$, $R_2(x_2)$. Then the governing equations for the problem in question can be written in the dimensionless form as

$$\frac{d^2\theta(\xi_1)}{d\xi_1^{\,2}} = \frac{1}{N_c}\left[\beta(\xi_1) - \int_{\xi_2=0}^{1} \beta(\xi_2)\, dF_{d\xi_1-d\xi_2} - \int_{\xi_3=0}^{1} \beta_3(\xi_3)\, dF_{d\xi_1-d\xi_3}\right] \quad \text{in } 0 \leq \xi_1 \leq 1 \tag{6-65a}$$

$$\theta(\xi_1) = 1 \quad \text{at } \xi_1 = 0 \tag{6-65b}$$

$$\frac{d\theta(\xi_1)}{d\xi_1} = 0 \quad \text{at } \xi_1 = 1 \tag{6-65c}$$

$$\beta(\xi_1) = \varepsilon\theta^4(\xi_1) + (1-\varepsilon)\int_{\xi_2=0}^{1} \beta(\xi_2)\, dF_{d\xi_1-d\xi_2} + (1-\varepsilon)\int_{\xi_3=0}^{1} \beta_3(\xi_3)\, dF_{d\xi_1-d\xi_3} \tag{6-66}$$

and

$$\beta_3(\xi_3) = \varepsilon + (1 - \varepsilon)\int_{\xi_1=0}^{1} \beta(\xi_1)\, dF_{d\xi_3 - d\xi_1} + (1 - \varepsilon)\int_{\xi_2=0}^{1} \beta(\xi_2)\, dF_{d\xi_3 - d\xi_2} \tag{6-67}$$

where we have defined

$$\beta \equiv \frac{R_1}{\bar{\sigma}T_b^4} = \frac{R_2}{\bar{\sigma}T_b^4}, \qquad \beta_3 \equiv \frac{R_3}{\bar{\sigma}T_b^4}, \qquad N_c \equiv \frac{kt}{L^2\bar{\sigma}T_b^3}$$
$$\theta \equiv \frac{T_1}{T_b}, \qquad \xi_1 \equiv \frac{x_1}{L}, \qquad \xi_2 \equiv \frac{x_2}{L}, \qquad \text{and} \qquad \xi_3 \equiv \frac{x_3}{b} \tag{6-68}$$

Equations 6-65, 6-66, and 6-67 are three simultaneous equations for the three unknowns $\theta(\xi_1)$, $\beta(\xi_1)$, and $\beta_3(\xi_3)$.

Frost and Eraslan [11] solved the problem of heat transfer by combined conduction and radiation, including the interaction between the base and adjacent fin surfaces, by using the approximate Galerkin method [17, pp. 258–309; 6, pp. 338–345]. Donovan and Rohrer [12] solved a similar problem by a numerical iterative method. In references 11 and 12 the fourth-power radiative heat transfer boundary condition was used at the fin tip in place of the boundary condition given by Eq. 6-65c. For an external environment at zero temperature the fourth-power radiation boundary condition is given as

$$-\frac{d\theta(\xi_1)}{d\xi_1} = \frac{\varepsilon t}{LN_c}\,\theta^4(\xi_1) \quad \text{at } \xi_1 = 1 \tag{6-69}$$

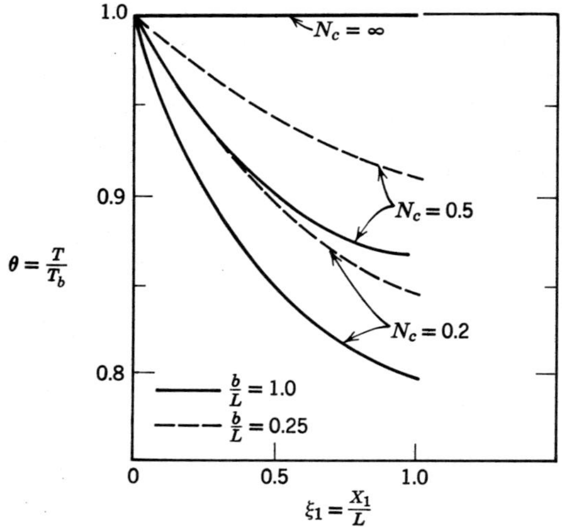

Fig. 6-7. The distribution of temperature at the fin plate. (From R. C. Donovan and W. M. Rohrer [12]).

Figure 6-7 shows the temperature distribution at the fin for two base widths and for several values of the conduction-to-radiation parameter when the boundary at $x = L$ is assumed to be insulated (i.e., Eq. 6-65c). For $N_c \to \infty$, the heat transfer is by pure conduction and the temperature profile in the fin is uniform since the boundary at $x_1 = L$ is insulated. For small values of N_c, the radiation becomes dominant and hence a temperature gradient is established along the fin. This temperature gradient is steeper for the larger values of b/L, which corresponds to increased interaction from the fin base.

REFERENCES

1. J. P. Gallinan and W. P. Berggren, "Some Radiator Design Criteria for Space Vehicles," *J. Heat Transfer*, **81C**, 237–248, 1959.
2. A. P. Fraas and M. N. Özişik, *Heat Exchanger Design*, John Wiley and Sons, New York, 1965.
3. J. G. Bartas and W. H. Sellers, "Radiation Fin Effectiveness," *J. Heat Transfer*, **82C**, 73–75, 1960.
4. E. N. Nilson and R. Curry, "The Minimum Weight Straight Fin of Triangular Profile Radiating to Space," *J. Aerospace Sci.*, **27**, 146, 1960.
5. E. R. G. Eckert, T. F. Irvine, Jr., and E. M. Sparrow, "Analytical Formulation for Radiating Fins with Mutual Irradiation," *Am. Rocket Soc.*, **30**, 644–646, 1960.
6. R. L. Chambers and E. V. Somers, "Radiation Fin Efficiency for One-Dimensional Heat Flow in a Circular Fin," *J. Heat Transfer*, **81C**, 327–329, 1959.
7. E. M. Sparrow, E. R. G. Eckert, and T. F. Irvine, "The Effectiveness of Radiating Fins with Mutual Irradiation," *J. Aerospace Sci.*, **28**, 763–772, 1961.
8. R. G. Hering, "Radiative Heat Exchange Between Conducting Plates with Specular Reflection," *J. Heat Transfer*, **88C**, 29–36, 1966.
9. E. M. Sparrow and E. R. G. Eckert, "Radiant Interaction Between Fin and Base Surfaces," *J. Heat Transfer*, **84C**, 12–18, 1962.
10. E. M. Sparrow, G. B. Miller, and V. K. Jonsson, "Radiating Effectiveness of Annular-Finned Space Radiators, Including Mutual Irradiation Between Radiator Elements," *J. Aerospace Sci.*, **29**, 1291–1299, 1962.
11. W. Frost and A. H. Eraslan, "An Iterative Method for Determining the Heat Transfer from a Fin with Radiative Interaction Between the Base and Adjacent Fin Surfaces," AIAA 3rd Thermophysics Conference, *AIAA Paper* No. 68-772, June 1968.
12. R. C. Donovan and W. M. Rohrer, "Radiative and Conductive Fins on a Plane Wall, Including Mutual Irradiation," *ASME Paper* No. 68-WA/HT-22, November 1969.
13. H. F. Mueller and N. D. Malmuth, "Temperature Distribution in Radiating Heat Shields by the Method of Singular Perturbations," *Intern. J. Heat Mass Transfer*, **8**, 915–920, 1965.
14. M. A. Heaslet and H. Lomax, "Numerical Predictions of Radiative Interchange Between Conducting Fins with Mutual Irradiations," *NASA Tech. Rept.* TR R-116, 1961.

15. H. Schlichting, *Boundary Layer Theory*, 6th ed., McGraw-Hill Book Co., New York, 1968.

16. C. L. Tien, "Approximate Solutions of Radiative Exchange Between Conducting Plates with Specular Reflection," *J. Heat Transfer*, **89C**, 119–120, 1967.

17. L. V. Kantorovich and V. I. Krylow, *Approximate Methods of Higher Analysis*, John Wiley and Sons, New York, 1964.

NOTES

[1] Equation 5-10a can be written for plate 1 as

$$q_1{}^r(x_1) = R_1(x_1) - \sum_{j=1}^{N} \int_{x_j} R_j(x_j)\, dF_{dx_1 - dx_j} \tag{1}$$

The problem in question involves only two zones, plates 1 and 2, having radiosities $R_1(x_1)$ and $R_2(x_2)$; then Eq. 1 simplifies to

$$q_1{}^r(x_1) = R_1(x_1) - \int_{x_2=0}^{L} R_2(x_2)\, dF_{dx_1 - dx_2} \tag{2}$$

[2] The reciprocal of N_c has also been used in the literature.

CHAPTER 7

Radiation and Convection in NonParticipating Media

In many engineering applications, such as in advanced types of power plants for nuclear rockets, high-speed flights, and re-entry vehicles, temperatures are so high that heat transfer by thermal radiation becomes important. In this chapter we consider the interaction of radiation with convection for the flow of transparent fluid (i.e., a fluid that does not absorb, emit, or scatter radiation). The treatment of simultaneous convection and radiation for a participating fluid will be presented in Chapters 13 and 14.

The presence of thermal radiation does not alter the standard equations of motion and energy for a transparent fluid; therefore the equations of motion and of energy given in the books by Schlichting [1], Kays [2], and Moore [3] are applicable in the formulation of heat transfer problems for transparent fluids with radiation. The coupling between the radiation and convection results only through the fourth-power thermal radiation boundary condition at the surface of the walls. However, one should distinguish between the situations in which the temperature is prescribed at the boundary surface and those in which the net heat flux is specified. In the former case the radiation and convection can be treated independently since the boundary surface temperature is fixed and radiation does not alter it. In the latter case, however, the presence of radiation at the boundary surface alters the surface temperature; hence convection and radiation are coupled.

Simultaneous convection and radiation for the flow of transparent fluid has been studied by several investigators. Cess [4, 5], Sparrow and Lin [6], and Lind and Cebeci [7] studied the effects of radiation for laminar boundary layer flow along a flat plate. Perlmutter and Siegel [8], Siegel and Perlmutter [9], Dussan and Irvine [10], Chen [11], and Siegel and Keshock [12] investigated the radiation effects for flow inside circular tubes; Keshock and Siegel [13], the radiation effects for flow between two heated parallel plates. The analyses in references 8, 9, 12, and 13 for flow inside channels are limited in that they require *a priori* knowledge of the heat transfer coefficient. Chen [11] and Dussan and Irvine [10] analyzed the problem without any *a priori* assumption regarding the heat transfer coefficient. However, Chen [11] introduced radiation effects by assuming that heat transfer to the gas at the tube wall is proportional to the fourth power of the wall temperature, but this is a physically unrealistic boundary condition since it neglects radiation incident on the surface from other surface element. Dussan and Irvine [10] calculated heat transfer by assuming linearized radiation and using an exponential kernel approximation. A more complete and realistic model that did not require *a priori* knowledge of the heat transfer coefficient was used by Thorsen [14] and Thorsen and Kanchanagom [15] to investigate the effects of radiation on heat transfer for flow inside circular tubes and by Liu and Thorsen [16] for flow between parallel channels.

7-1 LAMINAR BOUNDARY LAYER FLOW ALONG A FLAT PLATE WITH RADIATION BOUNDARY CONDITION

Consider the steady, laminar boundary layer flow of an incompressible, transparent fluid along a flat plate with a constant applied heat flux q_w Btu/hr ft^2 at the wall surface. The heat is dissipated from the surface of the plate by conduction to the fluid and by the fourth-power radiation law to an external environment at temperature T_e. The surface of the plate is opaque and gray and has uniform emissivity ε. The properties of the fluid are constant, the velocity u_∞ and temperature T_∞ of the external flow are uniform, and the flow velocities are sufficiently small so that the viscous energy dissipation can be neglected. Figure 7-1 shows the geometry and coordinates.

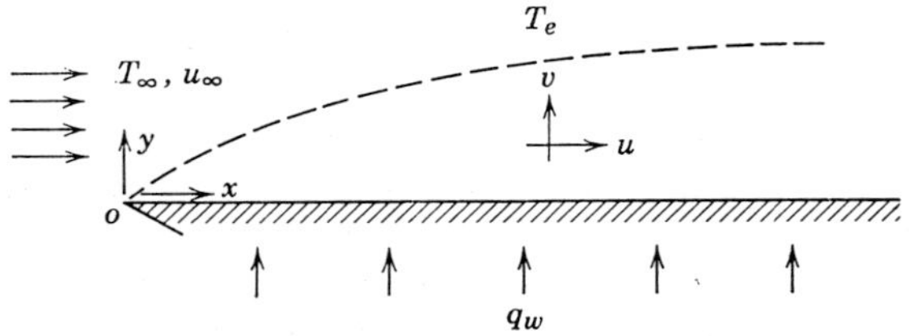

Fig. 7-1. Laminar boundary layer along a flat plate with radiation boundary condition.

The velocity problem for the boundary layer flow in question is given as [1, p. 125]

$$\frac{\partial u}{\partial x}+\frac{\partial v}{\partial y}=0 \qquad \text{(continuity)} \tag{7-1}$$

$$u\frac{\partial u}{\partial x}+v\frac{\partial u}{\partial y}=\nu\frac{\partial^2 u}{\partial y^2} \qquad (x \text{ momentum}) \tag{7-2}$$

with the boundary conditions

$$u=v=0 \quad \text{at } y=0 \tag{7-3a}$$

$$u=u_\infty \quad \text{at } y\to\infty \tag{7-3b}$$

where u and v are the x- and y-direction velocity components, respectively, and ν is the kinematic viscosity.

The temperature distribution in the boundary layer satisfies the following energy equation [1, p. 273]:

$$u\frac{\partial T}{\partial x}+v\frac{\partial T}{\partial y}=\alpha\frac{\partial^2 T}{\partial y^2} \tag{7-4}$$

with the boundary conditions[1]

$$q_w=-k\frac{\partial T}{\partial y}+\varepsilon(\bar{\sigma}T^4-\bar{\sigma}T_e^4) \quad \text{at } y=0 \tag{7-5a}$$

$$T=T_\infty \quad \text{at } y\to\infty \tag{7-5b}$$

where α is the thermal diffusivity and k the thermal conductivity of the fluid.

The velocity problem given by Eqs. 7-1 through 7-3 is not coupled to the temperature problem; hence it can be solved independently by standard techniques. A stream function $\psi(x, y)$ is defined as

$$u=\frac{\partial\psi(x,y)}{\partial y} \quad \text{and} \quad v=-\frac{\partial\psi(x,y)}{\partial x} \tag{7-6}$$

Then the continuity equation 7-1 is identically satisfied and the momentum equation is written in terms of the stream function. By introducing the well-known similarity variables $f(\eta)$ and η, defined as

$$\eta=y\sqrt{\frac{u_\infty}{\nu x}} \tag{7-7a}$$

$$f(\eta)=\frac{\psi(x,y)}{\sqrt{x\nu u_\infty}} \tag{7-7b}$$

The momentum equation and the boundary conditions for the velocity problem are transformed into the following ordinary differential equation

[1, p. 126]:

$$2f''' + ff'' = 0 \tag{7-8a}$$

with the boundary conditions

$$f = 0, f' = 0 \quad \text{at } \eta = 0 \tag{7-8b}$$

$$f' = 1 \quad \text{at } \eta \to \infty \tag{7-8c}$$

where primes denote differentiation with respect to η. The velocity components u and v are related to the similarity variables by

$$u = u_\infty f' \tag{7-9a}$$

$$v = \frac{1}{2}\sqrt{\frac{\nu u_\infty}{x}}\,(\eta f' - f) \tag{7-9b}$$

The numerical solution to the transformed velocity problem given by Eqs. 7-8 is available in the literature, and the functions f, f' are well tabulated as a function of η.

To solve the energy equation 7-4 with the boundary conditions given by Eqs. 7-5, Cess [4] utilized a technique similar to that employed by Jaeger [17]. Two independent variables ξ and η are defined as

$$\xi = \frac{\varepsilon\bar{\sigma}T_\infty{}^3}{k}\sqrt{\frac{\nu x}{u_\infty}} \tag{7-10a}$$

$$\eta = y\sqrt{\frac{u_\infty}{\nu x}} \tag{7-10b}$$

Here the variable η is the same as that defined by Eq. 7-7a for the velocity problem. When the transformation of Eqs. 7-10 and the velocity components of Eqs. 7-9 are introduced into the energy equation 7-4, the latter transforms into[2]

$$\frac{\partial^2 T}{\partial \eta^2} + \tfrac{1}{2}\,\mathrm{Pr}\, f\frac{\partial T}{\partial \eta} - \tfrac{1}{2}\,\mathrm{Pr}\,\frac{df}{d\eta}\,\xi\,\frac{\partial T}{\partial \xi} = 0 \tag{7-11}$$

where Pr is the Prandtl number.

To solve Eq. 7-11, a power-series expansion technique has been used. That is, the temperature function $T(\xi, \eta)$ is expanded in powers of ξ in the form

$$T(\xi, \eta) - T_\infty = T_\infty \sum_{n=1}^{\infty} a_n \theta_n(\eta)\xi^n \tag{7-12}$$

with the requirement that

$$\theta_1(0) = \theta_2(0) = \theta_3(0) = \cdots = 1 \tag{7-13}$$

where the coefficients a_n and the functions $\theta_n(\theta)$ are unknowns.

By substituting the expansion given by Eq. 7-12 into Eq. 7-11 and equating the coefficients of ξ^n to zero (for $a_n \neq 0$), it is found that the functions $\theta_n(\eta)$ constitute the solution of the following ordinary differential equation:

$$\theta_n'' + \tfrac{1}{2}\,\mathrm{Pr}\, f\theta_n' - \frac{n}{2}\,\mathrm{Pr}\, f'\theta_n = 0 \tag{7-14}$$

with the boundary conditions

$$\theta_n = 1 \quad \text{at } \eta = 0 \tag{7-15a}$$

$$\theta_n = 0 \quad \text{at } \eta \to \infty \tag{7-15b}$$

where primes denote differentiation with respect to η. The ordinary differential equation 7-14 with the boundary conditions of Eq. 7-15 can be solved numerically for a given value of the Prandtl number since the functions f and f' are available from the solution of the velocity problem.

When the functions $\theta_n(\eta)$ are known, the problem of determining the temperature distribution $T(\xi, \eta)$ in the boundary layer reduces to evaluating the unknown expansion coefficients a_n in Eq. 7-12. The boundary condition 7-5a can now be utilized to determine these expansion coefficients.

From Eq. 7-12 we write

$$T(\xi, 0) = T_\infty\left(1 + \sum_{n=1}^{\infty} a_n \xi^n\right) \tag{7-16}$$

and $\partial T/\partial y$ at the wall can be evaluated as

$$\begin{aligned}\left.\frac{\partial T(x, y)}{\partial y}\right|_{y=0} &= \left.\frac{\partial T}{\partial \eta}\right|_{\eta=0}\frac{d\eta}{dy} = \left[T_\infty \sum_{n=1}^{\infty} a_n \theta_n'(0)\xi^n\right]\sqrt{\frac{u_\infty}{\nu x}}\\ &= \left[T_\infty \sum_{n=1}^{\infty} a_n \theta_n'(0)\xi^n\right]\frac{1}{\xi}\,\frac{\varepsilon\bar{\sigma}T_\infty^{\;3}}{k}\\ &= \frac{\varepsilon\bar{\sigma}T_\infty^{\;4}}{k}\sum_{n=1}^{\infty} a_n\theta_n'(0)\xi^{n-1}\end{aligned} \tag{7-17}$$

Substituting Eqs. 7-16 and 7-17 into the boundary condition 7-5a and equating like powers of ξ, one obtains the desired relation for the determination of coefficients a_n. For example, by equating the constant terms (i.e., the coefficients of ξ^0) we obtain

$$a_1 = -\frac{1}{\theta_1'(0)}\left[\frac{q_w}{\varepsilon\bar{\sigma}T_\infty^{\;4}} + \left(\frac{T_e}{T_\infty}\right)^4 - 1\right] \tag{7-18a}$$

and, similarly, equating the coefficients of ξ yields

$$\frac{a_2}{a_1} = \frac{4}{\theta_2'(0)} \tag{7-18b}$$

Other coefficients are determined in a similar manner.

Knowing the functions $\theta_n(\eta)$ and the coefficients a_n, we can evaluate the distribution of temperature in the boundary layer from Eq. 7-12.

Results

In the problems of convective heat transfer the Nusselt number is a quantity of interest. The local Nusselt number Nu is defined as

$$\text{Nu} = -\frac{x}{T_w - T_\infty} \frac{\partial T}{\partial y}\bigg|_{y=0} \tag{7-19}$$

Substituting Eqs. 7-16 and 7-17 into Eq. 7-19 and in the resulting expression replacing $\varepsilon\bar{\sigma}T_\infty^{\ 4}/k$ by $\xi\sqrt{u_\infty/\nu x}$ according to Eq. 7-10a, we obtain

$$\frac{\text{Nu}}{\text{Re}^{1/2}} = -\frac{\sum_{n=1}^{\infty} a_n\theta_n'(0)\xi^n}{\sum_{n=1}^{\infty} a_n\xi^n} \tag{7-20a}$$

where the Reynolds number is defined as

$$\text{Re} = \frac{u_\infty x}{\nu} \tag{7-20b}$$

Dividing one series into the other, we can write Eq. 7-20a as

$$\frac{\text{Nu}}{\text{Re}^{1/2}} = -\theta_1'(0) - \frac{a_2}{a_1}[\theta_2'(0) - \theta_1'(0)]\xi - \cdots \tag{7-21}$$

Substituting a_2/a_1 from Eq. 7-18b into Eq. 7-21, we obtain

$$\frac{\text{Nu}}{\text{Re}^{1/2}} = -\theta_1'(0) - 4\left[1 - \frac{\theta_1'(0)}{\theta_2'(0)}\right]\xi - \cdots \tag{7-22}$$

The values of $\theta_1'(0)$ and $\theta_2'(0)$ have been evaluated by Donoughe and Livingood [18] and are given as

$$\theta_1'(0) = -0.4059 \qquad \text{and} \qquad \theta_2'(0) = -0.4803 \tag{7-23}$$

Then the local Nusselt number becomes

$$\frac{\text{Nu}}{\text{Re}^{1/2}} = 0.4059 - 0.620\xi - \cdots \tag{7-24}$$

The first term on the right-hand side of Eq. 7-24 is the local Nusselt number for laminar boundary layer along a flat plate with applied constant wall heat flux. The second term gives the first-order effect of radiation at the boundary surface on convective heat transfer. Cess [4] has shown that there is little use in evaluating higher-order terms in Eq. 7-24 because of slow convergence of the series.

7-2 FORCED CONVECTION INSIDE A CIRCULAR TUBE WITH RADIATION BOUNDARY CONDITION

Consider the steady, fully developed flow of a transparent fluid inside a circular tube having a uniformly applied heat flux q_w Btu/hr ft^2 at the wall surface. The fluid enters the tube at the origin of the axial coordinate $x = 0$ at a uniform temperature T_{g1} and is heated to an average exit temperature T_{g2} at $x = L$. Figure 7-2 shows the geometry and the coordinates. The heat supplied to the tube wall is dissipated from the inner surface by convection and radiation, while the outer surface is kept insulated. The two ends of the tube at $x = 0$ and $x = L$ are open to outside environments at temperatures T_1 and T_2, respectively. The inside surface of the tube is an opaque, gray, diffuse emitter and a diffuse reflector and has uniform emissivity ε. The Kirchhoff law is assumed to be valid.

The determination of the tube surface temperature and the gas temperature rise as functions of the axial distance along the tube is of interest in this problem. Here we present a simplified analysis of this problem by considering a radially averaged gas temperature and consequently by assuming that *a priori* knowledge of the heat transfer coefficient is available. Therefore limitation to such simplification should be recognized since there is a coupling through the nonlinear boundary condition, between radiation at the wall and conduction and convection in the fluid.

Let $T_g(x)$ be the radially averaged gas temperature, u_m the mean velocity, and h the convective heat transfer coefficient, which is assumed to be uniform along the entire length of the tube.

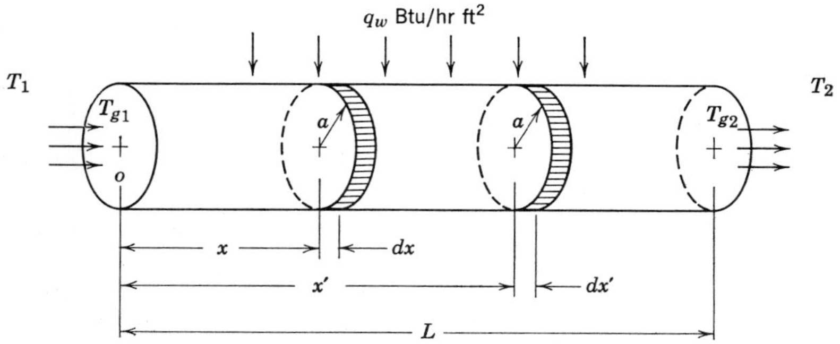

Fig. 7-2. Forced convection heat transfer inside a circular tube with radiation boundary condition.

Consider a cylindrical control volume of length dx and radius a about position x, as illustrated in Fig. 7-2. An energy balance equation for this control volume is written as

$$\rho u_m c_p \pi a^2 \frac{dT_g(x)}{dx}\, dx = h[T_w(x) - T_g(x)]2\pi a\, dx \tag{7-25}$$

or

$$\frac{dT_g(x)}{dx} = \frac{2h}{\rho u_m c_p a}[T_w(x) - T_g(x)] \tag{7-26a}$$

with the boundary condition

$$T_g(x) = T_{g1} \quad \text{at } x = 0 \tag{7-26b}$$

Equations 7-26 involve two unknowns: $T_g(x)$, the gas temperature; and $T_w(x)$, the tube wall temperature. To determine an additional relationship an energy balance equation is written at the tube wall surface by equating the externally supplied heat flux to the heat dissipation from the wall surface by convection and radiation, that is,

$$q_w = q^{cv}(x) + q^r(x) \tag{7-27}$$

where the convective heat flux is given by

$$q^{cv}(x) = h[T_w(x) - T_g(x)] \tag{7-28}$$

and heat transfer coefficient h is assumed to be specified. The radiation heat flux $q^r(x)$ is determined from Eq. 5-10a as

$$q^r(x) = R(x) - \left[\bar{\sigma}T_1^4 F(x) + \bar{\sigma}T_2^4 F(L-x) + \int_{x'=0}^{L} R(x')\, dF_{dx-dx',|x-x'|}\right] \tag{7-29}$$

where $R(x)$ = radiosity at the cylindrical surface,
$F(x)$ = diffuse view factor from the band (a, dx) at x to the opening at $x = 0$ at a distance x apart,
$F(L - x)$ = diffuse view factor from the band (a, dx) at x to the opening at $x = L$ at a distance $L - x$ apart,
$dF_{dx-dx',|x-x'|}$ = elemental diffuse view factor from the band (a, dx) at x to the band (a, dx) at x' at a distance $|x - x'|$ apart.

In Eq. 7-29, the first, second, and third terms inside the bracket are respectively the radiation from the openings at $x = 0$ and $x = L$ and from the entire inside surface of the tube to the band (a, dx) at x.

Finally, the relation for the radiosity function $R(x)$ is obtained from Eq. 5-10b as[3]

$$R(x) = \bar{\sigma} T_w^4(x) - \frac{1-\varepsilon}{\varepsilon} q^r(x) \tag{7-30a}$$

or, by substituting for $q^r(x)$ from Eqs. 7-27 and 7-28 into Eq. 7-30a, we obtain

$$R(x) = \bar{\sigma} T_w^4(x) - \frac{1-\varepsilon}{\varepsilon} \{q_w - h[T_w(x) - T_g(x)]\} \tag{7-30b}$$

Equations 7-26 through 7-30 give the complete mathematical formulation of the problem.

The Dimensionless Equations

The foregoing equations can be expressed in the dimensionless form by introducing the following dimensionless quantities:

$$\left.\begin{aligned}
\theta &\equiv \left(\frac{\bar{\sigma}}{q_w}\right)^{1/4} T, \text{ dimensionless temperature} \\
\beta &= \frac{R}{q_w}, \text{ dimensionless radiosity} \\
\mathrm{S} &\equiv \frac{4h}{\rho u_m c_p}, \text{ Stanton number} \\
h^* &\equiv \frac{h}{q_w}\left(\frac{q_w}{\bar{\sigma}}\right)^{1/4}, \text{ dimensionless heat transfer coefficient} \\
\xi &\equiv \frac{x}{2a}, \quad \xi' \equiv \frac{x'}{2a}, \quad \text{and} \quad \xi_L \equiv \frac{L}{2a}
\end{aligned}\right\} \tag{7-31}$$

Then Eqs. 7-26 become

$$\frac{d\theta_g(\xi)}{d\xi} = \mathrm{S}[\theta_w(\xi) - \theta_g(\xi)] \tag{7-32a}$$

with

$$\theta_g(\xi) = \theta_{g1} \quad \text{at } \xi = 0 \tag{7-32b}$$

Equations 7-27, 7-28, and 7-29 are combined as

$$\begin{aligned}
1 = h^*[\theta_w(\xi) - \theta_g(\xi)] + \beta(\xi) - \Bigg[&\theta_1^4 F(\xi) + \theta_2^4 F(\xi_L - \xi) \\
&+ \int_{\xi'=0}^{\xi} \beta(\xi')\, dF_{d\xi - d\xi', (\xi - \xi')} + \int_{\xi'=\xi}^{\xi_L} \beta(\xi')\, dF_{d\xi - d\xi', (\xi' - \xi)}\Bigg]
\end{aligned} \tag{7-33}$$

and Eq. 7-30b becomes

$$\beta(\xi) = \theta_w^4(\xi) - \frac{1-\varepsilon}{\varepsilon}\{1 - h^*[\theta_w(\xi) - \theta_g(\xi)]\} \tag{7-34}$$

Equations 7-32, 7-33, and 7-34 are three simultaneous equations for the three unknown functions $\theta_w(\xi)$, $\theta_g(\xi)$, and $\beta(\xi)$.

The view factors $F(\xi)$ and $F(\xi_L - \xi)$ in the above relations can be obtained from Eq. 5-84 as

$$F(z) = \frac{\frac{1}{2} + z^2}{\sqrt{1+z^2}} - z \tag{7-35}$$

where $z = \xi$ or $\xi_L - \xi$. The elemental view factor $dF_{d\xi-d\xi',|\xi'-\xi|}$ is obtained from Eq. 5-85 as

$$dF_{d\xi-d\xi',|z|} = \left[1 - |z|\frac{z^2 + \frac{3}{2}}{(z^2+1)^{3/2}}\right] d\xi' \tag{7-36}$$

where $z = \xi - \xi'$.

Cylinder with Black Walls

In the following analysis the inside surface of the cylinder is assumed to be a gray, diffuse emitter and a diffuse reflector. If the cylinder wall is assumed to be a *black surface*, we set in the above equations $\varepsilon = 1$. Then Eq. 7-34 simplifies to

$$\beta(\xi) = \theta_w^4(\xi) \tag{7-37}$$

Equations 7-32 remain unchanged, that is,

$$\frac{d\theta_g(\xi)}{d\xi} = S[\theta_w(\xi) - \theta_g(\xi)] \tag{7-38a}$$

$$\theta_g(\xi) = \theta_{g1} \quad \text{at } \xi = 0 \tag{7-38b}$$

Substituting Eq. 7-37 into Eq. 7-33, we obtain

$$1 = h^*[\theta_w(\xi) - \theta_g(\xi)] + \theta_w^4(\xi) - \left[\theta_1^4 F(\xi) + \theta_2^4 F(\xi_L - \xi) + \int_{\xi'=0}^{\xi} \theta_w^4(\xi')\, dF_{d\xi-d\xi',(\xi-\xi')} + \int_{\xi'=\xi}^{\xi_L} \theta_w^4(\xi')\, dF_{d\xi-d\xi',(\xi'-\xi)}\right] \tag{7-39}$$

Then the unknown temperature functions $\theta_w(\xi)$ and $\theta_g(\xi)$ are determined from the simultaneous solution of Eqs. 7-38 and 7-39. Although analytical solutions cannot be found to these equations, numerical solutions can certainly be obtained by using a high-speed digital computer. These equations were solved [8] by direct numerical integration for a cylinder with black walls. However, the utility of this method was restricted to short tube lengths (i.e., $\xi_L = 5$ and 10) because too many subdivisions were required to approximate the integral for long tubes, giving rise to a set of simultaneous equations too numerous to be solved accurately by the standard matrix technique. Therefore the exponential kernel approximation technique was used [8, 9] to treat the cases involving long tubes, and the accuracy was checked by comparing the results for short tubes with those obtained by direct numerical integration. The exponential kernel technique is based on the approximate representation of diffuse view factors by exponential functions. To illustrate the technique we consider now its application to the foregoing cylinder problem for black walls.

The diffuse view factors $F(z)$ and $dF_{d\xi-d\xi',|z|}$ of Eqs. 7-35 and 7-36 can be represented by an exponential function in the form [19]

$$F(z) = \frac{\frac{1}{2} + z^2}{\sqrt{1 + z^2}} - z \cong \frac{e^{-2z}}{2} \tag{7-40}$$

$$F_{d\xi-d\xi',z} = \left[1 - |z|\frac{z^2 + \frac{3}{2}}{(z^2 + 1)^{3/2}}\right] d\xi' \cong e^{-2z}\, d\xi' \tag{7-41}$$

The approximation of Eq. 7-41 was used first by Buckley [20] with good results.

Substituting Eqs. 7-40 and 7-41 into Eq. 7-39, we obtain

$$1 = h^*[\theta_w(\xi) - \theta_g(\xi)] + \theta_w^{\,4}(\xi) - \left[\tfrac{1}{2}\theta_1^{\,4}e^{-2\xi} + \tfrac{1}{2}\theta_2^{\,4}e^{-2(\xi_L-\xi)} + \int_{\xi'=0}^{\xi} \theta_w^{\,4}(\xi')e^{-2(\xi-\xi')}\, d\xi' + \int_{\xi'=\xi}^{\xi_L} \theta_w^{\,4}(\xi')e^{-2(\xi'-\xi)}\, d\xi'\right] \tag{7-42}$$

The integral equation 7-42 can be transformed into an ordinary differential equation by differentiating it with respect to ξ twice and eliminating from the resulting expression the integral term by means of Eq. 7-42 and the term $d^2\theta_g/d\xi^2$ by means of Eq. 7-38a. We obtain[4]

$$[h^* + 4\theta_w^{\,3}(\xi)]\frac{d^2\theta_w(\xi)}{d\xi^2} + 12\theta_w^{\,2}(\xi)\left(\frac{d\theta_w}{d\xi}\right)^2 - Sh^*\frac{d\theta_w}{d\xi} + h^*(S^2 - 4)\theta_w(\xi) = h^*(S^2 - 4)\theta_g(\xi) - 4 \tag{7-43}$$

Equation 7-43 is a second-order ordinary differential equation for the temperature function $\theta_w(\xi)$ and should be solved simultaneously with the ordinary differential equation 7-38 for the temperature function $\theta_g(\xi)$. The two boundary conditions needed for the solution of Eq. 7-43 are obtained by evaluating the original integral equation 7-42 at $\xi = 0$ and $\xi = \xi_L$. This yields

$$1 = h^*[\theta_w(0) - \theta_{g1}] + \theta_w{}^4(0) - \left[\tfrac{1}{2}\theta_1{}^4 + \tfrac{1}{2}\theta_2{}^4 e^{-2\xi_L} + \int_{\xi'=0}^{\xi_L} \theta_w{}^4(\xi') e^{-2\xi'}\, d\xi'\right]$$

$$\xi = 0 \quad (7\text{-}44a)$$

$$1 = h^*[\theta_w(\xi_L) - \theta_{g2}] + \theta_w{}^4(\xi_L) - \left[\tfrac{1}{2}\theta_1{}^4 e^{-2\xi_L} + \tfrac{1}{2}\theta_2{}^4 + \int_{\xi'=0}^{\xi_L} \theta_w{}^4(\xi') e^{-2(\xi_L-\xi')}\, d\xi'\right]$$

$$\xi = \xi_L \quad (7\text{-}44b)$$

Here we note that the external environment temperatures θ_1 and θ_2 do not appear in the differential equation 7-43 because they have been eliminated in transforming the integral equation into the differential equation, but they appear in the boundary conditions 7-44.

If Eq. 7-43 is to be solved with a forward integration scheme, the values of $\theta_w(\xi)$ and $d\theta_w(\xi)/d\xi$ at $\xi = 0$ are needed in order to start the forward integration. A trial value may be chosen for $\theta_w(0)$, and the value of $d\theta_w(0)/d\xi$ evaluated from

$$\frac{d\theta_w(0)}{d\xi} = \frac{h^*(S+2)[\theta_w(0) - \theta_{g1}] + 2[\theta_w{}^4(0) - \theta_1{}^4 - 1]}{h^* + 4\theta_w{}^3(0)} \quad (7\text{-}45)$$

Table 7-1 Comparison of Exponential Kernel Method with Direct Numerical Integration[a]

($h^* = 0.8$, $S = 0.01$, $\theta_{g1} = \theta_1 = 1.5$, $\theta_{g2} = \theta_2$, $\xi_L = 10$, $\varepsilon = 1$)

Axial Position ξ	Tube Wall Temperature $\theta_w(\xi)$ Direct Integration	Exponential Kernel	% Difference from Exponential Kernel Method
0	1.786	1.797	0.6
10/6	2.140	2.165	1.2
30/6	2.305	2.348	1.8
50/6	2.161	2.185	1.1
10	1.825	1.838	0.7

	Gas Temperature Rise $\theta_{g2} - \theta_{g1}$ Direct Integration	Exponential Kernel	% Difference from Exponential Kernel Method
	0.0645	0.0673	4.2

[a] From M. Perlmutter and R. Siegel [8].

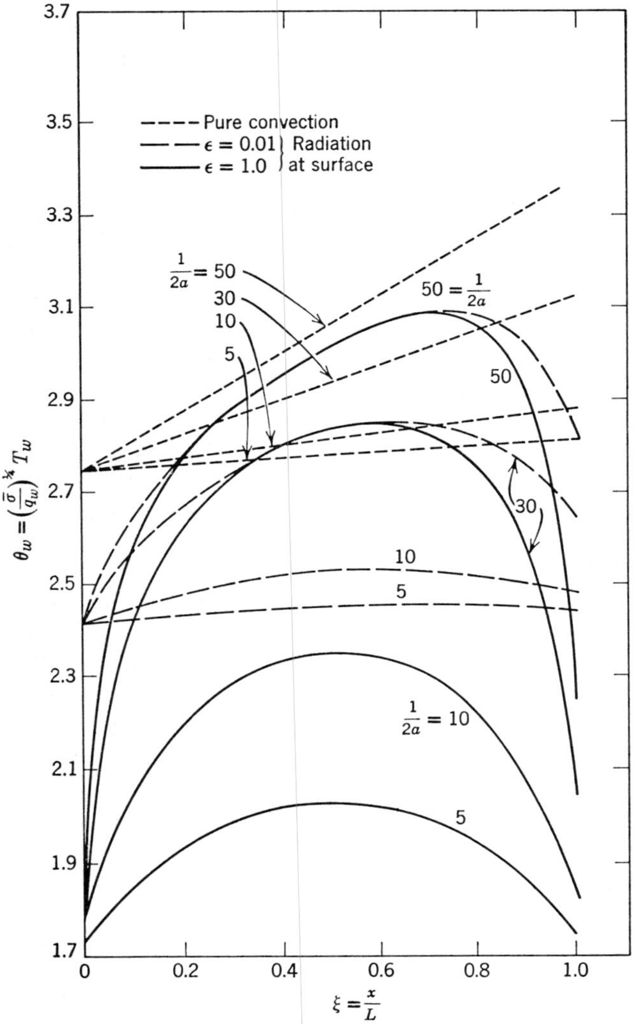

Fig. 7-3. Effects of tube length and wall emissivity on wall temperature distribution; $h^* = 0.8$, $S = 0.01$, $\theta_{g1} = \theta_1 = 1.5$, $\theta_{g2} = \theta_2$. (From R. Siegel and M. Perlmutter [9]).

This relation has been determined from Eq. 7-42 by differentiating it with respect to ξ, evaluating the resulting expression at $\xi = 0$, eliminating from it the integral terms by means of the boundary condition 7-44a, and eliminating the term $d\theta_g(0)/d\xi$ by means of Eq. 7-38a evaluated at $\xi = 0$.

When Eq. 7-45 is used, the boundary condition of Eq. 7-44a at $\xi = 0$ is automatically satisfied. Therefore all one needs in this integration is to satisfy the boundary condition 7-44b at $\xi = \xi_L$. If the correct value is chosen for $\theta_w(0)$ to start the integration, the resulting solution should satisfy the boundary condition 7-44b at $\xi = \xi_L$. This condition serves as a test to interpolate the correct value of $\theta_w(0)$ over the trial values.

Results and Discussion

Table 7-1 shows a comparison of the dimensionless tube wall temperature $\varepsilon_w(\xi)$ at several axial positions, obtained by direct integration and by exponential kernel approximation for a cylinder with black walls. Included in this table is the gas temperature rise $(\theta_{g2} - \theta_{g1})$ obtained by the two methods. It is apparent from these results that the exponential kernel approximation gives the tube wall temperature with an error less than 2 per cent and the gas temperature rise with an error less than 4 per cent for the case considered.

Figure 7-3 shows the effects of tube length and wall emissivity on the distribution of wall temperature for the case $h^* = 0.8$, $S = 0.01$, $\theta_{g1} = \theta_1 = 1.5$, and $\theta_{g2} = \theta_2$. Included in this figure is the tube wall temperature for pure forced convection (i.e., no radiation at the tube surface). It appears that as the convection becomes important the temperature distribution curves approach those for pure convection for long tubes. For short tubes, however, the radiative transfer from the tube wall to the external environment reduces the tube wall temperature. For long tubes the emissivity has an effect on tube wall temperature only near the inlet and the exit. As the emissivity is decreased from 1 to 0.01, the wall temperature rises toward that for pure forced convection because heat cannot be dissipated by radiation efficiently.

REFERENCES

1. H. Schlichting, *Boundary Layer Theory*, 6th ed., McGraw-Hill Book Co., New York, 1968.
2. W. M. Kays, *Convective Heat and Mass Transfer*, McGraw-Hill Book Co., New York, 1966.
3. F. K. Moore, *Theory of Laminar Flows*, Princeton University Press, Princeton, N.J., 1964.
4. R. D. Cess, "The Effect of Radiation upon Forced-Convection Heat Transfer," *Appl. Sci. Res.*, Section A, **10**, 430–438, 1962.
5. R. D. Cess, in *Advances in Heat Transfer*, Vol. 1, edited by T. F. Irvine and J. P. Hartnett, pp. 33–50, Academic Press, New York, 1964.
6. E. M. Sparrow and S. H. Lin, "Boundary Layers with Prescribed Heat Flux—Application to Simultaneous Convection and Radiation," *Intern. J. Heat Mass Transfer*, **8**, 437–448, 1965.
7. R. C. Lind and T. Cebeci, "Solution of the Equations of the Compressible Laminar Boundary Layer with Surface Radiation," Douglas Aircraft Co., Report No. DAC-33482, Los Angeles, Calif., December 1966.

8. M. Perlmutter and R. Siegel, "Heat Transfer by Combined Forced Convection and Thermal Radiation in a Heated Tube," *J. Heat Transfer*, **84C**, 301–311, 1962.
9. R. Siegel and M. Perlmutter, "Convective and Radiant Heat Transfer for Flow of a Transparent Gas in a Tube with Gray Wall," *Intern. J. Heat Mass Transfer*, **5**, 639–660, 1962.
10. B. I. Dussan and T. F. Irvine, "Laminar Heat Transfer in a Round Tube with Radiating Flux at the Outer Wall," *Proceedings of the Third International Heat Transfer Conference*, Chicago, Aug. 7–12, Vol. 5, pp. 184–189, 1966.
11. John C. Chen, "Laminar Heat Transfer in a Tube with Nonlinear Radiant Heat-Flux Boundary Condition," *Intern. J. Heat Mass Transfer*, **9**, 433–440, 1966.
12. R. Siegel and E. G. Keshock, "Wall Temperature in a Tube with Forced Convection, Internal Radiation Exchange and Axial Wall Conduction," *NASA Tech. Note* TN D-2116, March 1964.
13. E. G. Keshock and R. Siegel, "Combined Radiation and Convection in a Asymmetrically Heated Parallel Plate Flow Channel," *J. Heat Transfer*, **86C**, 341–350, 1964.
14. R. S. Thorsen, "Heat Transfer in a Tube with Forced Convection, Internal Radiation Exchange, Axial Wall Heat Conduction and Arbitrary Wall Heat Generation," *Intern. J. Heat Mass Transfer*, **12**, 1182–1187, 1969.
15. R. Thorsen and D. Kanchanagom, "The Influence of Internal Radiation Exchange, Arbitrary Wall Heat Generation and Wall Heat Conduction on Heat Transfer in Laminar and Turbulent Flows," *Proceedings of the Fourth International Heat Transfer Conference*, Paris, Vol. 3, Section R2.8, pp. 1–10, 1970.
16. S. T. Liu and R. S. Thorsen, "Combined Forced Convection and Radiation Heat Transfer in Asymmetrically Heated Parallel Plates," *Proceedings of the Heat Transfer and Fluid Mechanics Institute*, pp. 32–44, Stanford University Press, Palo Alto, Calif., 1970.
17. J. C. Jaeger, "Conduction of Heat in a Solid with a Power Law Heat Transfer at Its Surface," *Proc. Cambridge Phil. Soc.*, **46**, 634–641, 1950.
18. P. L. Donoughe and N. B. Livingood, "Exact Solutions of Laminar Boundary-Layer Equations with Constant Property Values for Porous Wall with Variable Temperature," *NASA Tech. Rept*. 1229, 1958.
19. C. M. Usiskin and R. Siegel, "Thermal Radiation from a Cylindrical Enclosure with Specified Heat Flux," *J. Heat Transfer*, **82C**, 369–374, 1960.
20. H. Buckley, "On the Radiation from the Inside of a Circular Cylinder, Part I," *Phil. Mag*., **6**, 447–457, 1928.

NOTES

[1] In writing the second term on the right-hand side of Eq. 7-5a we assumed that Kirchhoff's law is valid (i.e., emissivity = absorptivity). Otherwise this term must be replaced by

$$\bar{\sigma}(\varepsilon T^4 - \alpha T_e^{\,4})$$

where α and ε are the absorptivity and emissivity, respectively, of the surface.

[2] In the transformation from x, y variables to ξ, η variables, various derivatives can be evaluated as:

$$\frac{\partial}{\partial x} = \frac{\partial}{\partial \xi}\frac{d\xi}{dx} + \frac{\partial}{\partial \eta}\frac{d\eta}{ax} = \frac{\xi}{2x}\frac{\partial}{\partial \xi} - \frac{\eta}{2x}\frac{\partial}{\partial \eta} \tag{1}$$

$$\frac{\partial}{\partial y} = \frac{\partial}{\partial \xi}\frac{d\xi}{dy} + \frac{\partial}{\partial \eta}\frac{d\eta}{dy} = \sqrt{\frac{u_\infty}{\nu x}}\frac{\partial}{\partial \eta} \tag{2}$$

$$\frac{\partial^2}{\partial y^2} = \frac{u_\infty}{\nu x}\frac{\partial^2}{\partial \eta^2} \tag{3}$$

[3] Equation 5-10b for the one-dimensional case considered here is written as

$$q^r(x) = \frac{1}{\rho}\left[\varepsilon\bar{\sigma}T_w^{\,4}(x) - (1-\rho)R(x)\right] \tag{1}$$

Assuming that the Kirchhoff law is valid (i.e., $\varepsilon = 1 - \rho$), we can rearrange this relation as

$$R(x) = \bar{\sigma}T_w^{\,4}(x) - \frac{1-\varepsilon}{\varepsilon}\,q^r(x) \tag{2}$$

[4] Differentiation of Eq. 7-42 twice with respect to ξ yields

$$0 = h^*\left(\frac{d^2\theta_w}{d\xi^2} - \frac{d^2\theta_g}{d\xi^2}\right) + 12\theta_w^{\,2}\left(\frac{d\theta_w}{d\xi}\right)^2 + 4\frac{d^2\theta_w}{d\xi^2}\theta_w^{\,3}$$

$$- 4\left[\tfrac{1}{2}\theta_1^{\,4}e^{-2\xi} + \tfrac{1}{2}\theta_2^{\,4}e^{-2(\xi_L-\xi)} + \int_{\xi'=0}^{\xi}\theta_w^{\,4}(\xi')e^{-2(\xi-\xi')}\,d\xi'\right.$$

$$\left.+\int_{\xi'=\xi}^{\xi_L}\theta_w^{\,4}(\xi')e^{-2(\xi'-\xi)}\,d\xi' - \theta_w^{\,4}(\xi)\right] \tag{1}$$

The integral terms inside the bracket can be eliminated by means of Eq. 7-42, and we obtain

$$0 = h^*\left(\frac{d^2\theta_w}{d\xi^2} - \frac{d^2\theta_g}{d\xi^2}\right) + 12\theta_w\left(\frac{d\theta_w}{d\xi}\right)^2 + 4\frac{d^2\theta_w}{d\xi^2}\theta_w^{\,3} - 4\,h^*(\theta_w - \theta_g) + 4 \tag{2}$$

The term $d^2\theta_g/d\xi^2$ is evaluated from Eq. 7-38a as

$$\frac{d^2\theta_g}{d\xi^2} = \mathrm{S}\left[\frac{d\theta_w}{d\xi} - \mathrm{S}(\theta_w - \theta_g)\right] \tag{3}$$

Substituting Eq. 3 into Eq. 2 yields Eq. 7-43

CHAPTER 8

Basic Relations for Radiative Heat Transfer in Participating Media

In this chapter consideration is given to radiative heat transfer in a participating medium, that is, a medium that absorbs, emits, and scatters radiation. The equation of radiative transfer is derived, a formal integration of this equation is given, and formal solutions are obtained for the net radiative heat flux, the gradient of the net radiative heat flux, and the incident radiation for a plane-parallel medium. Models for nongray radiative heat transfer are described, and the reduction of problems with no azimuthal symmetry to problems with azimuthal symmetry is discussed.

8-1 EQUATION OF RADIATIVE TRANSFER

The angular distribution of radiation intensity $I_\nu(\mathbf{r}, \hat{\mathbf{\Omega}})$ satisfies the equation of radiative transfer. A number of different approaches have been applied to derive this equation.

Sampson [1, Chapter 2] derived it directly from the Boltzmann equation by treating radiation as photons; Chandrasekhar [2, pp. 8–9], Kourganoff [3, pp. 16–17], Sobolev [4, pp. 6–9], and Viskanta [5, pp. 29–36; 6, pp. 183–193] obtained derivations by adapting an Eulerian approach and writing a radiant energy balance for a differential volume along a beam of radiation. Weinberg and Wigner [7, p. 232] and Murray [8, pp. 242–245] gave derivations of the equivalent equation in neutron transport theory.

Consider an absorbing, emitting, scattering medium characterized by a spectral absorption coefficient κ_ν and spectral scattering coefficient σ_ν. A beam of monochromatic radiation of intensity $I_\nu(s, \hat{\mathbf{\Omega}}, t)$ travels in this medium in the direction $\hat{\mathbf{\Omega}}$ along a path s. The equation of radiative transfer can be derived conveniently by Eulerian approach. We choose the elemental volume as a right-angle cylinder of cross section dA and length ds about position s with the cylinder axis lying along s as illustrated in Fig. 8-1. Let $I_\nu(s, \hat{\mathbf{\Omega}}, t)$ be the intensity at s, and $I_\nu(s, \hat{\mathbf{\Omega}}, t) + dI_\nu$ the intensity at $s + ds$, where dI_ν represents the net increase in the spectral intensity. The quantity

$$dI_\nu(s, \hat{\mathbf{\Omega}}, t)\, dA\, d\Omega\, d\nu\, dt \tag{8-1a}$$

represents the difference in the radiative energy crossing the surfaces dA at $s + ds$ and s in the time interval dt about t and in the frequency interval $d\nu$ about ν, and contained in an element of solid angle $d\Omega$ about the direction $\hat{\mathbf{\Omega}}$.

Let W_ν denote the net gain of radiative energy by the beam in this volume element per unit volume, per unit time about t, per unit frequency about ν, and per unit solid angle about the direction $\hat{\mathbf{\Omega}}$. Then the quantity

$$W_\nu\, dA\, ds\, d\Omega\, d\nu\, dt \tag{8-1b}$$

represents the net gain of radiative energy by the beam contained in an element of solid angle $d\Omega$ about the direction $\hat{\mathbf{\Omega}}$, in the time interval dt, in the frequency interval $d\nu$, and in a cylindrical element of volume $dA\, ds$. By

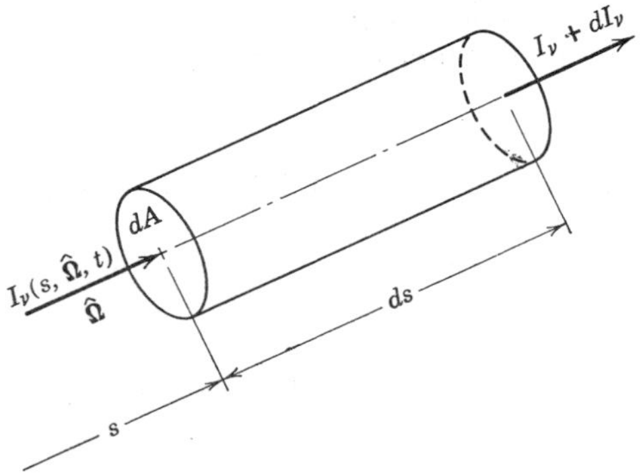

Fig. 8-1. Coordinates for derivation of equation of radiative transfer.

equating Eqs. 8-1a and 8-1b, we obtain

$$\frac{dI_\nu(s, \hat{\mathbf{\Omega}}, t)}{ds} = W_\nu \tag{8-1c}$$

If c denotes the speed of propagation of radiation in the medium, the distance ds traversed by the beam is given by

$$ds = c\,dt \tag{8-2}$$

Then Eq. 8-1c can be written in the form

$$\frac{1}{c}\frac{DI_\nu(s, \hat{\mathbf{\Omega}}, t)}{Dt} = W_\nu \tag{8-3}$$

where D/Dt is the substantial derivative for the element moving with a velocity c and is related to the partial derivatives with respect to time and space coordinates by

$$\frac{DI_\nu}{Dt} \equiv \frac{\partial I_\nu}{\partial t} + c\hat{\mathbf{\Omega}} \cdot \nabla I_\nu \tag{8-4a}$$

This relation can be written alternatively in the form

$$\frac{DI_\nu}{Dt} \equiv \frac{\partial I_\nu}{\partial t} + c\nabla \cdot (\hat{\mathbf{\Omega}} I_\nu) \tag{8-4b}$$

where we have utilized the vector identity

$$\nabla \cdot (\hat{\mathbf{\Omega}} I_\nu) = I_\nu \nabla \cdot \hat{\mathbf{\Omega}} + \hat{\mathbf{\Omega}} \cdot \nabla I_\nu = \hat{\mathbf{\Omega}} \cdot \nabla I_\nu$$

When the direction $\hat{\mathbf{\Omega}}$ is along the path s, we have $\nabla \cdot (\hat{\mathbf{\Omega}} I_\nu) = \partial I_\nu/\partial s$ and Eq. 8-4b can be written as[1]

$$\frac{DI_\nu}{Dt} \equiv \frac{\partial I_\nu}{\partial t} + c\frac{\partial I_\nu}{\partial s} \tag{8-4c}$$

where $\partial/\partial s$ denotes differentiation along the path s.

By utilizing the above relations for the substantial derivative, we can write Eq. 8-3 in the form

$$\frac{1}{c}\frac{\partial I_\nu}{\partial t} + \nabla \cdot (\hat{\mathbf{\Omega}} I_\nu) = W_\nu \tag{8-5a}$$

or

$$\frac{1}{c}\frac{\partial I_\nu}{\partial t} + \frac{\partial I_\nu}{\partial s} = W_\nu \tag{8-5b}$$

An explicit relation will now be derived for W_ν. For an absorbing, emitting, and scattering medium W_ν is composed of components to account for the gains and losses of radiant energy and it may be written formally as

$$W_\nu = W_{\text{emission}} - W_{\text{absorption}} + W_{\text{in-scattering}} - W_{\text{out-scattering}} \tag{8-6a}$$

The first term on the right-hand side represents the gain of radiant energy by the beam due to radiation emitted by the matter per unit time, volume, solid angle, and frequency and will be designated by the symbol $j_\nu^e(s, t)$. Hence

$$W_{\text{emission}} = j_\nu^e(s, t) \tag{8-6b}$$

The second term represents the loss of radiant energy from the beam due to the absorption of radiation by the matter per unit time, volume, solid angle, and frequency and is given by (see Eq. 1-57)

$$W_{\text{absorption}} = \kappa_\nu(s) I_\nu(s, \hat{\mathbf{\Omega}}, t) \tag{8-6c}$$

The third term is the gain of radiant energy by the beam due to radiation incident on the matter from all directions in the spherical space that is scattered by the matter in the direction $\hat{\mathbf{\Omega}}$ per unit time, volume, solid angle, and frequency. For purely coherent scattering and an isotropic medium it is given by[2] (see Eq. 1-66)

$$W_{\text{in-scattering}} = \frac{1}{4\pi} \sigma_\nu(s) \int_{4\pi} p(\hat{\mathbf{\Omega}}' \cdot \hat{\mathbf{\Omega}}) I_\nu(s, \hat{\mathbf{\Omega}}', t)\, d\Omega' \tag{8-6d}$$

where $\hat{\mathbf{\Omega}}' \cdot \hat{\mathbf{\Omega}} = \cos\theta_0$, and θ_0 is the angle between the incident and scattered rays.

The last term represents the loss of radiant energy from the beam due to scattering by the matter out of the direction $\hat{\mathbf{\Omega}}$ per unit time, volume, solid angle, and frequency, and is given by (see Eq. 1-60)

$$W_{\text{out-scattering}} = \sigma_\nu(s) I_\nu(s, \hat{\mathbf{\Omega}}, t) \tag{8-6e}$$

Substitution of Eqs. 8-6 into Eq. 8-5b yields the equation of radiative transfer in the form

$$\frac{1}{c}\frac{\partial I_\nu(s, \hat{\mathbf{\Omega}}, t)}{\partial t} + \frac{\partial I_\nu(s, \hat{\mathbf{\Omega}}, t)}{\partial s} + [\kappa_\nu(s) + \sigma_\nu(s)] I_\nu(s, \hat{\mathbf{\Omega}}, t)$$
$$= j_\nu^e(s, t) + \frac{1}{4\pi}\sigma_\nu(s) \int_{\Omega'=4\pi} p(\hat{\mathbf{\Omega}}' \cdot \hat{\mathbf{\Omega}}) I_\nu(s, \hat{\mathbf{\Omega}}', t)\, d\Omega' \tag{8-7}$$

For most engineering applications the term $(1/c)(\partial I_\nu/\partial t)$ in this equation can be neglected in comparison to other terms because of the large magnitude of the speed of propagation c. Then Eq. 8-7 simplifies to

$$\frac{dI_\nu(s, \hat{\mathbf{\Omega}}, t)}{ds} + [\kappa_\nu(s) + \sigma_\nu(s)] I_\nu(s, \hat{\mathbf{\Omega}}, t)$$
$$= j_\nu^e(s, t) + \frac{1}{4\pi}\sigma_\nu(s) \int_{\Omega'=4\pi} p(\hat{\mathbf{\Omega}}' \cdot \hat{\mathbf{\Omega}}) I_\nu(s, \hat{\mathbf{\Omega}}', t)\, d\Omega' \tag{8-8}$$

In this equation the intensity depends on time if the emission term $j_\nu{}^e(s, t)$ is a function of time. Therefore in Eq. 8-8 the time variable is considered merely as a parameter.

If we assume also that a local thermodynamic equilibrium (LTE) is established and that the Kirchhoff law is valid, the emission term $j_\nu{}^e(s)$ is related to the Planck function by (see Eq. 1-59)

$$j_\nu{}^e(s) = \kappa_\nu(s) I_{\nu b}(T) \tag{8-9}$$

Then Eq. 8-8 can be written in the form

$$\frac{1}{\beta_\nu(s)} \frac{dI_\nu(s, \hat{\mathbf{\Omega}})}{ds} + I_\nu(s, \hat{\mathbf{\Omega}}) = S_\nu(s) \tag{8-10a}$$

where we have defined

$$S_\nu(s) \equiv (1 - \omega_\nu) I_{\nu b}(T) + \frac{1}{4\pi} \omega_\nu \int_{4\pi} p(\hat{\mathbf{\Omega}}' \cdot \hat{\mathbf{\Omega}}) I_\nu(s, \hat{\mathbf{\Omega}}')\, d\Omega' \tag{8-10b}$$

$$\beta_\nu(s) \equiv \kappa_\nu(s) + \sigma_\nu(s) \tag{8-10c}$$

$$\omega_\nu \equiv \frac{\sigma_\nu(s)}{\beta_\nu(s)}, \qquad 1 - \omega_\nu \equiv \frac{\kappa_\nu(s)}{\beta_\nu(s)} \tag{8-10d}$$

$$I_{\nu b}(T) \equiv \frac{2h\nu^3}{c^2[\exp(h\nu/kT) - 1]} \tag{8-10e}$$

Here $S_\nu(s)$ is called the spectral *source function*, $\beta_\nu(s)$ the spectral *extinction coefficient*, and ω_ν the spectral *albedo*, which is the ratio of the scattering coefficient to the extinction coefficient. For simplicity we have omitted the time variable in the above relations.

We now examine some special cases of Eqs. 8-10.

(a) Gray Medium

When the radiative properties of the medium are independent of frequency, the integration of Eqs. 8-10 over all frequencies yields

$$\frac{1}{\beta(s)} \frac{dI(s, \hat{\mathbf{\Omega}})}{ds} + I(s, \hat{\mathbf{\Omega}}) = (1 - \omega) \frac{n^2 \bar{\sigma} T^4}{\pi} + \frac{\omega}{4\pi} \int_{4\pi} p(\hat{\mathbf{\Omega}}' \cdot \hat{\mathbf{\Omega}}) I(s, \hat{\mathbf{\Omega}}')\, d\Omega' \tag{8-11a}$$

where we have defined

$$I(s, \hat{\mathbf{\Omega}}) \equiv \int_{\nu=0}^{\infty} I_\nu(s, \hat{\mathbf{\Omega}})\, d\nu \tag{8-11b}$$

$$\int_{\nu=0}^{\infty} I_{\nu b}(T)\, d\nu \equiv I_b(T) = \frac{n^2 \bar{\sigma} T^4}{\pi} \tag{8-11c}$$

and n is the refractive index of the medium.

(b) Purely Scattering Medium

For a medium that does not absorb or emit radiation but only scatters it, we set $\omega_\nu = 1$ (i.e., $\kappa_\nu = 0$) and Eqs. 8-10 simplify to

$$\frac{1}{\beta_\nu(s)} \frac{dI_\nu(s, \hat{\mathbf{\Omega}})}{ds} + I_\nu(s, \hat{\mathbf{\Omega}}) = \frac{1}{4\pi} \int_{4\pi} p(\hat{\mathbf{\Omega}}' \cdot \hat{\mathbf{\Omega}}) I_\nu(s, \hat{\mathbf{\Omega}}')\, d\Omega' \tag{8-12}$$

In this equation the frequency dependence of the intensity is merely a parameter, since for any given frequency Eq. 8-12 is uncoupled from equations for other frequencies.

(c) Purely Absorbing and Emitting Medium

For a medium that absorbs and emits radiation but does not scatter it, we set $\omega_\nu = 0$ (i.e., $\sigma_\nu = 0$) in Eqs. 8-10 and obtain

$$\frac{1}{\kappa_\nu(s)} \frac{dI_\nu(s, \hat{\mathbf{\Omega}})}{ds} + I_\nu(s, \hat{\mathbf{\Omega}}) = I_{\nu b}(T) \tag{8-13}$$

(d) Diathermal (Nonparticipating) Medium

A nonabsorbing, nonemitting, and nonscattering medium is called a *diathermal* or a *nonparticipating* medium. For such a medium the absorption and scattering coefficients are zero. By setting $\kappa_\nu = 0$ and $\sigma_\nu = 0$, Eqs. 8-10 simplify to

$$\frac{dI_\nu}{ds} = 0 \qquad \text{or} \qquad \nabla \cdot (\hat{\mathbf{\Omega}} I_\nu) = 0 \tag{8-14}$$

This equation implies that the radiation intensity remains constant everywhere in a diathermal medium.

Divergence of Radiative Heat Flux Vector

The divergence of the net radiative heat flux vector $\nabla \cdot \mathbf{q}^r$ characterizes the net radiative energy emitted or absorbed by the matter per unit time and per unit volume over all frequencies because of radiation to and from the entire spherical space. To derive a relation defining $\nabla \cdot \mathbf{q}^r$ we consider the equation of radiative transfer (Eq. 8-10) in the form

$$\nabla \cdot [\hat{\mathbf{\Omega}} I_\nu(s, \hat{\mathbf{\Omega}})] + \beta_\nu I_\nu(s, \hat{\mathbf{\Omega}}) = \kappa_\nu I_{\nu b}(T) + \frac{\sigma_\nu}{4\pi} \int_{\Omega'=4\pi} p(\hat{\mathbf{\Omega}}' \cdot \hat{\mathbf{\Omega}}) I_\nu(s, \hat{\mathbf{\Omega}}')\, d\Omega' \tag{8-15}$$

We operate on both sides of this equation by the operator $\int_{4\pi} d\Omega$ and obtain

$$\nabla \cdot \mathbf{q}_\nu{}^r = 4\pi\kappa_\nu I_{\nu b}(T) - \kappa_\nu \int_{4\pi} I_\nu(s, \hat{\boldsymbol{\Omega}})\, d\Omega \tag{8-16a}$$

since

$$\mathbf{q}_\nu{}^r = \int_{4\pi} \hat{\boldsymbol{\Omega}} I_\nu(s, \hat{\boldsymbol{\Omega}})\, d\Omega \tag{8-16b}$$

$$\frac{1}{4\pi}\int_{4\pi} p(\hat{\boldsymbol{\Omega}}' \cdot \hat{\boldsymbol{\Omega}})\, d\Omega = 1 \tag{8-16c}$$

$$\int_{4\pi} d\Omega = 4\pi \tag{8-16d}$$

Integration of Eq. 8-16 over all frequencies yields the desired relation as

$$\nabla \cdot \mathbf{q}^r = 4\pi \int_{\nu=0}^{\infty} \kappa_\nu I_{\nu b}(T)\, d\nu - \int_{\nu=0}^{\infty} \kappa_\nu \left[\int_{4\pi} I_\nu(s, \hat{\boldsymbol{\Omega}}')\, d\Omega'\right] d\nu \tag{8-17a}$$

where we have defined

$$\mathbf{q}^r \equiv \int_{\nu=0}^{\infty} q_\nu{}^r\, d\nu \tag{8-17b}$$

The first and the second terms on the right-hand side of Eq. 8-18a represent, respectively, the radiation emitted and absorbed by the matter per unit time and per unit volume. Therefore $\nabla \cdot \mathbf{q}^r$ represents the net emission or the net absorption of radiation (i.e., per unit time and per unit volume) by the material, depending on whether it is positive or negative, respectively. Equation 8-18a is sometimes referred to as the *equation for the conservation of radiative energy*. We note that the scattering does not enter into this equation.

For a medium that contains distributed energy sources of strength g Btu/hr ft^3 and in which the energy transfer is by radiation only (i.e., convective and conductive modes of heat transfer are negligible), $\nabla \cdot \mathbf{q}^r$ should be equal to g; that is, the energy equation becomes[3]

$$\nabla \cdot \mathbf{q}^r = g \tag{8-18a}$$

where g may be a function of position and time. For the one-dimensional case, for example, this equation may be written as

$$\frac{dq^r(y)}{dy} = g(y) \tag{8-18b}$$

8-2 RADIATIVE EQUILIBRIUM

Consider a medium in which there are no internal energy sources or sinks, the energy transfer is by pure radiation (i.e., conduction and convection modes of heat transfer are negligible), and the steady-state temperature distribution is established. In such a medium the radiation absorbed by the matter at any location should be equal to the radiation emitted by the matter. This condition is equivalent to the requirement that the net emission (or net absorption) of radiation should be equal to zero everywhere in the medium and is called *radiative equilibrium;* the equation characterizing radiative equilibrium is obtained by equating the right-hand side of Eq. 8-17a equal to zero:

$$4\pi \int_{\nu=0}^{\infty} \kappa_\nu I_{\nu b}(T)\, d\nu = \int_{\nu=0}^{\infty} \kappa_\nu \left[\int_{\phi=0}^{2\pi} \int_{\mu=-1}^{1} I_\nu(s, \mu, \phi)\, d\mu\, d\phi \right] d\nu \tag{8-19}$$

or, alternatively,[4]

$$\nabla \cdot \mathbf{q}^r = 0 \tag{8-20}$$

Equation 8-19 or 8-20 constitutes the mathematical definition of radiative equilibrium.

When the intensity of radiation is independent of the azimuthal angle ϕ, Eq. 8-19 simplifies to

$$\int_{\nu=0}^{\infty} \kappa_\nu I_{\nu b}(T)\, d\nu = \frac{1}{2} \int_{\nu=0}^{\infty} \kappa_\nu \left[\int_{\mu=-1}^{1} I_\nu(s, \mu)\, d\mu \right] d\nu \tag{8-21}$$

For a gray medium with azimuthal symmetry Eq. 8-19 simplifies to

$$I_b(T) = \frac{1}{2} \int_{-1}^{1} I(s, \mu)\, d\mu \tag{8-22}$$

or

$$\frac{n^2 \bar{\sigma} T^4}{\pi} = \frac{1}{2} \int_{-1}^{1} I(s, \mu)\, d\mu \tag{8-23}$$

For the one-dimensional case with the oy axis chosen as the coordinate axis, the condition of radiative equilibrium given by Eq. 8-20 becomes

$$\frac{dq^r}{dy} = 0 \tag{8-24}$$

This relation implies that the net radiative heat flux in the y direction is constant at all depths along the oy axis. However, the constancy of the net radiative heat flux q^r can by no means be generalized to the constancy of the spectral radiative heat flux $q_\nu{}^r$; the spectral radiative heat flux may vary with depth even though q^r remains constant.

8-3 FORMAL INTEGRATION OF EQUATION OF RADIATIVE TRANSFER

Consider the equation of radiative transfer in the form

$$\frac{1}{\beta_\nu(s)} \frac{dI_\nu(s, \hat{\mathbf{\Omega}})}{ds} + I_\nu(s, \hat{\mathbf{\Omega}}) = S_\nu(s, \hat{\mathbf{\Omega}}) \tag{8-25}$$

where the spectral source function $S_\nu(s, \hat{\mathbf{\Omega}})$ is defined as

$$S_\nu(s, \hat{\mathbf{\Omega}}) \equiv (1 - \omega_\nu) I_{\nu b}(T) + \frac{1}{4\pi} \omega_\nu \int_{4\pi} p(\hat{\mathbf{\Omega}}' \cdot \hat{\mathbf{\Omega}}) I_\nu(s, \hat{\mathbf{\Omega}}')\, d\Omega' \tag{8-26}$$

Equation 8-25 is an integro- partial differential equation, because the directional derivative d/ds involves partial derivatives with respect to the space variables when it is written explicitly for a specific coordinate system,[5] and the intensity $I_\nu(s, \hat{\mathbf{\Omega}})$ appears under the integral sign in the source function. Therefore the solution of Eq. 8-25 is very complex even for one-dimensional cases. However, it is instructive to examine the formal integration of Eq. 8-25 along the path s in the direction $\hat{\mathbf{\Omega}}$ subject to a formal boundary condition

$$I_\nu(s, \hat{\mathbf{\Omega}}) = I_{0\nu} \quad \text{at } s = s_0 \tag{8-27}$$

Figure 8-2 shows the path considered.

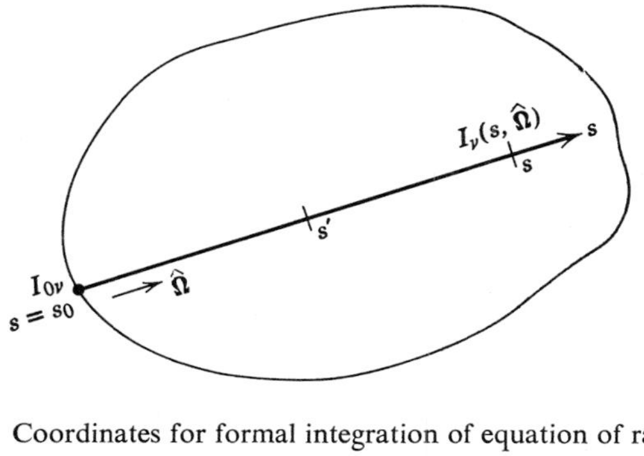

Fig. 8-2. Coordinates for formal integration of equation of radiative transfer.

The integration of Eq. 8-25 along the path s with the boundary condition 8-27 gives

$$I_\nu(s,\hat{\Omega}) = I_{0\nu}\exp\left[-\int_{s0}^{s}\beta_\nu(s')\,ds'\right] + \int_{s0}^{s}\beta_\nu(s')S_\nu(s',\hat{\Omega})\exp\left[-\int_{s'}^{s}\beta_\nu(s'')\,ds''\right]ds' \tag{8-28}$$

Equation 8-28 is called the integral form of the equation of radiative transfer, but it is *not* a solution in the real sense because the source function $S_\nu(s', \hat{\Omega})$ is dependent on the intensity. However, this integral form gives a good insight into the physical significance of the factors affecting the intensity at any point along the path s.

The first term on the right-hand side of Eq. 8-28 is the contribution of the radiation intensity originating from the boundary at $s = s_0$. The intensity $I_{0\nu}$ from the boundary surface diminishes as a result of absorption and out-scattering by a factor $\exp\,[-\int_{s_0}^{s}\beta_\nu(s')\,ds']$ when it reaches location s along the path of integration.

The second term on the right-hand side of Eq. 8-28 is the contribution of the source function $S_\nu(s, \hat{\Omega})$ distributed along the path from s_0 to s.

8-4 EQUATION OF RADIATIVE TRANSFER FOR PLANE-PARALLEL MEDIUM

Because it is very difficult to solve the equation of radiative transfer for the general three-dimensional case, one-dimensional symmetry is considered for most practical purposes. Here we present the simplification of the equation of radiative transfer for a plane-parallel system.

Consider a medium that is stratified in planes perpendicular to the oy axis so that the radiative properties are uniform over each layer. Let s be the length measured along an arbitrary direction $\hat{\Omega}$, and θ be the polar angle between the direction $\hat{\Omega}$ and the positive oy axis, as shown in Fig. 8-3. The directional derivative d/ds can be expressed in terms of the derivatives with respect to the space coordinate y as

$$\frac{d}{ds} = \frac{\partial}{\partial y}\frac{dy}{ds} = \mu\frac{\partial}{\partial y} \tag{8-29}$$

where μ is the cosine of the angle θ between the direction $\hat{\Omega}$ of the beam and the oy axis, that is,

$$\mu \equiv \cos\theta, \tag{8-30}$$

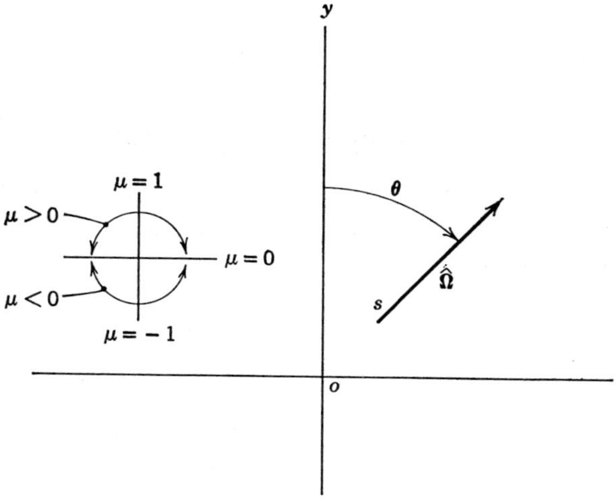

Fig. 8-3. Coordinates for plane-parallel system.

and the partial derivatives with respect to x and z vanish for a plane-parallel system. Then the equation of radiative transfer 8-10 becomes

$$\frac{\mu}{\beta_\nu}\frac{\partial I_\nu(y, \mu, \phi)}{dy} + I_\nu(y, \mu, \phi) = S_\nu(y, \mu, \phi) \tag{8-31a}$$

where

$$S_\nu(y, \mu, \phi) \equiv (1 - \omega_\nu)I_{\nu b}[T(y)] + \frac{\omega_\nu}{4\pi}\int_{\phi'=0}^{2\pi}\int_{\mu'=-1}^{1} p(\mu_0)I_\nu(y, \mu', \phi')\, d\mu'\, d\phi' \tag{8-31b}$$

and μ_0 is the cosine of the angle θ_0 between the directions of the incident and scattered rays, that is,

$$\hat{\mathbf{\Omega}}' \cdot \hat{\mathbf{\Omega}} = \cos\theta_0 \equiv \mu_0 \tag{8-32}$$

and μ_0 is given by the relation (see Eq. 1-65)

$$\mu_0 = \mu\mu' + \sqrt{1-\mu^2}\sqrt{1-\mu'^2}\cos(\phi - \phi') \tag{8-33}$$

In the problems of radiative heat transfer it is convenient to define an optical variable τ as[6]

$$d\tau \equiv \beta_\nu\, dy \qquad \text{or} \qquad \tau = \int_0^y \beta_\nu\, dy' \tag{8-34}$$

Then the equation of radiative transfer 8-31 becomes

$$\mu\frac{\partial I_\nu(\tau, \mu, \phi)}{\partial\tau} + I_\nu(\tau, \mu, \phi) = S_\nu(\tau, \mu, \phi) \tag{8-35a}$$

where

$$S_\nu(\tau, \mu, \phi) \equiv (1 - \omega_\nu)I_{\nu b}[T(\tau)] + \frac{\omega_\nu}{4\pi}\int_{\phi'=0}^{2\pi}\int_{\mu'=-1}^{1} p(\mu_0)I_\nu(\tau, \mu', \phi')\, d\mu'\, d\phi' \tag{8-35b}$$

Azimuthally Symmetric Radiation

If the boundary conditions for the equation of radiative transfer are azimuthally symmetric, the radiation intensity in the medium is independent of ϕ and Eq. 8-35 simplifies to

$$\mu \frac{\partial I_\nu(\tau, \mu)}{\partial \tau} + I_\nu(\tau, \mu) = (1 - \omega_\nu)I_{\nu b}[T(\tau)] + \frac{\omega_\nu}{4\pi}\int_{\mu'=-1}^{1} I_\nu(\tau, \mu')\int_{\phi'=0}^{2\pi} p(\mu_0)\, d\phi\, d\mu' \tag{8-36}$$

To perform the integration with respect to ϕ' we expand the phase function $p(\mu_0)$ in Legendre polynomials in the form[7] (see Eq. 2-55)

$$p(\mu_0) = \sum_{n=0}^{N} a_n P_n(\mu_0), \quad a_0 = 1 \tag{8-37}$$

where $P_n(\mu_0)$ is the Legendre polynomial of order n and argument μ_0, and we note that, when the argument μ_0 of Legendre polynomials $P_n(\mu_0)$ is related to μ and μ' as in Eq. 8-33, the following relation exists among Legendre polynomials [10, pp. 326–328]:

Grad. & Ryzhik p. 1015

$$P_n(\mu_0) = P_n(\mu)P_n(\mu') + 2\sum_{m=1}^{\infty} \frac{(n - m)!}{(n + m)!} P_n^{\,m}(\mu)P_n^{\,m}(\mu') \cos m(\phi - \phi') \tag{8-38}$$

where $P_n^{\,m}(\mu)$ is the associated Legendre functions. Then Eq. 8-37 becomes

$$p(\mu_0) = \sum_{n=0}^{N} a_n P_n(\mu)P_n(\mu') + 2\sum_{n=1}^{N}\sum_{m=1}^{n} a_n^{\,m} P_n^{\,m}(\mu)P_n^{\,m}(\mu') \cos m(\phi - \phi') \tag{8-39a}$$

where

$$a_n^{\,m} \equiv a_n \frac{(n - m)!}{(n + m)!}, \quad n = m, \ldots, N;\ 0 \le m \le N \tag{8-39b}$$

$$a_0 = 1$$

Integrating Eq. 8-39 over ϕ' from 0 to 2π, we obtain

$$\int_0^{2\pi} p(\mu_0)\, d\phi' = 2\pi \sum_{n=0}^{N} a_n P_n(\mu)P_n(\mu'), \quad a_0 = 1 \tag{8-40}$$

since the integration of $\cos m(\phi - \phi')$ over 2π is zero for integral values of m. Substitution of Eq. 8-40 into Eq. 8-36 yields the equation of radiative transfer for azimuthal symmetry as

$$\mu \frac{\partial I_\nu(\tau, \mu)}{\partial \tau} + I_\nu(\tau, \mu) = S_\nu(\tau, \mu) \tag{8-41}$$

where

$$S_\nu(\tau, \mu) \equiv (1 - \omega_\nu)I_{\nu b}[T(\tau)] + \frac{\omega_\nu}{2}\int_{-1}^{1} p(\mu, \mu')I_\nu(\tau, \mu')\, d\mu' \tag{8-42a}$$

$$p(\mu, \mu') \equiv \sum_{n=0}^{N} a_n P_n(\mu)P_n(\mu'), \quad a_0 = 1 \tag{8-42b}$$

and the phase function $p(\mu, \mu')$ is independent of the azimuthal angle. Clearly, several special cases can be obtained from the phase function given by Eq. 8-42b. The case $N = 0$ gives *isotropic* scattering, $N = 1$ yields the phase function for *linearly anisotropic* scattering, that is,

$$p(\mu, \mu') = 1 + a_1\mu\mu' \tag{8-43}$$

and $N = 2$ corresponds to the phase function for *second-degree anisotropic* scattering:

$$p(\mu, \mu') = 1 + a_1\mu\mu' + \tfrac{1}{4}a_2(3\mu^2 - 1)(3\mu'^2 - 1) \tag{8-44}$$

The phase function for the *Rayleigh* scattering is obtainable from Eq. 8-44 by setting $a_1 = 0$ and $a_2 = \frac{1}{2}$, that is,

$$p(\mu, \mu') = 1 + \tfrac{1}{8}(3\mu^2 - 1)(3\mu'^2 - 1) = \tfrac{3}{8}[3 - \mu^2 + (3\mu^2 - 1)\mu'^2] \tag{8-45}$$

The equivalence of this expression and the Rayleigh phase function given by Eq. 2-89 is shown by averaging the Rayleigh phase function of Eq. 2-89 over ϕ as

$$p(\mu, \mu') = \frac{1}{2\pi}\int_0^{2\pi} \tfrac{3}{4}(1 + \mu_0^2)\, d\phi' \tag{8-46a}$$

where

$$\mu_0 = \mu\mu' + \sqrt{1 - \mu^2}\sqrt{1 - \mu'^2}\cos(\phi - \phi') \tag{8-46b}$$

When this integration is performed, the terms involving $\cos(\phi - \phi')$ vanish and the result given by Eq. 8-45 is recovered.

8-5 EQUATION OF RADIATIVE TRANSFER FOR CYLINDRICAL AND SPHERICAL SYMMETRY

In this section we present the equation of radiative transfer for isotropic scattering for the cylindrical and spherical coordinate systems. These equations can be obtained from the equation of radiative transfer for isotropic scattering, Eq. 8-10, that is,

$$\frac{dI_\nu(s, \hat{\mathbf{\Omega}})}{ds} + \beta_\nu I_\nu(s, \hat{\mathbf{\Omega}}) = \kappa_\nu I_{\nu b}(T) + \frac{1}{4\pi}\sigma_\nu \int_{4\pi} I_\nu(s, \hat{\mathbf{\Omega}}')\, d\Omega' \tag{8-47}$$

by expressing the directional derivative d/ds in this equation in terms of the partial derivatives with respect to the space variables in the coordinate system considered. The transformation of d/ds into the spherical and cylindrical coordinates has been described by Weinberg and Wigner [7, pp. 273–276] for a related equation in neutron transport theory.

Spherical Symmetry

In a system with spherical symmetry the radiation intensity at any location depends on two variables: r, the radial distance from the origin; and μ, the cosine of the angle between the direction $\hat{\mathbf{\Omega}}$ of the beam and the extension of the radius vector $\mathbf{r}$. Then the directional derivative d/ds can be related to the partial derivatives with respect to r and μ by

$$\frac{d}{ds} = \frac{\partial}{\partial r}\frac{dr}{ds} + \frac{\partial}{\partial \mu}\frac{d\mu}{ds} \tag{8-48}$$

The derivatives dr/ds and $d\mu/ds$ can be evaluated by referring to the geometry shown in Fig. 8-4. A change in the position by a distance ds along the path s in the direction $\hat{\mathbf{\Omega}}$ will result in an increase dr in the radial position and in a

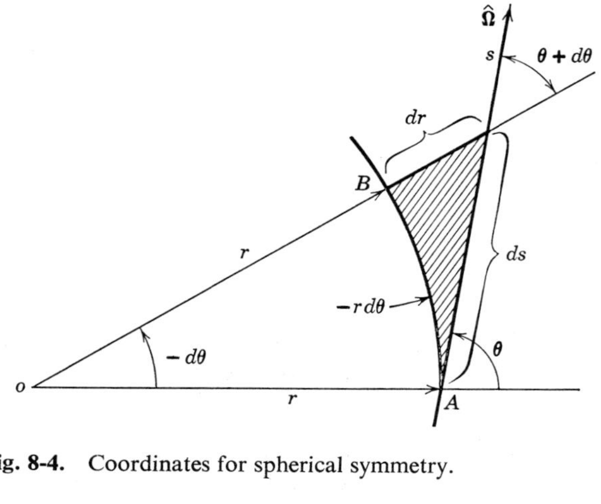

Fig. 8-4. Coordinates for spherical symmetry.

decrease by $d\theta$ in the angular position. Then, by referring to Fig. 8-4, we write

$$\frac{dr}{ds} = \cos\theta \equiv \mu \tag{8-49}$$

$$-\frac{d\theta}{ds} = \frac{\sin\theta}{r} \tag{8-50a}$$

or

$$\frac{d\mu}{ds} = \frac{\sin^2\theta}{r} = \frac{1-\mu^2}{r} \tag{8-50b}$$

since

$$d\mu = d(\cos\theta) = -\sin\theta\, d\theta \tag{8-51}$$

Substituting Eqs. 8-49 and 8-50b into Eq. 8-48, we obtain

$$\frac{d}{ds} = \mu\frac{\partial}{\partial r} + \frac{1-\mu^2}{r}\frac{\partial}{\partial \mu} \tag{8-52}$$

Then the equation of radiative transfer 8-47 for spherical symmetry becomes

$$\mu\frac{\partial I_\nu(r,\mu)}{\partial r} + \frac{1-\mu^2}{r}\frac{\partial I_\nu(r,\mu)}{\partial \mu} + \beta_\nu I_\nu(r,\mu) = \kappa_\nu I_{\nu b}(T) + \tfrac{1}{2}\sigma_\nu\int_{-1}^{1} I_\nu(r,\mu')\, d\mu' \tag{8-53}$$

Cylindrical Symmetry

Consider a cylindrical coordinate system with the z axis taken as the axis of the cylinder. We assume that the radiation intensity does not vary in the z direction. The radial position vector can be specified by the coordinates r, ϕ_r, and the direction $\hat{\boldsymbol{\Omega}}$ of the radiation intensity at that point can be specified by the coordinates θ, ϕ_Ω, where θ is the polar angle between the direction $\hat{\boldsymbol{\Omega}}$ and the z axis. The cylindrical symmetry requires that the radiation intensity remain invariant under rotation about the z axis. Then the radiation intensity depends on three variables: the radial distance r from the z axis, the polar angle θ, and the difference $\phi = \phi_\Omega - \phi_r$ of the aximuthal angles. Figure 8-5 illustrates the coordinates considered here. The directional derivative d/ds is related to the partial derivatives with respect to these variables by

$$\frac{d}{ds} = \frac{\partial}{\partial r}\frac{dr}{ds} + \frac{\partial}{\partial \phi}\frac{d\phi}{ds} \tag{8-54}$$

The derivatives dr/ds and $d\phi/ds$ can be evaluated. A change in the position by ds along the path s in the direction $\hat{\boldsymbol{\Omega}}$ will result in an increase by dr in the radial position and a decrease by $d\phi$ in the angle ϕ, but the angle θ will

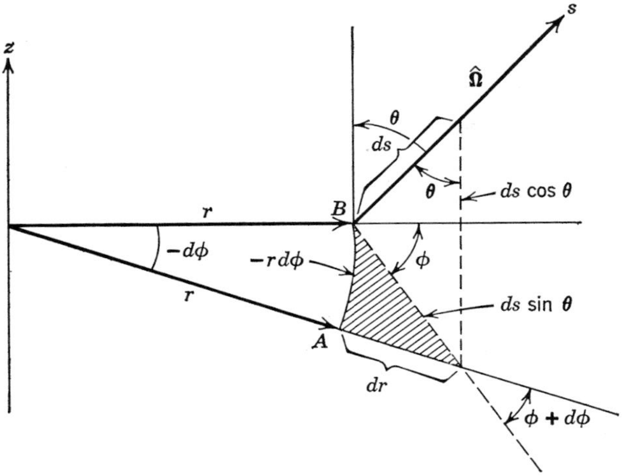

Fig. 8-5. Coordinates for cylindrical symmetry.

remain unchanged. We write

$$\frac{dr}{ds \sin \theta} = \cos \phi \tag{8-55}$$

$$-\frac{r\, d\phi}{ds \sin \theta} = \sin \phi \tag{8-56}$$

Substitution of Eqs. 8-55 and 8-56 into Eq. 8-54 yields

$$\frac{d}{ds} = \sin \theta \cos \phi \frac{\partial}{\partial r} - \frac{\sin \theta \sin \phi}{r} \frac{\partial}{\partial \phi} \tag{8-57}$$

Then the equation of radiative transfer 8-47 in cylindrical symmetry becomes

$$\sin \theta \left[\cos \phi \frac{\partial I_\nu(r, \theta, \phi)}{\partial r} - \frac{\sin \phi}{r} \frac{\partial I_\nu(r, \theta, \phi)}{\partial \phi} \right] + \beta_\nu I_\nu(r, \theta, \phi)$$
$$= \kappa_\nu I_{\nu b}(T) + \frac{1}{4\pi} \sigma_\nu \int_{\phi'=0}^{2\pi} \int_{\theta'=0}^{\pi} I_\nu(r, \theta', \phi') \sin \theta' \, d\theta' \, d\phi' \tag{8-58}$$

8-6 FORMAL SOLUTION OF EQUATION OF RADIATIVE TRANSFER IN PLANE-PARALLEL GEOMETRY WITH AZIMUTHAL SYMMETRY

In this section consideration is given to the formal solution of the equation of radiative transfer in plane-parallel geometry for azimuthal symmetry.

The treatment of problems with no azimuthal symmetry is discussed at the end of this chapter, where it is shown that the equation of radiative transfer with no azimuthal symmetry can be transformed into a set of equations of azimuthal symmetry.

Consider the equation of radiative transfer given in the form (see Eq. 8-41)

$$\mu \frac{\partial I_\nu(\tau, \mu)}{\partial \tau} + I_\nu(\tau, \mu) = S_\nu(\tau, \mu), \quad \text{in } 0 \le \tau \le \tau_0, \ -1 \le \mu \le 1 \tag{8-59a}$$

where

$$S_\nu(\tau, \mu) \equiv (1 - \omega_\nu) I_{\nu b}[T(\tau)] + \frac{\omega_\nu}{2} \int_{-1}^{1} p(\mu, \mu') I_\nu(\tau, \mu') \, d\mu' \tag{8-59b}$$

$$p(\mu, \mu') \equiv \sum_{n=0}^{N} a_n P_n(\mu) P_n(\mu'), \quad a_0 = 1 \tag{8-60a}$$

$$\omega_\nu \equiv \frac{\sigma_\nu}{\beta_\nu}, \qquad 1 - \omega_\nu \equiv \frac{\kappa_\nu}{\beta_\nu} \tag{8-60b}$$

and μ is the cosine of the angle between the direction of the radiation intensity and the positive τ axis.

The boundary conditions for Eq. 8-59 can be written formally as

$$I_\nu(\tau, \mu) \big|_{\tau=0} = I_\nu^+(0, \mu) \quad \text{for } 0 < \mu \le 1 \tag{8-61a}$$

$$I_\nu(\tau, \mu) \big|_{\tau=\tau_0} = I_\nu^-(\tau_0, \mu) \quad \text{for } -1 \le \mu < 0 \tag{8-61b}$$

In general, the boundary surface intensity functions $I_\nu^+(0, \mu)$, $\mu > 0$, and $I_\nu^-(\tau_0, \mu)$, $\mu < 0$, are not necessarily known at the onset of the problem. For example, for reflecting, opaque boundaries the functions on the right-hand side of Eqs. 8-61 will involve the radiation intensity from the interior of the medium that is reflected by the boundary surface; hence they are unknowns. On the other hand, for transparent boundaries with externally incident azimuthally symmetric radiation or for black boundaries at prescribed temperature these functions are known.

In seeking a formal solution of Eq. 8-59a it is convenient to separate the intensity $I_\nu(\tau, \mu)$ into a *forward component* $I_\nu^+(\tau, \mu)$, $\mu > 0$, and a *backward component* $I_\nu^-(\tau, \mu)$, $\mu < 0$. Then the equations for $I_\nu^+(\tau, \mu)$ and $I_\nu^-(\tau, \mu)$ and the boundary conditions are given as

$$\mu \frac{\partial I_\nu^+(\tau, \mu)}{\partial \tau} + I_\nu^+(\theta, \mu) = S_\nu(\tau, \mu), \quad \text{in } 0 \le \tau \le \tau_0, \ 0 < \mu \le 1 \tag{8-62a}$$

$$I_\nu^+(\tau, \mu)\big|_{\tau=0} = I_\nu^+(0, \mu), \quad \text{for } 0 < \mu \le 1 \tag{8-62b}$$

and

$$\mu \frac{\partial I_\nu^-(\tau, \mu)}{\partial \tau} + I_\nu^-(\tau, \mu) = S_\nu(\tau, \mu) \quad \text{in } 0 \le \tau \le \tau_0, \ -1 \le \mu < 0 \tag{8-63a}$$

$$I_\nu^-(\tau, \mu)\big|_{\tau=\tau_0} = I_\nu^-(\tau_0, \mu), \quad \text{for } -1 \le \mu < 0 \tag{8-63b}$$

These equations are coupled through the source function $S_\nu(\tau, \mu)$:

$$S_\nu(\tau, \mu) = (1 - \omega_\nu) I_{\nu b}[T(\tau)] + \frac{\omega_\nu}{2}\left[\int_0^1 p(\mu, \mu') I_\nu^+(\tau, \mu')\, d\mu' + \int_{-1}^0 p(\mu, \mu') I_\nu^-(\tau, \mu')\, d\mu'\right] \tag{8-64}$$

The geometry and coordinates for the problem considered are shown in Fig. 8-6. Equations 8-62 and 8-63 are two coupled integrodifferential equations the solution of which, for the general case, is not very straightforward. Therefore, before presenting the details of the solution of these equations, we summarize an outline of our analysis in order to provide an overall view of the procedure followed in the remaining sections of this chapter.

First, Eqs. 8-62 and 8-63 will be solved formally for the intensities $I_\nu^+(\tau, \mu)$, $\mu > 0$, and $I_\nu^-(\tau, \mu)$, $\mu < 0$. In most applications quantities such as the incident radiation $G(\tau)$, the net radiative heat flux $q^r(\tau)$, and the derivative of the net radiative heat flux $dq^r(\tau)/d\tau$ are of interest. Therefore, by utilizing these formal solutions for $I_\nu^+(\tau, \mu)$ and $I_\nu^-(\tau, \mu)$, formal relations will be obtained for $G(\tau)$, $q^r(\tau)$, and $dq^r(\tau)/d\tau$. It will be found that all these expressions will include the boundary surface intensity functions $I_\nu^+(0, \mu)$, $\mu > 0$, and $I_\nu^-(\tau_0, \mu)$, $\mu < 0$, and the source function $S_\nu(\tau, \mu)$, which are, in general, unknown. The next step in the analysis is, therefore, the determination of relations defining the boundary surface intensities and the source function. In Section 8-7 we discuss the boundary conditions appropriate to radiative heat transfer problems and in Section 8-8 present formal solutions for the boundary surface intensities. However, to evaluate the boundary surface intensities from these relations the source function $S_\nu(\tau, \mu)$ is required. To

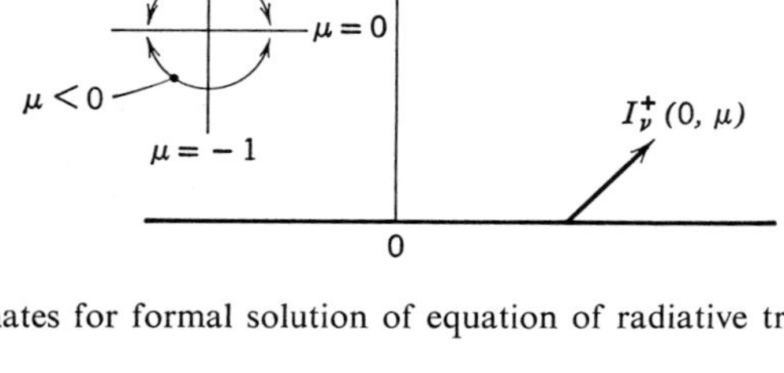

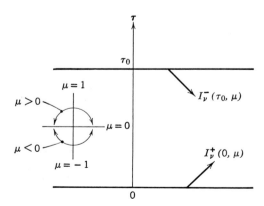

Fig. 8-6. Coordinates for formal solution of equation of radiative transfer for a plane-parallel slab.

complete the analysis we present in Section 8-9 the integral equation defining the source function.

Here we would like to point out that the formal relations for $G(\tau)$, $q^r(\tau)$, $dq^r(\tau)/d\tau$, and the boundary surface intensities that will be derived in the course of this analysis are frequently used in the radiative heat transfer applications considered in Chapters 11 through 14. Therefore, to provide a ready reference for such applications, we first derive the general relations and then present several special cases for $G(\tau)$, $q^r(\tau)$, $dq^r(\tau)/d\tau$, and the boundary surface intensities.

We now proceed to carry out the details of this analysis.

Formal Solution for Radiation Intensities

The formal solution for the forward component $I_\nu^+(\tau, \mu)$ is obtained from, Eqs. 8-62 as

$$I_\nu^+(\tau, \mu) = I_\nu^+(0, \mu)e^{-\tau/\mu} + \int_0^\tau \frac{1}{\mu} S_\nu(\tau', \mu)e^{-(\tau-\tau')/\mu}\, d\tau' \quad \text{for } \mu > 0 \quad (8\text{-}65)$$

and that for the backward component $I_\nu^-(\tau, \mu)$ is obtained from Eq. 8-63 as

$$I_\nu^-(\tau, \mu) = I_\nu^-(\tau_0, \mu)e^{(\tau_0-\tau)/\mu} - \int_\tau^{\tau_0} \frac{1}{\mu} S_\nu(\tau', \mu)e^{-(\tau-\tau')/\mu}\, d\tau' \quad \text{for } \mu < 0 \quad (8\text{-}66a)$$

or, by replacing μ by $-\mu$, Eq. 8-66a becomes

$$I_\nu^-(\tau, -\mu) = I_\nu^-(\tau_0, -\mu)e^{-(\tau_0-\tau)/\mu} + \int_\tau^{\tau_0} \frac{1}{\mu} S_\nu(\tau', -\mu)e^{-(\tau'-\tau)/\mu}\, d\tau' \quad \text{for } \mu > 0 \quad (8\text{-}66b)$$

In the above result, for example in Eq. 8-65, the first term on the right-hand side is the uncollided contribution of radiation from the boundary surface $\tau = 0$ that has penetrated to a depth τ without having undergone any scattering; the second term is the contribution of the source function in the region from $\tau = 0$ to τ to radiation intensity at τ. A similar interpretation can be given for the solution of Eq. 8-66.

The formal solutions given by Eqs. 8-65 and 8-66 are not solutions in the real sense because, in general, the source function and the boundary surface intensities depend on the radiation intensity from the medium; hence they are not immediately available at the onset of the problem.

Formal Solution for Incident Radiation

The spectral *incident radiation* $G_\nu(\tau)$ for azimuthally symmetric radiation is defined as

$$G_\nu(\tau) \equiv 2\pi \int_{-1}^{1} I_\nu(\tau, \mu)\, d\mu \quad (8\text{-}67)$$

Separating the intensity into a forward and a backward component, we can write Eq. 8-67 as

$$G_\nu(\tau) = 2\pi\left[\int_0^1 I_\nu^+(\tau, \mu)\, d\mu + \int_{-1}^0 I_\nu^-(\tau, \mu)\, d\mu\right]$$

$$= 2\pi\left[\int_0^1 I_\nu^+(\tau, \mu)\, d\mu + \int_0^1 I_\nu^-(\tau, -\mu)\, d\mu\right] \tag{8-68}$$

Substituting the intensities from Eqs. 8-65 and 8-66 into Eq. 8-68, we obtain the formal relation for the spectral incident radiation as

$$G_\nu(\tau) = 2\pi\left[\int_0^1 I_\nu^+(0, \mu)e^{-\tau/\mu}\, d\mu + \int_0^1 I_\nu^-(\tau_0, -\mu)e^{-(\tau_0-\tau)/\mu}\, d\mu\right.$$
$$+ \int_{\mu=0}^1 \int_{\tau'=0}^{\tau} \frac{1}{\mu} S_\nu(\tau', \mu)e^{-(\tau'-\tau)/\mu}\, d\tau'\, d\mu$$
$$\left. + \int_{\mu=0}^1 \int_{\tau'=\tau}^{\tau_0} \frac{1}{\mu} S_\nu(\tau', -\mu)e^{-(\tau'-\tau)/\mu}\, d\tau'\, d\mu\right] \tag{8-69}$$

We examine some special cases of Eq. 8-69.

(a) Isotropic Scattering

The source function $S_\nu(\tau, \mu)$ for isotropic scattering is independent of μ, and Eq. 8-69 simplifies to

$$G_\nu(\tau) = 2\pi\left[\int_0^1 I_\nu^+(0, \mu)e^{-\tau/\mu}\, d\mu + \int_0^1 I_\nu^-(\tau_0, -\mu)e^{-(\tau_0-\tau)/\mu}\, d\mu\right.$$
$$\left. + \int_{\tau'=0}^{\tau_0} S_\nu(\tau')E_1(|\tau - \tau'|)\, d\tau'\right] \tag{8-70}$$

where we have defined

$$\int_{\tau'=0}^{\tau_0} S_\nu(\tau')E_1(|\tau - \tau'|)\, d\tau'$$
$$\equiv \int_{\tau'=0}^{\tau_0} S_\nu(\tau')E_1(\tau - \tau')\, d\tau' + \int_{\tau'=\tau}^{\tau_0} S_\nu(\tau')E_1(\tau' - \tau)\, d\tau' \tag{8-71}$$

and $E_n(z)$ is the *exponential integral function*, defined as

$$E_n(z) = \int_0^1 \eta^{n-2}e^{-z/n}\, d\eta \tag{8-72}$$

A summary of the properties of the exponential integral functions is given in the Appendix. For a detailed discussion of the properties of exponential functions the reader should refer to the books by Chandrasekhar [2, pp. 373–374] and Kourganoff [3, pp. 253–260].

(b) Isotropic Scattering, Boundary Surface Intensities Independent of Direction

When the boundary surface intensities are independent of direction, Eq. 8-70 simplifies to

$$G_\nu(\tau) = 2\pi\left[I_\nu^+(0)E_2(\tau) + I_\nu^-(\tau_0)E_2(\tau_0 - \tau) + \int_0^{\tau_0} S_\nu(\tau')E_1(|\tau - \tau'|)\,d\tau'\right] \tag{8-73}$$

(c) Nonscattering Medium, Boundary Surface Intensities Independent of Direction

For a nonscattering medium we set $\omega_\nu = 0$ and the spectral source function (Eq. 8-59b) simplifies to

$$S_\nu(\tau) = I_{\nu b}[T(\tau)] \tag{8-74}$$

Substituting Eq. 8-74 into Eq. 8-69 and noting that the boundary surface intensities are independent of direction, we obtain

$$G_\nu(\tau) = 2\pi\left[I_\nu^+(0)E_2(\tau) + I_\nu^-(\tau_0)E_2(\tau_0 - \tau) + \int_0^{\tau_0} I_{\nu b}[T(\tau')]E_1(|\tau - \tau'|)\,d\tau'\right] \tag{8-75}$$

Formal Solution for Net Radiative Heat Flux

The net radiative heat flux $q^r(\tau)$ for azimuthally independent radiation in a plane-parallel medium is given as

$$q^r(\tau) = \int_{\nu=0}^{\infty} q_\nu^r(\tau)\,d\nu \tag{8-76}$$

where the spectral radiative heat flux is related to the radiation intensity by

$$q_\nu^r(\tau) = 2\pi\int_{\mu=-1}^{1} I_\nu(\tau, \mu)\mu\,d\mu \tag{8-77}$$

The net radiative heat flux $q^r(\tau)$ represents the net flow of radiant energy per unit time and per unit area perpendicular to the τ axis; $q^r(\tau)$ is considered positive when the net radiant energy is being transported in the direction of the positive τ axis.

Separating the intensity into a forward and a backward component, we can write Eq. 8-77 in the form

$$\begin{aligned} q_\nu^r(\tau) &= 2\pi\left[\int_{\mu=0}^{1} I_\nu^+(\tau, \mu)\mu\,d\mu + \int_{\mu=-1}^{0} I_\nu^-(\tau, \mu)\mu\,d\mu\right] \\ &= 2\pi\left[\int_{\mu=0}^{1} I_\nu^+(\tau, \mu)\mu\,d\mu - \int_{\mu=0}^{1} I_\nu^-(\tau, -\mu)\mu\,d\mu\right] \end{aligned} \tag{8-78}$$

Sometimes it is convenient to express the spectral radiative heat flux $q_\nu^r(\tau)$ as the algebraic sum of two *half-range fluxes* as

$$q_\nu^r(\tau) = q_\nu^+(\tau) - q_\nu^-(\tau) \tag{8-79}$$

By comparing Eqs. 8-78 and 8-79 we note that the forward and backward half-range fluxes are defined as

$$q_\nu^+(\tau) = 2\pi \int_{\mu=0}^{1} I_\nu^+(\tau, \mu)\mu \, d\mu \tag{8-80a}$$

$$q_\nu^-(\tau) = 2\pi \int_{\mu=0}^{1} I_\nu^-(\tau, -\mu)\mu \, d\mu \tag{8-80b}$$

Substituting the forward and backward intensities from Eqs. 8-65 and 8-66 into Eq. 8-78, we obtain the formal relation for the spectral net radiative heat flux as

$$\begin{aligned} q_\nu^r(\tau) = {} & 2\pi\left[\int_0^1 I_\nu^+(0, \mu)e^{-\tau/\mu}\mu \, d\mu + \int_0^1 \int_0^\tau S_\nu(\tau', \mu)e^{-(\tau-\tau')/\mu} \, d\tau' \, d\mu\right] \\ & - 2\pi\left[\int_0^1 I_\nu^-(\tau_0, -\mu)e^{-(\tau_0-\tau)/\mu}\mu \, d\mu + \int_0^1 \int_\tau^{\tau_0} S_\nu(\tau', -\mu)e^{-(\tau'-\tau)/\mu} \, d\tau' \, d\mu\right] \end{aligned} \tag{8-81}$$

We now examine some special cases of Eq. 8-81.

(a) Isotropic Scattering

For isotropic scattering the spectral source function is independent of μ and Eq. 8-81 simplifies to

$$\begin{aligned} q_\nu^r(\tau) = {} & 2\pi\left[\int_0^1 I_\nu^+(0, \mu)e^{-\tau/\mu}\mu \, d\mu + \int_0^\tau S_\nu(\tau')E_2(\tau - \tau') \, d\tau'\right] \\ & - 2\pi\left[\int_0^1 I_\nu^-(\tau_0, -\mu)e^{-(\tau_0-\tau)/\mu}\mu \, d\mu + \int_\tau^{\tau_0} S_\nu(\tau')E_2(\tau' - \tau) \, d\tau'\right] \end{aligned} \tag{8-82}$$

(b) Isotropic Scattering, Boundary Surface Intensities Independent of Direction

When the boundary surface intensities are independent of direction, Eq. 8-82 simplifies to

$$\begin{aligned} q_\nu^r(\tau) = {} & 2\pi\left[I_\nu^+(0)E_3(\tau) + \int_0^\tau S_\nu(\tau')E_2(\tau - \tau') \, d\tau'\right] \\ & - 2\pi\left[I_\nu^-(\tau)E_3(\tau_0 - \tau) + \int_\tau^{\tau_0} S_\nu(\tau')E_2(\tau' - \tau) \, d\tau'\right] \end{aligned} \tag{8-83}$$

(c) Nonscattering Medium, Boundary Surface Intensities Independent of Direction

For a nonscattering medium the spectral source function is given by Eq. 8-74. Substituting this relation into Eq. 8-81 and noting that the boundary surface intensities are independent of direction, we obtain

$$q_\nu^r(\tau) = 2\pi\left[I_\nu^+(0)E_3(\tau) + \int_0^\tau I_{\nu b}[T(\tau')]E_2(\tau - \tau')\,d\tau'\right]$$
$$- 2\pi\left[I_\nu^-(\tau_0)E_3(\tau_0 - \tau) + \int_\tau^{\tau_0} I_{\nu b}[T(\tau')]E_2(\tau' - \tau)\,d\tau'\right] \quad (8\text{-}84)$$

Formal Solution for $dq_\nu^r/d\tau$

In problems involving the interaction of radiation with conduction and convection in a participating medium the derivative of the net radiative heat flux, $dq^r/d\tau$, enters the energy equation. Although this quantity can be determined by differentiating with respect to τ the foregoing relations for $q^r(\tau)$, it is convenient to derive such relations in the straightforward manner now described.

Consider the relation for the divergence of the spectral radiative flux vector given by Eq. 8-16a:

$$\nabla \cdot \mathbf{q}_\nu^r = 4\pi\kappa_\nu I_{\nu b}(T) - \kappa_\nu G_\nu \quad (8\text{-}85)$$

where

$$G_\nu \equiv \int_{4\pi} I_\nu\, d\Omega'$$

For a plane-parallel medium with the planes of stratification perpendicular to the oy axis Eq. 8-85 simplifies to

$$\frac{dq_\nu^r(y)}{dy} = \kappa_\nu[4\pi I_{\nu b}(T) - G_\nu(y)] \quad (8\text{-}86)$$

In terms of the optical variable Eq. 8-86 is written as

$$\frac{dq_\nu^r(\tau)}{d\tau} = (1 - \omega_\nu)[4\pi I_{\nu b}(T) - G_\nu(\tau)] \quad (8\text{-}87)$$

where we have defined

$$G_\nu(\tau) = 2\pi\int_{-1}^{1} I_\nu(\tau, \mu)\, d\mu \quad (8\text{-}88a)$$

$$d\tau = \beta_\nu\, dy \quad (8\text{-}88b)$$

$$\omega_\nu = \frac{\sigma_\nu}{\beta_\nu} \quad (8\text{-}88c)$$

By substituting for $G_\nu(\tau)$ from Eq. 8-69 into Eq. 8-87, we obtain the formal solution for $dq_\nu^r(\tau)/d\tau$ for azimuthally independent radiation in a plane-parallel slab as

$$\frac{dq_\nu^r(\tau)}{d\tau} = (1-\omega_\nu)4\pi I_{\nu b}(T) - (1-\omega_\nu)2\pi\left[\int_0^1 I_\nu^+(0,\mu)e^{-\tau/\mu}\,d\mu + \int_0^1 I_\nu^-(\tau_0,-\mu)e^{-(\tau_0-\tau)/\mu}\,d\mu + \int_0^1\int_{\tau'=0}^{\tau}\frac{1}{\mu}S_\nu(\tau',\mu)e^{-(\tau-\tau')/\mu}\,d\tau'\,d\mu + \int_0^1\int_{\tau'=\tau}^{\tau_0}\frac{1}{\mu}S_\nu(\tau',-\mu)e^{-(\tau'-\tau)/\mu}\,d\tau'\,d\mu\right] \tag{8-89}$$

We examine now some special cases of Eq. 8-89.

(a) Isotropic Scattering

For isotropic scattering the spectral source function is independent of direction and Eq. 8-89 simplifies to

$$\frac{dq_\nu^r(\tau)}{d\tau} = (1-\omega_\nu)4\pi I_{\nu b}(T) - (1-\omega_\nu)2\pi\left[\int_0^1 I_\nu^+(0,\mu)e^{-\tau/\mu}\,d\mu + \int_0^1 I_\nu^-(\tau_0,-\mu)e^{-(\tau_0-\tau)/\mu}\,d\mu + \int_0^{\tau_0} S_\nu(\tau')E_1(|\tau-\tau'|)\,d\tau'\right] \tag{8-90}$$

where we have defined

$$\int_0^{\tau_0} S_\nu(\tau')E_1(|\tau-\tau'|)\,d\tau' \equiv \int_0^{\tau} S_\nu(\tau')E_1(\tau-\tau')\,d\tau' + \int_\tau^{\tau_0} S_\nu(\tau')E_1(\tau'-\tau)\,d\tau' \tag{8-91}$$

An alternative form of Eq. 8-90 is obtained by eliminating $(1-\omega_\nu)4\pi I_{\nu b}(T)$ by means of Eq. 8-122, given in the next section.[8] We obtain

$$\frac{dq_\nu^r(\tau)}{d\tau} = 4\pi S_\nu(\tau) - 2\pi\left[\int_0^1 I_\nu^+(0,\mu)e^{-\tau/\mu}\,d\mu + \int_0^1 I_\nu^-(\tau_0,-\mu)e^{-(\tau_0-\tau)/\mu}\,d\mu + \int_0^{\tau_0} S_\nu(\tau')E_1(|\tau-\tau'|)\,d\tau'\right] \tag{8-92}$$

(b) Isotropic Scattering, Boundary Surface Intensities Independent of Direction

When the boundary surface intensities are independent of direction, Eqs. 8-90 and 8-92, respectively, simplify to

$$\frac{dq_\nu^r(\tau)}{d\tau} = (1-\omega_\nu)4\pi I_{\nu b}(T) - (1-\omega_\nu)2\pi\left[I_\nu^+(0)E_2(\tau) + I_\nu^-(\tau_0)E_2(\tau_0-\tau) + \int_0^{\tau_0} S_\nu(\tau')E_1(|\tau-\tau'|)\,d\tau'\right] \tag{8-93}$$

and

$$\frac{dq_\nu^r(\tau)}{d\tau} = 4\pi S_\nu(\tau) - 2\pi\Big[I_\nu^+(0)E_2(\tau) + I_\nu^-(\tau_0)E_2(\tau_0 - \tau) + \int_0^{\tau_0} S_\nu(\tau')E_1(|\tau - \tau'|)\, d\tau'\Big] \tag{8-94}$$

(c) Nonscattering Medium, Boundary Surface Intensities Independent of Direction

For a nonscattering medium, setting $\omega_\nu = 0$ and noting that the boundary surface intensities are independent of direction, Eqs. 8-90 and 8-92 both simplify to

$$\frac{dq_\nu^r(\tau)}{d\tau} = 4\pi I_{\nu b}(T) - 2\pi\Big[I_\nu^+(0)E_2(\tau) + I_\nu^-(\tau_0)E_2(\tau_0 - \tau) + \int_0^{\tau_0} I_{\nu b}(T)E_1(|\tau - \tau'|)\, d\tau'\Big] \tag{8-95}$$

8-7 BOUNDARY CONDITIONS

In Section 8-6, for simplicity in the analysis, we denoted the boundary condition functions formally by $I_\nu^+(0, \mu)$, $\mu > 0$, and $I_\nu^-(\tau_0, \mu)$, $\mu < 0$, at the boundary surfaces $\tau = 0$ and $\tau = \tau_0$, respectively. In this section we present explicit relations for these boundary conditions for transparent boundaries and opaque boundaries with diffusely and specularly reflecting surfaces.

(a) Transparent Boundaries

When the boundary surfaces at $\tau = 0$ and $\tau = \tau_0$ are transparent and the external environment adjacent to the boundaries is a vacuum (i.e., does not interact with radiation), the boundary conditions for externally incident azimuthally symmetric radiation can be written as

$$I_\nu^+(0, \mu) = f_{1\nu}(\mu) \quad \text{for } \mu > 0 \tag{8-96a}$$

$$I_\nu^-(\tau_0, \mu) = f_{2\nu}(\mu) \quad \text{for } \mu < 0 \tag{8-96b}$$

where $f_{1\nu}(\mu)$ and $f_{2\nu}(\mu)$ are prescribed functions of μ. When the externally applied radiation is constant, Eqs. 8-96 simplify to

$$I_\nu^+(0) = f_{1\nu} \quad \text{for } \mu > 0 \tag{8-97a}$$

$$I_\nu^-(\tau_0) = f_{2\nu} \quad \text{for } \mu < 0 \tag{8-97b}$$

where $f_{1\nu}$ and $f_{2\nu}$ are constants.

(b) Black Boundaries

When the boundary surfaces at $\tau = 0$ and $\tau = \tau_0$ are both black surfaces kept at uniform temperatures T_1 and T_2, respectively, the intensity of spectral radiation from these surfaces into the medium is equal to the Planck function at the surface temperature. Then the boundary conditions are given as

$$I_\nu^+(0) = I_{\nu b}(T_1) \quad \text{for } \mu > 0 \tag{8-98a}$$

$$I_\nu^-(\tau_0) = I_{\nu b}(T_2) \quad \text{for } \mu < 0 \tag{8-98b}$$

where $I_{\nu b}$ is the Planck function, which is independent of direction. We note that the boundary conditions for black surfaces are similar to those given by Eqs. 8-97 for transparent boundaries with externally applied constant radiation.

(c) Opaque, Diffusely Emitting, Diffusely Reflecting Boundaries

When the boundary surfaces are opaque, diffuse emitters and diffuse reflectors, the intensity of radiation from the boundary surfaces is independent of direction and is composed of an emitted and a diffusely reflected component. Let T_1 and T_2 be the temperatures, $\varepsilon_{1\nu}$ and $\varepsilon_{2\nu}$ the spectral hemispherical emissivities, and $\rho_{1\nu}^d$ and $\rho_{2\nu}^d$ the spectral hemispherical diffuse reflectivities of the boundary surfaces at $\tau = 0$ and $\tau = \tau_0$, respectively.

The intensity of radiation, $I_\nu^+(0)$ for $0 < \mu \le 1$, leaving the boundary surface at $\tau = 0$ in the positive μ direction, is given as

$$I_\nu^+(0) = (\text{Emitted component}) + (\text{Diffusely reflected component}), \quad \mu > 0$$

$$= \varepsilon_{1\nu} I_{\nu b}(T_1) + \rho_{1\nu}^d \frac{\int_0^{2\pi}\int_{-1}^{0} I_\nu^-(0,\mu')\mu'\,d\mu'\,d\phi'}{\int_0^{2\pi}\int_{-1}^{0} \mu'\,d\mu'\,d\phi'}, \qquad \mu > 0$$

$$= \varepsilon_{1\nu} I_{\nu b}(T_1) + 2\rho_{1\nu}^d \int_0^1 I_\nu^-(0,-\mu')\mu'\,d\mu', \qquad \mu > 0 \tag{8-99a}$$

Similarly, the intensity of radiation, $I_\nu^-(\tau_0)$ for $-1 \le \mu \le 0$, leaving the boundary surface at $\tau = \tau_0$ in the negative μ direction is given as

$$I_\nu^-(\tau_0) = \varepsilon_{2\nu} I_{\nu b}(T_2) + \rho_{2\nu}^d \frac{\int_0^{2\pi}\int_0^1 I_\nu^+(\tau_0,\mu')\mu'\,d\mu'\,d\phi'}{\int_0^{2\pi}\int_0^1 \mu'\,d\mu'\,d\phi'}, \quad \mu < 0$$

$$= \varepsilon_{2\nu} I_{\nu b}(T_2) + 2\rho_{2\nu}^d \int_0^1 I_\nu^+(\tau_0,\mu')\mu'\,d\mu', \qquad \mu < 0 \tag{8-99b}$$

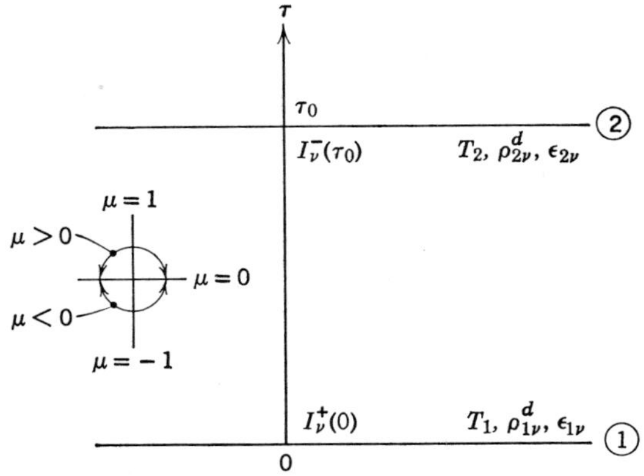

Fig. 8-7. Boundary conditions for a plane-parallel slab with diffusely emitting, diffusely reflecting boundaries.

Equations 8-99 are the boundary conditions desired for opaque, diffusely emitting and diffusely reflecting surfaces; Fig. 8-7 illustrates these boundary conditions.

When the Kirchhoff law is applicable, $\rho_{i\nu}^d$ can be replaced by $1 - \varepsilon_{i\nu}$, $i = 1$ or 2. For black surfaces, the boundary conditions of Eqs. 8-99 simplify to those given by Eqs. 8-98.

(d) Opaque, Diffusely Emitting, Purely Specularly Reflecting Boundaries

Consider opaque, diffusely emitting, purely specularly reflecting boundaries. Let T_1 and T_2 be the temperatures, $\varepsilon_{1\nu}$ and $\varepsilon_{2\nu}$ the spectral hemispherical emissivities, and $\rho_{1\nu}^s$ and $\rho_{2\nu}^s$ the spectral specular reflectivities at the boundary surfaces $\tau = 0$ and $\tau = \tau_0$, respectively.

The intensity of radiation, $I_\nu^+(0, \mu)$ for $0 < \mu \leq 1$, leaving the boundary surface at $\tau = 0$ in the positive μ direction is composed of an emitted and a specularly reflected component and is given as

$$I_\nu^+(0, \mu) = \varepsilon_{1\nu} I_{\nu b}(T_1) + \rho_{1\nu}^s I_\nu^-(0, -\mu), \quad \mu > 0 \tag{8-100a}$$

The intensity of radiation, $I_\nu^-(\tau_0, \mu)$ for $-1 \leq \mu < 0$, leaving the boundary surface at $\tau = \tau_0$ in the negative μ direction is determined in a similar manner and is given as

$$\begin{aligned} I_\nu^-(\tau_0, \mu) &= \varepsilon_{2\nu} I_{\nu b}(T_2) + \rho_{2\nu}^s I^+(\tau_0, -\mu), \quad && \mu < 0 \\ I_\nu^-(\tau_0, -\mu) &= \varepsilon_{2\nu} I_{\nu b}(T_2) + \rho_{2\nu}^s I_\nu^+(\tau_0, \mu), \quad && \mu > 0 \end{aligned} \tag{8-100b}$$

Equations 8-100 are the desired boundary conditions.

8-8 EQUATIONS FOR BOUNDARY SURFACE INTENSITIES

The formal solutions given in Section 8-6 involve the boundary surface intensities $I_\nu^+(0, \mu)$ and $I_\nu^-(\tau_0, -\mu)$, for $\mu > 0$, which are known only for the transparent and black boundaries and are unknown for the reflecting boundaries. In this section we shall derive the equations for the boundary surface intensities for a plane-parallel slab with reflecting boundaries and azimuthally independent radiation.

Slab with Opaque, Diffusely Emitting, Diffusely Reflecting Boundaries

Consider a plane-parallel slab of optical thickness τ_0 with opaque, diffusely emitting and diffusely reflecting boundaries. Let T_1 and T_2 be the temperatures, $\varepsilon_{1\nu}$ and $\varepsilon_{2\nu}$ the spectral hemispherical emissivities, and $\rho_{1\nu}^d$ and $\rho_{2\nu}^d$ the spectral hemispherical diffuse reflectivities at the boundary surfaces $\tau = 0$ and $\tau = \tau_0$, respectively. The boundary conditions for this case are given by Eqs. 8-99a and 8-99b, and the right-hand sides of these equations involve the intensity functions $I_\nu^-(0, -\mu')$ and $I_\nu^+(\tau_0, \mu')$, respectively. These intensity functions can be determined from the formal solutions given by Eqs. 8-65 and 8-66b in the following manner.

By setting $\tau = 0$ in Eq. 8-66b and noting that for diffuse reflection the intensity $I_\nu^-(\tau_0, -\mu)$ on the right-hand side of this equation is independent of direction, we obtain

$$I_\nu^-(0, -\mu) = I_\nu^-(\tau_0)e^{-\tau_0/\mu} + \int_0^{\tau_0} \frac{1}{\mu} S_\nu(\tau', -\mu)e^{-\tau'/\mu}\, d\tau', \quad \mu > 0 \qquad (8\text{-}101)$$

By setting $\tau = \tau_0$ in Eq. 8-65 and noting that for diffuse reflection the intensity $I_\nu^-(0, \mu)$ on the right-hand side of this equation is independent of direction, we obtain

$$I_\nu^+(\tau_0, \mu) = I_\nu^+(0)e^{-\tau_0/\mu} + \int_0^{\tau_0} \frac{1}{\mu} S_\nu(\tau', \mu)e^{-(\tau_0-\tau')/\mu}\, d\tau', \quad \mu > 0 \qquad (8\text{-}102)$$

By substituting Eqs. 8-101 and 8-102 on the right-hand sides of Eqs. 8-99a and 8-99b, respectively, we obtain

$$I_\nu^+(0) = \varepsilon_{1\nu} I_{\nu b}(T_1) + 2\rho_{1\nu}^d \Big[I_\nu^-(\tau_0)E_3(\tau_0) + \int_0^1 \int_0^{\tau_0} S_\nu(\tau', -\mu')e^{-\tau'/\mu'}\, d\tau'\, d\mu' \Big] \quad \text{for } \mu > 0 \qquad (8\text{-}103a)$$

$$I_\nu^-(\tau_0) = \varepsilon_{2\nu} I_{\nu b}(T_2) + 2\rho_{2\nu}^d \Big[I_\nu^+(0)E_3(\tau_0) + \int_0^1 \int_0^{\tau_0} S_\nu(\tau', \mu')e^{-(\tau_0-\tau')/\mu'}\, d\tau'\, d\mu' \Big] \quad \text{for } \mu < 0 \qquad (8\text{-}103b)$$

Equations 8-103a and 8-103b, two simultaneous equations for the boundary surface intensities $I_\nu^+(0)$ for $\mu > 0$ and $I_\nu^-(\tau_0)$ for $\mu < 0$, can be written compactly in the forms

$$I_\nu^+(0) = a_1 + b_1 I_\nu^-(\tau_0) \tag{8-104a}$$

$$I_\nu^-(\tau_0) = a_2 + b_2 I_\nu^+(0) \tag{8-104b}$$

where we have defined

$$a_1 \equiv \varepsilon_{1\nu} I_{\nu b}(T_1) + 2\rho_{1\nu}^d \int_0^1 \int_0^{\tau_0} S_\nu(\tau', -\mu') e^{-\tau'/\mu'}\, d\tau'\, d\mu' \tag{8-105a}$$

$$a_2 \equiv \varepsilon_{2\nu} I_{\nu b}(T_2) + 2\rho_{2\nu}^d \int_0^1 \int_0^{\tau_0} S_\nu(\tau', \mu') e^{-(\tau_0-\tau')/\mu'}\, d\tau'\, d\mu' \tag{8-105b}$$

$$b_1 \equiv 2\rho_{1\nu}^d E_3(\tau_0) \tag{8-105c}$$

$$b_2 \equiv 2\rho_{2\nu}^d E_3(\tau_0) \tag{8-105d}$$

A simultaneous solution of Eqs. 8-104 for the boundary surface intensities $I_\nu^+(0)$ and $I_\nu^-(\tau_0)$ yields

$$I_\nu^+(0) = \frac{a_1 + b_1 a_2}{1 - b_1 b_2}, \quad \mu > 0 \tag{8-106a}$$

$$I_\nu^-(\tau_0) = \frac{a_2 + b_2 a_1}{1 - b_1 b_2}, \quad \mu < 0 \tag{8-106b}$$

In Eqs. 8-106 the only unknown quantity is the spectral source function.

We now examine some special cases of Eqs. 8-103.

(a) Isotropically Scattering Medium

In this case the spectral source function $S_\nu(\tau, \mu)$ is independent of μ (see Eq. 8-42a for $p = 1$), and Eqs. 8-103 simplify to

$$I_\nu^+(0) = \varepsilon_{1\nu} I_{\nu b}(T_1) + 2\rho_{1\nu}^d \left[I_\nu^-(\tau_0) E_3(\tau_0) + \int_0^{\tau_0} S_\nu(\tau') E_2(\tau')\, d\tau' \right] \tag{8-107a}$$

$$I_\nu^-(\tau_0) = \varepsilon_{2\nu} I_{\nu b}(T_2) + 2\rho_{2\nu}^d \left[I_\nu^+(0) E_3(\tau_0) + \int_0^{\tau_0} S_\nu(\tau') E_2(\tau_0 - \tau')\, d\tau' \right] \tag{8-107b}$$

A simultaneous solution of Eqs. 8-107 for $I_\nu^+(0)$ and $I_\nu^-(\tau_0)$ can be expressed in the form

$$I_\nu^+(0) = \frac{a_1 + b_1 a_2}{1 - b_1 b_2} \tag{8-108a}$$

$$I_\nu^-(\tau_0) = \frac{a_2 + b_2 a_1}{1 - b_1 b_2} \tag{8-108b}$$

where we have defined

$$a_1 = \varepsilon_{1\nu} I_{\nu b}(T_1) + 2\rho_{1\nu}^d \int_0^{\tau_0} S_\nu(\tau') E_2(\tau')\, d\tau' \tag{8-108c}$$

$$a_2 = \varepsilon_{2\nu} I_{\nu b}(T_2) + 2\rho_{2\nu}^d \int_0^{\tau_0} S_\nu(\tau') E_2(\tau_0 - \tau')\, d\tau' \tag{8-108d}$$

$$b_1 = 2\rho_{1\nu}^d E_3(\tau_0) \tag{8-108e}$$

$$b_2 = 2\rho_{2\nu}^d E_3(\tau_0) \tag{8-108f}$$

(b) Nonscattering Medium

For a nonscattering medium the spectral source function $S_\nu(\tau)$ is given by

$$S_\nu(\tau) = I_{\nu b}[T(\tau)] \tag{8-109}$$

Then Eqs. 8-103 simplify to

$$I_\nu^+(0) = \varepsilon_{1\nu} I_{\nu b}(T_1) + 2\rho_{1\nu}^d \left\{ I_\nu^-(\tau_0) E_3(\tau_0) + \int_0^{\tau_0} I_{\nu b}[T(\tau')] E_2(\tau')\, d\tau' \right\} \tag{8-110a}$$

$$I_\nu^-(\tau_0) = \varepsilon_{2\nu} I_{\nu b}(T_2) + 2\rho_{2\nu}^d \left\{ I_\nu^+(0) E_3(\tau_0) + \int_0^{\tau_0} I_{\nu b}[T(\tau')] E_2(\tau_0 - \tau')\, d\tau' \right\} \tag{8-110b}$$

Then the solution of Eqs. 8-110 can be expressed in the same form as given by Eqs. 8-108 except that we replace in these equations $S_\nu(\tau)$ by the Planck function $I_{\nu b}[T(\tau)]$.

Slab with Opaque, Diffusely Emitting, Purely Specularly Reflecting Boundaries

Consider a plane-parallel slab of optical thickness τ_0 with opaque, diffusely emitting, purely specularly reflecting boundaries. Let T_1 and T_2 be the temperatures, $\varepsilon_{1\nu}$ and $\varepsilon_{2\nu}$ the spectral hemispherical emissivities, and $\rho_{1\nu}^s$ and $\rho_{2\nu}^s$ the spectral specular reflectivities at the boundary surfaces $\tau = 0$ and $\tau = \tau_0$, respectively. The boundary conditions for this problem are given by Eqs. 8-100a and 8-100b. The right-hand sides of these equations involve the intensity functions $I_\nu^-(0, -\mu)$ and $I_\nu^+(\tau_0, \mu)$, which can be determined from the formal solutions for the intensities given by Eqs. 8-65 and 8-66b in the following manner.

By setting $\tau = 0$ in Eq. 8-66b and $\tau = \tau_0$ in Eq. 8-65, we obtain respectively

$$I_\nu^-(0, -\mu) = I_\nu^-(\tau_0, -\mu) e^{-\tau_0/\mu} + \int_0^{\tau_0} \frac{1}{\mu} S_\nu(\tau', -\mu) e^{-\tau'/\mu}\, d\tau' \quad \text{for } \mu > 0 \tag{8-111}$$

$$I_\nu^+(\tau_0, \mu) = I_\nu^+(0, \mu) e^{-\tau_0/\mu} + \int_0^{\tau_0} \frac{1}{\mu} S_\nu(\tau', \mu) e^{-(\tau_0 - \tau')/\mu}\, d\tau' \quad \text{for } \mu > 0 \tag{8-112}$$

By substituting Eq. 8-111 on the right-hand side of Eq. 8-100a and Eq. 8-112 on the right-hand side of Eq. 8-100b, we obtain respectively

$$I_\nu^+(0,\mu) = \varepsilon_{1\nu} I_{\nu b}(T_1) + \rho_{1\nu}^s \left[I_\nu^-(\tau_0, -\mu) e^{-\tau_0/\mu} + \int_0^{\tau_0} \frac{1}{\mu} S_\nu(\tau', -\mu) e^{-\tau'/\mu}\, d\tau' \right] \quad \text{for } \mu > 0 \tag{8-113a}$$

$$I_\nu^-(\tau_0, -\mu) = \varepsilon_{2\nu} I_{\nu b}(T_2) + \rho_{2\nu}^s \left[I_\nu^+(0, \mu) e^{-\tau_0/\mu} + \int_0^{\tau_0} \frac{1}{\mu} S_\nu(\tau', \mu) e^{-(\tau_0-\tau')/\mu}\, d\tau' \right] \quad \text{for } \mu > 0 \tag{8-113b}$$

Equations 8-113 are two simultaneous equations for the determination of the boundary surface intensities $I_\nu^+(0,\mu)$ and $I_\nu^-(\tau_0, -\mu)$, for $\mu > 0$. These equations can be written compactly in the forms

$$I_\nu^+(0,\mu) = a_1 + b_1 I_\nu^-(\tau_0, -\mu), \quad \mu > 0 \tag{8-114a}$$

$$I_\nu^-(\tau_0, -\mu) = a_2 + b_2 I_\nu^+(0,\mu), \quad \mu > 0 \tag{8-114b}$$

where we have defined

$$a_1 \equiv \varepsilon_{1\nu} I_{\nu b}(T_1) + \rho_{1\nu}^s \int_0^{\tau_0} \frac{1}{\mu} S_\nu(\tau', -\mu) e^{-\tau'/\mu}\, d\tau' \tag{8-115a}$$

$$a_2 \equiv \varepsilon_{2\nu} I_{\nu b}(T_2) + \rho_{2\nu}^s \int_0^{\tau_0} \frac{1}{\mu} S_\nu(\tau', \mu) e^{-(\tau_0-\tau')/\mu}\, d\tau' \tag{8-115b}$$

$$b_1 \equiv \rho_{1\nu}^s e^{-\tau_0/\mu} \tag{8-115c}$$

$$b_2 \equiv \rho_{2\nu}^s e^{-\tau_0/\mu} \tag{8-115d}$$

Simultaneous solution of Eqs. 8-114 for $I_\nu^+(0,\mu)$ and $I_\nu^-(\tau_0, -\mu)$ yields

$$I_\nu^+(0,\mu) = \frac{a_1 + b_1 a_2}{1 - b_1 b_2}, \quad \mu > 0 \tag{8-116a}$$

$$I_\nu^-(\tau_0, -\mu) = \frac{a_2 + b_2 a_1}{1 - b_1 b_2}, \quad \mu > 0 \tag{8-116b}$$

where a_i and b_i, $i = 1$ or 2, were defined by Eqs. 8-115.

We examine now some special cases of Eqs. 8-113.

(a) Isotropically Scattering Medium

For an isotropically scattering medium the spectral source function is independent of direction and we replace $S_\nu(\tau', \mu)$ by $S_\nu(\tau')$ in Eqs. 8-113.

(b) Nonscattering Medium

For a nonscattering medium the spectral source function $S_\nu(\tau', \mu)$ in Eqs. 8-113 is replaced by the Planck function $I_{\nu b}[T(\tau')]$.

8-9 EQUATION FOR THE SOURCE FUNCTION

The formal solutions given in Section 8-6 and the equations for the boundary surface intensities given in Section 8-8 involve the spectral source function $S_\nu(\tau, \mu)$, which is unknown for most practical cases. In this section we derive the integral equation for the determination of the spectral source function for a plane-parallel slab with azimuthally independent radiation.

The spectral source function $S_\nu(\tau, \mu)$ was defined by Eq. 8-42a as

$$S_\nu(\tau, \mu) = (1 - \omega_\nu)I_{\nu b}[T(\tau)] + \tfrac{1}{2}\omega_\nu \int_{-1}^{1} p(\mu, \mu')I_\nu(\tau, \mu')\, d\mu' \qquad (8\text{-}117)$$

By separating the intensity into a forward and a backward component, we can write Eq. 8-117 in the form

$$S_\nu(\tau, \mu) = (1 - \omega_\nu)I_{\nu b}[T(\tau)] + \tfrac{1}{2}\omega_\nu \left[\int_0^1 p(\mu, \mu')I_\nu^+(\tau, \mu')\, d\mu' + \int_{-1}^{0} p(\mu, \mu')I_\nu^-(\tau, \mu')\, d\mu' \right] \qquad (8\text{-}118)$$

By replacing μ' with $-\mu'$ in the second integral on the right-hand side of Eq. 8-118 we obtain

$$S_\nu(\tau, \mu) = (1 - \omega_\nu)I_{\nu b}[T(\tau)] + \tfrac{1}{2}\omega_\nu \left[\int_0^1 p(\mu, \mu')I_\nu^+(\tau, \mu')\, d\mu' + \int_0^1 p(\mu, -\mu')I_\nu^-(\tau, -\mu')\, d\mu' \right] \qquad (8\text{-}119)$$

The intensity components $I_\nu^+(\tau, \mu')$ and $I_\nu^-(\tau, -\mu')$ on the right-hand side of Eq. 8-119 can be obtained from Eqs. 8-65 and 8-66b, respectively. Then Eq. 8-119 becomes

$$\begin{aligned} S_\nu(\tau, \mu) = {} & (1 - \omega_\nu)I_{\nu b}[T(\tau)] \\ & + \tfrac{1}{2}\omega_\nu \int_{\mu'=0}^{1} p(\mu, \mu') \\ & \times \left[I_\nu^+(0, \mu')e^{-\tau/\mu'} + \int_{\tau'=0}^{\tau} \frac{1}{\mu'} S_\nu(\tau', \mu')e^{-(\tau-\tau')/\mu'}\, d\tau' \right] d\mu' \\ & + \tfrac{1}{2}\omega_\nu \int_{\mu'=0}^{1} p(\mu, -\mu') \\ & \times \left[I_\nu^-(\tau_0, -\mu')e^{-(\tau_0-\tau)/\mu'} + \int_{\tau'=\tau}^{\tau_0} \frac{1}{\mu'} S_\nu(\tau', -\mu')e^{-(\tau'-\tau)/\mu'}\, d\tau' \right] d\mu' \end{aligned} \qquad (8\text{-}120)$$

Equation 8-120 provides an integral equation for the spectral source function $S_\nu(\tau, \mu)$, since relations are available for the boundary surface intensities $I_\nu^+(0, \mu')$ and $I_\nu^-(\tau_0, -\mu')$ in terms of this function.

We now examine some special cases of Eq. 8-120.

(a) Isotropically Scattering Medium

For an isotropically scattering medium the phase function is unity and Eq. 8-120 simplifies to

$$\begin{aligned} S_\nu(\tau) &= (1-\omega_\nu)I_{\nu b}[T(\tau)] \\ &\quad + \tfrac{1}{2}\omega_\nu\left[\int_{\mu=0}^{1} I_\nu^+(0,\mu)e^{-\tau/\mu}\,d\mu + \int_{\tau'=0}^{\tau} S_\nu(\tau')E_1(\tau-\tau')\,d\tau'\right] \\ &\quad + \tfrac{1}{2}\omega_\nu\left[\int_{\mu=0}^{1} I_\nu^-(\tau_0,-\mu)e^{-(\tau_0-\tau)/\mu}\,d\mu + \int_{\tau'=\tau}^{\tau_0} S_\nu(\tau')E_1(\tau'-\tau)\,d\tau'\right] \end{aligned} \tag{8-121}$$

which can be written more compactly as

$$\begin{aligned} S_\nu(\tau) &= (1-\omega_\nu)I_{\nu b}[T(\tau)] \\ &\quad + \tfrac{1}{2}\omega_\nu\left[\int_0^1 I_\nu^+(0,\mu)e^{-\tau/\mu}\,d\mu + \int_0^1 I_\nu^-(\tau_0,-\mu)e^{-(\tau_0-\tau)/\mu}\,d\mu \right. \\ &\quad \left. + \int_{\tau'=0}^{\tau_0} S_\nu(\tau')E_1(|\tau-\tau'|)\,d\tau'\right] \end{aligned} \tag{8-122}$$

(b) Isotropically Scattering Medium, Boundary Surface Intensities Independent of Direction

In this case Eq. 8-122 simplifies to

$$\begin{aligned} S_\nu(\tau) &= (1-\omega_\nu)I_{\nu b}[T(\tau)] \\ &\quad + \tfrac{1}{2}\omega_\nu\left[I_\nu^+(0)E_2(\tau) + I_\nu^-(\tau_0)E_2(\tau_0-\tau) + \int_{\tau'=0}^{\tau_0} S_\nu(\tau')E_1(|\tau-\tau'|)\,d\tau'\right] \end{aligned} \tag{8-123}$$

8-10 RADIATIVE EQUILIBRIUM IN A PLANE-PARALLEL SLAB—GRAY AND NONGRAY CASES

Applications of pure radiative heat transfer in participating media will be considered in greater detail in Chapter 11, and those of interactions with conduction and convection modes of heat transfer in Chapters 12 through 14. In this section and the following ones we present simple examples of radiative heat transfer in participating media in order to illustrate the applications of some of the formal relations derived previously and some of the methods of treatment of nongray radiative heat transfer.

The problem considered here is that of heat transfer by pure radiation in a plane-parallel slab in radiative equilibrium; that is, an absorbing, emitting, scattering plane-parallel slab of thickness L is in radiative equilibrium between two boundaries at $y = 0$ and $y = L$, which are kept at uniform temperatures T_1 and T_2, respectively. The boundaries are assumed to be opaque, diffuse reflectors and diffuse emitters and to have emissivities ε_1, ε_2 and reflectivities $\rho_1{}^d$, $\rho_2{}^d$. In this problem we are concerned with the determination of temperature distribution and the net radiative heat flux in the medium. We consider first the gray case and then extend the analysis to the nongray case.

Gray Medium

It is assumed that both the medium and the boundaries are gray. We considered in Section 8-2 the concept of radiative equilibrium and showed that for a plane-parallel geometry and gray assumption the equations defining the condition of radiative equilibrium are given as (Eqs. 8-23 and 8-24)

$$\frac{n^2\bar{\sigma}T^4(\tau)}{\pi} = \frac{1}{2}\int_{-1}^{1} I(\tau, \mu)\, d\mu \equiv \frac{1}{4\pi}\, G(\tau) \tag{8-124a}$$

$$\frac{dq^r}{dy} = 0 \qquad \text{or} \qquad \frac{dq^r}{d\tau} = 0 \tag{8-124b}$$

where τ is the optical variable, that is, $d\tau = \beta\, dy$, and $G(\tau)$ is the incident radiation. The first of these conditions provides a relation for the determination of temperature distribution in the medium provided that the angular distribution of radiation intensity $I(\tau, \mu)$ is known. The second condition implies that the net radiative heat flux q^r is constant everywhere in the medium for radiative equilibrium. We also note that q^r is related to the radiation intensity by (see Eq. 8-77)

$$q^r = 2\pi \int_{-1}^{1} I(\tau, \mu)\mu\, d\mu \tag{8-124c}$$

That is, the net radiative heat flux can readily be determined if the radiation intensity in the medium is known.

Now the problem is reduced to determining the angular distribution of radiation intensity $I(\tau, \mu)$ that satisfies the equation of radiative transfer

$$\mu \frac{\partial I(\tau, \mu)}{\partial \tau} + I(\tau, \mu) = (1 - \omega) I_b[T(\tau)] + \frac{\omega}{2}\int_{-1}^{1} I(\tau, \mu')\, d\mu', \quad 0 \le \tau \le \tau_0 \tag{8-125a}$$

with the boundary conditions (see Eqs. 8-99 for the gray case)

$$I^+(0) = \varepsilon_1 I_b(T_1) + 2\rho_1{}^d \int_0^1 I^-(0, -\mu')\mu'\, d\mu', \quad \mu > 0 \tag{8-125b}$$

$$I^-(\tau_0) = \varepsilon_2 I_b(T_2) + 2\rho_2{}^d \int_0^1 I^+(\tau_0, \mu')\mu'\, d\mu', \quad \mu < 0 \tag{8-125c}$$

where τ is the optical variable and $I_b(T)$ is the black-body radiation intensity, that is,

$$I_b(T) = \frac{n^2\bar{\sigma}T^4}{\pi} \tag{8-125d}$$

The integral term in the equation of radiative transfer 8-125a can be eliminated by means of Eq. 8-124a. We obtain

$$\mu \frac{\partial I(\tau, \mu)}{\partial \tau} + I(\tau, \mu) = I_b[T(\tau)] \quad \text{in } 0 \le \tau \le \tau_0 \tag{8-126}$$

We note that, for a gray medium in radiative equilibrium, the equation of radiative transfer 8-126 is equivalent to that for an absorbing, emitting nonscattering, gray medium.

If Eq. 8-126 is solved subject to the boundary conditions 8-125b and 8-125c and the radiation intensity $I(\tau, \mu)$ is determined, then the substitution of this relation for intensity into Eqs. 8-124a and 8-124c provides the necessary relations for the evaluation of temperature distribution and the net radiative heat flux in the medium. However, we have already treated in this chapter the formal solution of Eq. 8-126 subject to the boundary conditions considered here and have determined formal relations for $G(\tau)$ and $q^r(\tau)$. Therefore, without repeating these calculations, we can readily obtain the formal relations for $G(\tau)$ and $q^r(\tau)$ from Eqs. 8-73 and 8-84, respectively, as

$$G(\tau) = 2\pi\left[I^+(0)E_2(\tau) + I^-(\tau_0)E_2(\tau_0 - \tau) + \int_0^{\tau_0} I_b[T(\tau')]E_1(|\tau - \tau'|)\, d\tau'\right] \tag{8-127}$$

and

$$q^r(\tau) = 2\pi\left[I^+(0)E_3(\tau) + \int_0^{\tau} I_b[T(\tau')]E_2(\tau - \tau')\, d\tau'\right]$$
$$- 2\pi\left[I^-(\tau_0)E_3(\tau_0 - \tau) + \int_{\tau}^{\tau_0} I_b[T(\tau')]E_2(\tau' - \tau)\, d\tau'\right] \tag{8-128}$$

To obtain these relations we removed the subscript ν for the gray case and replaced the source function $S_\nu(\tau)$ by the black-body radiation intensity $I_b(T)$. Here $I^+(0)$ and $I^-(\tau_0)$ are the boundary surface intensities, for which formal solutions are obtained from Eq. 8-110 as

$$I^+(0) = \varepsilon_1 I_b(T_1) + 2\rho_1{}^d\left\{I^-(\tau_0)E_3(\tau_0) + \int_0^{\tau_0} I_b[T(\tau')]E_2(\tau')\,d\tau'\right\} \tag{8-129a}$$

$$I^-(\tau_0) = \varepsilon_2 I_b(T_2) + 2\rho_2{}^d\left\{I^+(0)E_3(\tau_0) + \int_0^{\tau_0} I_b[T(\tau')]E_2(\tau_0 - \tau')\,d\tau'\right\} \tag{8-129b}$$

where T_1 and T_2 are the temperatures of the boundary surfaces at $\tau = 0$ and $\tau = \tau_0$, respectively.

We now derive the relations for the determination of temperature distribution and the net radiative heat flux in the medium.

By combining Eq. 8-127 with the condition of radiative equilibrium given by Eq. 8-124a we obtain

$$n^2\bar{\sigma}T^4(\tau) = \frac{1}{2}\left[\pi I^+(0)E_2(\tau) + \pi I^-(\tau_0)E_2(\tau_0 - \tau) + \int_0^{\tau_0} n^2\bar{\sigma}T^4(\tau')E_1(|\tau - \tau'|)\,d\tau'\right] \tag{8-130a}$$

which can be rearranged as

$$\frac{n^2\bar{\sigma}T^4(\tau) - \pi I^-(\tau_0)}{\pi I^+(0) - \pi I^-(\tau_0)} = \frac{1}{2}\left[E_2(\tau) + \int_0^{\tau_0} \frac{n^2\bar{\sigma}T^4(\tau') - \pi I^-(\tau_0)}{\pi I^+(0) - \pi I^-(\tau_0)} E_1(|\tau - \tau'|)\,d\tau'\right] \tag{8-130b}$$

We now introduce a dimensionless function $\theta(\tau)$, defined as

$$\theta(\tau) \equiv \frac{n^2\bar{\sigma}T^4(\tau) - \pi I^-(\tau_0)}{\pi I^+(0) - \pi I^-(\tau_0)} \tag{8-131}$$

Substitution of the transformation given by Eq. 8-131 into Eq. 8-130b yields

$$\theta(\tau) = \frac{1}{2}\left[E_2(\tau) + \int_0^{\tau_0} \theta(\tau')E_1(|\tau - \tau'|)\,d\tau'\right] \tag{8-132}$$

Equation 8-132 is a singular Fredholm integral equation of the second kind for the dimensionless universal function $\theta(\tau)$.

The equation for the net radiative heat flux $q^r(\tau)$ (Eq. 8-128) can be rearranged as

$$\frac{q^r(\tau)}{\pi I^+(0) - \pi I^-(\tau_0)} = 2\left[E_3(\tau) + \int_0^{\tau} \frac{n^2\bar{\sigma}T^4(\tau') - \pi I^-(\tau_0)}{\pi I^+(0) - \pi I^-(\tau_0)} E_2(\tau - \tau')\,d\tau' - \int_\tau^{\tau_0} \frac{n^2\bar{\sigma}T^4(\tau') - \pi I^-(\tau_0)}{\pi I^+(0) - \pi I^-(\tau_0)} E_2(\tau' - \tau)\,d\tau'\right] \tag{8-133}$$

where we have replaced $I_b[T(\tau)]$ by $n^2\bar{\sigma}T^4(\tau)/\pi$. Substitution of the transformation given by Eq. 8-131 into Eq. 8-133 yields

$$\frac{q^r}{\pi I^+(0) - \pi I^-(\tau_0)} = 2\left[E_3(\tau) + \int_0^\tau \theta(\tau')E_2(\tau - \tau')\, d\tau' - \int_\tau^{\tau_0} \theta(\tau')E_2(\tau' - \tau)\, d\tau'\right] \equiv Q \quad \text{(8-134a)}$$

where Q is the dimensionless radiative heat flux. For radiative equilibrium the radiative heat flux q^r in the medium is independent of τ; then Q is a constant, and the relation defining Q can be simplified by evaluating it at $\tau = 0$. We obtain

$$Q = 1 - 2\int_0^{\tau_0} \theta(\tau')E_2(\tau')\, d\tau' \quad \text{(8-134b)}$$

Once the dimensionless universal function $\theta(\tau)$ is determined from the solution of Eq. 8-132, the dimensionless radiative heat flux constant Q is evaluated from Eq. 8-134b. The determination of the actual temperature distribution $T(\tau)$ and the net radiative heat flux q^r depends on the type of boundary conditions for the radiation problem. The simplest case involves black boundaries, for which $I^+(0)$ and $I^-(\tau_0)$ are known quantities, that is,

$$I^+(0) = \frac{n^2\bar{\sigma}T_1^4}{\pi} \quad \text{and} \quad I^-(\tau_0) = \frac{n^2\bar{\sigma}T_2^4}{\pi} \quad \text{(8-135)}$$

In this case the temperature distribution $T(\tau)$ is immediately available from Eq. 8-131 and the net radiative heat flux q^r from Eq. 8-134a. For diffusely reflecting and diffusely emitting boundaries, the boundary surface intensities $I^-(\tau_0)$ and $[I^+(0) - I^-(\tau_0)]$ appearing in Eqs. 8-131 and 8-134a can be related to the temperatures and emissivities of the boundary surfaces and q^r by simple relations. This subject will be considered further in Chapter 11 with specific examples, and numerical results will be discussed.

An Alternative Formulation for a Gray Medium

An alternative formulation of the equation of radiative transfer for the case of radiative equilibrium is obtained from Eq. 8-125a by eliminating in that equation the term $I_b[T(\tau)]$ by means of the relation given by Eq. 8-124a. We find

$$\mu\frac{\partial I(\tau, \mu)}{\partial \tau} + I(\tau, \mu) = \frac{1}{2}\int_{-1}^{1} I(\tau, \mu')\, d\mu' \quad \text{(8-136)}$$

The boundary conditions 8-125b and 8-125c can be written as

$$I(0) = \varepsilon_1\frac{n^2\bar{\sigma}T_1^4}{\pi} + 2\rho_1^d\int_0^1 I(0, -\mu')\mu'\, d\mu', \quad \mu > 0 \quad \text{(8-137a)}$$

$$I(\tau_0) = \varepsilon_2\frac{n^2\bar{\sigma}T_2^4}{\pi} + 2\rho_2^d\int_0^1 I(\tau_0, \mu')\mu'\, d\mu', \quad \mu < 0 \quad \text{(8-137b)}$$

Once the integral equation 8-136 is solved subject to the boundary conditions 8-137 and $I(\tau, \mu)$ is determined, the temperature distribution and the net radiative heat flux everywhere in the medium are evaluated from Eqs. 8-124a and 8-124c, respectively. The net radiative heat flux being constant in the medium (see Eq. 8-124b), it should be evaluated at only one location. The application of this alternative method and a discussion of the results will be presented in Chapter 11.

Nongray Medium

When the radiative properties are strongly dependent on the frequency, the nongray treatment of the problem is desirable. Unfortunately the nongray analysis for the general frequency-dependent case is very complex. Howell and Perlmutter [11] applied the Monte Carlo technique to solve nongray radiative heat transfer for a slab. Various models have been developed to simplify the characterization of the frequency dependence of radiative properties. For example, the absorption of radiation by carbon dioxide, water vapor, and glass is restricted to specific regions of the spectrum, and outside these regions the absorption is essentially zero. The *band model* illustrated in Fig. 8-8 can be used in such situations to characterize the frequency dependence of the absorption coefficient κ_ν. In this model κ_i is assumed to be uniform within each band $\Delta\nu_i$ and zero between two neighboring bands. A discussion of the band model is given by Bevans and Dunkle [12] and Edwards [13]. Sparrow and Cess [14, pp. 239–243] used a three-band model to represent the frequency dependence of the absorption coefficient for carbon dioxide, and Crosbie and Viskanta [15] employed a two-band model to solve a problem of nongray radiative heat transfer.

When the absorption spectrum consists of narrow bands or spectral lines,

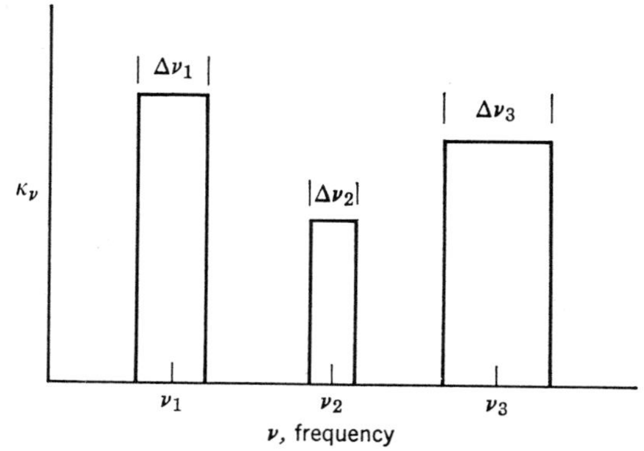

Fig. 8-8. The band model.

the *narrow-band model* can be used to simplify the nongray problem. This model assumes that the width of a band is sufficiently narrow so that the Planck function $I_{\nu b}(T)$ does not vary greatly across the band and can be taken out of the integral. For example, for a spectral line or a narrow band of width $\Delta\nu_i$, centered at frequency ν_i, the following approximation can be made:

$$\int_{\Delta\nu_i} \kappa_\nu I_{\nu b}(T)\, d\nu \cong I_{bi}(T) \int_{\Delta\nu_i} \kappa_\nu\, d\nu \tag{8-138}$$

where $I_{bi}(T)$ is the value of the Planck function at the center ν_i of the line or narrow band. If the band is broad, it can be divided into a number of sufficiently narrow bands so that each can be treated as a narrow band. Crosbie and Viskanta [16, 17] applied the narrow-band model to solve the problem of nongray radiative heat transfer for a slab exposed to a collimated external flux and for a slab with internal energy sources.

Figure 8-9 shows the so-called *picket-fence* models that may be used to characterize the frequency dependence of the absorption coefficient. The uniform picket-fence model consists of spectral lines all having identical height, width, and spacing and superimposed on a gray background. In most stellar spectra, for example, the spectral lines are seen against a continuous background. The generalized picket-fence model consists of lines (or narrow bands) having width, height, and spacing that are not uniform over the spectrum. The picket-fence model was used originally by Chandrasekhar [18] and has subsequently been employed by Siewert and Zweifel [19, 20], Kung and Sibulkin [21], and Reith, Siewert, and Özişik [22] in problems of pure radiative heat transfer, and by Greif [23] in the problem of interaction of radiation by conduction.

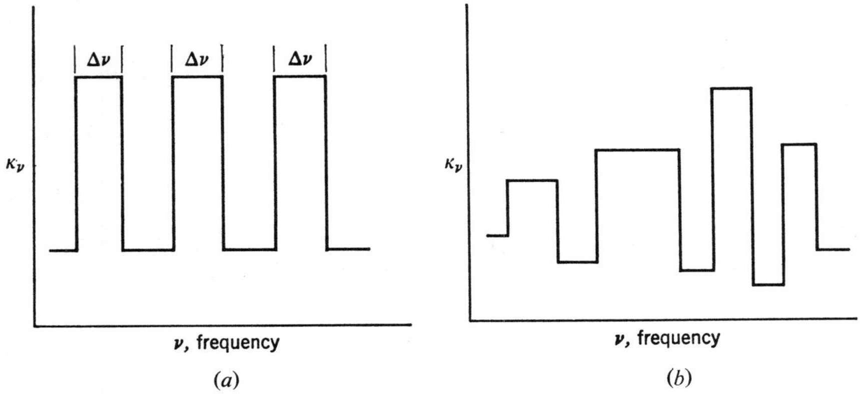

Fig. 8-9. Picket-fence models. (*a*) Uniform picket fence. (*b*) Generalized picket fence.

In this section we illustrate the application of the band model and the narrow-band model to the problem of radiative heat transfer in an absorbing, emitting, nongray slab in radiative equilibrium and formulate the equations for the determination of the temperature distribution and the net radiative heat flux in the medium.

General Formulation

The condition of radiative equilibrium is written as

$$\frac{dq^r}{dy} = 0 \tag{8-139a}$$

where[9]

$$q^r = \int_{\nu=0}^{\infty} q_\nu^{\,r}(y)\, d\nu \tag{8-139b}$$

$$q_\nu^{\,r}(y) = 2\pi \int_{-1}^{1} I_\nu(y, \mu)\mu\, d\mu \tag{8-139c}$$

For an absorbing, emitting, nongray, plane-parallel slab the equation of radiative transfer can be given as

$$\mu \frac{\partial I_\nu(y,\mu)}{\partial y} + \kappa_\nu I_\nu(y,\mu) = \kappa_\nu I_{\nu b}[T(y)], \quad 0 \le y \le L,\ -1 \le \mu \le 1 \tag{8-140}$$

Assuming that the boundaries at $y = 0$ and $y = L$ are diffusely emitting, diffusely reflecting surfaces at temperatures T_1 and T_2, respectively, the boundary conditions are given as (see Eqs. 8-99)

$$I_\nu^+(0) = \varepsilon_{1\nu} I_{\nu b}(T_1) + 2\rho_{1\nu}^d \int_0^1 I_\nu^-(0, -\mu')\mu'\, d\mu', \quad \mu > 0 \tag{8-141a}$$

$$I_\nu^-(L) = \varepsilon_{2\nu} I_{\nu b}(T_2) + 2\rho_{2\nu}^d \int_0^1 I_\nu^+(L, \mu')\mu'\, d\mu', \quad \mu < 0 \tag{8-141b}$$

where $I_{\nu b}(T)$ is the Planck function.

The spectral absorption coefficient κ_ν is assumed to depend on temperature in the form [15]

$$\kappa_\nu \equiv \kappa_\nu(T) = \alpha_\nu \xi(T) \tag{8-142a}$$

with

$$\alpha_\nu \le 1 \tag{8-142b}$$

where α_ν is a function of the frequency alone, and $\xi(T)$ of temperature alone.

We define the optical variable τ and the optical thickness τ_0 as

$$\tau \equiv \int_0^y \xi(T)\, dy \tag{8-143a}$$

$$\tau_0 \equiv \int_0^L \xi(T)\, dy \tag{8-143b}$$

Equations 8-140 and 8-141 can be written in terms of the optical variable as

$$\mu \frac{\partial I_\nu(\tau, \mu)}{\partial \tau} + \alpha_\nu I_\nu(\tau, \mu) = \alpha_\nu I_{\nu b}[T(\tau)], \quad 0 \leq \tau \leq \tau_0, \ -1 \leq \mu \leq 1 \qquad (8\text{-}144)$$

with the boundary conditions

$$I_\nu^+(0) = \varepsilon_{1\nu} I_{\nu b}(T_1) + 2\rho_{1\nu}^d \int_0^1 I_\nu^-(0, -\mu')\mu' \, d\mu', \quad \mu > 0 \qquad (8\text{-}145\text{a})$$

$$I_\nu^-(\tau_0) = \varepsilon_{2\nu} I_{\nu b}(T_2) + 2\rho_{2\nu}^d \int_0^1 I_\nu^+(\tau_0, \mu')\mu' \, d\mu', \quad \mu < 0 \qquad (8\text{-}145\text{b})$$

Equation 8-144 can be solved formally subject to the boundary conditions 8-145, and the equation for the spectral radiative heat flux $q_\nu^r(\tau)$ determined as (see Eq. 8-84)

$$q_\nu^r(\tau) = 2\pi\left\{I_\nu^+(0)E_3(\alpha_\nu\tau) + \int_0^\tau \alpha_\nu I_{\nu b}[T(\tau')]E_2[\alpha_\nu(\tau - \tau') \, d\tau'\right\}$$
$$- 2\pi\left\{I_\nu^-(\tau_0)E_3[\alpha_\nu(\tau_0 - \tau)] + \int_\tau^{\tau_0} \alpha_\nu I_{\nu b}[T(\tau')]E_2[\alpha_\nu(\tau' - \tau)] \, d\tau'\right\} \qquad (8\text{-}146)$$

and the derivative of $q_\nu^r(\tau)$ with respect to τ is given as (see Eq. 8-95)

$$\frac{dq_\nu^r(\tau)}{d\tau} = 4\pi\alpha_\nu I_{\nu b}[T(\tau)]$$
$$- 2\pi\left\{\alpha_\nu I_\nu^+(0)E_2(\alpha_\nu\tau) + \alpha_\nu I_\nu^-(\tau_0)E_2[\alpha_\nu(\tau_0 - \tau)]\right.$$
$$\left. + \int_0^{\tau_0} \alpha_\nu^2 I_{\nu b}[T(\tau')]E_1(\alpha_\nu |\tau - \tau'|) \, d\tau'\right\} \qquad (8\text{-}147)$$

The formal solutions for the boundary surface intensities $I_\nu^+(0)$ and $I_\nu^-(\tau_0)$ are given as (see Eqs. 8-110)

$$I_\nu^+(0) = \varepsilon_{1\nu} I_{\nu b}(T_1)$$
$$+ 2\rho_{1\nu}^d\left\{I_\nu^-(\tau_0)E_3(\alpha_\nu\tau_0) + \int_0^{\tau_0} \alpha_\nu I_{\nu b}[T(\tau')]E_2(\alpha_\nu\tau') \, d\tau'\right\} \qquad (8\text{-}148\text{a})$$

$$I_\nu^-(\tau_0) = \varepsilon_{2\nu} I_{\nu b}(T_2)$$
$$+ 2\rho_{2\nu}^d\left\{I_\nu^+(0)E_3(\alpha_\nu\tau_0) + \int_0^{\tau_0} \alpha_\nu I_{\nu b}[T(\tau')]E_2[\alpha_\nu(\tau_0 - \tau')] \, d\tau'\right\} \qquad (8\text{-}148\text{b})$$

and the condition of radiative equilibrium (Eq. 8-139) is written as

$$\int_{\nu=0}^\infty \frac{dq_\nu^r(\tau)}{d\tau} \, d\nu = 0 \qquad (8\text{-}149)$$

By substituting Eq. 8-147 into Eq. 8-149 we obtain

$$\int_{\nu=0}^{\infty} \alpha_\nu I_{\nu b}[T(\tau)]\, d\nu$$
$$= \frac{1}{2}\int_{\nu=0}^{\infty}\left\{\alpha_\nu I_\nu^+(0)E_2(\alpha_\nu\tau) + \alpha_\nu I_\nu^-(\tau_0)E_2[\alpha_\nu(\tau_0-\tau)]\right.$$
$$\left. + \int_0^{\tau_0}\alpha_\nu^2 I_{\nu b}[T(\tau')]E_1(\alpha_\nu\,|\tau-\tau'|)\, d\tau'\right\} d\nu \qquad (8\text{-}150)$$

Equation 8-150 with the boundary surface intensities given by Eqs. 8-148 provides an integral equation for the determination of temperature distribution in the slab but is very difficult to solve for the general frequency-dependent case considered here. If the frequency dependence of the absorption coefficient is approximated by a band model or a narrow-band model, however, Eq. 8-150 can be reduced to a form that is more tractable for computation purposes. We examine below the application of the band model and the narrow-band model to Eq. 8-150 by means of an approach adopted by Crosbie and Viskanta [15–17].

(a) Application of Band Model

Consider an absorption spectrum consisting of M bands each of width $\Delta\nu_i$ and positioned at frequency ν_i, $i = 1, 2, \ldots, M$. We assume that the parameter α_ν takes a value $0 < \alpha_\nu \leq 1$ within the band and vanishes between two neighboring bands. Noting that the integrals in Eq. 8-150 vanish when $\alpha_\nu = 0$, we can write Eq. 8-150 as

$$\sum_{i=1}^{M}\alpha_i\int_{\Delta\nu_i} I_{\nu b}[T(\tau)]\, d\nu$$
$$= \frac{1}{2}\sum_{i=1}^{M}\alpha_i E_2(\alpha_i\tau)\int_{\Delta\nu_i} I_\nu^+(0)\, d\nu$$
$$+ \frac{1}{2}\sum_{i=1}^{M}\alpha_i E_2[\alpha_i(\tau_0-\tau)]\int_{\Delta\nu_i} I_\nu^-(\tau_0)\, d\nu$$
$$+ \frac{1}{2}\int_0^{\tau_0}\sum_{i=1}^{M}\alpha_i^2 E_1(\alpha_i\,|\tau-\tau'|)\left\{\int_{\Delta\nu_i} I_{\nu b}[T(\tau')]\, d\nu\right\} d\tau' \qquad (8\text{-}151)$$

The formulation given by this equation is quite general. In order to illustrate its specific application we consider below a two-band model.

Two-Band Model. Equation 8-151 can be simplified further if it is assumed that the absorption spectrum consists of two regions, an absorbing and a transparent region, and that α_ν takes a value of unity in the absorbing region contained in the frequency band $\Delta\nu$ and a value of zero outside this region, that is,

$$\alpha_\nu = \begin{cases} 1 & \text{for } \nu \text{ in } \Delta\nu \\ 0 & \text{for } \nu \text{ outside } \Delta\nu \end{cases} \qquad (8\text{-}152)$$

The application of the two-band model given in Eq. 8-152 to Eq. 8-151 yields

$$\int_{\nu=0}^{\infty} \alpha_\nu I_{\nu b}[T(\tau)]\, d\nu = \tfrac{1}{2}E_2(\tau)\int_{\nu=0}^{\infty} \alpha_\nu I_\nu^+(0)\, d\nu$$
$$+ \tfrac{1}{2}E_2(\tau_0 - \tau)\int_{\nu=0}^{\infty} \alpha_\nu I_\nu^-(\tau_0)\, d\nu$$
$$+ \frac{1}{2}\int_{\tau'=0}^{\tau_0} E_1(|\tau - \tau'|)\left\{\int_{\nu=0}^{\infty} \alpha_\nu I_{\nu b}[T(\tau')]\, d\nu\right\} d\tau' \quad (8\text{-}153)$$

where α_ν is given by Eq. 8-152.

We define new functions as

$$f(\tau) \equiv \int_{\nu=0}^{\infty} \alpha_\nu I_{\nu b}[T(\tau)]\, d\nu \quad (8\text{-}154a)$$

$$f_1 \equiv \int_{\nu=0}^{\infty} \alpha_\nu I_\nu^+(0)\, d\nu \quad (8\text{-}154b)$$

$$f_2 \equiv \int_{\nu=0}^{\infty} \alpha_\nu I_\nu^-(\tau_0)\, d\nu \quad (8\text{-}154c)$$

Then Eq. 8-153 becomes

$$f(\tau) = \frac{1}{2}\left[f_1 E_2(\tau) + f_2 E_2(\tau_0 - \tau) + \int_{\tau'=0}^{\tau_0} f(\tau')E_1(|\tau - \tau'|)\, d\tau'\right] \quad (8\text{-}155a)$$

which can be rearranged as

$$\frac{f(\tau) - f_2}{f_1 - f_2} = \frac{1}{2}\left[E_2(\tau) + \int_{\tau'=0}^{\tau_0} \frac{f(\tau') - f_2}{f_1 - f_2} E_1(|\tau - \tau'|)\, d\tau'\right] \quad (8\text{-}155b)$$

Equation 8-155b defines the temperature distribution in the medium and is in a form similar to Eq. 8-130b. Therefore we introduce a dimensionless function $\theta(\tau)$, defined as

$$\theta(\tau) \equiv \frac{f(\tau) - f_2}{f_1 - f_2} \quad (8\text{-}156)$$

Then Eq. 8-155b becomes

$$\theta(\tau) = \frac{1}{2}\left[E_2(\tau) + \int_0^{\tau_0} \theta(\tau')E_1(|\tau - \tau'|)\, d\tau'\right] \quad (8\text{-}157)$$

This is an integral equation for the function $\theta(\tau)$ and is exactly the same as Eq. 8-132 for the gray case.

A relation will now be determined for the net radiative heat flux q^r. By integrating Eq. 8-146 over all frequencies and applying the two-band approximation defined by Eq. 8-152, we obtain the relation for the net radiative heat flux as

$$q^r = \pi\left[\int_{\text{outside }\Delta\nu} I_\nu^+(0)\,d\nu - \int_{\text{outside }\Delta\nu} I_\nu^-(\tau_0)\,d\nu\right]$$
$$+ 2\pi\left\{E_3(\tau)\int_{\Delta\nu} I_\nu^+(0)\,d\nu - E_3(\tau_0-\tau)\int_{\Delta\nu} I_\nu^-(\tau_0)\,d\nu\right.$$
$$+\int_0^\tau E_2(\tau-\tau')\int_{\Delta\nu} I_{\nu b}[T(\tau')]\,d\nu\,d\tau'$$
$$\left.-\int_\tau^{\tau_0} E_2(\tau'-\tau)\int_{\Delta\nu} I_{\nu b}[T(\tau')]\,d\nu\,d\tau'\right\} \tag{8-158a}$$

This equation can be written more compactly in the form

$$q^r = [\pi f_1^* - \pi f_2^*] + 2\pi\left\{f_1E_3(\tau) - f_2E_3(\tau_0-\tau)\right.$$
$$\left.+\int_0^\tau f(\tau')E_2(\tau-\tau')\,d\tau' - \int_\tau^{\tau_0} f(\tau')E_2(\tau'-\tau)\,d\tau'\right\} \tag{8-158b}$$

where f_1, f_2, and $f(\tau')$ were defined by Eqs. 8-154, and f_1^* and f_2^* are defined as

$$f_1^* \equiv \int_{\text{outside }\Delta\nu} I_\nu^+(0)\,d\nu = \int_{\nu=0}^\infty (1-\alpha_\nu)I_\nu^+(0)\,d\nu \tag{8-159a}$$

$$f_2^* \equiv \int_{\text{outside }\Delta\nu} I_\nu^-(\tau_0)\,d\nu = \int_{\nu=0}^\infty (1-\alpha_\nu)I_\nu^-(\tau_0)\,d\nu \tag{8-159b}$$

Equation 8-158b can be written in the form

$$\frac{q^r - \pi(f_1^* - f_2^*)}{\pi(f_1-f_2)} = 2\left[E_3(\tau) + \int_0^\tau \frac{f(\tau')-f_2}{f_1-f_2}E_2(\tau-\tau')\,d\tau'\right.$$
$$\left.-\int_\tau^{\tau_0}\frac{f(\tau')-f_2}{f_1-f_2}E_2(\tau'-\tau)\,d\tau'\right] \tag{8-160}$$

Substitution of the transformation given by Eq. 8-156 into Eq. 8-160 yields the following expression for the net radiative heat flux q^r:

$$\frac{q^r - \pi(f_1^* - f_2^*)}{\pi(f_1-f_2)} = Q \tag{8-161a}$$

where we have defined the dimensionless heat flux Q as

$$Q \equiv 2\left[E_3(\tau) + \int_0^\tau \theta(\tau')E_2(\tau-\tau')\,d\tau' - \int_\tau^{\tau_0}\theta(\tau')E_2(\tau'-\tau)\,d\tau'\right] \tag{8-161b}$$

The dimensionless heat flux Q is independent of τ since the net radiative heat flux q^r in the medium is constant for radiative equilibrium. Equations 8-161 are of the same form as Eqs. 8-134 for a gray medium.

For black boundaries the boundary surface intensities $I_\nu^+(0)$ and $I_\nu^-(\tau_0)$ are known, and the functions f_1, f_2, f_1^*, and f_2^* are immediately available. Therefore, once the function $\theta(\tau)$ is determined from the solution of the integral equation 8-157, the net radiative heat flux can be evaluated from Eq. 8-161. The determination of temperature distribution in the medium, however, requires the solution of the transcendental equation 8-156, in which the function $f(\tau)$ involves the temperature.

(b) Application of Narrow-Band Model

Equation 8-150 can be simplified by the application of the narrow-band model. Consider an absorption spectrum consisting of M narrow bands (or lines) each of width $\Delta\nu_i$ and centered at frequency ν_i, $i = 1, 2, \ldots, M$. To simplify the analysis we assume that the boundaries of the slab are black surfaces, that is, $I_{\nu i}^+(0) = I_{bi}(T_1)$ and $I_{\nu i}^-(\tau_0) = I_{bi}(T_2)$.

When the narrow-band approximation given by Eq. 8-138 is applied to Eq. 8-150, we obtain

$$\begin{aligned}\sum_{i=1}^{M} I_{bi}[T(\tau)] \int_{\Delta\nu_i} \alpha_\nu \, d\nu \\ = \frac{1}{2}\sum_{i=1}^{M} I_{bi}(T_1)\int_{\Delta\nu_i} \alpha_\nu E_2(\alpha_\nu \tau)\, d\nu \\ + \frac{1}{2}\sum_{i=1}^{M} I_{bi}(T_2)\int_{\Delta\nu_i} \alpha_\nu E_2[\alpha_\nu(\tau_0-\tau)]\, d\nu \\ + \frac{1}{2}\int_0^{\tau_0}\sum_{i=1}^{M} I_{bi}[T(\tau')]\left[\int_{\Delta\nu_i} \alpha_\nu^2 E_1(\alpha_\nu\,|\tau-\tau'|)\, d\nu\right] d\tau' \qquad (8\text{-}162)\end{aligned}$$

where $\alpha_i \leq 1$, $i = 1, 2, \ldots, M$, within the band, and the Planck function I_{bi} is evaluated at the center of the ith band or line.

Equation 8-162 can be written in the form

$$\begin{aligned}\sum_{i=1}^{M} I_{bi}[T(\tau)]\gamma_i = \frac{1}{2}\sum_{i=1}^{M} I_{bi}(T_1)\gamma_i K_2^i(\tau) + \frac{1}{2}\sum_{i=1}^{M} I_{bi}(T_2)\gamma_i K_2^i(\tau_0-\tau) \\ + \frac{1}{2}\int_0^{\tau_0}\sum_{i=1}^{M} I_{bi}[T(\tau')]\gamma_i K_1^i(|\tau-\tau'|)\, d\tau' \qquad (8\text{-}163)\end{aligned}$$

where we have defined

$$\gamma_i \equiv \int_{\Delta\nu_i} \alpha_\nu \, d\nu \qquad (8\text{-}164a)$$

$$K_1^i(z) \equiv \frac{1}{\gamma_i}\int_{\Delta\nu_i} \alpha_\nu^2 E_1(\alpha_\nu z)\, d\nu \qquad (8\text{-}164b)$$

$$K_2^i(z) \equiv \frac{1}{\gamma_i}\int_{\Delta\nu_i} \alpha_\nu E_2(\alpha_\nu z)\, d\nu \qquad (8\text{-}164c)$$

If we assume also that the absorption coefficient for each band (or line) is identical in shape and intensity and the profiles are not overlapping, the quantities γ_i, $K_1^i(z)$, and $K_2^i(z)$ become independent of i; then Eq. 8-163 simplifies to

$$F(\tau) = \frac{1}{2}\left[F_1 K_2(\tau) + F_2 K_2(\tau_0 - \tau) + \int_0^{\tau_0} F(\tau') K_1(|\tau - \tau'|)\, d\tau'\right] \tag{8-165}$$

where we have defined

$$F(\tau) \equiv \sum_{i=1}^{M} I_{bi}[T(\tau)] \tag{8-166a}$$

$$F_1 \equiv \sum_{i=1}^{M} I_{bi}(T_1) \tag{8-166b}$$

$$F_2 \equiv \sum_{i=1}^{M} I_{bi}(T_2) \tag{8-166c}$$

and

$$K_1(z) \equiv \frac{1}{\gamma}\int_{\Delta\nu} \alpha_\nu^2 E_1(\alpha_\nu z)\, d\nu \tag{8-167a}$$

$$K_2(z) \equiv \frac{1}{\gamma}\int_{\Delta\nu} \alpha_\nu E_2(\alpha_\nu z)\, d\nu \tag{8-167b}$$

$$\gamma \equiv \int_{\Delta\nu} \alpha_\nu\, d\nu \tag{8-167c}$$

It is convenient to introduce a dimensionless function $\theta^*(\tau)$, defined as

$$\theta^*(\tau) = \frac{F(\tau) - F_2}{F_1 - F_2} \tag{8-168}$$

Substitution of Eq. 8-168 into Eq. 8-165 yields the following integral equation for the function $\theta^*(\tau)$:

$$\theta^*(\tau) = \frac{1}{2}\left[K_2(\tau) + \int_0^{\tau_0} \theta^*(\tau') K_1(|\tau - \tau'|)\, d\tau'\right] \tag{8-169}$$

The integral equation 8-169 can be solved for the characteristic functions $K_1(z)$ and $K_2(z)$, defined by Eqs. 8-167. The mathematical properties of the $K_1(z)$ and $K_2(z)$ functions have been discussed by Crosbie and Viskanta [17].

The relation for the net radiative heat flux q^r can be determined from Eq. 8-146 by integrating this equation over all frequencies and applying the narrow-band approximation given by Eq. 8-138. If we assume that the boundaries are black, the absorption coefficient in each region is identical in shape and intensity, and the profiles are not overlapping, Eq. 8-146 yields

$$q^r = \bar{\sigma}(T_1^4 - T_2^4) + 2\pi\gamma[F_1 K_3(\tau) - F_2 K_3(\tau_0 - \tau)] + 2\pi\gamma\left[\int_0^{\tau} F(\tau')K_2(\tau - \tau')\,d\tau' - \int_{\tau}^{\tau_0} F(\tau')K_2(\tau' - \tau)\,d\tau'\right] \quad (8\text{-}170)$$

where F_1, F_2, and $F(\tau)$ were defined by Eqs. 8-166, γ and $K_2(z)$ were defined by Eqs. 8-167, and the function $K_3(z)$ is defined as

$$K_3(z) \equiv \frac{1}{\gamma}\int_{\Delta\nu} [E_3(\alpha_\nu z) - \tfrac{1}{2}]\,d\nu \quad (8\text{-}171)$$

We note that Eq. 8-170 is of the same form as Eq. 8-158b; therefore it can be rearranged as in Eqs. 8-161 and related to the function $\theta^*(\tau)$. We obtain

$$\frac{q^r - \bar{\sigma}(T_1^4 - T_2^4)}{\pi\gamma(F_1 - F_2)} = Q \quad (8\text{-}172)$$

where the dimensionless heat flux Q is defined as

$$Q \equiv 2\left[K_3(\tau) + \int_0^{\tau} \theta^*(\tau')E_2(\tau - \tau')\,d\tau' - \int_{\tau}^{\tau_0} \theta^*(\tau')E_2(\tau' - \tau)\,d\tau'\right] \quad (8\text{-}173)$$

Once the function $\theta^*(\tau)$ is known from the solution of the integral equation 8-169, Q is evaluated from Eq. 8-173 and the net radiative heat flux q^r from Eq. 8-172. The determination of temperature distribution in the medium requires the solution of the transcendental equation 8-168, in which the function $F(\tau)$ involves the temperature.

8-11 PLANE-PARALLEL SLAB WITH INTERNAL ENERGY SOURCES

In this section consideration is given to the problem of radiative heat transfer in a plane-parallel slab having thickness L and containing distributed energy sources of strength $g(y)$ Btu/hr ft³. It will be assumed that the medium absorbs and emits radiation and that the boundaries at $y = 0$ and $y = L$ are opaque, diffuse emitters and diffuse reflectors and are kept at uniform temperatures T_1 and T_2, respectively. We shall be concerned with the determination of the temperature distribution and the net radiative heat flux in the medium. Here we present the mathematical formulation of this problem for both gray and nongray cases.

Gray Medium

When the energy transfer is by pure radiation (i.e., conductive and convective modes of heat transfer are negligible), the conservation of energy equation in one dimension for a medium containing energy sources of strength $g(y)$ is given by (see Eq. 8-18b)

$$\frac{dq^r(y)}{dy} = g(y) \qquad \text{or} \qquad \frac{dq^r(\tau)}{d\tau} = \frac{g(\tau)}{\kappa} \tag{8-174}$$

where $d\tau = \kappa\, dy$ is the optical variable, and $q^r(\tau)$ is related to the radiation intensity $I(\tau, \mu)$ by

$$q^r(\tau) = 2\pi \int_{-1}^{1} I(\tau, \mu)\mu\, d\mu \tag{8-175}$$

and the radiation intensity $I(\tau, \mu)$ is to be determined from the solution of the equation of radiative transfer. For an absorbing and emitting gray slab with diffusely emitting, diffusely reflecting boundaries the equation of radiative transfer and the boundary conditions are given as

$$\mu \frac{\partial I(\tau, \mu)}{\partial \tau} + I(\tau, \mu) = I_b[T(\tau)] \quad \text{in } 0 \le \tau \le \tau_0,\ -1 \le \mu \le 1 \tag{8-176a}$$

with (see Eqs. 8-125b and 8-125c)

$$I^+(0) = \varepsilon_1 I_b(T_1) + 2\rho_1{}^d \int_0^1 I(0, -\mu')\mu'\, d\mu', \quad \mu > 0 \tag{8-176b}$$

$$I^-(\tau_0) = \varepsilon_2 I_b(T_2) + 2\rho_2{}^d \int_0^1 I^+(\tau_0, \mu')\mu'\, d\mu', \quad \mu < 0 \tag{8-176c}$$

where $\tau_0 = \kappa L$ is the optical thickness of the slab.

A relation defining the temperature distribution in the medium is obtained by operating on Eq. 8-176a by the operator $2\pi \int_{-1}^{1} d\mu$. We find

$$\frac{dq^r(\tau)}{d\tau} + G(\tau) = 4\pi I_b[T(\tau)] \equiv 4n^2\bar{\sigma}T^4(\tau) \tag{8-177a}$$

and by substituting for $dq^r(\tau)/d\tau$ from Eq. 8-174 we obtain

$$n^2\bar{\sigma}T^4(\tau) = \frac{g(\tau)}{4\kappa} + \tfrac{1}{4}G(\tau) \tag{8-177b}$$

where $G(\tau)$ is the incident radiation:

$$G(\tau) = 2\pi \int_{-1}^{1} I(\tau, \mu)\, d\mu \tag{8-177c}$$

Therefore, once the angular distribution of radiation intensity $I(\tau, \mu)$ is known, the net radiative heat flux and the temperature distribution everywhere in the medium are obtained from Eqs. 8-175 and 8-177b, respectively. The radiative heat transfer problem considered here (Eqs. 8-176) is exactly the same as that considered previously for radiative equilibrium (i. e., Eqs. 8-126, 8-125b and 8-125c). Therefore substitution of $G(\tau)$ from Eq. 8-127 into Eq. 8-177b yields

$$n^2\bar{\sigma}T^4(\tau) = \frac{g(\tau)}{4\kappa} + \frac{1}{2}\left[\pi I^+(0)E_2(\tau) + \pi I^-(\tau_0)E_2(\tau_0 - \tau) + \int_0^{\tau_0} n^2\bar{\sigma}T^4(\tau')E_1(|\tau - \tau'|)\,d\tau'\right] \tag{8-178}$$

This integral equation defines the distribution of temperature within the slab; for the case of no internal generation it simplifies to Eq. 8-130a. The solution of Eq. 8-178 for constant g has been considered by Heaslet and Warming [24, 25], who reduced this equation into two simpler Fredholm integral equations of the second kind.

Constant Energy Generation

To simplify the analysis the energy generation in the medium is assumed to be uniform (i.e., g = constant). We rearrange Eq. 8-178 in the form

$$\frac{n^2\bar{\sigma}T^4(\tau) - \pi I^-(\tau_0)}{\pi I^+(0) - \pi I^-(\tau_0)} = \frac{1}{4}\frac{g/\kappa}{\pi I^+(0) - \pi I^-(\tau_0)} + \frac{1}{2}\left[E_2(\tau) + \int_0^{\tau_0} \frac{n^2\bar{\sigma}T^4(\tau') - \pi I^-(\tau_0)}{\pi I^+(0) - \pi I^-(\tau_0)} E_1(|\tau - \tau'|)\,d\tau'\right] \tag{8-179}$$

and introduce a transformation that relates the temperature distribution to the functions $\theta(\tau)$ and $\theta_g(\tau)$ in the form

$$\frac{n^2\bar{\sigma}T^4(\tau) - \pi I^-(\tau_0)}{\pi I^+(0) - \pi I^-(\tau_0)} = \theta(\tau) + \frac{g/\kappa}{\pi I^+(0) - \pi I^-(\tau_0)}\theta_g(\tau) \tag{8-180}$$

If we choose the function $\theta(\tau)$ to satisfy the following integral equation (see Eq. 8-132):

$$\theta(\tau) = \frac{1}{2}\left[E_2(\tau) + \int_0^{\tau_0} \theta(\tau')E_1(|\tau - \tau'|)\,d\tau'\right] \tag{8-181}$$

then the transformation 8-180 will satisfy the integral equation 8-179 if the function $\theta_g(\tau)$ is the solution of the following integral equation:

$$\theta_g(\tau) = \frac{1}{4} + \frac{1}{2}\int_0^{\tau_0} \theta_g(\tau')E_1(|\tau - \tau'|)\,d\tau' \tag{8-182}$$

The validity of Eq. 8-182 is readily verified by substituting the transformation 8-180 into Eq. 8-179 and utilizing Eq. 8-181.

Thus we have reduced the integral equation 8-178 (or 8-179) defining the temperature distribution in the medium into two simpler integral equations for the universal functions $\theta(\tau)$ and $\theta_g(\tau)$. The function $\theta(\tau)$ is exactly the same as that defined by Eq. 8-132.

The relation for the net radiative heat flux $q^r(\tau)$ in the medium is of exactly the same form as given by Eq. 8-128 and can be rearranged as Eq. 8-133, that is,

$$\frac{q^r(\tau)}{\pi I^+(0) - \pi I^-(\tau_0)} = 2\left[E_3(\tau) + \int_0^\tau \frac{n^2\bar{\sigma}T^4(\tau') - \pi I^-(\tau_0)}{\pi I^+(0) - \pi I^-(\tau_0)} E_2(\tau - \tau')\, d\tau' - \int_\tau^{\tau_0} \frac{n^2\bar{\sigma}T^4(\tau') - \pi I^-(\tau_0)}{\pi I^+(0) - \pi I^-(\tau_0)} E_2(\tau' - \tau)\, d\tau'\right] \tag{8-183}$$

Now, if we substitute the transformation 8-180 into Eq. 8-183, the expression for the net radiative heat flux can be written in the form

$$\frac{q^r(\tau)}{\pi[I^+(0) - I^-(\tau_0)]} = Q(\tau) + \frac{g/\kappa}{\pi[I^+(0) - I^-(\tau_0)]} Q_g(\tau) \tag{8-184}$$

where we have defined the functions $Q(\tau)$ and $Q_g(\tau)$ as

$$Q(\tau) \equiv 2\left[E_3(\tau) + \int_0^\tau \theta(\tau')E_2(\tau - \tau')\, d\tau' - \int_\tau^{\tau_0} \theta(\tau')E_2(\tau' - \tau)\, d\tau'\right] \tag{8-185}$$

$$Q_g(\tau) \equiv 2\left[\int_0^\tau \theta_g(\tau')E_2(\tau - \tau')\, d\tau' - \int_\tau^{\tau_0} \theta_g(\tau')E_2(\tau' - \tau)\, d\tau'\right] \tag{8-186}$$

We note that the function $Q(\tau)$ is exactly the same as that defined by Eq. 8-134a.

For black boundaries the boundary surface intensities $I^+(0)$ and $I^-(\tau_0)$ are known. Once the functions $\theta(\tau)$ and $\theta_g(\tau)$ are determined from the solution of the integral equations 8-181 and 8-182, respectively, the distribution of temperature is evaluated from Eq. 8-180. The functions $Q(\tau)$ and $Q_g(\tau)$ are evaluated from Eqs. 8-185 and 8-186, respectively, and the net radiative heat flux $q^r(\tau)$ is determined from Eq. 8-184.

Nongray Medium

We now illustrate the application of the band model, the narrow-band model, and the picket-fence model to the radiative heat transfer problem for the nongray case. For simplicity we assume that the boundaries are black; hence the boundary surface intensities $I^+(0)$, $\mu > 0$, and $I^-(\tau_0)$, $\mu < 0$, are known quantities. For the band model and the narrow-band model the equations for the temperature distribution and the net radiative heat flux can be derived by an analysis similar to that described in the Section 8-10. Therefore for these models the derivations will not be repeated; only the resulting equations will be given here. In the case of the picket-fence model, however, the formulation will be considered in detail.

(a) Two-Band Model

For a two-band model, the equation that defines the temperature distribution in the medium is given as (see Eq. 8-155a)

$$f(\tau) = \frac{g(\tau)}{4\pi\xi(T)} + \frac{1}{2}\left[f_1 E_2(\tau) + f_2 E_2(\tau_0 - \tau) + \int_{\tau'=0}^{\tau_0} f(\tau')E_1(|\tau - \tau'|)\, d\tau'\right] \tag{8-187a}$$

which can be rearranged as

$$\frac{f(\tau) - f_2}{f_1 - f_2} = \frac{g(\tau)}{4\pi\xi(T)(f_1 - f_2)} + \frac{1}{2}\left[E_2(\tau) + \int_{\tau'=0}^{\tau_0} \frac{f(\tau') - f_2}{f_1 - f_2} E_1(|\tau - \tau'|)\, d\tau'\right] \tag{8-187b}$$

where $f(\tau)$, f_1, and f_2 were defined by Eqs. 8-154, and

$$\kappa_\nu \equiv \kappa_\nu(T) = \alpha_\nu \xi(T) \tag{8-187c}$$

$$\tau \equiv \int_0^y \xi(T)\, dy \tag{8-187d}$$

We note that for no energy generation Eq. 8-187b reduces to Eq. 8-155b.

Assuming that $g(\tau)/\xi(T)$ is constant, we can relate the temperature distribution function $f(\tau)$ to the universal functions $\theta(\tau)$ and $\theta_g(\tau)$ by (see Eq. 8-180)

$$\frac{f(\tau) - f_2}{f_1 - f_2} = \theta(\tau) + \frac{g}{\pi(f_1 - f_2)\xi}\theta_g(\tau) \tag{8-188}$$

where the functions $\theta(\tau)$ and $\theta_g(\tau)$ are the solutions of the integral equations 8-181 and 8-182, respectively.

(b) Narrow Band Model

If we assume that each narrow band (or line) is identical in shape and intensity, profiles are not overlapping, and $g(\tau)/\xi(T)$ is constant, the application of the narrow-band model yields the following equation for the temperature distribution:

$$F(\tau) = \frac{g}{4\pi\gamma\xi} + \frac{1}{2}\left[F_1 K_2(\tau) + F_2 K_2(\tau_0 - \tau) + \int_{\tau'=0}^{\tau_0} F(\tau')K_1(|\tau - \tau'|)\, d\tau'\right] \tag{8-189}$$

where $\kappa_\nu = \alpha_\nu \xi$, $d\tau = \xi\, dy$, and other quantities were defined by Eqs. 8-166 and 8-167. We note that, for no internal energy generation, Eq. 8-189 simplifies to Eq. 8-165.

The temperature distribution function $F(\tau)$ can be related to new dimensionless functions $\theta^*(\tau)$ and $\theta_g^*(\tau)$ by

$$\frac{F(\tau) - F_2}{F_1 - F_2} = \theta^*(\tau) + \frac{g}{\pi(F_1 - F_2)\gamma\xi}\,\theta_g^*(\tau) \tag{8-190a}$$

where the functions $\theta^*(\tau)$ and $\theta_g^*(\tau)$ are the solutions of the following integral equations:

$$\theta^*(\tau) = \frac{1}{2}\left[K_2(\tau) + \int_0^{\tau_0} \theta^*(\tau')K_1(|\tau - \tau'|)\, d\tau'\right] \tag{8-190b}$$

$$\theta_g^*(\tau) = \frac{1}{4} + \frac{1}{2}\int_0^{\tau_0} \theta_g^*(\tau')K_1(|\tau - \tau'|)\, d\tau' \tag{8-190c}$$

and $K_1(\tau)$ and $K_2(\tau)$ were defined by Eqs. 8-167.

(c) Picket-Fence Model

The radiative heat transfer for an absorbing, emitting, nongray medium with internal energy sources will now be formulated by using a generalized picket-fence model.

We assume that the energy spectrum is divided into frequency bands $\Delta\nu_i$, $i = 1, 2, \ldots, N$, so that over each frequency band the absorption coefficient is uniform. Then the conservation of energy equation becomes

$$\frac{dq^r(y)}{dy} = g(y) \tag{8-191a}$$

where

$$q^r(y) = \int_{\nu=0}^{\infty} q_\nu^r(y)\, d\nu \tag{8-191b}$$

$$q_\nu^r(y) = 2\pi\int_{-1}^{1} I_\nu(y, \mu)\mu\, d\mu \tag{8-191c}$$

which can be written as

$$q^r(y) = \sum_{i=1}^{N} q_i^r(y) \tag{8-192a}$$

$$q_i^r(y) = 2\pi\int_{-1}^{1} I_i(y, \mu)\mu\, d\mu \tag{8-192b}$$

$$I_i(y, \mu) = \int_{\Delta\nu i} I_\nu(y, \mu)\, d\nu \tag{8-192c}$$

We now consider the equation of radiative transfer in the form

$$\mu \frac{\partial I_\nu(y,\mu)}{\partial y} + \kappa_\nu I_\nu(y,\mu) = \kappa_\nu I_\nu[T(y)] \quad \text{in } 0 \le y \le L \qquad \text{(8-193)}$$

subject to appropriate boundary conditions at $y = 0$ and $y = L$.

By integrating Eq. 8-193 over a frequency band $\Delta\nu_i$ and noting that the absorption coefficient is uniform over $\Delta\nu_i$ we obtain

$$\mu \frac{\partial I_i(y,\mu)}{\partial y} + \kappa_i I_i(y,\mu) = \kappa_i f_i \frac{\bar{\sigma}T^4(y)}{\pi}, \quad i = 1, 2, \ldots, N \qquad \text{(8-194)}$$

where we have defined

$$f_i \equiv \frac{I_{bi}[T(y)]}{\bar{\sigma}T^4(y)/\pi} \qquad \text{(8-195a)}$$

$$I_{bi}[T(y)] \equiv \int_{\Delta\nu i} I_{b\nu}[T(y)]\, d\nu \qquad \text{(8-195b)}$$

$$I_i(y,\mu) \equiv \int_{\Delta\nu i} I_\nu(y,\mu)\, d\mu \qquad \text{(8-195c)}$$

and κ_i is the mean absorption coefficient in the band $\Delta\nu_i$.

We operate on both sides of Eq. 8-194 by the operator

$$2\pi \int_{-1}^{1} d\mu \qquad \text{(8-196)}$$

and obtain

$$\frac{dq_i^r(y)}{dy} + \kappa_i 2\pi \int_{-1}^{1} I_i(y,\mu)\, d\mu = 4\pi\kappa_i f_i \frac{\bar{\sigma}T^4(y)}{\pi}, \quad i = 1, 2, \ldots, N \qquad \text{(8-197)}$$

where we have utilized Eq. 8-192b.

We sum up Eqs. 8-197 for $i = 1, 2, \ldots, N$ and utilize Eq. 8-192a to obtain

$$\frac{dq^r(y)}{dy} + 2\pi \sum_{j=1}^{N} \kappa_j \int_{-1}^{1} I_j(y,\mu)\, d\mu = 4\pi \frac{\bar{\sigma}T^4(y)}{\pi} \sum_{j=1}^{N} \kappa_j f_j \qquad \text{(8-198)}$$

We eliminate $dq^r(y)/dy$ from Eq. 8-198 by means of the conservation of energy equation 8-191a and obtain

$$\frac{\bar{\sigma}T^4(y)}{\pi} = \frac{1}{\sum_{j=1}^{N} \kappa_j f_j} \frac{g(y)}{4\pi} + \sum_{j=1}^{N} \frac{\kappa_j}{2\sum_{s=1}^{N} \kappa_s f_s} \int_{-1}^{1} I_j(y,\mu)\, d\mu \qquad \text{(8-199)}$$

Elimination of $\bar{\sigma}T^4(y)/\pi$ from Eq. 8-194 by means of Eq. 8-199 yields

$$\mu \frac{\partial I_i(y,\mu)}{\partial y} + \kappa_i I_i(y,\mu) = \frac{\kappa_i f_i}{\sum_{j=1}^{N} \kappa_j f_j} \frac{g(y)}{4\pi} + \sum_{j=1}^{N} \frac{\kappa_i f_i \kappa_j}{2\sum_{s=1}^{N} \kappa_s f_s} \int_{-1}^{1} I_j(y,\mu)\, d\mu$$

$$i = 1, 2, \ldots, N \qquad \text{(8-200)}$$

Equation 8-200 provides N simultaneous integrodifferential equations for the N unknown intensity functions $I_i(y, \mu)$, $i = 1, 2, \ldots, N$. Once these intensities have been determined, the distribution of temperature in the medium is evaluated from Eq. 8-199, and the net radiative heat flux from Eqs. 8-192.

(d) Two-Group Picket-Fence Model

As a special case of the generalized picket-fence model we consider here a two-group picket-fence model, that is, the absorption coefficient takes only two values, κ_1 and κ_2, over the entire energy spectrum. Then Eq. 8-200 simplifies to

$$\mu \frac{\partial I_i(y, \mu)}{\partial y} + \kappa_i I_i(y, \mu) = \frac{\kappa_i f_i}{\kappa_1 f_1 + \kappa_2 f_2} \frac{g(y)}{4\pi} + \sum_{j=1}^{2} \frac{\kappa_i f_i \kappa_j}{2(\kappa_1 f_1 + \kappa_2 f_2)} \int_{-1}^{1} I_j(y, \mu)\, d\mu, \quad i = 1 \text{ or } 2 \tag{8-201}$$

which provides two coupled integral equations for the intensities $I_1(y, \mu)$ and $I_2(y, \mu)$. Once these intensity functions have been determined from the solution of these equations subject to appropriate boundary conditions, the distribution of temperature is evaluated from Eq. 8-199 and the net radiative heat flux from Eq. 8-192 by setting $N = 2$.

If we further assume that one of the κ_i is zero, say $\kappa_2 = 0$ and κ_1 is finite, then the two-group picket-fence model reduces to the two-band model discussed previously and Eqs. 8-201 becomes uncoupled, that is,

$$\mu \frac{\partial I_1(y, \mu)}{\partial y} + \kappa_1 I_1(y, \mu) = \frac{g(y)}{4\pi} + \tfrac{1}{2}\kappa_1 \int_{-1}^{1} I_1(y, \mu)\, d\mu \tag{8-202a}$$

$$\mu \frac{\partial I_2(y, \mu)}{\partial y} = 0 \tag{8-202b}$$

Defining the optical variable as $d\tau = \kappa_1\, dy$, we obtain for Eqs. 8-202

$$\mu \frac{\partial I_1(\tau, \mu)}{\partial \tau} + I_1(\tau, \mu) = \frac{g(\tau)}{4\pi\kappa_1} + \frac{1}{2} \int_{-1}^{1} I_1(\tau, \mu)\, d\mu \tag{8-203a}$$

$$\frac{\partial I_2(\tau, \mu)}{\partial \tau} = 0 \tag{8-203b}$$

Equation 8-203a is a gray problem for the intensity function $I_1(\tau, \mu)$, and the solution of Eq. 8-203b for $I_2(\tau, \mu)$ is very simple.

8-12 PROBLEMS WITH NO AZIMUTHAL SYMMETRY

In the preceding sections we considered the formal solution of the equation of radiative transfer in a plane-parallel medium with azimuthal symmetry. In the case of isotropic scattering the problem of radiative heat transfer in a plane-parallel medium with no azimuthal symmetry can readily be transformed into a problem with azimuthal symmetry. For an anisotropically scattering medium, if the phase function is postulated to be expanded in a series of Legendre polynomials as in Eq. 8-37, the azimuthally dependent problem can be transformed into a set of azimuthally independent problems by expanding the intensity $I(\tau, \mu, \phi)$ in a Fourier series in the variable ϕ. Chu et al. [26] and Evans, Chu, and Churchill [27], for example, have expanded the intensity in the form

$$I(\tau, \mu, \phi) = \sum_{m=0}^{\infty} I_m(\tau, \mu) \cos m(\phi - \phi_0) \tag{8-204}$$

as suggested by Chandrasekhar [2, p. 23] to solve the problem of radiative heat transfer in an anisotropic half-space and a finite slab irradiated externally by obliquely incident rays.

We present now the transformation of the azimuthally dependent problem of radiative transfer into a problem with azimuthal symmetry.

Isotropic Scattering

Consider an isotropically scattering, plane-parallel slab having transparent boundaries irradiated externally at the surfaces $\tau = 0$ and $\tau = \tau_0$ with obliquely incident rays. The equation of radiative transfer and the boundary conditions are given as

$$\mu \frac{\partial I(\tau, \mu, \phi)}{\partial \tau} + I(\tau, \mu, \phi) = g(\tau) + \frac{\omega}{4\pi} \int_{\phi'=0}^{2\pi} \int_{\mu'=-1}^{1} I(\tau, \mu', \phi')\, d\mu'\, d\phi',$$

$$\text{in } 0 \leq \tau \leq \tau_0, \ -1 \leq \mu \leq 1 \tag{8-205a}$$

$$I(\tau, \mu, \phi)\big|_{\tau=0} = F_1(\mu, \phi), \quad \mu > 0 \tag{8-205b}$$

$$I(\tau, \mu, \phi)\big|_{\tau=\tau_0} = F_2(\mu, \phi), \quad \mu < 0 \tag{8-205c}$$

where the inhomogeneous term $g(\tau)$ and the boundary conditions $F_1(\mu, \phi)$ and $F_2(\mu, \phi)$ are prescribed functions.

We now consider the following auxiliary problem:

$$\mu \frac{\partial I_0(\tau, \mu, \phi)}{\partial \tau} + I_0(\tau, \mu, \phi) = g(\tau) \quad \text{in } 0 \leq \tau \leq \tau_0, \ -1 \leq \mu \leq 1 \tag{8-206a}$$

with the boundary conditions

$$I_0(\tau, \mu, \phi)|_{\tau=0} = F_1(\mu, \phi), \quad \mu > 0 \tag{8-206b}$$

$$I_0(\tau, \mu, \phi)|_{\tau=\tau_0} = F_2(\mu, \phi), \quad \mu < 0 \tag{8-206c}$$

Here the intensity $I_0(\tau, \mu, \phi)$ characterizes the unscattered component of intensity for the problem given by Eqs. 8-205.

Equations 8-206 can readily be solved by separating the intensity $I_0(\tau, \mu, \phi)$ into a forward component $I_0^+(\tau, \mu, \phi)$, $\mu > 0$, and a backward component $I_0^-(\tau, \mu, \phi)$, $\mu < 0$. We obtain (see Eqs. 8-65 and 8-66a)

$$I_0^+(\tau, \mu, \phi) = F_1(\mu, \phi)e^{-\tau/\mu} + \int_0^{\tau} g(\tau') \frac{e^{-(\tau-\tau')/\mu}}{\mu} \, d\tau', \qquad \mu > 0 \tag{8-207a}$$

$$I_0^-(\tau, \mu, \phi) = F_2(\mu, \phi)e^{(\tau_0-\tau)/\mu} - \int_{\tau}^{\tau_0} g(\tau') \frac{e^{(\tau'-\tau)/\mu}}{\mu} \, d\tau', \quad \mu < 0 \tag{8-207b}$$

The quantity

$$\int_{\phi'=0}^{2\pi} \int_{\mu'=-1}^{1} I_0(\tau, \mu', \phi') \, d\mu' \, d\phi'$$

can be evaluated as (see Eq. 8-68)

$$\begin{aligned} G_0(\tau) &\equiv \int_{\phi'=0}^{2\pi} \int_{\mu'=-1}^{1} I_0(\tau, \mu', \phi') \, d\mu' \, d\phi' \\ &= \int_{\phi'=0}^{2\pi} \int_{\mu'=0}^{1} [I_0^+(\tau, \mu', \phi') + I_0^-(\tau, -\mu', \phi')] \, d\mu' \, d\phi' \\ &= \int_{\phi'=0}^{2\pi} \int_{\mu'=0}^{1} [F_1(\mu', \phi')e^{-\tau/\mu'} + F_2(-\mu', \phi')e^{-(\tau_0-\tau)/\mu'}] \, d\mu' \, d\phi' \\ &\quad + 2\pi\left[\int_{\tau'=0}^{\tau} g(\tau')E_1(\tau - \tau') \, d\tau' + \int_{\tau'=\tau}^{\tau_0} g(\tau')E_1(\tau' - \tau) \, d\tau'\right] \end{aligned} \tag{8-208}$$

We assume that the intensity $I(\tau, \mu, \phi)$ for the problem expressed by Eqs. 8-205 can be represented as a sum of the unscattered radiation intensity $I_0(\tau, \mu, \phi)$ defined by Eqs. 8-206 and a scattered intensity $I_1(\tau, \mu)$ in the form

$$I(\tau, \mu, \phi) = I_0(\tau, \mu, \phi) + I_1(\tau, \mu) \tag{8-209}$$

By substituting Eq. 8-209 into Eqs. 8-205 and utilizing Eqs. 8-206 and 8-208 it can be shown that the intensity $I_1(\tau, \mu)$ is the solution of the following azimuthally independent problem:

$$\mu \frac{\partial I_1(\tau, \mu)}{\partial \tau} + I_1(\tau, \mu) = \frac{\omega}{4\pi} G_0(\tau) + \frac{\omega}{2} \int_{\mu'=-1}^{1} I_1(\tau, \mu') \, d\mu'$$

$$\text{in } 0 \le \tau \le \tau_0, \ -1 \le \mu \le 1 \tag{8-210a}$$

with the boundary conditions

$$I_1(\tau, \mu)|_{\tau=0} = 0, \quad \mu > 0 \tag{8-210b}$$

$$I_1(\tau, \mu)|_{\tau=\tau_0} = 0, \quad \mu < 0 \tag{8-210c}$$

where $G_0(\tau)$ is a known function given by Eq. 8-208. Thus we have transformed the azimuthally dependent problem (Eqs. 8-205) into the solution of a problem with azimuthal symmetry, given by Eqs. 8-210.

Anisotropic Scattering

Consider an anisotropically scattering plane-parallel slab having transparent boundaries subjected to an obliquely incident external radiation at the boundary $\tau = 0$. For simplicity we assume no internal energy sources and no externally incident radiation at the boundary $\tau = \tau_0$. The equation of radiative transfer and the boundary conditions are given as

$$\mu \frac{\partial I(\tau, \mu, \phi)}{\partial \tau} + I(\tau, \mu, \phi) = \frac{\omega}{4\pi} \int_{\phi'=0}^{2\pi} \int_{\mu'=-1}^{1} p(\mu_0) I(\tau, \mu', \phi')\, d\mu'\, d\phi'$$

$$\text{in } 0 \le \tau \le \tau_0, \ -1 \le \mu \le 1 \tag{8-211a}$$

$$I(\tau, \mu, \phi)|_{\tau=0} = F_1(\mu, \phi), \quad \mu > 0 \tag{8-211b}$$

$$I(\tau, \mu, \phi)|_{\tau=\tau_0} = 0, \qquad \mu < 0 \tag{8-211c}$$

where $F_1(\mu, \phi)$ is a prescribed function, $p(\mu_0)$ is the phase function, and μ_0 is the cosine of the angle between the incident and scattered rays, given by

$$\mu_0 = \mu\mu' + \sqrt{1 - \mu^2}\sqrt{1 - \mu'^2} \cos(\phi - \phi') \tag{8-212}$$

We now consider the following auxiliary problem:

$$\mu \frac{\partial I_0(\tau, \mu, \phi)}{\partial \tau} + I_0(\tau, \mu, \phi) = 0 \quad \text{in } 0 \le \tau \le \tau_0, \ -1 \le \mu \le 1 \tag{8-213a}$$

with the boundary conditions

$$I_0(\tau, \mu, \phi)|_{\tau=0} = F_1(\mu, \phi), \quad \mu > 0 \tag{8-213b}$$

$$I_0(\tau, \mu, \phi)|_{\tau=\tau_0} = 0, \qquad \mu < 0 \tag{8-213c}$$

Here the function $I_0(\tau, \mu, \phi)$ characterizes the unscattered part of the solution $I(\tau, \mu, \phi)$ of the problem given by Eqs. 8-211. The solution of Eqs. 8-213 for the forward component $I_0^+(\tau, \mu, \phi)$ and the backward component $I_0^-(\tau, \mu, \phi)$ is immediately obtained as

$$I_0^+(\tau, \mu, \phi) = F_1(\mu, \phi) e^{-\tau/\mu}, \quad \mu > 0 \tag{8-214a}$$

$$I_0^-(\tau, \mu, \phi) = 0, \qquad \mu < 0 \tag{8-214b}$$

Then the quantity

$$\int_{\phi'=0}^{2\pi} \int_{\mu'=-1}^{1} p(\mu_0) I_0(\tau, \mu', \phi') \, d\mu' \, d\phi'$$

is given as

$$G(\tau) \equiv \int_{\phi'=0}^{2\pi} \int_{\mu'=-1}^{1} p(\mu_0) I_0(\tau, \mu', \phi') \, d\mu' \, d\phi'$$
$$= \int_{\phi'=0}^{2\pi} \int_{\mu'=0}^{1} p(\mu_0) F_1(\mu', \phi') e^{-\tau/\mu'} \, d\mu' \, d\phi' \tag{8-215}$$

We assume that the solution $I(\tau, \mu, \phi)$ of Eqs. 8-211 can be separated into an unscattered and a scattered component in the form

$$I(\tau, \mu, \phi) = I_0(\tau, \mu, \phi) + I_1(\tau, \mu, \phi) \tag{8-216}$$

where $I_0(\tau, \mu, \phi)$ is the solution of Eqs. 8-213. Substituting Eq. 8-216 into Eqs. 8-211 and utilizing Eqs. 8-213 and 8-215, we can show that the function $I_1(\tau, \mu, \phi)$ is the solution of the following problem:

$$\mu \frac{\partial I_1(\tau, \mu, \phi)}{\partial \tau} + I_1(\tau, \mu, \phi) = \frac{\omega}{4\pi} \int_{\phi'=0}^{2\pi} \int_{\mu'=0}^{1} p(\mu_0) F_1(\mu', \phi') e^{-\tau/\mu'} d\mu' \, d\phi'$$
$$+ \frac{\omega}{4\pi} \int_{\phi'=0}^{2\pi} \int_{\mu'=-1}^{1} p(\mu_0) I_1(\tau, \mu', \phi') \, d\mu' \, d\phi' \quad \text{in } 0 \le \tau \le \tau_0, \; -1 \le \mu \le 1 \tag{8-217a}$$

with the boundary conditions

$$I_1(\tau, \mu, \phi)|_{\tau=0} = 0, \quad \mu > 0 \tag{8-217b}$$
$$I_1(\tau, \mu, \phi)|_{\tau=\tau_0} = 0, \quad \mu < 0 \tag{8-217c}$$

In general, the azimuthally dependent problem given in Eq. 8-217 can be separated into a set of problems with azimuthal symmetry by assuming that the phase function $p(\mu_0)$ can be expanded by Legendre polynomials as given by Eq. 8-39 and the intensity can be expanded in a Fourier series in the form

$$I_1(\tau, \mu, \phi) = \sum_{k=1}^{\infty} [I_s^k(\tau, \mu) \sin k\phi + I_c^k(\tau, \mu) \cos k\phi] \tag{8-218}$$

By substituting these expansions into Eqs. 8-217 and collecting the coefficients of $\cos k\phi$ and $\sin k\phi$, we obtain a set of equations with homogeneous boundary conditions for the functions $I_s^k(\tau, \mu)$ and $I_c^k(\tau, \mu)$.

For simplicity in the analysis we consider below the separation of Eq. 8-217 into a set of problems with azimuthal symmetry for a special case of function $F_1(\mu, \phi)$ that is represented by delta functions.

Obliquely Incident Parallel Plane Rays

Consider the function $F_1(\mu, \phi)$, given in the form

$$F_1(\mu, \phi) = f_1\, \delta(\mu_1 - \mu)\, \delta(\phi_1 - \phi) \tag{8-219}$$

where $\delta(x)$ is the delta function and f_1 is a constant. The boundary condition given by Eq. 8-219 characterizes external parallel plane radiation of intensity f_1 incident on the boundary surface $\tau = 0$ at an angle $\theta_1 = \cos^{-1} \mu_1$ with the τ axis and at an azimuthal angle ϕ_1.

We assume that the phase function $p(\mu_0)$ can be expanded in terms of Legendre polynomials in the form (see Eq. 8-37)

$$p(\mu_0) = \sum_{n=0}^{N} a_n P_n(\mu_0), \quad a_0 = 1 \tag{8-220}$$

When μ_0 is given by Eq. 8-212, the phase function can be expressed in the form

$$p(\mu_0) = \sum_{m=0}^{N} \sum_{n=m}^{N} (2 - \delta_{0m}) a_n^{(m)} P_n^{\ m}(\mu) P_n^{\ m}(\mu') \cos m(\phi - \phi') \tag{8-221a}$$

where

$$a_n^{(m)} \equiv a_n \frac{(n - m)!}{(n + m)!}, \quad n = m, \ldots, N; 0 \le m \le N \tag{8-221b}$$

$$\delta_{0m} = \begin{cases} 1 & \text{for } m = 0 \\ 0 & \text{otherwise} \end{cases} \tag{8-221c}$$

Here Eq. 8-221 is obtained from Eq. 8-39 by inverting the order of summation on the right-hand side of that equation.

We now expand the intensity $I_1(\tau, \mu, \phi)$ in a Fourier series in the form [2, p. 23]

$$I_1(\tau, \mu, \phi) = \sum_{k=0}^{\infty} I^k(\tau, \mu) \cos k(\phi_1 - \phi) \tag{8-222}$$

Substitution of expansions 8-221 and 8-222 into Eq. 8-217a yields

$$\begin{aligned}
&\sum_{k=0}^{\infty} \cos k(\phi_1 - \phi)\left[\mu \frac{\partial I^k(\tau, \mu)}{\partial \tau} + I^k(\tau, \mu)\right] \\
&= \sum_{m=0}^{N} \cos m(\phi_1 - \phi)\left[\frac{\omega}{4\pi} f_1 e^{-\tau/\mu_1} \sum_{n=m}^{N} (2 - \delta_{0m}) a_n^{(m)} P_n^{\ m}(\mu) P_n^{\ m}(\mu_1)\right] \\
&+ \sum_{k=0}^{\infty} \sum_{m=0}^{N} \sum_{n=m}^{N} \frac{\omega}{4\pi} a_n^{(m)} P_n^{\ m}(\mu) \int_{\mu'=-1}^{1} P_n^{\ m}(\mu') I^k(\tau, \mu')\, d\mu' \\
&\times \left[(2 - \delta_{0m}) \int_{\phi'=0}^{2\pi} \cos k(\phi_1 - \phi') \cos m(\phi - \phi')\, d\phi'\right]
\end{aligned} \tag{8-223}$$

The integral over ϕ' on the right-hand side can be performed as

$$\int_{\phi'=0}^{2\pi} \cos k(\phi_1 - \phi') \cos m(\phi - \phi')\, d\phi' = \delta_{mk}\pi[\delta_{0m} + \cos m(\phi_1 - \phi)] \tag{8-224}$$

where

$$\delta_{mk} = \begin{cases} 1 & \text{for } m = k \\ 0 & \text{otherwise} \end{cases}$$

When Eq. 8-224 is substituted into Eq. 8-223, the summation over k on the right-hand side of Eq. 8-223 drops out. In the resulting expression we replace superscript m by k for convenience, split up the summation on the left-hand side into two parts, one from $k = 0$ to N and the other from $k = N + 1$ to ∞, and obtain

$$\begin{aligned} &\sum_{k=0}^{N} \cos k(\phi_1 - \phi)\left[\mu \frac{\partial I^k(\tau, \mu)}{\partial \tau} + I^k(\tau, \mu)\right] \\ &\quad + \sum_{k=N+1}^{\infty} \cos k(\phi_1 - \phi)\left[\mu \frac{\partial I^k(\tau, \mu)}{\partial \tau} + I^k(\tau, \mu)\right] \\ &= \sum_{k=0}^{N} \cos k(\phi_1 - \phi)\left[\frac{\omega}{4\pi} f_1 e^{-\tau/\mu_1} \sum_{n=k}^{N} (2 - \delta_{0k}) a_n^{(k)} P_n^{\,k}(\mu) P_n^{\,k}(\mu_1)\right. \\ &\quad + \sum_{k=0}^{N} \cos k(\phi_1 - \phi)\left[\frac{\omega}{2} \sum_{n=k}^{N} a_n^{(k)} P_n^{\,k}(\mu) \int_{\mu'=-1}^{1} P_n^{\,k}(\mu') I^k(\tau, \mu')\, d\mu'\right] \end{aligned} \tag{8-225}$$

In writing Eq. 8-225 we utilized the following relation:

$$(2 - \delta_{0k})[\delta_{0k} + \cos k(\phi_1 - \phi)] = 2 \cos k(\phi_1 - \phi), \quad k = 0, 1, 2, \ldots \tag{8-226}$$

Equation 8-225 can be separated into a set of simpler equations for the functions $I^k(\tau, \mu)$ because of the linear independence of $\cos k(\phi_1 - \phi)$. By collecting the coefficients of $\cos k(\phi_1 - \phi)$ we obtain

$$\mu \frac{\partial I^k(\tau, \mu)}{\partial \tau} + I^k(\tau, \mu) = 0, \quad k > N \tag{8-227}$$

$$\begin{aligned} \mu \frac{\partial I^k(\tau, \mu)}{\partial \tau} + I^k(\tau, \mu) &= \frac{\omega}{4\pi} f_1 e^{-\tau/\mu_1} \sum_{n=k}^{N} (2 - \delta_{0k}) a_n^{(k)} P_n^{\,k}(\mu) P_n^{\,k}(\mu_1) \\ &\quad + \frac{\omega}{2} \sum_{n=k}^{N} a_n^{(k)} P_n^{\,k}(\mu) \int_{\mu'=-1}^{1} P_n^{\,k}(\mu') I^k(\tau, \mu')\, d\mu', \quad k \le N \end{aligned} \tag{8-228}$$

where $k = 0, 1, \ldots, N$ and $n = k, k + 1, \ldots, N$.

The boundary conditions for these equations are given as

$$I^k(\tau, \mu)|_{\tau=0} = 0, \quad \mu > 0 \tag{8-229a}$$

$$I^k(\tau, \mu)|_{\tau=\tau_0} = 0, \quad \mu < 0 \tag{8-229b}$$

Equations 8-227 have trivial solutions since both the equations and the boundary conditions are homogeneous; hence they are not needed. The solution of Eqs. 8-228 subject to the boundary conditions 8-229 yields the intensity functions $I^k(\tau, \mu)$, $k = 0, 1, 2, \ldots, N$. Thus we have transformed the azimuthally dependent equation 8-217 to a system of $N + 1$ simultaneous integrodifferential equations 8-228 in azimuthal symmetry.

The case $N = 0$ characterizes isotropic scattering, and Eqs. 8-228 reduce to

$$\mu \frac{\partial I(\tau, \mu)}{\partial \tau} + I(\tau, \mu) = \frac{\omega}{4\pi} f_1 e^{-\tau/\mu_1} + \frac{\omega}{2} \int_{\mu'=-1}^{1} I(\tau, \mu')\, d\mu' \tag{8-230}$$

where we have omitted the superscript 0 for simplicity. This result is obtainable from Eq. 8-210a by setting in that equation $g(\tau) = 0$, $F_2(-\mu, \phi) = 0$, and $F_1(\mu, \phi) = f_1\, \delta(\mu_1 - \mu)$.

REFERENCES

1. D. H. Sampson, *Radiative Contributions to Energy and Momentum Transport in a Gas*, Interscience Publishers, New York, 1965.
2. S. Chandrasekhar, *Radiative Transfer*, Oxford University Press, London, 1950; also Dover Publications, New York, 1960.
3. V. Kourganoff, *Basic Methods in Transfer Problems*, Dover Publications, New York, 1963.
4. V. V. Sobolev, *A Treatise on Radiative Transfer*, D. Van Nostrand Company, Princeton, N.J., 1963.
5. R. Viskanta, *Heat Transfer in Thermal Radiation Absorbing and Scattering Media*, ANL-6170, Argonne National Laboratory, Argonne, Ill., 1960.
6. R. Viskanta, "Radiation Transfer and Interaction of Convection with Radiation Heat Transfer," in *Advances in Heat Transfer*, edited by T. F. Irvine and J. P. Hartnett, Academic Press, New York, 1966.
7. A. M. Weinberg and E. P. Wigner, *The Physical Theory of Neutron Chain Reactions*, The University of Chicago Press, Chicago, Ill., 1958.
8. R. L. Murray, *Nuclear Reactor Physics*, Prentice-Hall, Englewood Cliffs, N.J., 1957.
9. V. A. Ambartsumyan, *Theoretical Astrophysics*, Pergamon Press, New York, 1958.
10. E. T. Whittaker and G. N. Watson, *A Course in Modern Analysis*, Cambridge University Press, London, 1931.
11. J. R. Howell and M. Perlmutter, "Monte Carlo Solution of Radiant Heat Transfer in a Nongrey Nonisothermal Gas with Temperature Dependent Properties," *A.I.Ch.E. J.*, **10**, 562–567, 1964.
12. J. T. Bevans and R. V. Dunkle, "Radiant Interchange Within an Enclosure, Parts I and II," *J. Heat Transfer*, **82C**, 1–19, 1960.

13. D. K. Edwards, "Radiation Interchange in a Nongray Enclosure Containing an Isothermal Carbon Dioxide-Nitrogen Gas Mixture," *J. Heat Transfer*, **84C**, 1–11, 1962.

14. E. M. Sparrow and R. D. Cess, *Radiation Heat Transfer*, Brooks/Cole Publishing Co., Belmont, Calif., 1966.

15. A. L. Crosbie and R. Viskanta, "The Exact Solution to a Simple Nongray Radiative Transfer Problem," *J. Quant. Spectry. Radiative Transfer*, **9**, 553–568, 1969.

16. A. L. Crosbie and R. Viskanta, "Nongray Radiative Transfer in a Planar Medium Exposed to a Collimated Flux," *J. Quant. Spectry. Radiative Transfer*, **10**, 465–485, 1970.

17. A. L. Crosbie and R. Viskanta, "Effect of Band or Line Shape on the Radiative Transfer in a Nongray Planar Medium," *J. Quant. Spectry. Radiative Transfer*, **10**, 487–509, 1970.

18. S. Chandrasekhar, "The Radiative Equilibrium of the Outer Layers of a Star with a Special Reference to the Blanketing Effects of the Reversing Layer," *Monthly Notices Roy Astron. Soc.*, **96**, 21–42, 1935.

19. C. E. Siewert and P. F. Zweifel, "An Exact Solution of Equations of Radiative Transfer for Local Thermodynamic Equilibrium in the Non-Gray Case: Picket-Fence Approximation," *Ann. Phys. (N.Y.)*, **36**, 61–85, 1966.

20. C. E. Siewert and P. F. Zweifel, "Radiative Transfer, II," *J. Math. Phys.*, **7**, 2092–2102, 1966.

21. H. C. Kung and M. Sibulkin, "Radiative Transfer in a Nongray Gas Between Parallel Walls," *J. Quant. Spectry. Radiative Transfer*, **9**, 1447–1461, 1969.

22. R. J. Reith, C. E. Siewert, and M. N. Özişik, "Non-grey Radiative Heat Transfer in Conservative Plane-Parallel Media with Reflecting Boundaries," *J. Quant. Spectry. Radiative Transfer*, **11**, 1441–1462, 1971.

23. Ralph Greif, "Energy Transfer by Radiation and Conduction with Variable Gas Properties," *Intern. J. Heat Mass Transfer*, **7**, 891–900, 1964.

24. M. A. Heaslet and R. F. Warming, "Radiative Transport and Wall Temperature Slip in an Absorbing Planar Medium," *Intern. J. Heat Mass Transfer*, **8**, 979–994, 1965.

25. M. A. Heaslet and R. F. Warming, "Radiative Transport in an Absorbing Planar Medium, II: Prediction of Radiative Source Functions," *Intern. J. Heat Mass Transfer*, **10**, 1413–1427, 1967.

26. C. M. Chu, J. A. Leacock, J. C. Chen, and S. W. Churchill, "Numerical Solutions for Multiple Anisotropic Scattering," in *Electromagnetic Scattering*, edited by M. Kerker, pp. 567–582, Macmillan Co., New York, 1963.

27. L. B. Evans, C. M. Chu, and S. W. Churchill, "The Effect of Anisotropic Scattering on Radiant Transport," *J. Heat Transfer*, **87C**, 381–387, 1965.

NOTES

[1] Equation 8-4c can also be derived directly from the definition of the substantial derivative, that is,

$$\frac{DI_\nu}{Dt} = \frac{\partial I_\nu}{\partial t}\frac{dt}{dt} + \frac{\partial I_\nu}{\partial s}\frac{ds}{dt} = \frac{\partial I_\nu}{\partial t} + c\frac{\partial I_\nu}{\partial s}$$

since $c = ds/dt$.

[2] For noncoherent scattering there is redistribution of the frequency of the scattered radiation, and the in-scattering term is given as

$$\frac{1}{4\pi}\int_{\Omega'=4\pi}\int_{\nu'=0}^{\infty} \sigma_{\nu'}(s)p(\nu' \to \nu; \hat{\mathbf{\Omega}}' \cdot \hat{\mathbf{\Omega}})I_{\nu'}(s, \hat{\mathbf{\Omega}}', t)\, d\nu'\, d\Omega' \tag{1}$$

where the phase function is normalized so that

$$\frac{1}{4\pi}\int_{\Omega'=4\pi}\int_{\nu=0}^{\infty} p(\nu' \to \nu; \hat{\mathbf{\Omega}}' \cdot \hat{\mathbf{\Omega}})\, d\nu\, d\Omega' = 1 \tag{2}$$

Equations 1 and 2 reduce to those for purely coherent scattering if we let

$$p(\nu' \to \nu; \hat{\mathbf{\Omega}}' \cdot \hat{\mathbf{\Omega}}) = p(\hat{\mathbf{\Omega}}' \cdot \hat{\mathbf{\Omega}})\delta(\nu' - \nu) \tag{3}$$

where δ is the Dirac delta function. Substitution of Eq. 3 into Eqs. 1 and 2, respectively, yields

$$\frac{1}{4\pi}\sigma_\nu(s)\int_{\Omega'=4\pi} p(\hat{\mathbf{\Omega}}' \cdot \hat{\mathbf{\Omega}})I_\nu(s, \hat{\mathbf{\Omega}}', t)\, d\Omega' \tag{4}$$

$$\frac{1}{4\pi}\int_{\Omega=4\pi} p(\hat{\mathbf{\Omega}}' \cdot \hat{\mathbf{\Omega}})\, d\Omega' = 1 \tag{5}$$

For further discussion of noncoherent scattering the reader should refer to the text by Ambartsumyan [9].

[3] Equation 8-18 is analogous to the steady-state heat conduction equation with internal energy sources, that is,

$$\nabla \cdot \mathbf{q}^c = g$$

where $\mathbf{q}^c$ is the conductive heat flux vector.

[4] Equation 8-20 is analogous to the steady-state heat conduction equation with no energy generation term, that is,

$$\nabla \cdot \mathbf{q}^c = 0$$

where $\mathbf{q}^c$ is the conductive heat flux vector.

[5] In the rectangular coordinate system, for example, Eq. 8-25 can be written as

$$\frac{1}{\beta_\nu}\left(l\frac{\partial}{\partial x} + m\frac{\partial}{\partial y} + n\frac{\partial}{\partial z}\right)I_\nu(x, y, z; l, m, n) + I_\nu = S_\nu(x, y, z; l, m, n)$$

where l, m, and n are the direction cosines of the vector $\hat{\mathbf{\Omega}}$, that is, the cosines of the angles between the direction $\hat{\mathbf{\Omega}}$ and the ox, oy, or oz axis, respectively.

[6] In the frequency-dependent problems the optical variable is a function of frequency; hence it should be written as τ_ν. However, for simplicity we shall omit the subscript ν.

[7] Legendre polynomials are given as

$$P_0(x) = 1 \qquad P_2(x) = \tfrac{1}{2}(3x^2 - 1) \qquad P_4(x) = \tfrac{1}{8}(35x^4 - 30x^2 + 3)$$

$$P_1(x) = x \qquad P_3(x) = \tfrac{1}{2}(5x^3 - 3x) \qquad \text{etc.}$$

8 The equation for the spectral source function (Eq. 8-122) is given as

$$S_\nu(\tau) = (1 - \omega_\nu)I_{\nu b}(T) + \frac{\omega_\nu}{2}\left[\int_0^1 I_\nu^+(0, \mu)e^{-\tau/\mu}\, d\mu + \int_0^1 I_\nu^-(\tau_0, -\mu)e^{-(\tau_0-\tau)/\mu}\, d\mu\right.$$

$$\left. + \int_0^{\tau_0} S(\tau')E_1(|\tau - \tau'|)\, d\tau'\right]$$

9 Equation 8-139a implies that for radiative equilibrium the net radiative heat flux is constant everywhere in the medium, that is,

$$q^r = \int_{\nu=0}^{\infty} q_\nu^{\,r}(y)\, d\nu = \text{constant}$$

but the spectral radiative heat flux $q_\nu^{\,r}(y)$ is not necessarily constant.

CHAPTER 9

Approximate Methods in the Solution of the Equation of Radiative Transfer

The mathematical difficulties of working with the integrodifferential equations have resulted in a number of approximate methods of solving the equation of radiative transfer. The *opticially thin limit* and *optically thick limit* (diffusion or Rosseland) approximations involve simplifications based on the optical thickness of the medium. In the *Eddington* and *Schuster-Schwarzchild* approximations a simplifying assumption is introduced in regard to the angular distribution of radiation intensity. In the *exponential kernel approximation* the exponential integral functions in the formal solutions are represented by exponentials. The *spherical harmonics method*, the *moment method*, and the *discrete ordinate method* are more elaborate schemes that may provide higher-order approximation.

In this chapter we describe briefly various approximate methods for solving the equation of radiative transfer. The reader should refer also to the publications by Viskanta [1], Chandrasekhar [2], Kourganoff [3], Woolley and Stibbs [4], and Sparrow and Cess [5] on this subject. The approximate methods are useful in that they provide various simple means of solving the complicated problem of radiative heat transfer, but their utility is restricted in that the accuracy of an approximate solution cannot be assessed until it is compared with the exact solution. Therefore we shall consider in Chapters 11, 12, and 13 the application and the accuracy of some of the approximate methods described here for the solution of problems of pure radiative heat transfer and the interaction of radiation with conduction and convection.

9-1 OPTICALLY THIN LIMIT APPROXIMATION

The optically thin limit approximation is based on the consideration that the optical thickness τ_0 of the medium is very small (i.e., $\tau_0 \ll 1$). Then the following simplifications can be made in the exponential integral functions and the exponentials:

$$E_2(\tau) = 1 - 0(\tau) \tag{9-1a}$$

$$E_3(\tau) = \tfrac{1}{2} - \tau + 0(\tau^2) \tag{9-1b}$$

$$e^{-\tau} = 1 - \tau + 0(\tau^2) \tag{9-1c}$$

for $\tau \leq \tau_0$. When these approximations are introduced into the formal solutions given in Chapter 8, relatively simple expressions are obtained for the source function, the boundary surface intensities, the net radiative heat flux, and other quantities. The application of the optically thin limit approximation is illustrated below with specific examples.

Equation for the Source Function

The integral equation for the spectral source function in isotropic scattering and azimuthal symmetry is given by Eq. 8-123. For an optically thin medium (i.e., $\tau_0 \ll 1$), by introducing the approximations given by Eqs. 9-1 and neglecting the terms of the order of τ_0, Eq. 8-123 simplifies to

$$S_\nu(\tau) = (1 - \omega_\nu)I_{\nu b}[T(\tau)] + \tfrac{1}{2}\omega_\nu[I_\nu^+(0) + I_\nu^-(\tau_0)] \tag{9-2}$$

The absence of the integral term in this equation implies that there is no attenuation of radiation originating from the medium itself. This is physically meaningful because the medium is optically so thin that it does not have sufficient thickness for self-attenuation of radiation.

Equations for Boundary Surface Intensities

Consider Eqs. 8-103a and 8-103b for the boundary surface intensities for an isotropically scattering, plane-parallel slab with diffusely reflecting boundaries. For an optically thin medium, by introducing the approximation given by Eq. 9-1b and neglecting the terms of the order of τ_0, Eqs. 8-103a and 8-103b simplify, respectively, to

$$I_\nu^+(0) = \varepsilon_{1\nu} I_{\nu b}(T_1) + \rho_{1\nu}^d I_\nu^-(\tau_0), \quad \mu > 0 \tag{9-3a}$$

$$I_\nu^-(\tau_0) = \varepsilon_{2\nu} I_{\nu b}(T_2) + \rho_{2\nu}^d I_\nu^+(0), \quad \mu < 0 \tag{9-3b}$$

When these equations are solved simultaneously for the boundary surface intensities $I_\nu^+(0)$ and $I_\nu^-(\tau_0)$, we obtain

$$I_\nu^+(0) = \frac{\varepsilon_{1\nu} I_{\nu b}(T_1) + \rho_{1\nu}\varepsilon_{2\nu} I_{\nu b}(T_2)}{1 - \rho_{1\nu}\rho_{2\nu}} \tag{9-4a}$$

$$I_\nu^-(\tau_0) = \frac{\varepsilon_{2\nu} I_{\nu b}(T_2) + \rho_{2\nu}\varepsilon_{1\nu} I_{\nu b}(T_1)}{1 - \rho_{1\nu}\rho_{2\nu}} \tag{9-4b}$$

where we have omitted the superscript d from ρ^d for simplicity.

Equation for the Spectral Net Radiative Heat Flux

The spectral net radiative heat flux (Eq. 8-83) for an isotropically scattering, plane-parallel slab with boundary surface intensities independent of direction simplifies to

$$q_\nu^r(\tau) = 2\pi\left\{I_\nu^+(0)(\tfrac{1}{2} - \tau) + \int_0^\tau S_\nu(\tau')\,d\tau' - I_\nu^-(\tau_0)[\tfrac{1}{2} - (\tau_0 - \tau)] - \int_\tau^{\tau_0} S_\nu(\tau')\,d\tau'\right\} \tag{9-5}$$

Here we have retained the terms of the order of τ_0; hence this equation is correct to the order of τ_0. If the terms of the order of τ_0 are neglected, Eq. 9-5 simplifies to

$$q_\nu^r = \pi[I_\nu^+(0) - I_\nu^-(\tau_0)] \tag{9-6a}$$

If the boundary surfaces are opaque and $\rho_{1\nu} = 1 - \varepsilon_{1\nu}$, $\rho_{2\nu} = 1 - \varepsilon_{2\nu}$, the substitution of Eqs. 9-4 into Eq. 9-6a yields

$$q_\nu^r = \frac{\pi[I_{\nu b}(T_1) - I_{\nu b}(T_2)]}{[(1/\varepsilon_{1\nu}) + (1/\varepsilon_{2\nu}) - 1]} \tag{9-6b}$$

which is the standard expression for the spectral net radiative heat flux between two opaque parallel plates separated by a transparent medium.

Equation for $dq_\nu^r/d\tau$

In Eqs. 8-93 and 8-94 we gave two alternative expressions for $dq_\nu^r(\tau)/d\tau$ for an isotropically scattering, plane-parallel slab with azimuthally symmetric radiation. By introducing the optically thin limit approximation and neglecting the terms of the order of τ_0 these equations simplify respectively to

$$\frac{dq_\nu^r(\tau)}{d\tau} = (1 - \omega_\nu)[4\pi I_{\nu b}(T) - 2\pi I_\nu^+(0) - 2\pi I_\nu^-(\tau_0)] \tag{9-7}$$

and

$$\frac{dq_\nu^r(\tau)}{d\tau} = 4\pi S_\nu(\tau) - 2\pi I_\nu^+(0) - 2\pi I_\nu^-(\tau_0) \tag{9-8}$$

Here we note that Eq. 9-8 is obtainable from Eq. 9-5 by differentiating the latter equation with respect to τ; if the source term given by Eq. 9-2 is substituted into Eq. 9-8, we obtain Eq. 9-7.

By substituting the boundary surface intensities given by Eqs. 9-4a and 9-4b into Eq. 9-7 we obtain

$$\frac{dq_\nu^r(\tau)}{d\tau} = (1 - \omega_\nu)\Bigg\{4\pi I_{\nu b}[T(\tau)] - 2\pi \times \frac{\varepsilon_{1\nu} I_{\nu b}(T_1) + \varepsilon_{2\nu} I_{\nu b}(T_2) + \varepsilon_{1\nu}\rho_{2\nu} I_{\nu b}(T_1) + \varepsilon_{2\nu}\rho_{1\nu} I_{\nu b}(T_2)}{1 - \rho_{1\nu}\rho_{2\nu}}\Bigg\} \tag{9-9}$$

In the foregoing relations, $dq_\nu^r/d\tau$ for the optically thin limit approximation is independent of the scattering coefficient because $d\tau = \beta_\nu\, dy$ and $1 - \omega_\nu = \kappa_\nu/\beta_\nu$; hence β_ν cancels out.

9-2 OPTICALLY THICK LIMIT (ROSSELAND OR DIFFUSION) APPROXIMATION

A medium is said to be optically thick if the radiation photon mean free path (i.e., the reciprocal of the extinction coefficient) is very small compared with the characteristic dimension of the medium. This approximation, which is also known as the Rosseland or the diffusion approximation, was derived originally by Rosseland [6]. Its derivation can be found, in addition to the original work, in several other references, for example, the article by Viskanta [1, pp. 208–209]. The principal advantage of this approximation is that it provides a very simple relation for the net radiative heat flux. We present now the simplification of the equation for the spectral radiative heat flux under the optically thick limit approximation.

Consider the relations for the spectral radiative heat flux $q_\nu^r(\tau)$ and the spectral source function $S_\nu(\tau)$, given by Eqs. 8-82 and 8-122, respectively, in the forms

$$q_\nu^r(\tau) = 2\pi\left[\int_0^1 I_\nu^+(0,\mu)e^{-\tau/\mu}\mu\, d\mu + \int_0^\tau S_\nu(\tau')E_2(\tau - \tau')\, d\tau'\right] - 2\pi\left[\int_0^1 I_\nu^-(\tau_0, -\mu)e^{-(\tau_0-\tau)/\mu}\mu\, d\mu + \int_\tau^{\tau_0} S_\nu(\tau')E_2(\tau' - \tau)\, d\tau'\right] \tag{9-10}$$

and

$$S_\nu(\tau) = (1 - \omega_\nu) I_{\nu b}[T(\tau)] + \tfrac{1}{2}\omega_\nu \left[\int_0^1 I_\nu^+(0, \mu) e^{-\tau/\mu} \, d\mu \right.$$

$$\left. + \int_0^1 I_\nu^-(\tau_0, -\mu) e^{-(\tau_0 - \tau)/\mu} \, d\mu + \int_{\tau'=0}^{\tau_0} S_\nu(\tau') E_1(|\tau - \tau'|) \, d\tau' \right] \quad (9\text{-}11)$$

We start the analysis by expanding the source function $S_\nu(\tau')$ in a Taylor series about τ:

$$S_\nu(\tau') = S_\nu(\tau) + (\tau' - \tau) \left. \frac{dS_\nu(\tau')}{d\tau'} \right|_\tau + \frac{1}{2!} (\tau' - \tau)^2 \left. \frac{d^2 S_\nu(\tau')}{d\tau'^2} \right|_\tau + \cdots \quad (9\text{-}12)$$

For an optically thick medium τ, τ_0, and $\tau_0 - \tau$ are very large except in the regions close to the boundaries. Therefore we consider regions away from the boundaries and assume that

$$\tau, \tau_0, \text{ and } \tau_0 - \tau \gg 1 \quad (9\text{-}13)$$

For large values of τ, the exponential integral function and the exponential function can be simplified as

$$e^{-\tau} \to 0, \qquad E_n(\tau) \to 0, \qquad \tau^n E_n(\tau) \to 0 \quad \text{for } \tau \to \infty, n = 1, 2, 3, \cdots \quad (9\text{-}14)$$

The Taylor-series expansion (Eq. 9-12) is substituted into Eqs. 9-10 and 9-11, the integrals with respect to τ' are evaluated by parts, and the resulting expressions are simplified by applying the approximations given in Eq. 9-14. We obtain

$$S_\nu(\tau) = I_{\nu b}[T(\tau)] \quad (9\text{-}15)$$

$$q_\nu^r(\tau) = -\frac{4\pi}{3} \frac{dS_\nu(\tau)}{d\tau} = -\frac{4\pi}{3} \frac{dI_{\nu b}[T(\tau)]}{d\tau} \quad (9\text{-}16)$$

which are valid for an optically thick medium in regions away (optically) from the boundaries.

The spectral net radiative heat flux $q_\nu^r(\tau)$, given by Eq. 9-16, is called the optically thick limit (Rosseland or diffusion) approximation for the spectral net radiative heat flux.

The total net radiative heat flux under the optically thick limit approximation is given by

$$q^r(\tau) = -\frac{4\pi}{3} \int_{\nu=0}^{\infty} \frac{dI_{\nu b}[T(\tau)]}{d\tau} \, d\nu \quad (9\text{-}17)$$

or

$$q^r(y) = -\frac{4\pi}{3} \int_{\nu=0}^{\infty} \frac{1}{\beta_\nu} \frac{dI_{\nu b}[T(y)]}{dy} \, d\nu \quad (9\text{-}18)$$

The derivative of $I_{\nu b}[T(y)]$ with respect to y can be written as

$$\frac{dI_{\nu b}(T)}{dy} = \frac{dI_{\nu b}(T)}{dI_b(T)} \frac{dI_b(T)}{dy} \tag{9-19}$$

Substituting Eq. 9-19 into Eq. 9-18, we obtain

$$q^r(y) = -\frac{4\pi}{3} \frac{dI_b(T)}{dy} \int_{\nu=0}^{\infty} \frac{1}{\beta_\nu} \frac{dI_{\nu b}(T)}{dI_b(T)} \, d\nu \tag{9-20}$$

We now define the *Rosseland mean extinction coefficient* β_R as (see Eq. 1-56a)

$$\frac{1}{\beta_R} \equiv \int_{\nu=0}^{\infty} \frac{1}{\beta_\nu} \frac{dI_{\nu b}(T)}{dI_b(T)} \, d\nu \tag{9-21}$$

Then Eq. 9-20 is written as

$$q^r(y) = -\frac{4\pi}{3\beta_R} \frac{dI_b(T)}{dy} \tag{9-22}$$

or in the form

$$q^r(y) = -\frac{4\bar{\sigma}}{3\beta_R} \frac{d(n^2T^4)}{dy} \tag{9-23}$$

since

$$I_b(T) = \frac{n^2\bar{\sigma}T^4}{\pi} \tag{9-24}$$

where n is the refractive index of the medium.

For constant n, Eq. 9-23 can be written as

$$q^r(y) = -k_r \frac{dT}{dy} \tag{9-25a}$$

where

$$k_r \equiv \frac{16n^2\bar{\sigma}T^3}{3\beta_R} \tag{9-25b}$$

The coefficient k_r is called the *radiative conductivity*, a terminology analogous to *thermal conductivity* in heat conduction. Equation 9-25a has exactly the same form as the expression for the conductive heat flux; therefore the optically thick limit approximation characterizes the transport of radiation as a diffusion process.

Equation 9-23 (or 9-25) is called the *Rosseland approximation* or the *diffusion approximation* for the radiative heat flux. The Rosseland mean extinction coefficient β_R, defined by Eq. 9-21, can be evaluated by utilizing the fractional function of the second kind as described in Chapter 1 (see Eq. 1-56b).

For the case of a three-dimensional temperature field the derivation of the diffusion approximation by Taylor-series expansion is given by Deissler [7]. The three-dimensional equivalent of Eq. 9-16 for the spectral net radiative heat flux vector $\mathbf{q}_\nu{}^r$ is given as

$$\mathbf{q}_\nu{}^r = -\frac{4\pi}{3\beta_\nu} \nabla I_{\nu b}(T) \tag{9-26}$$

and the three-dimensional equivalent of Eq. 9-22 for the net radiative heat flux vector $\mathbf{q}^r$ as

$$\mathbf{q}^r = -\frac{4\pi}{3\beta_{\mathrm{R}}} \nabla I_b(T) \tag{9-27a}$$

or

$$\mathbf{q}^r = -\frac{4\bar{\sigma}}{3\beta_{\mathrm{R}}} \nabla(n^2T^4) \tag{9-27b}$$

The limitation to the use of diffusion approximation should be recognized. It is valid in the interior of a medium but should not be employed near the boundaries, where assumption 9-13 does not hold. It can not provide a complete description of the physical situation near the boundaries since it does not include any terms for radiation from the boundary surfaces. However, the boundary surface effects are negligible in the interior of an optically thick region since radiation from the boundaries is attenuated before reaching the interior.

For the optically thick limit approximation, the effect of the absorption and scattering characteristics of a medium on radiative heat flux is expressed only through the extinction coefficient β_{R}. For a purely absorbing and emitting medium (i.e., $\sigma = 0$), β_{R} is replaced by the *Rosseland mean absorption coefficient* κ_{R}. For a purely scattering medium, however, the temperature loses its physical significance with respect to radiative heat transfer.

Care must be exercised in applying the diffusion approximation. Significant error can be introduced into the prediction of net radiative heat flux unless the medium is optically thick or the thickness of the slab is more than several radiation photon mean free paths, that is, $\tau_0 \equiv \int_0^L \beta_\nu \, dy \gg 1$.

9-3 A MODIFIED DIFFUSION APPROXIMATION TO DETERMINE NET RADIATIVE HEAT FLUX IN A MEDIUM IN RADIATIVE EQUILIBRIUM

The determination of the net radiative heat flux in a participating medium in radiative equilibrium is of considerable interest in engineering applications. The diffusion approximation provides a simple relation for the net radiative heat flux, but its appicability is restricted to a medium with a thickness more than several radiation photon mean free paths. Shorin [8] introduced the temperature jump condition at the boundaries in order to derive a simple expression for the net radiative heat flux that would be sufficiently accurate for both small and large values of the optical thickness. However, his results for large optical thickness differed from the correct limit because the Schuster-Schwarzchild approximation was used in the analysis instead of the Rosseland approximation. Deissler [7] utilized the temperature jump boundary condition at the walls and applied a Taylor-series expansion for the distribution of the Planck function in the medium in his analysis to extend the applicability of diffusion approximation to the prediction of the net radiative heat flux in a medium in radiative equilibrium at lower values of the optical thickness. Before presenting Deissler's analysis we give a brief description of the temperature jump condition at the walls.

Consider a plane-parallel, gray slab of finite optical thickness τ_0 in radiative equilibrium between two parallel black boundaries at $\tau = 0$ and $\tau = \tau_0$ that are kept at uniform temperatures T_1 and T_2 $(T_2 > T_1)$, respectively. Let $\theta(\tau)$ be the dimensionless temperature distribution in the medium, defined as $\theta(\tau) = [\bar{\sigma}T^4(\tau) - \bar{\sigma}T_1^{\,4}]/(\bar{\sigma}T_2^{\,4} - \bar{\sigma}T_1^{\,4})$. Usiskin and Sparrow [9] and Heaslet and Warming [10] determined this temperature distribution in the slab by solving the problem exactly, and Fig. 9-1 shows the results of the calculations, plotted as a function of τ/τ_0 for several different values of

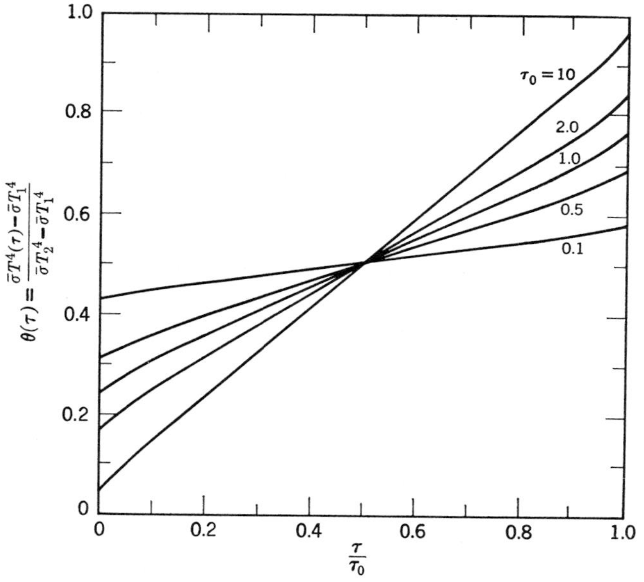

Fig. 9-1. Dimensionless temperature distribution for different optical thicknesses. (From M. A. Heaslet and R. F. Warming [10].)

the optical thickness τ_0. An examination of this figure reveals that there exists a temperature discontinuity (i.e., a step or jump) at the boundaries for all finite values of the optical thickness. However, as τ_0 approaches infinity, the temperature of the medium adjacent to the boundary and the boundary surface temperature become continuous.[1]

The physical situation giving rise to the temperature jump at the boundary can be envisioned with the following consideration. The net radiative energy flux passing through a plane in the immediate vicinity of the boundary surface is composed of two half-fluxes, one coming directly from the boundary surface and the other from the interior region of the medium, which on the average corresponds to the radiation mean free path away from the boundary surface. Therefore the average temperature of the net radiative energy flux passing through a plane in the immediate vicinity of the boundary surface will lie between the wall temperature and the temperature a mean free path away from the surface, giving rise to a temperature jump at the boundaries for finite values of the radiation photon mean free path (i.e., $1/\beta$). When the radiation photon mean free path approaches zero, corresponding to optical thickness $\tau_0 = \beta L$ approaching infinity, the temperature becomes continuous at the wall. This is analogous to the temperature jump that occurs at the wall in heat transfer in rarefied gases.

We now present a brief account of Deissler's analysis to predict the net radiative heat flux in a plane-parallel slab and two concentric cylinders in radiative equilibrium.

Plane-Parallel Slab

Consider a plane-parallel, nongray slab of optical thickness τ_0 between two diffusely reflecting, diffusely emitting, opaque parallel boundaries, as illustrated in Fig. 9-2. The boundary surfaces at $\tau = 0$ and $\tau = \tau_0$ are kept at uniform temperatures T_1 and T_2 and have spectral emissivities $\varepsilon_{1\nu}$ and $\varepsilon_{2\nu}$, respectively. The energy transfer is by pure radiation (i.e., conduction and convection are negligible), there are no energy sources or sinks in the medium, and the steady-state conditions are established. We shall derive equations

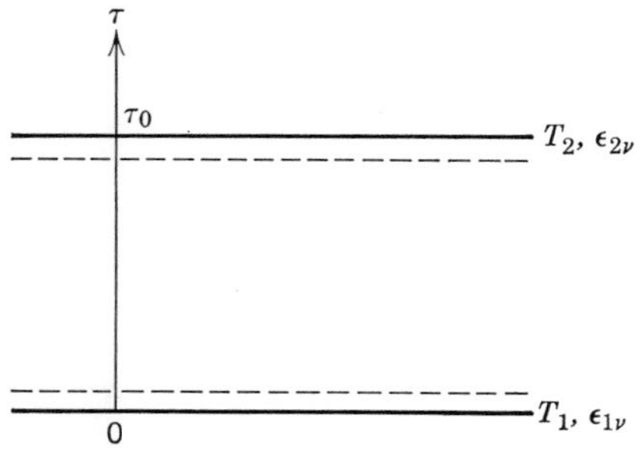

Fig. 9-2. Coordinates for extension of diffusion approximation.

for the temperature jump at the boundaries and the net radiative heat flux in the medium.

The spectral radiative heat flux $q_\nu{}^r(\tau)$ in a plane-parallel slab with boundary surface intensities independent of direction is given by Eq. 8-83. By setting $\tau = 0$ in this equation we obtain

$$q_\nu{}^r(0) = \pi I_\nu^+(0) - 2\pi\left[I_\nu^-(\tau_0)E_3(\tau_0) + \int_0^{\tau_0} S_\nu(\tau')E_2(\tau')\,d\tau'\right] \tag{9-28}$$

The boundary surface intensity function $I_\nu^+(0)$ is given by Eq. 8-107a as

$$I_\nu^+(0) = \varepsilon_{1\nu}I_{\nu b}(T_1) + 2(1-\varepsilon_{1\nu})\left[I_\nu^-(\tau_0)E_3(\tau_0) + \int_0^{\tau_0} S_\nu(\tau')E_2(\tau')\,d\tau'\right] \tag{9-29}$$

where we have replaced $\rho_{1\nu}^d$ by $1 - \varepsilon_{1\nu}$. Eliminating $I_\nu^+(0)$ between Eqs. 9-28 and 9-29 we obtain

$$q_\nu{}^r(0) = \pi\varepsilon_{1\nu}\left[I_{\nu b}(T_1) - 2I_\nu^-(\tau)E_3(\tau_0) - 2\int_0^{\tau_0} S_\nu(\tau')E_2(\tau')\,d\tau'\right] \tag{9-30}$$

For an optically thick medium (i.e., τ_0 is large) we have

$$E_3(\tau_0) \to 0 \tag{9-31}$$

and by Eqs. 9-15 and 9-16 we write

$$S_\nu(\tau) = I_{\nu b}[T(\tau)] \equiv I_{\nu b}(\tau) \tag{9-32}$$

$$q_\nu{}^r(0) = -\frac{4\pi}{3}\frac{dI_{\nu b}(\tau)}{d\tau}\bigg|_{\tau=0} \tag{9-33}$$

Substitution of Eqs. 9-31 through 9-33 into Eq. 9-30 yields

$$-\frac{4\pi}{3}\frac{dI_{\nu b}(\tau)}{d\tau}\bigg|_{\tau=0} = \pi\varepsilon_{1\nu}\left[I_{\nu b}(T_1) - 2\int_0^{\tau_0} I_{\nu b}(\tau')E_2(\tau')\,d\tau'\right] \tag{9-34}$$

The Planck function $I_{\nu b}(\tau')$ can be expanded in a Taylor series about $\tau' = 0$ as

$$I_{\nu b}(\tau') = I_{\nu b}(0) + \tau'\left(\frac{dI_{\nu b}(\tau)}{d\tau}\right)_{\tau=0} + \frac{\tau'^2}{2!}\left(\frac{d^2I_{\nu b}(\tau)}{d\tau^2}\right)_{\tau=0} + \cdots \tag{9-35}$$

Substituting Eq. 9-35 into Eq. 9-34, we obtain

$$\begin{aligned}-\frac{4\pi}{3}\frac{dI_{\nu b}(\tau)}{d\tau}\bigg|_0 = \pi\varepsilon_{1\nu}\bigg[I_{\nu b}(T_1) &- 2I_{\nu b}(0)\int_0^{\tau_0} E_2(\tau')\,d\tau' \\ &- 2\frac{dI_{\nu b}(\tau)}{d\tau}\bigg|_0\int_0^{\tau_0}\tau' E_2(\tau')\,d\tau' \\ &- 2\frac{1}{2!}\frac{d^2I_{\nu b}(\tau)}{d\tau^2}\bigg|_0\int_0^{\tau_0}\tau'^2 E_2(\tau')\,d\tau' + \cdots\end{aligned} \tag{9-36}$$

For large values of τ_0 various integrals in Eq. 9-36 are evaluated as

$$\int_0^\infty E_2(\tau')\,d\tau' = \tfrac{1}{2}, \qquad \int_0^\infty \tau' E_2(\tau')\,d\tau' = \tfrac{1}{3}, \qquad \int_0^\infty \tau'^2 E_2(\tau')\,d\tau' = \tfrac{1}{2} \tag{9-37}$$

Then Eq. 9-36 simplifies to

$$-\frac{4\pi}{3}\frac{dI_{\nu b}(\tau)}{d\tau}\bigg|_0 = \pi\varepsilon_{1\nu}\left[I_{\nu b}(T_1) - I_{\nu b}(0) - \frac{2}{3}\frac{dI_{\nu b}(\tau)}{d\tau}\bigg|_0 - \frac{1}{2}\frac{d^2 I_{\nu b}(\tau)}{d\tau^2}\bigg|_0 - \cdots\right] \tag{9-38}$$

Neglecting the second- and higher-order terms and rearranging, we obtain for Eq. 9-38

$$\pi[I_{\nu b}(T_1) - I_{\nu b}(0)] = -\left(\frac{1}{\varepsilon_{1\nu}} - \frac{1}{2}\right)\frac{4\pi}{3}\frac{dI_{\nu b}(\tau)}{d\tau}\bigg|_0 \tag{9-39a}$$

or

$$\pi[I_{\nu b}(T_1) - I_{\nu b}(0)] = \left(\frac{1}{\varepsilon_{1\nu}} - \frac{1}{2}\right) q_\nu^r(0) \tag{9-39b}$$

Similarly, the corresponding relation for the boundary surface at $\tau = \tau_0$ is obtained as

$$\pi[I_{\nu b}(\tau_0) - I_{\nu b}(T_2)] = -\left(\frac{1}{\varepsilon_{2\nu}} - \frac{1}{2}\right)\frac{4\pi}{3}\frac{dI_{\nu b}(\tau)}{d\tau}\bigg|_{\tau_0} \tag{9-40a}$$

or

$$\pi[I_{\nu b}(\tau_0) - I_{\nu b}(T_2)] = \left(\frac{1}{\varepsilon_{2\nu}} - \frac{1}{2}\right) q_\nu^r(\tau_0) \tag{9-40b}$$

Equations 9-39 and 9-40 are for a single frequency ν; integration of these equations over all frequencies for constant properties yields respectively

$$n^2[\bar{\sigma}T_1^4 - \bar{\sigma}T^4(0)] = \left(\frac{1}{\varepsilon_1} - \frac{1}{2}\right) q^r \tag{9-41}$$

$$n^2[\bar{\sigma}T^4(\tau_0) - \bar{\sigma}T_2^4] = \left(\frac{1}{\varepsilon_2} - \frac{1}{2}\right) q^r \tag{9-42}$$

since

$$\int_{\nu=0}^\infty I_{\nu b}(\tau)\,d\nu = I_b(\tau) = \frac{n^2\bar{\sigma}T^4(\tau)}{\pi}$$

Here $T(0)$ and $T(\tau_0)$ denote the temperatures in the medium adjacent to the boundaries at $\tau = 0$ and $\tau = \tau_0$, respectively, T_1 and T_2 are the boundary surface temperatures, and the net radiative heat flux q^r is constant everywhere in the medium. Equations 9-41 and 9-42 will now be utilized to determine the net radiative heat flux, q^r.

For an optically thick medium, the net radiative heat flux is given by the diffusion approximation as (see Eq. 9-22)

$$q^r = -\frac{4\pi}{3}\frac{dI_b(\tau)}{d\tau} \tag{9-43}$$

Integration of Eq. 9-43 from $\tau = 0$ to $\tau = \tau_0$ for q^r as constant yields

$$\tau_0 q^r = -\frac{4\pi}{3}[I_b(\tau_0) - I_b(0)] = -\tfrac{4}{3}n^2[\bar{\sigma}T^4(\tau_0) - \bar{\sigma}T^4(0)] \tag{9-44}$$

From Eqs. 9-41 and 9-42 we have

$$n^2[\bar{\sigma}T^4(\tau_0) - \bar{\sigma}T^4(0)] = n^2(\bar{\sigma}T_2^4 - \bar{\sigma}T_1^4) + \left(\frac{1}{\varepsilon_1} + \frac{1}{\varepsilon_2} - 1\right)q^r \tag{9-45}$$

and from Eqs. 9-44 and 9-45 we obtain the net radiative heat flux q^r as

$$q^r = \frac{n^2(\bar{\sigma}T_1^4 - \bar{\sigma}T_2^4)}{\frac{3}{4}\tau_0 + (1/\varepsilon_1) + (1/\varepsilon_2) - 1} \tag{9-46}$$

Knowing the net radiative heat flux q^r, we can evaluate the temperature jumps at the boundary surfaces $\tau = 0$ and $\tau = \tau_0$ from Eqs. 9-41 and 9-42, respectively.

For a nonparticipating medium we have $\beta = 0$; hence $\tau_0 = 0$, and Eq. 9-46 yields the correct limit for the net radiative heat flux between two diffusely reflecting, diffusely emitting, parallel plates separated by a nonparticipating medium.

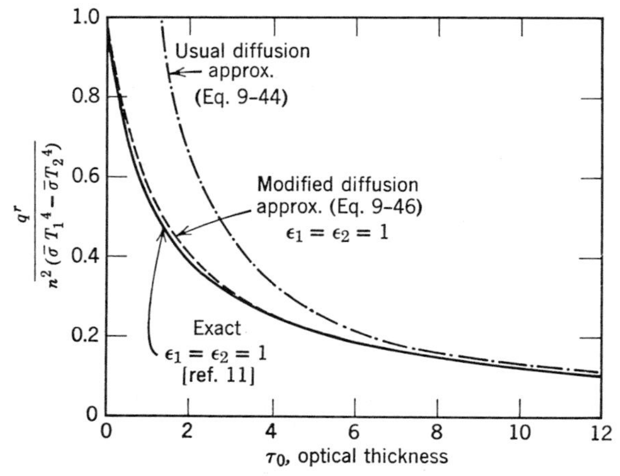

Fig. 9-3. Comparison of diffusion and modified diffusion approximations with the exact solution.

Figure 9-3 shows a comparison of the net radiative heat flux $(q^r/n^2) \times (\bar{\sigma}T_1^4 - \bar{\sigma}T_2^4)$, evaluated from Eq. 9-46 with the exact solution of Hottel [11], for $\varepsilon_1 = \varepsilon_2 = 1$.

Equation 9-46 agrees closely with the exact solution for all values of the optical thickness τ_0. Included in this figure is the prediction of the net radiative heat flux from the result of the usual diffusion approximation given by Eq. 9-44. It is apparent that the diffusion approximation breaks down for small values of τ_0.

Gray Medium Between Two Concentric Cylinders

Consider a gray medium that contains no energy sources or sinks, bounded between two concentric, coaxial, long cylinders of inner and outer radii r_1 and r_2, respectively. The inner and outer cylinders are opaque, diffuse emitters and diffuse reflectors, are kept at uniform temperatures T_1 and T_2, and have emissivities ε_1 and ε_2. The energy transfer is by pure radiation. A modified diffusion approximation will now be used to determine the net radiative heat flux in the medium and the temperature jump conditions at the boundaries. For simplicity we consider $n = 1$.

Let $q^r(\tau)$ be the net radiative heat flux at any radial location r. The total net radiative heat transfer Q^r in the radial direction per unit length of the cylinder is

$$Q^r = 2\pi r q^r(r) = \text{constant} \tag{9-47}$$

Then the temperature jump conditions given by Eqs. 9-41 and 9-42 when applied to the boundaries at $r = r_1$ and $r = r_2$, respectively, yield

$$\bar{\sigma}T_1^4 - \bar{\sigma}T^4(r_1) = \left(\frac{1}{\varepsilon_1} - \frac{1}{2}\right)\frac{Q^r}{2\pi r_1} \tag{9-48}$$

$$\bar{\sigma}T^4(r_2) - \bar{\sigma}T_2^4 = \left(\frac{1}{\varepsilon_2} - \frac{1}{2}\right)\frac{Q^r}{2\pi r_2} \tag{9-49}$$

The diffusion approximation 9-43 for a cylinder with $q^r(\tau)$ as given by Eq. 9-47 becomes

$$\pi \frac{dI_b(r)}{dr} = -\frac{3\beta Q^r}{8\pi}\frac{1}{r} \tag{9-50}$$

Integration of Eq. 9-50 from $r = r_1$ to $r = r_2$ for Q^r as constant yields

$$\pi[I_b(r_2) - I_b(r_1)] = -\frac{3\beta Q^r}{8\pi}\ln\left(\frac{r_2}{r_1}\right) \tag{9-51a}$$

or

$$\bar{\sigma}T^4(r_2) - \bar{\sigma}T^4(r_1) = -\frac{3\beta Q^r}{8\pi}\ln\left(\frac{r_2}{r_1}\right) \tag{9-51b}$$

From Eqs. 9-48 and 9-49 we have

$$\bar{\sigma}T^4(r_2) - \bar{\sigma}T^4(r_1) = \bar{\sigma}T_2^{\,4} - \bar{\sigma}T_1^{\,4} + \frac{Q^r}{2\pi}\left[\frac{1}{r_1}\left(\frac{1}{\varepsilon_1} - \frac{1}{2}\right) + \frac{1}{r_2}\left(\frac{1}{\varepsilon_2} - \frac{1}{2}\right)\right] \quad (9\text{-}52)$$

Eliminating $\bar{\sigma}T^4(r_2) - \bar{\sigma}T^4(r_1)$ between Eqs. 9-51b and 9-52, we obtain the following relation for the total radiative heat flow rate Q^r:

$$\frac{\bar{\sigma}T_1^{\,4} - \bar{\sigma}T_2^{\,4}}{Q^r} = \frac{1}{2\pi}\left[\frac{3\beta}{4}\ln\left(\frac{r_2}{r_1}\right) + \frac{1}{r_1}\left(\frac{1}{\varepsilon_1} - \frac{1}{2}\right) + \frac{1}{r_2}\left(\frac{1}{\varepsilon_2} - \frac{1}{2}\right)\right] \quad (9\text{-}53)$$

Knowing Q^r, we can evaluate the temperature jumps at the boundaries $r = r_1$ and $r = r_2$ from Eqs. 9-48 and 9-49 respectively.

The net radiative heat flux at the inner boundary surface can be determined by setting $r = r_1$ in Eq. 9-47, that is,

$$Q^r = 2\pi r_1 q^r(r_1) \quad (9\text{-}54)$$

and then, substituting Eq. 9-54 into Eq. 9-53, we obtain

$$\frac{\bar{\sigma}T_1^{\,4} - \bar{\sigma}T_2^{\,4}}{q^r(r_1)} = \frac{3\beta}{4} r_1 \ln\left(\frac{r_2}{r_1}\right) + \left(\frac{1}{\varepsilon_1} - \frac{1}{2}\right) + \frac{r_1}{r_2}\left(\frac{1}{\varepsilon_2} - \frac{1}{2}\right) \quad (9\text{-}55)$$

If the usual diffusion approximation is used, the net radiative heat flux $q^r(r_1)$ at the inner boundary surface is obtained from Eqs. 9-51b and 9-54 as

$$\frac{\bar{\sigma}T^4(r_1) - \bar{\sigma}T^4(r_2)}{q^r(r_1)} = \frac{3\beta}{4} r_1 \ln\left(\frac{r_2}{r_1}\right) \quad (9\text{-}56)$$

Figure 9-4 shows a comparison of $q^r(r_1)/(\bar{\sigma}T_1^{\,4} - \bar{\sigma}T_2^{\,4})$, evaluated from the modified diffusion approximation of Eq. 9-55 and the usual diffusion approximation of Eq. 9-56, with the Monte Carlo solution of Perlmutter and Howell [12] for an absorbing and emitting medium. The result from the modified diffusion approximation agrees with the exact solution reasonably well for large values of $\kappa(r_2 - r_1)$; the agreement is also good for small values of $\kappa(r_2 - r_1)$ when the ratio r_2/r_1 is not too large. In all cases, the improvement in the evaluation of the radiative heat flux over the usual diffusion approximation is considerable.

For a nonparticipating medium between the cylinders we set $\beta = 0$, and Eq. 9-55 simplifies to

$$\frac{\bar{\sigma}T_1^{\,4} - \bar{\sigma}T_2^{\,4}}{q^r(r_1)} = \left(\frac{1}{\varepsilon_1} - \frac{1}{2}\right) + \frac{r_1}{r_2}\left(\frac{1}{\varepsilon_2} - \frac{1}{2}\right) \quad (9\text{-}57)$$

This equation yields the correct limit for radiative interchange between two concentric cylinders separated by a nonparticipating medium only when $r_1/r_2 = 1$.

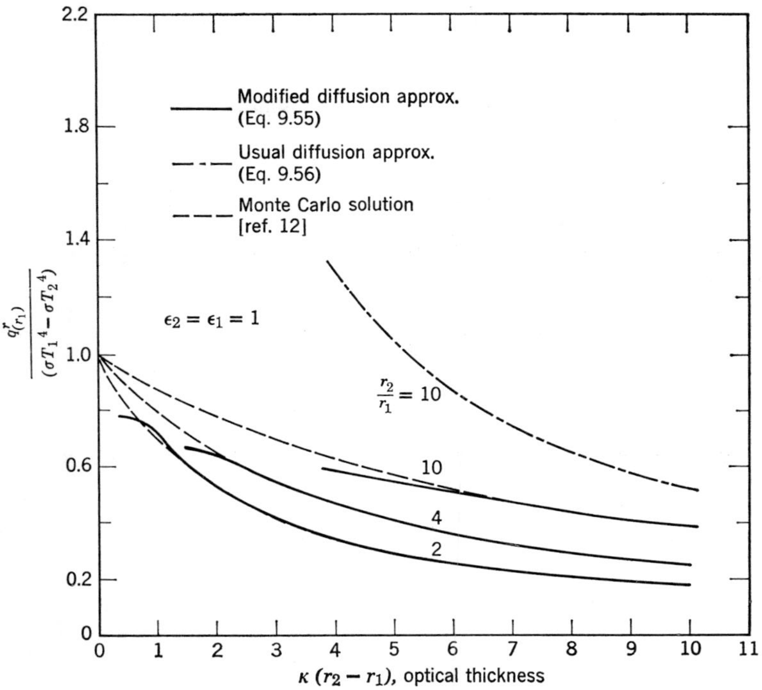

Fig. 9-4. Comparison of diffusion and modified diffusion approximations with the Monte Carlo solution for a concentric cylinder. (From R. G. Deissler [7].)

9-4 EDDINGTON APPROXIMATION

One of the earliest approximations developed to obtain an approximate solution to the equation of radiative transfer is due to Eddington [13, pp. 321–324]. The basis of this approach involves an approximation to the angular distribution of radiation intensity such that the integrodifferential equation of radiative transfer is transformed into an ordinary differential equation. The derivation of the Eddington approximation can be found, in addition to the original work, in the publications by Viskanta [1, pp. 204–205] and Woolley and Stibbs [4]. We present now a brief account of the Eddington approximation.

Consider the equation of radiative transfer for a plane-parallel medium in the form

$$\mu \frac{\partial I(\tau, \mu)}{\partial \tau} + I(\tau, \mu) = (1 - \omega) I_b(\tau) + \frac{\omega}{2} \int_{-1}^{1} I(\tau, \mu')\, d\mu' \qquad (9\text{-}58)$$

We operate on both sides of this equation by the operator $2\pi \int_{-1}^{1} d\mu$ and obtain

$$\frac{dq^r(\tau)}{d\tau} + G(\tau) = 4\pi(1-\omega)I_b(\tau) + \omega G(\tau)$$

or

$$\frac{dq^r(\tau)}{d\tau} = (1-\omega)[4\pi I_b(\tau) - G(\tau)] \tag{9-59}$$

where we have defined the net radiative heat flux and the incident radiation, respectively, as

$$q^r(\tau) = 2\pi \int_{-1}^{1} \mu I(\tau, \mu)\, d\mu \tag{9-60}$$

$$G(\tau) = 2\pi \int_{-1}^{1} I(\tau, \mu)\, d\mu \tag{9-61}$$

We now operate on both sides of Eq. 9-58 by the operator $2\pi \int_{-1}^{1} \mu\, d\mu$ and obtain

$$c\frac{dP^r(\tau)}{d\tau} = -q^r(\tau) \tag{9-62}$$

where we have defined the radiation pressure $P^r(\tau)$ as

$$P^r(\tau) = \frac{1}{c} 2\pi \int_{-1}^{1} \mu^2 I(\tau, \mu)\, d\mu \tag{9-63}$$

and c is the speed of propagation of radiation in the medium.

Up to this point our analysis has been exact; Eqs. 9-59 and 9-62 are two coupled equations, but they involve three unknowns: $q^r(\tau)$, $G(\tau)$, and $P^r(\tau)$. An additional relation was obtained by Eddington by introducing an approximation for the angular distribution of the radiation intensity.

By separating the intensity into a forward component $I^+(\tau, \mu)$, $\mu \in (0, 1)$, and a backward component $I^-(\tau, \mu)$, $\mu \in (-1, 0)$, we write

$$G(\tau) = 2\pi\left[\int_{-1}^{0} I^-(\tau, \mu)\, d\mu + \int_{0}^{1} I^+(\tau, \mu)\, d\mu\right] \tag{9-64}$$

$$q^r(\tau) = 2\pi\left[\int_{-1}^{0} \mu I^-(\tau, \mu)\, d\mu + \int_{0}^{1} \mu I^+(\tau, \mu)\, d\mu\right] \tag{9-65}$$

$$P^r(\tau) = \frac{1}{c} 2\pi\left[\int_{-1}^{0} \mu^2 I^-(\tau, \mu)\, d\mu + \int_{0}^{1} \mu^2 I^+(\tau, \mu)\, d\mu\right] \tag{9-66}$$

If it is assumed that the intensity components $I^+(\tau, \mu)$ and $I^-(\tau, \mu)$ are independent of direction, that is,

$$I^+(\tau, \mu) = I^+(\tau), \quad 0 < \mu \leq 1 \tag{9-67a}$$

$$I^-(\tau, \mu) = I^-(\tau), \quad -1 \leq \mu < 0 \tag{9-67b}$$

Eqs. 9-64 through 9-66 simplify to

$$G(\tau) = 2\pi[I^+(\tau) + I^-(\tau)] \tag{9-68}$$

$$q^r(\tau) = \pi[I^+(\tau) - I^-(\tau)] \tag{9-69}$$

$$P^r(\tau) = \frac{2\pi}{3c}[I^+(\tau) + I^-(\tau)] \tag{9-70}$$

Any two of these three relations can be used to determine an additional relation for use with Eqs. 9-59 and 9-62. For example, from Eqs. 9-68 and 9-70 we obtain

$$P^r(\tau) = \frac{1}{3c}G(\tau) \tag{9-71}$$

which relates $P^r(\tau)$ to $G(\tau)$. Equations 9-59, 9-62, and 9-71 are three relations for the three unknowns $q^r(\tau)$, $G(\tau)$, and $P^r(\tau)$; they can be combined to yield a single differential equation for any one of these unknowns.

For example, $P^r(\tau)$ is eliminated between Eqs. 9-62 and 9-71 to yield

$$\frac{1}{3}\frac{dG(\tau)}{d\tau} = -q^r(\tau) \tag{9-72}$$

When Eq. 9-59 is differentiated and in the resulting expression $dG/d\tau$ is eliminated by means of Eq. 9-72, the following ordinary differential is obtained for the net radiative heat flux $q^r(\tau)$:

$$\frac{d^2q^r(\tau)}{d\tau^2} = (1-\omega)\left[4\pi\frac{dI_b(\tau)}{d\tau} + 3q^r(\tau)\right] \tag{9-73a}$$

which is called the Eddington approximation. Alternatively, if $dq^r/d\tau$ is eliminated from Eq. 9-59 by means of Eq. 9-72, the following ordinary differential equation is obtained for $G(\tau)$:

$$\frac{d^2G(\tau)}{d\tau^2} = 3(1-\omega)[G(\tau) - 4\pi I_b(\tau)] \tag{9-73b}$$

These equations are valid within an optically thick medium, but not as accurate near the boundaries. It will be seen later in this chapter that the Eddington approximation given by Eqs. 9-73 is exactly the same as the P_1 approximation given by Eqs. 9-120 and 9-121. In fact, Eq. 9-73a reduces to the usual diffusion approximation 9-43 if we neglect in this equation the second-order term, that is,

$$q^r(\tau) = -\frac{4\pi}{3}\frac{dI_b(\tau)}{d\tau} \tag{9-74}$$

Pomraning [14] extended the Eddington approximation. His numerical calculations for simple problems in which exact results are known indicate that the modified approximation gives substantial improvement over the Eddington results.

Several applications of the Eddington approximation can be found in the literature on the problems of interaction of radiation with conduction and convection. Goody [15] applied the Eddington approximation to the problem of simultaneous radiation and natural convection for an absorbing and emitting fluid between two horizontal parallel plates heated below, and Viskanta and Grosh [16] applied it to a problem of simultaneous conduction and radiation.

To solve the Eddington approximation given by Eq. 9-73b two boundary conditions are needed. Since this equation is similar to that obtained with the P_1 approximation, we shall delay the discussion of boundary conditions to Section 9-7, where the Mark and Marshak boundary conditions are presented for more general situations. Some applications of the Eddington approximation will be given in Chapter 11.

9-5 SCHUSTER-SCHWARZCHILD APPROXIMATION

The integrodifferential equation of radiative transfer for an isotropically scattering, plane-parallel medium can be transformed into a pair of ordinary differential equations by the approximation introduced by Schuster [17] and Schwarzchild [18]. The Schuster-Schwarzchild approximation is discussed in several publications, for example, those by Viskanta [1, pp. 203–204], Chandrasekhar [2, p. 55], and Sobolev [19, p. 42]. We present below the derivation of this approximation.

The radiation field is divided into a forward and a backward radiation stream, j^+ for $0 < \mu \leq 1$ and j^- for $-1 \leq \mu < 0$, respectively, defined as

$$j^+ \equiv \int_0^1 I^+(\tau, \mu')\, d\mu' \quad \text{for } 0 < \mu \leq 1 \tag{9-75a}$$

$$j^- \equiv \int_{-1}^0 I^-(\tau, \mu')\, d\mu' \quad \text{for } -1 \leq \mu < 0 \tag{9-75b}$$

By utilizing these definitions the equations of radiative transfer for the forward and backward intensities can be written as

$$\mu \frac{\partial I^+(\tau, \mu)}{\partial \tau} + I^+(\tau, \mu) = (1 - \omega) I_b(T) + \frac{\omega}{2}(j^+ + j^-), \quad \mu > 0 \tag{9-76a}$$

$$\mu \frac{\partial I^-(\tau, \mu)}{\partial \tau} + I^-(\tau, \mu) = (1 - \omega) I_b(T) + \frac{\omega}{2}(j^+ + j^-), \quad \mu < 0 \tag{9-76b}$$

We operate on Eq. 9-76a by the operator $\int_0^1 d\mu$, and on Eq. 9-76b by the operator $\int_{-1}^0 d\mu$ and obtain respectively

$$\frac{d}{d\tau}\left[\int_0^1 \mu I^+(\tau, \mu)\, d\mu\right] + j^+ = (1-\omega)I_b(T) + \frac{\omega}{2}(j^+ + j^-) \quad \text{(9-77a)}$$

$$\frac{d}{d\tau}\left[\int_{-1}^0 \mu I^-(\tau, \mu)\, d\mu\right] + j^- = (1-\omega)I_b(T) + \frac{\omega}{2}(j^+ + j^-) \quad \text{(9-77b)}$$

So far Eqs. 9-77 are exact. The Schuster-Schwarzchild approximation is now introduced as

$$\int_0^1 \mu I^+(\tau, \mu)\, d\mu \cong \frac{1}{2}\int_0^1 I^+(\tau, \mu)\, d\mu = \tfrac{1}{2}j^+ \quad \text{(9-78a)}$$

$$\int_{-1}^0 \mu I^-(\tau, \mu)\, d\mu \cong -\frac{1}{2}\int_{-1}^0 I^-(\tau, \mu)\, d\mu = -\tfrac{1}{2}j^- \quad \text{(9-78b)}$$

When these approximations are substituted into Eqs. 9-77, we obtain

$$\frac{1}{2}\frac{dj^+}{d\tau} + j^+ = (1-\omega)I_b(T) + \frac{\omega}{2}(j^+ + j^-) \quad \text{(9-79a)}$$

$$-\frac{1}{2}\frac{dj^-}{d\tau} + j^- = (1-\omega)I_b(T) + \frac{\omega}{2}(j^+ + j^-) \quad \text{(9-79b)}$$

Equations 9-79 are two coupled ordinary differential equations for the functions j^+ and j^-. Once j^+ and j^- are determined, the net radiative heat flux is evaluated from

$$\begin{aligned} q^r(\tau) &= 2\pi\int_{-1}^1 \mu I(\tau, \mu)\, d\mu = 2\pi\left[\int_0^1 \mu I^+(\tau, \mu)\, d\mu + \int_{-1}^0 \mu I^-(\tau, \mu)\, d\mu\right] \\ &= \pi(j^+ - j^-) \end{aligned} \quad \text{(9-80)}$$

The gradient of the radiative heat flux, $dq^r/d\tau$, is obtained by adding Eqs. 9-79a and 9-79b and utilizing the result of Eq. 9-80. We find

$$\frac{1}{2\pi}\frac{dq^r(\tau)}{d\tau} = (1-\omega)[2I_b(T) - (j^+ + j^-)] \quad \text{(9-81)}$$

The Boundary Conditions

To solve Eqs. 9-79 two boundary conditions are needed. To illustrate the construction of the boundary conditions we consider a plane-parallel slab between two diffusely reflecting, diffusely emitting opaque boundaries at $\tau = 0$ and $\tau = \tau_0$, which are kept at temperatures T_1 and T_2 and have emissivities ε_1 and ε_2, respectively. The boundary conditions for such a case have already been given as (see Eqs. 8-99)

$$I^+(0) = \varepsilon_1 \frac{n^2\bar{\sigma}T_1^4}{\pi} - 2(1-\varepsilon_1)\int_{-1}^{0} I^-(0, \mu')\mu' \, d\mu' \quad \mu > 0 \qquad (9\text{-}82a)$$

$$I^-(\tau_0) = \varepsilon_2 \frac{n^2\bar{\sigma}T_2^4}{\pi} + 2(1-\varepsilon_2)\int_{0}^{1} I^+(\tau_0, \mu')\mu' \, d\mu' \quad \mu < 0 \qquad (9\text{-}82b)$$

where we have removed the frequency dependence in Eqs. 8-99 and replaced ρ by $1 - \varepsilon$.

The approximation 9-78 is introduced on the right-hand sides of Eqs. 9-82 to yield

$$I^+(0) = \varepsilon_1 \frac{n^2\bar{\sigma}T_1^4}{\pi} + (1-\varepsilon_1)j^-(0) \qquad (9\text{-}83a)$$

$$I^-(\tau_0) = \varepsilon_2 \frac{n^2\bar{\sigma}T_2^4}{\pi} + (1-\varepsilon_2)j^+(\tau_0) \qquad (9\text{-}83b)$$

When Eq. 9-83a is operated on by the operator $\int_0^1 d\mu$ and Eq. 9-83 by the operator $\int_{-1}^0 d\mu$, we obtain, respectively,

$$j^+(0) = \varepsilon_1 \frac{n^2\bar{\sigma}T_1^4}{\pi} + (1-\varepsilon_1)j^-(0) \qquad (9\text{-}84a)$$

$$j^-(\tau_0) = \varepsilon_2 \frac{n^2\bar{\sigma}T_2^4}{\pi} + (1-\varepsilon_2)j^+(\tau_0) \qquad (9\text{-}84b)$$

Equations 9-84 provide two boundary conditions for the two coupled ordinary differential equations 9-79.

The Schuster-Schwarzchild approximation has been utilized for solving problems of radiative heat transfer by Larkin and Churchill [20] and Chen [21].

9-6 EXPONENTIAL KERNEL APPROXIMATIONS

In Chapter 8 we presented formal solutions for the net radiative heat flux and for its derivative in a plane-parallel geometry. These expressions can be simplified significantly if the exponential integral functions entering them are represented by exponentials. Krook [22] describes a general procedure to approximate the $E_2(\tau)$ function in terms of exponentials in the form

$$E_2(\tau) \simeq \sum_{j=1}^{n} a_j e^{-b_j\tau} \qquad (9\text{-}85)$$

where the coefficients a_j and b_j are determined by the moment method.

We consider a one-term representation of $E_2(\tau)$ in the form

$$E_2(\tau) \cong ae^{-b\tau} \tag{9-86}$$

Then $E_3(\tau)$ is given as

$$E_3(\tau) = -\int E_2(\tau')\, d\tau' = \frac{a}{b}\, e^{-b\tau} \tag{9-87}$$

Lick [23] determined the coefficients a and b by requiring that the areas and the first moments of the exponential and exponential integral over the range $\tau = 0$ to $\tau = \infty$ be equal; he obtained

$$E_2(\tau) \cong \tfrac{3}{4}e^{-(3/2)\tau} \tag{9-88}$$

$$E_3(\tau) \cong \tfrac{1}{2}e^{-(3/2)\tau} \tag{9-89}$$

Vincenti and Baldwin [24] proposed a representation of $E_2(\tau)$ in the form

$$E_2(\tau) \cong 0.813e^{-1.562\tau} \tag{9-90}$$

An alternative representation of $E_2(\tau)$ is given in the form

$$E_2(\tau) \cong e^{-\sqrt{3}\tau} \tag{9-91}$$

It can readily be verified that Eq. 9-91 characterizes the Eddington approximation. To prove this we consider the expression for the net radiative heat flux for an absorbing and emitting medium given by Eq. 8-84 as

$$q^r(\tau) = 2\pi[I^+(0)E_3(\tau) - I^-(\tau_0)E_3(\tau_0 - \tau)] + 2\pi\left[\int_0^\tau I_b(\tau')E_2(\tau - \tau')\, d\tau' - \int_\tau^{\tau_0} I_b(\tau')E_2(\tau' - \tau)\, d\tau'\right] \tag{9-92}$$

We substitute in Eq. 9-92 the representation given by Eqs. 9-86 and 9-87 and obtain

$$q^r(\tau) = 2\pi\frac{a}{b}\,[I^+(0)e^{-b\tau} - I^-(\tau_0)e^{-b(\tau_0-\tau)}] + 2\pi a\left[\int_0^\tau I_b(\tau')e^{-b(\tau-\tau')}\, d\tau' - \int_\tau^{\tau_0} I_b(\tau')e^{-b(\tau'-\tau)}\, d\tau'\right] \tag{9-93}$$

When Eq. 9-93 is differentiated twice with respect to τ, and in the resulting expression the integral terms and the boundary surface intensities are eliminated by means of Eq. 9-93, we obtain

$$\frac{d^2q^r(\tau)}{d\tau^2} = 4\pi a\frac{dI_b(\tau)}{d\tau} + b^2q^r(\tau) \tag{9-94}$$

The Eddington approximation 9-73, for $\omega = 0$, becomes

$$\frac{d^2q^r(\tau)}{d\tau^2} = 4\pi\frac{dI_b(\tau)}{d\tau} + 3q^r(\tau) \tag{9-95}$$

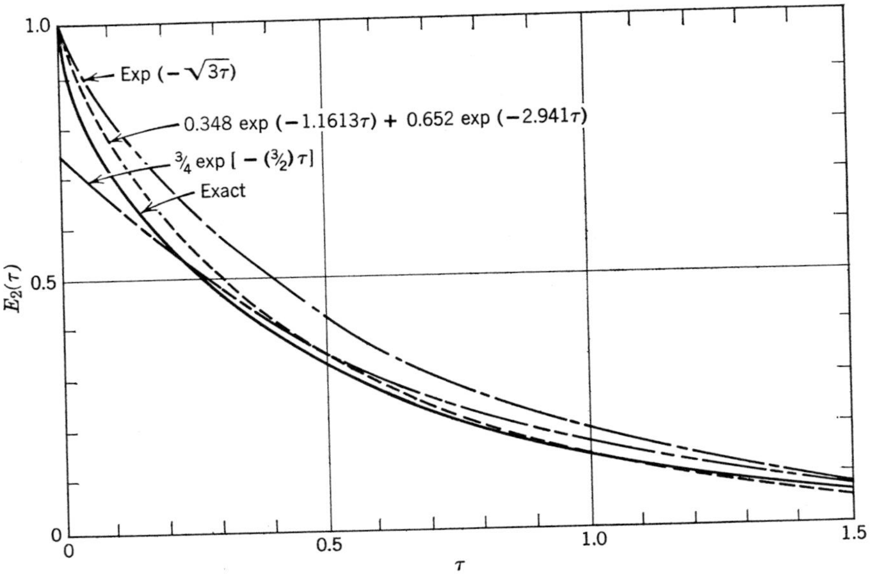

Fig. 9-5. Comparison of exact and approximate values of $E_2(\tau)$.

Equation 9-94 is identical to Eq. 9-95 if we choose

$$a = 1 \qquad \text{and} \qquad b = \sqrt{3} \tag{9-96}$$

which correspond to the coefficients in the exponential approximation 9-91.

Murty [25] gives a two-term exponential approximation for the $E_2(\tau)$ function in the form

$$E_2(\tau) \cong 0.348e^{-1.1613\tau} + 0.652e^{-2.941\tau} \tag{9-97}$$

Figure 9-5 shows a comparison of various approximations for $E_2(\tau)$ with the exact value this function.

9-7 SPHERICAL HARMONICS METHOD

The method of *spherical harmonics* provides a means to obtain a higher-order approximate solution to the equation of radiative transfer at the expense of additional labor and calculations. This method was first suggested by Jeans [26] in connection with the problem of radiative transfer in stars. A general description of the method of spherical harmonics can be found in the books by Kourganoff [3, pp. 90–101] for radiative transfer applications, and by Davison [27, pp. 116–145] and Murray [28, pp. 247–259] for neutron transport applications.

Consider the equation of radiative transfer for a plane-parallel, gray medium with azimuthal symmetry, given in the form

$$\mu \frac{\partial I(\tau, \mu)}{\partial \tau} + I(\tau, \mu) = (1 - \omega)I_b(T) + \frac{\omega}{2}\int_{-1}^{1} p(\mu, \mu')I(\tau, \mu')\,d\mu' \quad (9\text{-}98)$$

where the phase function $p(\mu, \mu')$ is assumed to be represented in a series of Legendre polynomials as[2]

$$p(\mu, \mu') = \sum_{n=0}^{\infty}(2n + 1)f_n P_n(\mu)P_n(\mu'), \quad f_0 = 1 \quad (9\text{-}99)$$

We now assume that the intensity of radiation $I(\tau, \mu)$ can be expanded in a series of Legendre polynomials in the form

$$I(\tau, \mu) = \sum_{m=0}^{\infty} \frac{2m + 1}{4\pi} P_m(\mu)\Psi_m(\tau) \quad (9\text{-}100)$$

If the functions $\Psi_m(\tau)$ are known, the radiation intensity can be determined from Eq. 9-100. Therefore in the following analysis we shall be concerned with the determination of functions $\Psi_m(\tau)$.

The expansions 9-99 and 9-100 are substituted into Eq. 9-98 to yield

$$\sum_{m=0}^{\infty} \frac{2m + 1}{4\pi} P_m(\mu)\left[\mu \frac{d\Psi_m(\tau)}{d\tau} + \Psi_m(\tau)\right]$$
$$= (1 - \omega)I_b(T) + \frac{\omega}{2}\sum_{m=0}^{\infty}\sum_{n=0}^{\infty}(2n + 1)f_n P_n(\mu)\frac{2m + 1}{4\pi}\Psi_m(\tau)$$
$$\times \int_{\mu'=-1}^{1} P_n(\mu')P_m(\mu')\,d\mu' \quad (9\text{-}101a)$$

which simplifies to

$$\sum_{m=0}^{\infty} \frac{2m + 1}{4\pi} P_m(\mu)\left[\mu \frac{d\Psi_m(\tau)}{d\tau} + \Psi_m(\tau)\right]$$
$$= (1 - \omega)I_b(T) + \omega\sum_{m=0}^{\infty}\frac{2m + 1}{4\pi} f_m P_m(\mu)\Psi_m(\tau) \quad (9\text{-}101b)$$

To obtain Eq. 9-101b we utilized the orthogonality of Legendre polynomials, that is,

$$\int_{-1}^{1} P_n(\mu')P_m(\mu')\,d\mu' = \begin{cases} 0 & \text{for } m \neq n \\ \dfrac{2}{2m + 1} & \text{for } m = n \end{cases} \quad (9\text{-}102)$$

We consider the recurrence formula for the Legendre polynomials given as [29, p. 308]

$$\mu P_m(\mu) = \frac{mP_{m-1}(\mu) + (m + 1)P_{m+1}(\mu)}{2m + 1} \quad (9\text{-}103)$$

Substitution of $\mu P_m(\mu)$ from Eq. 9-103 into Eq. 9-101b yields

$$\sum_{m=0}^{\infty}\left[m\,\frac{d\Psi_m(\tau)}{d\tau}\,P_{m-1}(\mu) + (m+1)\,\frac{d\Psi_m(\tau)}{d\tau}\,P_{m+1}(\mu) + (2m+1)(1-\omega f_m)\Psi_m(\tau)P_m(\mu) - 4\pi(1-\omega)I_b(T)\,\delta_{0m}\right] = 0 \quad (9\text{-}104)$$

where

$$\delta_{0m} = \begin{cases} 1 & \text{for } m = 0 \\ 0 & \text{otherwise} \end{cases}$$

The series given in Eq. 9-104 can be rearranged as

$$\sum_{m=0}^{\infty}\left[(m+1)\,\frac{d\Psi_{m+1}(\tau)}{d\tau} + m\,\frac{d\Psi_{m-1}(\tau)}{d\tau} + (2m+1)(1-\omega f_m)\Psi_m(\tau) - 4\pi(1-\omega)I_b(T)\,\delta_{0m}\right]P_m(\mu) = 0 \quad (9\text{-}105)$$

If Eq. 9-105 should be valid for all μ, then the coefficients of $P_m(\mu)$ must vanish identically. This requirement results in the following set of coupled ordinary differential equations for the functions $\Psi_m(\tau)$, $m = 0, 1, 2, \ldots$:

$$(m+1)\Psi'_{m+1} + m\Psi'_{m-1} + (2m+1)(1-\omega f_m)\Psi_m = 4\pi(1-\omega)I_b(T)\,\delta_{0m}, \quad m = 0, 1, 2, \ldots \quad (9\text{-}106)$$

where $f_0 = 1$ and the prime denotes differentiation with respect to τ.

For isotropic scattering we set in Eq. 9-106 all f_m equal to zero except f_0, which is equal to unity.

Equations 9-106 are an infinite set of coupled ordinary differential equations for the functions $\Psi_m(\tau)$. In practice, however, only a finite number of equations $m = N$ are considered and the term Ψ'_{N+1} is neglected.

The resulting system of equations is written as

$$\left.\begin{aligned}
&\Psi'_1 + (1-\omega)\Psi_0 = 4\pi(1-\omega)I_b(T) \\
&2\Psi'_2 + \Psi'_0 + 3(1-\omega f_1)\Psi_1 = 0 \\
&3\Psi'_3 + 2\Psi'_1 + 5(1-\omega f_2)\Psi_2 = 0 \\
&\cdots\cdots\cdots\cdots\cdots\cdots\cdots\cdots\cdots\cdots \\
&N\Psi'_N + (N-1)\Psi'_{N-2} + (2N-1)(1-\omega f_{N-1})\Psi_{N-1} = 0 \\
&N\Psi'_{N-1} + (2N+1)(1-\omega f_N)\Psi_N = 0
\end{aligned}\right\} \quad (9\text{-}107)$$

System 9-107 provides $N + 1$ simultaneous, linear, ordinary differential equations for the $N + 1$ unknown functions $\Psi_0, \Psi_1, \ldots, \Psi_N$ and is called the P_N *approximation*.

The desired solution of the system of equations 9-107 can be written as a linear sum of the solution to the homogeneous part of these equations and a particular solution; the latter cannot be determined, however, until the black-body intensity function, $I_b(T) = n^2\bar{\sigma}T^4(\tau)/\pi$, is specified. We seek the solution of the homogeneous part of system 9-107 in the form

$$\Psi_m^H(\tau) = g_m e^{k\tau}, \quad m = 0, 1, 2, \ldots, N \tag{9-108}$$

where the g_m are arbitrary constants and the k are the exponents that are yet to be determined.

Substitution of Eq. 9-108 into the homogeneous part of Eq. 9-107 (or 9-106) yields the following $N + 1$ simultaneous, homogeneous, algebraic equations for the coefficients g_m:

$$k[(m + 1)g_{m+1} + mg_{m-1}] + (2m + 1)(1 - \omega f_m)g_m = 0 \tag{9-109a}$$

where $m = 0, 1, 2, \ldots, N, f_0 = 1$, and $g_{N+1} = 0$.

For isotropic scattering we set

$$f_m = \delta_{0m} = \begin{cases} 1 & \text{for } m = 0 \\ 0 & \text{otherwise} \end{cases}$$

and Eq. 9-109a simplifies to

$$k[(m + 1)g_{m+1} + mg_{m-1}] + (2m + 1)(1 - \omega\delta_{0m})g_m = 0 \tag{9-109b}$$

If the system of homogeneous algebraic equations 9-109 should have a nontrivial solution, the determinant of the coefficients must vanish. In the case of *isotropic scattering* this requirement yields

$$\begin{vmatrix} (1-\omega) & k & 0 & 0 & 0 & \cdots & 0 & 0 & 0 & 0 \\ k & 3 & 2k & 0 & 0 & \cdots & 0 & 0 & 0 & 0 \\ 0 & 2k & 5 & 3k & 0 & \cdots & 0 & 0 & 0 & 0 \\ 0 & 0 & 3k & 7 & 4k & \cdots & 0 & 0 & 0 & 0 \\ \cdot & \cdot & \cdot & \cdot & \cdot & \cdots & \cdot & \cdot & \cdot & \cdot \\ \cdot & \cdot & \cdot & \cdot & \cdot & \cdots & \cdot & \cdot & \cdot & \cdot \\ 0 & 0 & 0 & 0 & 0 & \cdots & (N-2)k & (2N-3) & (N-1)k & 0 \\ 0 & 0 & 0 & 0 & 0 & \cdots & 0 & (N-1)k & (2N-1) & Nk \\ 0 & 0 & 0 & 0 & 0 & \cdots & 0 & 0 & Nk & (2N+1) \end{vmatrix} = 0 \tag{9-110}$$

The solution of Eq. 9-110 gives the permissible values of k_i for each value of ω. Then, for each value of k_i, a set of $g_m(k_i)$'s, $m = 0, 1, 2, \ldots, N$, is determined from Eq. 9-109b, and the solution to the homogeneous part of

Eq. 9-107, for isotropic scattering, can be written in the form

$$\Psi_m^H(\tau) = \sum_{i=0}^{N} \bar{A}_i g_m(k_i) e^{k_i \tau}, \quad m = 0, 1, \ldots, N \tag{9-111}$$

The complete solution is given as the sum of the homogeneous solution and a particular solution in the form

$$\psi_m(\tau) = \Psi_m^H(\tau) + \Psi^P \tag{9-112}$$

where the particular solution Ψ^P depends on the black-body radiation intensity function $I_b[T(\tau)]$. The unknown coefficients $\bar{A}_i$ associated with the homogeneous solution are determined by constraining the complete solution to meet the boundary conditions for the problem. Once the functions $\Psi_m(\tau)$ are known, the distribution of the radiation intensity is determined from Eq. 9-100.

Davison [27, pp. 118–126] discusses an alternative representation of the homogeneous solution of Eq. 9-111 in terms of the auxiliary functions $H_m(k_i)$ in the form

$$\Psi_m{}^H(\tau) = \sum_{i=0}^{N} A_i H_m(k_i) e^{k_i \tau} \tag{9-113}$$

where A_i are the expansion coefficients, and the auxiliary function $H_n(k_i)$ is defined by

$$H_n(k) = (-1)^n \left\{ P_n\left(\frac{1}{k}\right) - \frac{\omega}{k}\left[Q_0\left(\frac{1}{k}\right) - Q_n\left(\frac{1}{k}\right)\right]\right\} \tag{9-114}$$

Here P_n and Q_n are the Legendre polynomials and the Legendre functions of the second kind, respectively. The reader may refer to the book by Davison [27, pp. 119–121] for a tabulation of the permissible values of k_i and the corresponding values of $H_n(k_i)$, $n = 0, 1, \ldots, N$, for various values of ω and N.

P_1 Approximation

As a special case we examine below the P_1 approximation for isotropic scattering. This approximation is obtained immediately from Eqs. 9-107 by setting $N = 1$ and $f_m = \delta_{0m}$ and by neglecting the term $d\Psi_2(\tau)/d\tau$, that is,

$$\Psi_1'(\tau) + (1 - \omega)\Psi_0(\tau) = 4\pi(1 - \omega) I_b(T) \tag{9-115a}$$

$$\Psi_0'(\tau) + 3\Psi_1(\tau) = 0 \tag{9-115b}$$

The physical significance of the functions $\Psi_0(\tau)$ and $\Psi_1(\tau)$ is envisioned by recalling the expansion of intensity as in Eq. 9-100, that is,

$$I(\tau, \mu) = \sum_{m=0}^{\infty} \frac{2m + 1}{4\pi} P_m(\mu) \Psi_m(\tau) \tag{9-116}$$

By utilizing the orthogonality of the Legendre polynomials as given by Eq. 9-102, it can be shown from Eq. 9-116 that the function $\Psi_m(\tau)$ is related to the intensity by

$$\Psi_m(\tau) = 2\pi \int_{-1}^{1} P_m(\mu') I(\tau, \mu')\, d\mu' \tag{9-117}$$

Then we have

$$\Psi_0(\tau) = 2\pi \int_{-1}^{1} I(\tau, \mu)\, d\mu \equiv G(\tau) = \text{incident radiation} \tag{9-118a}$$

$$\Psi_1(\tau) = 2\pi \int_{-1}^{1} \mu I(\tau, \mu)\, d\mu \equiv q^r(\tau) = \text{net radiative heat flux} \tag{9-118b}$$

By replacing $\Psi_0(\tau)$ and $\Psi_1(\tau)$ by $G(\tau)$ and $q^r(\tau)$, respectively, we can write Eqs. 9-115 as

$$\frac{dq^r(\tau)}{d\tau} + (1 - \omega)G(\tau) = 4\pi(1 - \omega)I_b(T) \tag{9-119a}$$

$$\frac{dG(\tau)}{d\tau} + 3q^r(\tau) = 0 \tag{9-119b}$$

where

$$I_b(T) \equiv \frac{n^2 \bar{\sigma} T^4(\tau)}{\pi}$$

Equations 9-119 can be combined to yield a single second-degree ordinary differential equation for $G(\tau)$ or $q^r(\tau)$ as

$$\frac{d^2G(\tau)}{d\tau^2} = 3(1 - \omega)[G(\tau) - 4\pi I_b(T)] \tag{9-120}$$

$$\frac{d^2q^r(\tau)}{d\tau^2} = (1 - \omega)\left[3q^r(\tau) + 4\pi \frac{dI_b(T)}{d\tau}\right] \tag{9-121}$$

Here we note that Eq. 9-121 is exactly the same as the Eddington approximation given by Eq. 9-73a.

Once $G(\tau)$ is determined from the solution of Eq. 9-120 subject to a set of appropriate boundary conditions, the radiation intensity $I(\tau, \mu)$ is determined from Eq. 9-116 as

$$I(\tau, \mu) = \frac{1}{4\pi}[P_0(\mu)\Psi_0(\tau) + 3P_1(\mu)\Psi_1(\tau)] \tag{9-122}$$

or

$$I(\tau, \mu) = \frac{1}{4\pi}[G(\tau) + 3\mu q^r(\tau)] = \frac{1}{4\pi}\left[G(\tau) - \mu \frac{dG(\tau)}{d\tau}\right] \tag{9-123}$$

since $q^r(\tau)$ is related to $dG(\tau)/d\tau$ by Eq. 9-119b.

Optically Thick Medium

For an optically thick medium (i.e., $\tau \gg 1$) the left-hand side of Eq. 9-121 is negligible and we obtain

$$q^r(\tau) = -\frac{4\pi}{3}\frac{dI_b(T)}{d\tau} \tag{9-124}$$

which is the same as the Rosseland approximation given by Eq. 9-22.

Boundary Conditions for P_N Approximation

Mark [30] and Marshak [31] proposed two different schemes of approximations to represent the boundary conditions for the spherical harmonics method in neutron transport theory. A discussion of Mark's and Marshak's boundary conditions is given, in addition to the original references, in the book by Davison [27, pp. 129–137]. Here we present a brief description of these two boundary conditions.

Mark's Boundary Condition

Consider a slab of optical thickness τ_o with boundary conditions prescribed as

$$I(0, \mu) = f_1(\mu) \quad \text{for } \mu > 0 \tag{9-125}$$

$$I(\tau_0, \mu) = f_2(\mu) \quad \text{for } \mu < 0 \tag{9-126}$$

where $f(\mu)$ is some known function of μ.

Mark suggested that these boundary conditions can be used with the P_N approximation for certain values of μ_i, which are taken to be the roots of the Legendre polynomial $P_{N+1}(\mu)$, that is,

$$P_{N+1}(\mu) = 0 \tag{9-127}$$

Then, boundary conditions 9-125 and 9-126 can be represented as

$$I(0, \mu_i) = f_1(\mu_i), \quad \mu > 0 \tag{9-128a}$$

$$I(\tau_0, \mu_i) = f_2(\mu_i), \quad \mu < 0 \tag{9-128b}$$

where the μ_i are the positive roots of Eq. 9-127.

The limitation to the choice of the functions $f(\mu)$ for use in Mark's boundary condition should be recognized, since the boundary surface intensities $I(0, \mu)$, $\mu > 0$, and $I(\tau_0, \mu)$, $\mu < 0$, are represented by the truncated series 9-116. If $f(\mu)$ is a highly singular function, such as the one that contains the delta function of μ, the convergence of the series is extremely slow. If $f(\mu)$ is a smooth function, the convergence does not pose any problem.

To illustrate the application with the P_1 approximation we consider the following boundary conditions:

$$I(0, \mu) = \frac{n^2\bar{\sigma}T_1^{\,4}}{\pi} \quad \text{for } \mu > 0 \tag{9-129a}$$

$$I(\tau_0, -\mu) = \frac{n^2\bar{\sigma}T_2^{\,4}}{\pi} \quad \text{for } \mu > 0 \tag{9-129b}$$

For the P_1 approximation, these boundary conditions should be satisfied at the values of μ_i that are taken to be the positive roots of

$$P_2(\mu_i) = 3\mu_i^{\,2} - 1 = 0 \tag{9-130a}$$

or

$$\mu_i = \frac{1}{\sqrt{3}} \tag{9-130b}$$

The intensity $I(\tau, \mu)$ is related to $G(\tau)$ by

$$I(\tau, \mu) = \frac{1}{4\pi}\left[G(\tau) - \mu\frac{dG(\tau)}{d\tau}\right] \tag{9-131}$$

Then, from Eqs. 9-129, 9-130, and 9-131, boundary conditions 9-129 are represented in the form

$$\frac{1}{4\pi}\left[G(\tau) - \frac{1}{\sqrt{3}}\frac{dG(\tau)}{d\tau}\right]_{\tau=0} = \frac{n^2\bar{\sigma}T_1^{\,4}}{\pi} \tag{9-132a}$$

$$\frac{1}{4\pi}\left[G(\tau) + \frac{1}{\sqrt{3}}\frac{dG(\tau)}{d\tau}\right]_{\tau=\tau_0} = \frac{n^2\bar{\sigma}T_2^{\,4}}{\pi} \tag{9-132b}$$

Marshak's Boundary Condition

An alternative approach was proposed by Marshak [31] to represent the boundary condition for use with the P_N approximation. Consider the boundary conditions given in the form

$$I(0, \mu) = f_1(\mu) \quad \text{for } \mu > 0 \tag{9-133a}$$

$$I(\tau_0, \mu) = f_2(\mu) \quad \text{for } \mu < 0 \tag{9-133b}$$

where $f(\mu)$ is some known function of μ. Marshak's representation of these boundary conditions for use with the P_N approximation is given in the form

$$\int_0^1 I(0, \mu)\mu^{2i-1}\,d\mu = \int_0^1 f_1(\mu)\mu^{2i-1}\,d\mu \quad \text{for } \mu > 0 \tag{9-134a}$$

and

$$\int_{-1}^0 I(\tau_0, \mu)\mu^{2i-1}\,d\mu = \int_{-1}^0 f_2(\mu)\mu^{2i-1}\,d\mu \quad \text{for } \mu < 0 \tag{9-134b}$$

where $i = 1, 2, 3, \ldots, \frac{1}{2}(N + 1)$.

We illustrate now the application of the Marshak boundary condition with the P_1 approximation. Consider the boundary conditions

$$I(0, \mu) = \varepsilon_1 \frac{\bar{\sigma} T_1^4}{\pi} + \rho_1 I(0, -\mu), \quad \mu > 0 \tag{9-135a}$$

$$I(\tau_0, -\mu) = \varepsilon_2 \frac{\bar{\sigma} T_2^4}{\pi} + \rho_2 I(\tau_0, \mu), \quad \mu > 0 \tag{9-135b}$$

The intensity is related to $G(\tau)$ by

$$I(\tau, \mu) = \frac{1}{4\pi}\left[G(\tau) - \mu \frac{dG(\tau)}{d\tau}\right] \tag{9-136}$$

The application of the Marshak boundary condition to Eqs. 9-135 with $I(\tau, \mu)$ as given by Eq. 9-136 yields

$$\frac{1}{4\pi}\int_0^1 \left[G(\tau) - \mu \frac{dG(\tau)}{d\tau}\right]_{\tau=0} \mu \, d\mu$$
$$= \varepsilon_1 \frac{n^2 \bar{\sigma} T_1^4}{\pi} \int_0^1 \mu \, d\mu + \frac{\rho_1}{4\pi}\int_0^1 \left[G(\tau) + \mu \frac{dG(\tau)}{d\tau}\right]_{\tau=0} \mu \, d\mu \tag{9-137a}$$

$$\frac{1}{4\pi}\int_0^1 \left[G(\tau) + \mu \frac{dG(\tau)}{d\tau}\right]_{\tau=\tau_0} \mu \, d\mu$$
$$= \varepsilon_2 \frac{n^2 \bar{\sigma} T_2^4}{\pi} \int_0^1 \mu \, d\mu + \frac{\rho_2}{4\pi}\int_0^1 \left[G(\tau) - \mu \frac{dG(\tau)}{d\tau}\right] \mu \, d\mu \tag{9-137b}$$

Performing the integrations, we obtain for Eqs. 9-137

$$(1 - \rho_1)G(0) - \tfrac{2}{3}(1 + \rho_1)\frac{dG(0)}{d\tau} = 4\varepsilon_1 n^2 \bar{\sigma} T_1^4 \tag{9-138a}$$

$$(1 - \rho_2)G(\tau_0) + \tfrac{2}{3}(1 + \rho_2)\frac{dG(\tau_0)}{d\tau} = 4\varepsilon_2 n^2 \bar{\sigma} T_2^4 \tag{9-138b}$$

In comparing the accuracy of these two types of boundary conditions it was stated in Davison's book [27, p. 134] that in low-order P_N approximations Marshak's boundary conditions would give better results, but in high-order P_N approximations Mark's boundary conditions would be more accurate. However, more recent numerical work by Pellaud [32] and Schmidt and Gelbard [33] indicates that Marshak's boundary conditions are generally better than Mark's boundary conditions in all P_N approximations.

Another type of boundary condition was obtained independently by Federighi [34] for the general P_N approximation for slabs, spheres, and cylinders and by Pomraning [35] for the P_3 approximation. These authors obtained for the Milne problem in neutron transport theory better results than those obtained with the Marshak conditions. Recently Canosa [36] compared the Federighi and the Marshak boundary conditions in the solution of a neutron critical length problem; he concluded that to obtain exact values with the Marshak condition was difficult because the convergence became very slow for the higher P_N calculations, whereas the Federighi conditions converged to the exact values for relatively low-order P_N computations.

9-8 MOMENT METHOD

The *moment method* described by Krook [22] and the *discrete ordinate method* discussed by Chandrasekhar [2] and Kourganoff [3] provide means to obtain higher-order approximate solutions to the equation of radiative transfer. However, it has been shown by Krook [22] that the moment method, the discrete ordinate method, and the spherical harmonics method are closely related and completely equivalent.

We describe here the derivation of the lowest-order moment method and compare it with the P_1 approximation in order to show the equivalence of the results.

We consider the equation of radiative transfer in the form

$$\frac{dI(s, \hat{\mathbf{\Omega}})}{ds} + \beta I(s, \hat{\mathbf{\Omega}}) = \kappa I_b(T) + \frac{\sigma}{4\pi} \int_{4\pi} I(s, \hat{\mathbf{\Omega}}')\, d\Omega' \tag{9-139}$$

where s is the path measured along the direction of propagation $\hat{\mathbf{\Omega}}$. The directional derivative d/ds in the rectangular coordinate system is related to the partial derivatives with respect to x, y, and z by

$$\frac{d}{ds} = l \frac{\partial}{\partial x} + m \frac{\partial}{\partial y} + n \frac{\partial}{\partial z} \tag{9-140}$$

where l, m, and n are the direction cosines of the direction vector $\hat{\mathbf{\Omega}}$ with the ox, oy, and oz axes, respectively, and are given as

$$l = \sin\theta \cos\phi, \qquad m = \sin\theta \sin\phi, \qquad \text{and} \qquad n = \cos\theta \tag{9-141}$$

Here θ is the polar angle measured from the positive z axis, and ϕ the azimuthal angle. Substitution of Eqs. 9-140 and 9-141 into Eq. 9-139 yields

$$\left(l \frac{\partial I}{\partial x} + m \frac{\partial I}{\partial y} + n \frac{\partial I}{\partial z}\right) + \beta I = \kappa I_b(T) + \frac{\sigma}{4\pi} \int_{4\pi} I\, d\Omega \tag{9-142}$$

where $I \equiv I(x, y, z; l, m, n)$.

The intensity is represented in the form

$$I(x, y, z; l, m, n) = I_0(x, y, z) + a(x, y, z)l + b(x, y, z)m + c(x, y, z)n \tag{9-143a}$$

which is written more compactly as

$$I = I_0 + al + bm + cn \tag{9-143b}$$

where I_0, a, b, and c are functions of position only and are to be determined.

We substitute Eq. 9-143 into Eq. 9-142, integrate the resulting expression over all solid angles (i.e., $\Omega = 4\pi$), and obtain[3]

$$\frac{\partial a}{\partial x} + \frac{\partial b}{\partial y} + \frac{\partial c}{\partial z} = 3\kappa[I_b(T) - I_0] \tag{9-144}$$

Equation 9-143b is substituted into Eq. 9-142, the resulting expression is multiplied successively by l, m, and n, and for each of these cases the resulting expression is integrated over a solid angle 4π. We obtain

$$a = -\frac{1}{\beta}\frac{\partial I_0}{\partial x} \tag{9-145a}$$

$$b = -\frac{1}{\beta}\frac{\partial I_0}{\partial y} \tag{9-145b}$$

$$c = -\frac{1}{\beta}\frac{\partial I_0}{\partial z} \tag{9-145c}$$

Thus we have related the coefficients a, b, and c to the function I_0. Substitution of Eqs. 9-145 into Eq. 9-143 yields the following expression for the radiation intensity $I(x, y, z; l, m, n)$ in terms of the function $I_0(x, y, z)$:

$$I(x, y, z; l, m, n) = I_0(x, y, z) - \frac{1}{\beta}\left(l\frac{\partial I_0}{\partial x} + m\frac{\partial I_0}{\partial y} + n\frac{\partial I_0}{\partial z}\right) \tag{9-146}$$

Thus, if the function $I_0(x, y, z)$ is known, the distribution of radiation intensity is determined from Eq. 9-146.

Substitution of Eq. 9-145 into Eq. 9-144 yields the following partial differential equation for $I_0(x, y, z)$:

$$\frac{1}{\beta}\nabla^2 I_0(x, y, z) = 3\kappa[I_0(x, y, z) - I_b(T)] \tag{9-147}$$

where

$$\nabla^2 \equiv \frac{\partial^2}{\partial x^2} + \frac{\partial^2}{\partial y^2} + \frac{\partial^2}{\partial z^2}$$

The function $I_0(x, y, z)$ is related to the incident radiation G by integrating I over all solid angles. We obtain

$$G = \int_{4\pi} I \, d\Omega = \int_{\phi=0}^{2\pi} \int_{\theta=0}^{\pi} (I_0 + al + bm + cn) \sin\theta \, d\theta \, d\phi = 4\pi I_0 \quad \text{(9-148a)}$$

or

$$I_0 = \frac{G}{4\pi} \quad \text{(9-148b)}$$

Substitution of Eq. 9-148b into Eq. 9-147 yields

$$\frac{1}{\beta} \nabla^2 G(x, y, z) = 3\kappa[G(x, y, z) - 4\pi I_b(T)] \quad \text{(9-149)}$$

For the one-dimensional, plane-parallel medium Eq. 9-149 simplifies to

$$\frac{1}{\beta} \frac{d^2G(z)}{dz^2} = 3\kappa[G(z) - 4\pi I_b(T)] \quad \text{(9-150a)}$$

or

$$\frac{d^2G}{d\tau^2} = 3(1 - \omega)[G - 4\pi I_b(T)] \quad \text{(9-150b)}$$

where we have defined

$$d\tau = \beta \, dz \qquad \text{and} \qquad 1 - \omega = \frac{\kappa}{\beta}$$

We note that Eq. 9-150b is exactly the same as Eq. 9-120 obtained by the P_1 approximation.

In the problems of interaction of radiation with conduction and convection the energy equation includes the radiative heat flux term $\nabla \cdot \mathbf{q}^r$, which can be related to I_0 or G as now described.

$$\nabla \cdot \mathbf{q}^r = \int_{4\pi} \nabla \cdot \hat{\mathbf{\Omega}} I \, d\Omega \quad \text{(9-151)}$$

where

$$\nabla \equiv \hat{\mathbf{i}} \frac{\partial}{\partial x} + \hat{\mathbf{j}} \frac{\partial}{\partial y} + \hat{\mathbf{k}} \frac{\partial}{\partial z} \quad \text{(9-152a)}$$

$$\hat{\mathbf{\Omega}} \equiv \hat{\mathbf{i}} l + \hat{\mathbf{j}} m + \hat{\mathbf{k}} n, \qquad I = I_0 + al + bm + cn \quad \text{(9-152b)}$$

Substitution of Eqs. 9-152 into 9-151 yields

$$\nabla \cdot \mathbf{q}^r = \frac{4\pi}{3} \left(\frac{\partial a}{\partial x} + \frac{\partial b}{\partial y} + \frac{\partial c}{\partial z} \right) \quad \text{(9-153)}$$

By utilizing Eqs. 9-145 and assuming constant β, we can write Eq. 9-153 as

$$\nabla \cdot \mathbf{q}^r = -\frac{4\pi}{3\beta}\left(\frac{\partial^2 I_0}{\partial x^2} + \frac{\partial^2 I_0}{\partial y^2} + \frac{\partial^2 I_0}{\partial z^2}\right) = -\frac{4\pi}{3\beta}\nabla^2 I_0 \tag{9-154}$$

and in view of Eq. 9-148b this relation becomes

$$\nabla \cdot \mathbf{q}^r = -\frac{1}{3\beta}\nabla^2 G \tag{9-155}$$

We note that for the one-dimensional case Eq. 9-155 reduces to the expression given by Eq. 9-119b, obtained by the P_1 approximation.

The moments method has been applied by Edwards and Bobco [37], Bobco [38], and Cheng [39] in the solution of the problem of radiative heat transfer.

REFERENCES

1. R. Viskanta, "Radiation Transfer and Interaction of Convection with Radiation Heat Transfer," in *Advances in Heat Transfer*, edited by T. F. Irvine and J. P. Hartnett, Vol. 3, pp. 175–251, Academic Press, New York, 1966.
2. S. Chandrasekhar, *Radiative Transfer*, Oxford University Press, London, 1950; also Dover Publications, New York, 1960.
3. V. Kourganoff, *Basic Methods in Transfer Problems*, Dover Publications, New York, 1963.
4. R. v. d. R. Woolley and D. W. N. Stibbs, *The Outer Layers of a Star*, Oxford University Press, London, 1953.
5. E. M. Sparrow and R. D. Cess, *Radiation Heat Transfer*, Brooks/Cole Publishing Co., Belmont, Calif., 1966.
6. S. Rosseland, *Theoretical Astrophysics*, Oxford University Press, London, 1936.
7. R. G. Deissler, "Diffusion Approximation for Thermal Radiation in Gases with Jump Boundary Conditions," *J. Heat Transfer*, **86C**, 240–246, 1964.
8. S. N. Shorin, "Heat Exchange by Radiation in the Presence of an Absorbing Medium," *Izv. Akad. Nauk SSSR*, No. 3, pp. 389–406, 1951; abridged translation in *Engineer's Digest*, Vol. 12, pp. 324–328, 1951.
9. C. M. Usiskin and E. M. Sparrow, "Thermal Radiation Between Parallel Plates Separated by an Absorbing-Emitting Nonisothermal Gas," *Intern. J. Heat Mass Transfer*, **1**, 28–36, 1960.
10. M. A. Heaslet and R. F. Warming, "Radiative Transfer and Wall Temperature Slip in an Absorbing Planar Medium," *Intern. J. Heat Mass Transfer*, **8**, 979–994, 1965.
11. H. C. Hottel, "Radiation as a Diffusion Process," *Intern. J. Heat Mass Transfer*, **5**, 82–83, 1962.
12. M. Perlmutter and J. R. Howell, "Radiant Transfer through a Gray Gas Between Concentric Cylinders Using Monte Carlo," *J. Heat Transfer*, **86C**, 169–179, 1964.

13. A. S. Eddington, *The Internal Constitution of Stars*, Cambridge University Press, London, 1926; also Dover Publications, New York, 1960.

14. G. C. Pomraning, "An Extension of the Eddington Approximation," *J. Quant. Spectry Radiative Transfer*, **9**, 407–422, 1969.

15. R. M. Goody, "The Influence of Radiative Transfer on Cellular Convection," *J. Fluid Mech.*, **1**, 424–435, 1956.

16. R. Viskanta and R. J. Grosh, "Heat Transfer by Simultaneous Conduction and Radiation in an Absorbing Medium," *J. Heat Transfer*, **84C**, 63–72, 1962.

17. A. Schuster, "Radiation Through a Foggy Atmosphere," *Astrophys. J.*, **21**, 1–22, 1905.

18. K. Schwarzchild, "Ueber das Gleichgewicht der Sonneatmosphare," *Akad. Wiss. Gottingen, Math.-Phys. Kl. Nachr.*, **1**, 41–53, 1906.

19. V. V. Sobolev, *A Treatise on Radiative Transfer*, D. van Nostrand Co., Princeton, N.J., 1963.

20. B. K. Larkin and S. W. Churchill, "Heat Transfer by Radiation through Porous Insulations," *A.I.Ch.E. J.*, **5**, pp. 467–474, 1959.

21. J. C. Chen, "Simultaneous Radiative and Conductive Heat Transfer in an Absorbing, Emitting and Scattering Medium in Slug Flow," *A.I.Ch.E. J.*, **10**, 253–259, 1964.

22. Max Krook, "On the Solution of Equation of Transfer, I," *Astrophys. J.*, **122**, 488–497, 1955.

23. W. Lick, "Energy Transfer by Radiation and Conduction," *Proceedings of the Heat Transfer and Fluid Mechanics Institute*, pp. 14–26, Stanford University Press, Palo Alto, Calif., 1963.

24. W. G. Vincenti and B. S. Baldwin, Jr., "Effects of Thermal Radiation on the Propagation of Plane Acoustic Waves," *J. Fluid Mech.*, **12**, pp. 449–477, 1962.

25. S. S. Murty, "Approximations on Angular Distribution of Thermal Radiation," *Intern. J. Heat Mass Transfer*, **8**, 1203–1208, 1965.

26. J. H. Jeans, "The Equations of Radiative Transfer of Energy," *Monthly Notices Roy. Astron. Soc.*, **78**, 28–36, 1917.

27. B. Davison, *Neutron Transport Theory*, Oxford University Press, London, 1958.

28. R. L. Murray, *Nuclear Reactor Physics*, Prentice-Hall, Englewood Cliffs, N.J., 1957.

29. E. T. Whittaker and G. N. Watson, *A Course of Modern Analysis*, 4th ed., Cambridge University Press, London, 1965.

30. J. C. Mark, *The Spherical Harmonics Method*, Parts I and II. National Research Council of Canada, Atomic Energy Reports No. MT 92 (1944) and MT 97 (1945).

31. R. E. Marshak, "Note on the Spherical Harmonics Method as Applied to the Milne Problem for a Sphere," *Phys. Rev.*, **71**, 443–446, 1947.

32. Bruno Pellaud, "Numerical Comparison of Different Types of Vacuum Boundary Conditions for the P_N Approximation," *Trans. Am. Nucl. Soc.*, **9**, 434–435, 1966.

33. E. Schmidt and E. M. Gelbard, "A Double P_N Method for Spheres and Cylinders," *Trans. Am. Nucl. Soc.*, **9**, 432–433, 1966.

34. F. D. Federighi, "Vacuum Boundary Conditions for Spherical Harmonics Methods," *Nucleonics*, **6**, 277–285, 1964.

35. G. C. Pomraning, "An Improved Three-Surface Boundary Condition for the P-3 Approximation," *Nucl. Sci. Eng.*, **18**, 528–530, 1964.

36. Jose Canosa, "A Comparison of Federighi's with Marshak's Boundary Conditions," *Nucl. Sci. Eng.*, **43,** 349–350, 1971.

37. R. H. Edwards and R. P. Bobco, "Radiant Heat Transfer from Isothermal Dispersions with Isotropic Scattering," *ASME Paper* No. 67-HT-8, 1967.

38. R. P. Bobco, "Directional Emissivities from a Two-Dimensional, Absorbing-Scattering Medium: the Semi-Infinite Slab," *ASME Paper* No. 67-HT-12, 1967.

39. Ping Cheng, "Two-Dimensional Radiating Gas Flow by a Moment Method," *AIAA J.*, **2,** 1662–1664, 1964.

NOTES

[1] Here we have considered energy transfer by pure radiation. If the energy transfer were by simultaneous radiation and conduction, the temperature would be continuous at the wall for any value of the optical thickness τ_0 because of the requirement of the conduction boundary condition.

[2] In Chapter 8 we expanded $p(\mu, \mu')$ in the form (see Eq. 8-42b)

$$p(\mu, \mu') = \sum_{n=0}^{\infty} a_n P_n(\mu) P_n(\mu'), \quad a_0 = 1$$

By comparing this expansion with Eq. 9-99 we note that

$$a_n = (2n + 1) f_n$$

[3] When Eq. 9-143b is substituted into Eq. 9-142, the integral on the right-hand side of the latter equation is evaluated as

$$\int_{4\pi} I\, d\Omega = \int_{\phi=0}^{2\pi} \int_{\theta=0}^{\pi} (I_0 + al + bm + cn) \sin\theta\, d\theta\, d\phi = 4\pi I_0 \tag{1}$$

To obtain this result we utilized the relations given by Eq. 9-141 for l, m, and n and noted that the integration of $\sin\phi$, $\cos\phi$ over 2π and the integration of $\sin\theta$, $\cos\theta$ over π vanish. Then Eq. 9-142 becomes

$$\left(l\frac{\partial I_0}{\partial x} + l^2\frac{\partial a}{\partial x} + lm\frac{\partial b}{\partial x} + ln\frac{\partial c}{\partial x}\right) + \left(m\frac{\partial I_0}{\partial y} + ml\frac{\partial a}{\partial y} + m^2\frac{\partial b}{\partial y} + mn\frac{\partial c}{\partial y}\right)$$
$$+ \left(n\frac{\partial I_0}{\partial z} + nl\frac{\partial a}{\partial z} + nm\frac{\partial b}{\partial z} + n^2\frac{\partial c}{\partial z}\right) + \beta(I_0 + al + bm + cn)$$
$$= \kappa I_b(T) + \sigma I_0 \tag{2}$$

When Eq. 2 is integrated over all solid angles, the terms involving lm, ln, mn, l, m, and n vanish, and the integration of l^2, m^2, and n^2 each contribute $4\pi/3$. Then Eq. 2 becomes

$$\frac{4\pi}{3}\left(\frac{\partial a}{\partial x} + \frac{\partial b}{\partial y} + \frac{\partial c}{\partial z}\right) + 4\pi\beta I_0 = 4\pi\kappa I_b(T) + 4\pi\sigma I_0 \tag{3}$$

or

$$\frac{\partial a}{\partial x} + \frac{\partial b}{\partial y} + \frac{\partial c}{\partial z} = \tfrac{4}{3}\kappa[I_b(T) - I_0] \tag{4}$$

which is the same as Eq. 9-144.

CHAPTER 10

Case's Normal-Mode Expansion Technique in the Solution of the Equation of Radiative Transfer

The normal-mode expansion technique, introduced by Case [1] in 1960 to solve rigorously the one-dimensional, plane-parallel neutron transport equation, also provides a systematic approach to one-dimensional, plane-parallel, radiative transfer problems. In this method the desired solution of the equation of radiative transfer is written as a linear sum of the normal modes (i.e., eigenfunctions) of the homogeneous part of the equation of radiative transfer and a particular solution appropriate to the inhomogeneous term of interest. The unknown expansion coefficients associated with the homogeneous solution are then determined by constraining the complete solution to satisfy the boundary conditions for the problem and by utilizing the orthogonality of the normal modes and various normalization integrals. The approach is analogous to the classical orthogonal expansion technique.

In this chapter we present an introductory treatment of the Case technique as applied to the solution of the equation of radiative transfer in one-dimensional, plane-parallel geometry for a gray medium with isotropic scattering. The normal modes of the homogeneous equation are given, the orthogonality properties of the normal modes and various normalization integrals are presented, and the representation of an arbitrary function in terms of the normal modes is described. The reader should refer to the book by Case and Zweifel [2] for a comprehensive treatment of the theory and application of this method and for a rather broad survey of the original papers on the subject. Several recent works on the Case technique have been collected by Inönü and Zweifel [3]. The technique has been extended by Mika [4] and Shure and Natelson [5] to include the case of anisotropic scattering. Several useful orthogonality relations for normal modes are given by Kuščer, McCormick, and Summerfield [6] and by McCormick and Kuščer [7, 8]. We do not intend to cite here numerous references on the application of the Case technique in the field of neutron transport theory but shall mention some of the pertinent references in the area of radiative transfer.

Ferziger and Simmons [9], Kriese and Siewert [10], Özişik and Siewert [11], and Beach, Özişik, and Siewert [12] applied this technique to solve radiative heat transfer in a plane-parallel gray slab. Case's method has been used by Siewert and Zweifel [13, 14] to construct the normal-mode solution to the set of N coupled integral equations encountered in the treatment of nongray radiative heat transfer in the picket-fence model. Simmons and Ferziger [15], Siewert and Özişik [16], and Reith, Siewert, and Özişik [17] solved nongray radiative heat transfer problems appropriate to the picket-fence model.

In the following analysis we are concerned only with the treatment of a *one-dimensional equation of radiative transfer in plane-parallel geometry for a gray medium and isotropic scattering* with the purpose of introducing this new and powerful technique to those interested in the heat transfer area. The extension of this method to higher-order anisotropy, nongray cases in the picket-fence model, and multidimensional and other coordinate systems involves more complicated analysis and therefore is not treated here. The reader interested in such applications should refer to the original work in these areas. An excellent review of all the work performed until 1972 on the normal mode expansion technique in neutron transport theory was given recently by McCormick and Kuščer [18].

10-1 NORMAL MODES OF THE HOMOGENEOUS EQUATION IN ISOTROPIC SCATTERING

Consider the equation of radiative transfer for a one-dimensional, plane-parallel, isotropically scattering, gray medium, given as

$$\mu \frac{\partial I(\tau, \mu)}{\partial \tau} + I(\tau, \mu) = (1 - \omega) \frac{n^2 \bar{\sigma} T^4(\tau)}{\pi} + \frac{\omega}{2} \int_{-1}^{1} I(\tau, \mu')\, d\mu' \quad (10\text{-}1)$$

where $I(\tau, \mu)$ is the radiation intensity, ω is the single scattering albedo, n is the refractive index, $T(\tau)$ is the temperature, τ is the optical variable, and μ is the cosine of the angle between the direction of radiation intensity and the positive τ axis.

The complete solution of this equation can be written in the form

$$I(\tau, \mu) = \psi(\tau, \mu) + I_p(\tau, \mu) \tag{10-2}$$

where $I_p(\tau, \mu)$ is a particular solution of Eq. 10-1 (i.e., a solution that satisfies the inhomogeneous equation 10-1 but not necessarily its boundary conditions), and its determination depends on the type of inhomogeneous term. At the end of this chapter we discuss the determination of a particular solution of Eq. 10-1 for different types of inhomogeneous terms; however, at this point in the analysis we assume that a particular solution is available and $I_p(\tau, \mu)$ is a known function. The function $\psi(\tau, \mu)$ is the solution of the homogeneous part of Eq. 10-1, that is, $\psi(\tau, \mu)$ satisfies the following equation:

$$\mu \frac{\partial \psi(\tau, \mu)}{\partial \tau} + \psi(\tau, \mu) = \frac{\omega}{2} \int_{-1}^{1} \psi(\tau, \mu')\, d\mu' \tag{10-3}$$

Once the normal modes (eigenfunctions) and the eigenvalues of this homogeneous equation are determined, the complete solution for $\psi(\tau, \mu)$ is constructed by the superposition of all permissible elementary solutions. The expansion coefficients associated with the homogeneous solution can then be determined by requiring that the complete solution given formally by Eq. 10-2 satisfy the boundary conditions for the problem. Therefore the first step in the analysis is the determination of the normal modes and the eigenvalues of the homogeneous equation 10-3; following Case [1], we present now the determination of these eigenvalues and eigenfunctions.

The function $\psi(\tau, \mu)$ is separated in the form

$$\psi(\tau, \mu) = e^{-\tau/\eta} \phi(\eta, \mu) \tag{10-4}$$

where $\phi(\eta, \mu)$ is the normal mode, and η is the eigenvalue of the homogeneous equation. Substitution of Eq. 10-4 into the integral equation 10-3 yields

$$\left(1 - \frac{\mu}{\eta}\right) \phi(\eta, \mu) = \frac{\omega}{2} \int_{-1}^{1} \phi(\eta, \mu')\, d\mu' \tag{10-5}$$

We note that the normal mode $\phi(\eta, \mu)$ is arbitrary within a multiplication constant, since Eq. 10-5 is homogeneous. This arbitrariness can be removed by normalizing $\phi(\eta, \mu)$ as

$$\int_{-1}^{1} \phi(\eta, \mu')\, d\mu' = 1 \tag{10-6}$$

provided that this integral never vanishes. It has been shown [2, p. 60] that the integral in Eq. 10-6 indeed does not vanish; hence the above normalization is always valid. By introducing this normalization condition into Eq. 10-5 we obtain

$$\left(1 - \frac{\mu}{\eta}\right)\phi(\eta, \mu) = \frac{\omega}{2} \tag{10-7}$$

Here we assume that the values of ω lie in the range 0 to 1, which covers all the cases of interest in radiative heat transfer[1]; μ takes values from -1 to 1.

To solve Eq. 10-7 for $\phi(\eta, \mu)$ we consider the following two cases separately: (a) η does not lie between -1 and 1, and (b) η lies between -1 and 1. The reason for this is that the latter case involves singularity at $\eta = \mu$ and hence its solution requires special treatment.

(a) η Does Not Lie Between -1 and 1

To distinguish the normal modes and the eigenvalues for this case, the notation $\eta = \eta_0$ will be used whenever η is not contained in $(-1, 1)$. The solution of Eq. 10-7 yields the *discrete normal modes* $\phi(\eta_0, \mu)$ of the homogeneous equation 10-3 as

$$\phi(\eta_0, \mu) = \frac{\omega\eta_0}{2}\frac{1}{\eta_0 - \mu}, \quad \eta_0 \notin (-1, 1) \tag{10-8}$$

A dispersion relation is now determined to evaluate the *discrete eigenvalues* η_0. The integration of Eq. 10-8 over μ from $\mu = -1$ to 1 and the application of the normalization condition 10-6 gives

$$\Lambda(\eta_0) \equiv 1 - \frac{\omega\eta_0}{2}\int_{-1}^{1}\frac{1}{\eta_0 - \mu}\,d\mu = 0 \tag{10-9a}$$

The integration of this equation yields

$$\Lambda(\eta_0) \equiv 1 - \frac{\omega\eta_0}{2}\ln\left[\frac{1 + (1/\eta_0)}{1 - (1/\eta_0)}\right] = 0 \tag{10-9b}$$

which can be written alternatively as

$$\Lambda(\eta_0) \equiv 1 - \omega\eta_0 \tanh^{-1}\left(\frac{1}{\eta_0}\right) = 0 \tag{10-9c}$$

Since

$$2\tanh^{-1} x = \ln\left(\frac{1 + x}{1 - x}\right) \quad \text{for } |x| < 1$$

Equation 10-9 is the desired dispersion relation for the determination of discrete eigenvalues η_0. Case [1, 2] has shown that for a given ω, when $0 \leq \omega < 1$, this dispersion relation has two roots which are real and appear in pairs, that is, $\pm\eta_0$. Then the corresponding two discrete normal modes $\phi(\pm\eta_0, \mu)$ are

$$\phi(\pm\eta_0, \mu) = \frac{\omega\eta_0}{2} \frac{1}{\eta_0 \mp \mu}, \quad \omega \neq 1 \tag{10-10}$$

and the two discrete solutions of the homogeneous equation 10-3 are given as

$$\psi_1(\tau, \mu) = e^{-\tau/\eta_0}\phi(\eta_0, \mu), \quad \omega \neq 1 \tag{10-11a}$$

$$\psi_2(\tau, \mu) = e^{\tau/\eta_0}\phi(-\eta_0, \mu), \quad \omega \neq 1 \tag{10-11b}$$

For $\omega = 1$, the discrete eigenvalues $\pm\eta_0$ both become infinite and the two discrete normal modes $\phi(\pm\eta_0, \mu)$ degenerate into a single one:

$$\lim_{\eta_0 \to \infty} \phi(\pm\eta_0, \mu) = \lim_{\eta_0 \to \infty} \frac{\omega\eta_0}{2} \frac{1}{\eta_0 \mp \mu} = \frac{1}{2} \tag{10-12}$$

However, it has been shown [2, p. 298; 3, p. 46] that the two discrete solutions of Eq. 10-3, for $\omega = 1$, are

$$\psi_1(\tau, \mu) = \tfrac{1}{2}, \quad \omega = 1 \tag{10-13a}$$

$$\psi_2(\tau, \mu) = \tfrac{1}{2}(\tau - \mu), \quad \omega = 1 \tag{10-13b}$$

For a given ω the discrete eigenvalues η_0 are determined from the dispersion relation 10-9. For the cases $\omega \ll 1$ and $\omega \to 1$, simple explicit relations can be obtained from Eq.10-9 to evaluate η_0 approximately. As ω approaches zero, η_0 approaches unity; therefore, for small values of ω, Eq. 10-9b is simplified as

$$1 - \frac{\omega}{2} \ln\left[\frac{2}{1 - (1/\eta_0)}\right] = 0, \quad \omega \ll 1 \tag{10-14a}$$

or

$$\ln\left[\frac{1}{2}\left(1 - \frac{1}{\eta_0}\right)\right] = -\frac{2}{\omega}, \quad \omega \ll 1 \tag{10-14b}$$

or

$$\frac{1}{\eta_0} = 1 - 2e^{-2/\omega}, \quad \omega \ll 1 \tag{10-14c}$$

Equation 10-14c is the desired relation to approximate η_0 for small values of ω.

As ω approaches unity, η_0 becomes large and $1/\eta_0$ becomes very small. Then, by expanding $\tanh^{-1}(1/\eta_0)$ in powers of $1/\eta_0$ and retaining only the first two terms of the expansion, we can simplify Eq. 10-9c to

$$1 - \omega\eta_0\left[\frac{1}{\eta_0} + \frac{1}{3}\left(\frac{1}{\eta_0}\right)^2\right] = 0, \quad \omega \to 1 \tag{10-15a}$$

By setting $\omega = 1$ in the term involving $1/\eta_0^2$, we obtain

$$\frac{1}{\eta_0} = \sqrt{3(1 - \omega)}, \quad \omega \to 1 \tag{10-15b}$$

which gives η_0 for the values of ω near unity. Alternatively, the following relation can be written from Eq. 10-15b:

$$\lim_{\omega \to 1} \eta_0^2(1 - \omega) = \tfrac{1}{3} \tag{10-15c}$$

Table 10-1 gives the values of the discrete eigenvalues η_0 for $0 < \omega < 1$.

(b) η Lies Between -1 and 1

When η takes values between -1 and 1, Eq. 10-7 is singular for $\eta = \mu$; in this case the most general solution of Eq. 10-7 is given as [1, 2]

$$\phi(\eta, \mu) = \frac{\omega\eta}{2} \frac{P}{\eta - \mu} + \lambda(\eta)\, \delta(\eta - \mu), \quad \eta \in (-1, 1) \tag{10-16}$$

where $\delta(\eta - \mu)$ is the Dirac delta function, $\lambda(\eta)$ is an arbitrary function to be determined, and the symbol P denotes that, when Eq. 10-16 is integrated over η or μ, the integration involving the term $1/(\eta - \mu)$ should be evaluated in the Cauchy principal-value integral sense.[2]

To determine the function $\lambda(\eta)$ we integrate Eq. 10-16 over μ from -1 to 1, utilize the normalization condition given by Eq. 10-6, and obtain

$$\lambda(\eta) = 1 - \frac{\omega\eta}{2} \int_{-1}^{1} P \frac{d\mu}{\eta - \mu} \tag{10-17a}$$

This relation establishes the function $\lambda(\eta)$ for use in Eq. 10-16. Then any value of η in the interval between -1 and 1 is an eigenvalue, and the function $\phi(\eta, \mu)$ as defined by Eq. 10-16 is called the *continuum normal mode* since η takes all values between -1 and 1 continuously.

The principal-value integration in Eq. 10-17a is performed to yield

$$\lambda(\eta) = 1 - \tfrac{1}{2}\omega\eta \ln \left(\frac{1 + \eta}{1 - \eta}\right) \tag{10-17b}$$

which is written alternatively as

$$\lambda(\eta) = 1 - \omega\eta \tanh^{-1} \eta \tag{10-17c}$$

Complete Solution of the Homogeneous Equation

The complete solution of the homogeneous integral equation 10-3 is obtained by taking a linear sum of the two discrete solutions for $\eta_0 \notin (-1, 1)$

Table 10-1 Discrete Eignevalues η_0 for Isotropic Scattering

ω	η_0	ω	η_0	ω	η_0
0.15	1.000003	0.41	1.01662	0.71	1.22123
0.16	1.000007	0.42	1.01884	0.72	1.23662
0.17	1.000016	0.43	1.02126	0.73	1.25305
0.18	1.000030	0.44	1.02338	0.74	1.27063
0.19	1.000054	0.45	1.02671	0.75	1.28946
0.20	1.000091	0.46	1.02977	0.76	1.30967
0.21	1.000146	0.47	1.03305	0.77	1.33141
0.22	1.000226	0.48	1.03657	0.78	1.35485
0.23	1.000336	0.49	1.04034	0.79	1.38018
0.24	1.000483	0.50	1.04438	0.80	1.40763
0.25	1.000675	0.51	1.04868	0.81	1.43748
0.26	1.000920	0.52	1.05328	0.82	1.47006
0.27	1.001225	0.53	1.05817	0.83	1.50576
0.28	1.001600	0.54	1.06338	0.84	1.54506
0.29	1.002053	0.55	1.06891	0.85	1.58855
0.30	1.002593	0.56	0.07479	0.86	1.63697
0.31	1.003227	0.57	1.08103	0.87	1.69124
0.32	1.003965	0.58	1.08765	0.88	1.75256
0.33	1.004815	0.59	1.09468	0.89	1.82251
0.34	1.005785	0.60	1.10213	0.90	1.90320
0.35	1.006883	0.61	1.11003	0.91	1.99754
0.36	1.008118	0.62	1.11840	0.92	2.10969
0.37	1.009498	0.63	1.12729	0.93	2.24582
0.38	1.011031	0.64	1.13671	0.94	2.41560
0.39	1.012724	0.65	1.14670	0.95	2.63514
0.40	1.014586	0.66	1.15730	0.96	2.93402
		0.67	1.16857	0.97	3.37403
		0.68	1.18053	0.98	4.11552
		0.69	1.19326	0.99	5.79672
		0.70	1.20680	1.00	∞

and all possible continuum solutions for $\eta \in (-1, 1)$. We consider the cases for $\omega < 1$ and $\omega = 1$ separately.

(a) $\omega < 1$

The complete solution of Eq. 10-3 can be written in the form

$$\psi(\tau, \mu) = A(\eta_0)\phi(\eta_0, \mu)e^{-\tau/\eta_0} + A(-\eta_0)\phi(-\eta_0, \mu)e^{\tau/\eta_0} + \int_{-1}^{1} A(\eta)\phi(\eta, \mu)e^{-\tau/\eta}\, d\eta \quad \text{for } \omega < 1 \qquad (10\text{-}18a)$$

where $A(\eta_0)$, $A(-\eta_0)$ and $A(\eta)$ are arbitrary expansion coefficients. The solution given in this form is useful for the problems of an infinite medium (i.e., $-\infty < \tau < \infty$) that require the expansion of a function in terms of the normal modes over the full-range of μ (i.e., $-1 \le \mu \le 1$). However, in the treatment of problems of a half-space or a slab, an arbitrary function is generally expanded in terms of the normal modes over the half-range of μ. For such cases it is desirable to separate the integral into two parts and write Eq. 10-18a in the alternative form as

$$\psi(\tau, \mu) = A(\eta_0)\phi(\eta_0, \mu)e^{-\tau/\eta_0} + A(-\eta_0)\phi(-\eta_0, \mu)e^{\tau/\eta_0} + \int_0^1 A(\eta)\phi(\eta, \mu)e^{-\tau/\eta}\, d\eta + \int_0^1 A(-\eta)\phi(-\eta, \mu)e^{\tau/\eta}\, d\eta \quad \text{for } \omega < 1 \tag{10-18b}$$

where $A(\eta_0)$, $A(-\eta_0)$, $A(\eta)$, and $A(-\eta)$ are arbitrary expansion coefficients. For convenience, we summarize below the normal modes defined previously:

$$\phi(\pm\eta_0, \mu) = \frac{\omega\eta_0}{2}\frac{1}{\eta_0 \mp \mu}, \qquad \eta_0 \notin (-1, 1) \tag{10-18c}$$

$$\phi(\eta, \mu) = \frac{\omega\eta}{2}\frac{P}{\eta - \mu} + \lambda(\eta)\,\delta(\eta - \mu), \quad \eta \in (-1, 1) \tag{10-18d}$$

where

$$\lambda(\eta) = 1 - \omega\eta \tanh^{-1}\eta \tag{10-18e}$$

and the discrete eigenvalues η_0 are the two zeros of the dispersion function

$$\Lambda(\eta_0) \equiv 1 - \omega\eta_0 \tanh^{-1}\frac{1}{\eta_0} \tag{10-18f}$$

We observe the following symmetry property of the normal modes:

$$\phi(\xi, \mu) = \phi(-\xi, -\mu) \tag{10-19a}$$

$$\phi(-\xi, \mu) = \phi(\xi, -\mu) \tag{10-19b}$$

where

$$\xi = \eta_0 \qquad \text{or} \qquad \eta \in (-1, 1)$$

(b) $\omega = 1$

The complete solution of Eq. 10-3 is obtained by taking a linear sum of the discrete solutions given by Eqs. 10-13 and the continuum solutions. We obtain

$$\psi(\tau, \mu) = A\tfrac{1}{2} + B\tfrac{1}{2}(\tau - \mu) + \int_{-1}^1 A(\eta)\phi(\eta, \mu)e^{-\tau/\eta}\, d\eta, \quad \omega = 1 \tag{10-20a}$$

where A, B and $A(\eta)$ are arbitrary expansion coefficients. The solution given in this form is useful for the problems of an infinite medium. However, in the treatment of problems of a half-space or a slab, an arbitrary function is generally expanded in terms of the normal modes over the half-range of μ. For such cases it is desirable to separate the integral into two parts and write Eq. 10-20a in the form

$$\psi(\tau, \mu) = A\tfrac{1}{2} + B\tfrac{1}{2}(\tau - \mu) + \int_0^1 A(\eta)\phi(\eta, \mu)e^{-\tau/\eta}\, d\eta$$
$$+ \int_0^1 A(-\eta)\phi(-\eta, \mu)e^{\tau/\eta}\, d\eta, \quad \omega = 1 \quad (10\text{-}20b)$$

where A, B, $A(\eta)$, and $A(-\eta)$ are arbitrary expansion coefficients. The normal mode $\phi(\eta, \mu)$ is obtained from Eqs. 10-18d and 10-18e by setting $\omega = 1$.

10-2 EXPANSION OF AN ARBITRARY FUNCTION IN TERMS OF NORMAL MODES

In the solution of the equation of radiative transfer by the normal-mode expansion technique it is generally required that an arbitrary function $f(\mu)$ be expanded in terms of the normal modes of the homogeneous equation over the full range of μ (i.e., $-1 \leq \mu \leq 1$) or over the half-range of μ (i.e., $0 \leq \mu \leq 1$). The validity of such expansions is ensured by the full-range and half-range completeness theorems proved by Case. We present now a statement of these completeness theorems; the reader should refer to the original publications [1, 2] for their proof.

Full-Range Completeness Theorem

A suitably well-behaved function[3] $f(\mu)$, defined in the full range of $\mu(-1 \leq \mu \leq 1)$, can be expanded in terms of the normal modes of the homogeneous equation 10-3 in the form

$$f(\mu) = A(\eta_0)\phi(\eta_0, \mu) + A(-\eta_0)\phi(-\eta_0, \mu) + \int_{-1}^1 A(\eta)\phi(\eta, \mu)\, d\eta$$
$$\text{for } \mu \in (-1, 1), \quad \omega < 1 \quad (10\text{-}21a)$$

and

$$f(\mu) = A\tfrac{1}{2} + B\tfrac{1}{2}\mu + \int_{-1}^1 A(\eta)\phi(\eta, \mu)\, d\eta \quad \text{for } \mu \in (-1, 1), \quad \omega = 1 \quad (10\text{-}21b)$$

where $A(\eta_0)$, $A(-\eta_0)$, $A(\eta)$, A, and B are the expansion coefficients.

Half-Range Completeness Theorem

A suitably well-behaved function $f(\mu)$, defined over the positive half-range of $\mu (0 \leq \mu \leq 1)$, can be expanded in terms of the normal modes in the form

$$f(\mu) = A(\eta_0)\phi(\eta_0, \mu) + \int_0^1 A(\eta)\phi(\eta, \mu)\, d\eta \quad \text{for } \mu \in (0, 1), \quad \omega < 1 \tag{10-22a}$$

and

$$f(\mu) = A\tfrac{1}{2} + \int_0^1 A(\eta)\phi(\eta, \mu)\, d\eta \quad \text{for } \mu \in (0, 1), \quad \omega = 1 \tag{10-22b}$$

If the function $f(\mu)$ were defined in the negative half-range of μ (i.e., $-1 \leq \mu \leq 0$), the expansion would be given in the form

$$f(\mu) = A(-\eta_0)\phi(-\eta_0, \mu) + \int_{-1}^0 A(\eta)\phi(\eta, \mu)\, d\eta \quad \text{for } \mu \in (-1, 0), \quad \omega < 1 \tag{10-22c}$$

10-3 THE ORTHOGONALITY OF NORMAL MODES

The orthogonality of normal modes constitutes the basis for determining the unknown coefficients associated with the expansion of an arbitrary function in terms of the normal modes. The procedure is analogous to the application of the orthogonality of eigenfunctions in the classical orthogonal expansion technique. In this section we present the orthogonality of normal modes for the full-range and half-range expansions.

Full-Range Orthogonality

The discrete normal modes $\phi(\pm\eta_0, \mu)$ and the continuum normal modes $\phi(\eta, \mu)$ are orthogonal over the full range of $\mu (-1 \leq \mu \leq 1)$ with respect to the weight function μ. This orthogonality relation is given as

$$\int_{-1}^1 \mu\phi(\xi, \mu)\phi(\xi', \mu)\, d\mu = 0 \quad \text{for } \xi \neq \xi' \tag{10-23}$$

where ξ and ξ' may be the discrete eigenvalues $\pm\eta_0$ or may take values between -1 and 1 as in the case of continuum eigenvalues.

To prove the above orthogonality relation we write Eq. 10-7 for ξ and ξ' as

$$\left(1 - \frac{\mu}{\xi}\right)\phi(\xi, \mu) = \frac{\omega}{2} \tag{10-24a}$$

$$\left(1 - \frac{\mu}{\xi'}\right)\phi(\xi', \mu) = \frac{\omega}{2} \tag{10-24b}$$

We multiply the first equation by $\phi(\xi', \mu)$ and the second equation by $\phi(\xi, \mu)$, subtract the resulting expressions, integrate the final expression over μ from -1 to 1, utilize the normalization condition 10-6, and obtain

$$\left(\frac{1}{\xi'} - \frac{1}{\xi}\right)\int_{-1}^{1} \mu\phi(\xi', \mu)\phi(\xi, \mu)\, d\mu = 0 \tag{10-25}$$

If Eq. 10-25 should be valid for $\xi' \neq \xi$, the following relation must be satisfied:

$$\int_{-1}^{1} \mu\phi(\xi', \mu)\phi(\xi,\mu)\, d\mu = 0 \quad \text{for } \xi' \neq \xi \tag{10-26}$$

where ξ, ξ' takes all eigenvalues, that is, $\xi, \xi' = \eta_0$ or $\eta \in (-1, 1)$. Equation 10-26 is the desired orthogonality relation for the normal modes over the full range of μ.

Half-Range Orthogonality

Kuščer, McCormick, and Summerfield [6] have shown that the discrete normal modes $\phi(\pm\eta_0, \mu)$ and the continuum normal modes $\phi(\eta, \mu)$ are orthogonal with respect to the weight function $W(\mu)$ over the half-range of μ. A proof of this orthogonality relation and the evaluation of the weight function are given, in addition to the original reference, in the book by Case and Zweifel [2, pp. 82–86]. Here we present only the resulting orthogonality relations.

For the positive half-range of μ (i.e., $0 \leq \mu \leq 1$) the orthogonality relation is given as

$$\int_{0}^{1} W(\mu)\phi(\xi, \mu)\phi(\xi', \mu)\, d\mu = 0 \quad \text{for } \xi \neq \xi', \quad \omega < 1 \tag{10-27a}$$

where

$$\xi, \xi' = \eta_0 \qquad \text{or} \qquad \eta \in (0, 1) \tag{10-27b}$$

and the weight function $W(\mu)$ is defined as[4]

$$W(\mu) = (\eta_0 - \mu)\gamma(\mu), \quad \omega < 1 \tag{10-28}$$

For $\omega = 1$ we have $\eta_0 \to \infty$. In this case, by dividing both sides of Eq. 10-27a by η_0 and letting $\eta_0 \to \infty$, we note that the weight function $W(\mu)$ simplifies to $\gamma(\mu)$. The function $\gamma(\mu)$ is related to Chandrasekhar's [19] $H(\mu)$ function for isotropic scattering by the relation [7]

$$\gamma(\mu) = \frac{\omega\mu}{2} \frac{H(\mu)}{(1 - \omega)^{1/2}(\eta_0 - \mu)}, \quad 0 \leq \mu \leq 1, \quad \omega < 1 \tag{10-29a}$$

or it is related to Case's $X(z)$ function by the relation [2, p. 131]

$$\gamma(\mu) = \frac{\omega\mu}{2} \frac{1}{(\eta_0^2 - \mu^2)(1 - \omega)X(-\mu)}, \quad 0 \leq \mu \leq 1, \quad \omega < 1 \tag{10-29b}$$

For $\omega = 1$ we have $\eta_0 \to \infty$ and $\eta_0^2(1-\omega) = \frac{1}{3}$. Then Eqs. 10-29a and 10-29b, respectively, simplify to

$$\gamma(\mu) = \frac{\sqrt{3}}{2}\mu H(\mu), \quad 0 \le \mu \le 1, \quad \omega = 1 \tag{10-29c}$$

$$\gamma(\mu) = \frac{3}{2}\frac{\mu}{X(-\mu)}, \quad 0 \le \mu \le 1, \quad \omega = 1 \tag{10-29d}$$

We present now a brief discussion of $H(z)$ and $X(z)$ functions.

Chandrasekhar's $H(z)$ Function

Among various integral relations satisfied by the $H(z)$ function we consider the one given in the form [19, p. 104]

$$H(z) = 1 + zH(z)\int_0^1 \frac{f(\mu)}{z+\mu} H(\mu)\, d\mu \tag{10-30a}$$

where $f(\mu)$ is called the *characteristic function*, the form of which depends on the problem considered.

An alternative form of the integral equation satisfied by the $H(z)$ function is given as [19, pp. 107 and 123]

$$\frac{1}{H(z)} = \left[1 - 2\int_0^1 f(\mu)\, d\mu\right]^{1/2} + \int_0^1 \frac{\mu f(\mu)}{z+\mu} H(\mu)\, d\mu \tag{10-30b}$$

The *characteristic function $f(\mu)$ for isotropic scattering* is given as

$$f(\mu) = \frac{\omega}{2} \tag{10-31}$$

Then Eqs. 10-30a and 10-30b simplify respectively to

$$H(z) = 1 + \frac{\omega}{2} zH(z)\int_0^1 \frac{H(\mu)}{z+\mu}\, d\mu \tag{10-32a}$$

and

$$\frac{1}{H(z)} = (1-\omega)^{1/2} + \frac{\omega}{2}\int_0^1 \frac{\mu H(\mu)}{z+\mu}\, d\mu \tag{10-32b}$$

For a purely scattering medium we set $\omega = 1$. Equation 10-32 can be used to evaluate the $H(z)$ function numerically.

The reader may refer to the books by Chandrasekhar [19, Chapter V] and Kourganoff [20, Chapter VI] for further discussion of the properties of the $H(z)$ function.

Case's $X(z)$ Function

The $X(z)$ function has been defined by Case as a function that is analytic and nonvanishing in the complex plane cut from 0 to 1, and vanishes as $-1/z$ at infinity, that is,

$$\lim_{z\to\infty} zX(z) = -1$$

A number of identities define the $X(z)$ function. One such identity is given as

$$X(z) = \int_0^1 \frac{\gamma(\mu)}{\mu - z}\, d\mu \tag{10-33}$$

Substituting $\gamma(\mu)$ from Eq. 10-29b into Eq. 10-33, we obtain

$$X(z) = \frac{\omega}{2(1-\omega)} \int_0^1 \frac{\mu}{(\eta_0^2 - \mu^2)(\mu - z)X(-\mu)}\, d\mu, \quad \omega < 1 \tag{10-34a}$$

which can be used to evaluate the $X(z)$ function numerically.

For large values of z, noting that the $X(z)$ function vanishes as $-1/z$, the following identity is obtained from Eq. 10-34a:

$$1 = \frac{\omega}{2(1-\omega)} \int_0^1 \frac{\mu}{(\eta_0^2 - \mu^2)X(-\mu)}\, d\mu, \quad \omega < 1 \tag{10-34b}$$

For $\omega = 1$, Eq. 10-34a can be simplified by utilizing the limiting relation

$$\lim_{\omega\to 1} \eta_0^2(1-\omega) = \tfrac{1}{3}$$

We obtain

$$X(z) = \frac{3}{2} \int_0^1 \frac{\mu}{(\mu - z)X(-\mu)}\, d\mu, \qquad \omega = 1 \tag{10-34c}$$

A relation relating $X(z)$ and $X(-z)$ is given by

$$X(z)X(-z) = \frac{\Lambda(z)}{(1-\omega)(\eta_0^2 - z^2)}, \quad \omega < 1 \tag{10-35a}$$

where

$$\Lambda(z) = 1 - \frac{\omega z}{2} \int_{-1}^1 \frac{1}{z - \mu}\, d\mu = 1 - \omega z \tanh^{-1}\frac{1}{z} \tag{10-35b}$$

The value of $X(z)X(-z)$ for $z \to \eta_0$ is determined as

$$\begin{aligned} X(\eta_0)X(-\eta_0) &= \frac{1}{1-\omega} \lim_{z\to\eta_0} \frac{\Lambda(z)}{\eta_0^2 - z^2} = \frac{1}{1-\omega} \frac{\left.\dfrac{\partial \Lambda(z)}{\partial z}\right|_{z=\eta_0}}{-2\eta_0} \\ &= -\frac{N(\eta_0)}{\omega(1-\omega)\eta_0^3} \end{aligned} \tag{10-35c}$$

where

$$N(\eta_0) = \tfrac{1}{2}\omega\eta_0^{\,3}\left(\frac{\omega}{\eta_0^{\,2} - 1} - \frac{1}{\eta_0^{\,2}}\right) \tag{10-35d}$$

since

$$\left.\frac{\partial\Lambda(z)}{\partial z}\right|_{z=\eta_0} = \frac{2}{\omega\eta_0^{\,2}}\,N(\eta_0) \tag{10-35e}$$

The value of $X(0)$ is immediately obtainable from Eqs. 10-35a by setting $z = 0$, that is,

$$X^2(0) = \frac{1}{\eta_0^{\,2}(1-\omega)}, \qquad X(0) = \frac{1}{\eta_0\sqrt{1-\omega}} \quad \text{for } \omega < 1 \tag{10-36a}$$

For $\omega = 1$, by noting that

$$\lim_{\omega\to 1} \eta_0^{\,2}(1-\omega) = \tfrac{1}{3}$$

Eq. 10-36a simplifies to

$$X^2(0) = 3, \qquad X(0) = \sqrt{3} \quad \text{for } \omega = 1 \tag{10-36b}$$

By multiplying both sides of Eq. 10-33 by z, then letting z become infinite, and utilizing

$$\lim_{z\to\infty} zX(z) = -1$$

we obtain

$$1 = \int_0^1 \gamma(\mu)\, d\mu \tag{10-37}$$

which is valid for $0 \le \omega \le 1$.

Relation Between $H(z)$ and $X(z)$ Functions

The $H(z)$ and $X(z)$ functions, in fact, are related to each other.

For isotropic scattering the relation between the $H(z)$ and $X(z)$ functions is given as [5]

$$H(z) = \frac{X(0)}{(1 + z/\eta_0)X(-z)} \tag{10-38a}$$

For $\omega < 1$, substituting the value of $X(0)$ from Eq. 10-36a into Eq. 10-38a yields

$$H(z) = \frac{1}{(1-\omega)^{1/2}(\eta_0 + z)X(-z)}, \quad \omega < 1 \tag{10-38b}$$

For $\omega = 1$, we have $\eta_0 \to \infty$ and $X(0) = \sqrt{3}$; then Eq. 10-38a simplifies to

$$H(z) = \frac{\sqrt{3}}{X(-z)}, \quad \omega = 1 \tag{10-38c}$$

For $z = \eta_0$, Eq. 10-38b becomes

$$H(\eta_0) = \frac{1}{2\eta_0(1-\omega)^{1/2}X(-\eta_0)}, \quad \omega < 1 \tag{10-38d}$$

Substituting $X(-\eta_0)$ from Eq. 10-35c into Eq. 10-38d, we obtain

$$H(\eta_0) = -\frac{\omega(1-\omega)^{1/2}\eta_0^2}{2N(\eta_0)}X(\eta_0), \quad \omega < 1 \tag{10-38e}$$

where $N(\eta_0)$ is given by Eq. 10-35d.

10-4 NORMALIZATION INTEGRALS

The expansion coefficients associated with the representation of an arbitrary function in terms of the normal modes can be determined by utilizing the orthogonality property of the normal modes and various normalization integrals. In this section we present the normalization integrals for the discrete and continuum normal modes and isotropic scattering. The cases involving the full range of $\mu(-1 \leq \mu \leq 1)$ and the half-range of $\mu(0 \leq \mu \leq 1)$ will be considered separately.

(a) Full Range Normalization Integrals

Consider the normalization integral for the discrete normal modes defined over the full range of μ as

$$\int_{-1}^{1} \mu\phi^2(\pm\eta_0, \mu)\, d\mu \equiv \pm N(\eta_0) \tag{10-39}$$

where the discrete normal modes were defined by Eq. 10-10, that is,

$$\phi(\pm\eta_0, \mu) = \frac{\omega\eta_0}{2}\frac{1}{\eta_0 \mp \mu}, \quad \omega < 1 \tag{10-40}$$

The integral in Eq. 10-39a is readily evaluated to yield[5]

$$N(\eta_0) = \tfrac{1}{2}\omega\eta_0^3\left(\frac{\omega}{\eta_0^2 - 1} - \frac{1}{\eta_0^2}\right), \quad \omega < 1 \tag{10-41}$$

The normalization integral $N(\eta)$ over the full range of μ for the continuum normal mode may be defined as

$$\int_{-1}^{1} \mu\phi(\eta', \mu)\phi(\eta, \mu)\, d\mu = N(\eta)\,\delta(\eta - \eta') \tag{10-42}$$

where $N(\eta)$ has been computed by Case [1] and is given as[6]

$$N(\eta) = \eta\left[\lambda^2(\eta) + \left(\frac{\pi\omega\eta}{2}\right)^2\right] \tag{10-43a}$$

$$= \frac{\eta}{g(\omega, \eta)} \tag{10-43b}$$

where

$$\lambda(\eta) \equiv 1 - \omega\eta \tanh^{-1} \eta \tag{10-43c}$$

The function $g(\omega, \eta)$ has been tabulated by Case, de Hoffmann, and Placzek [22].

Here we should like to point out that the normalization integral as defined by Eq. 10-42 may be considered as an abbreviation for the statement that, if we expand a function $f(\mu)$ in terms of $\phi(\eta, \mu)$ in the form

$$f(\mu) = \int_{-1}^{1} A(\eta)\phi(\eta, \mu)\, d\eta \tag{10-44a}$$

then the unknown expansion coefficient $A(\eta)$ will be given by

$$A(\eta) = \frac{1}{N(\eta)} \int_{-1}^{1} \mu\phi(\eta, \mu) f(\mu)\, d\mu \tag{10-44b}$$

(b) Half-Range Normalization Integrals

Kuščer, McCormick, and Summerfield [6] evaluated the normalization integrals over the half-range of μ for both discrete and continuum normal modes. The normalization integral $N(\eta_0)$ for the discrete normal mode over the positive half-range of μ (i.e., $0 \leq \mu \leq 1$) is defined as

$$\int_0^1 W(\mu)\phi^2(\eta_0, \mu)\, d\mu \equiv N(\eta_0) \tag{10-45a}$$

where

$$N(\eta_0) = -\left(\frac{\omega\eta_0}{2}\right)^2 X(\eta_0) \tag{10-45b}$$

The above result can readily be proved by substituting the explicit relations for $W(\mu)$ and $\phi(\eta_0, \mu)$:

$$W(\mu) = (\eta_0 - \mu)\gamma(\mu) \qquad \text{and} \qquad \phi(\eta_0, \mu) = \frac{\omega\eta_0}{2} \frac{1}{\eta_0 - \mu}$$

and utilizing the definition of the $X(z)$ function given by Eq. 10-33.

The normalization integral for the continuum normal modes over the positive half-range of μ may be defined as

$$\int_0^1 W(\mu)\phi(\eta', \mu)\phi(\eta, \mu)\, d\mu = W(\eta)\frac{N(\eta)}{\eta}\,\delta(\eta - \eta'),$$

$$0 \le \eta,\quad \eta' \le 1,\quad \omega < 1 \qquad (10\text{-}46a)$$

where various terms on the right-hand side are given by [6][7]

$$W(\eta) = (\eta_0 - \eta)\gamma(\eta) \qquad (10\text{-}46b)$$

$$\frac{N(\eta)}{\eta} = \lambda^2(\eta) + \left(\frac{\pi\omega\eta}{2}\right)^2 = \frac{1}{g(\omega, \eta)} \qquad (10\text{-}46c)$$

$$\lambda(\eta) = 1 - \omega\eta \tanh^{-1}\eta \qquad (10\text{-}46d)$$

and the function $\gamma(\eta)$ can be evaluated from Eq. 10-29a or 10-29b.

For $\omega = 1$, we have $\eta_0 \to \infty$. In this case, dividing both sides of Eq. 10-46a by η_0 and letting $\eta_0 \to \infty$, we obtain

$$\int_0^1 \gamma(n)\phi(\eta', \mu)\phi(\eta, \mu)\, d\mu = \gamma(\eta)\frac{N(\eta)}{\eta}\,\delta(\eta - \eta')$$

$$0 \le \eta, \eta' \le 1,\quad \omega = 1 \qquad (10\text{-}47a)$$

where

$$\frac{N(\eta)}{\eta} = \left[\lambda^2(\eta) + \left(\frac{\pi\eta}{2}\right)^2\right] = \frac{1}{g(1, \eta)} \qquad (10\text{-}47b)$$

$$\lambda(\eta) = 1 - \eta \tanh^{-1}\eta \qquad (10\text{-}47c)$$

and $\gamma(\eta)$ is given by Eq. 10-29c or 10-29d.

Numerical Values of $H(\mu)$, $X(-\mu)$, and $g(\omega, \mu)$ Functions

The functions $H(\mu)$, $X(-\mu)$, and $g(\omega, \mu)$ for isotropic scattering have been tabulated by Chandrasekhar [19, p. 125], Mendelson of Case and Zweifel [2], and Case, de Hoffmann, and Placzek [22, p. 73], respectively; also they are given by Lii [23] and Stibbs and Weir [24]. Tables 10-2, 10-3, and 10-4, obtained from reference 23, give the functions $H(\mu)$, $X(-\mu)$, and $g(\omega, \mu)$ for isotropic scattering for μ from 0 to 1.0 (with 0.05) and for several different values of ω.

10-5 DETERMINATION OF EXPANSION COEFFICIENTS

In this section we illustrate the application of the orthogonality of the normal modes and various normalization integrals to determine the expansion

Table 10-2 $H(\mu)$ **Function for Isotropic Scattering**

μ	$\omega = 0.1$	$\omega = 0.2$	$\omega = 0.3$	$\omega = 0.4$	$\omega = 0.5$	$\omega = 0.6$	$\omega = 0.7$	$\omega = 0.8$	$\omega = 0.9$	$\omega = 1.0$
0.0	1.000000	1.000000	1.000000	1.000000	1.000000	1.000000	1.000000	1.000000	1.000000	1.000000
0.05	1.007809	1.016053	1.024805	1.034163	1.044265	1.055317	1.067655	1.081915	1.099678	1.136575
0.10	1.012378	1.025618	1.039875	1.055360	1.072369	1.091349	1.113032	1.138808	1.172143	1.247350
0.15	1.015842	1.032943	1.051546	1.071980	1.094710	1.120442	1.150344	1.186640	1.234918	1.350834
0.20	1.018644	1.038917	1.061147	1.085781	1.113461	1.145165	1.182516	1.228639	1.291434	1.450351
0.25	1.020993	1.043955	1.069299	1.097592	1.129653	1.166733	1.210934	1.266322	1.343271	1.547326
0.30	1.023006	1.048296	1.076365	1.107899	1.143890	1.185868	1.236419	1.300588	1.391350	1.642522
0.35	1.024759	1.052095	1.082581	1.117017	1.156569	1.203044	1.259517	1.332034	1.436280	1.736404
0.40	1.026306	1.055459	1.088110	1.125169	1.167972	1.218600	1.280619	1.361090	1.478496	1.829276
0.45	1.027685	1.058467	1.093072	1.132519	1.178307	1.232789	1.300019	1.388081	1.518327	1.921350
0.50	1.028922	1.061177	1.097559	1.139192	1.187735	1.245807	1.317945	1.413263	1.556034	2.012779
0.55	1.030042	1.063634	1.101641	1.145285	1.196381	1.257809	1.334582	1.436842	1.591826	2.103677
0.60	1.031060	1.065875	1.105375	1.150877	1.204348	1.268920	1.350079	1.458989	1.625880	2.194133
0.65	1.031991	1.067930	1.108806	1.156031	1.211718	1.279245	1.364562	1.479848	1.658345	2.284214
0.70	1.032846	1.069820	1.111971	1.160800	1.218560	1.288871	1.378136	1.499540	1.689348	2.373975
0.75	1.033635	1.071567	1.114903	1.165228	1.224933	1.297871	1.390889	1.518169	1.719001	2.463460
0.80	1.034365	1.073187	1.117627	1.169352	1.230885	1.306308	1.402901	1.535828	1.747402	2.552704
0.85	1.035043	1.074694	1.120165	1.173204	1.236460	1.314235	1.414237	1.552596	1.774638	2.641739
0.90	1.035674	1.076099	1.122537	1.176811	1.241694	1.321701	1.424957	1.568543	1.800787	2.730588
0.95	1.036264	1.077413	1.124758	1.180196	1.246618	1.328746	1.435111	1.583732	1.825920	2.819272
1.00	1.036816	1.078645	1.126844	1.183381	1.251260	1.335407	1.444746	1.598220	1.850099	2.907811

Table 10-3 $X(-\mu)$ Function for Isotropic Scattering

μ	$\omega = 0.1$	$\omega = 0.2$	$\omega = 0.3$	$\omega = 0.4$	$\omega = 0.5$	$\omega = 0.6$	$\omega = 0.7$	$\omega = 0.8$	$\omega = 0.9$	$\omega = 1.0$
0.0	1.054093	1.117932	1.192138	1.272435	1.354115	1.434618	1.512873	1.588529	1.661554	1.732051
0.05	0.996119	1.047880	1.108024	1.172613	1.237472	1.300424	1.360633	1.417893	1.472267	1.523921
0.10	0.946549	0.990925	1.042449	1.097515	1.152391	1.205187	1.255224	1.302383	1.346772	1.388584
0.15	0.902308	0.941123	0.986158	1.034109	1.081615	1.127017	1.169754	1.209764	1.247181	1.282209
0.20	0.862333	0.896727	0.936606	0.978935	1.020672	1.060345	1.097485	1.132071	1.164250	1.194228
0.25	0.825936	0.856706	0.892364	0.930111	0.967180	1.002257	1.034946	1.065251	1.093330	1.119383
0.30	0.792606	0.820347	0.852477	0.886411	0.919620	0.950922	0.979979	1.006816	1.031594	1.054507
0.35	0.761944	0.787113	0.816251	0.846962	0.876924	0.905071	0.931110	0.955083	0.977147	0.997493
0.40	0.733624	0.756584	0.783153	0.811105	0.838302	0.863775	0.887271	0.908841	0.928640	0.946851
0.45	0.707377	0.728421	0.752763	0.778330	0.803147	0.826328	0.847654	0.867182	0.885064	0.901476
0.50	0.682975	0.702344	0.724739	0.748227	0.770976	0.792175	0.811630	0.829406	0.855649	0.860527
0.55	0.660225	0.678118	0.698801	0.720462	0.741401	0.760870	0.778701	0.794958	0.809787	0.823344
0.60	0.638962	0.655547	0.674712	0.694759	0.714102	0.732053	0.748461	0.763393	0.776989	0.789401
0.65	0.619040	0.634460	0.652273	0.670886	0.688814	0.705423	0.720576	0.734344	0.746861	0.758270
0.70	0.600336	0.614713	0.631316	0.648646	0.665314	0.680729	0.694770	0.707509	0.719072	0.729599
0.75	0.582738	0.596177	0.611692	0.627871	0.643410	0.657759	0.670810	0.682633	0.693351	0.703097
0.80	0.566151	0.578742	0.593276	0.608417	0.622940	0.636332	0.648496	0.659501	0.669466	0.678516
0.85	0.550489	0.562312	0.575956	0.590158	0.603764	0.616294	0.627660	0.637930	0.647220	0.655648
0.90	0.535676	0.546800	0.559635	0.572985	0.585759	0.597508	0.608154	0.617763	0.626445	0.634314
0.95	0.521644	0.532130	0.544227	0.556800	0.568818	0.579859	0.589852	0.598862	0.606995	0.614361
1.00	0.508332	0.518235	0.529657	0.541520	0.552848	0.563244	0.572643	0.581110	0.588745	0.595655

Table 10-4 $g(\omega, \mu)$ Function for Isotropic Scattering

$$g(\omega, \mu) = \frac{1}{(1 - \omega\mu \tanh^{-1} \mu)^2 + [(\pi/2)\omega\mu]^2}$$

μ	$\omega = 0.1$	$\omega = 0.2$	$\omega = 0.3$	$\omega = 0.4$	$\omega = 0.5$	$\omega = 0.6$	$\omega = 0.7$	$\omega = 0.8$	$\omega = 0.9$	$\omega = 1.0$
0.0	1.000000	1.000000	1.000000	1.000000	1.000000	1.000000	1.000000	1.000000	1.000000	1.000000
0.05	1.000439	1.000754	1.000946	1.001015	1.000959	1.000780	1.000478	1.000051	0.999502	0.998831
0.10	1.001762	1.003032	1.003805	1.004079	1.003855	1.003131	1.001911	1.000198	0.997997	0.995314
0.15	1.003990	1.006874	1.008634	1.009257	1.008739	1.007084	1.004303	1.000414	0.995444	0.989425
0.20	1.007157	1.012356	1.015535	1.016656	1.015704	1.012692	1.007657	1.000657	0.991776	0.981116
0.25	0.011314	0.019586	1.024659	1.026437	1.024885	1.020033	1.011974	1.000860	0.986895	0.970324
0.30	0.016533	1.028716	1.036217	1.038826	1.036467	1.029208	1.017254	1.000930	0.980664	0.956959
0.35	1.022908	1.039952	1.050495	1.054125	1.050697	1.040349	1.023484	1.000735	0.972905	0.940906
0.40	1.030563	1.053567	1.067875	1.072740	1.067902	1.053620	1.030639	1.000095	0.963383	0.922015
0.45	1.039664	1.069927	1.088873	1.095214	1.088506	1.069219	1.038661	0.998763	0.951798	0.900096
0.50	1.050429	1.089526	1.114189	1.122280	1.113068	1.087383	1.047444	0.996396	0.947756	0.874907
0.55	1.063153	1.113042	1.144794	1.154945	1.142327	1.108388	1.056795	0.992511	0.920748	0.846135
0.60	1.078243	1.141430	1.182065	1.194621	1.177276	1.132534	1.066373	0.986423	0.900094	0.813371
0.65	1.096280	1.176083	1.228029	1.243357	1.219267	1.160114	1.075579	0.977127	0.874880	0.776067
0.70	1.118125	1.219125	1.285812	1.304241	1.270177	1.191317	1.083337	0.963119	0.843836	0.733466
0.75	1.145133	1.274000	1.360533	1.382188	1.332652	1.225965	1.087675	0.942055	0.805131	0.684478
0.80	1.179610	1.346788	1.461329	1.485572	1.410425	1.262782	1.084798	0.910092	0.755983	0.627430
0.85	1.225969	1.449658	1.606660	1.630111	1.508395	1.297165	1.066868	0.860446	0.691798	0.559508
0.90	1.294427	1.612462	1.842025	1.849536	1.629844	1.313164	1.015674	0.779605	0.603895	0.475236
0.95	1.419421	1.945044	2.331852	2.228849	1.743429	1.244393	0.878172	0.633330	0.470747	0.360398
1.00	0.0	0.0	0.0	0.0	0.0	0.0	0.0	0.0	0.0	0.0

coefficients associated with the representation of a suitably well-behaved function in terms of the normal modes. The full-range and the half-range expansions will be considered separately.

(a) The Full-Range Expansion

Consider a function $f(\mu)$ defined over the full range of μ expanded in terms of the normal modes in the form (see Eq. 10-21a)

$$f(\mu) = A(\eta_0)\phi(\eta_0, \mu) + A(-\eta_0)\phi(-\eta_0, \mu) + \int_{-1}^{1} A(\eta)\phi(\eta, \mu)\, d\eta$$

$$\mu \in (-1, 1), \quad \omega < 1 \qquad (10\text{-}48)$$

where $A(\eta_0)$, $A(-\eta_0)$, and $A(\eta)$ are the expansion coefficients.

To determine the discrete expansion coefficient $A(\eta_0)$ we operate on both sides of Eq. 10-48 by the operator

$$\int_{-1}^{1} \mu\phi(\eta_0, \mu)\, d\mu$$

and utilize the full-range orthogonality condition 10-23 to obtain

$$\int_{-1}^{1} \mu\phi(\eta_0, \mu) f(\mu)\, d\mu = A(\eta_0)\int_{-1}^{1} \mu\phi^2(\eta_0, \mu)\, d\mu$$

or

$$A(\eta_0) = \frac{1}{N(\eta_0)}\int_{-1}^{1} \mu\phi(\eta_0, \mu) f(\mu)\, d\mu \qquad (10\text{-}49a)$$

where the normalization integral $N(\eta_0)$ is given by Eq. 10-41 as

$$N(\eta_0) \equiv \int_{-1}^{1} \mu\phi^2(\eta_0, \mu)\, d\mu = \tfrac{1}{2}\omega\eta_0^{\,3}\left(\frac{\omega}{\eta_0^{\,2} - 1} - \frac{1}{\eta_0^{\,2}}\right), \quad \omega < 1 \qquad (10\text{-}49b)$$

The discrete coefficient $A(-\eta_0)$ is determined by operating on both sides of Eq. 10-48 with the operator $\int_{-1}^{1} \mu\phi(-\eta_0, \mu)\, d\mu$ and by utilizing the full-range orthogonality condition 10-23:

$$A(-\eta_0) = -\frac{1}{N(\eta_0)}\int_{-1}^{1} \mu\phi(-\eta_0, \mu) f(\mu)\, d\mu \qquad (10\text{-}50)$$

where the normalization integral $N(\eta_0)$ is given by Eq. 10-49b.

To evaluate the continuum coefficient $A(\eta)$ we operate on both sides of Eq. 10-48 by the operator $\int_{-1}^{1} \mu\phi(\eta', \mu)\, d\mu$, utilize the full-range orthogonality condition 10-23, and obtain

$$\int_{-1}^{1} \mu\phi(\eta', \mu) f(\mu)\, d\mu = \int_{-1}^{1} \mu\phi(\eta', \mu)\left[\int_{-1}^{1} A(\eta)\phi(\eta, \mu)\, d\eta\right] d\mu \qquad (10\text{-}51a)$$

The order of integration on the right-hand side of Eq. 10-51a is important because of the singular nature of the continuum normal modes $\phi(\xi, \mu)$. However, it has been shown [2, p. 69] by the application of the Poincaré-Bertrand formula that the order of integration can be changed; then Eq. 10-51a becomes

$$\int_{-1}^{1} \mu\phi(\eta', \mu) f(\mu)\, d\mu = \int_{-1}^{1} A(\eta) \left[\int_{-1}^{1} \mu\phi(\eta, \mu)\phi(\eta', \mu)\, d\mu \right] d\eta \quad (10\text{-}51\text{b})$$

The integral inside the bracket is evaluated by Eq. 10-42 to yield

$$\int_{-1}^{1} \mu\phi(\eta', \mu) f(\mu)\, d\mu = \int_{-1}^{1} A(\eta) N(\eta)\, \delta(\eta - \eta')\, d\eta \quad (10\text{-}51\text{c})$$

By performing the integration on the right-hand side and interchanging η and η' in the resulting expression, the continuum coefficient $A(\eta)$ is obtained as

$$A(\eta) = \frac{1}{N(\eta)} \int_{-1}^{1} \mu\phi(\eta, \mu) f(\mu)\, d\mu \quad (10\text{-}52\text{a})$$

where $N(\eta)$ is given by Eqs. 10-43, that is,

$$N(\eta) = \eta \left[\lambda^2(\eta) + \left(\frac{\pi\omega\eta}{2} \right)^2 \right] \quad (10\text{-}52\text{b})$$

$$\lambda(\eta) = 1 - \omega\eta \tanh^{-1} \eta \quad (10\text{-}52\text{c})$$

(b) The Half-Range Expansion

Consider a function $f(\mu)$ defined over the positive half-range of μ expanded in terms of the normal modes in the form (see Eq. 10-22a)

$$f(\mu) = A(\eta_0)\phi(\eta_0, \mu) + \int_0^1 A(\eta)\phi(\eta, \mu)\, d\eta, \quad \mu \in (0, 1), \quad \omega < 1 \quad (10\text{-}53)$$

where $A(\eta_0)$ and $A(\eta)$ are the unknown expansion coefficients.

To determine the discrete coefficient $A(\eta_0)$, we operate on both sides of Eq. 10-53 by the operator $\int_0^1 W(\mu)\phi(\eta_0, \mu)\, d\mu$, utilize the half-range orthogonality condition 10-27, and obtain

$$A(\eta_0) = \frac{1}{N(\eta_0)} \int_0^1 W(\mu)\phi(\eta_0, \mu) f(\mu)\, d\mu \quad (10\text{-}54\text{a})$$

where the normalization integral is given by Eq. 10-45 as

$$N(\eta_0) \equiv \int_0^1 W(\mu)\phi^2(\eta_0, \mu)\, d\mu = -\left(\frac{\omega\eta_0}{2} \right)^2 X(\eta_0) \quad (10\text{-}54\text{b})$$

To determine the continuum coefficient $A(\eta)$, we operate on both sides of Eq. 10-53 by the operator $\int_0^1 W(\mu)\phi(\eta', \mu)\, d\mu$, utilize the half-range orthogonality condition 10-27, and obtain

$$\int_0^1 W(\mu)\phi(\eta', \mu)f(\mu)\, d\mu = \int_0^1 W(\mu)\phi(\eta', \mu)\left[\int_0^1 A(\eta)\phi(\eta, \mu)\, d\eta\right] d\mu \quad (10\text{-}55a)$$

It has been shown in reference 6 that inversion of the order of integration on the right-hand side is permissible; therefore Eq. 10-55a can be written as

$$\int_0^1 W(\mu)\phi(\eta', \mu)f(\mu)\, d\mu = \int_0^1 A(\eta)\left[\int_0^1 W(\mu)\phi(\eta, \mu)\phi(\eta', \mu)\, d\mu\right] d\eta \quad (10\text{-}55b)$$

When the integral inside the bracket is evaluated by Eq. 10-46, we obtain

$$\int_0^1 W(\mu)\phi(\eta', \mu)f(\mu)\, d\mu = \int_0^1 A(\eta)W(\eta)\frac{N(\eta)}{\eta}\, \delta(\eta - \eta')\, d\eta \quad (10\text{-}55c)$$

Evaluating the integral on the right-hand side and interchanging η and η' in the resulting expression, we obtain the continuum coefficient as

$$A(\eta) = \frac{\eta}{W(\eta)N(\eta)}\int_0^1 W(\mu)\phi(\eta, \mu)f(\mu)\, d\mu \quad (10\text{-}56a)$$

or

$$A(\eta) = \frac{g(\omega, \eta)}{W(\eta)}\int_0^1 W(\mu)\phi(\eta, \mu)f(\mu)\, d\mu \quad (10\text{-}56b)$$

where $W(\eta)$, $N(\eta)$, and $g(\omega, \eta)$ have already been defined.

(c) The Degenerate Case

The discrete normal modes become degenerate for $\omega = 1$. Consider the representation of an arbitrary function $f(\mu)$ over the half-range of $\mu(0 \leq \mu \leq 1)$ in the form

$$f(\mu) = A\tfrac{1}{2} + \int_0^1 A(\eta)\phi(\eta, \mu)\, d\eta, \quad \mu \in (0, 1) \quad (10\text{-}57)$$

The discrete coefficient A is determined by operating on both sides of Eq. 10-57 by the operator $\int_0^1 \gamma(\mu)\, d\mu$, and noting that $\int_0^1 \gamma(\mu)\, d\mu = 1$ and $\int_0^1 \gamma(\mu)\phi(\eta, \mu)\, d\mu = 0$ (see Eqs. 10-37 and 10-87). We obtain

$$A\tfrac{1}{2} = \int_0^1 \gamma(\mu)f(\mu)\, d\mu \quad (10\text{-}58)$$

To determine the continuum coefficient $A(\eta)$ we operate on both sides of Eq. 10-57 by the operator $\int_0^1 \gamma(\mu)\phi(\eta', \mu)\, d\mu$ and obtain

$$\int_0^1 \gamma(\mu)\phi(\eta', \mu) f(\mu)\, d\mu = \int_0^1 \gamma(\mu)\phi(\eta', \mu)\left[\int_0^1 A(\eta)\phi(\eta, \mu)\, d\eta\right] d\mu \tag{10-59a}$$

The order of integration can be changed to yield

$$\int_0^1 \gamma(\mu)\phi(\eta', \mu) f(\mu)\, d\mu = \int_0^1 A(\eta)\left[\int_0^1 \gamma(\mu)\phi(\eta', \mu)\phi(\eta, \mu)\, d\mu\right] d\eta \tag{10-59b}$$

The integral inside the bracket is evaluated by Eq. 10-47a; then

$$\int_0^1 \gamma(\mu)\phi(\eta', \mu) f(\mu)\, d\mu = \int_0^1 A(\eta)\gamma(\eta)\frac{N(\eta)}{\eta}\,\delta(\eta - \eta')\, d\eta \tag{10-59c}$$

Performing the integration on the right-hand side and interchanging η and η' in the resulting expression, we obtain the continuum coefficient as

$$A(\eta) = \frac{\eta}{\gamma(\eta)N(\eta)}\int_0^1 \gamma(\mu)\phi(\eta, \mu) f(\mu)\, d\mu \tag{10-59d}$$

where $N(\eta)$ is defined by Eq. 10-47b.

10-6 A SUMMARY OF INTEGRALS INVOLVING NORMAL MODES

In the solution of problems of radiative heat transfer by the normal-mode expansion technique, various functions of the normal modes are to be integrated over the full-range or the half-range of μ. We summarize now various normalization integrals, orthogonality relations, and the results of several useful integrations involving normal modes in isotropic scattering. The reader should refer to the original publications [1, 2, 6, 25] for derivations and more comprehensive tabulations.

(a) Full Range of $\mu(-1 \le \mu \le 1)$

The normalization integrals are given for the discrete normal modes as

$$\int_{-1}^1 \mu\phi^2(\pm\eta_0, \mu)\, d\mu \equiv \pm N(\eta_0) \tag{10-60a}$$

where

$$N(\eta_0) = \tfrac{1}{2}\omega\eta_0^{\,3}\left(\frac{\omega}{\eta_0^{\,2} - 1} - \frac{1}{\eta_0^{\,2}}\right), \quad \omega < 1 \tag{10-60b}$$

and for the continuum normal modes as

$$\int_{-1}^{1} \mu\phi(\eta, \mu)\phi(\eta', \mu)\, d\mu = N(\eta)\, \delta(\eta - \eta') \tag{10-61a}$$

where

$$N(\eta) = \eta\left[(1 - \omega\eta \tanh^{-1} \eta)^2 + \left(\frac{\pi\omega\eta}{2}\right)^2\right] = \frac{1}{g(\omega, \eta)}, \quad \omega \leq 1 \tag{10-61b}$$

The orthogonality relations are given as

$$\int_{-1}^{1} \mu\phi(\eta, \mu)\phi(\pm\eta_0, \mu)\, d\mu = 0 \tag{10-62a}$$

$$\int_{-1}^{1} \mu\phi(\eta_0, \mu)\phi(-\eta_0, \mu)\, d\mu = 0 \tag{10-62b}$$

Other useful integrals for $\omega < 1$ include [25, p. 33][8]

$$\int_{-1}^{1} \mu\phi(\pm\eta_0, \mu)\, d\mu = \pm\eta_0(1 - \omega) \tag{10-63a}$$

$$\int_{-1}^{1} \mu^2\phi(\pm\eta_0, \mu)\, d\mu = \eta_0^2(1 - \omega) \tag{10-63b}$$

$$\int_{-1}^{1} \mu\phi(\eta, \mu)\, d\mu = \eta(1 - \omega) \tag{10-64a}$$

$$\int_{-1}^{1} \mu^2\phi(\eta, \mu)\, d\mu = \eta^2(1 - \omega) \tag{10-64b}$$

For $\omega = 1$, Eqs. 10-63 simplify, respectively, to

$$\int_{-1}^{1} \mu\, d\mu = 0 \qquad 10\text{-}65a)$$

$$\int_{-1}^{1} \mu^2\, d\mu = \tfrac{2}{3} \tag{10-65b}$$

since

$$\lim_{\omega\to 1} \phi(\pm\eta_0, \mu) = \tfrac{1}{2} \qquad \text{and} \qquad \lim_{\omega\to 1} \eta_0^2(1 - \omega) = \tfrac{1}{3}$$

and Eqs. 1–64 simplify to

$$\int_{-1}^{1} \mu\phi(\eta, \mu)\, d\mu = 0 \tag{10-66a}$$

$$\int_{-1}^{1} \mu^2\phi(\eta, \mu)\, d\mu = 0 \tag{10-66b}$$

(b) Half-Range of $\mu (0 \leq \mu \leq 1)$

The normalization integrals are given for the discrete normal modes as

$$\int_0^1 W(\mu)\phi^2(\eta_0, \mu)\, d\mu = N(\eta_0) \tag{10-67}$$

where

$$N(\eta_0) = -\left(\frac{\omega\eta_0}{2}\right)^2 X(\eta_0), \quad \omega < 1 \tag{10-68a}$$

$$W(\mu) = (\eta_0 - \mu)\gamma(\mu), \qquad \omega < 1 \tag{10-68b}$$

and for the continuum normal modes as

$$\int_0^1 W(\mu)\phi(\eta', \mu)\phi(\eta, \mu)\, d\mu = W(\mu)\frac{N(\eta)}{\eta}\delta(\eta - \eta'), \quad 0 \leq \eta, \eta' \leq 1 \tag{10-69a}$$

where

$$\frac{N(\eta)}{\eta} = (1 - \omega\eta \tanh^{-1}\eta)^2 + \left(\frac{\pi\omega\eta}{2}\right)^2 = \frac{1}{g(\omega, \eta)}, \quad \omega \leq 1 \tag{10-69b}$$

Other useful relations for the case $\omega < 1$ and $0 \leq \eta$, $\eta' \leq 1$, are given as follows:

$$\int_0^1 W(\mu)\phi(\eta_0, \mu)\phi(\eta, \mu)\, d\mu = 0 \tag{10-70}$$

$$\int_0^1 W(\mu)\phi(-\eta_0, \mu)\phi(\eta, \mu)\, d\mu = \omega\eta\eta_0 X(-\eta_0)\phi(-\eta_0, \eta) \tag{10-71}$$

$$\int_0^1 W(\mu)\phi(\pm\eta_0, \mu)\phi(\eta_0, \mu)\, d\mu = \mp\left(\frac{\omega\eta_0}{2}\right)^2 X(\pm\eta_0) \tag{10-72}$$

$$\int_0^1 W(\mu)\phi(-\eta, \mu)\phi(\eta_0, \mu)\, d\mu = \tfrac{1}{4}\omega^2\eta\eta_0 X(-\eta) \tag{10-73}$$

$$\int_0^1 W(\mu)\phi(-\eta, \mu)\phi(\eta', \mu)\, d\mu = \tfrac{1}{2}\omega\eta'(\eta_0 + \eta)\phi(-\eta, \eta')X(-\eta) \tag{10-74}$$

$$\int_0^1 W(\mu)\phi(\eta_0, \mu)\, d\mu = \tfrac{1}{2}\omega\eta_0 \tag{10-75}$$

$$\int_0^1 \mu W(\mu)\phi(\eta_0, \mu)\, d\mu = \tfrac{1}{2}\omega\eta_0\gamma^{(1)} \tag{10-76}$$

$$\int_0^1 \mu^2 W(\mu)\phi(\eta_0, \mu)\, d\mu = \tfrac{1}{2}\omega\eta_0\gamma^{(2)} \tag{10-77}$$

$$\int_0^1 W(\mu)\phi(\eta, \mu)\, d\mu = \tfrac{1}{2}\omega\eta \tag{10-78}$$

$$\int_0^1 \mu W(\mu)\phi(\eta, \mu)\, d\mu = \tfrac{1}{2}\omega\eta[\gamma^{(1)} + \eta - \eta_0] \tag{10-79}$$

$$\int_0^1 \mu^2 W(\mu)\phi(\eta, \mu)\, d\mu = \tfrac{1}{2}\omega\eta[\gamma^{(2)} + (\eta - \eta_0)(\gamma^{(1)} + \eta)] \tag{10-80}$$

where we have defined the moments $\gamma^{(n)}$ of the $\gamma(\mu)$ function as

$$\gamma^{(n)} \equiv \int_0^1 \mu^n \gamma(\mu)\, d\mu, \quad n = 0, 1, 2, 3, \ldots \tag{10-81}$$

and (see Eq. 10-37)

$$\gamma^{(0)} \equiv \int_0^1 \gamma(\mu)\, d\mu = 1$$

The function $\gamma(\mu)$ is related to the $H(\mu)$ and $X(-\mu)$ functions by Eqs. 10-29. Table 10-5 gives the numerical values of the moments $\gamma^{(n)}$ for several different values of ω.[9]

The foregoing equations can be written in alternative form by noting that $W(\mu)\phi(\eta_0, \mu) = \frac{1}{2}\omega\eta_0\gamma(\mu)$. For example, Eqs. 10-70, 10-72, and 10-73, respectively, become

$$\int_0^1 \gamma(\mu)\phi(\eta, \mu)\, d\mu = 0 \tag{10-82}$$

$$\int_0^1 \gamma(\mu)\phi(\pm\eta_0, \mu)\, d\mu = \mp\tfrac{1}{2}\omega\eta_0 X(\pm\eta_0) \tag{10-83}$$

$$\int_0^1 \gamma(\mu)\phi(-\eta, \mu)\, d\mu = \tfrac{1}{2}\omega\eta X(-\eta) \tag{10-84}$$

For $\omega = 1$, the above relations are simplified by dividing both sides of the equation by η_0 and then letting $\omega \to 1$. For example, by noting that $\eta_0 \to \infty$, $\eta_0 X(-\eta_0) \to 1$, and $\phi(\eta_0, \mu) \to \frac{1}{2}$ for $\omega = 1$, Eq. 10-67 yields

$$\int_0^1 \gamma(\mu)\, d\mu = 1 \tag{10-85}$$

Similarly, Eq. 10-69 simplifies to

$$\int_0^1 \gamma(\mu)\phi(\eta, \mu)\phi(\eta', \mu)\, d\mu = \gamma(\eta)\frac{N(\eta)}{\eta}\,\delta(\eta - \eta'), \quad 0 \le \eta, \eta' \le 1, \omega = 1 \tag{10-86a}$$

where

$$\frac{N(\eta)}{\eta} = \left[(1 - \eta \tanh^{-1} \eta)^2 + \left(\frac{\pi\eta}{2}\right)^2\right] = \frac{1}{g(1, \eta)} \tag{10-86b}$$

Table 10-5 Moments $\gamma^{(n)}$ of $\gamma(\mu)$ Function for Isotropic Scattering

$$\gamma^{(n)} = \int_0^1 \mu^n \gamma(\mu)\, d\mu$$

n	$\omega = 0.1$	$\omega = 0.2$	$\omega = 0.3$	$\omega = 0.4$	$\omega = 0.5$	$\omega = 0.6$	$\omega = 0.7$	$\omega = 0.8$	$\omega = 0.9$	$\omega = 1.0$
0	1.000000	1.000000	1.000000	1.000000	1.000000	1.000000	1.000000	1.000000	1.000000	1.000000
1	0.972825	0.940482	0.903427	0.865808	0.831018	0.800151	0.773129	0.749501	0.728760	0.710446
2	0.954676	0.900678	0.839258	0.778393	0.723996	0.677468	0.638164	0.604926	0.576640	0.552367
3	0.941051	0.870784	0.791395	0.714378	0.647545	0.592131	0.546693	0.509312	0.478286	0.452262
4	0.930146	0.846854	0.753340	0.664340	0.589098	0.528380	0.479852	0.440847	0.409142	0.383039
5	0.921054	0.826908	0.721829	0.623557	0.542409	0.478485	0.428526	0.389169	0.357741	0.332267
6	0.913260	0.809811	0.694991	0.589334	0.503942	0.438122	0.387695	0.348663	0.317968	0.293417
7	0.906438	0.794853	0.671656	0.559990	0.471514	0.404653	0.354339	0.315997	0.286246	0.262721
8	0.900374	0.781560	0.651041	0.534409	0.443684	0.376361	0.326516	0.289058	0.260336	0.237852
9	0.894915	0.769598	0.632601	0.511810	0.419459	0.352073	0.302918	0.266441	0.238765	0.217290
10	0.889952	0.758727	0.615936	0.491632	0.398125	0.330956	0.282625	0.247168	0.220521	0.200006

Other relations for $\omega = 1, 0 \leq \eta, \eta' \leq 1$, include the following:

$$\int_0^1 \gamma(\mu)\phi(\eta, \mu)\, d\mu = 0 \tag{10-87}$$

$$\int_0^1 \mu\gamma(\mu)\phi(\eta, \mu)\, d\mu = -\frac{\eta}{2} \tag{10-88}$$

$$\int_0^1 \mu^2\gamma(\mu)\phi(\eta, \mu)\, d\mu = -\frac{\eta}{2}\,[\gamma^{(1)} + \eta] \tag{10-89}$$

$$\int_0^1 \gamma(\mu)\phi(-\eta, \mu)\, d\mu = \frac{\eta}{2}\, X(-\eta) \tag{10-90}$$

$$\int_0^1 \gamma(\mu)\phi(-\eta, \mu)\phi(\eta', \mu)\, d\mu = \frac{\eta'}{2}\,\phi(-\eta, \eta')X(-\eta) = \frac{\eta\eta'}{4}\frac{1}{\eta + \eta'}\, X(-\eta) \tag{10-91}$$

10-7 APPLICATION TO RADIATIVE HEAT TRANSFER FOR $\omega < 1$

To illustrate the application of the normal-mode expansion technique for the case $\omega < 1$, we consider radiative heat transfer in an absorbing, emitting, isotropically scattering, plane-parallel, semi-infinite $(0 \leq \tau < \infty)$ gray medium. For simplicity we assume that the boundary surface at $\tau = 0$ is transparent and irradiated by an externally incident radiation assumed to be azimuthally symmetrical. The equation of radiative transfer is given as

$$\mu \frac{\partial I(\tau, \mu)}{\partial \tau} + I(\tau, \mu) = (1 - \omega)\frac{\bar{\sigma}T^4(\tau)}{\pi} + \frac{\omega}{2}\int_{-1}^{1} I(\tau, \mu')\, d\mu'$$

$$0 \leq \tau < \infty,\ -1 \leq \mu \leq 1,\ \omega < 1 \tag{10-92}$$

with the boundary conditions:

$$I(\tau, \mu)|_{\tau=0} = F(\mu), \quad \mu > 0 \tag{10-93}$$

and at $\tau \to \infty$ the solution converges to a particular solution $I_p(\tau, \mu)$ of Eq. 10-92.

Here we assume that the distribution of temperature $T(\tau)$ in the medium is prescribed and that a particular solution $I_p(\tau, \mu)$ of Eq. 10-92 is available for the inhomogeneous term in question.

The complete solution of Eq. 10-92 can be written as a linear sum of the normal modes of the homogeneous equation and a particular solution in the form (see: Eq. 10-18b)

$$I(\tau,\mu) = A(\eta_0)\phi(\eta_0,\mu)e^{-\tau/\eta_0} + A(-\eta_0)\phi(-\eta_0,\mu)e^{\tau/\eta_0} + \int_0^1 A(\eta)\phi(\eta,\mu)e^{-\tau/\eta}\,d\eta$$
$$+ \int_0^1 A(-\eta)\phi(-\eta,\mu)e^{\tau/\eta}\,d\eta + I_p(\tau,\mu) \quad (10\text{-}94)$$

where $A(\eta_0)$, $A(-\eta_0)$, $A(\eta)$, and $A(-\eta)$ are the arbitrary expansion coefficients to be determined.

If the solution given by Eq. 10-94 should satisfy the boundary condition at infinity, the part of the solution that diverges at infinity must be excluded; then Eq. 10-94 simplifies to

$$I(\tau, \mu) = A(\eta_0)\phi(\eta_0, \mu)e^{-\tau/\eta_0} + \int_0^1 A(\eta)\phi(\eta, \mu)e^{-\tau/\eta}\,d\eta + I_p(\tau, \mu) \quad (10\text{-}95)$$

which satisfies both Eq. 10-92 and the boundary condition at infinity. The expansion coefficients $A(\eta_0)$ and $A(\eta)$ can be determined by constraining the solution 10-95 to satisfy the boundary condition 10-93, and by utilizing the orthogonality of normal modes and various normalization integrals. The application of boundary condition 10-93 yields

$$f(0, \mu) = A(\eta_0)\phi(\eta_0, \mu) + \int_0^1 A(\eta)\phi(\eta, \mu)\,d\eta, \quad \mu \in (0, 1) \quad (10\text{-}96a)$$

where we have defined

$$f(0, \mu) \equiv F(\mu) - I_p(0, \mu), \quad \mu \in (0, 1) \quad (10\text{-}96b)$$

Here the function $f(0, \mu)$ is considered to be known, since $F(\mu)$ is prescribed and a particular solution of Eq. 10-92 is assumed to be available.

Equation 10-96a is a representation of a function $f(0, \mu)$, defined over the positive half-range of μ in terms of the normal modes; the validity of such an expansion is ensured by the half-range completeness theorem stated previously. Then the espansion coefficients $A(\eta_0)$ and $A(\eta)$ can be determined by utilizing the orthogonality property of the normal modes over the half-range of μ and various normalization integrals. We note that Eq. 10-96a is of exactly the same form as Eq. 10-53; therefore the expansion coefficients $A(\eta_0)$ and $A(\eta)$ are immediately obtained from Eqs. 10-54 and 10-56, respectively. The coefficient $A(\eta_0)$ is given as

$$A(\eta_0) = \frac{1}{N(\eta_0)} \int_0^1 W(\mu)\phi(\eta_0, \mu)f(0, \mu)\,d\mu \quad (10\text{-}97a)$$

where

$$N(\eta_0) = -\left(\frac{\omega\eta_0}{2}\right)^2 X(\eta_0) \quad (10\text{-}97b)$$

$$W(\mu) = (\eta_0 - \mu)\gamma(\mu) \quad (10\text{-}97c)$$

and $A(\eta)$ is given as

$$A(\eta) = \frac{\eta}{W(\eta)N(\eta)} \int_0^1 W(\mu)\phi(\eta, \mu) f(0, \mu)\, d\mu \tag{10-98a}$$

where

$$\frac{N(\eta)}{\eta} = \left[(1 - \omega\eta \tanh^{-1} \eta)^2 + \left(\frac{\pi\omega\eta}{2}\right)^2\right] = \frac{1}{g(\omega, \eta)} \tag{10-98b}$$

Once the expansion coefficients $A(\eta_0)$ and $A(\eta)$ have been determined, the intensity of radiation $I(\tau, \mu)$ everywhere in the medium is available from Eq. 10-95. Other physical quantities of interest such as the incident radiation $G(\tau)$ and the net radiative heat flux $q^r(\tau)$ are determined from their definitions. The incident radiation is given as

$$\begin{aligned} G(\tau) &= 2\pi \int_{-1}^1 I(\tau, \mu)\, d\mu \\ &= 2\pi\left[A(\eta_0)e^{-\tau/\eta_0} + \int_0^1 A(\eta)e^{-\tau/\eta}\, d\eta + \int_{-1}^1 I_p(\tau, \mu)\, d\mu\right] \end{aligned} \tag{10-99}$$

since

$$\int_{-1}^1 \phi(\xi, \mu)\, d\mu = 1 \quad \text{for } \xi = \eta_0 \text{ or } \eta \in (0, 1)$$

and the net radiative heat flux as

$$\begin{aligned} q^r(\tau) &= 2\pi \int_{-1}^1 \mu I(\tau, \mu)\, d\mu \\ &= 2\pi(1 - \omega)\left[A(\eta_0)\eta_0 e^{-\tau/\eta_0} + \int_0^1 A(\eta)\eta e^{-\tau/\eta}\, d\eta + \frac{1}{1 - \omega}\int_{-1}^1 \mu I_p(\tau, \mu)\, d\mu\right] \end{aligned} \tag{10-100}$$

since

$$\int_{-1}^1 \mu\phi(\xi, \mu)\, d\mu = \xi(1 - \omega) \quad \text{for } \xi = \eta_0 \text{ or } \eta \in (0, 1)$$

Another physical quantity of interest is the angular distribution of the emergent radiation intensity $I_e(0, \mu)$, $\mu \in (-1, 0)$, at the boundary surface $\tau = 0$. This quantity can be evaluated from the solution 10-95 by setting in that equation $\tau = 0$ and evaluating it for $\mu < 0$. We obtain

$$I_e(0, \mu) = A(\eta_0)\phi(\eta_0, \mu) + \int_0^1 A(\eta)\phi(\eta, \mu)\, d\eta + I_p(0, \mu), \quad \mu < 0 \tag{10-101a}$$

or, alternatively,

$$I_e(0, -\mu) = A(\eta_0)\phi(\eta_0, -\mu) + \int_0^1 A(\eta)\phi(\eta, -\mu)\, d\eta + I_p(0, -\mu), \quad \mu > 0 \tag{10-101b}$$

The exit distribution function $I_e(0, \mu)$, $\mu \in (-1, 0)$, should be distinguished from the boundary condition function $I(0, \mu)$, $\mu \in (0, 1)$. We note that the integral in Eqs. 10-101 is nonsingular.

A Simple Method to Determine the Emergent Intensity

The emergent radiation intensity can be evaluated from Eq. 10-101 provided that the expansion coefficients $A(\eta_0)$ and $A(\eta)$ have been determined as described above. An alternative, simple method described by McCormick and Kuščer [7] may be used to evaluate the emergent radiation intensity without determining the expansion coefficients. We present now a brief discussion of this approach.

Consider Eq. 10-96a, that is,

$$f(0, \mu) = A(\eta_0)\phi(\eta_0, \mu) + \int_0^1 A(\eta)\phi(\eta, \mu)\, d\eta, \quad \mu > 0,\ 0 \leq \eta \leq 1 \quad (10\text{-}102)$$

We operate on both sides of this equation by the operator

$$\int_0^1 W(\mu)\phi(-\mu', \mu)\, d\mu, \quad \mu' > 0 \quad (10\text{-}103)$$

and obtain

$$\int_0^1 W(\mu)\phi(-\mu', \mu) f(0, \mu)\, d\mu = A(\eta_0)\int_0^1 W(\mu)\phi(-\mu', \mu)\phi(\eta_0, \mu)\, d\mu$$
$$+ \int_0^1 A(\eta)\left[\int_0^1 W(\mu)\phi(-\mu', \mu)\phi(\eta, \mu)\, d\mu\right] d\eta, \quad \mu' > 0 \quad (10\text{-}104)$$

The integrals with respect to μ on the right-hand side of Eq. 10-104 can be evaluated by means of the relations given by Eqs. 10-73 and 10-74, that is,

$$\int_0^1 W(\mu)\phi(-\mu', \mu)\phi(\eta_0, \mu)\, d\mu = \tfrac{1}{4}\omega^2\mu'\eta_0 X(-\mu')$$
$$= \tfrac{1}{2}\omega\mu'(\eta_0 + \mu')X(-\mu')\phi(\eta_0, -\mu') \quad (10\text{-}105)$$

$$\int_0^1 W(\mu)\phi(-\mu', \mu)\phi(\eta, \mu)\, d\mu = \tfrac{1}{2}\omega\eta(\eta_0 + \mu')X(-\mu')\phi(-\mu', \eta)$$
$$= \tfrac{1}{2}\omega\mu'(\eta_0 + \mu')X(-\mu')\phi(\eta, -\mu') \quad (10\text{-}106)$$

Substitution of Eqs. 10-105 and 10-106 into Eq. 10-104 yields

$$\int_0^1 W(\mu)\phi(-\mu', \mu) f(0, \mu)\, d\mu$$
$$= \tfrac{1}{2}\omega\mu'(\eta_0 + \mu')X(-\mu')\left[A(\eta_0)\phi(\eta_0, -\mu') + \int_0^1 A(\eta)\phi(\eta, -\mu')\, d\eta\right], \quad \mu' > 0 \quad (10\text{-}107)$$

Interchanging μ and μ', we obtain

$$\frac{2}{\omega\mu(\eta_0+\mu)X(-\mu)}\int_0^1 W(\mu')\phi(-\mu,\mu')f(0,\mu')\,d\mu'$$

$$= A(\eta_0)\phi(\eta_0,-\mu)+\int_0^1 A(\eta)\phi(\eta,-\mu)\,d\eta,\quad \mu>0 \quad (10\text{-}108)$$

and, replacing μ by $-\mu$, we have

$$-\frac{2}{\omega\mu(\eta_0-\mu)X(\mu)}\int_0^1 W(\mu')\phi(\mu,\mu')f(0,\mu')\,d\mu'$$

$$= A(\eta_0)\phi(\eta_0,\mu)+\int_0^1 A(\eta)\phi(\eta,\mu)\,d\eta,\quad \mu<0 \quad (10\text{-}109)$$

Equation 10-109 provides the desired relation to eliminate the expansion coefficients from the right-hand side of Eq. 10-101a. Substituting Eq. 10-109 into Eq. 10-101a, we obtain

$$I_e(0,\mu) = I_p(0,\mu) - \frac{2}{\omega\mu(\eta_0-\mu)X(\mu)}\int_0^1 W(\mu')\phi(\mu,\mu')f(0,\mu')\,d\mu',\quad \mu<0 \quad (10\text{-}110a)$$

Equation 10-110a is a simple expression for the emergent radiation intensity at $\tau = 0$. The integral in Eq. 10-110a is nonsingular.

For computational purposes it may be convenient to rearrange Eq. 10-110a in the form

$$I_e(0,\mu) = I_p(0,\mu) - \frac{1}{X(\mu)}\int_0^1 \gamma(\mu')\left(\frac{1}{\eta_0-\mu}+\frac{1}{\mu-\mu'}\right)f(0,\mu')\,d\mu',\quad \mu<0 \quad (10\text{-}110b)$$

Since

$$W(\mu') = (\eta_0-\mu')\gamma(\mu') \quad (10\text{-}111)$$

$$\phi(\mu,\mu') = \tfrac{1}{2}\omega\mu\frac{1}{\mu-\mu'} \quad (10\text{-}112)$$

$$\frac{\eta_0-\mu'}{(\eta_0-\mu)(\mu-\mu')} = \frac{1}{\eta_0-\mu}+\frac{1}{\mu-\mu'} \quad (10\text{-}113)$$

We examine now special cases of the above half-space problem.

(a) Semi-infinite Half-Space Externally Irradiated by Isotropic Radiation

Consideration will be given to a situation in which a semi-infinite medium at a uniform temperature T_0 is irradiated externally through the transparent boundary at $\tau = 0$ by an isotropic radiation of intensity I_0. The mathematical formulation of this problem is given as

$$\mu \frac{\partial I(\tau, \mu)}{\partial \tau} + I(\tau, \mu) = (1 - \omega) \frac{\bar{\sigma} T_0^4}{\pi} + \frac{\omega}{2} \int_{-1}^{1} I(\tau, \mu') \, d\mu'$$

$$\text{for } 0 \leq \tau \leq \infty, \; -1 \leq \mu \leq 1, \; \omega < 1 \qquad (10\text{-}114)$$

with the boundary conditions:

$$I(0, \mu) = I_0, \quad \mu > 0$$

and at $\tau \to \infty$ the solution converges to a particular solution $I_p(\tau, \mu)$ of the equation of radiative transfer.

A particular solution of Eq. 10-114 is given as (see Table 8-1)

$$I_p(\tau, \mu) = \frac{\bar{\sigma} T_0^4}{\pi} \qquad (10\text{-}115)$$

Then the complete solution of this problem becomes (see Eq. 10-95)

$$I(\tau, \mu) = A(\eta_0)\phi(\eta_0, \mu)e^{-\tau/\eta_0} + \int_0^1 A(\eta)\phi(\eta, \mu)e^{-\tau/\eta} \, d\eta + \frac{\bar{\sigma} T_0^4}{\pi} \qquad (10\text{-}116)$$

The expansion coefficients $A(\eta_0)$ and $A(\eta)$ are immediately determined from Eqs. 10-97 and 10-98, respectively, by setting $f(0, \mu)$ as (see Eq. 10-96b)

$$f(0) \equiv I_0 - \frac{\bar{\sigma} T_0^4}{\pi} = \text{constant} \qquad (10\text{-}117)$$

Then the discrete expansion coefficient $A(\eta_0)$ becomes

$$A(\eta_0) = -\left(\frac{2}{\omega \eta_0}\right)^2 \frac{f(0)}{X(\eta_0)} \int_0^1 W(\mu)\phi(\eta_0, \mu) \, d\mu = -\frac{2f(0)}{\omega \eta_0 X(\eta_0)} \qquad (10\text{-}118)$$

Here we have utilized Eq. 10-75 to evaluate the integral in Eq. 10-118.

The continuum coefficient $A(\eta)$ becomes

$$A(\eta) = \frac{g(\omega, \eta) f(0)}{W(\eta)} \int_0^1 W(\mu)\phi(\eta, \mu) \, d\mu = \frac{\omega \eta g(\omega, \eta) f(0)}{2W(\eta)} \qquad (10\text{-}119)$$

where we have utilized Eq. 10-78 to evaluate the integral in Eq. 10-119.

Knowing the expansion coefficients, we can determine the radiation intensity $I(\tau, \mu)$, the incident radiation $G(\tau, \mu)$, and the net radiative heat flux $q^r(\tau)$ anywhere in the medium from Eqs. 10-116, 10-99, and 10-100, respectively. For example, the net radiative heat flux is given as

$$q^r(\tau) = 2\pi(1 - \omega) f(0) \left[-\frac{2}{\omega X(\eta_0)} e^{-\tau/\eta_0} + \frac{\omega}{2} \int_0^1 \frac{\eta^2 g(\omega, \eta)}{W(\eta)} e^{-\tau/\eta} \, d\eta \right] \qquad (10\text{-}120)$$

The angular distribution of the emergent radiation intensity can be evaluated from Eq. 10-101 or Eq. 10-110b; the latter expression is used here since it is simpler. We obtain

$$I_e(0, \mu) = I_p(0, \mu) - \frac{f(0)}{X(\mu)}\left[\frac{1}{\eta_0 - \mu}\int_0^1 \gamma(\mu')\, d\mu' + \int_0^1 \frac{\gamma(\mu')}{\mu - \mu'}\, d\mu'\right] \tag{10-121a}$$

When the integrals are evaluated by means of Eqs. 10-37 and 10-33, we obtain

$$I_e(0, \mu) = I_p(0, \mu) - \frac{f(0)}{(\eta_0 - \mu)X(\mu)} + f(0), \quad \mu < 0 \tag{10-121b}$$

or

$$I_e(0, \mu) = I_0 - \frac{1}{(\eta_0 - \mu)X(\mu)}\left(I_0 - \frac{\bar{\sigma} T_0^4}{\pi}\right), \quad \mu < 0 \tag{10-121c}$$

(b) Semi-infinite Half-Space Externally Irradiated by Parallel Rays

Consideration will be given to radiative heat transfer in a semi-infinite medium having a uniform temperature T_0 and a transparent boundary at $\tau = 0$. The medium is irradiated by parallel rays obliquely incident on the boundary surface at $\tau = 0$ at an angle $\theta = \cos^{-1} \mu_0$ with the positive τ axis. The mathematical formulation of this problem can be given as

$$\mu \frac{\partial I(\tau, \mu)}{\partial \tau} + I(\tau, \mu) = (1 - \omega)\frac{\bar{\sigma} T_0^4}{\pi} + \frac{\omega}{2}\int_{-1}^{1} I(\tau, \mu')\, d\mu'$$

$$\text{for } 0 \le \tau < \infty,\ -1 \le \mu \le 1,\ \omega < 1 \tag{10-122}$$

with the boundary conditions:

$$I(0, \mu) = I_0\, \delta(\mu - \mu_0), \quad \mu > 0 \tag{10-123}$$

and at $\tau \to \infty$ the solution converges to a particular solution $I_p(\tau, \mu)$ of the equation of radiative transfer.

The complete solution of this problem is given as

$$I(\tau, \mu) = A(\eta_0)\phi(\eta_0, \mu)e^{-\tau/\eta_0} + \int_0^1 A(\eta)\phi(\eta, \mu)e^{-\tau/\eta}\, d\eta + \frac{\bar{\sigma} T_0^4}{\pi} \tag{10-124}$$

The expansion coefficients $A(\eta_0)$ and $A(\eta)$ are immediately determined from Eqs. 10-97 and 10-98, respectively, by setting in these relations

$$f(0, \mu) \equiv I_0\, \delta(\mu - \mu_0) - \frac{\sigma T_0^4}{\pi}, \quad \mu > 0 \tag{10-125}$$

We obtain

$$A(\eta_0) = -\left(\frac{2}{\omega\eta_0}\right)^2 \frac{1}{X(\eta_0)} \int_0^1 W(\mu)\phi(\eta_0, \mu)\left[I_0\,\delta(\mu - \mu_0) - \frac{\bar{\sigma}T_0^4}{\pi}\right] d\mu$$

$$= \frac{2}{\omega\eta_0 X(\eta_0)}\left[-I_0\gamma(\mu_0) + \frac{\bar{\sigma}T_0^4}{\pi}\right] \tag{10-126}$$

$$A(\eta) = \frac{g(\omega, \eta)}{W(\eta)} \int_0^1 W(\mu)\phi(\eta, \mu)\left[I_0\,\delta(\mu - \mu_0) - \frac{\bar{\sigma}T_0^4}{\pi}\right] d\mu$$

$$= \frac{g(\omega, \eta)}{W(\eta)}\left[W(\mu_0)\phi(\eta, \mu_0)I_0 - \frac{\omega\eta}{2}\frac{\bar{\sigma}T_0^4}{\pi}\right] \tag{10-127}$$

Knowing the expansion coefficients, we can evaluate the radiation intensity $I(\tau, \mu)$, the incident radiation $G(\tau)$, and the net radiative heat flux $q^r(\tau)$ from Eqs. 10–124, 10-99, and 10-100, respectively:

$$I(\tau, \mu) = A(\eta_0)\phi(\eta_0, \mu)e^{-\tau/\eta_0} + \int_0^1 A(\eta)\phi(\eta, \mu)e^{-\tau/\eta}\, d\eta + \frac{\bar{\sigma}T_0^4}{\pi} \tag{10-128}$$

$$G(\tau) = 2\pi\left[A(\eta_0)e^{-\tau/\eta_0} + \int_0^1 A(\eta)e^{-\tau/\eta}\, d\eta + \frac{2\bar{\sigma}T_0^4}{\pi}\right] \tag{10-129}$$

$$q^r(\tau) = 2\pi(1 - \omega)\left[A(\eta_0)\eta_0 e^{-\tau/\eta_0} + \int_0^1 A(\eta)\eta e^{-\tau/\eta}\, d\eta\right] \tag{10-130}$$

The radiation intensity, for example, is given explicitly as

$$I(\tau, \mu) = \frac{2}{\omega\eta_0 X(\eta_0)}\left[-I_0\gamma(\mu_0) + \frac{\bar{\sigma}T_0^4}{\pi}\right]\phi(\eta_0, \mu)e^{-\tau/\eta_0} + \int_0^1 \frac{g(\omega, \eta)}{W(\eta)} \times \left[W(\mu_0)\phi(\eta, \mu_0)I_0 - \frac{\omega\eta}{2}\frac{\bar{\sigma}T_0^4}{\pi}\right]\phi(\eta, \mu)e^{-\tau/\eta}\, d\eta + \frac{\bar{\sigma}T_0^4}{\pi} \tag{10-131}$$

In many engineering applications the angular distribution of emergent radiation intensity is of interest. This quantity is determined by setting in Eq. 10-131 $\tau = 0$ and evaluating it for $\mu < 0$. In this case we note that the integral in Eq. 10-131 is nonsingular.

10-8 PARTICULAR SOLUTIONS OF EQUATION OF RADIATIVE TRANSFER IN PLANE-PARALLEL GEOMETRY FOR ISOTROPIC SCATTERING

Consider the equation of radiative transfer in plane-parallel geometry for azimuthally symmetric radiation in the form

$$\mu\frac{\partial I(\tau, \mu)}{\partial \tau} + I(\tau, \mu) = H(\tau) + \frac{\omega}{2}\int_{-1}^1 I(\tau, \mu')\, d\mu' \tag{10-132}$$

Three cases of practical interest can be obtained from this equation.

1. $H(\tau) = (1 - \omega)[n^2\bar{\sigma}T^4(\tau)/\pi]$ with $0 < \omega < 1$ characterizes absorbing, emitting, isotropically scattering medium.

2. $H(\tau) = 0$ with $\omega = 1$ characterizes a purely scattering medium or a gray medium in radiative equilibrium (see Eq. 8-136).

3. $H(\tau) = g(\tau)/4\pi\kappa$ with $\omega = 1$ characterizes an absorbing and emitting medium with internal energy sources (see Eq. 8-203a).

In seeking the solution of Eq. 10-132 a frequently invoked approach is to write the solution as a sum of a particular solution $I_p(\tau, \mu)$ (i.e., a solution that satisfies the original inhomogeneous equation but not necessarily its boundary conditions) and a homogeneous solution $\psi(\tau, \mu)$ in the form

$$I(\tau, \mu) = \psi(\tau, \mu) + I_p(\tau, \mu) \tag{10-133}$$

where $\psi(\tau, \mu)$ is the solution of the homogeneous part of Eq. 10-132, that is,

$$\mu \frac{\partial \psi(\tau, \mu)}{\partial \tau} + \psi(\tau, \mu) = \frac{\omega}{2} \int_{-1}^{1} \psi(\tau, \mu')\, d\mu' \tag{10-134}$$

Several particular solutions of the inhomogeneous equation 10-132 have been determined by Lundquist and Horak [26], Hans-Dieter Freund [27], and Özişik and Siewert [28]. We present in Table 10-6 particular solutions of Eq. 10-132 for several different types of the inhomogeneous term $H(\tau)$ for $0 < \omega \le 1$.

To illustrate the approach for the derivation of the results shown in Table 10-6 we consider now the determination of a particular solution of Eq. 10-132 for an inhomogeneous term of the form

$$H(\tau) = e^{-\tau/\eta}, \quad |\eta| > 1 \tag{10-135}$$

for the case $0 < \omega < 1$ which is given in this table.

We assume a particular solution in the form

$$I_p(\tau, \mu) = F(\eta, \mu)e^{-\tau/\eta} \tag{10-136}$$

where $F(\eta, \mu)$ is to be determined. Substitution of Eqs. 10-135 and 10-136 into Eq. 10-132 yields

$$\mu F(\eta, \mu)\left(-\frac{1}{\eta}\right)e^{-\tau/\eta} + F(\eta, \mu)e^{-\tau/\eta} = e^{-\tau/\eta} + \frac{\omega}{2} e^{-\tau/\eta} \int_{-1} F(\eta, \mu')\, d\mu'$$

or

$$\left(1 - \frac{\mu}{\eta}\right) F(\eta, \mu) = 1 + \frac{\omega}{2} \int_{-1}^{1} F(\eta, \mu')\, d\mu' \tag{10-137}$$

This equation can be written in the form

$$F(\eta, \mu) = \frac{\eta}{\eta - \mu}\left[1 + \frac{\omega}{2} K(\eta)\right] \tag{10-138a}$$

Table 10-6 Particular Solutions of Equation of Radiative Transfer in Plane-Parallel Geometry for Isotropic Scattering[a]

$H(\tau)$	$I_p(\tau, \mu)$, Particular Solution[b]	Remarks
1	$(1-\omega)^{-1}$	$\omega \neq 1$
	$-\frac{3}{2}\tau^2 + 3\tau\mu - 3\mu^2$	$\omega = 1$
τ	$(1-\omega)^{-1}(\tau - \mu)$	$\omega \neq 1$
	$-\frac{1}{2}\tau^3 + \frac{3}{2}\tau^2\mu - 3\tau\mu^2 + 3\mu^3$	$\omega = 1$
τ^2	$(1-\omega)^{-1}\left(\tau^2 - 2\tau\mu + 2\mu^2 + \frac{2}{3}\frac{\omega}{1-\omega}\right)$	$\omega \neq 1$
	$-\frac{1}{4}\tau^4 + \tau^3\mu - 3\tau^2\mu^2 + 6\tau\mu^3 - 6\mu^4 - \frac{6}{5}(-\frac{3}{2}\tau^2 + 3\tau\mu - 3\mu^2)$	$\omega = 1$
τ^3	$(1-\omega)^{-1}\left[\tau^3 - 3\tau^2\mu + 6\tau\mu^2 - 6\mu^3 + \frac{2\omega}{1-\omega}(\tau - \mu)\right]$	$\omega \neq 1$
	$-\frac{3}{20}\tau^5 + \frac{3}{4}\tau^4\mu - 3\tau^2\mu^2 + 9\tau^2\mu^3 - 18\tau\mu^4 + 18\mu^5 - \frac{18}{5}(-\frac{1}{2}\tau^3 + \frac{3}{2}\tau^2\mu - 3\tau\mu^2 + 3\mu^3)$	$\omega = 1$
$e^{-\tau/\eta}$	$-\frac{2}{\omega}\,\delta(\eta - \mu)e^{-\tau/\eta}$	$\lvert\eta\rvert < 1$
	$\left(1 - \omega\eta \tanh^{-1}\frac{1}{\eta}\right)^{-1} \frac{\eta}{\eta - \mu} e^{-\tau/\eta}$	$\lvert\eta\rvert > 1$ and $\left(1 - \omega\eta \tanh^{-1}\frac{1}{\eta}\right) \neq 0$

[a] From M. N. Özişik and C. E. Siewert [28].
[b] Particular solution $I_p^{(n)}(\tau, \mu)$ for an inhomogeneous term $H(\tau) = A\tau^n$ for $\omega \neq 1$ is given by Lundquist and Horak [26] as

$$I_p^{(n)}(\tau, \mu) = \frac{A}{1-\omega} \sum_{r=0,1,2,\ldots}^{n} (-1)^r \frac{n!}{(n-r)!} \tau^{n-r}\mu^r + \frac{\omega}{1-\omega} \times \sum_{s=1,2,\ldots}^{\substack{(n-1)/2 \ (n \text{ odd}) \\ n/2 \ (n \text{ even})}} \frac{n!}{(n-2s)!(2s+1)} I_p^{(n-2s)}(\tau, \mu)$$

Table 10-6 (continued)

$H(\tau)$	$I_p(\tau, \mu)$, Particular Solution	Remarks
$\cos\frac{\tau}{\eta}$	$\left(1 - \omega\eta \tan^{-1}\frac{1}{\eta}\right)^{-1} (\eta^2 + \mu^2)^{-1}$ $\times \left(\eta^2 \cos\frac{\tau}{\eta} + \eta\mu \sin\frac{\tau}{\eta}\right)$	$\left(1 - \omega\eta \tan^{-1}\frac{1}{\eta}\right) \neq 0$
$\sin\frac{\tau}{\eta}$	$-\left(1 - \omega\eta \tan^{-1}\frac{1}{\eta}\right)^{-1} (\eta^2 + \mu^2)^{-1}$ $\times \left(\eta\mu \cos\frac{\tau}{\eta} - \eta^2 \sin\frac{\tau}{\eta}\right)$	$\left(1 - \omega\eta \tan^{-1}\frac{1}{\eta}\right) \neq 0$
$\int_{-1}^{1} A(\eta)e^{-\tau/\eta}\, d\eta$	$-\frac{2}{\omega} A(\eta)e^{-\tau/\eta}$	
$E_N(\tau)$	$-\frac{2}{\omega} \mu^{N-2}e^{-\tau/\mu}$	$0 < \mu < 1$
	0	$-1 < \mu < 0$

where we have defined

$$K(\eta) \equiv \int_{-1}^{1} F(\eta, \mu')\, d\mu' \tag{10-138b}$$

We operate on both sides of Eq. 10-138a by the operator $\int_{-1}^{1} d\mu$ and obtain

$$K(\eta) = \left[1 + \frac{\omega}{2} K(\eta)\right] \eta \int_{-1}^{1} \frac{1}{\eta - \mu}\, d\mu \tag{10-139}$$

The integral on the right-hand side can be evaluated for $|\eta| > 1$, and we find[10]

$$K(\eta) = \left[1 + \frac{\omega}{2} K(\eta)\right] \eta \left(2 \tanh^{-1}\frac{1}{\eta}\right) \tag{10-140a}$$

and solving for $K(\eta)$ yields

$$K(\eta) = \frac{2\eta \tanh^{-1}(1/\eta)}{1 - \omega\eta \tanh^{-1}(1/\eta)} \tag{10-140b}$$

Substituting $K(\eta)$ into Eq. 10-138a gives

$$F(\eta, \mu) = \frac{\eta}{\eta - \mu} \frac{1}{1 - \omega\eta \tanh^{-1}(1/\eta)} \tag{10-141}$$

Substitution of Eq. 10-141 into Eq. 10-136 gives the particular solution of Eq. 10-132 for an inhomogeneous term of the form $e^{-\tau/\eta}$ as

$$I_p(\tau, \mu) = \frac{\eta}{\eta - \mu} \frac{1}{1 - \omega\eta \tanh^{-1}(1/\eta)} e^{-\tau/\eta} \quad \text{for } 0 < \omega < 1, \ |\eta| > 1 \tag{10-142}$$

which is given in Table 10-6. The validity of the particular solutions given in Table 10-6 can be readily verified by direct substitution into Eq. 10-132.

REFERENCES

1a. K. M. Case, "Elementary Solutions of the Transport Equation and Their Applications," *Ann. Phys. (N.Y.)*, **9,** 1–23, 1960.

1b. K. M. Case, "Recent Developments in Neutron Transport Theory," *Michigan Memorial Phoenix Project, Lectures Presented at the Neutron Physics Conference,* University of Michigan, Ann Arbor, Mich., June 12–17, 1961.

2. K. M. Case and P. F. Zweifel, *Linear Transport Theory*, Addison-Wesley Publishing Co., Reading, Mass., 1967.

3. E. Inönü and P. F. Zweifel, Eds., *Developments in Transport Theory*, Academic Press, New York, 1967.

4. J. R. Mika, "Neutron Transport with Anisotropic Scattering," *Nucl. Sci. Eng.*, **11,** 415–427, 1961.

5. F. Shure and M. Natelson, "Anisotropic Scattering in Half-Space Transport Problems," *Ann. Phys. (N.Y.)*, **26,** 274–291, 1964.

6. I. Kuščer, N. J. McCormick, and G. C. Summerfield, "Orthogonality of Case's Eigenfunctions in One-Speed Transport Theory," *Ann. Phys. (N.Y.)*, **30,** 411–421, 1964.

7. N. J. McCormick and I. Kuščer, "Half-Space Neutron Transport with Linearly Anisotropic Scattering," *J. Math. Phys.*, **6,** 1939–1945, 1965.

8. N. J. McCormick and I. Kuščer, "Bi-orthogonality Relations for Solving Half-Space Transport Problems," *J. Math. Phys.*, **7,** 2036–2045, 1966.

9. J. H. Ferziger and G. M. Simmons, "Application of Case's Method to Plane-Parallel Radiative Transfer," *Intern. J. Heat Mass Transfer*, **9,** 987–992, 1966.

10. J. T. Kriese and C. E. Siewert, "Radiative Transfer in a Conservative Finite Slab with an Internal Source," *Intern. J. Heat Mass Transfer*, **13,** 1349–1357, 1970.

11. M. N. Özişik and C. E. Siewert, "On the Normal-Mode Expansion Technique for Radiative Transfer in a Scattering, Absorbing and Emitting Slab with Specularly Reflecting Boundaries," *Intern. J. Heat Mass Transfer*, **12,** 611–620, 1969.

12. H. L. Beach, M. N. Özişik, and C. E. Siewert, "Radiative Transfer in Linearly Anisotropic-Scattering, Conservative and Non-conservative Slabs with Reflective Boundaries," *Intern. J. Heat Mass Transfer*, **14,** 1551–1565, 1971.

13. C. E. Siewert and P. F. Zweifel, "An Exact Solution of Equations of Radiative Transfer for Local Thermodynamic Equilibrium in the Non-gray Case: Picket Fence Approximation," *Ann. Phys. (N.Y.)*, **36**, 61–85, 1966.

14. C. E. Siewert and P. F. Zweifel, "Radiative Transfer, II," *J. Math. Phys.*, **7**, 2092–2102, 1966.

15. G. M. Simmons and J. H. Ferziger, "Non-gray Radiative Heat Transfer Between Parallel Plates," *Intern. J. Heat Mass Transfer*, **11**, 1611–1620, 1968.

16. C. E. Seiwert and M. N. Özişik, "An Exact Solution in the Theory of Line Formation," *Monthly Notices Roy. Astron. Soc.*, **146**, 351–360, 1969.

17. R. J. Reith, Jr., C. E. Siewert, and M. N. Özişik, "Non-gray Radiative Heat Transfer in Conservative Plane-Parallel Media with Reflecting Boundaries," *J. Quant. Spectry. Radiative Transfer*, **11**, 1441–1462, 1971.

18. N. J. McCormick and I. Kuščer, "Singular Eigenfunction Expansions in Neutron Transport Theory," in *Advances in Nuclear Science and Technology*, Vol. 7, edited by E. J. Henley and J. Lewins (in press.)

19. S. Chandrasekhar, *Radiative Transfer*, Oxford University Press, London, 1950; also Dover Publications, New York, 1960.

20. V. Kourganoff, *Basic Methods in Transfer Problems*, Dover Publications, New York, 1963.

21. N. I. Muskhelishvili, *Singular Integral Equations*, Noordhoff, Groningen, Holland, 1953.

22. K. M. Case, F. de Hoffmann, and G. Placzek, *Introduction to the Theory of Neutron Diffusion*, Vol. 1, Los Alamos Scientific Laboratory, Los Alamos, N.M., 1953.

23. C. C. Lii, "Normal-Mode Expansion Technique in Radiative Heat Transfer," Ph.D. Dissertation, Mechanical and Aerospace Engineering Department, North Carolina State University, Raleigh, N.C., 1972.

24. D. W. N. Stibbs and R. E. Weir, "On the *H*-Function for Isotropic Scattering," *Monthly Notices Roy. Astron. Soc.*, **119**, No. 5, 512–525, 1959.

25. N. J. McCormick, "One Speed Neutron Transport Problems in Plane Geometry," Ph.D. Dissertation, Nuclear Engineering Department, University of Michigan, Ann Arbor, Mich., 1965.

26. C. A. Lundquist and H. G. Horak, "The Transfer of Radiation by an Emitting Atmosphere, IV," *Astrophys. J.*, **121**, 175–182, 1955.

27. Hans-Dieter Freund, "Particular Solutions of the Unhomogeneous, One-Velocity Boltzmann Equation in Plane Geometry," *Atomkernenergie*, **14**, 222, 1969.

28. M. N. Özişik and C. E. Siewert, "Several Particular Solutions of One-Speed Transport Equation," *Nucl. Sci. Eng.*, **40**, 491–494, 1970.

NOTES

[1] The case $\omega > 1$ is of interest in certain problems in the nuclear reactor theory.

[2] The Cauchy principal-value integration of a function $f(x)$ from $x = a$ to $x = b$, in which function $f(x)$ becomes infinite at a point $x = c$ inside the range of integration (a, b), is defined as

$$\int_a^b Pf(x)\,dx = \lim_{\varepsilon\to 0}\left[\int_a^{c-\varepsilon} f(x)\,dx + \int_{c+\varepsilon}^b f(x)\,dx\right]$$

when the limit exists. For example, in the integral $\int_{-1}^{2}(dx/x)$, the function $1/x$ is infinite at $x = 0$. The Cauchy principal value of this integral is evaluated as

$$\int_{-1}^{2} P\frac{dx}{x} = \lim_{\varepsilon\to 0}\left(\int_{-1}^{-\varepsilon}\frac{dx}{x} + \int_{\varepsilon}^{2}\frac{dx}{x}\right) = \ln 2$$

Hence the limit exists.

[3] A discussion of a suitably well-behaved function is given in Appendix G of reference 2.

[4] If the negative half-range of μ (i.e., $-1 \le \mu \le 0$) had been considered, the weight function would be given as

$$W(\mu) = (\eta_0 + \mu)\gamma(\mu)$$

[5] The normalization integral $N(\eta_0)$ can be evaluated as follows:

$$N(\eta_0) = \int_{-1}^{1} \mu\phi^2(\eta_0, \mu)\, d\mu = \left(\frac{\omega\eta_0}{2}\right)^2 \int_{-1}^{1} \frac{\mu}{(\eta_0 - \mu)^2}\, d\mu$$

$$= \left(\frac{\omega\eta_0}{2}\right)^2 \int_{-1}^{1} \frac{1}{\eta_0 - \mu}\left(\frac{\eta_0}{\eta_0 - \mu} - 1\right) d\mu$$

$$= \frac{\omega\eta_0}{2}\left[\frac{\omega\eta_0^2}{2}\int_{-1}^{1}\frac{d\mu}{(\eta_0 - \mu)^2} - \frac{\omega\eta_0}{2}\int_{-1}^{1}\frac{d\mu}{\eta_0 - \mu}\right] \tag{1}$$

From the normalization condition 10-6 we have

$$\frac{\omega\eta_0}{2}\int_{-1}^{1}\frac{d\mu}{\eta_0 - \mu} = 1 \tag{2}$$

Substituting Eq. 2 into Eq. 1, we obtain

$$N(\eta_0) = \frac{\omega\eta_0}{2}\left[\frac{\omega\eta_0^2}{2}\int_{-1}^{1}\frac{d\mu}{(\eta_0 - \mu)^2} - 1\right]$$

$$= \frac{\omega\eta_0}{2}\left(\frac{\omega\eta_0^2}{\eta_0^2 - 1} - 1\right) = \frac{\omega\eta_0^3}{2}\left(\frac{\omega}{\eta_0^2 - 1} - \frac{1}{\eta_0^2}\right) \tag{3}$$

which is the desired result.

[6] The evaluation of the integral in Eq. 10-42 requires some discussion. The continuum normal modes appearing in this equation are given as (see Eq. 10-16)

$$\phi(\xi, \mu) = \frac{\omega\xi}{2}\frac{P}{\xi - \mu} + \lambda(\xi)\,\delta(\xi - \mu), \quad \xi = \eta \text{ or } \eta' \in (-1, 1) \tag{1}$$

Substituting these normal modes into Eq. 10-42 and performing the multiplication yield four separate terms. The integrals in three of these terms can be evaluated readily. The fourth term, however, involves an integral in the form

$$\tfrac{1}{4}\omega^2\eta\eta'\int_{-1}^{1}\frac{P}{\eta - \mu}\,\frac{P}{\eta' - \mu}\,\mu\, d\mu \tag{2}$$

This integral can be evaluated by making use of the Poincaré-Bertrand formula given by

Muskhelishvili [21, p. 56] as

$$\int_L P\frac{1}{\eta-\mu}\left[\int_L P\frac{f(\eta',\mu)}{\eta'-\mu}d\eta'\right]d\mu = \int_L P\,d\eta'\int_L P\frac{f(\eta',\mu)}{(\eta-\mu)(\eta'-\mu)}\,d\mu + \pi^2 f(\eta,\eta)$$

$$= \int_L \frac{d\eta'}{\eta-\eta'}\int_L Pf(\eta',\mu) \times \left(\frac{1}{\eta'-\mu}-\frac{1}{\eta-\mu}\right)d\mu + \pi^2 f(\eta,\eta) \quad (3)$$

A mnemonic form of Eq. 3 may be written as

$$\frac{P}{\eta-\mu}\frac{P}{\eta'-\mu} = \frac{1}{\eta-\eta'}\left(\frac{P}{\eta'-\mu}-\frac{P}{\eta-\mu}\right) + \pi^2\,\delta(\eta-\mu)\,\delta(\eta'-\mu) \quad (4)$$

When all the integrals in Eq. 10-42 are evaluated and the results are combined, one obtains the right-hand side of Eq. 10-42 with $N(\eta)$ as given by Eq. 10-43a.

[7] The integral in Eq. 10-46a is evaluated by making use of the Poincaré-Bertrand formula [21] discussed in note 6.

[8] The results given by Eqs. 10-63, for example, can be derived as follows. Consider the discrete normal modes rearranged in the form

$$(\eta_0 \mp \mu)\phi(\pm\eta_0,\mu) = \frac{\omega\eta_0}{2} \quad (1)$$

(a) By operating on both sides of Eq. 1 by the operator $\int_{-1}^{1} d\mu$ we obtain

$$\eta_0\int_{-1}^{1}\phi(\pm\eta_0,\mu)\,d\mu \mp \int_{-1}^{1}\mu\phi(\pm\eta_0,\mu)\,d\mu = \omega\eta_0 \quad (2)$$

By the normalization condition $\int_{-1}^{1}\phi(\pm\eta_0,\mu)\,d\mu = 1$, Eq. 2 becomes

$$\int_{-1}^{1}\mu\phi(\pm\eta_0,\mu)\,d\mu = \pm\eta_0(1-\omega) \quad (3)$$

which is the result given by Eq. 10-63a.

(b) By operating on both sides of Eq. 1 by the operator $\int_{-1}^{1}\mu\,d\mu$ we obtain

$$\eta_0\int_{-1}^{1}\mu\phi(\pm\eta_0,\mu)\,d\mu \mp \int_{-1}^{1}\mu^2\phi(\pm\eta_0,\mu)\,d\mu = 0 \quad (4)$$

Substitution of Eq. 3 into Eq. 4 yields

$$\int_{-1}^{1}\mu^2\phi(\pm\eta_0,\mu)\,d\mu = \eta_0^{\,2}(1-\omega) \quad (5)$$

which is the result given by Eq. 10-63b.

[9] The relation defining $\gamma^{(n)}$ is obtained from Eqs. 10-81 and 10-29a as

$$\gamma^{(n)} = \frac{\omega}{2\sqrt{1-\omega}}\int_0^1 \mu^{n+1}H(\mu)\frac{d\mu}{\eta_0-\mu} \quad (1)$$

This relation can be used to evaluate $\gamma^{(n)}$. However, for small values of ω difficulty may arise in numerical computations since η_0 approaches unity as ω approaches zero. To alleviate this difficulty Eq. 1 is written in an alternative form as [23]

$$\gamma^{(n)} = \frac{{\eta_0}^n}{\sqrt{1-\omega}} - \frac{\omega}{2\sqrt{1-\omega}}({\eta_0}^n H_0 + \eta_0^{n-1} H_1 + \cdots + H_n) \tag{2a}$$

where

$$H_n \equiv \int_0^1 \mu^n H(\mu)\, d\mu \tag{2b}$$

and $H(\mu)$ is Chandrasekhar's $H(\mu)$ function for isotropic scattering. To obtain Eq. 2 the following relations are utilized:

$$\frac{\mu^{n+1}}{\eta_0 - \mu} = \frac{\eta_0^{n+1}}{\eta_0 - \mu} - ({\eta^n}_0 + \eta_0^{n-1}\mu + \eta_0^{n-2}\mu^2 + \cdots + \eta_0\mu^{n-1} + \mu^n) \tag{3}$$

where

$$\frac{\omega\eta_0}{2}\int_0^1 H(\mu)\frac{d\mu}{\eta_0 - \mu} = 1 \tag{4}$$

[10] For $|\eta| < 1$ the integral in Eq. 10-139 is the principal-value integral and its evaluation requires special consideration.

CHAPTER 11

Pure Radiative Heat Transfer in Participating Media

There are several applications of radiative heat transfer in participating media in which conduction and convection are negligible. For example, in heat transfer through porous materials such as fibers, powders, and many others used as lightweight insulators for both low- and high-temperature applications, radiation is the dominant mode of energy transfer. These insulating materials possess a large amount of void space, and some of them have voids in the form of dispersed bubbles. As radiation is transmitted through such a medium, it is absorbed, scattered, and reradiated at the surface of the fibrous materials or at the surface of the bubbles.

In rocket applications, for example, when aluminized solid propellants are burned, the exhaust gas contains a significant number of micron-size particles which scatter radiation. Therefore heat transfer in the flow of highly turbulent gas at high temperatures and high speeds containing scattering particles is characterized as a radiative problem in an absorbing, emitting, scattering medium.

For the field of astrophysics, the theory and applications of radiative transfer are given in the books by Chandrasekhar [1], Kourganoff [2], Sobolev [3], and Busbridge [4], as well as in numerous other publications that we do not intend to cite here.

The literature on engineering applications of radiative heat transfer is ever growing. Usiskin and Sparrow [5] and Meghreblian [6] investigated radiative heat transfer for an absorbing, emitting, plane-parallel gas layer between two emitting, parallel, black plates. Howell and Perlmutter [7, 8] used the Monte Carlo method to solve similar problems with reflecting boundaries. Viskanta and Grosh [9], Love and Grosh [10], and Hsia and Love [11] solved numerically radiative heat transfer in an absorbing, emitting, scattering slab. Cess and Sotak [12] used the exponential kernel approximation, and Edwards and Bobco [13] and Bobco [14] applied the method of moments to solve approximately the radiative heat transfer in a slab.

Heaslet and Warming [15, 16] used the methods and tabulated the functions developed by Chandrasekhar [1] to obtain highly accurate solutions to radiative heat transfer in an absorbing, emitting slab. Crosbie and Viskanta [17–19] applied a two-band model and narrow-band model to solve radiative heat transfer in an absorbing, emitting slab.

Recently, the normal-mode expansion technique has been applied to obtain rigorous solutions to radiative heat transfer problems in absorbing, emitting, scattering, plane-parallel media. This method, originally developed by Case [20] for treating one-dimensional neutron transport problems, is described in the book by Case and Zweifel [21], and developments in regard to its application have been compiled by Inönü and Zweifel [22]. Seiwert and Zweifel [23, 24] applied this technique to solve the nongray, radiative equilibrium problem in a semi-infinite medium. Ferziger and Simmons [25] and Simmons and Ferziger [26] treated radiative equilibrium for a gray and a nongray slab. Heaslet and Warming [27] combined the probabilistic method of Sobolev [3] and the method of normal modes for radiative heat transfer in a plane-parallel slab. Özişik and Siewert [28] and Beach, Özişik, and Siewert [29] applied the method of normal modes to radiative heat transfer in an isotropic and linearly anisotropic scattering slab with reflecting boundaries. Kreise and Siewert [30] solved the problem of radiative transfer for a conservative slab with internal energy sources. Reith, Siewert, and Özişik [31], Siewert and McCormick [32], and Siewert and Özişik [33] treated nongray radiative heat transfer problems.

In this chapter we present the application of various methods to solve radiative heat transfer problems in plane-parallel geometry.

11-1 TEMPERATURE DISTRIBUTION AND RADIATIVE HEAT FLUX IN A PLANE-PARALLEL SLAB IN RADIATIVE EQUILIBRIUM

In this section consideration will be given to the problem of radiative equilibrium in a plane-parallel, gray slab between two diffusely emitting, diffusely reflecting, opaque, gray boundaries. The boundary surfaces at $\tau = 0$ and $\tau = \tau_0$ are kept at temperatures T_1 and T_2 and have emissivities

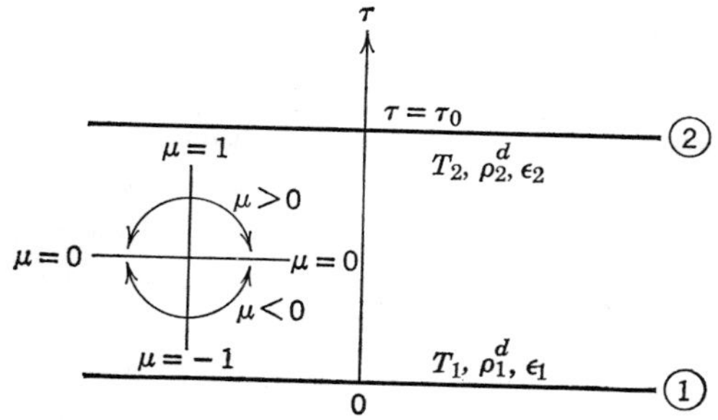

Fig. 11-1. Absorbing, emitting slab in radiative equilibrium.

ε_1 and ε_2 and diffuse reflectivities $\rho_1{}^d$ and $\rho_2{}^d$, respectively. Figure 11-1 shows the geometry and coordinates. The distribution of temperature and the net radiative heat flux in the medium will be determined.

The equation of radiative transfer and the boundary conditions are given as (see Eqs. 8-126, 8-125b, and 8-125c)

$$\mu \frac{\partial(\tau, \mu)}{\partial \tau} + I(\tau, \mu) = \frac{n^2 \bar{\sigma} T^4(\tau)}{\pi}, \quad \text{in} \quad 0 \leq \tau \leq \tau_0, \ -1 \leq \mu \leq 1 \tag{11-1}$$

$$I^+(0) = \varepsilon_1 \frac{n^2 \bar{\sigma} T_1{}^4}{\pi} + 2\rho_1{}^d \int_0^1 I^-(0, -\mu')\mu' \, d\mu', \quad \mu > 0 \tag{11-2a}$$

$$I^-(\tau_0) = \varepsilon_2 \frac{n^2 \bar{\sigma} T_2{}^4}{\pi} + 2\rho_2{}^d \int_0^1 I^+(\tau_0, \mu')\mu' \, d\mu', \quad \mu < 0 \tag{11-2b}$$

The formal solution of this problem was considered in Chapter 8, and the distribution of temperature $T(\tau)$ in the medium related to the universal function $\theta(\tau)$ by the relation (see Eq. 8-131)

$$\frac{n^2 \bar{\sigma} T^4(\tau) - \pi I^-(\tau_0)}{\pi I^+(0) - \pi I^-(\tau_0)} = \theta(\tau) \tag{11-3}$$

where the function $\theta(\tau)$ satisfies the following integral equation (see Eq. 8-132)

$$\theta(\tau) = \frac{1}{2}\left[E_2(\tau) + \int_0^{\tau_0} \theta(\tau') E_1(|\tau - \tau'|) \, d\tau'\right] \tag{11-4}$$

The net radiative heat flux q_r in the medium is given by (see Eqs. 8-134)

$$\frac{q^r}{\pi I^+(0) - \pi I^-(\tau_0)} = Q \tag{11-5}$$

where the heat flux constant Q is related to the function $\theta(\tau)$ by

$$Q = 1 - 2\int_0^{\tau_0} \theta(\tau')E_2(\tau')\,d\tau' \tag{11-6}$$

The integral equation 11-4 has been solved by Usiskin and Sparrow [5], using the method of successive iterations, and by Viskanta and Grosh [9], using the method of undetermined coefficients. Heaslet and Warming [15] have shown that the temperature distribution function $\theta(\tau)$ and the heat flux constant Q can be evaluated very accurately by means of the methods and tabulated X and Y functions of Chandrasekhar [1]. Figure 11-2 shows a

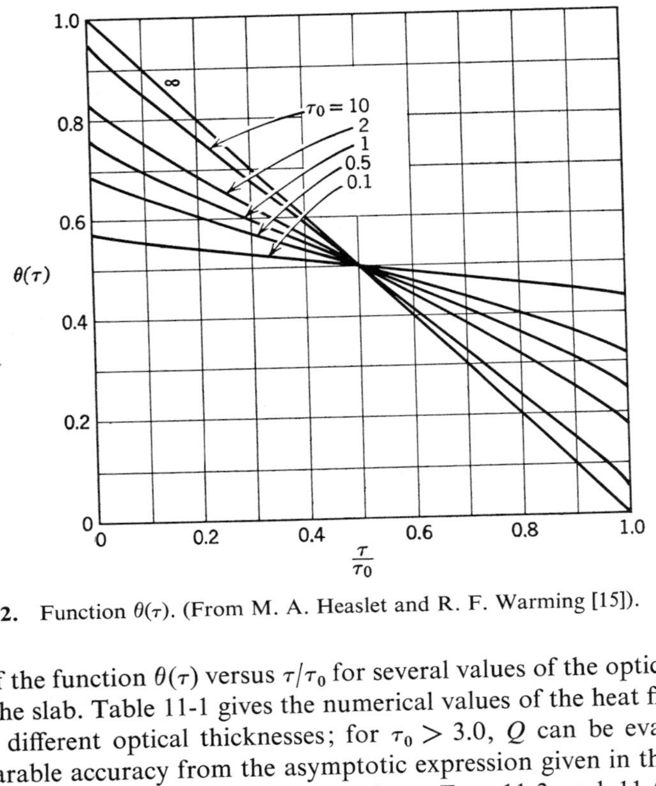

Fig. 11-2. Function $\theta(\tau)$. (From M. A. Heaslet and R. F. Warming [15]).

plot of the function $\theta(\tau)$ versus τ/τ_0 for several values of the optical thickness τ_0 of the slab. Table 11-1 gives the numerical values of the heat flux constant Q for different optical thicknesses; for $\tau_0 > 3.0$, Q can be evaluated with comparable accuracy from the asymptotic expression given in this table.

The determination of $T(\tau)$ and q^r from Eqs. 11-3 and 11-5 requires a knowledge of boundary surface intensities. For black boundaries these intensities $I^+(0)$ and $I^-(\tau_0)$ are known, that is,

$$I^+(0) = \frac{n^2\bar{\sigma}T_1^{\,4}}{\pi} \quad \text{and} \quad I^-(\tau_0) = \frac{n^2\bar{\sigma}T_2^{\,4}}{\pi} \tag{11-7}$$

Table 11-1 Numerical Values of Q for Different Optical Thicknesses[a]

τ_0	Q
0.1	0.9157
0.2	0.8491
0.3	0.7934
0.4	0.7458
0.5	0.7040
0.6	0.6672
0.8	0.6046
1.0	0.5532
1.5	0.4572
2.0	0.3900
2.5	0.3401
3.0	0.3016

[a] From M. A. Heaslet and R. F. Warming [15].
[b] For $\tau_0 \gg 1$, $Q = 4/3(\gamma + \tau_0)$, where $\gamma = 1.24089$.

Hence the temperature distribution $T(\tau)$ and the net radiative heat flux q^r can be determined from Eqs. 11-3 and 11-5, respectively, since the functions $\theta(\tau)$ and Q and the boundary surface intensities $I^+(0)$ and $I^-(\tau_0)$ are all known. However, for diffusely reflecting boundaries $I^+(0)$ and $I^-(\tau_0)$ are not immediately available, although they can be determined as now described.

Determination of $I^+(0)$ and $I^-(\tau_0)$

The boundary conditions 11-2 can be written in terms of the half-range fluxes as

$$q^+(0) = \varepsilon_1 n^2 \bar{\sigma} T_1^4 + (1 - \varepsilon_1) q^-(0) \tag{11-8a}$$

$$q^-(\tau_0) = \varepsilon_2 n^2 \bar{\sigma} T_2^4 + (1 - \varepsilon_2) q^+(\tau_0) \tag{11-8b}$$

where we have defined the half-range fluxes $q^-(0)$ and $q^+(\tau_0)$ as (see Eqs. 8-80)

$$q^-(0) = 2\pi \int_{\mu'=0}^{1} I^-(0, -\mu')\mu' \, d\mu' \tag{11-9a}$$

$$q^+(\tau_0) = 2\pi \int_{\mu'=0}^{1} I^+(\tau_0, \mu')\mu' \, d\mu' \tag{11-9b}$$

and the half-range fluxes $q^+(0)$ and $q^-(\tau_0)$ as

$$q^+(0) = \pi I^+(0) \tag{11-10a}$$

$$q^-(\tau_0) = \pi I^-(\tau_0) \tag{11-10b}$$

and have assumed that Kirchhoff's law is valid, hence expressed the reflectivities in terms of the emissivities.

The net radiative heat flux at the boundaries can be related to the half-range fluxes by (see Eq. 8-79)

$$q^r(0) = q^+(0) - q^-(0) \tag{11-11a}$$

$$q^r(\tau_0) = q^+(\tau_0) - q^-(\tau_0) \tag{11-11b}$$

For radiative equilibrium the net radiative heat flux q^r is constant everywhere in the medium; then we have

$$q^r = q^r(0) = q^r(\tau_0) \tag{11-12}$$

In Eq. 11-11a we substitute q^r for $q^r_{(0)}$ and eliminate $q^-_{(0)}$ by means of Eq. 11-8a. Similarly, in Eq. 11-11b we substitute q^r for $q^r_{(\tau_0)}$ and eliminate $q^+_{(\tau_0)}$ by Eq. 11-8b. Then Eqs. 11-11a and 11-11b respectively become

$$q^+(0) = n^2\bar{\sigma}T_1^4 - \left(\frac{1}{\varepsilon_1} - 1\right)q^r \tag{11-13a}$$

$$q^-(\tau_0) = n_1^2\bar{\sigma}T_2^4 + \left(\frac{1}{\varepsilon_2} - 1\right)q^r \tag{11-13b}$$

Equations 11-10 and 11-13 provide the desired relations for the determination of boundary surface intensities $I^-(\tau_0)$ and $[I^+(0) - I^-(\tau_0),]$ appearing in Eqs. 11-3 and 11-5, respectively. We obtain

$$\pi I^-(\tau_0) = n^2\bar{\sigma}T_2^4 + \left(\frac{1}{\varepsilon_2} - 1\right)q^r \tag{11-14a}$$

$$\pi I^+(0) - \pi I^-(\tau_0) = n^2\bar{\sigma}(T_1^4 - T_2^4) - \left(\frac{1}{\varepsilon_1} + \frac{1}{\varepsilon_2} - 2\right)q^r \tag{11-14b}$$

To evaluate the boundary surface intensities from Eqs. 11-14 a knowledge of the net radiative heat flux q^r is needed.

Relation for q^r

An explicit relation is now determined for the net radiative heat flux q^r in terms of the boundary surface temperatures, emissivities, and heat flux constant Q.

By substituting Eq. 11-14b into Eq. 11-5 and solving for q^r we obtain

$$q^r = n^2\bar{\sigma}(T_1^4 - T_2^4)\frac{Q}{1 + [(1/\varepsilon_1) + (1/\varepsilon_2) - 2]Q} \tag{11-15}$$

where the heat flux constant Q is available in Table 11-1 for several optical thicknesses τ_0 of the slab. Now knowing q^r, the boundary surface intensities $I^-_{(\tau_0)}$ and $I^+_{(0)}$ are evaluated from Eqs. 11-14.

For black boundaries Eq. 11-15 simplifies to

$$q^r = n^2\bar{\sigma}(T_1{}^4 - T_2{}^4)Q \quad \text{for } \varepsilon_1 = \varepsilon_2 = 1 \tag{11-16}$$

For a nonparticipating medium we have $\tau_0 = 0$, $n = 1$, and $Q = 1$ (see Eq. 11-6); then Eq. 11-15 simplifies to

$$q^r = \frac{\bar{\sigma}(T_1{}^4 - T_2{}^4)}{(1/\varepsilon_1) + (1/\varepsilon_2) - 1} \quad \text{for } \tau_0 = 0 \tag{11-17}$$

which is the expression commonly given in heat transfer books for net radiative heat flux between two parallel, infinitely large, diffusely emitting, diffusely reflecting surfaces separated by a nonparticipating medium.

Relation for $T(\tau)$

An explicit relation can be determined for the temperature distribution $T(\tau)$ in the medium from Eq. 11-3 by substituting in that equation the relations for surface intensities $\pi I^-(\tau_0)$ and $[\pi I^-(0) - \pi I^+(\tau_0)]$ as obtained from Eqs. 11-14a and 11-5, respectively. We find

$$\frac{n^2\bar{\sigma}T^4(\tau) - \{n^2\bar{\sigma}T_2{}^4 + [(1/\varepsilon_2) - 1]q^r\}}{q^r/Q} = \theta(\tau)$$

or

$$n^2\bar{\sigma}[T^4(\tau) - T_2{}^4] = \frac{q^r}{Q}\left[\theta(\tau) + \left(\frac{1}{\varepsilon_2} - 1\right)Q\right] \tag{11-18}$$

Substituting q^r/Q from Eq. 11-15 into Eq. 11-18, we obtain the temperature distribution in the slab as

$$\frac{T^4(\tau) - T_2{}^4}{T_1{}^4 - T_2{}^4} = \frac{\theta(\tau) + [(1/\varepsilon_2) - 1]Q}{1 + [(1/\varepsilon_1) + (1/\varepsilon_2) - 2]Q} \tag{11-19}$$

For black boundaries Eq. 11-19 simplifies to

$$\frac{T^4(\tau) - T_2{}^4}{T_1^4 - T_2^4} = \theta(\tau) \quad \text{for } \varepsilon_1 = \varepsilon_2 = 1 \tag{11-20}$$

Equation 11-20 implies that the function $\theta(\tau)$ is the dimensionless temperature distribution in an absorbing and emitting slab in radiative equilibrium between two black boundaries. Referring to Fig. 11-2, we note that a temperature discontinuity (i.e., jump) exists between the wall temperature and the temperature of the medium at the immediate vicinity of the walls for all values of τ_0 except $\tau_0 \to \infty$. The reason for this temperature discontinuity was discussed in Chapter 9.

11-2 TEMPERATURE DISTRIBUTION AND RADIATIVE HEAT FLUX IN A PLANE-PARALLEL SLAB WITH UNIFORMLY DISTRIBUTED INTERNAL ENERGY SOURCES

In this section we consider radiative transfer in an absorbing, emitting medium having uniformly distributed internal energy sources and bounded between two parallel black boundaries at $\tau = 0$ and $\tau = \tau_0$, which are kept at uniform temperatures T_1 and T_2, respectively. The temperature distribution and the net radiative heat flux in the medium will be determined for both gray and nongray cases.

Gray Medium

We summarize below the governing equations and present a discussion of the solution, since the formulation of this problem and its formal solution were given in Section 8-11.

The conservation of energy equation is given as (see Eq. 8-18b)

$$\frac{dq^r(\tau)}{d\tau} = \frac{g}{\kappa} \tag{11-21}$$

where g is the energy generation rate per unit time and per unit volume and is assumed to be constant. The net radiative heat flux $q^r(\tau)$ is related to the radiation intensity $I(\tau, \mu)$ by

$$q^r(\tau) = 2\pi \int_{-1}^{1} I(\tau, \mu)\mu \, d\mu \tag{11-22}$$

and the radiation intensity satisfies the equation of radiative transfer:

$$\mu \frac{\partial I(\tau, \mu)}{\partial \tau} + I(\tau, \mu) = \frac{\bar{\sigma} T^4(\tau)}{\pi} \quad \text{in } 0 \le \tau \le \tau_0,\ -1 \le \mu \le 1 \tag{11-23}$$

with the boundary conditions

$$I(\tau, \mu)\big|_{\tau=0} = \frac{n^2 \bar{\sigma} T_1^4}{\pi}, \quad \mu > 0 \tag{11-24a}$$

$$I(\tau, \mu)\big|_{\tau=\tau_0} = \frac{n^2 \bar{\sigma} T_2^4}{\pi}, \quad \mu < 0 \tag{11-24b}$$

The temperature distribution $T(\tau)$ in the medium is related to the universal functions $\theta(\tau)$ and $\theta_g(\tau)$ by (see Eq. 8-180)

$$\frac{T^4(\tau) - T_2^4}{T_1^4 - T_2^4} = \theta(\tau) + \frac{g}{n^2\kappa\bar{\sigma}(T_1^4 - T_2^4)}\theta_g(\tau) \tag{11-25}$$

where the functions $\theta(\tau)$ and $\theta_g(\tau)$ are the solutions of the singular integral equations 8-181 and 8-182, respectively. The functions $\theta(\tau)$ and $\theta_g(\tau)$ are plotted in Figs. 11-2 and 11-3, respectively, against τ/τ_0 for several values of the optical thickness τ_0.

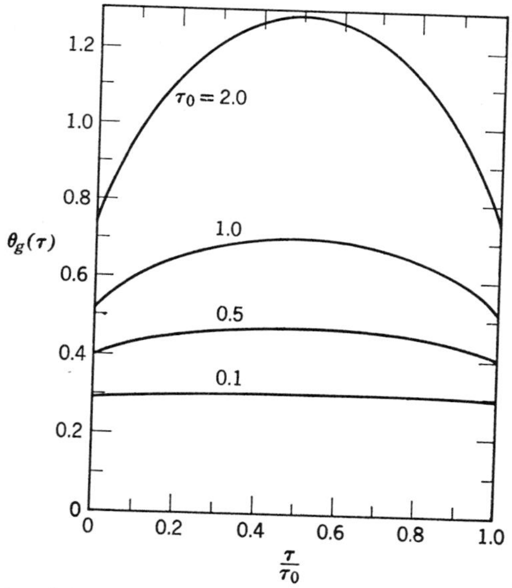

Fig. 11-3. Function $\theta_g(\tau)$. (From M. A. Heaslet and R. F. Warming [15].)

The net radiative heat flux $q^r(\tau)$ in the medium is given in the form (see Eq. 8-184)

$$\frac{q^r(\tau)}{n^2\bar{\sigma}(T_1^4 - T_2^4)} = Q(\tau) + \frac{g}{n^2\kappa\bar{\sigma}(T_1^4 - T_2^4)}Q_g(\tau) \tag{11-26}$$

where the functions $Q(\tau)$ and $Q_g(\tau)$ are defined by Eqs. 8-185 and 8-186, respectively. In many engineering applications the net radiative heat flux at the boundaries is of interest; for the boundary at $\tau = 0$ Eq. 11-26 yields

$$\frac{q^r(0)}{n^2\bar{\sigma}(T_1^4 - T_2^4)} = Q(0) + \frac{g}{n^2\kappa\bar{\sigma}(T_1^4 - T_2^4)}Q_g(0) \tag{11-27}$$

where $Q(0)$ and $Q_g(0)$ are obtained from Eqs. 8-185 and 8-186, respectively, by setting $\tau = 0$, that is,

$$Q(0) = 1 - 2\int_0^{\tau_0} \theta(\tau')E_2(\tau')\,d\tau' \tag{11-28}$$

$$Q_g(0) = -2\int_0^{\tau_0} \theta_g(\tau')E_2(\tau')\,d\tau' \tag{11-29}$$

and the functions $\theta(\tau)$ and $\theta_g(\tau)$ have already been defined.

Knowing the net radiative heat flux $q^r(0)$ at the wall $\tau = 0$, we can immediately determine the net radiative heat flux $q^r(\tau_0)$ at the wall $\tau = \tau_0$ by integrating Eq. 11-21 from $\tau = 0$ to $\tau = \tau_0$. We obtain

$$q^r(\tau_0) = q^r(0) + \frac{g}{\kappa}\tau_0 \tag{11-30}$$

Heaslet and Warming [15] showed that the functions $\theta(\tau)$ and $\theta_g(\tau)$ can be expressed in terms of the well-tabulated $X(\mu, \tau_0)$ and $Y(\mu, \tau_0)$ functions for isotropic scattering[1] and hence computed with a high degree of accuracy. If $\alpha_n(\tau_0)$ and $\beta_n(\tau_0)$ denote the nth moments of the X and Y functions, respectively, that is,

$$\alpha_n(\tau_0) \equiv \int_0^1 X(\mu, \tau_0)\mu^n\,d\mu \tag{11-31a}$$

$$\beta_n(\tau_0) \equiv \int_0^1 Y(\mu, \tau_0)\mu^n\,d\mu \tag{11-31b}$$

then the functions θ, θ_g, Q, and Q_g for $\tau = 0$ are evaluated from [15]

$$\theta(0) = \tfrac{1}{2}\alpha_0(\tau_0), \qquad \theta(\tau_0) = \tfrac{1}{2}\beta_0(\tau_0) \tag{11-32a}$$

$$\theta_g(0) = \theta_g(\tau_0) = \frac{1}{4\beta_0(\tau_0)} \tag{11-32b}$$

$$Q(0) = \beta_0(\tau_0)[\alpha_1(\tau_0) + \beta_1(\tau_0)] \tag{11-33a}$$

$$Q_g(0) = \tfrac{1}{2}\tau_0 \tag{11-33b}$$

Table 11-2 gives the numerical values of α_0, $1/4\beta_0$, $\alpha_1 + \beta_1$, and $\beta(\alpha_1 + \beta_1)$ for values of τ_0 from 0.1 to 3.0. For values of $\tau_0 > 3$ they can be evaluated from the asymptotic expressions given at the bottom of the table.

Nongray Medium

Crosbie and Viskanta [17, 19] applied both a two-band model and a narrow-band model to investigate the nongray effects on radiative heat transfer

Table 11-2 Numerical Values of α_0, $1/4\beta_0$, $\alpha_1 + \beta_1$, and $\beta_0(\alpha_1 + \beta_1)$

τ_0	α_0	$1/4\beta_0$	$\alpha_1 + \beta_1$	$\beta_0(\alpha_1 + \beta_1)$
0.1	1.1419	0.2914	1.0672	0.9157
0.2	1.2228	0.3217	1.0926	0.8491
0.3	1.2838	0.3491	1.1080	0.7934
0.4	1.3331	0.3749	1.1185	0.7458
0.5	1.3746	0.3998	1.1259	0.7040
0.6	1.4103	0.4240	1.1316	0.6672
0.8	1.4692	0.4711	1.1392	0.6046
1.0	1.5163	0.5170	1.1440	0.5532
1.5	1.6024	0.6289	1.1501	0.4572
2.0	1.6615	0.7388	1.1525	0.3900
2.5	1.7051	0.8480	1.1538	0.3401
3.0	1.7386	0.9568	1.1542	0.3016
$\tau_0 \gg 1$	$2 - \dfrac{2/\sqrt{3}}{\gamma + \tau_0}$	$\dfrac{\sqrt{3}}{8}(\gamma + \tau_0)$	$\dfrac{2}{\sqrt{3}}$	$\dfrac{4/3}{\gamma + \tau_0}$

where $\gamma = 1.42089$

[a] From M. M. Heaslet and R. F. Warming [15].

for a plane-parallel slab with uniformly distributed internal energy sources. A two-band model will be considered here.

For simplicity in the analysis it is assumed that the boundaries of the slab at $\tau = 0$ and $\tau = \tau_0$ are black surfaces, kept at temperatures T_1 and T_2, respectively. When the internal energy source is uniform and κ_ν is independent of temperature, the distribution of temperature $f(\tau)$ in the medium for a two-band model is given in the form (see Eq. 8-188)

$$\frac{f(\tau) - f_2}{f_1 - f_2} = \theta(\tau) + \frac{g}{\pi(f_1 - f_2)\xi}\,\theta_g(\tau) \tag{11-34}$$

where we have defined (see Eqs. 8-154)

$$f(\tau) \equiv \int_{\nu=0}^{\infty} \alpha_\nu I_{\nu b}[T(\tau)]\, d\nu \tag{11-35a}$$

$$f_i \equiv \int_{\nu=0}^{\infty} \alpha_\nu I_{\nu b}(T_i)\, d\nu, \quad i = 1 \text{ or } 2 \tag{11-35b}$$

Also, $\theta(\tau)$ and $\theta_g(\tau)$ are the universal functions, $\kappa_\nu = \alpha_\nu \xi$, and α_ν takes only two values, zero and unity.

For computational purposes it is convenient to nondimensionalize the functions $f(\tau)$, f_1, and f_2 with respect to the reference intensity $\bar{\sigma}T_2^4/\pi$. Then Eq. 11-34 becomes

$$\frac{F(t) - F_2(1)}{F_1(t_1) - F_2(1)} = \theta(\tau) + \frac{g}{[F_1(T_1) - F_2(1)]\xi\bar{\sigma}T_2^4}\,\theta_g(\tau) \tag{11-36}$$

where we have defined[2]

$$t \equiv \frac{T(\tau)}{T_2}, \qquad t_1 \equiv \frac{T_1}{T_2} \tag{11-37}$$

$$F(t) \equiv t^4 \sum_{i=1}^{N} \alpha_i \left(\frac{15}{\pi^4}\int_{\frac{z_{i-1}}{t}}^{\frac{z_i}{t}} \frac{x^3}{e^x - 1}\,dx\right) \tag{11-38a}$$

$$F_1(t_1) \equiv t_1^4 \sum_{i=1}^{N} \alpha_i \left(\frac{15}{\pi^4}\int_{\frac{z_{i-1}}{t_1}}^{\frac{z_i}{t_1}} \frac{x^3}{e^x - 1}\,dx\right) \tag{11-38b}$$

$$F_2(1) \equiv \sum_{i=1}^{N} \alpha_i \left(\frac{15}{\pi^4}\int_{z_{i-1}}^{z_i} \frac{x^3}{e^r - 1}\,dx\right) \tag{11-38c}$$

$$z_i \equiv \frac{h\nu_i}{kT_2}$$

Here h and k are the Planck and Boltzmann constants, respectively, and α_i is the value of α_ν in the frequency interval between ν_{i-1} and ν_i.

Equation 11-36 has been solved numerically [17] by the algorithm known as *regula falsi* [36, p. 87] for several different arrangements of a two-band model, and the results are compared with the solution from the gray analysis for the case of no generation, that is, $g = 0$. Figure 11-4 shows a comparison of temperature distribution in slabs of optical thickness $\tau_0 = 1$ and $\tau_0 \to \infty$ for two different arrangements of two-group band models (model A and model B) and a gray model. We note that the temperature distribution for model A is lower than that for the gray solution and approaches the gray solution value as the cut-off point z_c become infinite. Conversely, the temperature distribution for model B is higher than that for the gray solution and approaches the gray solution value as the cut-off point z_c approaches zero.

11-3 THE RADIATIVE HEAT FLUX IN AN ABSORBING, EMITTING SLAB AT A PRESCRIBED TEMPERATURE

In this section we examine radiative heat transfer in an absorbing, emitting, nonscattering, gray medium having a prescribed temperature and bounded

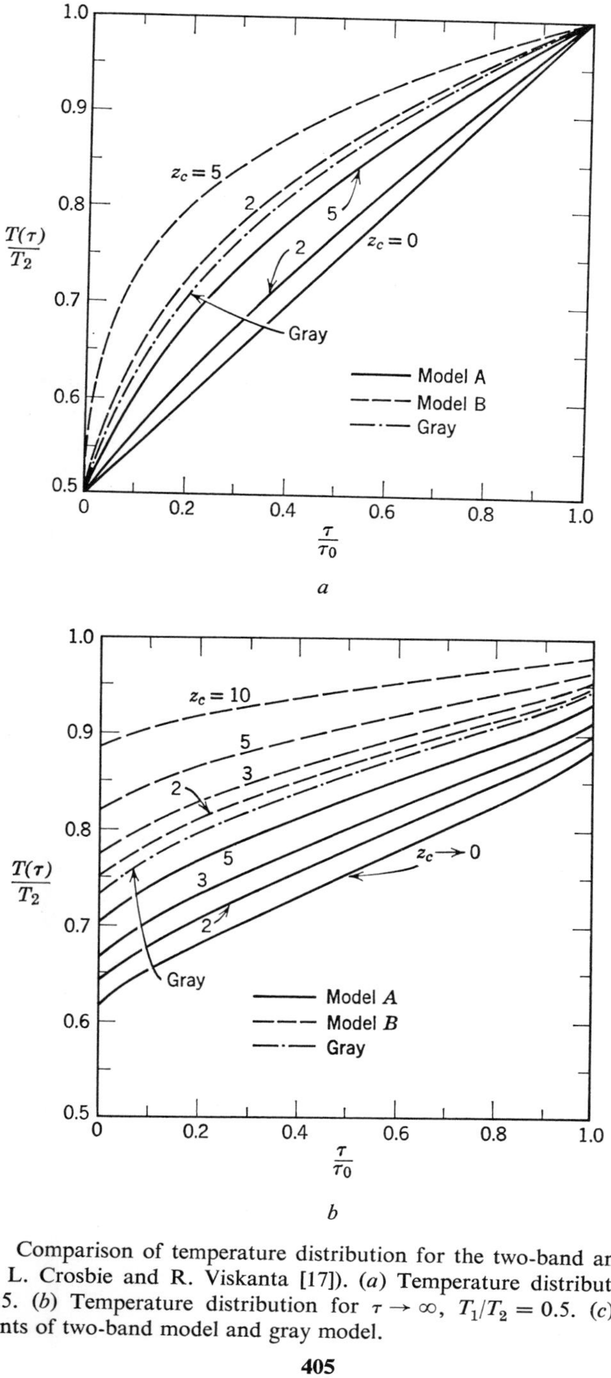

Fig. 11-4. Comparison of temperature distribution for the two-band and gray models. (From A. L. Crosbie and R. Viskanta [17]). (*a*) Temperature distribution for $\tau_0 = 1$, $T_1/T_2 = 0.5$. (*b*) Temperature distribution for $\tau \to \infty$, $T_1/T_2 = 0.5$. (*c*) Two different arrangements of two-band model and gray model.

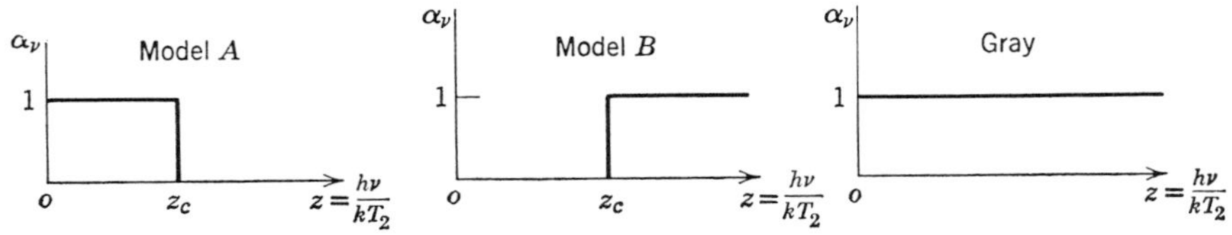

c

Fig. 11-4. (*Continued*)

between two parallel boundaries. In a physical situation, radiative heat transfer from an absorbing, emitting gas flowing at high temperature and high velocity between two parallel plates may be approximated by such a problem. Figure 11-5 shows the geometry and the coordinates; we assume that the boundaries at $\tau = 0$ and $\tau = \tau_0$ are opaque, gray, diffuse emitters, and diffuse reflectors, have emissivities ε_1 and ε_2 and reflectivities ρ_1 and ρ_2, and are kept at uniform temperatures T_1 and T_2, respectively. The medium between the boundaries is maintained at a prescribed temperature $T(\tau)$. The net radiative heat flux in the medium will be determined.

The mathematical formulation of this problem is given as

$$\mu \frac{\partial I(\tau, \mu)}{\partial \tau} + I(\tau, \mu) = \frac{n^2 \bar{\sigma} T^4(\tau)}{\pi} \quad \text{in } 0 \le \tau \le \tau_0,\ -1 \le \mu \le 1 \tag{11-39}$$

with the boundary conditions (see Eqs. 8-99)

$$I^+(0) = \varepsilon_1 \frac{n^2 \bar{\sigma} T_1^4}{\pi} + 2\rho_1 \int_0^1 I^-(0, -\mu')\mu' \, d\mu', \quad \mu > 0 \tag{11-40a}$$

$$I^-(\tau_0) = \varepsilon_2 \frac{n^2 \bar{\sigma} T_2^4}{\pi} + 2\rho_2 \int_0^1 I^+(\tau_0, \mu')\mu' \, d\mu', \quad \mu < 0 \tag{11-40b}$$

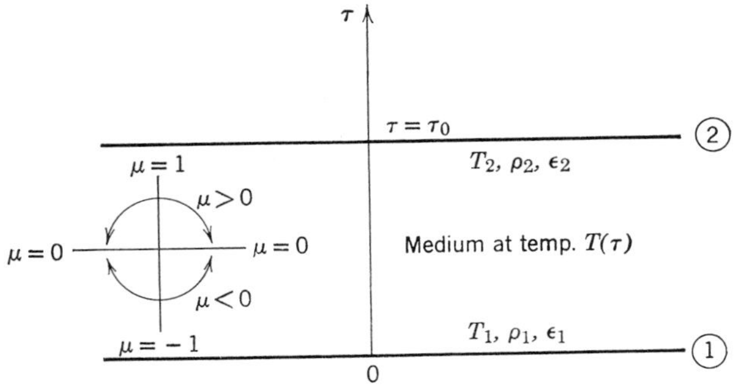

Fig. 11-5. Absorbing emitting slab at temperature $T(\tau)$ with reflecting boundaries.

The formal solution of Eq. 11-39 was considered in Chapter 8; the net radiative heat flux $q^r(\tau)$ is immediately obtained from Eq. 8-84 by integrating that equation over all frequencies:

$$q^r(\tau) = 2\pi[I^+(0)E_3(\tau) - I^-(\tau_0)E_3(\tau_0 - \tau)]$$
$$+ 2\pi\left\{\int_0^\tau I_b[T(\tau')]E_2(\tau - \tau')\, d\tau' - \int_\tau^{\tau_0} I_b[T(\tau')]E_2(\tau' - \tau)\, d\tau'\right\} \quad (11\text{-}41)$$

where

$$I_b[T(\tau)] \equiv \frac{n^2\bar{\sigma}T(\tau)}{\pi}$$

To evaluate the net radiative heat flux from Eq. 11-41, the boundary surface intensities $I^+(0)$ and $I^-(\tau_0)$ are needed. We consider the determination of the boundary surface intensities for (a) black boundaries and (b) diffusely reflecting, diffusely emitting boundaries.

(a) Black Boundaries

For black boundaries the boundary surface intensities are immediately available from the relations

$$I^+(0) = \frac{n^2\bar{\sigma}T_1^{\,4}}{\pi} \quad (11\text{-}42a)$$

$$I^-(\tau_0) = \frac{n^2\bar{\sigma}T_2^{\,4}}{\pi} \quad (11\text{-}42b)$$

substituting Eqs. 11-42 into Eq. 11-41, we obtain the net radiative heat flux in the slab.

(b) Diffusely Reflecting Boundaries

For diffusely reflecting and diffusely emitting boundaries the equations for the boundary surface intensities $I^+(0)$ and $I^-(\tau_0)$ are determined from Eqs. 8-110 by integrating those equations over all frequencies. We obtain

$$I^+(0) = \varepsilon_1 I_b(T_1) + 2\rho_1\left\{I^-(\tau_0)E_3(\tau_0) + \int_0^{\tau_0} I_b[T(\tau')]E_2(\tau')\, d\tau'\right\} \quad (11\text{-}43a)$$

$$I^-(\tau_0) = \varepsilon_2 I_b(T_2) + 2\rho_2\left\{I^+(0)E_3(\tau_0) + \int_0^{\tau_0} I_b[T(\tau')]E_2(\tau_0 - \tau')\, d\tau'\right\} \quad (11\text{-}43b)$$

A simultaneous solution of Eqs. 11-43 yields the desired relations for the boundary surface intensities. Alternatively, these results can be obtained from Eqs. 8-108 by omitting the frequency dependence in those equations and replacing the source function $S(\tau)$ by $I_b[T(\tau)]$ for absorbing and emitting media. We obtain

$$I^+(0) = \frac{\varepsilon_1 I_b(T_1) + 2\rho_1 E_3(\tau_0)\varepsilon_2 I_b(T_2) + 2\rho_1[A + 2\rho_2 E_3(\tau_0)B]}{1 - 4\rho_1\rho_2 E_3^2(\tau_0)} \tag{11-44a}$$

$$I^-(\tau_0) = \frac{\varepsilon_2 I_b(T_2) + 2\rho_2 E_3(\tau_0)\varepsilon_1 I_b(T_1) + 2\rho_2[B + 2\rho_1 E_3(\tau_0)A]}{1 - 4\rho_1\rho_2 E_3^2(\tau_0)} \tag{11-44b}$$

where we have defined

$$A \equiv \int_0^{\tau_0} I_b[T(\tau')]E_2(\tau')\,d\tau' \tag{11-45a}$$

$$B \equiv \int_0^{\tau_0} I_b[T(\tau')]E_2(\tau - \tau')\,d\tau' \tag{11-45b}$$

$$I_b(T_i) = \frac{n^2\bar{\sigma}T_i^4}{\pi}, \quad i = 1 \text{ or } 2 \tag{11-45c}$$

$$I_b[T(\tau)] = \frac{n^2\bar{\sigma}T^4(\tau)}{\pi} \tag{11-45d}$$

Substituting Eqs. 11-44 into Eq. 11-41, we obtain the following relation for the net radiative heat flux $q^r(\tau)$ in an absorbing, emitting slab with diffusely emitting, diffusely reflecting opaque boundaries:[3]

$$\begin{aligned} q^r(\tau) = {} & 2\pi E_3(\tau)\frac{\varepsilon_1 I_b(T_1) + 2\rho_1 E_3(\tau_0)\varepsilon_2 I_b(T_2) + 2\rho_1[A + 2\rho_2 E_3(\tau_0)B]}{1 - 4\rho_1\rho_2 E_3^2(\tau_0)} \\ & - 2\pi E_3(\tau_0 - \tau)\frac{\varepsilon_2 I_b(T_2) + 2\rho_2 E_3(\tau_1)\varepsilon_1 I_b(T_1) + 2\rho_2[B + 2\rho_1 E_3(\tau_0)A]}{1 - 4\rho_1\rho_2 E_3^2(\tau_0)} \\ & + 2\pi\int_0^{\tau} I_b[T(\tau')]E_2(\tau - \tau')\,d\tau' - 2\pi\int_{\tau}^{\tau_0} I_b[T(\tau')]E_2(\tau' - \tau)\,d\tau' \end{aligned} \tag{11-46}$$

We now examine some special cases of Eq. 11-46.

Medium at Uniform Temperature: When the medium is at a uniform temperature T_0 we have

$$I_b[T(\tau)] = I_b(T_0) = \frac{n^2\bar{\sigma}T_0^4}{\pi} = \text{constant} \tag{11-47}$$

Then various integrals in Eq. 11-46 can be evaluated as

$$\int_0^{\tau} I_b(T_0)E_2(\tau - \tau')\,d\tau' = I_b(T_0)[\tfrac{1}{2} - E_3(\tau)] \tag{11-48a}$$

$$\int_{\tau}^{\tau_0} I_b(T_0)E_2(\tau' - \tau)\,d\tau' = I_b(T_0)[\tfrac{1}{2} - E_3(\tau_0 - \tau)] \tag{11-48b}$$

$$A = B = I_b(T_0)[\tfrac{1}{2} - E_3(\tau_0)] \tag{11-48c}$$

and Eq. 11-46 becomes

$$\begin{aligned} q^r(\tau) &= 2\pi E_3(\tau) \\ &\times \frac{\varepsilon_1 I_b(T_1) + 2\rho_1 E_3(\tau_0)\varepsilon_2 I_b(T_2) - I_b(T_0)[(1-\rho_1) + 2\rho_1(1-\rho_2)E_3(\tau_0)]}{1 - 4\rho_1\rho_2 E_3^2(\tau_0)} \\ &- 2\pi E_3(\tau_0 - \tau) \\ &\times \frac{\varepsilon_2 I_b(T_2) + 2\rho_2 E_3(\tau_0)\varepsilon_1 I_b(T_1) - I_b(T_0)[(1-\rho_2) + 2\rho_2(1-\rho_1)E_3(\tau_0)]}{1 - 4\rho_1\rho_2 E_3^2(\tau_0)} \end{aligned} \tag{11-49}$$

If we assume further that the boundaries are black, we have $\rho_1 = \rho_2 = 0$, $\varepsilon_1 = \varepsilon_2 = 1$, and Eq. 11-49 simplifies to

$$q^r(\tau) = 2\pi E_3(\tau)[I_b(T_1) - I_b(T_0)] - 2\pi E_3(\tau_0 - \tau)[I_b(T_2) - I_b(T_0)] \tag{11-50}$$

Nonparticipating Medium: For a nonparticipating (i.e., transparent) medium between two opaque, diffusely reflecting, diffusely emitting parallel boundaries we set $\kappa = 0$. Then

$$\tau = \tau_0 = 0, \qquad E_3(\tau) = E_3(\tau_0) = E_3(0) = \tfrac{1}{2} \tag{11-51}$$

Substituting these results into Eq. 11-46 and replacing ρ_1 and ρ_2 by $1 - \varepsilon_1$ and $1 - \varepsilon_2$, respectively, we obtain

$$q^r = \frac{\pi[I_b(T_1) - I_b(T_2)]}{(1/\varepsilon_1) + (1/\varepsilon_2) - 1} = \frac{\bar{\sigma}T_1^4 - \bar{\sigma}T_2^4}{(1/\varepsilon_1) + (1/\varepsilon_2) - 1} \tag{11-52}$$

This is the expression found in the standard textbooks on heat transfer for the net radiative heat flux between two parallel, gray, diffusely reflecting, diffusely emitting infinite plates separated by a nonparticipating medium.

11-4 ABSORBING, EMITTING, ISOTROPICALLY SCATTERING SLAB AT A PRESCRIBED TEMPERATURE—SOLUTION WITH P_1-APPROXIMATION

In this section we apply the P_1 approximation to determine the angular distribution of radiation intensity and the net radiative heat flux for an absorbing, emitting, isotropically scattering gray slab at uniform temperature T_0. The boundary surfaces 1 and 2 are positioned at $\tau = 0$ and $\tau = \tau_0$ and are kept at uniform temperatures T_1 and T_2, respectively. It is assumed that the surfaces are gray, diffuse emitters with emissitivites ε_1 and ε_2, and the reflectivities are expressed as the sum of the diffuse and specular reflectivity components in the following form: $\rho_i = \rho_i^s + \rho_i^d$, $i = 1$ or 2. The mathematical formulation of the radiation problem here considered is given as

$$\mu \frac{\partial I(\tau, \mu)}{\partial \tau} + I(\tau, \mu) = (1 - \omega) \frac{\bar{\sigma} T_0^4}{\pi} + \frac{\omega}{2} \int_{-1}^{1} I(\tau, \mu')\, d\mu' \quad \text{in } 0 \leq \tau \leq \tau_0,\ -1 \leq \mu \leq 1 \tag{11-53a}$$

with the boundary conditions

$$I(0, \mu) = \varepsilon_1 \frac{\bar{\sigma} T_1^4}{\pi} + \rho_1^s I(0, -\mu) + 2\rho_1^d \int_0^1 I(0, -\mu')\mu'\, d\mu', \quad \mu > 0 \tag{11-53b}$$

$$I(\tau_0, -\mu) = \varepsilon_2 \frac{\bar{\sigma} T_2^4}{\pi} + \rho_2^s I(\tau_0, \mu) + 2\rho_2^d \int_0^1 I(\tau_0, \mu')\mu'\, d\mu', \quad \mu > 0 \tag{11-53c}$$

and for simplicity we have taken $n = 1$. Several special cases can be obtained from this problem. For example:

$\rho_1^d = \rho_2^d = 0$ corresponds to purely specularly reflecting boundaries,
$\rho_1^s = \rho_2^s = 0$ to purely diffusely reflecting boundaries,
$\rho_1^s = \rho_2^s = \rho_1^d = \rho_2^d = \varepsilon_1 = \varepsilon_2 = 0$ to transparent boundaries,
and so on.

This problem will be solved with the P_1 approximation, and the angular distribution of radiation intensity and the net radiation heat flux in the medium will be determined.

The P_1 approximations for the net radiative heat flux $q^r(\tau)$ and the radiation intensity $I(\tau, \mu)$ are given in the form (see Eqs. 9-119b and 9-123)

$$q^r(\tau) = -\frac{1}{3} \frac{dG(\tau)}{d\tau} \tag{11-54a}$$

$$I(\tau, \mu) = \frac{1}{4\pi} \left[G(\tau) - \mu \frac{dG(\tau)}{d\tau} \right] \tag{11-54b}$$

where $G(\tau)$ is the incident radiation. The equation of radiative transfer 11-53a is transformed into an ordinary differential equation for $G(\tau)$ as

(see Eq. 9-120 or 9-150b)

$$\frac{d^2G(\tau)}{d\tau^2} - K^2G(\tau) = -K^2 4\bar{\sigma}T_0^4 \quad \text{in } 0 \leq \tau \leq \tau_0 \tag{11-55a}$$

where we have defined

$$K^2 \equiv 3(1-\omega), \quad \omega < 1 \tag{11-55b}$$

When Marshak's approach is applied, the boundary conditions 11-53b and 11-53c are transformed, respectively, to[4]

$$\left[a_1 G(\tau) - \tfrac{2}{3}b_1 \frac{dG(\tau)}{d\tau}\right]_{\tau=0} = 4\varepsilon_1\bar{\sigma}T_1^4 \tag{11-55c}$$

$$\left[a_2 G(\tau) + \tfrac{2}{3}b_2 \frac{dG(\tau)}{d\tau}\right]_{\tau=\tau_0} = 4\varepsilon_2\bar{\sigma}T_2^4 \tag{11-55d}$$

where we have defined

$$a_i \equiv 1 - \rho_i^s - \rho_i^d, \quad i = 1 \text{ or } 2 \tag{11-55e}$$

$$b_i \equiv 1 + \rho_i^s + \rho_i^d, \quad i = 1 \text{ or } 2 \tag{11-55f}$$

It is convenient to separate the general problem given by Eqs. 11-55 into simpler problems by introducing new dependent variables as

$$G(\tau) = G_0(\tau) + G_1(\tau) + G_2(\tau) \tag{11-56}$$

where the functions $G_i(\tau)$, $i = 0, 1, 2$, are the solutions of three simple problems:

$$\frac{d^2G_i(\tau)}{d\tau^2} - K^2G_i(\tau) = -\delta_{0i}K^2 4\bar{\sigma}T_0^4 \quad \text{in } 0 \leq \tau \leq \tau_0 \tag{11-57a}$$

$$\left[a_1 G_i(\tau) - \tfrac{2}{3}b_1 \frac{dG_i(\tau)}{d\tau}\right]_{\tau=0} = \delta_{1i}4\varepsilon_1\bar{\sigma}T_1^4 \tag{11-57b}$$

$$\left[a_2 G_i(\tau) + \tfrac{2}{3}b_2 \frac{dG_i(\tau)}{d\tau}\right]_{\tau=\tau_0} = \delta_{2i}4\varepsilon_2\bar{\sigma}T_2^4 \tag{11-57c}$$

where $i = 0$, 1, or 2.

We now introduce new dimensionless functions $\psi_i(\tau)$, $i = 0$, 1, or 2, defined as

$$G(\tau) = 4\bar{\sigma}T_0^4\psi_0(\tau) + 4\varepsilon_1\bar{\sigma}T_1^4\psi_1(\tau) + 4\varepsilon_2\bar{\sigma}T_2^4\psi_2(\tau) \tag{11-58}$$

Then the subproblems given by Eqs. 11-57 are replaced by the following three simple subproblems:

$$\frac{d^2\psi_i(\tau)}{d\tau^2} - K^2\psi_i(\tau) = -\delta_{0i}K^2 \quad \text{in } 0 \leq \tau \leq \tau_0 \tag{11-59a}$$

$$\left[a_1\psi_i(\tau) - \tfrac{2}{3}b_1\frac{d\psi_i(\tau)}{d\tau}\right]_{\tau=0} = \delta_{1i} \tag{11-59b}$$

$$\left[a_2\psi_i(\tau) + \tfrac{2}{3}b_2\frac{d\psi_i(\tau)}{d\tau}\right]_{\tau=\tau_0} = \delta_{2i} \tag{11-59c}$$

where $i = 0, 1$, or 2, and

$$\delta_{ij} = \begin{cases} 1, & i = j \\ 0, & \text{otherwise} \end{cases}$$

Once the functions $\psi_i(\tau)$, $i = 0, 1$, or 2, are determined from the solution of Eqs. 11-59, the incident radiation $G(\tau)$, the radiation intensity $I(\tau, \mu)$, and the net radiative heat flux $q^r(\tau)$ anywhere in the medium are evaluated from

$$G(\tau) = 4\sum_{i=0}^{2} \varepsilon_i \bar{\sigma} T_i^4 \psi_i(\tau) \tag{11-60a}$$

$$I(\tau, \mu) = \frac{1}{\pi}\sum_{i=0}^{2} \varepsilon_i \bar{\sigma} T_i^4 \left[\psi_i(\tau) - \mu\frac{d\psi_i(\tau)}{d\tau}\right] \tag{11-60b}$$

$$q^r(\tau) = -\frac{4}{3}\sum_{i=0}^{2} \varepsilon_i \bar{\sigma} T_i^4 \frac{d\psi_i(\tau)}{d\tau} \tag{11-60c}$$

where $\varepsilon_0 = 1$, and $i = 0, 1$, or 2.

The physical significance of the functions $\psi_i(\tau)$ are as follows. The function $\psi_0(\tau)$ characterizes a subproblem in which the medium is at a uniform temperature and the boundary surfaces are at zero temperature. The function $\psi_1(\tau)$ characterizes a subproblem in which the boundary surface at $\tau = 0$ is at a constant temperature while the boundary surface at $\tau = \tau_0$ and the medium are at zero temperature. The function $\psi_2(\tau)$ has an interpretation similar to that of function $\psi_1(\tau)$, with, however, the boundary surface temperatures interchanged.

Conservative Medium

For a purely scattering medium (i.e., a conservative medium) we have $\omega = 1$; then $K = 0$ and the subproblems given by Eqs. 11-59 simplify to

$$\frac{d^2\psi_i(\tau)}{d\tau^2} = 0 \quad \text{in } 0 \leq \tau \leq \tau_0 \tag{11-61a}$$

$$\left[a_1\psi_i(\tau) - \tfrac{2}{3}b_1\frac{d\psi_i(\tau)}{d\tau}\right]_{\tau=0} = \delta_{1i} \tag{11-61b}$$

$$\left[a_2\psi_i(\tau) + \tfrac{2}{3}b_2\frac{d\psi_i(\tau)}{d\tau}\right]_{\tau=\tau_0} = \delta_{2i} \tag{11-61c}$$

where $i = 1$ or 2.

Once the functions $\psi_i(\tau)$, $i = 1$ or 2, are determined from the solution of Eqs. 11-61, the incident radiation $G(\tau)$, the radiation intensity $I(\tau, \mu)$, and the net radiative heat flux in the medium are evaluated from

$$G(\tau) = 4 \sum_{i=1}^{2} \varepsilon_i \bar{\sigma} T_i^4 \psi_i(\tau) \tag{11-62a}$$

$$I(\tau, \mu) = \frac{1}{\pi} \sum_{i=1}^{2} \varepsilon_i \bar{\sigma} T_i^4 \left[\psi_i(\tau) - \mu \frac{d\psi_i(\tau)}{d\tau} \right] \tag{11-62b}$$

$$q^r = -\frac{4}{3} \sum_{i=1}^{2} \varepsilon_i \bar{\sigma} T_i^4 \frac{d\psi_i}{d\tau} \tag{11-62c}$$

The net radiative heat flux q^r for a conservative medium is independent of τ. The constancy of q^r is readily shown by noting that the solution of Eq. 11-61 is of the form $\psi_i(\tau) = c_1 + c_2\tau$ and the radiative flux q^r is proportional to $d\psi_i/d\tau$.

The solutions of Eqs. 11-59 and 11-61 are straightforward and need not be given here. However, to illustrate the application of the P_1 approximation we consider a problem with transparent boundaries.

Slab with Transparent Boundaries

Consider an absorbing, emitting, isotropically scattering slab at a uniform temperature T_0, with transparent boundaries at $\tau = 0$ and $\tau = \tau_0$, and no externally incident radiation. The angular distribution of the radiation intensity and the net radiative heat flux will be determined using the P_1 approximation.

For transparent boundaries Eqs. 11-57 are simplified by setting $\rho_j^s = \rho_j^d = \varepsilon_j = 0$, $j = 1$ or 2 (i.e., $a_1 = a_2 = b_1 = b_2 = 1$). We obtain

$$\frac{d^2G_0(\tau)}{d\tau^2} - K^2 G_0(\tau) = -K^2 4\bar{\sigma} T_0^4 \quad \text{in } 0 \le \tau \le \tau_0 \tag{11-63a}$$

$$\left[G_0(\tau) - \frac{2}{3} \frac{dG_0(\tau)}{d\tau} \right]_{\tau=0} = 0 \tag{11-63b}$$

$$\left[G_0(\tau) + \frac{2}{3} \frac{dG_0(\tau)}{d\tau} \right]_{\tau=\tau_0} = 0 \tag{11-63c}$$

The problems characterized by the functions $G_1(\tau)$ and $G_2(\tau)$ have trivial solutions and hence are not included here. The problem given by Eqs. 11-63 possesses symmetry about $\tau = \tau_0/2$; therefore it is simpler to consider the solution in the region $0 \le \tau \le \tau_0/2$ with the symmetry boundary condition at $\tau = \tau_0/2$. We replace Eqs. 11-63 with the following problem:

$$\frac{d^2\psi_0(\tau)}{d\tau^2} - K^2\psi_0(\tau) = -K^2 \quad \text{in } 0 \leq \tau \leq \frac{\tau_0}{2} \tag{11-64a}$$

$$\left[\psi_0(\tau) - \frac{2}{3}\frac{d\psi_0(\tau)}{d\tau}\right]_{\tau=0} = 0 \tag{11-64b}$$

$$\left.\frac{d\psi_0(\tau)}{d\tau}\right|_{\tau=\tau_0/2} = 0 \tag{11-64c}$$

where we have defined

$$\psi_0(\tau) \equiv \frac{G_0(\tau)}{4\bar{\sigma}T_0^4} \tag{11-64d}$$

The solution of Eqs. 11-64 is given as

$$\psi_0(\tau) = 1 - \frac{\cosh K[(\tau_0/2) - \tau]}{\cosh (K\tau_0/2) + \frac{2}{3}K \sinh (K\tau_0/2)} \tag{11-65}$$

Then the radiation intensity $I(\tau, \mu)$ and the net radiative heat flux $q^r(\tau)$ in the medium are given by (see Eqs. 11-60b and 11-60c)

$$\begin{aligned} I(\tau, \mu) &= \frac{\bar{\sigma}T_0^4}{\pi}\left[\psi_0(\tau) - \mu\frac{d\psi_0(\tau)}{d\tau}\right] \\ &= \frac{\bar{\sigma}T_0^4}{\pi}\left[1 - \frac{\cosh K[(\tau_0/2) - \tau] + \mu K \sinh K[(\tau_0/2) - \tau]}{\cosh (K\tau_0/2) + \frac{2}{3}K \sinh (K\tau_0/2)}\right] \end{aligned} \tag{11-66a}$$

$$q^r(\tau) = -\tfrac{4}{3}\bar{\sigma}T_0^4\frac{d\psi_0(\tau)}{d\tau} = -\tfrac{4}{3}\bar{\sigma}T_0^4\frac{K \sinh K[(\tau_0/2) - \tau]}{\cosh (K\tau_0/2) + \frac{2}{3}K \sinh (K\tau_0/2)} \tag{11-66b}$$

At the boundary surface $\tau = 0$, Eqs. 11-66 simplify to

$$I(0, \mu) = \frac{\bar{\sigma}T_0^4}{\pi}\frac{K \tanh (K\tau_0/2)}{1 + \frac{2}{3}K \tanh (K\tau_0/2)}\left(\tfrac{2}{3} - \mu\right) \tag{11-67a}$$

$$q^r(0) = -\bar{\sigma}T_0^4\frac{2K}{K + \frac{3}{2}\coth (K\tau_0/2)} \tag{11-67b}$$

The net radiative flux given by Eq. 11-67b also characterizes the emergent radiative flux at the boundary surface $\tau = 0$ since there is no externally incident radiation. The angular distribution of the emergent radiation intensity can be determined from Eq. 11-67a.

For a semi-infinite medium, $\tau_0 \to \infty$, Eqs. 11-67 simplify to

$$I(0, \mu) = \frac{\bar{\sigma} T_0^4}{\pi} \frac{K}{1 + \frac{2}{3}K} (\tfrac{2}{3} - \mu) \tag{11-68a}$$

$$q^r(0) = -\bar{\sigma} T_0^4 \frac{2K}{K + \frac{2}{3}} \tag{11-68b}$$

Limitations to the angular distribution of the emergent radiation intensity as given by Eqs. 11-67a and 11-68a should be recognized. These equations show an inward radiation intensity at the boundary surface $\tau = 0$ despite the fact that there is no externally incident radiation in the directions $0 \leq \mu \leq 1$ at the surface $\tau = 0$. Edwards and Bobco [13] compared the exit distribution of the radiation intensity obtained from the lowest-order moment method (i.e., equivalent to the P_1 approximation) with Chandrasekhar's [1, p. 77] exact solution for a semi-infinite medium. There was a large difference between the exact and approximate solutions for the angular distribution of the emergent radiation intensity, but the agreement between the exact and approximate solutions for the emergent radiative flux was good.

Accuracy of P_1-Approximation to Predict Radiative Flux

To illustrate the accuracy of the P_1 approximation to predict the net radiative heat flux we consider the net radiative fluxes $q_0{}^r(0)$ and $q_1{}^r(0)$ at the boundary surface $\tau = 0$, obtained from the solution of Eqs. 11-59 for $i = 0$ and $i = 1$, respectively, with $\omega \neq 1$. We obtain

$$Q_0(0) \equiv \frac{q_0{}^r(0)}{\bar{\sigma} T_0^4} = -\frac{4K}{3D}(-a_2 + a_2 \cosh K\tau_0 + \tfrac{2}{3} b_2 K \sinh K\tau_0) \tag{11-69a}$$

$$Q_1(0) \equiv \frac{q_1{}^r(0)}{\bar{\sigma} T_1^4} = -\frac{4\varepsilon_1 K}{3a_1 D}(a_2 \cosh K\tau_0 + \tfrac{2}{3} b_2 K \sinh K\tau_0) \tag{11-69b}$$

where

$$D \equiv \left(a_2 + \frac{4b_1 b_2 K^2}{9a_1}\right) \sinh K\tau_0 + \tfrac{2}{3}K\left(b_2 + \frac{b_1 a_2}{a_1}\right) \cosh K\tau_0$$

In the case of a conservative medium (i.e., $\omega = 1$) the net radiative heat flux $q_1{}^r$ obtained from the solution of Eqs. 11-61 for $i = 1$, is given as

$$Q_1 \equiv \frac{q_1{}^r}{\bar{\sigma} T_1^4} = \tfrac{4}{3}\varepsilon_1 \frac{1}{a_1\tau_0 + \frac{2}{3}[b_1 + (a_1 b_2/a_2)]} \tag{11-69c}$$

We note that for a conservative medium the net radiative heat flux is constant everywhere in the medium.

Table 11-3 shows a comparison of the dimensionless radiative heat fluxes

Table 11-3 Accuracy of P_1 Approximation to Predict Radiative Flux for a Slab with Opaque Boundaries

(a) Nonconservative Medium, $\omega < 1$

ω	τ_0	ρ_1^d	ρ_1^s	ρ_2^d	ρ_2^s	$-Q_0(0)$ P_1 Approximation	$-Q_0(0)$ Exact[a]	$\left\lvert \frac{Q_{\text{Approx}} - Q_{\text{Exact}}}{Q_{\text{Exact}}} \right\rvert \times 100$
0.5	0.1	0	0	1	0	0.1810	0.1736	4.3
0.5	0.1	0	0	0.5	0	0.1376	0.1316	4.6
0.5	0.1	0	0	0	0	0.0951	0.0911	4.4
0.5	1.0	0	0	1	0	0.8142	0.7572	7.5
0.5	1.0	0	0	0.5	0	0.7106	0.6510	9.2
0.5	1.0	0	0	0	0	0.6165	0.5591	10.2
0.5	10.0	0	0	1	0	0.8989	0.8535	5.3
0.5	10.0	0	0	0.5	0	0.8989	0.8534	5.4
0	0.1	0	0	0	0	0.1814	0.1674	8.3
0	1.0	0	0	0	0	0.8935	0.7806	14.4
0	10.0	0	0	0	0	1.072	1.0000	7.2
						$Q_1(0)$ P_1 Approximation	$Q_1(0)$ Exact[a]	
0.5	0.1	0	0	1	0	0.1810	0.1736	4.3
0.5	0.1	0	0	0.5	0	0.5839	0.5753	1.5
0.5	0.1	0	0	0	0	0.9779	0.9616	1.7
0.5	1.0	0	0	1	0	0.8142	0.7572	7.5
0.5	1.0	0	0	0.5	0	0.8632	0.8154	5.9
0.5	1.0	0	0	0	0	0.9076	0.8658	4.8
0.5	10.0	0	0	1	0	0.8989	0.8535	5.3
0.5	10.0	0	0	0.5	0	0.8989	0.8535	5.3

(b) Conservative Medium, $\omega = 1$; $\rho^s = \rho_2^s = 0$

τ_0	ρ_1^d	ρ_2^d	Q_1 P_1 Approximation	Q_1 Exact[a]	$\left\lvert \frac{Q_{\text{Approx}} - Q_{\text{Exact}}}{Q_{\text{Exact}}} \right\rvert \times 100$
0.1	0.5	0	0.4819	0.4780	0.82
0.1	0	0	0.9302	0.9157	1.59
1.0	0.5	0	0.3636	0.3562	2.07
1.0	0	0	0.5714	0.5534	0.26
10.0	0.5	0	0.1052	0.1045	0.69
10.0	0	0	0.1176	0.1167	0.77

[a] From J. L. Beach, M. N. Özişik, and C. E. Siewert [29].

given by Eqs. 11-69, obtained by the P_1 approximation, with the exact solution provided by Beach, Özişik, and Siewert [29]. The P_1 approximation appears to overestimate the net radiative heat flux for all the cases considered here, although its accuracy is reasonably good for conservative media. However, for nonconservative media the accuracy is not as good; the magnitude of the error depends on the optical thickness, the value of ω, and the reflectivity of the boundary surfaces.

11-5 ABSORBING, EMITTING, SCATTERING SLAB AT PRESCRIBED TEMPERATURE—NUMERICAL SOLUTION USING THE GAUSSIAN QUADRATURE

In this section we examine the method of solution of the equation of radiative transfer numerically by the Gaussian quadrature scheme and the determination of net radiative heat flux for an absorbing, emitting, anisotropically scattering, plane-parallel, gray slab kept at a prescribed temperature $T(\tau)$ between two diffusely reflecting, diffusely emitting, opaque, gray boundaries. The geometry and coordinates for the problem are similar to those shown in Fig. 11-5. The boundary surfaces at $\tau = 0$ and $\tau = \tau_0$ are kept at uniform temperatures T_1 and T_2 and have emissivities ε_1 and ε_2 and diffuse reflectivities ρ_1 and ρ_2, respectively. The mathematical formulation of this problem is given as

$$\mu \frac{\partial I(\tau,\mu)}{\partial \tau} + I(\tau, \mu) = (1 - \omega) I_b[T(\tau)] + \frac{\omega}{2} \int_{-1}^{1} p(\mu, \mu') I(\tau, \mu')\, d\mu'$$

$$\text{in } 0 \le \tau \le \tau_0, \ -1 \le \mu \le 1 \quad (11\text{-}70)$$

with the boundary conditions (see Eqs. 8-99 for the gray case)

$$I(0, \mu) = \varepsilon_1 I_b(T_1) + 2\rho_1 \int_0^1 I(0, -\mu')\mu'\, d\mu', \quad \mu > 0 \quad (11\text{-}71a)$$

$$I(\tau_0, \mu) = \varepsilon_2 I_b(T_2) + 2\rho_2 \int_0^1 I(\tau_0, \mu')\mu'\, d\mu', \quad \mu < 0 \quad (11\text{-}71b)$$

For simplicity we have omitted the superscripts + and − from the boundary surface intensities.

Love and Grosh [10] reduced the above equations to a set of simultaneous first-order linear differential equations by approximating the integrals by the Gaussian quadrature scheme and solved them for a constant inhomogeneous term (i.e., uniform temperature distribution in the medium). Hsia and Love [11] utilized a similar approach to solve them for a linear temperature distribution in the medium. To illustrate this approach we consider the transformation of the above integrodifferential equation into a set of ordinary differential equations by using a Gaussian quadrature scheme and discuss the method of solution of the resulting equations.

We separate the intensity into a forward component $I(\tau, \mu)$, $\mu \in (0, 1)$, and a backward component $I(\tau, \mu)$, $\mu \in (-1, 0)$, as discussed in Chapter 8, and write the integrodifferential equation 11-70 in the forms (see Eqs. 8-62a, 8-63a, 8-64)

$$\mu \frac{\partial I(\tau, \mu)}{\partial \tau} + I(\tau, \mu) = (1 - \omega) I_b[T(\tau)] + \frac{\omega}{2}\left[\int_0^1 p(\mu, \mu') I(\tau, \mu')\, d\mu' + \int_0^1 p(\mu, -\mu') I(\tau, -\mu')\, d\mu'\right]$$

$$\text{in } 0 \leq \tau \leq \tau_0,\ \mu > 0 \qquad (11\text{-}72)$$

and[5]

$$-\mu \frac{\partial I(\tau, -\mu)}{\partial \tau} + I(\tau, -\mu) = (1 - \omega) I_b[T(\tau)] + \frac{\omega}{2}\left[\int_0^1 p(\mu, -\mu') I(\tau, \mu')\, d\mu' + \int_0^1 p(\mu, \mu') I(\tau, -\mu')\, d\mu'\right]$$

$$\text{in } 0 \leq \tau \leq \tau_0,\ \mu > 0 \qquad (11\text{-}73)$$

We note that Eqs. 11-72 and the boundary conditions 11-71 are valid in the positive half-range of μ, that is, $\mu \in (0, 1)$, and the two intensity components are distinguished by the notation $I(\tau, \mu)$ and $I(\tau, -\mu)$.

The integral terms in the above equations can be approximated by a summation using the double Gaussian quadrature formula given in the form

$$\int_0^1 f(\tau, \mu')\, d\mu' \cong \sum_{j=1}^{N} a_j f(\tau, \mu_j), \quad \mu > 0 \qquad (11\text{-}74)$$

where a_j = the weight factor specified by the Gaussian quadrature formula,
μ_j = discrete values of μ_j specified by the Gaussian quadrature formula.

A comprehensive tabulation of the weight factors a_j is available in the *Handbook of Mathematical Functions* [37, pp. 916–922], and an improved Gaussian quadrature scheme is given by Kronrod [38].

The integrodifferential equations 11-72 and 11-73 are transformed by the application of an N-point Gaussian quadrature formula into $2N$ coupled ordinary differential equations for the intensity functions $I(\tau, \mu_i)$ and

$I(\tau, -\mu_i)$, $i = 1, 2, \ldots, N$. We obtain

$$\mu_i \frac{dI(\tau, \mu_i)}{d\tau} + I(\tau, \mu_i) = (1 - \omega)I_b[T(\tau)] + \frac{\omega}{2} \sum_{j=1}^{N} a_j[p(\mu_i, \mu_j)I(\tau, \mu_j) + p(\mu_i, -\mu_j)I(\tau, -\mu_j)], \quad \mu_i \in (0, 1) \tag{11-75a}$$

$$-\mu_i \frac{dI(\tau, -\mu_i)}{d\tau} + I(\tau, -\mu_i) = (1 - \omega)I_b[T(\tau)] + \frac{\omega}{2} \sum_{j=1}^{N} a_j[p(\mu_i, -\mu_j)I(\tau, \mu_j) + p(\mu_i, \mu_j)I(\tau, -\mu_j)], \quad \mu_i \in (0, 1) \tag{11-75b}$$

where $i = 1, 2, \ldots, N$.

Assuming that the boundary conditions 11-71 are satisfied at each discrete value of μ_i and applying the Gaussian quadrature formula to transform the integrals into summations, we obtain from Eqs. 11-71 the following set of $2N$ boundary conditions:

$$I(0, \mu_i) = \varepsilon_1 I_b(T_1) + 2\rho_1 \sum_{j=1}^{N} a_j \mu_j I(0, -\mu_j), \quad \mu_i \in (0, 1) \tag{11-76a}$$

$$I(\tau_0, -\mu_i) = \varepsilon_2 I_b(T_2) + 2\rho_2 \sum_{j=1}^{N} a_j \mu_j I(\tau_0, \mu_j), \quad \mu_i \in (0, 1) \tag{11-76b}$$

where $i = 1, 2, \ldots, N$.

Equations 11-75 with boundary conditions 11-76 provide a system of $2N$ simultaneous ordinary differential equations for the $2N$ unknown intensities $I(\tau, \mu_i)$ and $I(\tau, -\mu_i)$, $i = 1, 2, \ldots, N$. Once these equations are solved and the intensities are determined, the net radiative heat flux $q^r(\tau)$ is evaluated from

$$q^r(\tau) = 2\pi \int_{-1}^{1} I(\tau, \mu)\mu \, d\mu = 2\pi\left[\int_0^1 I(\tau, \mu)\mu \, d\mu - \int_0^1 I(\tau, -\mu)\mu \, d\mu\right]$$

$$= 2\pi \sum_{j=1}^{N} a_j \mu_j [I(\tau, \mu_j) - I(\tau, -\mu_j)] \tag{11-77}$$

Then the net radiative heat flux at the boundary surface $\tau = 0$ is given by

$$q^r(0) = 2\pi \sum_{j=1}^{N} a_j \mu_j [I(0, \mu_j) - I(0, -\mu_j)] \tag{11-78}$$

Method of Solution

Equations 11-75 can be rearranged as

$$\frac{dI(\tau, \mu_i)}{d\tau} - \sum_{j=1}^{N} \alpha_{ij} I(\tau, \mu_j) - \sum_{j=1}^{N} \beta_{ij} I(\tau, -\mu_j) = \frac{1}{\mu_i}(1 - \omega)I_b[T(\tau)] \tag{11-79a}$$

$$\frac{dI(\tau, -\mu_i)}{d\tau} + \sum_{j=1}^{N} \beta_{ij} I(\tau, \mu_j) + \sum_{j=1}^{N} \alpha_{ij} I(\tau, -\mu_j) = -\frac{1}{\mu_i}(1 - \omega)I_b[T(\tau)] \tag{11-79b}$$

where $\mu_i \in (0, 1)$, $i = 1\ 2, \ldots, N$, and

$$\alpha_{ij} \equiv \frac{1}{\mu_i} \frac{\omega}{2} a_j p(\mu_i, \mu_j) - \frac{\delta_{ij}}{\mu_i} \tag{11-80a}$$

$$\beta_{ij} \equiv \frac{1}{\mu_i} \frac{\omega}{2} a_j p(\mu_i, -\mu_j) \tag{11-80b}$$

$$\delta_{ij} = \begin{cases} 1 & \text{for } i = j \\ 0 & \text{otherwise} \end{cases} \tag{11-80c}$$

Equations 11-79 are $2N$ simultaneous, linear, ordinary differential equations for the $2N$ unknown intensity functions $I(\tau, \mu_i)$ and $I(\tau, -\mu_i)$, $i = 1, 2, \ldots, N$, which should be solved subject to the $2N$ boundary conditions given by Eqs. 11-76.

We assume that the solution to the homogeneous part of Eqs. 11-79 can be written in the form

$$I(\tau, \mu_i) = g_i(k)e^{k\tau}, \qquad I(\tau, -\mu_i) = g_i^*(k)e^{k\tau} \tag{11-81}$$

where $i = 1, 2, \ldots, N$.

When these solutions are substituted into the homogeneous part of Eqs. 11-79, we obtain a set of $2N$ linear, homogeneous, algebraic equations in $g_i(k)$ and $g_i^*(k)$ with k as a parameter. The permissible values of k are determined from the requirement that the determinant of the coefficients of the $g_i(k)$ and $g_i^*(k)$ vanish if the resulting system of algebraic homogeneous equations has a nontrivial solution. Once the k_j are known, the algebraic equations are solved for each value of k_j, $j = 1, 2, \ldots, 2N$, and the corresponding values of $g_i(k_j)$ and $g_i^*(k_j)$, $i = 1, 2, \ldots, N$, are determined. The complete solution of Eqs. 11-79 is written as a linear sum of the homogeneous solutions and a particular solution I_p in the forms

$$I(\tau, \mu_i) = \sum_{j=1}^{N} c_j g_i(k_j)e^{k_j\tau} + I_p \tag{11-82a}$$

$$I(\tau, -\mu_i) = \sum_{j=1}^{N} c_j g_i^*(k_j)e^{k_j\tau} + I_p \tag{11-82b}$$

where the c_j are $2N$ integration constants that must be determined by the application of $2N$ boundary conditions. For an isothermal medium a particular solution is readily determined, and for a nonisothermal medium it can be developed as described in reference 11.

11-6 ABSORBING, EMITTING, ISOTROPICALLY SCATTERING SLAB AT PRESCRIBED TEMPERATURE—SOLUTION BY NORMAL-MODE EXPANSION TECHNIQUE FOR ω < 1

In this section we illustrate the application of the normal-mode expansion technique to solve the equation of radiative transfer and determine the angular distribution of radiation intensity and the net radiative heat flux for an absorbing, emitting, isotropically scattering, plane-parallel, gray slab kept at a prescribed temperature $T(\tau)$ between two specularly reflecting, diffusely emitting, opaque, gray boundaries. The boundary surfaces at $\tau = 0$ and $\tau = \tau_0$ are at uniform temperatures T_1 and T_2 and have emissivities ε_1 and ε_2 and specular reflectivities $\rho_1{}^s$ and $\rho_2{}^s$ respectively. The geometry and the coordinates are similar to those shown in Fig. 11-5. The mathematical formulation of the problem is given as

$$\mu \frac{\partial I(\tau, \mu)}{\partial \tau} + I(\tau, \mu) = (1 - \omega) \frac{\bar{\sigma} T^4(\tau)}{\pi} + \frac{\omega}{2} \int_{-1}^{1} I(\tau, \mu')\, d\mu'$$

$$\text{in } 0 \le \tau \le \tau_0,\ -1 \le \mu \le 1 \qquad (11\text{-}83)$$

with the boundary conditions (see Eqs. 8-100 for the gray case)

$$I(0, \mu) = \varepsilon_1 \frac{\bar{\sigma} T_1{}^4}{\pi} + \rho_1{}^s I(0, -\mu), \quad \mu > 0 \qquad (11\text{-}84a)$$

$$I(\tau_0, \mu) = \varepsilon_2 \frac{\bar{\sigma} T_2{}^4}{\pi} + \rho_2{}^s I(\tau_0, -\mu), \quad \mu < 0$$

or

$$I(\tau_0, -\mu) = \varepsilon_2 \frac{\bar{\sigma} T_2{}^4}{\pi} + \rho_2{}^s I(\tau_0, \mu), \quad \mu > 0 \qquad (11\text{-}84b)$$

For convenience in the analysis we write Eqs. 11-83 and 11-84 in the form

$$\mu \frac{\partial I(\tau, \mu)}{\partial \tau} + I(\tau, \mu) = s(\tau) + \frac{\omega}{2} \int_{-1}^{1} I(\tau, \mu')\, d\mu'$$

$$\text{in } 0 \le \tau \le \tau_0,\ -1 \le \mu \le 1 \qquad (11\text{-}85)$$

with the boundary conditions

$$I(0, \mu) = a_1 + b_1 I(0, -\mu), \quad \mu > 0 \qquad (11\text{-}86a)$$

$$I(\tau_0, -\mu) = a_2 + b_2 I(\tau_0, \mu), \quad \mu > 0 \qquad (11\text{-}86b)$$

where we have defined

$$s(\tau) \equiv (1 - \omega) \frac{\bar{\sigma} T^4(\tau)}{\pi} \qquad (11\text{-}87a)$$

$$a_i = \varepsilon_i \frac{\bar{\sigma} T_i^4}{\pi}, \qquad b_i = \rho_i^s, \quad i = 1 \text{ or } 2 \qquad (11\text{-}87b)$$

The complete solution of Eq. 11-85 can be written as a linear sum of the elementary solutions of the homogeneous part of this equation and a particular solution $I_p(\tau, \mu)$ in the form (see Eq. 10-18b for the homogeneous solution)

$$I(\tau, \mu) = A(\eta_0)\phi(\eta_0, \mu)e^{-\tau/\eta_0} + A(-\eta_0)\phi(-\eta_0, \mu)e^{\tau/\eta_0} + \int_0^1 A(\eta)\phi(\eta, \mu)e^{-\tau/\eta}\, d\eta + \int_0^1 A(-\eta)\phi(-\eta, \mu)e^{\tau/\eta}\, d\eta + I_p(\tau, \mu) \tag{11-88}$$

where the discrete normal modes are defined as (see Eq. 10-18c)

$$\phi(\pm\eta_0, \mu) = \frac{\omega\eta_0}{2}\frac{1}{\eta_0 \mp \mu}, \quad \eta_0 \notin (-1, 1) \tag{11-89a}$$

and the continuum normal modes as (see Eqs. 10-18d and 10-18e)

$$\phi(\eta, \mu) = \frac{\omega\eta}{2}\frac{P}{\eta - \mu} + (1 - \omega\eta \tanh^{-1}\eta)\,\delta(\eta - \mu), \quad \eta \in (-1, 1) \tag{11-89b}$$

Here P denotes that the integration of $1/(\eta - \mu)$ over η or μ should be performed in the Cauchy principal-value integral sense, and the discrete eigenvalues $\pm\eta_0$ are the two zeros of the dispersion relation

$$\Lambda(\eta_0) \equiv 1 - \omega\eta_0 \tanh^{-1}\frac{1}{\eta_0} = 0 \tag{11-89c}$$

Equation 11-88 involves four unknown expansion coefficients $A(\eta_0)$, $A(-\eta_0)$, $A(\eta)$, and $A(-\eta)$, which must be determined by constraining the general solution given by Eq. 11-88 to satisfy the boundary conditions 11-86.

Once a particular solution $I_p(\tau, \mu)$ of Eq. 11-85 is found and the expansion coefficients are determined, the angular distribution of radiation intensity $I(\tau, \mu)$ is given by Eq. 11-88, and the incident radiation $G(\tau)$ and the net radiative heat flux $q^r(\tau)$ anywhere in the medium are evaluated respectively from

$$\begin{aligned} G(\tau) &= 2\pi \int_{-1}^{1} I(\tau, \mu)\, d\mu \\ &= 2\pi\Big[A(\eta_0)e^{-\tau/\eta_0} + A(-\eta_0)e^{\tau/\eta_0} + \int_0^1 A(\eta)e^{-\tau/\eta}\, d\eta \\ &\quad + \int_0^1 A(-\eta)e^{\tau/\eta}\, d\eta + \int_{-1}^{1} I_p(\tau, \mu)\, d\mu\Big] \end{aligned} \tag{11-90}$$

$$\begin{aligned} q^r(\tau) &= 2\pi \int_{-1}^{1} \mu I(\tau, \mu)\, d\mu \\ &= 2\pi(1 - \omega)\Big[A(\eta_0)\eta_0 e^{-\tau/\eta_0} - A(-\eta_0)\eta_0 e^{\tau/\eta_0} + \int_0^1 A(\eta)\eta e^{-\tau/\eta}\, d\eta \\ &\quad - \int_0^1 A(-\eta)\eta e^{\tau/\eta}\, d\eta + \frac{1}{1 - \omega}\int_{-1}^{1} I_p(\tau, \mu)\mu\, d\mu\Big] \end{aligned} \tag{11-91}$$

At this point we assume that a particular solution of Eq. 11-85 can be found for the considered inhomogeneous term $s(\tau)$ (see Table 10-6), and thus

$I_p(\tau, \mu)$ is a known function. We now proceed to the determination of the expansion coefficients.

Substitution of solution 11-88 into the boundary conditions 11-86a and 11-86b, respectively, yields

$$[a_1 + b_1 I_p(0, -\mu) - I_p(0, \mu)] + [b_1 A(\eta_0) - A(-\eta_0)]\phi(-\eta_0, \mu) + \int_0^1 [b_1 A(\eta) - A(-\eta)]\phi(-\eta, \mu)\, d\eta$$
$$= [A(\eta_0) - b_1 A(-\eta_0)]\phi(\eta_0, \mu) + \int_0^1 [A(\eta) - b_1 A(-\eta)]\phi(\eta, \mu)\, d\eta, \qquad \mu > 0 \quad (11\text{-}92)$$

$$[a_2 + b_2 I_p(\tau_0, \mu) - I_p(\tau_0, -\mu)] + [-e^{-\tau_0/\eta_0} A(\eta_0) + b_2 e^{\tau_0/\eta_0} A(-\eta_0)]\phi(-\eta_0, \mu) + \int_0^1 [-e^{-\tau_0/\eta} A(\eta) + b_2 e^{\tau_0/\eta} A(-\eta)]\phi(-\eta, \mu)\, d\eta$$
$$= [-b_2 e^{-\tau_0/\eta_0} A(\eta_0) + e^{\tau_0/\eta_0} A(-\eta_0)]\phi(\eta_0, \mu) + \int_0^1 [-b_2 e^{-\tau_0/\eta} A(\eta) + e^{-\tau_0/\eta} A(-\eta)]\phi(\eta, \mu)\, d\eta, \qquad \mu > 0 \quad (11\text{-}93)$$

To obtain Eqs. 11-92 and 11-93 we utilized the reciprocity relation of normal modes

$$\phi(\xi, -\mu) = \phi(-\xi, \mu) \qquad (11\text{-}94a)$$

$$\phi(-\xi, -\mu) = \phi(\xi, \mu) \qquad (11\text{-}94b)$$

where

$$\xi = \pm\eta_0 \text{ or } \eta \in (0, 1)$$

We note that the right-hand sides of Eqs. 11-92 and 11-93 are two half-range expansions similar to the one given by Eq. 10-22a. By the half-range completeness theorem stated in Chapter 10 these expansions are sufficiently general to represent an arbitrary function (i.e., the left-hand sides of these equations) defined in the interval $\mu \in (0, 1)$. The expansion coefficients appearing in these equations can be isolated by utilizing the orthogonality property of the normal modes and various normalization integrals now described.

Isolation of Discrete Coefficients

To isolate the discrete coefficients $A(\eta_0)$ and $A(-\eta_0)$, we operate on both sides of Eqs. 11-92 and 11-93 by the operator

$$\int_0^1 \phi(\eta_0, \mu) W(\mu)\, d\mu \qquad (11\text{-}95a)$$

where the weight function $W(\mu)$ is given as (see Eq. 10-28)

$$W(\mu) = (\eta_0 - \mu)\gamma(\mu) \qquad (11\text{-}95b)$$

Then Eqs. 11-92 and 11-93, respectively, yield[6]

$$\int_0^1 [a_1 + b_1 I_p(0, -\mu) - I_p(0, \mu)]\phi(\eta_0, \mu)W(\mu)\, d\mu + [b_1 A(\eta_0) - A(-\eta_0)](\tfrac{1}{2}\omega\eta_0)^2 X(-\eta_0) + \int_0^1 [b_1 A(\eta) - A(-\eta)]\tfrac{1}{4}\omega^2\eta_0\eta X(-\eta)\, d\eta = -[A(\eta_0) - b_1 A(-\eta_0)](\tfrac{1}{2}\omega\eta_0)^2 X(\eta_0) \tag{11-96}$$

$$\int_0^1 [a_2 + b_2 I_p(\tau_0, \mu) - I_p(\tau_0, -\mu)]\phi(\eta_0, \mu)W(\mu)\, d\mu + [-e^{-\tau_0/\eta_0}A(\eta_0) + b_2 e^{\tau_0/\eta_0}A(-\eta_0)](\tfrac{1}{2}\omega\eta_0)^2 X(-\eta_0) + \int_0^1 [-e^{-\tau_0/\eta}A(\eta) + b_2 e^{\tau_0/\eta}A(-\eta)]\tfrac{1}{4}\omega^2\eta_0\eta X(-\eta)\, d\eta = -[-b_2 e^{-\tau_0/\eta_0}A(\eta_0) + e^{\tau_0/\eta_0}A(-\eta_0)](\tfrac{1}{2}\omega\eta_0)^2 X(\eta_0) \tag{11-97}$$

Thus we have isolated $[A(\eta_0) - b_1 A(-\eta_0)]$ and $[-b_2 e^{-\tau_0/\eta_0}A(\eta_0) + e^{\tau_0/\eta_0}A(-\eta_0)]$ by Eqs. 11-96 and 11-97, respectively.

Isolation of Continuum Coefficients

To isolate the continuum coefficients we operate on both sides of Eqs. 11-92 and 11-93 by the operator

$$\int_0^1 \phi(\eta', \mu)W(\mu)\, d\mu \tag{11-98a}$$

where

$$W(\mu) = (\eta_0 - \mu)\gamma(\mu) \tag{11-98b}$$

After performing various normalization integrals[7] and interchanging η and η' in the resulting expressions, we obtain for Eqs. 11-92 and 11-93, respectively,

$$\int_0^1 [a_1 + b_1 I_p(0, -\mu) - I_p(0, \mu)]\phi(\eta, \mu)W(\mu)\, d\mu + [b_1 A(\eta_0) - A(-\eta_0)]\omega\eta\eta_0 X(-\eta_0)\phi(-\eta_0, \eta) + \int_0^1 [b_1 A(\eta') - A(-\eta')]\tfrac{1}{2}\omega\eta(\eta_0 + \eta')X(-\eta')\phi(-\eta', \eta)\, d\eta' = [A(\eta) - b_1 A(-\eta)]\frac{W(\eta)}{g(\omega, \eta)} \tag{11-99}$$

$$\int_0^1 [a_2 + b_2 I_p(\tau_0, \mu) - I_p(\tau_0, -\mu)\phi(\eta, \mu)W(\mu)\, d\mu + [-e^{-\tau_0/\eta_0}A(\eta_0) + b_2 e^{\tau_0/\eta_0}A(-\eta_0)]\omega\eta\eta_0 X(-\eta_0)\phi(-\eta_0, \eta) + \int_0^1 [-e^{-\tau_0/\eta}A(\eta) + b_2 e^{\tau_0/\eta}A(-\eta)]\tfrac{1}{2}\omega\eta(\eta_0 + \eta')X(-\eta')\phi(-\eta', \eta)\, d\eta' = [-b_2 e^{-\tau_0/\eta}A(\eta) + e^{\tau_0/\eta}A(-\eta)]\frac{W(\eta)}{g(\omega, \eta)} \tag{11-100}$$

Thus we have isolated $[A(\eta) - b_1 A(-\eta)]$ and $[-b_2 e^{-\tau_0/\eta} A(\eta) + e^{\tau_0/\eta} A(-\eta)]$ by Eqs. 11-99 and 11-100, respectively.

Equations 11-96, 11-97, 11-99, and 11-100 provide four relations for the determination of four unknown expansion coefficients, $A(\eta_0)$, $A(-\eta_0)$, $A(\eta)$, and $A(-\eta)$. We note that two of these equations, 11-99 and 11-100, are coupled Fredholm type integral equations for the continuum coefficients $A(\eta)$ and $A(-\eta)$, but the other two equations are not integral equations. These four equations can be written more compactly in the matrix form as

$$\mathbf{M}\mathbf{A}(\eta_0) = \mathbf{G}(\eta_0) + \int_0^1 \mathbf{B}(\eta_0)\mathbf{A}(\eta') K_0(\eta')\, d\eta' \tag{11-101}$$

$$\mathbf{M}(\eta)\mathbf{A}(\eta) = \mathbf{G}(\eta) + \mathbf{B}(\eta_0)\mathbf{A}(\eta_0) K_1(\eta) + \int_0^1 \mathbf{B}(\eta')\mathbf{A}(\eta') K(\eta', \eta)\, d\eta', \quad \eta \in (0, 1) \tag{11-102}$$

where various matrices have been defined as

$$\mathbf{A}(\eta) \equiv \begin{bmatrix} A(\eta_0) \\ A(-\eta_0) \end{bmatrix}, \qquad \mathbf{A}(\eta) \equiv \begin{bmatrix} A(\eta) \\ A(-\eta) \end{bmatrix} \tag{11-103a}$$

$$\mathbf{M} \equiv \begin{bmatrix} b_1 e^{-2z_0/\eta_0} - 1 & b_1 - e^{-2z_0/\eta_0} \\ (b_2 - e^{-2z_0/\eta_0}) e^{-\tau_0/\eta_0} & (b_2 e^{-2z_0/\eta_0} - 1) e^{\tau_0/\eta_0} \end{bmatrix} \tag{11-103b}$$

$$\mathbf{M}(\eta) \equiv \begin{bmatrix} b_1 & -1 \\ -b_2 e^{-\tau_0/\eta} & e^{\tau_0/\eta} \end{bmatrix} \tag{11-103c}$$

$$\mathbf{B}(\eta_0) \equiv \begin{bmatrix} b_1 & -1 \\ -e^{-\tau_0/\eta_0} & b_2 e^{\tau_0/\eta_0} \end{bmatrix}, \qquad \mathbf{B}(\eta) \equiv \begin{bmatrix} b_1 & -1 \\ -e^{-\tau_0/\eta} & b_2 e^{\tau_0/\eta} \end{bmatrix} \tag{11-103d}$$

$$\mathbf{G}(\eta_0) \equiv \begin{bmatrix} g_1(\eta_0) \\ g_2(\eta_0) \end{bmatrix}, \qquad \mathbf{G}(\eta) \equiv \begin{bmatrix} g_1(\eta) \\ g_2(\eta) \end{bmatrix} \tag{11-103e}$$

$$g_1(\eta_0) \equiv \left(\frac{2}{\omega\eta_0}\right)^2 \frac{1}{X(\eta_0)} \int_0^1 [a_1 + b_1 I_p(0, -\mu) - I_p(0, \mu)] W(\mu) \phi(\eta_0, \mu)\, d\mu \tag{11-104a}$$

$$g_2(\eta_0) \equiv \left(\frac{2}{\omega\eta_0}\right)^2 \frac{1}{X(\eta_0)} \int_0^1 [a_2 + b_2 I_p(\tau_0, \mu) - I_p(\tau_0, -\mu)] W(\mu) \phi(\eta_0, \mu)\, d\mu \tag{11-104b}$$

$$g_1(\eta) \equiv \frac{g(\omega, \eta)}{W(\eta)} \int_0^1 [a_1 + b_1 I_p(0, -\mu) - I_p(0, \mu)] W(\mu) \phi(\eta, \mu)\, d\mu \tag{11-104c}$$

$$g_2(\eta) \equiv \frac{g(\omega, \eta)}{W(\eta)} \int_0^1 [a_2 + b_2 I_p(\tau_0, \mu) - I_p(\tau_0, -\mu)] W(\mu) \phi(\eta, \mu)\, d\mu \tag{11-104d}$$

and

$$e^{-2z_0/\eta_0} = -\frac{X(-\eta_0)}{X(\eta_0)} \tag{11-105a}$$

Here z_0 is the Milne problem extrapolated end point, which can be evaluated from the relation [21, p. 142]

$$z_0 = \frac{\omega\eta_0}{2}\int_0^1 g(\omega, \eta)\left(1 + \frac{\omega\mu^2}{1-\mu^2}\right)\tanh^{-1}\left(\frac{\mu}{\eta_0}\right)d\mu \tag{11-105b}$$

Various coefficients in Eqs. 11-101 and 11-102 have been defined as

$$K_0(\eta') \equiv \frac{\eta' X(-\eta')}{\eta_0 X(\eta_0)} \tag{11-106a}$$

$$\begin{aligned} K_1(\eta) &\equiv \frac{g(\omega, \eta)}{W(\eta)}\,\omega\eta\eta_0 X(-\eta_0)\phi(-\eta_0, \eta) \\ &= \omega(1-\omega)\eta_0^{\,2} g(\omega, \eta)X(-\eta)X(-\eta_0) \end{aligned} \tag{11-106b}$$

$$\begin{aligned} K(\eta, \eta') &\equiv \frac{g(\omega, \eta)}{W(\eta)}\,\frac{\omega n}{2}(\eta_0 + \eta')X(-\eta')\phi(-\eta', \eta) \\ &= \frac{\omega(1-\omega)\eta'}{2}\,\frac{(\eta_0+\eta)(\eta_0+\eta')}{\eta+\eta'}\,g(\omega, \eta)X(-\eta)X(-\eta') \end{aligned} \tag{11-106c}$$

The function $g(\omega, \eta)$ can be evaluated from (see Eq. 10-69b)

$$\frac{1}{g(\omega, \eta)} = (1 - \omega\eta\tanh^{-1}\eta)^2 + \left(\frac{\pi\omega\eta}{2}\right)^2 \tag{11-107}$$

and Case's $X(-\eta)$ function is related to Chandrasekhar's $H(\eta)$ function by (see Eq. 10-38b)

$$X(-\eta) = \frac{1}{(1-\omega)^{1/2}(\eta_0 + \eta)H(\eta)}, \quad \omega < 1 \tag{11-108}$$

Once a particular solution $I_p(\tau, \mu)$ of Eq. 11-85 is found for a prescribed source term $s(\tau)$, Eqs. 11-101 and 11-102 are solved and the four expansion coefficients are determined. Knowing the expansion coefficients, we can determine the radiation intensity $I(\tau, \mu)$, the incident radiation $G(\tau)$, the net radiative heat flux $q^r(\tau)$ anywhere in the medium from Eqs. 11-88, 11-90 and 11-91, respectively.

Superposition of Elementary Solutions

The general problem given by Eqs. 11-83 and 11-84 contains many parameters and thus includes many special cases. The linearity of the governing equations permits the construction of the general solution by the superposition of elementary solutions. Here we present the superposition for a uniform temperature distribution in the medium, that is,

$$T(\tau) = T_0 = \text{constant} \tag{11-109}$$

In this case it can be shown that the solution $I(\tau, \mu)$ of Eqs. 11-83 and 11-84 can be constructed by the superposition of the functions $\psi_i(\tau, \mu)$, $i = 0, 1, 2$, defined as

$$I(\tau, \mu) = \bar{\sigma} T_0^4 \psi_0(\tau, \mu) + \bar{\sigma} T_1^4 \psi_1(\tau, \mu) + \bar{\sigma} T_2^4 \psi_2(\tau, \mu) \tag{11-110}$$

where the functions $\psi_i(\tau, \mu)$ are the solutions of three simple problems:

$$\mu \frac{\partial \psi_0(\tau, \mu)}{\partial \tau} + \psi_0(\tau, \mu) = \frac{1-\omega}{\pi} + \frac{\omega}{2} \int_{-1}^{1} \psi_0(\tau, \mu')\, d\mu', \quad 0 \le \tau \le \tau_0 \tag{11-111a}$$

$$\psi_0(0, \mu) = \rho_1^s \psi_0(0, -\mu), \quad \mu > 0 \tag{11-111b}$$

$$\psi_0(\tau_0, -\mu) = \rho_2^s \psi_0(\tau_0, \mu), \quad \mu > 0 \tag{11-111c}$$

and

$$\mu \frac{\partial \psi_i(\tau, \mu)}{\partial \tau} + \psi_i(\tau, \mu) = \frac{\omega}{2} \int_{-1}^{1} \psi_i(\tau, \mu')\, d\mu', \quad 0 \le \tau \le \tau_0 \tag{11-112a}$$

$$\psi_i(0, \mu) = \delta_{1i} \frac{\varepsilon_1}{\pi} + \rho_1^s \psi_i(0, -\mu), \quad \mu > 0 \tag{11-112b}$$

$$\psi_i(\tau_0, -\mu) = \delta_{2i} \frac{\varepsilon_2}{\pi} + \rho_2^s \psi_i(\tau_0, \mu), \quad \mu > 0 \tag{11-112c}$$

where $i = 1$ or 2, and δ_{ij} is the Kronecker delta.

Then the net radiative heat flux $q^r(\tau)$ in the medium is given by

$$q^r(\tau) = \bar{\sigma}[T_0^4 Q_0(\tau) + T_1^4 Q_1(\tau) + T_2^4 Q_2(\tau)] \tag{11-113a}$$

where

$$Q_i(\tau) = 2\pi \int_{-1}^{1} \psi_i(\tau, \mu) \mu\, d\mu, \quad i = 0, 1, 2 \tag{11-113b}$$

In most engineering applications the net radiative heat flux at the boundaries is of interest. For example, the net radiative heat flux at $\tau = 0$ is determined from

$$q^r(0) = \bar{\sigma}[T_0^4 Q_0(0) + T_1^4 Q_1(0) + T_2^4 Q_2(0)] \tag{11-114}$$

A close scrutiny of the subproblems given by Eqs. 11-111 and 11-112 reveals that to evaluate the net radiative heat flux one needs to solve the problem defining function $\psi_0(\tau)$ and only one of the problems defining functions $\psi_1(\tau)$ and $\psi_2(\tau)$. Suppose that the subproblems defining functions $\psi_0(\tau)$ and $\psi_1(\tau)$ are to be solved; then Eq. 11-114 for the wall heat flux can be written as

$$q^r(0) = \bar{\sigma}[T_0^4 Q_0(0) + T_1^4 Q_1(0) - T_2^4 Q_1^*(\tau_0)] \tag{11-115}$$

where the heat flux function $Q_1^*(\tau_0)$ is obtained from the solution of the subproblem defining function $\psi_1(\tau, \mu)$, however, by interchanging in that problem the radiative properties of surfaces 1 and 2 and evaluating the solution for $\tau = \tau_0$.

Table 11-4 shows the numerical values of the heat flux functions $Q_0(0)$, $Q_1(0)$, and $Q_1^*(\tau_0)$ appearing in Eq. 11-115 for purely diffusely reflecting boundaries, for three different optical thicknesses and for $\omega = 0$ and $\omega = 0.5$.

Table 11-4 Heat Flux Functions $Q_0(0)$, $Q_1(0)$, and $Q_1^*(\tau_0)$ for a Slab at Constant Temperature and with Diffusely Reflecting Boundaries[a]

Boundary at $\tau = 0$		Boundary at $\tau = \tau_0$		$\tau_0 = 0.1$		$\tau_0 = 1.0$		$\tau_0 = 10.0$	
ε_1	ρ_1^d	ε_2	ρ_2^d	$\omega = 0$	$\omega = 0.5$	$\omega = 0$	$\omega = 0.5$	$\omega = 0$	$\omega = 0.5$
					$-Q_0(0)$				
1.0	0.0	0.0	1.0	0.3068	0.1736	0.9519	0.7572	1.0000	0.8535
1.0	0.0	0.5	0.5	0.2371	0.1316	0.8662	0.6510	1.0000	0.8535
1.0	0.0	1.0	0.0	0.1674	0.0911	0.7806	0.5591	1.0000	0.8535
					$Q_1(0)$				
1.0	0.0	0.0	1.0	0.3068	0.1736	0.9519	0.7572	1.0000	0.8535
1.0	0.0	0.5	0.5	0.6534	0.5753	0.9759	0.8154	1.0000	0.8535
1.0	0.0	1.0	0.0	1.0000	0.9616	1.0000	0.8658	1.0000	0.8535
					$Q_1^*(\tau_0)$				
1.0	0.0	0.0	1.0	0.0	0.0	0.0	0.0	0.0	0.0
1.0	0.0	0.5	0.5	0.4163	0.4437	0.1097	0.1644	0.0	0.0
1.0	0.0	1.0	0.0	0.8326	0.8704	0.2194	0.3067	0.0	0.0

[a] From H. L. Beach, M. N. Özişik, and C. E. Siewert [29].

11-7 SLAB WITH DISTRIBUTED INTERNAL ENERGY SOURCES—SOLUTION BY NORMAL-MODE EXPANSION TECHNIQUE

To illustrate the application of the normal-mode expansion technique for the case $\omega = 1$, we consider radiative heat transfer in a gray, plane-parallel slab having distributed internal energy sources and bounded between two black boundaries at $\tau = 0$ and $\tau = \tau_0$, which are kept at temperatures T_1 and T_2, respectively. In this problem we are concerned with the determination of temperature distribution and the net radiative heat flux in the medium.

The conservation of energy equation is given as (see Eq. 8-18b)

$$\frac{dq^r(y)}{dy} = g(y) \qquad \text{or} \qquad \frac{dq^r(\tau)}{d\tau} = \frac{g(\tau)}{\beta} \tag{11-116}$$

where $g(\tau)$ is the interal energy source (i.e., the energy per unit volume and time) and τ is the optical variable, defined as $d\tau = \beta\, dy$. The net radiative heat flux $q^r(\tau)$ is related to the radiation intensity $I(\tau, \mu)$ by

$$q^r(\tau, \mu) = 2\pi \int_{-1}^{1} I(\tau, \mu)\mu \, d\mu \tag{11-117}$$

and for a gray medium the radiation intensity $I(\tau, \mu)$ satisfies the equation of radiative transfer

$$\mu \frac{\partial I(\tau, \mu)}{\partial \tau} + I(\tau, \mu) = (1 - \omega) \frac{n^2 \bar{\sigma} T^4(\tau)}{\pi} + \frac{\omega}{2} \int_{-1}^{1} I(\tau, \mu')\, d\mu'$$

$$\text{in } 0 \leq \tau \leq \tau_0, \; -1 \leq \mu \leq 1 \tag{11-118}$$

and for black boundaries the boundary conditions are given as

$$I(0, \mu) = \frac{n^2 \bar{\sigma} T_1^{\,4}}{\pi} \equiv f_1, \quad \mu > 0 \tag{11-119}$$

$$I(\tau_0, \mu) = \frac{n^2 \bar{\sigma} T_2^{\,4}}{\pi} \equiv f_2, \quad \mu < 0 \tag{11-120}$$

Now, by utilizing the energy equation 11-116, we shall transform the equation of radiative transfer, Eq. 11-118, into a form appropriate for $\omega = 1$. We operate on both sides of Eq. 11-118 by the operator $2\pi \int_{-1}^{1} d\mu$, utilize the relations given by Eqs. 11-116 and 11-117, and obtain

$$\frac{n^2 \bar{\sigma} T^4(\tau)}{\pi} = \frac{g(\tau)}{4\pi\beta} + \frac{1}{2} \int_{-1}^{1} I(\tau, \mu')\, d\mu' \tag{11-121}$$

Substituting Eq. 11-121 into Eq. 11-118, we eliminate the temperature term and find

$$\mu \frac{\partial I(\tau, \mu)}{\partial \tau} + I(\tau, \mu) = \frac{g(\tau)}{4\pi\beta} + \frac{1}{2} \int_{-1}^{1} I(\tau, \mu')\, d\mu'$$

$$\text{in } 0 \leq \tau \leq \tau_0, \; -1 \leq \mu \leq 1 \tag{11-122}$$

with the boundary conditions

$$I(0, \mu) = f_1, \quad \mu > 0 \tag{11-123a}$$

$$I(\tau_0, \mu) = f_2, \quad \mu < 0 \tag{11-123b}$$

Equation 11-122 is now of the same form as the equation of radiative transfer for $\omega = 1$ and has a prescribed inhomogeneous source term. The radiative heat transfer problem given by Eqs. 11-122 and 11-123 has been solved by the application of the normal-mode expansion technique by Kriese and Siewert [30], Ferziger and Simmons [25], and Beach, Özişik, and Siewert [29] for different boundary conditions. Once the angular distribution of radiation intensity $I(\tau, \mu)$ is determined, the temperature distribution is evaluated from Eq. 11-121 and the net radiative heat flux from Eq. 11-117. We present now the solution of Eq. 11-122 subject to the boundary conditions 11-123 by the normal-mode expansion technique.

The solution of Eq. 11-122 is written as a linear sum of the elementary solutions of the homogeneous part of this equation and a particular solution in the form (see Eq. 10-20b for the homogeneous solution)

$$I(\tau, \mu) = A + B(\tau - \mu) + \int_0^1 A(\eta)\phi(\eta, \mu)e^{-\tau/\eta}\, d\eta + \int_0^1 A(-\eta)\phi(-\eta, \mu)e^{\tau/\eta}\, d\eta + I_p(\tau, \mu) \quad (11\text{-}124)$$

where it is assumed that a particular solution $I_p(\tau, \mu)$ can be found, depending on the type of the inhomogeneous term $g(\tau)/4\pi\beta$, and the continuum normal mode is given by (see Eqs. 10-18d and 10-18e for $\omega = 1$)

$$\phi(\eta, \mu) = \frac{\eta}{2}\frac{P}{\eta - \mu} + (1 - \eta \tanh^{-1} \eta)\, \delta(\eta - \mu), \quad \eta \in (-1, 1) \quad (11\text{-}125)$$

The problem is now reduced to determining the unknown expansion coefficients A, B, $A(\eta)$, and $A(-\eta)$; this can be done by constraining the solution 11-124 to meet the boundary conditions given by Eqs. 11-123.

Once a particular solution is found and the expansion coefficients are determined, the angular distribution of radiation intensity $I(\tau, \mu)$ is given by Eq. 11-124, and the net radiative heat flux is evaluated by means of Eq. 11-117, which becomes[8]

$$q^r(\tau) = 2\pi\left[-\tfrac{2}{3}B + \int_{-1}^{1} I_p(\tau, \mu)\mu\, d\mu\right] \quad (11\text{-}126)$$

and the temperature distribution is evaluated from Eq. 11-121, which becomes

$$\frac{n^2\bar{\sigma}T^4(\tau)}{\pi} = \frac{g(\tau)}{4\pi\kappa} + \frac{1}{2}\left[2A + 2B + \int_0^1 A(\eta)e^{-\tau/\eta}\, d\eta + \int_0^1 A(-\eta)e^{\tau/\eta}\, d\eta + \int_{-1}^{1} I_p(\tau, \mu)\, d\mu\right] \quad (11\text{-}127)$$

We now consider the determination of the expansion coefficients.

Substitution of solution 11-124 into the boundary conditions 11-123a and 11-123b, respectively, yields

$$f_1 - I_p(0, \mu) + \mu B - \int_0^1 A(-\eta)\phi(-\eta, \mu)\, d\eta$$
$$= A + \int_0^1 A(\eta)\phi(\eta, \mu)\, d\eta, \quad \mu > 0 \quad (11\text{-}128)$$

$$f_2 - I_p(\tau_0, -\mu) - \mu B - \int_0^1 A(\eta)e^{-\tau_0/\eta}\phi(-\eta, \mu)\, d\eta$$
$$= (A + B\tau_0) + \int_0^1 A(-\eta)e^{\tau_0/\eta}\phi(\eta, \mu)\, d\eta, \quad \mu > 0 \quad (11\text{-}129)$$

To obtain Eqs. 11-128 and 11-129 we utilized the reciprocity relations

$$\phi(\eta, -\mu) = \phi(-\eta, \mu) \quad (11\text{-}130a)$$
$$\phi(-\eta, -\mu) = \phi(\eta, \mu) \quad (11\text{-}130b)$$

We note that the right-hand sides of Eqs. 11-128 and 11-129 are of the same form as the half-range expansion given by Eq. 10-22b for the case $\omega = 1$; by the half-range expansion theorem stated in Chapter 10 this expansion is sufficiently general to represent an arbitrary function (i.e., the left-hand side of Eqs. 11-128 and 11-129) defined in the interval $\mu \in (0, 1)$. By utilizing the orthogonality of normal modes and various normalization integrals given in Chapter 10 for the case $\omega = 1$, the expansion coefficients can be isolated from these equations as now described.

Isolation of Discrete Coefficients

To isolate the discrete coefficients A and B, we operate on both sides of Eqs. 11-128 and 11-129 by the operator

$$\int_0^1 \gamma(\mu)\, d\mu \quad (11\text{-}131)$$

and obtain respectively[9]

$$f_1 - \int_0^1 I_p(0, \mu)\gamma(\mu)\, d\mu + B\gamma^{(1)} - \int_0^1 A(-\eta')\frac{\eta' X(-\eta')}{2}\, d\eta' = A \quad (11\text{-}132)$$

$$f_2 - \int_0^1 I_p(\tau_0, -\mu)\gamma(\mu)\, d\mu - B\gamma^{(1)} - \int_0^1 A(\eta')e^{-\tau_0/\eta'}\frac{\eta' X(-\eta')}{2}\, d\eta' = A + B\tau_0$$
$$(11\text{-}133)$$

Thus we have isolated A and $A + B\tau_0$ on the right-hand side of Eqs. 11-132 and 11-133, respectively.

Isolation of Continuum Coefficients

To isolate the continuum coefficients $A(\eta)$ and $A(-\eta)$ we operate on both sides of Eqs. 11-128 and 11-129 by the operator

$$\int_0^1 \gamma(\mu)\phi(\eta', \mu)\, d\mu \tag{11-134}$$

After performing various normalization integrals and interchanging η and η' in the resulting expressions, Eqs. 11-128 and 11-129 yield respectively[10]

$$-\int_0^1 I_p(0, \mu)\gamma(\mu)\phi(\eta, \mu)\, d\mu - B\frac{\eta}{2} - \int_0^1 A(-\eta')\frac{\eta\eta'}{4(\eta + \eta')}X(-\eta')\, d\eta' = A(\eta)\frac{\gamma(\eta)}{g(1, \eta)} \tag{11-135}$$

$$-\int_0^1 I_p(\tau_0, -\mu)\gamma(\mu)\phi(\eta, \mu)\, d\mu + B\frac{\eta}{2} - \int_0^1 A(\eta')e^{-\tau_0/\eta'}\frac{\eta\eta'}{4(\eta + \eta')}X(-\eta')\, d\eta' = A(-\eta)e^{\tau_0/\eta}\frac{\gamma(\eta)}{g(1, \eta)} \tag{11-136}$$

Thus we have isolated $A(\eta)$ and $A(-\eta)$ on the right-hand side of Eqs. 11-135 and 11-136, respectively.

The function $g(1, \eta)$ can be evaluated from (see Eq. 10-86b)

$$\frac{1}{g(1, \eta)} = (1 - \eta \tanh^{-1}\eta)^2 + \left(\frac{\pi\eta}{2}\right)^2 \tag{11-137}$$

The $X(-\eta)$ and $\gamma(\eta)$ functions are related to Chandrasekhar's $H(\mu)$ function by (see Eqs. 10-38c and 10-29c)

$$X(-\eta) = \frac{\sqrt{3}}{H(\eta)}, \qquad \omega = 1 \tag{11-138}$$

$$\gamma(\eta) = \frac{\sqrt{3}}{3}\eta H(\eta), \quad \omega = 1 \tag{11-139}$$

and $\gamma^{(1)}$ is given as

$$\gamma^{(1)} \equiv \int_0^1 \mu\gamma(\mu)\, d\mu = \frac{\sqrt{3}}{2}\int_0^1 \mu^2 H(\mu)\, d\mu = 0.71044\ldots \tag{11-140}$$

Equations 11-132, 11-133, 11-135, and 11-136 are four equations for the determination of four unknown expansion coefficients, A, B, $A(\eta)$, and $A(-\eta)$. To summarize the results we rearrange these equations as

$$A - \gamma^{(1)}B = \left[f_1 - \int_0^1 I_p(0,\mu)\gamma(\mu)\,d\mu\right] - \frac{1}{2}\int_0^1 \eta' X(-\eta')A(-\eta')\,d\eta' \tag{11-141}$$

$$A + (\gamma^{(1)} + \tau_0)B = \left[f_2 - \int_0^1 I_p(\tau_0, -\mu)\gamma(\mu)\,d\mu\right] - \frac{1}{2}\int_0^1 \eta' X(-\eta')e^{-\tau_0/\eta'}A(\eta')\,d\eta' \tag{11-142}$$

$$A(\eta) = -\frac{g(1,\eta)}{\gamma(\eta)}\left[\int_0^1 I_p(0,\mu)\gamma(\mu)\phi(\eta,\mu)\,d\mu + \frac{\eta}{2}B + \frac{\eta}{4}\int_0^1 \frac{\eta' X(-\eta')}{\eta + \eta'}A(-\eta')\,d\eta'\right] \tag{11-143}$$

$$A(-\eta)e^{\tau_0/\eta} = -\frac{g(1,\eta)}{\gamma(\eta)}\left[\int_0^1 I_p(\tau_0, -\mu)\gamma(\mu)\phi(\eta,\mu)\,d\mu - \frac{\eta}{2}B + \frac{\eta}{4}\int_0^1 \frac{\eta' X(-\eta')}{\eta + \eta'}e^{-\tau_0/\eta'}A(\eta')\,d\eta'\right] \tag{11-144}$$

If a particular solution $I_p(\tau, \mu)$ of Eq. 11-122 is found for a specified inhomogeneous term of this equation, Eqs. 11-141 through 11-144 can be solved numerically by an iterative procedure and the expansion coefficients determined to any desired degree of accuracy. Knowing the expansion coefficients, we can evaluate the radiation intensity $I(\tau, \mu)$, the net radiative heat flux $q^r(\tau)$, and the temperature $T(\tau)$ anywhere in the medium from Eqs. 11-124, 11-126, and 11-127, respectively.

For simplicity in the analysis we considered above only the black boundaries. This approach can readily be extended to diffusely reflecting and specularly reflecting boundaries.

Analytical Approximations

Analytical approximations can also be obtained from Eqs. 11-141 through 11-144 that will yield the expansion coefficients with sufficient accuracy. Ferziger and Simmons [25], Özişik and Siewert [28], and Kriese and Siewert [30] obtained such approximations. For the conservative case ($\omega = 1$) the lowest-order approximation was better than the result obtained with the classical diffusion theory, whereas the second-order approximation was highly accurate. For the nonconservative ($\omega < 1$) medium, however, Beach, Özişik, and Siewert [29] showed that such approximations were not so accurate; hence we did not consider them in Section 11-6.

(a) First-Order Approximation

A first-order approximation is obtained by neglecting in Eqs. 11-141 through 11-144 continuum coefficients entirely, that is,

$$A^{(1)}(\eta) = A^{(1)}(-\eta) = 0 \tag{11-145}$$

Then the discrete coefficients are immediately obtained from

$$A^{(1)} - \gamma^{(1)}B^{(1)} = f_1 - \int_0^1 I_p(0, \mu)\gamma(\mu)\, d\mu \tag{11-146a}$$

$$A^{(1)} + [\gamma^{(1)} + \tau_0]B^{(1)} = f_2 - \int_0^1 I_p(\tau_0, -\mu)\gamma(\mu)\, d\mu \tag{11-146b}$$

where superscript (1) in the expansion coefficients denotes the order of approximation.

(b) Second-Order Approximations

A second-order approximation to the continuum coefficients is obtained from Eqs. 11-143 and 11-144 by neglecting the integals entirely and by using the first-order approximation for the discrete coefficients. We obtain

$$A^{(2)}(\eta) = -\frac{g(1, \eta)}{\gamma(\eta)}\left[\int_0^1 I_p(0, \mu)\gamma(\mu)\phi(\mu, \eta)\, d\mu + \frac{\eta}{2} B^{(1)}\right] \tag{11-147a}$$

$$A^{(2)}(-\eta)e^{\tau_0/\eta} = -\frac{g(1, \eta)}{\gamma(\eta)}\left[\int_0^1 I_p(\tau_0, -\mu)\gamma(\mu)\phi(\eta, \mu)\, d\mu - \frac{\eta}{2} B^{(1)}\right] \tag{11-147b}$$

A second-order approximation to the discrete coefficients is obtained from Eqs. 11-141 and 11-142 by using in these equations the second-order continuum coefficients. We obtain

$$A^{(2)} - \gamma^{(1)}B^{(2)} = \left[f_1 - \int_0^1 I_p(0, \mu)\gamma(\mu)\, d\mu\right] - \frac{1}{2}\int_0^1 \eta' X(-\eta')A^{(2)}(-\eta')\, d\eta' \tag{11-148a}$$

$$A^{(2)} + [\gamma^{(1)} + \tau_0]B^{(2)}$$
$$= \left[f_2 - \int_0^1 I_p(\tau_0, -\mu)\gamma(\mu) d\mu\right] - \frac{1}{2}\int_0^1 \eta' X(-\eta')e^{-\tau_0/\eta'}A^{(2)}(\eta')\, d\eta' \tag{11-148b}$$

Discussion of Results

The net radiative heat flux in the medium is a quantity of interest in most engineering applications. If a first-order approximation is used, a first-order approximation for the net radiative heat flux from Eqs. 11-126 and 11-146 becomes

$$q^{(1)r}(\tau) = 2\pi\left[-\tfrac{2}{3}B^{(1)} + \int_{-1}^{1} I_p(\tau,\mu)\mu\, d\mu\right] \tag{11-149a}$$

where

$$B^{(1)} = \frac{1}{2\gamma^{(1)} + \tau_0}\left\{(f_2 - f_1) + \int_0^1 [I_p(0,\mu) - I_p(\tau_0,-\mu)]\gamma(\mu)\, d\mu\right\} \tag{11-149b}$$

Assuming a constant energy generation rate, that is, $g(\tau) = g_0 =$ constant, we immediately obtain a particular solution of Eq. 11-122 from Table 10-6 as

$$I_p(\tau,\mu) = -\frac{3g_0}{4\pi\beta}\left(\tfrac{1}{2}\tau^2 - \tau\mu + \mu^2\right) \tag{11-150}$$

Then various terms in Eqs. 11-149 become

$$I_p(0,\mu) = -\frac{3g_0}{4\pi\beta}\mu^2 \tag{11-151a}$$

$$I_p(\tau_0,-\mu) = -\frac{3g_0}{3\pi\beta}\left(\tfrac{1}{2}\tau_0^2 + \tau_0\mu + \mu^2\right) \tag{11-151b}$$

$$I_p(0,\mu) - I_p(\tau_0,-\mu) = \frac{3g_0}{4\pi\beta}\left(\tfrac{1}{2}\tau_0^2 + \tau_0\mu\right) \tag{11-151c}$$

$$\int_0^1 [I_p(0,\mu) - I_p(\tau_0,-\mu)]\gamma(\mu)\, d\mu = \frac{3g_0}{4\pi\beta}\left[\tfrac{1}{2}\tau_0^2 + \tau_0\gamma^{(1)}\right] \tag{11-151d}$$

and

$$\int_0^1 I_p(\tau,\mu)\mu\, d\mu = \frac{g(\tau)}{2\pi\beta} \tag{11-151e}$$

Substitution of Eqs. 11-151 into Eqs. 11-149 yields a first-order approximation for the net radiative heat flux for the case of constant internal generation as

$$q^{(1)r}(\tau) = -\frac{4\pi}{3}\frac{f_2 - f_1}{2\gamma^{(1)} + \tau_0} - \frac{g}{\beta}\left(\frac{\tau_0}{2} - \tau\right) \tag{11-152}$$

and for no internal generation Eq. 11-152 simplifies to

$$\frac{q^{(1)r}}{n^2(\bar{\sigma}T_1^4 - \bar{\sigma}T_2^4)} = \frac{4}{3[2\gamma^{(1)} + \tau_0]} \cong \frac{4}{3(1.4209 + \tau_0)} \tag{11-153}$$

since

$$f_i \equiv \frac{n^2\bar{\sigma}T_i^4}{\pi}, \quad i = 1 \text{ or } 2 \text{ and } \gamma^{(1)} = 0.71044\ldots$$

Table 11-5 Accuracy of the First-Order and Second-Order Approximations to Predict Net Radiative Heat Flux for $g = 0$, $\omega = 1$

τ_0	$\left\lvert \dfrac{q^r_{\text{exact}} - q^r_{\text{approx}}}{q^r_{\text{exact}}} \right\rvert \times 100$ First-Order Approximation	Second-Order Approximation	$\left\lvert \dfrac{q^r_{\text{exact}}}{n^2\bar{\sigma}T_1^4 - \bar{\sigma}T_2^4} \right\rvert$
0.1	4.26	0.317	0.9157
1.0	0.48	0.0044	0.5534
10.0	0	0	0.1167

Table 11-5 shows the accuracy of the first-order and the second-order approximations to predict the net radiative heat flux for the case of no internal energy generation (i.e., $g_0 = 0$), together with the exact values q^r. For an optical thickness $\tau_0 = 0.1$ the first-order approximation underestimates the net radiative heat flux by about 4 per cent; the second-order approximation, by only about 0.3 per cent. The maximum error occurs for $\tau_0 = 0$ and is less than 6 per cent for the first-order approximation.

11-8 REFLECTIVITY AND TRANSMISSIVITY OF SEMITRANSPARENT MEDIA

The reflection and the transmission of radiation by semitransparent media are affected by the absorption and scattering characteristics of the material below the surface. The radiative properties of fibrous materials, powders, foam, wood, paint, refractory oxides, a thick layer of glass, and many others are therefore treated as bulk phenomena. A number of attempts have been made to describe the radiative properties of a plane-parallel, semitransparent medium in terms of the absorption and scattering characteristics and the optical thickness of the material from the solution of the equation of radiative transfer. In such studies it is generally assumed that the medium absorbs and scatters radiation but does not reradiate it. There are many physical situations in which this assumption is justified: when the absorbed radiation is dissipated to the surrounding fluid almost entirely by conduction, and when light is transmitted through a transparent layer containing embedded particulates that scatter and absorb radiation, such as the particulate matter contained in the atmosphere, the reradiation is neglected.

The radiative properties of semitransparent materials have been determined by various investigators from the solution of the equation of radiative transfer by both approximate and exact methods. Horak and Chandrasekhar [39] solved exactly the diffuse reflection by a semi-infinite atmosphere, and Pitts [40] applied the Eddington approximation to study the reflection and transmission of light through unexposed photographic emulsions. Chu and Churchill [41] transformed the equation of radiative transfer into a set of simultaneous ordinary differential equations to calculate the transmission of radiation through a slab of finite thickness. We do not cite here several other approximate solutions that are available in the literature. The accuracy of an approximate solution is not known unless it is compared with the exact result. Chandrasekhar [1] treated exactly the reflection and transmission of radiation by a plane-parallel medium of arbitrary thickness and with transparent boundaries for the cases of isotropic, linearly anisotropic, and Rayleigh scattering. Chu et al. [42] utilized the methods developed by Chandrasekhar to determine numerically the reflection of an obliquely incident radiation from a semi-infinite medium for a phase function represented by an arbitrarily truncated series in the form

$$p(\mu) = \frac{1}{4\pi}\left[1 + \sum_{n=1}^{N} b_n P_n(\mu)\right] \tag{11-154}$$

A few representative cases corresponding to finite coefficients b_n for $N = 0, 1, 2, 3,$ and 4 have been considered in these calculations. Evans, Chu, and Churchill [43] and Hottel et al. [44] utilized an approach similar to that in reference 42 to calculate the reflectivity and transmissivity of a slab of finite thickness. The former investigators [43] considered an obliquely incident radiation and transparent boundaries, whereas the latter workers [44] studied an azimuthally symmetric radiation but included the effects of Fresnel reflection at the boundaries. Giovanelli [45] calculated the reflectivity of a semi-infinite medium with a transparent boundary for the cases of isotropic scattering and linearly anisotropic scattering according to a phase function

$$p(\mu) = 1 + b\mu \tag{11-155}$$

by utilizing the technique and the relevant H functions given by Chandrasekhar. Busbridge and Orchard [46] computed the reflectivity and transmissivity of a plane-parallel, conservative (i.e., $\omega = 1$) slab with transparent boundaries for linearly anisotropic scattering according to the phase function given by Eq. 11-155. Progelhof and Throne [47] applied the Monte Carlo method to determine the reflectivity and absorptivity of a cylindrical body under diffuse incident radiation. Recently Lii and Özişik [48] determined the transmissivity and hemispherical reflectivity of an absorbing, isotropically scattering slab of finite thickness with a reflecting boundary by solving the equation of radiative transfer exactly with the normal-mode expansion technique.

In this section we present some of the computed results for the reflectivity of an absorbing, scattering, but not reradiating plane-parallel, semitransparent medium with transparent boundaries for the cases of isotropic scattering and linearly anisotropic scattering, and illustrate the effects of the optical thickness τ_0, the albedo ω, and the linear anisotropy factor b on the reflectivity.

Table 11-6 contains reflectivity data computed by Giovanelli [45] from the exact results for a semi-infinite medium with a transparent boundary for the cases of isotropic scattering and for linearly anisotropic scattering according to a phase function

$$p(\mu) = 1 + \mu \tag{11-156}$$

for several different values of ω from 1 to 0. In this table the *hemispherical reflectivity*, as described in Chapter 1, characterizes a uniformly diffused

Table 11-6 Hemispherical and Normal-Hemispherical Reflectivity for a Semi-infinite Medium for Isotropic Scattering and for a Phase Function[a] $p(\mu) = 1 + \mu$

	Normal-Hemispherical Reflectivity		Hemispherical Reflectivity	
ω	Isotropic	$1 + \mu$	Isotropic	$1 + \mu$
1.000	1.00000	1.00000	1.00000	1.00000
0.999	0.91285	0.89367	0.92971	0.91446
0.995	0.81705	0.77877	0.84985	0.81945
0.990	0.75275	0.70270	0.79457	0.75482
0.975	0.64092	0.57344	0.69501	0.64140
0.950	0.53555	0.45552	0.59667	0.53311
0.925	0.46655	0.38104	0 52965	0 46172
0.900	0.41495	0.32712	0.47802	0.40825
0.85	0.33966	. . .	0.40017	. . .
0.80	0.28526	0.20015	0.34187	0.27406
0.7	0.20867	0.13286	0.25655	0.19626
0.6	0.15541	0.09065	0.19471	0.14318
0.5	0.11521	0.06192	0.14653	0.10411
0.4	0.08336	0.04147	0.10734	0.07394
0.3	0.05721	0.02638	0.07445	0.04986
0.2	0.03524	0.01513	0.04626	0.03018
0.1	0.01639	0.00649	0.02170	0.01382
0	0.00000	0.00000	0.00000	0.00000

[a] From R. G. Giovanelli [45].

Table 11-7 Normal-Hemispherical Reflectivity of a Semi-infinite Medium for a Phase Function[a] $p(\mu) = 1 + b\mu$

	b				
ω	0	0.25	0.5	0.75	1.00
1.000	1.00000	1.00000	1.00000	1.00000	1.00000
0.999	0.91285	0.9090	0.9046	0.8996	0.89367
0.995	0.81705	0.8094	0.8006	0.7906	0.77877
0.990	0.75275	0.7427	0.7312	0.7180	0.70270
0.975	0.64092	0.6273	0.6116	0.5940	0.57344
0.950	0.53555	0.5191	0.5006	0.4796	0.45552
0.925	0.46655	0.4488	0.4289	0.4075	0.38104
0.900	0.41495	0.3965	0.3760	0.3531	0.32712
0.85	0.33966	0.3209	0.3003	0.2772	0.2516
0.80	0.28526	0.2668	0.2468	0.2246	0.20015
0.7	0.20867	0.1919	0.1738	0.1542	0.13286
0.6	0.15541	0.1408	0.1252	0.1085	0.09094
0.5	0.11521	0.1029	0.0900	0.0764	0.06192
0.4	0.08336	0.0735	0.0633	0.0526	0.04147
0.3	0.05721	0.0498	0.0423	0.0344	0.02638
0.2	0.03524	0.0304	0.0254	0.0203	0.01513
0.1	0.01639	0.0140	0.0115	0.0106	0.00649
0	0.00000	0.00000	0.00000	0.00000	0.00000

[a] From R. G. Giovanelli [45].

radiation incident on the surface from all directions in the space above it and collection of the reflected radiation over the entire hemispherical space. The *normal-hemispherical reflectivity* refers to a collimated beam of constant radiative flux incident normally on the surface and collection of the reflected radiation over the entire hemispherical space.

Table 11-7 gives the normal-hemispherical reflectivity of a semi-infinite medium ($0 \leq \tau < \infty$) having a transparent boundary at $\tau = 0$ for a phase function $p(\mu) = 1 + b$, $b = 0$, 0.25, 0.50, 0.75, and 1.00.

Table 11-8 shows the hemispherical and the normal-hemispherical reflectivity computed by Busbridge and Orchard [46] from both the exact solution and the Eddington (or P_1) approximation for a conservative (i.e., $\omega = 1$), plane-parallel slab having optical thickness τ_0 and transparent boundaries for scattering according to a phase function $p(\mu) = 1 + b\mu$. Since there is no absorption in a conservative medium, the sum of the reflectivity and the transmissivity of the slab must be equal to unity. Then the transmissivity of the slab is obtained by subtracting from unity the reflectivity given in Table 11-8. An examination of the reflectivity data shown

Table 11-8 Normal-Hemispherical Reflectivity of a Conservative (i.e., $\omega = 1$) Slab for Scattering with a Phase Function[a] $p(\mu) = 1 + b\mu$

Optical Thickness	Normal-Hemispherical Reflectivity		Hemispherical Reflectivity	
τ_0	Eddington	Exact	Eddington	Exact
		$b = 0.0$		
1.0	0.2857	0.3413	0.4286	0.4466
2.0	0.5000	0.5175	0.6000	0.6099
3.0	0.6154	0.6225	0.6923	0.6984
4.0	0.6875	0.6909	0.7500	0.7540
5.0	0.7368	0.7387	0.7895	0.7923
6.0	0.7727	0.7738	0.8182	0.8203
7.0	0.8000	0.8007	0.8400	0.8417
8.0	0.8214	0.8218	0.8571	0.8585
9.0	0.8387	0.8389	0.8710	0.8721
10.0	0.8529	0.8530	0.8824	0.8833
		$b = 0.5$		
1.0	0.2308	0.2924	0.3846	0.4055
2.0	0.4444	0.4654	0.5556	0.5678
3.0	0.5652	0.5743	0.6522	0.6599
4.0	0.6429	0.6476	0.7143	0.7195
5.0	0.6970	0.6997	0.7576	0.7614
6.0	0.7368	0.7368	0.7895	0.7923
7.0	0.7674	0.7686	0.8140	0.8162
8.0	0.7917	0.7924	0.8333	0.8351
9.0	0.8113	0.8118	0.8491	0.8505
10.0	0.8276	0.8279	0.8621	0.8633
		$b = 1.0$		
1.0	0.1667	0.2355	0.3333	0.3577
2.0	0.3750	0.4006	0.5000	0.5154
3.0	0.5000	0.5120	0.6000	0.6102
4.0	0.5833	0.5901	0.6667	0.6738
5.0	0.6429	0.6471	0.7143	0.7195
6.0	0.6875	0.6904	0.7500	0.7540
7.0	0.7222	0.7242	0.7778	0.7810
8.0	0.7500	0.7514	0.8000	0.8026
9.0	0.7727	0.7738	0.8182	0.8203
10.0	0.7917	0.7924	0.8333	0.8351

[a] From I. W. Busbridge and S. E. Orchard [46].

Table 11-9a Hemispherical Reflectivity of a Nonconservative (i.e., $\omega < 1$) Slab for Isotropic Scattering and with a Reflecting Boundary at $\tau = \tau_0$ and a Transparent Boundary at $\tau = 0$[a] (Radiation is incident at $\tau = 0$.)

ω	Wall Reflectivity at τ_0		$\tau_0 = 0.1$		$\tau_0 = 0.5$		$\tau_0 = 1$		$\tau_0 = 2$		$\tau_0 = 5$		$\tau_0 = 15$		$\tau_0 = 30$	
	ρ^s	ρ^d	Exact	P_1 App.	Exact	P_1 App.	Exact	P_1 App.	Exact	P_1 App.	Exact	P_1 App.	Exact	P_1 App.	Exact	P_1 App.
0.995	0	0	0.0838	0.0693	0.2932	0.2702	0.4412	0.4233	0.5988	0.5890	0.7636	0.7607	0.8438	0.8429	0.8497	0.8489
	0.5	0	0.5213	0.5170	0.5834	0.5739	0.6363	0.6270	0.7040	0.6975	0.7930	0.7904	0.8456	0.8447	0.8497	0.8489
	1.0	0	0.9980	0.9980	0.9901	0.9901	0.9803	0.9803	0.9617	0.9615	0.9151	0.9147	0.8568	0.8560	0.8500	0.8492
	0	1.0	0.9980	0.9980	0.9901	0.9901	0.9803	0.9803	0.9617	0.9615	0.9151	0.9147	0.8568	0.8560	0.8500	0.8492
0.975	0	0	0.0318	0.0673	0.2831	0.2601	0.4206	0.4028	0.5575	0.5476	0.6715	0.6671	0.6949	0.6911	0.6950	0.6912
	0.5	0	0.5170	0.5126	0.5642	0.5541	0.6014	0.5913	0.6440	0.6365	0.6856	0.6812	0.6950	0.6912	0.6950	0.6912
	1.0	0	0.9901	0.9901	0.9519	0.9515	0.9080	0.9070	0.8354	0.8331	0.7274	0.7236	0.6952	0.6914	0.6950	0.6912
	0	1.0	0.9901	0.9901	0.9518	0.9515	0.9078	0.9070	0.8351	0.8331	0.7274	0.7236	0.6952	0.6914	0.6950	0.4912
0.950	0	0	0.0793	0.0648	0.2708	0.2478	0.3964	0.3785	0.5121	0.5014	0.5886	0.5815	0.5967	0.5896	0.5967	0.5896
	0.5	0	0.5117	0.5072	0.5412	0.5303	0.5615	0.5500	0.5813	0.5717	0.5952	0.5879	0.5967	0.5896	0.5967	0.5896
	1.0	0	0.9804	0.9802	0.9074	0.9059	0.8296	0.8260	0.7193	0.7127	0.6106	0.6030	0.5967	0.5896	0.5967	0.5896
	0	1.0	0.9803	0.9802	0.9070	0.9059	0.8290	0.8260	0.7187	0.7127	0.6105	0.6030	0.5967	0.5896	0.5967	0.5896
0.9	0	0	0.0744	0.0599	0.2475	0.2242	0.3527	0.3337	0.4376	0.4233	0.4763	0.4635	0.4780	0.4650		
	0.5	0	0.5014	0.4964	0.4988	0.4854	0.4924	0.4769	0.4837	0.4690	0.4783	0.4652	0.4780	0.4650		
	1.0	0	0.9614	0.9608	0.8275	0.8222	0.7027	0.6914	0.5638	0.5484	0.4819	0.4683	0.4780	0.4650		
	0	1.0	0.9612	0.9608	0.8262	0.8222	0.7009	0.6919	0.5626	0.5484	0.4818	0.4683	0.4780	0.4650		
0.8	0	0	0.0649	0.0501	0.2056	0.1804	0.2806	0.2567	0.3280	0.3059	0.3417	0.3188	0.3419	0.3189		
	0.5	0	0.4814	0.4755	0.4253	0.4054	0.3846	0.3594	0.3527	0.3276	0.3420	0.3190	0.3419	0.3189		
	1.0	0	0.9253	0.9232	0.6964	0.6799	0.5252	0.4976	0.3879	0.3589	0.3425	0.3193	0.3419	0.3189		
	0	1.0	0.9246	0.9232	0.6926	0.6799	0.5208	0.4976	0.3859	0.3589	0.3425	0.3193	0.3419	0.3189		
0.7	0	0	0.0558	0.0406	0.1690	0.1407	0.2221	0.1928	0.2506	0.2203	0.2565	0.2251	0.2566	0.2252		
	0.5	0	0.4624	0.4551	0.3638	0.3360	0.3039	0.2688	0.2657	0.2318	0.2566	0.2252	0.2566	0.2252		
	1.0	0	0.8914	0.8871	0.5933	0.5633	0.4057	0.3629	0.2850	0.2464	0.2568	0.2252	0.2566	0.2252		
	0	1.0	0.8900	0.8871	0.5865	0.5633	0.3992	0.3629	0.2827	0.2464	0.2567	0.2252	0.2566	0.2252		

Table11-9a *(Continued)*.

	Wall Reflectivity at τ_0		$\tau_0 = 0.1$		$\tau_0 = 0.5$		$\tau_0 = 1$		$\tau_0 = 2$		$\tau_0 = 5$		$\tau_0 = 15$		$\tau_0 = 30$	
ω	ρ^s	ρ^d	Exact	P_1 App.	Exact	P_1 App.	Exact	P_1 App.	Exact	P_1 App.	Exact	P_1 App.	Exact	P_1 App.	Exact	P_1 App.
0.6	0	0	0.0470	0.0313	0.1365	0.1046	0.1743	0.1388	0.1919	0.1540	0.1947	0.1559	0.1947	0.1559		
	0.5	0	0.4444	0.4353	0.3115	0.2753	0.2408	0.1963	0.2021	0.1604	0.1948	0.1559	0.1947	0.1559		
	1.0	0	0.8597	0.8524	0.5099	0.4660	0.3188	0.2631	0.2141	0.1681	0.1948	0.1559	0.1947	0.1559		
	0	1.0	0.8574	0.8524	0.5002	0.4660	0 3107	0.2631	0.2118	0.1681	0.1948	0.1559	0.1947	0.1559		
0.5	0	0	0.0384	0.0221	0.1077	0.0716	0.1342	0.0924	0.1451	0.1003	0.1465	0.1010	0.1466	0.1010		
	0.5	0	0.4272	0.4161	0.2664	0.2217	0.1898	0.1368	0.1526	0.1041	0.1466	0.1010	0.1466	0.1010		
	1.0	0	0.8299	0.8190	0.4409	0.3835	0.2521	0.1857	0.1608	0.1084	0.1466	0.1010	0.1466	0.1010		
	0	1.0	0.8264	0.8190	0.4284	0.3835	0.2428	0.1857	0.1585	0.1084	0.1466	0.1010	0.1466	0.1010		
0.4	0	0	0.0302	0.0132	0.0818	0.0412	0.0999	0.0519	0.1066	0.0555	0.1073	0.0557	0.1073	0.0557		
	0.5	0	0.4107	0.3974	0.2271	0.1740	0.1475	0.0868	0.1122	0.0579	0.1073	0.0557	0.1073	0.0557		
	1.0	0	0.8018	0.7869	0.3828	0.3126	0.1990	0.1236	0.1183	0.0604	0.1074	0.0557	0.1073	0.0557		
	0	1.0	0.7971	0.7869	0.3678	0.3126	0.1888	0.1236	0.1161	0.0604	0.1074	0.0557	0.1073	0.0557		
0.3	0	0	0.0223	0.0043	0.0584	0.0132	0.0701	0.0163	0.0741	0.0172	0.0745	0.0173	0.0745	0.0173		
	0.5	0	0.3951	0.3793	0.1926	0.1313	0.1117	0.0441	0 0786	0.0188	0.0745	0.0173	0.0745	0.0173		
	1.0	0	0.7753	0.7559	0.3332	0.2510	0.1555	0.0723	0.0833	0.0203	0.0745	0.0173	0.0745	0.0173		
	0	1.0	0.7693	0.7559	0.3160	0.2510	0.1445	0.0723	0.0812	0.0203	0.0745	0.0173	0.0745	0.0173		
0.2	0	0	0.0146		0.0372		0.0439		0.0461		0.0463					
	0.5	0	0.3801	0.3617	0.1619	0.0928	0.0808	0.0070	0.0498		0.0463					
	1.0	0	0.7502	0.7260	0.2902	0.1970	0.1188	0.0290	0.0537		0.0463					
	0	1.0	0.7427	0.7260	0.2709	0.1970	0.1074	0.0290	0.0516		0.0463					
0.1	0	0	0.0072		0.0178		0.0207		0.0216		0.0217					
	0.5	0	0.3656	0.3445	0.1344	0.0579	0.0539		0.0248		0.0217					
	1.0	0	0.7262	0.6972	0.2525	0.1491	0.0874		0.0280		0.0217					
	0	1.0	0.7168	0.6972	0.2312	0.1491	0.0756		0.0261		0.0217					

[a] From C. C. Lii and M. N. Özişik [48].

Table 11-9b Transmissivity of a Nonconservative (i.e., $\omega < 1$) Slab for Isotropic Scattering and with a Partially Reflecting Boundary at $\tau = \tau_0$ and a Transparent Boundary at $\tau = 0$[a]
(Radiation is incident at $\tau = 0$.)

ω	Wall Reflectivity at τ_0		$\tau_0 = 0.1$		$\tau_0 = 0.5$		$\tau_0 = 1$		$\tau_0 = 2$		$\tau_0 = 5$		$\tau_0 = 15$		$\tau_0 = 30$	
	ρ^s	ρ^d	Exact	P_1 App.	Exact	P_1 App.	Exact	P_1 App.	Exact	P_1 App.	Exact	P_1 App.	Exact	P_1 App.	Exact	P_1 App.
0.995	0	0	0.9152	0.9297	0.7018	0.6248	0.5488	0.5668	0.3815	0.3913	0.1892	0.1920	0.0450	0.0453	0.0071	0.0070
	0.5	0	0.4772	0.4815	0.4096	0.4190	0.3504	0.3595	0.2710	0.2773	0.1525	0.1549	0.0388	0.0391	0.0061	0.0061
0.975	0	0	0.9132	0.9277	0.6924	0.7153	0.5313	0.5487	0.3505	0.3594	0.1255	0.1370	0.0087	0.0086	0.0001	0.0001
	0.5	0	0.4777	0.4800	0.4017	0.4110	0.3346	0.3435	0.2417	0.2474	0.1016	0.1028	0.0067	0.0066	0.0001	0.0001
0.95	0	0	0.9108	0.9253	0.6812	0.7036	0.5110	0.5274	0.3175	0.3247	0.0950	0.0948	0.0021	0.0020		
	0.5	0	0.4738	0.4781	0.3923	0.4015	0.3169	0.3253	0.2122	0.2166	0.0670	0.0668	0.0015	0.0014		
0.9	0	0	0.9060	0.9204	0.6599	0.6812	0.4747	0.4885	0.2656	0.2686	0.0534	0.0507	0.0002	0.0002		
	0.5	0	0.4700	0.4744	0.3748	0.3836	0.2868	0.2932	0.1689	0.1703	0.0349	0.0330	0.0002	0.0001		
0.8	0	0	0.8965	0.9107	0.6220	0.6398	0.4162	0.4232	0.1973	0.1917	0.0229	0.0187				
	0.5	0	0.4628	0.4670	0.3450	0.3516	0.2405	0.2427	0.1172	0.1132	0.0137	0.0111				
0.7	0	0	0.8875	0.9012	0.5891	0.6026	0.3712	0.3704	0.1551	0.1425	0.0124	0.0083				
	0.5	0	0 4560	0.4599	0.3202	0.3241	0.2075	0.2050	0.0880	0.0800	0.0070	0.0047				
0.6	0	0	0.8788	0.8919	0.5603	0.5688	0.3355	0.3272	0.1269	0.1091	0.0077	0.0041				
	0.5	0	0.4495	0.4530	0.2993	0.3001	0.1827	0.1758	0.0697	0.0591	0.0042	0.0022				
0.5	0	0	0.8704	0.8828	0.5350	0.5381	0.3067	0.2911	0.1071	0.0855	0.0053	0.0022				
	0.5	0	0.4433	0.4463	0.2816	0.2791	0.1635	0.1526	0.0573	0.0450	0.0028	0.0011				
0.4	0	0	0.8624	0.8738	0.5125	0.5101	0.2830	0.2607	0.0925	0.0681	0.0039	0.0012				
	0.5	0	0.4374	0.4398	0.2663	0.2604	0.1483	0.1338	0.0486	0.0350	0.0020	0.0006				
0.3	0	0	0.8546	0.8650	0.4925	0.4844	0.2631	0.2347	0.0814	0.0551	0.0030	0.0007				
	0.5	0	0.4318	0.4335	0.2529	0.2438	0.1359	0.1183	0.0420	0.0278	0.0016	0.0004				
0.2	0	0	0.8470	0.8564	0.4744	0.4607	0.2463	0.2124	0.0728	0.0451	0.0024	0.0004				
	0.5	0	0.4264	0.4273	0.2412	0.2289	0.1256	0.1054	0.0372	0.0224	0.0012	0.0002				
0.1	0	0	0.8394	0.8480	0.4578	0.4390	0.2317	0.1930	0.0658	0.0373	0.0020	0.0002				
	0.5	0	0.4210	0.4213	0.2307	0.2155	0.1169	0.0944	0.0332	0.0182	0.0010	0.0010				

[a] From C. C. Lii and M. N. Özişik [48].

in this table reveals that the accuracy of the Eddington approximation is not so good for small optical thicknesses τ_0 and larger values of the anisotropy factor b.

Tables 11-9a and 11-9b show the hemispherical reflectivity and transmissivity, respectively, computed by Lii and Özişik [48] from both the exact solution and the P_1 approximation for a nonconservative (i.e., $\omega < 1$), plane-parallel slab of finite optical thickness τ_0, with a reflecting boundary and for isotropic scattering. An isotropic radiation is incident externally on the boundary at $\tau = 0$, which is taken to be transparent, while the other boundary at $\tau = \tau_0$ is assumed to have either a specular reflectivity ρ^s or a diffuse reflectivity ρ^d. The exact results show that the hemispherical reflectivity of the slab is slightly higher with a specularly reflecting boundary than with a diffusely reflecting boundary; for optical thicknesses of 15 and larger the hemispherical reflectivity of the slab is almost equal to that of a semi-infinite medium. The results with the P_1 approximation, however, do not distinguish between the types of reflectivity (i.e., specular or diffuse) at the boundary. The accuracy of the P_1 approximation is poor for smaller values of ω; for some cases ($\omega \leq 0.2$) the results have even become negative and hence meaningless. However, for ω close to unity and large optical thicknesses the P_1 approximation gives reasonably good results. Table 11-9b shows the transmissivity of the slab for isotropic radiation externally incident on the boundary at $\tau = 0$. The absorptivity of the slab can be determined from the results given in Tables 11-9a and 11-9b since the sum of the reflectivity, transmissivity, and absorptivity should be equal to unity.

REFERENCES

1. S. Chandrasekhar, *Radiative Transfer*, Oxford University Press, London, 1950; also Dover Publications, New York, 1960.
2. V. Kourganoff, *Basic Methods in Transfer Problems*, Oxford University Press, London, 1952; also Dover Publications, New York, 1963.
3. V. V. Sobolev, *A Treatise on Radiative Transfer*, D. Van Nostrand Company, Princeton, N.J., 1963.
4. I. W. Busbridge, *The Mathematics of Radiative Transfer*, Cambridge University Press, London, 1960.
5. C. M. Usiskin and E. M. Sparrow, "Thermal Radiation Between Parallel Plates Separated by an Absorbing-Emitting Nonisothermal Gas," *Intern. J. Heat Mass Transfer*, **1**, 28–36, 1960.
6. R. V. Meghreblian, "An Approximate Analytical Solution for the Radiative Exchange Between Two Flat Surfaces Separated by an Absorbing Gas," *Intern. J. Heat Mass Transfer*, **5**, 1051–1052, 1962.
7. J. R. Howell and M. Perlmutter, "Monte Carlo Solution of Thermal Transfer through Radiant Media Between Gray Walls," *J. Heat Transfer*, **86C**, 117–122, 1964.

8. J. R. Howell and M. Perlmutter, "Monte Carlo Solution of Radiant Heat Transfer in a Nongrey Nonisothermal Gas with Temperature Dependent Properties," *A.I.Ch.E. J.*, **10**, 562–567, 1964.

9. R. Viskanta and R. J. Grosh, "Heat Transfer in a Thermal Radiation Absorbing and Scattering Medium," *International Developments in Heat Transfer*, Part IV, American Society of Mechanical Engineers, pp. 820–828, 1961.

10. T. J. Love and R. J. Grosh, "Radiative Heat Transfer in Absorbing, Emitting, and Scattering Media," *J. Heat Transfer*, **87C**, 161–166, 1965.

11. H. M. Hsia and T. J. Love, "Radiative Heat Transfer Between Parallel Plates Separated by a Nonisothermal Medium with Anisotropic Scattering," *J. Heat Transfer*, **89C**, 197–203, 1967.

12. R. D. Cess and E. Sotak, "Radiation Heat Transfer in an Absorbing Medium Bounded by a Specular Reflector," *Z. Angew. Math. Phys.*, **15**, 642–647, 1964.

13. R. H. Edwards and R. P. Bobco, "Radiant Heat Transfer from Isothermal Dispersions with Isotropic Scattering," *ASME Paper* No. 67-HT-8, 1967.

14. R. Bobco, "Directional Emissivities from a Two-Dimensional Absorbing-Scattering Medium: the Semi-Infinite Slab," *ASME Paper* No. 67-HT-12, 1967.

15. M. A. Heaslet and R. F. Warming, "Radiative Transport and Wall Temperature Slip in an Absorbing Planar Medium," *Intern. J. Heat Mass Transfer*, **8**, 979–994, 1965.

16. M. A. Heaslet and R. F. Warming, "Radiative Transport in an Absorbing Planar Medium, II: Prediction of Radiative Source Functions," *Intern. J. Heat Mass Transfer*, **10**, 1413–1427, 1967.

17. A. L. Crosbie and R. Viskanta, "The Exact Solution to a Simple Nongray Radiative Transfer Problem," *J. Quant. Spectry. Radiative Transfer*, **9**, 553–568, 1969.

18. A. L. Crosbie and R. Viskanta, "Nongray Radiative Transfer in a Planar Medium Exposed to a Collimated Flux," *J. Quant. Spectry. Radiative Transfer*, **10**, 465–485, 1970.

19. A. L. Crosbie and R. Viskanta, "Effect of Band or Line Shape on the Radiative Transfer in a Nongray Planar Medium," *J. Quant. Spectry. Radiative Transfer*, **10** 487–509, 1970.

20. K. M. Case, "Elementary Solutions of the Transport Equation and Their Applications," *Ann. Phys. (N.Y.)*, **9**, 1–23, 1960.

21. K. M. Case and P. F. Zweifel, *Linear Transport Theory*, Addison-Wesley Publishing Co., Reading, Mass., 1967.

22. E. Inönü and P. F. Zweifel, *Developments in Transport Theory*, Academic Press, New York, 1967.

23. C. E. Siewert and P. F. Zweifel, "An Exact Solution of Equations of Radiative Transfer for Local Thermodynamic Equilibrium in the Non-Gray Case: Picket Fence Approximation," *Ann. Phys. (N.Y.)*, **36**, 61–85, 1966.

24. C. E. Siewert and P. F. Zweifel, "Radiative Transfer, II," *J. Math. Phys.*, **7**, 2092–2102, 1966.

25. J. H. Ferziger and G. M. Simmons, "Application of Case's Method to Plane Parallel Radiative Transfer," *Intern J. Heat Mass Transfer*, **9**, 987-992, 1966.

26. G. M. Simmons and J. H. Ferziger, "Non-Gray Radiative Heat Transfer Between Parallel Plates," *Intern. J. Heat Mass Transfer*, **11**, 1611–1620, 1968.

27. M. A. Heaslet and R. F. Warming, "Radiative Source-Function Predictions for

Finite and Semi-infinite Non-conservative Atmospheres," *Astrophys. Space Sci.*, **1,** 460–498, 1968.

28. M. N. Özişik and C. E. Siewert, "On the Normal-Mode Expansion Technique for Radiative Transfer in a Scattering, Absorbing and Emitting Slab with Specularly Reflecting Boundaries," *Intern. J. Heat Transfer*, **12,** 611–620, 1969.

29. H. L. Beach, M. N. Özişik, and C. E. Siewert, "Radiative Transfer in Linearly Anisotropic-Scattering, Conservative and Non-conservative Slabs with Reflective Boundaries," *Intern. J. Heat Mass Transfer*, **14,** 1551–1565, 1971.

30. J. T. Kriese and C. E. Siewert, "Radiative Transfer in a Conservative Finite Slab with an Internal Source," *Intern. J. Heat Mass Transfer*, **13,** 1349–1357, 1970.

31. R. J. Reith, C. E. Siewert, and M. N. Özişik, "Non-grey Radiative Heat Transfer in Conservative Plane-Parallel Media with Reflecting Boundaries," *J. Quant. Spectry. Radiative Transfer*, **11,** 1441–1462, 1971.

32. C. E. Siewert and N. J. McCormick, "On an Exact Solution in the Theory of Line Formation in Stellar Atmospheres," *Astrophys. J.*, **149,** 649–653, 1967.

33. C. E. Siewert and M. N. Özişik, "An Exact Solution in the Theory of Line Formation," *Monthly Notices Roy. Astron. Soc.*, **146,** 351–360, 1969.

34. S. Chandrasekhar, D. Elbert, and A. Franklin, "The *X*- and *Y*-Functions for Isotropic Scattering, I," *Astrophys. J.*, **115,** 244–268, 1952.

35. S. Chandrasekhar and D. Elbert, "The *X*- and *Y*-Functions for Isotropic Scattering, II," *Astrophys. J.*, **115,** 269–278, 1952.

36. P. Henrici, *Elements of Numerical Analysis*, John Wiley and Sons, New York, 1964.

37. M. Abronowitz and Irene A. Stegun, Editors, *Handbook of Mathematical Functions*, National Bureau of Standards, Applied Mathematics Series No. 55, U.S. Department of Commerce, Washington, D.C., 1964; also Dover Publications, New York, 1964.

38. A. S. Kronrod, *Nodes and Weights of Quadrature Formulas*, Consultants Bureau New York, 1965.

39. H. G. Horak and S. Chandrasekhar, "Diffuse Reflection by a Semi-infinite Atmosphere," *Astrophys. J.*, **134,** 45–56, 1961.

40. E. Pitts, "The Application of Radiative Transfer Theory to Scattering Effects in Unexposed Photographic Emulsions," *Proc. Phys. Soc.*, **67B,** 105–119, 1954.

41. Chiao-Min Chu and S. W. Churchill, "Numerical Solution of Problems of Multiple Scattering of Electromagnatic Radiation," *J. Phys. Chem.*, **59,** 855–863, 1955.

42. C. M. Chu, J. A. Leacock, J. C. Chen, and S. W. Churchill, "Numerical Solutions for Multiple, Anisotropic Scattering," in *Electromagnetic Scattering*, edited by M. Kerker, pp. 567–582, Macmillan Co., New York, 1963.

43. L. B. Evans, C. M. Chu, and S. W. Churchill, "The Effects of Anisotropic Scattering on Radiant Transport," *J. Heat Transfer*, **87,** 381–387, 1965.

44. H. C. Hottel, A. F. Sarofim, L. B. Evans, and I. A. Vasolos, "Radiative Transfer in Anisotropically Scattered Media: Allowance for Fresnel Reflection at the Boundaries, *ASME Paper* No. 67-HT-19, September 1967.

45. R. G. Giovanelli, "Reflection by Semi-infinite Diffusers," *Opt. Acta*, **2,** 153–162, 1955.

46. I. W. Busbridge and S. E. Orchard, "Reflection and Transmission of Light by a Thick Atmosphere According to a Phase Function: $1 + x \cos \theta$," *Astrophys. J.*, **149,** 655–664, 1967.

47. R. C. Progelhof and J. L. Throne, "Determination of the Radiation Properties of a

Semi-transparent Cylindrical Body Using the Monte Carlo Method," *ASME Paper* No. 70-SWA/HT-13, 1970.

48. C. C. Lii and M. N. Özişik, "Hemispherical Reflectivity of an Absorbing, Isotropically Scattering Slab with Reflecting Boundary," Intern. J. Heat Mass Transfer, (in press).

NOTES

[1] The $X(\mu, \tau_0)$ and $Y(\mu, \tau_0)$ functions for isotropic scattering have been defined as

$$X(\mu, \tau_0) = 1 + \int_0^{\tau_0} \phi(\xi) e^{-\xi/\mu}\, d\xi$$

$$Y(\mu, \tau_0) = e^{-\tau_0/\mu} + \int_0^{\tau_0} \phi(\tau_0 - \xi) e^{-\xi/\mu}\, d\xi$$

where the function $\phi(\tau)$ can be calculated as a solution of the following integral equation

$$\phi(\tau) = \tfrac{1}{2} E_1(\tau) + \tfrac{1}{2} \int_0^{\tau_0} \phi(\xi) E_1(|\tau - \xi|)\, d\xi$$

The X and Y functions as defined above have been calculated by Chandrasekhar, Elbert, and Franklin [34] and Chandrasekhar and Elbert [35].

[2] The expression for $F(t)$ given by Eq. 11-38a is derived as follows. By definition we have

$$F(t) \equiv \frac{f(\tau)}{\bar{\sigma} T_2^4/\pi} = \frac{f(\tau)}{I_b(T_2)} = \frac{\int_{\nu=0}^{\infty} \alpha_\nu I_{\nu b}[T(\tau)]\, d\nu}{\int_{\nu=0}^{\infty} I_{\nu b}[T_2]\, d\nu} \tag{1}$$

and $I_{\nu b}(T)$ is given by Eq. 1-44a as

$$I_{\nu b}(T) = \frac{2h\nu^3}{c^2[\exp(h\nu/kT) - 1]} \tag{2}$$

Substituting Eq. 2 into Eq. 1, we obtain

$$F(t) = \left(\frac{T}{T_2}\right)^4 \frac{\int^{\infty} \alpha_\nu [x^3/(e^x - 1)]\, dx}{\int_{x=0}^{\infty} x^3/(e^x - 1)\, dx} \tag{3}$$

where we have defined $x \equiv h\nu/kT$. When the integral in the denominator is performed, Eq. 3 becomes

$$F(t) = t^4 \frac{15}{\pi^4} \int_{x=0}^{\infty} \alpha_\nu \frac{x^3}{e^x - 1}\, dx \tag{4}$$

since

$$\int_{x=0}^{\infty} \frac{x^3}{e^x - 1}\, dx = \frac{15}{\pi^4} \qquad \text{and} \qquad \frac{T}{T_2} \equiv t$$

If the entire frequency spectrum is divided into bands $\Delta\nu_i$, $i = 1, 2, \ldots, N$, Eq. 4

can be written as

$$F(t) = t^4 \frac{15}{\pi^4} \sum_{i=1}^{N} \alpha_i \int_{z_{i-1}/t}^{z_i/t} \frac{x^3}{e^x - 1}\, dx \tag{5}$$

since

$$x_i = \frac{h\nu_i}{kT} = \left(\frac{h\nu_i}{kT_2}\right)\left(\frac{T_2}{T}\right) = \frac{z_i}{t}$$

where $z_i \equiv h\nu_i/kT_2$.

Equation 5 is the same as Eq. 11-38a.

[3] In the problems of interaction of radiation with conduction or convection the radiative flux term $dq^r(\tau)/d\tau$ appears in the energy equation. The relation for $dq^r(\tau)/d\tau$ is determined by differentiating Eq. 11-46 with respect to τ; alternatively, it can be obtained directly from Eq. 8-95.

$$\frac{dq^r(\tau)}{d\tau} = 4\pi I_b[T(\tau)] - 2\pi\Big[I^+(0)E_2(\tau) + I^-(\tau_0)E_2(\tau_0 - \tau) + \int_{\tau'=0}^{\tau_0} I_b[T(\tau')]E_1(|\tau - \tau'|)\, d\tau'\Big] \tag{1}$$

By substituting the boundary surface intensities $I^+(0)$ and $I^-(\tau_0)$ from Eqs. 11-44a and 11-44b into Eq. 1, we obtain

$$\begin{aligned}\frac{dq^r(\tau)}{d\tau} &= 4\pi I_b[T(\tau)] \\ &\quad - 2\pi E_2(\tau)\frac{\varepsilon_1 I_b(T_1) + 2\rho_1 E_3(\tau_0)\varepsilon_2 I_b(T_2) + 2\rho_1[A + 2\rho_2 E_3(\tau_0)B]}{1 - 4\rho_1\rho_2 E_3^2(\tau_0)} \\ &\quad - 2\pi E_2(\tau_0 - \tau)\frac{\varepsilon_2 I_b(T_2) + 2\rho_2 E_3(\tau_0)\varepsilon_1 I_b(T_1) + 2\rho_2[B + 2\rho_1 E_3(\tau_0)A]}{1 - 4\rho_1\rho_2 E_3^2(\tau_0)} \\ &\quad - 2\pi\int_{\tau'=0}^{\tau_0} I_b[T(\tau')]E_1(|\tau - \tau'|)\, d\tau'\end{aligned} \tag{2}$$

where A and B were defined by Eqs. 11-45.

[4] The Marshak boundary condition for the P_1 approximation is given as (see Eqs. 9-134 for $i = 1$)

$$\int_0^1 I(0, \mu)\mu\, d\mu = \int_0^1 f_1(\mu)\mu\, d\mu, \qquad \mu > 0 \tag{1a}$$

$$\int_0^1 I(\tau_0, -\mu)\mu\, d\mu = \int_0^1 f_2(-\mu)\mu\, d\mu, \quad \mu > 0 \tag{1b}$$

where $f_1(\mu)$, $\mu > 0$, and $f_2(\mu)$, $\mu < 0$, are the boundary condition functions. Introducing Eq. 11-54b into the boundary condition 11-53b and applying the Marshak boundary condition 1a, we obtain

$$\frac{1}{4\pi}\int_0^1\left[G(\tau)-\mu\frac{dG(\tau)}{d\tau}\right]_{\tau=0}\mu\,d\mu=\varepsilon_1\frac{\bar{\sigma}T_1^{\,4}}{\pi}\int_0^1\mu\,d\mu$$

$$+\frac{1}{4\pi}\rho_1^{\,s}\int_0^1\left[G(\tau)+\mu\frac{dG(\tau)}{d\tau}\right]_{\tau=0}\mu\,d\mu$$

$$+\frac{1}{4\pi}2\rho_1^{\,d}\int_{\mu=0}^1\mu\,d\mu\int_{\mu'=0}^1\left[G(\tau)+\mu'\frac{dG(\tau)}{d\tau}\right]_{\tau=0}\mu'\,d\mu' \tag{2a}$$

After performing the integrations Eq. 2a simplifies to

$$\left[(1-\rho_1^{\,s}-\rho_1^{\,d})G(\tau)-\tfrac{2}{3}(1+\rho_1^{\,s}+\rho_1^{\,d})\frac{dG(\tau)}{d\tau}\right]_{\tau=0}=4\varepsilon_1\bar{\sigma}T_1^{\,4} \tag{2b}$$

Similarly, the boundary condition 11-53c becomes

$$\left[(1-\rho_2^{\,s}-\rho_2^{\,d})G(\tau)+\tfrac{2}{3}(1+\rho_2^{\,s}+\rho_2^{\,d})\frac{dG(\tau)}{d\tau}\right]_{\tau=\tau_0}=4\varepsilon_2\bar{\sigma}T_2^{\,4} \tag{3}$$

[5] Equation 11-73 is obtained from the following equation:

$$\mu\frac{\partial I(\tau,\mu)}{\partial\tau}+I(\tau,\mu)=(1-\omega)I_b[T(\tau)]+\frac{\omega}{2}\left[\int_0^1 p(\mu,\mu')I(\tau,\mu')\,d\mu'\right.$$

$$\left.+\int_0^1 p(\mu,-\mu')I(\tau,-\mu')\,d\mu'\right]\quad\text{in } 0\le\tau\le\tau_0,\ \mu<0 \tag{1}$$

by replacing in this equation μ by $-\mu$, and by utilizing the following reciprocity relation for the phase function:

$$p(-\mu,\mu')=p(\mu,-\mu') \tag{2a}$$

$$p(-\mu,-\mu')=p(\mu,\mu') \tag{2b}$$

[6] The following normalization integrals given in Chapter 10 were used to obtain Eqs. 11-96 and 11-97:

$$\int_0^1 W(\mu)\phi(\eta,\mu)\phi(\eta_0,\mu)\,d\mu=0 \tag{10-70}$$

$$\int_0^1 W(\mu)\phi(\pm\eta_0,\mu)\phi(\eta_0,\mu)\,d\mu=\mp(\tfrac{1}{2}\omega\eta_0)^2X(\pm\eta_0) \tag{10-72}$$

$$\int_0^1 W(\mu)\phi(-\eta,\mu)\phi(\eta_0,\mu)\,d\mu=\tfrac{1}{4}\omega^2\eta\eta_0X(-\eta) \tag{10-73}$$

[7] The following normalization integrals given in Chapter 10 were used to obtain Eqs.

11-99 and 11-100:

$$\int_0^1 W(\mu)\phi(\eta, \mu)\phi(\eta', \mu)\, d\mu = \frac{W(\eta)}{g(\omega, \eta)}\, \delta(\eta - \eta') \tag{10-69}$$

$$\int_0^1 W(\mu)\phi(\eta, \mu)\phi(\eta_0, \mu)\, d\mu = 0 \tag{10-70}$$

$$\int_0^1 W(\mu)\phi(-\eta_0, \mu)\phi(\eta, \mu)\, d\mu = \omega\eta\eta_0 X(-\eta_0)\phi(-\eta_0, \eta) \tag{10-71}$$

$$\int_0^1 W(\mu)\phi(-\eta, \mu)\phi(\eta', \mu)\, d\mu = \tfrac{1}{2}\omega\eta'(\eta_0 + \eta)X(-\eta)\phi(-\eta, \eta') \tag{10-74}$$

[8] To obtain the result given by Eq. 11-126 we considered the following integral (see Eq. 10-64a)

$$\int_{-1}^1 \mu\phi(\eta, \mu)\, d\mu = \eta(1 - \omega)$$

and noted that it vanishes for $\omega = 1$.

[9] The following normalization integrals given in Chapter 10 were used to obtain Eqs. 11-132 and 11-133:

$$\int_0^1 \gamma(\mu)\, d\mu = 1 \tag{10-81}$$

$$\int_0^1 \mu\gamma(\mu)\, d\mu \equiv \gamma^{(1)} \tag{10-81'}$$

$$\int_0^1 \gamma(\mu)\phi(-\eta, \mu)\, d\mu = \frac{\eta X(-\eta)}{2} \tag{10-90}$$

$$\int_0^1 \gamma(\mu)\phi(\eta, \mu)\, d\mu = 0 \tag{10-87}$$

[10] The following normalization integrals were used to obtain Eqs. 11-136 and 11-137:

$$\int_0^1 \gamma(\mu)\phi(\eta', \mu)\, d\mu = 0 \tag{10-87}$$

$$\int_0^1 \mu\gamma(\mu)\phi(\eta', \mu)\, d\mu = -\frac{\eta'}{2} \tag{10-88}$$

$$\int_0^1 \gamma(\mu)\phi(-\eta, \mu)\phi(\eta', \mu)\, d\mu = \frac{\eta'\eta}{4(\eta' + \eta)}\, X(-\eta) \tag{10-91}$$

$$\int_0^1 \gamma(\mu)\phi(\eta, \mu)\phi(\eta', \mu)\, d\mu = \frac{\gamma(\eta)}{g(1, \eta)}\, \delta(\eta - \eta') \tag{10-86}$$

where $0 \le \eta, \eta' \le 1$.

CHAPTER 12

Conduction with Radiation in a Participating Medium

There are several engineering applications of simultaneous conduction and radiation in a participating medium. For example, in heat transfer at sufficiently high temperatures through porous insulating materials such as fibers, powders, and foams, thermal radiation is equally as important as conduction. In heat transfer through solids that are semitransparent to infrared radiation, internal radiant heat exchange between the layers at different temperatures may become of the same order of magnitude as conductive heat transfer if the temperature is high. In such situations a separate calculation of conductive and radiative heat fluxes without any consideration of the interaction between them may introduce error in the heat transfer results.

A rigorous analytical treatment of the problem of simultaneous conduction and radiation is difficult because of the coupling between the equations of energy and of radiative transfer. Viskanta and Grosh [1, 2] and Viskanta [3] investigated the interaction of radiation with conduction, and an extensive bibliography of the work in related areas was given by Viskanta and Grosh [4]. Lick [5, 6] applied various approximate methods such as the boundary layer type of analysis, the diffusion approximation, the picket-fence model, and the linearization of temperature to solve the problem of steady and transient energy transfer by simultaneous conduction and radiation. Greif [7] considered a problem similar to that treated by Lick [5], with, however, temperature and frequency-dependent properties; he used the picket-fence model. Wang and Tien [8, 9] studied the range of validity of various limiting analysis, and Chang [10, 11] treated the problem of interaction of radiation with conduction by introducing an approximate technique to solve the radiation part of the problem. Timmons and Mingle [12] used the method of quasi linearization [13] to solve numerically the steady-state simultaneous conduction and radiation. The unsteady combined conduction and radiation problem for an absorbing, emitting, isotropically scattering plane-parallel slab was treated by Hazzak and Beck [14] with a differential method, and by Lii and Özişik [15a] exactly, using the normal-mode expansion technique for the radiation part of the problem. Viskanta and Merriam [16] considered simultaneous conduction and radiation for concentric spheres separated by a participating medium, and Tarshis, O'Hara, and Viskanta [17] solved the problem for two absorbing media in intimate contact. The problem of radiative heat transfer within an absorbing, emitting, and scattering ablating body was solved by Kadanoff [18] with the diffusion approximation, and by Boles and Özişik [19] using a rigorous approach.

In this chapter we present a general formulation of the problem of simultaneous conduction and radiation, illustrate with simple examples the approximate and exact methods of solution, and discuss the results of solution.

12-1 FORMULATION OF SIMULTANEOUS CONDUCTION AND RADIATION

The energy equation for simultaneous conduction and radiation in a participating medium is given as

$$-\nabla \cdot (\mathbf{q}^c + \mathbf{q}^r) + h(\mathbf{r}, t) = \rho c_p \frac{\partial T(\mathbf{r}, t)}{\partial t} \tag{12-1}$$

where c_p and ρ are the specific heat and density of the medium, $h(\mathbf{r}, t)$ is the energy generation rate per unit volume per unit time, and $\mathbf{q}^c$ and $\mathbf{q}^r$ are the conductive and radiative flux vectors, respectively. For a homogeneous, isotropic medium the conductive heat flux vector is given by the Fourier law as

$$\mathbf{q}^c = -k\nabla T(\mathbf{r}, t) \tag{12-2}$$

where k is the thermal conductivity of the medium. The radiative flux vector is related to the radiation intensity by (see Eq. 1-74)

$$\mathbf{q}^r = \int_{4\pi} \hat{\mathbf{\Omega}} I(\mathbf{r}, \hat{\mathbf{\Omega}}, t)\, d\Omega \tag{12-3a}$$

$$I(\mathbf{r}, \hat{\mathbf{\Omega}}, t) = \int_{\nu=0}^{\infty} I_\nu(\mathbf{r}, \hat{\mathbf{\Omega}}, t)\, d\nu \tag{12-3b}$$

where the spectral radiation intensity $I_\nu(\mathbf{r}, \hat{\mathbf{\Omega}}, t)$ satisfies the equation of radiative transfer. The energy equation and the equation of radiative transfer are coupled because the latter involves the temperature $T(\mathbf{r}, t)$. Therefore the equation of radiative transfer cannot be solved until the temperature distribution in the medium is obtained from the solution of the energy equation. The energy equation involves the net radiative heat flux, which should be determined from the solution of the equation of radiative transfer.

In the following analysis we focus our attention on the formulation and solution of the simultaneous conduction and radiation problems in the plane-parallel geometry since the solution in the general three-dimensional geometry is extremely difficult.

One-Dimensional Case

We consider the time-dependent simultaneous conduction and radiation in the y direction in a plane-parallel, semitransparent slab. It is assumed that the slab is stratified in planes perpendicular to the y direction. Figure 12-1 shows the geometry and the coordinates. Let L be the thickness of the slab, and $h(y, t)$ be the energy generation per unit volume and per unit time. The energy equation 12-1 simplifies to

$$\frac{\partial}{\partial y}\left[k\frac{\partial T(y,t)}{\partial y}\right] - \frac{\partial q^r}{\partial y} + h(y,t) = \rho c_p \frac{\partial T(y,t)}{\partial t} \quad \text{in } 0 \le y \le L \tag{12-4}$$

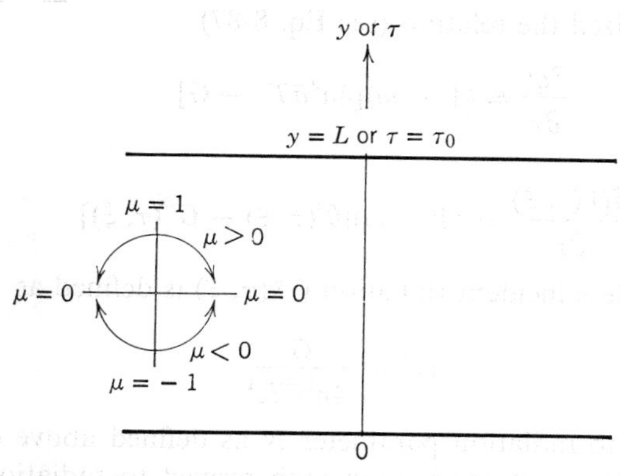

Fig. 12-1. Coordinates for combined conduction and radiation in a plane-parallel slab.

Assuming constant properties, we can write this equation in the dimensionless form as

$$\frac{\partial^2 \theta(\tau, \xi)}{\partial \tau^2} - \frac{1}{N}\frac{\partial Q^r(\theta, \xi)}{\partial \tau} + H(\tau, \xi) = \frac{\partial \theta(\tau, \xi)}{\partial \xi} \quad \text{in } 0 \leq \tau \leq \tau_0 \tag{12-5}$$

where we have defined various dimensionless quantities:

$$H(\tau, \xi) \equiv \frac{h}{kT_r\beta^2} = \text{dimensionless generation rate} \tag{12-6a}$$

$$N \equiv \frac{k\beta}{4n^2\bar{\sigma}T_r^3} = \text{conduction-to-radiation parameter} \tag{12-6b}$$

$$Q^r(\tau, \xi) \equiv \frac{q^r}{4n^2\bar{\sigma}T_r^4} = \text{dimensionless net radiative heat flux} \tag{12-6c}$$

$$\theta(\tau, \xi) \equiv \frac{T}{T_r} = \text{dimensionless temperature} \tag{12-6d}$$

$$\xi \equiv \frac{k}{\rho c_p}\beta^2 t = \text{dimensionless time variable} \tag{12-6e}$$

$$d\tau = \beta\, dy \text{ or } \tau = \beta y = \text{optical variable} \tag{12-6f}$$

$$\tau_0 = \beta L = \text{optical thickness} \tag{12-6g}$$

and T_r is a reference temperature.

Equation 12-5 can be written in the alternative form as

$$\frac{\partial^2 \theta(\tau, \xi)}{\partial \tau^2} - \frac{1-\omega}{N}[\theta^4(\tau, \xi) - G^*(\tau, \xi)] + H(\tau, \xi) = \frac{\partial \theta(\tau, \xi)}{\partial \xi} \quad \text{in } 0 \leq \tau \leq \tau_0 \tag{12-7}$$

where we have utilized the relation (see Eq. 8-87)

$$\frac{\partial q^r}{\partial \tau} = (1-\omega)[4n^2\bar{\sigma}T \ - G] \tag{12-8a}$$

or

$$\frac{\partial Q^r(\tau, \xi)}{\partial \tau} = (1-\omega)[\theta^4(\tau, \xi) - G^*(\tau, \xi)] \tag{12-8b}$$

and the dimensionless incident radiation $G^*(\tau, \xi)$ is defined as

$$G^* = \frac{G}{4n^2\bar{\sigma}T_r^4} \tag{12-8c}$$

The conduction-to-radiation parameter N as defined above characterizes the relative importance of conduction with respect to radiation. The large value of N corresponds to the conduction-dominating case, and the small value to the radiation-dominating case. When N is infinite, the radiation term drops out and Eqs. 12-5 and 12-7 both reduce to the standard conduction equation. For a purely scattering medium, that is, $\omega = 1$, the radiation does not interact with conduction and the temperature profile in the medium is that of pure conduction. This is apparent from Eq. 12-7, in which the radiation term drops out and the equation reduces to that of pure conduction for $\omega = 1$. For a nonscattering medium we set $\omega = 0$ and replace the extinction coefficient β by the absorption coefficient κ in the relations defining the optical variable and the parameter N.

The physical significance of the optical thickness τ_0 of the slab is better envisioned if it is written in the form

$$\tau_0 = \frac{L}{1/\beta} = \frac{\text{thickness of slab}}{\text{mean free path (or mean penetration length) of radiation photon}} \tag{12-9}$$

That is, τ_0 characterizes the number of mean free paths of radiation photon in the medium.

The dimensionless parameter N is a ratio of the conductive to radiative heat fluxes in the sense that, if $l \equiv 1/\beta$ designates a length, then N can be written as

$$N = \frac{kT/l}{4n^2\bar{\sigma}T^4} \equiv \frac{\text{conductive flux}}{\text{radiative flux}} \tag{12-10a}$$

For an optically thick ($\tau_0 \gg 1$) medium the radiative heat flux may be characterized by $16n^2\bar{\sigma}T^4/3\tau_0$ (i.e., see Eq. 9-25) and the conductive heat flux by kT/L. Then the ratio of the conductive to radiative heat fluxes for an optically thick medium may be given as

$$\left(\frac{\text{Conductive flux}}{\text{Radiative flux}}\right)_{\substack{\text{opt.}\\ \text{thick}}} = \frac{kT/L}{16n^2\bar{\sigma}T^4/3\tau_0} = \frac{3}{4}\frac{k\beta}{4n^2\bar{\sigma}T^3} = \tfrac{3}{4}N \quad \text{for } \tau_0 \ll 1 \tag{12-10b}$$

For an optically thin ($\tau_0 \ll 1$) medium the radiative heat flux may be characterized by $2\tau_0 n^2\bar{\sigma}T^4$ [this is obtainable from Eq. 8-84 by considering it for the gray case, omitting radiation from boundaries, setting $I_b = n^2\bar{\sigma}T^4/\pi =$ constant, evaluating at $\tau = \tau_0$, performing the integration, and then replacing $E_3(\tau_0)$ by $\frac{1}{2} - \tau_0$ for $\tau_0 \ll 1$] and the conductive heat flux is again taken as

kT/L. With this consideration we find

$$\left(\frac{\text{Conductive flux}}{\text{Radiative flux}}\right)_{\substack{\text{opt.}\\ \text{thin}}} = \frac{kT/L}{2\tau_0 n^2 \bar{\sigma} T^4} = \frac{2}{\tau_0^2}\frac{k\beta}{4n^2\bar{\sigma}T^3} = \frac{2}{\tau_0^2} N \quad \text{for } \tau_0 \ll 1 \tag{12-10c}$$

It is apparent from these results that, although the parameter N characterizes the relative importance of conduction in regard to radiation in general, the optical thickness τ_0 also plays an important role in this matter. For an optically thin medium the ratio N/τ_0^2 is important, but not N alone. For example, for the optically thin case the radiative heat flux tends to be negligible in comparison to the conductive heat flux from the medium, even for small values of N, since $\tau_0 \ll 1$.

Figure 12-2, obtained from Cess [20], shows the values of the conduction-to-radiation parameter N for water vapor, gaseous ammonia, and carbon dioxide at 1 atm pressure as a function of temperature. In this figure N is defined by setting $n = 1$ and replacing β by κ, because the refractive index for gases is close to unity and the scattering of thermal radiation by gas molecules is negligible unless the gas contains scattering particles. We note that the parameter N for gases is a property of the material and depends strongly on temperature, as shown in Fig. 12-2.

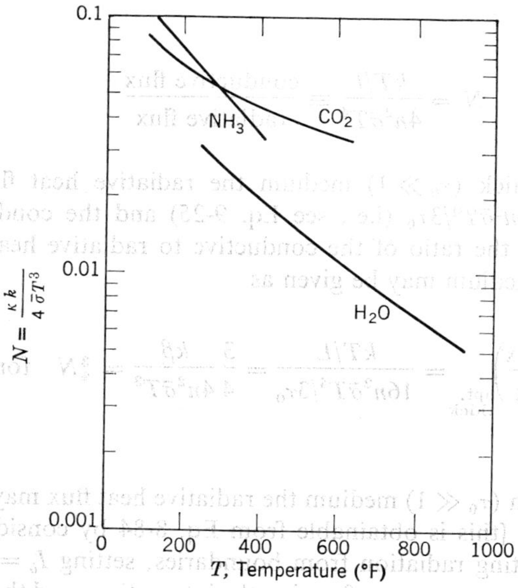

Fig. 12-2. Values of N for ammonia, carbon dioxide, and water vapor at 1 atm pressure. (From R. D. Cess [20].)

The dimensionless net radiative heat flux $Q^r(\tau, \xi)$ and the dimensionless incident radiation $G^*(\tau, \xi)$ in Eqs. 12-5 and 12-7 are related to the radiation intensity $I(\tau, \mu, \xi)$ by

$$Q^r(\tau, \xi) = \frac{2\pi \int_{-1}^{1} I(\tau, \mu, \xi)\mu \, d\mu}{4n^2\bar{\sigma}T_r^4} \equiv \frac{1}{2}\int_{-1}^{1} \psi(\tau, \mu, \xi)\mu \, d\mu \tag{12-11a}$$

$$G^*(\tau, \xi) = \frac{2\pi \int_{-1}^{1} I(\tau, \mu, \xi) \, d\mu}{4n^2\bar{\sigma}T_r^4} \equiv \frac{1}{2}\int_{-1}^{1} \psi(\tau, \mu, \xi) \, d\mu \tag{12-11b}$$

where the dimensionless radiation intensity $\psi(\tau, \mu, \xi)$ is defined as

$$\psi(\tau, \mu, \xi) \equiv \frac{I(\tau, \mu, \xi)}{(n^2\bar{\sigma}T_r^4/\pi)} \tag{12-11c}$$

For an absorbing, emitting, scattering, plane-parallel medium stratified in planes perpendicular to the 0τ axis, the function $\psi(\tau, \mu, \xi)$ satisfies the equation of radiative transfer given in the form

$$\mu \frac{\partial\psi(\tau, \mu, \xi)}{\partial\tau} + \psi(\tau, \mu, \xi) = (1 - \omega)\theta^4(\tau, \xi) + \frac{\omega}{2}\int_{-1}^{1} p(\mu, \mu')\psi(\tau, \mu', \xi) \, d\mu'$$

$$\text{in } 0 \le \tau \le \tau_0, \ -1 \le \mu \le 1 \tag{12-12}$$

where $\theta(\tau, \xi)$ is the dimensionless temperature distribution in the medium and should be obtained from the solution of the energy equation. Here we note that the equation of radiative transfer 12-12 is a steady-state equation even though the energy equation 12-5 (or 12-7) is time dependent; the reason for this, as discussed in Chapter 8, is that the radiation travels with the speed of light and hence the transients associated with the propagation of radiation are neglected. The intensity $\psi(\tau, \mu, \xi)$ depends on ξ because the temperature $\theta(\tau, \xi)$ depends on ξ; therefore the time variable enters the equation of radiative transfer merely as a parameter.

For the steady-state temperature distribution $\theta(\tau)$ the energy equation 12-5 and its alternative form, Eq. 12-7, respectively, simplify to

$$\frac{d^2\theta(\tau)}{d\tau^2} - \frac{1}{N}\frac{dQ^r(\tau)}{d\tau} + H(\tau) = 0 \tag{12-13}$$

and

$$\frac{d^2\theta(\tau)}{d\tau^2} - \frac{1-\omega}{N}[\theta^4(\tau) - G^*(\tau)] + H(\tau) = 0 \tag{12-14}$$

Net Heat Flux

In most engineering applications the net heat flux in the medium is of interest. Once the temperature distribution is determined from a simultaneous solution of the energy equation and the equation of radiative transfer, the net heat flux q in the y direction is obtained by summing the conductive and radiative heat fluxes:

$$q = -k\frac{\partial T}{\partial y} + q^r \tag{12-15}$$

This expression is written in the dimensionless form as

$$Q(\tau, \xi) = \frac{q}{k\beta T_r} = -\frac{\partial \theta(\tau, \xi)}{\partial \tau} + \frac{1}{N} Q^r(\tau, \xi) \tag{12-16}$$

For large values of N (i.e., $N \to \infty$) the radiation term vanishes and Eq. 12-16 reduces to that of pure conduction.

12-2 CONDUCTION AND RADIATION IN AN OPTICALLY THICK SLAB

In this section we examine the determination of the steady-state temperature distribution and the net heat flux for combined conduction and radiation in a slab by applying the optically thick limit approximation. We assume that the slab is optically thick (i.e., $\beta L = \tau_0 \gg 1$) and gray, that it has black boundaries of $\tau = 0$ and $\tau = \tau_0$, which are kept at uniform temperatures T_1 and T_2, respectively, and that there is an energy generation at a constant rate h per unit volume and per unit time.

The energy equation is given as (see Eq. 12-13)

$$\frac{d}{d\tau}\left[\frac{d\theta(\tau)}{d\tau} - \frac{1}{N} Q^r(\tau)\right] + H = 0 \quad \text{in } 0 \le \tau \le \tau_0 \tag{12-17}$$

with the boundary conditions

$$\theta(\tau) = \frac{T_1}{T_r} \equiv \theta_1 \quad \text{at } \tau = 0 \tag{12-18a}$$

$$\theta(\tau) = \frac{T_2}{T_r} \equiv \theta_2 \quad \text{at } \tau = \tau_0 \tag{12-18b}$$

where T_r is a reference temperature, and various dimensionless quantities have already been defined (see Eqs. 12-6). The net radiative heat flux for the optically thick limit approximation is given as (see Eq. 9-25)

$$q^r(\tau) = -\frac{16n^2\bar{\sigma}}{3} T^3(\tau)\frac{dT(\tau)}{d\tau} \tag{12-19}$$

or

$$Q^r = -\tfrac{4}{3}\theta^3(\tau)\frac{d\theta(\tau)}{d\tau} \tag{12-20}$$

Introducing Eq. 12-20 into the energy equation 12-17, we obtain

$$\frac{d}{d\tau}\left[\left(1 + \frac{4}{3N}\theta^3\right)\frac{d\theta}{d\tau}\right] + H = 0 \quad \text{in } 0 \le \tau \le \tau_0 \tag{12-21}$$

which can be written in the form

$$\frac{d}{d\eta}\left[k_e(\theta)\frac{d\theta}{d\eta}\right] = -H\tau_0^{\,2} \quad \text{in } 0 \le \eta \le 1 \tag{12-22}$$

with the boundary conditions

$$\theta = \theta_1 \quad \text{at } \eta = 0 \tag{12-23a}$$

$$\theta = \theta_2 \quad \text{at } \eta = 1 \tag{12-23b}$$

where we have defined the new independent variable η and the effective thermal conductivity $k_e(\theta)$ as

$$\eta \equiv \frac{\tau}{\tau_0} \tag{12-24}$$

$$k_e(\theta) \equiv 1 + \frac{4}{3N}\theta^3 \tag{12-25}$$

Equation 12-21 is equivalent to the steady-state heat conduction equation with temperature-dependent thermal conductivity and constant energy generation rate. To solve Eq. 12-22 a new dependent variable $u(\theta)$ is defined as [21, p. 353]

$$u(\theta) = \int_0^\theta k_e(\theta')\, d\theta' \tag{12-26}$$

Then Eq. 12-22 and the boundary conditions 12-23 are respectively transformed to

$$\frac{d^2u}{d\eta^2} = -H\tau_0^{\,2} \quad \text{in } 0 \le \eta \le 1 \tag{12-27}$$

with the boundary conditions

$$u = \int_0^{\theta_1} k_e(\theta')\, d\theta' = \theta_1 + \frac{1}{3N}\theta_1^{\,4} \equiv u_1 \quad \text{at } \eta = 0 \tag{12-28a}$$

$$u = \int_0^{\theta_2} k_e(\theta')\, d\theta' = \theta_2 + \frac{1}{3N}\theta_2^{\,4} \equiv u_2 \quad \text{at } \eta = 1 \tag{12-28b}$$

The solution of Eq. 12-27 subject to the boundary conditions 12-28 is given as

$$u = u_1(1-\eta) + u_2\eta + \frac{H\tau_0^{\,2}}{2}\eta(1-\eta) \tag{12-29a}$$

$$u = \theta_1(1-\eta) + \theta_2\eta + \frac{1}{3N}\left[(1-\eta)\theta_1^{\,4} + \eta\theta_2^{\,4}\right] + \frac{H\tau_0^{\,2}}{2}\eta(1-\eta) \tag{12-29b}$$

When Eq. 12-25 is substituted into Eq. 12-26, the transformation from u to θ becomes

$$u = \theta + \frac{1}{3N}\theta^4 \tag{12-30}$$

and from Eqs. 12-29b and 12-30, the relation defining the temperature distribution $\theta(\eta)$ in the slab is obtained as

$$\theta + \frac{1}{3N}\theta^4 = (1-\eta)\theta_1 + \eta\theta_2 + \frac{1}{3N}[(1-\eta)\theta_1^4 + \eta\theta_2^4] + \frac{H\tau_0^2}{2}\eta(1-\eta) \tag{12-31}$$

The relation for the net heat flux $q(\tau)$ in the medium is obtained from Eqs. 12-16 and 12-20 as

$$\frac{q(\tau)}{k\beta T_r} = -\left[1 + \frac{4}{3N}\theta^3(\tau)\right]\frac{d\theta(\tau)}{d\tau} \tag{12-32}$$

By changing the independent variable from τ to η with the transformation given by Eq. 12-24, we obtain

$$\frac{g(\eta)L}{kT_r} = -\left[1 + \frac{4}{3N}\theta^3(\eta)\right]\frac{d\theta(\eta)}{d\eta} \tag{12-33}$$

since $\tau_0 = \beta L$.

Finally, when $d\theta/d\eta$ is determined from Eq. 12-31 and substituted into Eq. 12-33, the net heat flux becomes

$$\frac{q(\eta)L}{kT_r} = (\theta_1 - \theta_2) + \frac{1}{3N}(\theta_1^4 - \theta_2^4) + (\eta - \tfrac{1}{2})H\tau_0^2 \tag{12-34}$$

Discussion of Results

The mathematical formulation of the one-dimensional conduction and radiation problem given by Eqs. 12-22 and 12-23 is equivalent to that considered by Viskanta and Grosh [22] for the problem of energy transport in Couette flow with thermal radiation. The energy generation term in the present formulation is equivalent to the viscous energy dissipation term in the Couette flow problem. Therefore we present some of the numerical results obtained by Viskanta and Grosh [22] for the case when dissipation can be neglected (i.e., $H = 0$).

Figure 12-3 shows a comparison of the temperature distribution in a slab for simultaneous conduction and radiation, obtained by using the optically thick approximation and the exact formulation for constant thermal conductivity and no energy generation (i.e., $H = 0$). The temperature T_2 of the boundary surface at $\tau = \tau_0$ is used as the reference temperature (i.e.,

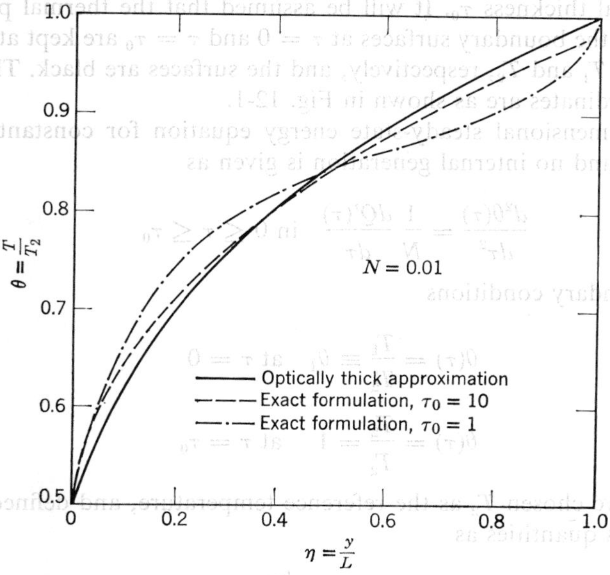

Fig. 12-3. Comparison of temperature distribution in a slab for the optically thick approximation and the exact formulation. (From R. Viskanta and R. J. Grosh [22].)

$T_r = T_2$). The results given in this figure are for $N = 0.01$, $\theta_1 = 0.5$, and $\theta_2 = 1.0$. The value of the parameter $N = 0.01$ characterizes a situation in which energy transport by radiation is dominant.

The exact solution for $\tau_0 = 10$ represents an optically thick medium more closely than the exact solution for $\tau_0 = 1$; therefore the temperature distribution obtained with the optically thick limit approximation agrees better with the exact formulation for $\tau_0 = 10$. However, the temperature gradients at the wall predicted by the optically thick limit approximation deviate considerably from those obtained by the exact solution; this may cause considerable error in heat transfer calculations at the wall. This discrepancy is to be expected since the optically thick limit approximation breaks down near the boundaries, and thus the utility of this approximation, especially when radiation effects are strong, is very limited.

12-3 CONDUCTION AND RADIATION IN AN ABSORBING AND EMITTING SLAB—Exact Treatment

Consideration will be given to the exact analysis of steady-state, combined conduction and radiation in an absorbing, emitting, nonscattering, gray slab of optical thickness τ_0. It will be assumed that the thermal properties are constant, the boundary surfaces at $\tau = 0$ and $\tau = \tau_0$ are kept at uniform temperatures T_1 and T_2, respectively, and the surfaces are black. The geometry and coordinates are as shown in Fig. 12-1.

The one-dimensional steady-state energy equation for constant thermal conductivity and no internal generation is given as

$$\frac{d^2\theta(\tau)}{d\tau^2} = \frac{1}{N}\frac{dQ^r(\tau)}{d\tau} \quad \text{in } 0 \le \tau \le \tau_0 \tag{12-35}$$

with the boundary conditions

$$\theta(\tau) = \frac{T_1}{T_2} \equiv \theta_1 \quad \text{at } \tau = 0 \tag{12-36a}$$

$$\theta(\tau) = \frac{T_2}{T_2} = 1 \quad \text{at } \tau = \tau_0 \tag{12-36b}$$

where we have chosen T_2 as the reference temperature, and defined various dimensionless quantities as

$$N \equiv \frac{k\kappa}{4n^2\bar{\sigma}T_2^{\,3}} \tag{12-37a}$$

$$Q^r(\tau) = \frac{q^r(\tau)}{4n^2\bar{\sigma}T_2^{\,4}} \tag{12-37b}$$

$$d\tau = \kappa\, dy, \qquad \tau_0 = \kappa L \tag{12-37c}$$

The radiative flux term $dQ^r/d\tau$ for an absorbing and emitting slab when the boundary surface intensities are independent of direction is obtained from Eq. 8-95 as

$$\frac{dq^r(\tau)}{d\tau} = 4n^2\bar{\sigma}T^4(\tau) - 2\Bigg[n^2\bar{\sigma}T_1^{\,4}E_2(\tau) + n^2\bar{\sigma}T_2^{\,4}E_2(\tau_0 - \tau) + \int_0^{\tau_0} n^2\bar{\sigma}T^4(\tau')E_1(|\tau - \tau'|)\, d\tau'\Bigg] \tag{12-38a}$$

or

$$\frac{dQ^r(\tau)}{d\tau} = \theta^4(\tau) - \frac{1}{2}\Bigg[\theta_1^{\,4}E_2(\tau) + E_2(\tau_0 - \tau) + \int_0^{\tau_0} \theta^4(\tau')E_1(|\tau - \tau'|)\, d\tau'\Bigg] \tag{12-38b}$$

where the index of refraction n is assumed to be constant.

The mathematical formulation of the problem is now complete. The temperature distribution in the medium is obtained from the solution of the energy equation 12-35 subject to the boundary conditions 12-36 and with $dQ^r/d\tau$ as given by Eq. 12-38b. Once the temperature distribution in the medium is known, the net heat flux $q(\tau)$ is evaluated from

$$\frac{q(\tau)}{k\kappa T_2} = -\frac{d\theta(\tau)}{d\tau} + \frac{1}{N} Q^r(\tau) \tag{12-39}$$

where the net radiative heat flux $Q^r(\tau)$ is obtainable from Eq. 8-84, that is,

$$q^r(\tau) = 2\left[n^2\bar{\sigma}T_1^4E_3(\tau) + \int_0^\tau n^2\bar{\sigma}T^4(\tau')E_2(\tau - \tau')\,d\tau'\right] - 2\left[n^2\bar{\sigma}T_2^4E_3(\tau_0 - \tau) + \int_\tau^{\tau_0} n^2\bar{\sigma}T^4(\tau')E_2(\tau' - \tau)\,d\tau'\right] \tag{12-40a}$$

or

$$Q^r(\tau) = \frac{1}{2}\left[\theta_1^4E_3(\tau) + \int_0^\tau \theta^4(\tau')E_2(\tau - \tau')\,d\tau'\right] - \frac{1}{2}\left[E_3(\tau_0 - \tau) + \int_\tau^{\tau_0} \theta^4(\tau')E_2(\tau' - \tau)\,d\tau'\right] \tag{12-40b}$$

Discussion of Results

It is highly unlikely that an analytical solution can be found to the problem of simultaneous conduction and radiation described above. However, the problem can be solved numerically with a high-speed digital computer. Viskanta and Grosh [1] transformed integrodifferential equations 12-35 and 12-38b into a nonlinear Fredholm type integral equation for $\theta(\tau)$ by integrating with respect to τ from 0 to τ. The integration constants were evaluated by utilizing the boundary conditions 12-36. The resulting integral equation was solved numerically by an iterative scheme.

Figure 12-4 shows the distribution of temperature $\theta(\tau)$ in the slab as a function of position for optical thicknesses $\tau_0 = 0.1$ and 1.0 for several values of the parameter N. The boundary conditions for these calculations were $\theta_1 = 0.5$ and $\theta_2 = 1.0$. In both the left and the right figure the curves for $N \geq 10$ are indistinguishable from those for pure conduction (i.e., $N = \infty$). As the value of the parameter N is decreased, the difference between the temperature profiles with and without radiation increases. For a given value of N the difference is larger with $\tau_0 = 1.0$ than with $\tau_0 = 0.1$. For example, for $N = 0.01$ the maximum difference between the temperature profiles with and without radiation is only about 3 per cent for $\tau_0 = 0.1$.

The case $N = 0$ corresponds to energy transfer by pure radiation. The temperature gradients at the cold wall are greater than those for pure conduction; however, at the hot wall the temperature gradients can be larger or smaller than those for pure conduction, depending on the value of the parameter N.

12-4 CONDUCTION AND RADIATION IN AN ABSORBING, EMITTING AND SCATTERING SLAB—Exact Treatment

Consideration will be given to the exact treatment of steady-state, simultaneous conduction and radiation in an absorbing, emitting, isotropically scattering, gray slab of optical thickness τ_0. The boundaries at $\tau = 0$ and $\tau = \tau_0$ are opaque, gray, diffusely emitting, diffusely reflecting surfaces kept at uniform temperatures T_1 and T_2, respectively. Figure 12-1 shows the geometry and coordinates under consideration. In this section we present two different approaches to solve the radiation part of the problem. In method 1 we apply the techniques described in Chapter 8; in method 2 we utilize the normal-mode expansion technique described in Chapter 10.

Method 1

The steady-state energy equation for simultaneous conduction and radiation, assuming constant thermal conductivity and no energy generation, may be

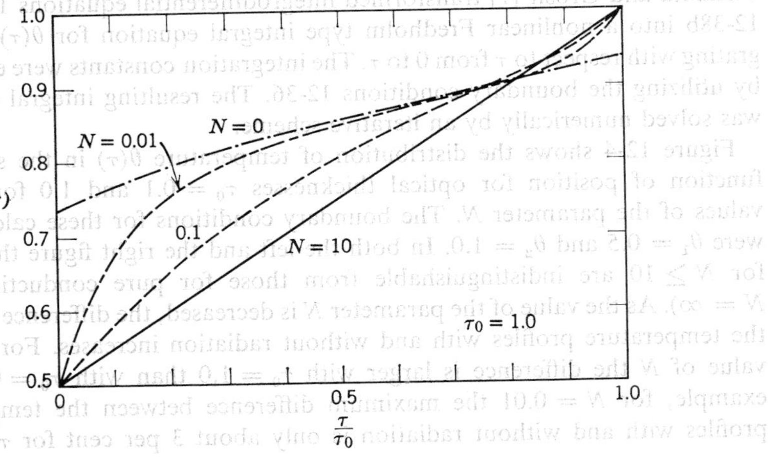

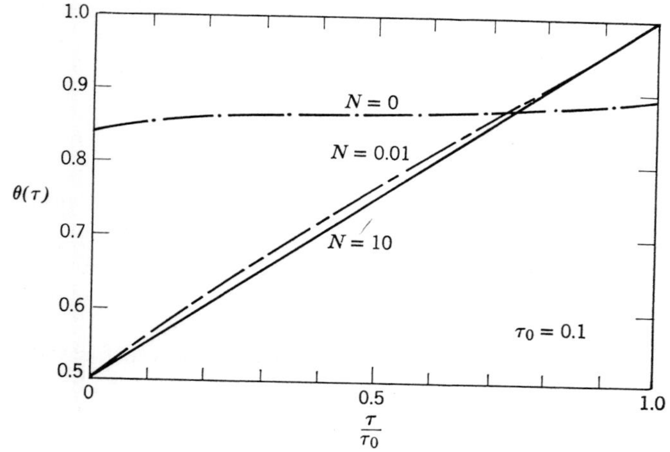

Fig. 12-4. Temperature distribution for combined conduction and radiation in an absorbing, emitting slab; $\theta_1 = 0.5$, $\theta_2 = 1.0$, $E_1 = E_2 - 1.0$. (From R. Viskanta and R. J. Grosh [1].)

written as (see Eq. 12-14)

$$\frac{d^2\theta(\tau)}{d\tau^2} = \frac{1-\omega}{N}\left[\theta^4(\tau) - G^*(\tau)\right] \quad \text{in } 0 \le \tau \le \tau_0 \tag{12-41}$$

with the boundary conditions

$$\theta(\tau) = 1 \qquad \text{at } \tau = 0 \tag{12-42a}$$

$$\theta(\tau) = \frac{T_2}{T_1} \equiv \theta_2 \quad \text{at } \tau = \tau_0 \tag{12-42b}$$

where $\theta(\tau)$ and N are defined as in Eqs. 12-6 by setting $T_r = T_1$.

The dimensionless incident radiation $G^*(\tau)$ is related to the radiation intensity $I(\tau, \mu)$ by

$$G^*(\tau) = \frac{2\pi \int_{-1}^{1} I(\tau, \mu')\, d\mu'}{4n^2\bar{\sigma}T_1^{\,4}} \equiv \frac{1}{2}\int_{-1}^{1} \psi(\tau, \mu')\, d\mu' \tag{12-43}$$

The function $\psi(\tau, \mu)$ satisfies the equation of radiative transfer

$$\mu \frac{\partial \psi(\tau, \mu)}{\partial \tau} + \psi(\tau, \mu) = (1-\omega)\theta^4(\tau) + \frac{\omega}{2}\int_{-1}^{1} \psi(\tau, \mu')\, d\mu' \quad \text{in } 0 \le \tau \le \tau_0 \tag{12-44}$$

with the boundary conditions

$$\psi^+(0) = \varepsilon_1 + 2(1 - \varepsilon_1)\int_0^1 \psi^-(0, -\mu')\mu'\, d\mu', \qquad \mu > 0 \qquad (12\text{-}45a)$$

$$\psi^-(\tau_0) = \varepsilon_2\theta_2^{\,4} + 2(1 - \varepsilon_2)\int_0^1 \psi^+(\tau_0, \mu')\mu'\, d\mu', \quad \mu < 0 \qquad (12\text{-}45b)$$

where we have assumed that the Kirchhoff law is valid and have replaced the reflectivity ρ by $1 - \varepsilon$ for the opaque boundaries.

The conduction and radiation problems given by Eqs. 12-41 and 12-44, respectively, are coupled through the relation given by Eq. 12-43 and should be solved simultaneously.

The radiation problem given by Eq. 12-44 can be solved formally by the method described in Chapter 8, and the relation for $G^*(\tau)$ is obtained as[1]

$$G^*(\tau) = \frac{1}{2}\Big\{\psi^+(0)E_2(\tau) + \psi^-(\tau_0)E_2(\tau_0 - \tau) + \int_0^{\tau_0}[(1 - \omega)\theta^4(\tau') + \omega G^*(\tau')]E_1(|\tau - \tau'|)\, d\tau'\Big\} \qquad (12\text{-}46)$$

where $\psi^+(0)$ and $\psi^-(\tau_0)$ are the dimensionless boundary surface intensities. For diffusely reflecting, diffusely emitting boundaries a formal solution for $\psi^+(0)$ and $\psi^-(\tau_0)$ is immediately obtained from Eqs. 8-108 as

$$\psi^+(0) = \frac{A_1 + b_1 A_2}{1 - b_1 b_2} \qquad (12\text{-}47)$$

$$\psi^-(\tau_0) = \frac{A_2 + b_2 A_1}{1 - b_1 b_2} \qquad (12\text{-}48)$$

where

$$A_1 \equiv \varepsilon_1 + 2(1 - \varepsilon_1)\int_0^{\tau_0}[(1 - \omega)\theta^4(\tau') + \omega G^*(\tau')]E_2(\tau')\, d\tau' \qquad (12\text{-}49a)$$

$$A_2 \equiv \varepsilon_2\theta_2^{\,4} + 2(1 - \varepsilon_2)\int_0^{\tau_0}[(1 - \omega)\theta^4(\tau') + \omega G^*(\tau')]E_2(\tau_0 - \tau')\, d\tau' \qquad (12\text{-}49b)$$

$$b_1 \equiv 2(1 - \varepsilon_1)E_3(\tau_0) \qquad (12\text{-}49c)$$

$$b_2 \equiv 2(1 - \varepsilon_2)E_3(\tau_0) \qquad (12\text{-}49d)$$

The above formulation can now be summarized. Equations 12-41, 12-46, 12-47, and 12-48 provide four independent relations for the four unknown functions $\theta(\tau)$, $G^*(\tau)$, $\psi^+(0)$, and $\psi^-(\tau_0)$. A direct iterative scheme may be used to solve these equations. The differential equation 12-41 is integrated subject to the boundary conditions 12-42a and 12-42b, and a nonlinear integral equation is obtained for $\theta(\tau)$. This integral equation for $\theta(\tau)$ and the equations for $G^*(\tau)$, $\psi^+(0)$, and $\psi^-(\tau_0)$ are then solved by iteration. Once the functions $\theta(\tau)$, $G^*(\tau)$, $\psi^+(0)$, and $\psi^-(\tau_0)$ are determined, the dimensionless net radiative heat flux $Q^r(\tau)$ anywhere in the medium is obtained from[2]

$$Q^r(\tau) \equiv \frac{q^r(\tau)}{4n^2\bar{\sigma}T_1^4} = \frac{1}{2}\Big\{\psi^+(0)E_3(\tau) - \psi^-(\tau_0)E_3(\tau_0 - \tau)$$

$$+ \int_0^\tau [(1-\omega)\theta^4(\tau') + \omega G^*(\tau')]E_2(\tau - \tau')\,d\tau'$$

$$- \int_\tau^{\tau_0} [(1-\omega)\theta^4(\tau') + \omega G^*(\tau')]E_2(\tau' - \tau)\,d\tau'\Big\} \qquad (12\text{-}50)$$

and the net heat flux q is evaluated from

$$\frac{q}{k\beta T_1} = -\frac{d\theta(\tau)}{d\tau} + \frac{1}{N}Q^r(\tau) \qquad (12\text{-}51)$$

Viskanta [3] determined the temperature distribution, the net radiative heat flux, and the total heat flux in the medium for combined conduction and radiation with a similar approach.

Method 2

The energy equation for combined conduction and radiation, assuming constant thermal conductivity and no internal energy generation, is written in the form (see Eq. 12-13)

$$\frac{d^2\theta(\tau)}{d\tau^2} - \frac{1}{N}\frac{dQ^r(\tau)}{d\tau} = 0 \quad \text{in } 0 \le \tau \le \tau_0 \qquad (12\text{-}52)$$

with the boundary conditions

$$\theta(\tau) = 1 \quad \text{at } \tau = 0 \qquad (12\text{-}53a)$$

$$\theta(\tau) = \theta_2 \quad \text{at } \tau = \tau_0 \qquad (12\text{-}53b)$$

The dimensionless net radiative heat flux $Q^r(\tau)$ is related to the radiation intensity $I(\tau, \mu)$ by

$$Q^r(\tau) \equiv \frac{q^r(\tau)}{4n^2\bar{\sigma}T_1^4} = \frac{2\pi\int_{-1}^{1} I(\tau,\mu')\mu'\,d\mu'}{4n^2\bar{\sigma}T_1^4} \equiv \frac{1}{2}\int_{-1}^{1}\psi(\tau,\mu')\mu'\,d\mu' \qquad (12\text{-}54a)$$

and the dimensionless radiation intensity $\psi(\tau, \mu)$ is defined as

$$\psi(\tau, \mu) \equiv \frac{I(\tau, \mu)}{n^2\bar{\sigma}T_1^4/\pi} \tag{12-54b}$$

The function $\psi(\tau, \mu)$ satisfies the equation of radiative transfer:

$$\mu \frac{\partial\psi(\tau, \mu)}{\partial\tau} + \psi(\tau, \mu) = (1 - \omega)\theta^4(\tau) + \frac{\omega}{2}\int_{-1}^{1}\psi(\tau, \mu')\,d\mu' \tag{12-55}$$

with the boundary conditions

$$\psi(0) = \varepsilon_1 + 2(1 - \varepsilon_1)\int_0^1 \psi(0, -\mu')\mu'\,d\mu', \qquad \mu > 0 \tag{12-56a}$$

$$\psi(\tau_0) = \varepsilon_2\theta_2^4 + 2(1 - \varepsilon_2)\int_0^1 \psi(\tau_0, \mu')\mu'\,d\mu', \quad \mu < 0 \tag{12-56b}$$

The heat conduction problem (Eqs. 12-52 and 12-53) and the radiation problem (Eqs. 12-55 and 12-56) are coupled through Eq. 12-54; they should be solved simultaneously.

The normal-mode expansion technique will now be applied to solve the radiation part of the problem. The complete solution of the equation of radiative transfer 12-55 is written as a sum of the solutions to the homogeneous part of this equation and a particular solution $\psi_P(\tau, \mu)$ in the form

$$\begin{aligned}\psi(\tau, \mu) = {} & A(\eta_0)\phi(\eta_0, \mu)e^{-\tau/\eta_0} + A(-\eta_0)\phi(-\eta_0, \mu)e^{\tau/\eta_0} \\ & + \int_0^1 A(\eta)\phi(\eta, \mu)e^{-\tau/\eta}\,d\eta \\ & + \int_0^1 A(-\eta)\phi(-\eta, \mu)e^{\tau/\eta}\,d\eta + \psi_P(\tau, \mu)\end{aligned} \tag{12-57}$$

where the discrete normal modes $\phi(\pm\eta_0, \mu)$ and the continuum normal modes $\phi(\pm\eta, \mu)$ are defined as (see Eqs. 10-18c and d)

$$\phi(\pm\eta_0, \mu) = \frac{\omega\eta_0}{2}\frac{1}{\eta_0 \mp \mu}, \qquad \eta_0 \notin (-1, 1) \tag{12-58a}$$

$$\phi(\pm\eta, \mu) = \frac{\omega\eta}{2}\frac{P}{\eta \mp \mu} + \lambda(\eta)\,\delta(\eta \mp \mu), \quad \eta \in (-1, 1) \tag{12-58b}$$

$$\lambda(\eta) = 1 - \omega\eta \tanh^{-1}\eta \tag{12-58c}$$

and the discrete eigenvalues $\pm\eta_0$ are the two zeros of the dispersion function:

$$\Lambda(\eta_0) \equiv 1 - \omega\eta_0 \tanh^{-1}\frac{1}{\eta_0} = 0 \tag{12-58d}$$

Here P in Eq. 12-58b is a mnemonic symbol used to denote that the ensuing integral is to be evaluated in the Cauchy principle-value sense, and $\delta(x)$ is the Dirac delta function. A particular solution $\psi_P(\tau, \mu)$ of the equation of radiative transfer 12-55 may be found if the temperature function $\theta^4(\tau)$ is known; however, the temperature distribution $\theta(\tau)$ is not known until the energy equation 12-52 is solved. Therefore, for the purpose of determining a particular solution, it is assumed that an initial guess is available for the temperature distribution $\theta^0(\tau)$ and that the function $[\theta^0(\tau)]^4$, defined in the interval $0 \le \tau \le \tau_0$, is represented by a polynomial in the optical variable τ in the form

$$[\theta^0(\tau)]^4 = \sum_{m=0}^{M} a_m \tau^m \quad \text{in } 0 \le \tau \le \tau_0 \tag{12-59}$$

The coefficients a_m are determined by taking a finite number of terms in the expansion. A particular solution of the equation of radiative transfer for an inhomogeneous term of the form τ^m, $m = 0, 1, 2, \ldots$, can be obtained from the general expression given in Table 10-6.

Therefore, if $\psi_{P,m}(\tau, \mu)$ denotes a particular solution for an inhomogeneous term $(1 - \omega)a_m\tau^m$, then a particular solution $\psi_P(\tau, \mu)$ for an inhomogeneous term $(1 - \omega)\sum_{m=0}^{M} a_m\tau^m$ is given as

$$\psi_P(\tau, \mu) = \sum_{m=0}^{M} \psi_{P,m}(\tau, \mu) \tag{12-60}$$

Knowing a particular solution $\psi_P(\tau, \mu)$, we can determine the expansion coefficients $A(\pm\eta_0)$ and $A(\pm\eta)$ by constraining the solution 12-57 to meet the boundary conditions 12-56 and by utilizing the orthogonality property of the normal modes and various normalization integrals as discussed in Chapters 10 and 11.

Once a particular solution is found for a given guess as to the distribution of temperature and the expansion coefficients $A(\pm\eta_0)$ and $A(\pm\eta)$ are determined, the corresponding dimensionless net radiative heat flux $Q^r(\tau)$ anywhere in the medium is evaluated from (see Eqs. 12-54a and 12-57)

$$Q^r(\tau) = \tfrac{1}{2}(1 - \omega)\Big[\eta_0 A(\eta_0)e^{-\tau/\eta_0} - \eta_0 A(-\eta_0)e^{\tau/\eta_0} + \int_0^1 \eta A(\eta)e^{-\tau/\eta}\, d\eta - \int_0^1 \eta A(-\eta)e^{\tau/\eta}\, d\eta + \frac{1}{1-\omega}\int_{-1}^{1} \psi_P(\tau, \mu)\mu\, d\mu\Big] \tag{12-61}$$

Therefore, for a given guess as to the distribution of temperature, $Q^r(\tau)$ is considered a known function; then integration of the energy equation 12-52 subject to the boundary conditions 12-53 yields

$$\theta(\tau) = 1 + \frac{\tau}{\tau_0}(\theta_2 - 1) + \frac{1}{N}\Big[\int_0^{\tau} Q^r(\tau')\, d\tau' - \frac{\tau}{\tau_0}\int_0^{\tau_0} Q^r(\tau')\, d\tau'\Big] \tag{12-62}$$

Then the temperature distribution $\theta(\tau)$ is evaluated from Eqs. 12-62 and 12-61 by an iterative scheme as now described.

An initial guess is made for the temperature distribution $\theta^0(\tau)$, the fourth power of this temperature distribution $[\theta^0(\tau)]^4$ is represented by a polynomial expansion in τ^m as in Eq. 12-59, and the coefficients a_m are determined by taking a finite number of terms in the series. A particular solution $\psi_P(\tau, \mu)$ of the equation of radiative transfer is found, the expansion coefficients $A(\pm\eta_0)$ and $A(\pm\eta)$ are determined, and the resulting net radiative heat flux function $Q^r(\tau)$ is introduced into Eq. 12-62 to obtain a first approximation for $\theta^1(\tau)$. This first approximation for the temperature distribution is used to find a second approximation, and so on. The calculations are repeated until the solution sufficiently converges.

Lii and Özişik [15a] applied the normal-mode expansion technique as described above and solved the simultaneous conduction and radiation problem in a slab for both the steady-state and the time-dependent cases.

Discussion of Results

The independent parameters that affect the steady-state temperature distribution in a slab for combined conduction and radiation include the conduction-to-radiation parameter N, the albedo ω, the optical thickness of the

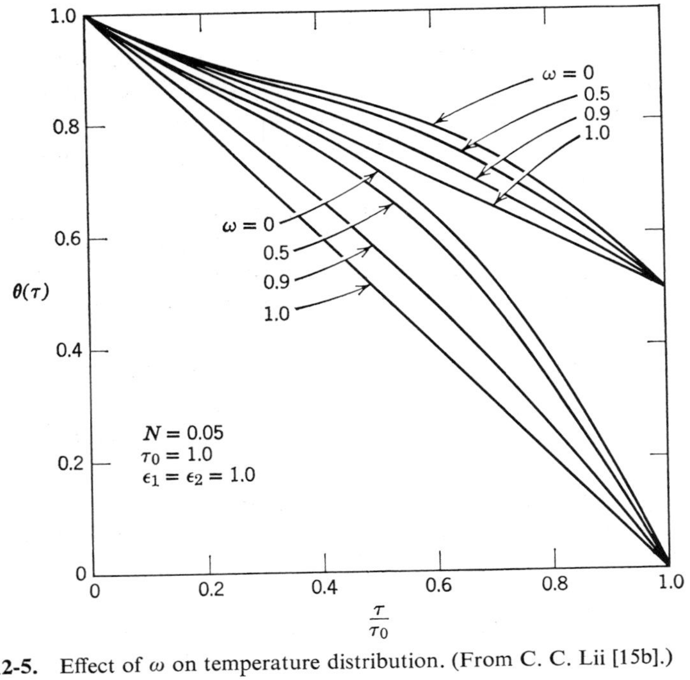

Fig. 12-5. Effect of ω on temperature distribution. (From C. C. Lii [15b].)

slab τ_0, the boundary surface emissivities ε_1 and ε_2, and the boundary surface temperature θ_2.

Figure 12-5 shows the effects of albedo ω on the temperature distribution in a slab with black walls for $N = 0.05$, $\tau_0 = 1.0$, $\theta_2 = 0$, and $\theta_2 = 0.5$. The two limiting cases $\omega = 0$ and $\omega = 1$ characterize, respectively, a purely absorbing, emitting medium and a purely scattering medium. The interaction of radiation with conduction is greatest for $\omega = 0$. The radiation has no effect on the temperature distribution, hence the temperature distribution is the same as that for pure conduction, when $\omega = 1.0$.

Figure 12-6 shows the effect of emissivity on the temperature distribution in a slab with diffusely emitting, diffusely reflecting opaque walls for $N = 0.1$, $\omega = 0.5$, and $\tau_0 = 1.0$.

The effect of the parameter N on the temperature distribution was discussed in Section 12-3 for the case $\omega = 0$, in which the effects are maximum. When $\omega \neq 0$ the parameter N has a similar effect on the temperature distribution; therefore it is not included here.

Table 12-1 shows the effects of the parameters N and ω on the conductive, radiative, and total heat fluxes in a slab with black walls for $\tau_0 = 1$, $\theta_1 = 1$,

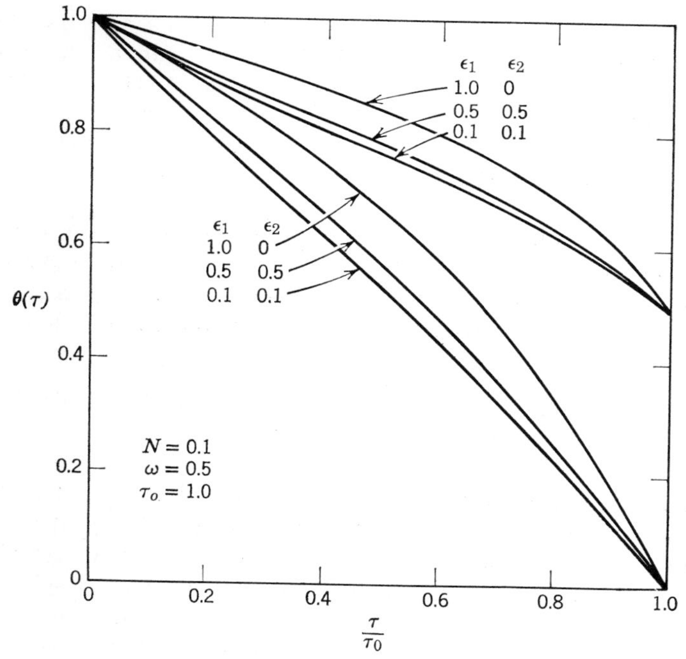

Fig. 12-6. Effect of emissivity on temperature distribution. (From C. C. Lii [15b].)

Table 12-1 Effect of Parameters N and ω on Conductive, Radiative, and Total Heat Fluxes[a]

$(\theta_1 = 1,\ \theta_2 = 0,\ \tau_0 = 1,\ \varepsilon_1 = \varepsilon_2 = 1)$

		$-\frac{\partial\theta}{\partial\tau}$			$\frac{Q^r}{N}$			Total Heat Flux q:
N	ω	$\frac{\tau}{\tau_0} = 0$	0.5	1.0	$\frac{\tau}{\tau_0} = 0$	0.5	1.0	$\frac{q}{k\beta T_1} = -\frac{\partial\theta}{\partial\tau} + \frac{Q^r}{N}$
0.5	0	0.9396	0.9879	1.1447	0.3585	0.3102	0.1534	1.2981
	0.5	0.9491	0.9930	1.0983	0.3392	0.2954	0.1900	1.2884
	0.9	0.9798	0.9985	1.0305	0.2995	0.2903	0.2488	1.2793
	1.0	1.0	1.0	1.0	0.2767	0.2767	0.2767	1.2767
0.1	0	0.8520	0.8799	1.6707	1.6651	1.6372	0.8464	2.5171
	0.5	0.8513	0.9386	1.4366	1.5924	1.5051	1.0071	2.4437
	0.9	0.9084	0.9901	1.1486	1.4896	1.4080	1.2494	2.3980
	1.0	1.0	1.0	1.0	1.3835	1.3835	1.3835	2.3835
0.05	0	0.8986	0.7159	2.2187	3.1544	3.3371	1.8344	4.0530
	0.5	0.7908	0.8314	1.8689	3.0937	3.0531	2.0156	3.8845
	0.9	0.8382	0.9745	1.2877	2.9608	2.8246	2.5114	3.7991
	1.0	1.0	1.0	1.0	2.7670	2.7670	2.7670	3.7670

[a] From C. C. Lii and M. N. Özişik [15a].

and $\theta_2 = 0$. The conductive and radiative heat fluxes vary with position, but for given values of ω and N the total net heat flux q is constant everywhere in the medium. Only for a purely scattering medium (i.e., $\omega = 1$) are the conductive and radiative heat fluxes constant everywhere in the slab.

12-5 ABLATION AND RADIATION IN AN ABSORBING, EMITTING, AND SCATTERING MEDIUM

One method of absorbing a large amount of externally applied heat flux at the surface of a body is to allow the material to ablate, that is, the material is removed from the surface partially by vaporization and partially by flowing away of the melt layer. For example, during high-speed re-entry conditions into the earth's atmosphere the aerodynamic heating is so high that it is removed by the ablation of the heat shield. When the ablating material is opaque to radiation, the standard techniques of heat conduction analysis may be used to determine the temperature distribution in the ablating body. However, when the ablating material is semitransparent and the temperature is high, energy transfer within the body from the hotter to the colder regions by thermal radiation becomes important. In such situations the determination of temperature distribution in the medium requires a simultaneous solution of the equations for conductive and radiative heat transfer.

During the ablation of a semitransparent, glassy material such as quartz or Pyrex the temperature at the heated surface may become very high. For example, temperatures as high as 2700°K have been reported during the ablation of quartz [23, 24]. At such high temperatures all glassy materials are liquids and have viscosities that are very strong functions of temperature [18]. Therefore, during the ablation of a glassy material, there exists a very thin liquid layer in the high-temperature region near the surface and most flow occurs within this liquid layer. At greater depths from the heated surface the body becomes cooler and the liquid layer gradually changes into solid. The determination of temperature distribution within an ablating glassy material is important in the prediction of the amount of material that is removed partially by the flowing away of the liquid layer and partially by vaporization. The reader should refer to the paper by Bethe and Adams [25] for a discussion of the determination of the surface temperature and the amount of material removed by ablation.

In this section consideration is given to the determination of the temperature distribution within an ablating, semitransparent body resulting from an externally applied heat flux at the boundary surface. For generality in the analysis we assume that the medium scatters radiation isotropically as well as absorbs and emits it. Figure 12-7 shows the geometry and the coordinate system. The medium, which is semi-infinite ($0 < x < \infty$) in extent, is ablating as a result of externally applied heating at the gas-liquid interface. The interface temperature T_0 is the maximum temperature in the steady-state ablation process, and the temperature decreases within the ablating body with the distance from the interface. The radiation emitted from the interior of the medium that impinges upon the liquid-air interface is partly transmitted and partly reflected by the interface, which is assumed to be an ideal surface that reflects radiation specularly. If the ablation has been proceeding at a constant rate for some time and the transients have passed, the temperature distribution within the medium becomes a function of the distance from the interface but not of time; this is called steady-state ablation. The problem of determining the temperature profile in an absorbing, emitting, isotropically scattering material for steady-state ablation has been studied by Kadanoff [18] and Boles and Özişik [19]. The former investigator used an

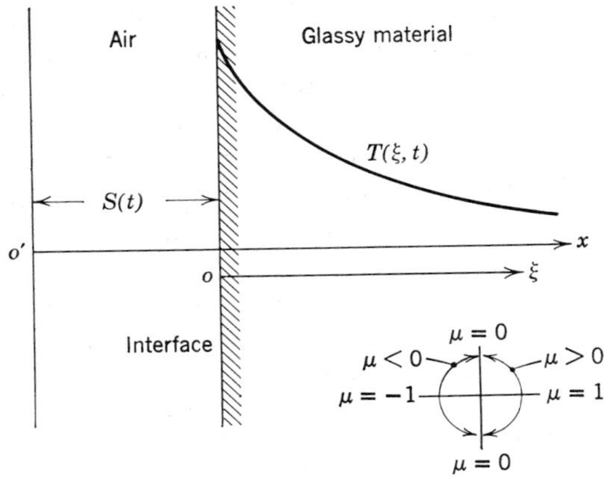

Fig. 12-7. Geometry and coordinates for the ablation of a glassy material.

approximate technique to solve the radiation part of the problem, and his results are applicable to cases with large amounts of scattering. The latter investigators treated the radiation part of the problem exactly with the normal-mode expansion technique, and their solution is applicable for all values, from 0 to 1, of the single scattering albedo ω.

We present now the mathematical formulation of the problem of temperature distribution in an ablating body, describe the method of solution for the case of steady-state ablation, and discuss the results.

The one-dimensional energy equation for combined conduction and radiation is given as (see Eq. 12-4)

$$\frac{\partial}{\partial x}\left(k\frac{\partial T}{\partial x}\right) - \frac{\partial q^r}{\partial x} = \rho c_p \frac{\partial T}{\partial t}, \quad S(t) \le x < \infty,\ t > 0 \qquad (12\text{-}63)$$

where $S(t)$ is the location of the air-liquid interface. The moving boundary can be eliminated by transforming

$$\xi(x, t) \equiv \xi = x - S(t) \qquad (12\text{-}64a)$$

and defining for convenience

$$\frac{dS(t)}{dt} \equiv v \qquad (12\text{-}64b)$$

where ξ is the coordinate measured from the air-liquid interface, and v is the ablation velocity. By introducing the transformation given by Eqs. 12-64 into the energy equation 12-63 we obtain[3]

$$\frac{\partial}{\partial \xi}\left(k\frac{\partial T}{\partial \xi} + v\rho c_p T - q^r\right) = \rho c_p \frac{\partial T}{\partial t}, \quad 0 \le \xi < \infty,\ t > 0 \quad (12\text{-}65)$$

Equation 12-65 is written in the dimensionless form as

$$\frac{\partial}{\partial \tau}\left(\frac{\partial \theta}{\partial \tau} + Z\theta - \frac{1}{N} Q^r\right) = \frac{\partial \theta}{\partial t^*}, \quad 0 \le \tau < \infty, \ t^* > 0 \tag{12-66}$$

where various dimensionless quantities are defined as

$$N \equiv \frac{k\beta}{4n^2\bar{\sigma}T_0^3} = \text{conduction-to-radiation parameter} \tag{12-67a}$$

$$Q^r \equiv \frac{q^r(\tau)}{4n^2\bar{\sigma}T_0^4} = \text{dimensionless net radiative heat flux} \tag{12-67b}$$

$$t^* \equiv \frac{k\beta^2 t}{\rho c_p} = \text{dimensionless time} \tag{12-67c}$$

$$Z \equiv \frac{v}{\alpha\beta} = \frac{1/\beta}{\alpha/v} \tag{12-67d}$$

$$\theta \equiv \frac{T}{T_0} \tag{12-67e}$$

$$\alpha = \frac{k}{\rho c_p} \tag{12-67f}$$

$$d\tau = \beta \, d\xi \tag{12-67g}$$

Here we have chosen the air-liquid interface temperature T_0, which is the maximum temperature in the ablation problem considered, as the reference temperature and have assumed constant properties. The parameter Z is a dimensionless radiation mean free path; hence it characterizes a dimensionless distance that a radiation photon will penetrate into the medium before being absorbed. The larger the value of Z, the larger is the depth of penetration.

Steady-State Ablation

If it is assumed that the ablation has been proceeding at a constant rate for a long time (i.e., v is constant), the temperature distribution in the medium is considered to be a function of the distance from the air-liquid interface but not of time. Then the energy equation 12-66 simplifies to

$$\frac{d}{d\tau}\left(\frac{d\theta}{d\tau} + Z\theta - \frac{1}{N} Q^r\right) = 0 \quad \text{in } 0 \le \tau < \infty \tag{12-68}$$

where Z is constant. The integration of this equation once for Q^r = constant and $d\theta/d\tau$ vanishes at $\tau \to \infty$ yields

$$\frac{d\theta}{d\tau} + Z\theta - \frac{1}{N} Q^r = 0 \quad \text{in } \tau \ge 0 \tag{12-69}$$

The boundary condition is taken as the temperature equal to the air-liquid interface temperature T_0 at $\tau = 0$, that is,

$$\theta = 1 \quad \text{at } \tau = 0 \tag{12-70}$$

The dimensionless net radiative heat flux term $Q^r(\tau)$ appearing in this equation is related to the radiation intensity by

$$Q^r(\tau) = \frac{2\pi \int_{-1}^{1} I(\tau, \mu')\mu' \, d\mu'}{4n^2\bar{\sigma}T_0^4} \equiv \frac{1}{2}\int_{-1}^{1} \psi(\tau, \mu')\mu' \, d\mu' \tag{12-71}$$

If it is assumed that the ablating body is gray, is stratified in planes perpendicular to the τ axis, and absorbs, emits, and isotropically scatters radiation, the dimensionless radiation intensity $\psi(\tau, \mu)$ satisfies the equation of radiative transfer, given in the form

$$\mu \frac{\partial \psi(\tau, \mu)}{\partial \tau} + \psi(\tau, \mu) = (1 - \omega)\theta^4(\tau) + \frac{\omega}{2}\int_{-1}^{1} \psi(\tau, \mu') \, d\mu'$$

$$\text{in } 0 \leq \tau < \infty, \; -1 \leq \mu \leq 1 \tag{12-72}$$

where ω is the single scattering albedo, and μ is the cosine of the angle between the direction of radiation and the positive τ axis. The air-liquid interface at $\tau = 0$ is assumed to be a nonemitting, specularly reflecting surface. Then the boundary condition for Eq. 12-72 at $\tau = 0$ is given as

$$\psi(0, \mu) = \rho^s \psi(0, -\mu), \quad \mu > 0 \tag{12-73}$$

while $\psi(\tau, \mu)$ equals to a particular solution of Eq. 12-72 at $\tau \to \infty$.

The solution of Eq. 12-72 is written as a linear sum of the solutions to the homogeneous part of this equation and a particular solution $\psi_P(\tau, \mu)$ in the form

$$\psi(\tau, \mu) = A(\eta_0)\phi(\eta_0, \mu)e^{-\tau/\eta_0} + \int_0^1 A(\eta)\phi(\eta, \mu)e^{-\tau/\eta} \, d\eta + \psi_p(\tau, \mu) \tag{12-74}$$

where we have omitted the parts of the homogeneous solution that diverge at infinity; hence Eq. 12-74 satisfies the boundary condition at $\tau \to \infty$. The discrete normal mode $\phi(\eta_0, \mu)$ and the continuum normal mode $\phi(\eta, \mu)$ were defined in Chapter 10 (see Eqs. 10-18).

The solution given by Eq. 12-74 contains two unknown expansion coefficients $A(\eta_0)$ and $A(\eta)$. Assuming that a particular solution $\psi_P(\tau, \mu)$ of the equation of radiative transfer is available, we can determine these coefficients, as described in Chapters 10 and 11, by constraining the solution 12-74 to meet the boundary condition at $\tau = 0$ and by utilizing the orthogonality of the normal modes and various normalization integrals. However, a particular solution cannot be determined until the fourth-power temperature $\theta^4(\tau)$ appearing in Eq. 12-72 is known. To circumvent this difficulty we assume that an initial guess is available for the temperature distribution $\theta(\tau)$ in the medium and that the function $\theta^4(\tau)$ is represented in a truncated cosine series in the form

$$\theta^4(\tau) = \sum_{m=0}^{M} B_m \cos \frac{m\pi\tau}{\tau_0} \quad \text{in } 0 \leq \tau \leq \tau_0 \tag{12-75}$$

where $\tau = \tau_0$ is the location beyond which $\theta^4(\tau)$ is less than a prescribed minimum; thus $\theta^4(\tau)$ is assumed to be negligible in the region $\tau > \tau_0$. In the present analysis this minimum is taken to be 10^{-4}. Once the function $\theta^4(\tau)$ is known, the coefficients B_m are readily determined by taking a finite number of terms in the expansion.

When $\theta^4(\tau)$ is represented as in Eq. 12-75, a particular solution of the equation of radiative transfer 12-72 for an inhomogeneous term $(1 - \omega)\theta^4(\tau)$ is obtained from Table 10-6 as

$$\psi_p(\tau, \mu) = (1 - \omega) \sum_{m=0}^{M} B_m \frac{(\tau_0/m\pi)[(\tau_0/m\pi) \cos (m\pi\tau/\tau_0) + \mu \sin (m\pi\tau/\tau_0)]}{[1 - \omega(\tau_0/m\pi) \tan^{-1} (m\pi/\tau_0)][(\tau_0/m\pi)^2 + \mu^2]} \tag{12-76}$$

Once a particular solution is available for a given guess as to temperature distribution and the corresponding expansion coefficients $A(\eta_0)$ and $A(\eta)$ are obtained, the dimensionless net radiative heat flux $Q^r(\tau)$ is determined from Eqs. 12-71 and 12-74 as

$$Q^r(\tau) = \tfrac{1}{2}(1 - \omega)\left[\eta_0 A(\eta_0) e^{-\tau/\eta_0} + \int_0^1 \eta A(\eta) e^{-\tau/\eta}\, d\eta + \frac{1}{1 - \omega} \int_{-1}^{1} \psi_p(\tau, \mu)\mu\, d\mu\right] \tag{12-77}$$

Substitution of Eq. 12-76 into Eq. 12-77 yields

$$Q^r(\tau) = \tfrac{1}{2}(1 - \omega)\left[\eta_0 A(\eta_0) e^{-\tau/\eta_0} + \int_0^1 \eta A(\eta) e^{-\tau/\eta}\, d\eta \right.$$
$$\left. + \sum_{m=0}^{M} 2B_m \frac{\tau_0}{m\pi} \frac{1 - (\tau_0/m\pi) \tan^{-1} (m\pi/\tau_0)}{1 - \omega(\tau_0/m\pi) \tan^{-1} (m\pi/\tau_0)} \sin (m\pi\tau/\tau_0)\right] \tag{12-78}$$

Method of Solution

The problem involves four independent parameters: N, Z, ρ^s, and ω. Once the values of these parameters are fixed and a guess as to distribution is available for $\theta(\tau)$, the function $\theta^4(\tau)$ is represented in a truncated cosine series as in Eq. 12-75 and the coefficients B_m are determined. Then a particular solution of the equation of radiative transfer is obtained from Eq. 12-76, and the corresponding expansion coefficients $A(\eta_0)$ and $A(\eta)$ are determined by an approach described in Chapters 10 and 11. Knowing $A(\eta_0)$, $A(\eta)$, and B_m, we can evaluate the dimensionless net radiative heat flux $Q^r(\tau)$ from Eq. 12-78. By treating $Q^r(\tau)$ as a prescribed function, the differential equation 12-69 subject to the boundary condition 12-70 is integrated numerically by a Runge-Kutta method and a first approximation is obtained for the temperature profile $\theta(\tau)$. This first approximation is used to obtain a second approximation, and the second approximation to obtain a third approximation, and so forth. The calculations are repeated until the solution converges to a prescribed criterion.

Boles and Özişik [19], using the above approach and a Runge-Kutta method that could readily correct the initial guess for the slope of the temperature at the boundary at $\tau = 0$, determined the distribution of the steady-state temperature and the net radiative heat flux in the ablating body.

Discussion of Results

Figure 12-8 shows the effects of the conduction-to-radiation parameter N on the distribution of temperature and the net radiative heat flux in an ablating body for $\omega = 0.9$, $\rho^s = 0.7$, and $Z = 0.05$. The temperature profile for the case of no radiation, that is, $\theta = e^{-z\tau}$, is also included in this figure. The temperature profile approaches that of pure conduction for large values of N. Decreasing N (i.e., increasing radiation) lowers the temperature in the region near the interface but raises it in the interior of the medium.

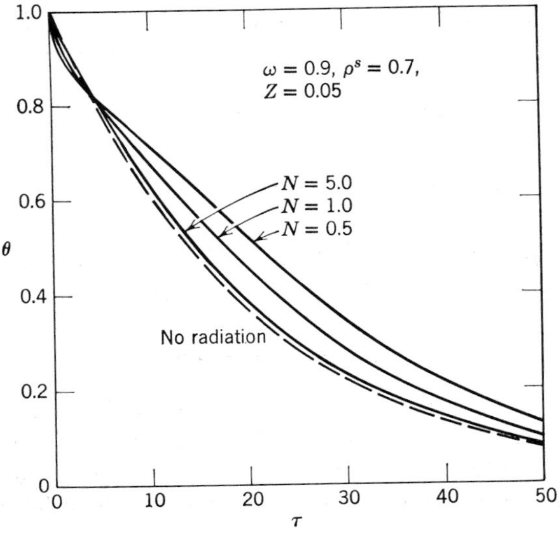

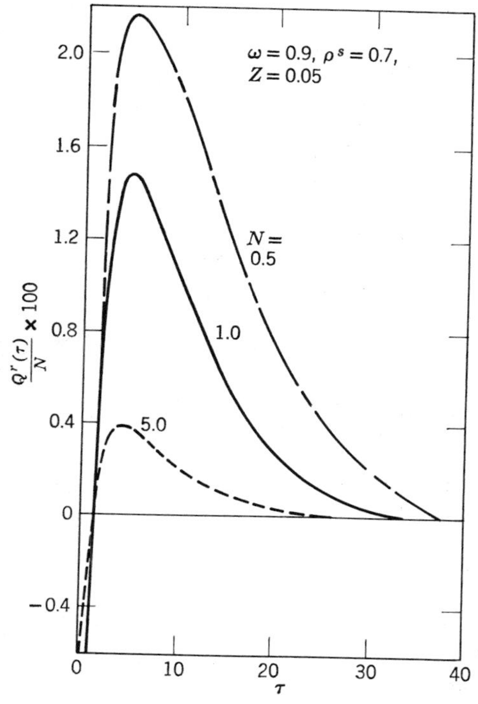

Fig. 12-8. Effect of conduction-to-radiation parameter N on temperature profile and radiative heat flux. (From M. A. Boles and M. N. Özişik [19].)

The reason for this is better envisoned by considering the distribution of the net radiative heat flux. Near the interface the net radiative heat flux is negative, indicating that the radiative energy flows from the medium into the external environment, thus reducing the temperature in this region below that of pure conduction as radiation effects are increased. The net radiative heat flux is positive in the interior of the body, indicating that the radiative energy flows in that direction; therefore the temperature in the interior region is higher than that for pure conduction. We also note that Q^r/N increases as N decreases.

Figure 12-9 shows the effects of the single scattering albedo ω on temperature distribution in the medium. The temperature profile for the case $\omega = 1$ corresponds to that of no radiation, since there is no radiation interaction in a purely scattering medium. Decreasing ω lowers the temperature in the region close to the interface and raises it in the interior of the medium. This result is consistent with the fact that the interaction of radiation with conduction increases with decreasing ω.

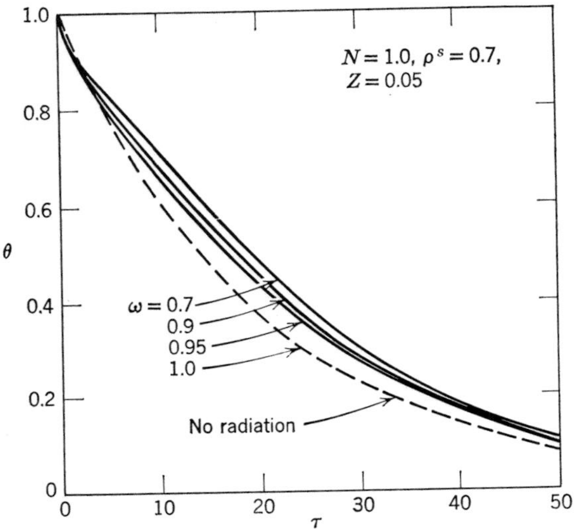

Fig. 12-9. Effect of single scattering albedo ω on temperature profile and radiative heat flux. (From M. A. Boles and M. N. Özişik [19].)

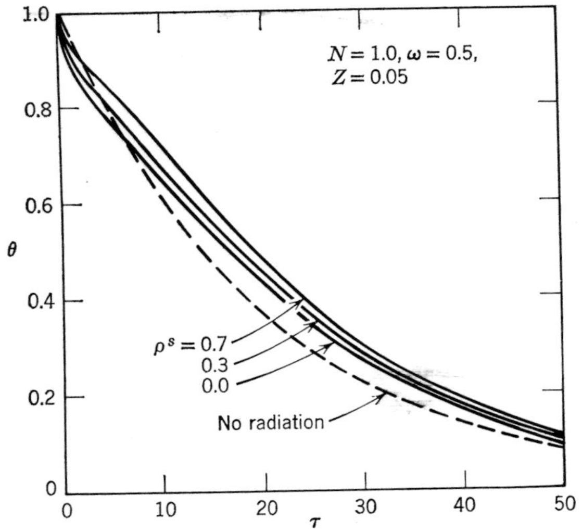

Fig. 12-10. Effect of reflectivity ρ^s on temperature profile. (From M. A. Boles and M. N. Özişik [19].)

Figure 12-10 shows the effects of the reflectivity of the air-liquid interface on the temperature distribution. Increasing the reflectivity decreases the radiative energy flow from the medium into the external surrounding through the interface and hence raises the temperature in the medium above that for $\rho^s = 0$, as shown in this figure.

12-6 EXPERIMENTAL WORK

A wealth of theoretical work has been reported in the literature on the prediction of temperature distribution and heat flux in a one-dimensional plane-parallel slab for combined conduction and radiation, but experimental verification of the theory is very limited because of the difficulties inherent in performing controlled experiments at high temperatures.

Nishimura, Hasatani, and Sugiyama [26] conducted experiments and compared the results with the calculated values. The agreement between the experiment and theory was fairly good for large values of N and τ_0 (i.e.,

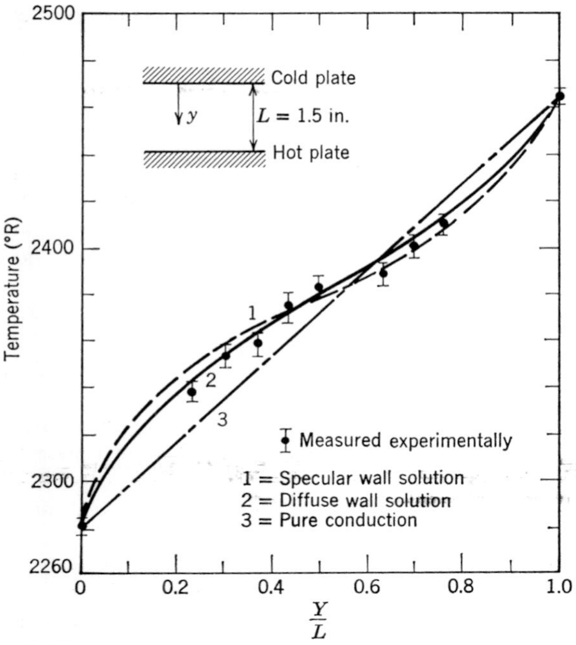

Fig. 12-11. A comparison of the measured and the predicted temperature distribution for combined conduction and radiation in a slab of molten glass. (From N. D. Eryou and L. R. Glicksman [28].)

$N \cong 0.5$, $\tau_0 = 15$ and $N \cong 0.9$, $\tau_0 = 25$) but was poor for small values (i.e., $N \cong 0.06$, $\tau_0 = 1.8$ and $N = 0.12$, $\tau_0 = 3$). However, some of the disagreement was attributed to the unsatisfactory temperature measurement when the contribution of radiation was great.

Finch, Noland, and Moeller [27] measured the steady-state heat flux through glass plates at temperatures up to 1000°F and compared the results with theoretical predictions. The agreement was not very good.

Eryou and Glicksman [28] measured the one-dimensional temperature profile and heat flux within a slab of molten glass contained between two parallel, platinum-lined ceramic plates. The temperatures of the plates were varied from a maximum of 2635°R to a minimum of 2240°R, and an optical method was used to measure temperatures. The experimental data were compared with the calculated results for both specularly and diffusely reflecting boundary conditions. In these calculations the frequency dependence of radiation properties was treated as stepwise variation. Figure 12-11 shows a comparison of the measured temperature profile with the values predicted for diffuse and specular wall conditions. The experimental results appear to fall between the specular and diffuse predictions. The heat flux data, however, were found to agree better with the diffuse model.

REFERENCES

1. R. Viskanta and R. J. Grosh, "Heat Transfer by Simultaneous Conduction and Radiation in an Absorbing Medium," *J. Heat Transfer*, **84C**, 63–72, 1963.
2. R. Viskanta and R. J. Grosh, "Effect of Surface Emissivity on Heat Transfer by Simultaneous Conduction and Radiation," *Intern. J. Heat Mass Transfer*, **5**, 729–734, 1962.
3. R. Viskanta, "Heat Transfer by Conduction and Radiation in Absorbing and Scattering Materials," *J. Heat Transfer*, **87C**, 143–150, 1965.
4. R. Viskanta and R. J. Grosh, "Recent Advances in Radiant Heat Transfer," *Appl. Mech. Rev.*, **17**, 91–100, 1964.
5. W. Lick, "Energy Transfer by Radiation and Conduction," *Proceedings of the Heat Transfer and Fluid Mechanics Institute*, pp. 14–26, Stanford University Press, Palo Alto, Calif., 1963.
6. W. Lick, "Transient Energy Transfer by Radiation and Conduction," *Intern. J. Heat Mass Transfer*, **8**, 119–127, 1965.
7. R. Greif, "Energy Transfer by Radiation and Conduction with Variable Gas Properties," *Intern. J. Heat Mass Transfer*, **7**, 891–900, 1964.
8. L. S. Wang and C. L. Tien, "Study of the Interaction Between Radiation and Conduction by a Differential Method," *Proceedings of the Third International Heat Transfer Conference*, Chicago, Vol. 5, pp. 190–199, 1966.
9. L. S. Wang and C. L. Tien, "A Study of Various Limits in Radiation Heat-Transfer Problems," *Intern. J. Heat Mass Transfer*, **10**, 1327–1338, 1967.

10. Yan-Po Chang, "A Potential Treatment of Energy Transfer by Conduction, Radiation, and Convection," *AIAA J.*, **5**, 1024–1026, 1967.

11. Yan-Po Chang, "A Potential Treatment of Energy Transfer in a Conducting, Absorbing, and Emitting Medium," *ASME Paper* No. 67-WA/HT-40, 1967.

12. D. H. Timmons and J. O. Mingle, "Simultaneous Radiation and Conduction with Specular Reflection," IAAA 6th Aerospace Sciences Meeting, Paper No. 68–28, New York, January 1968.

13. R. E. Bellman and R. E. Kalaba, *Quasilinearization and Non-linear Boundary Value Problems*, American Elsevier Publishing Co., New York, 1965.

14. A. S. Hazzak and J. V. Beck, "Unsteady Combined Conduction-Radiation Energy Transfer Using a Rigorous Differential Method," *Intern. J. Heat Mass Transfer*, **13**, 517–522, 1970.

15a. C. C. Lii and M. N. Özişik, "Transient Radiation and Conduction in an Absorbing, Emitting, Scattering Slab with Reflective Boundaries," *Intern. J. Heat Mass Transfer*, **15**, 1175–1179, 1972.

15b. C. C. Lii and M. N. Özişik, unpublished work.

16. R. Viskanta and R. L. Merriam, "Heat Transfer by Combined Conduction and Radiation Between Concentric Spheres Separated by Radiating Medium," *J. Heat Transfer*, **90C**, 248–256, 1968.

17. L. A. Tarshis, S. O'Hara and R. Viskanta, "Heat Transfer by Simultaneous Conduction and Radiation for Two Absorbing Media in Intimate Contact," *Intern. J. Heat Mass Transfer*, **12**, 333–347, 1969.

18. Leo P. Kadanoff, "Radiative Transport Within an Ablating Body," *J. Heat Transfer*, **83C**, 215–225, 1961.

19. M. Boles and M. N. Özişik, "Simultaneous Ablation and Radiation in an Absorbing, Emitting, and Isotropically Scattering Medium," *J. Quant. Spectry. Radiative Transfer*, **12**, 839–847, 1972.

20. R. D. Cess, "The Interaction of Thermal Radiation in Boundary Layer Heat Transfer," *Proceedings of the Third International Heat Transfer Conference*, Chicago, Vol. 5, pp. 129–137, 1966.

21. M. N. Özişik, *Boundary Value Problems of Heat Conduction*, International Textbook Co., Scranton, Penn., 1968.

22. R. Viskanta and R. J. Grosh, "Temperature Distribution in Couette Flow with Radiation," *Am. Rocket Soc. J.*, **31**, 839–840, 1961.

23. Mac C. Adams, W. E. Powers, and S. Georgiev, "An Experimental and Theoretical Study of Quartz Ablation at the Stagnation Point," *AVCO-Everett Res. Lab. Res. Rept*. No. 57, June 1959.

24. H. Hidalgo, "The Ablation of Glassy Materials for Laminar and Turbulent Heating on Blunted Bodies of Revolution," *AVCO-Everett Res. Lab. Res. Rept*. No. 62, June 1959.

25. H. A. Bethe and Mac C. Adams, "A Theory for the Ablation of Glassy Materials," *J. Aerospace Sci*., **26**, 321–350, 1959.

26. M. Nishimura, M. Hasatani and S. Sugiyama, "Simultaneous Heat Transfer by Radiation and Conduction: High-Temperature One-Dimensional Heat Transfer in Molten Glass," *Intern. Chem. Eng*., **8**, 739–745, 1968.

27. H. Finch, M. Noland, and C. Moeller, "Experimental Verification of the Analyses and Computer Programs Concerning Heat Transfer through Semi-transparent Materials," *Tech. Rept. AFFSL TR*-65-136, July 1965.

28. N. D. Eryou and L. R. Glicksman, "An Experimental and Analytical Study of Radiative and Conductive Heat Transfer in Molten Glass," *ASME Paper* No. 70-WA/HT-10, January 1971.

NOTES

[1] From Eq. 8-73 we have

$$G(\tau) = 2\pi[I^+(0)E_2(\tau) + I^-(\tau_0)E_2(\tau_0 - \tau) + \int_0^{\tau_0} S(\tau')E_1(|\tau - \tau'|)\,d\tau'] \tag{1a}$$

where $S(\tau)$ for isotropic scattering is defined as

$$S(\tau) \equiv (1 - \omega)\frac{n^2\bar{\sigma}T^4(\tau)}{\pi} + \frac{\omega}{2}\int_{-1}^{1} I(\tau, \mu')\,d\mu' = (1 - \omega)\frac{n^2\bar{\sigma}T^4(\tau)}{\pi} + \frac{\omega}{4\pi}G(\tau) \tag{1b}$$

We define the dimensionless quantities as

$$G^*(\tau) \equiv \frac{G(\tau)}{4n^2\bar{\sigma}T_1^{\,4}}, \qquad \psi^+(0) \equiv \frac{I^+(0)}{n^2\bar{\sigma}T_1^{\,4}/\pi}, \qquad \psi^-(\tau_0) \equiv \frac{I^-(\tau_0)}{n^2\bar{\sigma}T_1^{\,4}/\pi} \tag{2}$$

Then Eqs. 1 yield

$$G^*(\tau) = \tfrac{1}{2}\left\{\psi^+(0)E_2(\tau) + \psi^-(\tau_0)E_2(\tau_0 - \tau) + \int_0^{\tau_0} [(1 - \omega)\theta^4(\tau') + \omega G^*(\tau')] \cdot E_1(|\tau - \tau'|)\,d\tau'\right\} \tag{3}$$

[2] From Eq. 8-83 we have

$$q^r(\tau) = 2\pi\left[I^+(0)E_3(\tau) - I^-(\tau_0)E_3(\tau_0 - \tau) + \int_0^{\tau} S(\tau')E_2(\tau - \tau')\,d\tau' - \int_{\tau}^{\tau_0} S(\tau')E_2(\tau' - \tau)\,d\tau'\right] \tag{1a}$$

where

$$S(\tau) = (1 - \omega)\frac{n^2\bar{\sigma}T^4(\tau)}{\pi} + \frac{\omega}{4\pi}G(\tau) \tag{1b}$$

By dividing both sides of these equations by $4n^2\bar{\sigma}T_1^{\,4}$, we obtain the result given by Eq. 12-50.

[3] The transformation of various partial derivatives from the x, t variables to the ξ, t variables is given as:

$$\frac{\partial \xi}{\partial x} = 1, \qquad \frac{\partial \xi}{\partial t} = -\frac{dS}{dt} \equiv -v \tag{1}$$

$$\frac{\partial}{\partial x} = \frac{\partial}{\partial \xi}\cancelto{1}{\frac{\partial \xi}{\partial x}} + \frac{\partial}{\partial t}\cancelto{0}{\frac{dt}{dx}} = \frac{\partial}{\partial \xi} \tag{2}$$

$$\frac{\partial}{\partial t} = \frac{\partial}{\partial \xi}\frac{\partial \xi}{\partial t} + \frac{\partial}{\partial t}\frac{dt}{dt} = -v\frac{\partial}{\partial \xi} + \frac{\partial}{\partial t} \tag{3}$$

CHAPTER 13

Boundary Layer Flow with Radiation in a Participating Medium

In this chapter consideration will be given to the effects of thermal radiation on heat transfer and temperature distribution in the boundary layer flow of a participating fluid. At high temperatures the presence of thermal radiation alters the distribution of temperature in the boundary layer, which in turn affects the heat transfer at the wall. In such situations a simultaneous treatment of the convective and the radiative heat transfer is necessary. Goulard and Goulard [1] were among the early investigators who studied the coupling between convective and radiative heat transfer for one-dimensional Couette flow. It was shown that calculation based on the assumption of no interaction between convection and radiation overestimated the heat transfer.

A general treatment of heat transfer by combined convection and radiation in the boundary layer flow of a participating fluid leads to a set of partial differential and integrodifferential equations which must be solved simultaneously. The mathematical difficulties involved in handling the resulting system of complex equations have prompted various investigators to seek approximate methods of solving the radiation part of the problem. The optically thick limit approximation has been used by several investigators because this makes it possible to treat the problem by the conventional similarity approach. The optically thin limit and the exponential kernel approximations also result in significant simplifications in the analysis.

Viskanta and Grosh [2], Novotny and Yang [3], Rumynskii [4], Goulard [5], and Kim, Özişik, and Mulligan [6] applied the optically thick limit approximation to study the interaction of radiation with the boundary layer flow of a gray gas. Koh and De Silva [7], Cess [8], Tabaczynski and Kennedy [9], Cess [10], and Smith and Hassan [11] used the optically thin limit approximation. Pai and Tsao [12], Pai and Scaglione [13], and Oliver and McFadden [14] utilized, respectively, the exponential kernel approximation, optically thick limit approximation, and an iterative type of solution. Boles and Özişik [15a] and Boles [15b] applied the normal-mode expansion technique to treat exactly the radiation part of the problem and included the effects of scattering. Pai and Hsieh [16] and Hsieh and Pai [17] applied the perturbation theory to analyze the equilibrium and nonequilibrium radiative heat transfer, respectively, in supersonic steady flow.

Anderson [18], Olstad [19, 20], Anderson [21], Dirling, Rigdon, and Thomas [22], and Kim, Özişik and Mulligan [23] studied the effects of thermal radiation on the stagnation flow heat transfer. Elliott et al. [24] solved numerically the problem of a turbulent, optically thin boundary layer.

Cess [25] and Arpaci [26] utilized the singular perturbation technique and an approximate integral method, respectively, to investigate the effects of radiation on heat transfer in laminar free convection from a vertical plate, and Cheng and Özişik [27] used the normal-mode expansion technique for the exact treatment of this problem and included the scattering effects.

In the following sections we present the fundamental equations of fluid dynamics and energy for a radiating gas, describe the boundary layer simplifications, illustrate with representative examples the mathematical formulation of simultaneous convection and radiation in boundary layer flow, and discuss the methods of solution and the results. Since in the discussion of radiative heat transfer our attention will be focused more on the general solution of boundary layer equations, we present the boundary layer simplifications and similar solutions only for the two-dimensional steady boundary layer with radiative heat transfer. However, the transformed two-dimensional boundary layer equations are presented in general form so that several special cases are readily obtainable from them. For simplicity in the analysis only gray problems and laminar flow conditions are considered. The extension to nongray cases would require merely a nongray treatment of the radiation part of the problem.

13-1 EQUATIONS OF FLUID DYNAMICS AND ENERGY FOR A RADIATING FLUID

The equations of fluid dynamics and energy for a nonradiating fluid are given in the standard textbooks on the subject, for example, Schlichting [28].

Similar equations for a radiating gas are readily obtainable from those for the nonradiating case by introducing appropriate terms to account for the effects of radiation. Goulard [29], Viskanta [30], Pai [31], Vincenti and Kruger [32], and Pai [33] have given such equations. In this chapter we present the equations of continuity, momentum, and energy for a radiating gas.

Continuity Equation

The standard equation of continuity for a nonradiating fluid is applicable also to a radiating fluid, since change of mass due to radiation is negligible. Then the equation of continuity is given in the indicial notation as

$$\frac{\partial \rho}{\partial t} + \frac{\partial (\rho u_i)}{\partial x_i} = 0 \tag{13-1}$$

For an incompressible fluid this equation simplifies to

$$\frac{\partial u_i}{\partial x_i} = 0 \tag{13-2}$$

Momentum Equations

When radiation is present, it exerts pressure and hence introduces an additional rate of change of momentum into the system. In this case, analogous to that of the stress tensor in fluid dynamics, one considers that stresses are exerted across three mutually perpendicular planes and that each stress is characterized by the three components; then a radiation stress tensor is defined as [34]

$$p_{ij}^{\ r} \equiv \begin{vmatrix} p_{11}^{\ r} & p_{12}^{\ r} & p_{13}^{\ r} \\ p_{21}^{\ r} & p_{22}^{\ r} & p_{23}^{\ r} \\ p_{31}^{\ r} & p_{32}^{\ r} & p_{33}^{\ r} \end{vmatrix} = \frac{1}{c} \int_{4\pi} n_i n_j I \, d\Omega, \quad i, j = 1, 2, 3 \tag{13-3a}$$

where I is the total radiation intensity, that is,

$$I = \int_{\nu=0}^{\infty} I_\nu(\mathbf{r}, \hat{\mathbf{\Omega}}) \, d\nu \tag{13-3b}$$

c is the speed of propagation of radiation, and n_i and n_j are the direction cosines. The radiation stress tensor is symmetrical since $p_{ij}^{\ r} = p_{ji}^{\ r}$. Here we note that each diagonal term in Eq. 13-3 is similar to the normal radiation pressure considered in Eq. 1-83, that is,

$$p^r(\mathbf{r}) = \frac{1}{c} \int_{\nu=0}^{\infty} \left[\int_{\phi=0}^{2\pi} \int_{\mu=-1}^{1} I_\nu(\mathbf{r}, \mu, \phi) \mu^2 \, d\mu \, d\phi \right] d\nu \tag{13-4}$$

The arithmetic mean of the three diagonal terms is related to the radiation energy density u^r by

$$\tfrac{1}{3}(p_{11}{}^r + p_{22}{}^r + p_{33}{}^r) = \frac{1}{3c}\int_{4\pi} n_i n_i I \, d\Omega = \frac{1}{3c}\int_{4\pi} I \, d\Omega = \frac{u^r}{3} \tag{13-5}$$

since the sum of the squares of the direction cosines n_i is unity.[1]

In the case of an isotropic field of radiation (i.e., the radiation intensity I is independent of direction), the radiation stess tensor (Eq. 13-3) simplifies to[2]

$$p_{ij}{}^r = \frac{1}{c} I \int_{4\pi} n_i n_j \, d\Omega = \frac{4\pi}{3c} I \, \delta_{ij} = \tfrac{1}{3} u^r \, \delta_{ij} = \begin{vmatrix} \frac{u^r}{3} & 0 & 0 \\ 0 & \frac{u^r}{3} & 0 \\ 0 & 0 & \frac{u^r}{3} \end{vmatrix} \tag{13-6}$$

The foregoing definitions of radiation stress tensor are analogous to those of the radiation pressure tensor in fluid mechanics. To complete the analogy, the radiation stress tensor may be expressed in two parts in the form

$$p_{ij}{}^r = -p^r \delta_{ij} + \tau_{ij}{}^r \tag{13-7}$$

where p^r is the mean radiation pressure for an isotropic field of radiation and $\tau_{ij}{}^r$ is the nonisotropic part. For a black-body radiation field in a gas at temperature T, the mean radiation pressure p^r is given as

$$p^r = \frac{4\pi}{3c} \bar{\sigma} T^4 \tag{13-8}$$

According to this relation the radiation pressure is very small unless the temperature is of the order of 10^5 °K or higher.

Viscosity also arises in a radiation field. Jeans [35] gave the radiative viscosity μ^r as

$$\mu^r = \frac{4\bar{\sigma}T^4}{3c^2\kappa} \tag{13-9}$$

Later analysis by Hazlehurst and Sargent [36] showed that the radiation viscosity is twice that given by Eq. 13-9; it is apparent from the latter equation that the radiation viscosity is negligible unless the temperature is extremely high.

The ratio of the radiation pressure p^r as given by Eq. 13-8 to the gas pressure p is called the *radiation pressure number* R^p, defined as

$$R^p = \frac{\text{radiation pressure}}{\text{gas pressure}} = \frac{4\pi\bar{\sigma}T^4}{3cp} = \text{radiation pressure number} \tag{13-10}$$

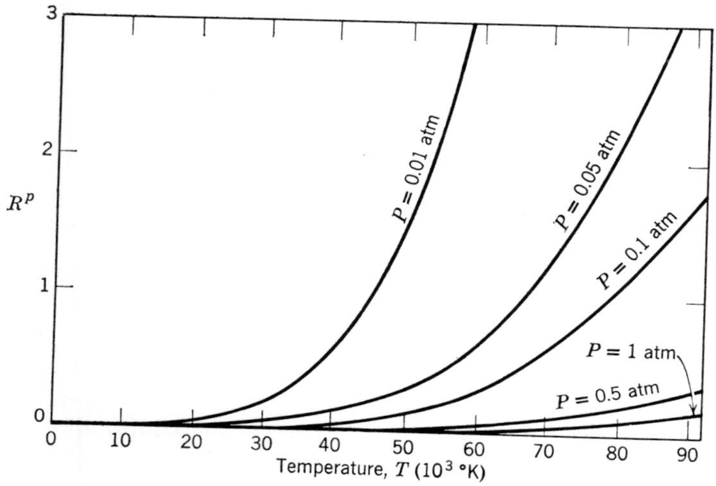

Fig. 13-1. Variation of radiation pressure number R^P with pressure and temperature. (From S. I. Pai [33].)

This number is useful to characterize the importance of radiation stresses in the momentum equation. Figure 13-1 shows some typical variation of R^p with temperature and pressure; R^p is very small unless the temperature is very high or the gas pressure is very small. Therefore at the temperatures and pressure encountered in most engineering applications R^p is very small and the radiation effects are negligible. Then the momentum equation for a radiating gas is taken to be the same as that for a nonradiating gas, and in the indicial notation it is given as[3]

$$\rho\left(\frac{\partial u_i}{\partial t} + u_j \frac{\partial u_i}{\partial x_j}\right) = F_i - \frac{\partial p}{\partial x_i} + \frac{\partial \tau_{ij}}{\partial x_j} \tag{13-11a}$$

where u_i and u_j are the velocity components, F_i is the body force per unit volume, p is the hydrodynamic pressure, and τ_{ij} is the viscous stress tensor, given as

$$\tau_{ij} = \mu\left(\frac{\partial u_i}{\partial x_j} + \frac{\partial u_j}{\partial x_i} - \frac{2}{3}\frac{\partial u_k}{\partial x_k}\delta_{ij}\right), \quad i, j, k = 1, 2, 3 \tag{13-11b}$$

For constant ρ (i.e., incompressible fluid) Eqs. 13-11 can be simplified by means of the continuity equation $\partial u_k/\partial x_k = 0$ to yield

$$\rho\left(\frac{\partial u_i}{\partial t} + u_j \frac{\partial u_i}{\partial x_j}\right) = F_i - \frac{\partial p}{\partial x_i} + \frac{\partial}{\partial x_j}\left[\mu\left(\frac{\partial u_i}{\partial x_j} + \frac{\partial u_j}{\partial x_i}\right)\right], \quad i, j = 1, 2, 3 \tag{13-12a}$$

For constant ρ and constant μ Eqs. 13-11 becomes

$$\rho\left(\frac{\partial u_i}{\partial t} + u_j \frac{\partial u_i}{\partial x_j}\right) = F_i - \frac{\partial p}{\partial x_i} + \mu \frac{\partial^2 u_i}{\partial x_j \, \partial x_j}, \quad i, j = 1, 2, 3 \quad \text{(13-12b)}$$

since $\partial u_j / \partial x_j = 0$ by the continuity equation.

Energy Equation

In the presence of radiation the energy equation involves additional terms to account for the effects of the radiation energy density u^r, the radiation stress tensor $p_{ij}^{\ r}$, and the radiative heat flux vector $q_i^{\ r}$. These quantities enter the energy equation in a manner analogous to the way in which the internal energy of the fluid, the hydrodynamic stress tensor, and the conductive heat flux vector enter the energy equation. However, the contribution of the radiation energy density and the radiation stress tensor in the energy equation is negligible when the radiation pressure number R^p is negligible. Therefore for most engineering applications only the contribution of the radiative heat flux is important, and the energy equation for a radiating fluid may be given as[4]

$$\rho \frac{Dh}{Dt} = -\frac{\partial}{\partial x_i}(q_i^{\ c} + q_i^{\ r}) + S + \frac{Dp}{Dt} + \mu\Phi \qquad \text{(13-13)}$$

where $q_i^{\ c}$ and $q_i^{\ r}$ are the conductive and radiative heat fluxes, respectively, h is the enthalpy of the fluid per unit mass, S is the external energy input per unit volume and per unit time, Φ is the viscous energy dissipation function, and D/Dt is the substantial derivative, defined as

$$\frac{D}{Dt} \equiv \frac{\partial}{\partial t} + u_i \frac{\partial}{\partial x_i} \qquad \text{(13-14)}$$

In Eq. 13-13 the first term on the right-hand side represents the energy transfer by conduction and radiation, the second term is the energy generation within the volume (if there is any), the third term is the work done on the fluid by the pressure forces (i.e., the work of compression), and the last term is the energy dissipation by viscous stresses.

For a perfect gas the energy equation 13-13 may be written as

$$\rho c_p \frac{DT}{Dt} = -\frac{\partial}{\partial x_i}(q_i^{\ c} + q_i^{\ r}) + S + \frac{Dp}{Dt} + \mu\Phi \qquad \text{(13-15)}$$

where the conductive heat flux vector $q_i^{\ c}$ is given by the Fourier law as

$$q_i^{\ c} = -k \frac{\partial T}{\partial x_i} = -k \nabla T \qquad \text{(13-16)}$$

The viscous energy dissipation function Φ depends on the nature of the fluid; for a Newtonian fluid it is given as [28, p. 254]

$$\Phi = 2\left[\left(\frac{\partial u_1}{\partial x_1}\right)^2 + \left(\frac{\partial u_2}{\partial x_2}\right)^2 + \left(\frac{\partial u_3}{\partial x_3}\right)^2\right] + \left(\frac{\partial u_2}{\partial x_1} + \frac{\partial u_1}{\partial x_2}\right)^2$$
$$+ \left(\frac{\partial u_3}{\partial x_2} + \frac{\partial u_2}{\partial x_3}\right)^2 + \left(\frac{\partial u_1}{\partial x_3} + \frac{\partial u_3}{\partial x_1}\right)^2 - \frac{2}{3}\left(\frac{\partial u_1}{\partial x_1} + \frac{\partial u_2}{\partial x_2} + \frac{\partial u_3}{\partial x_3}\right)^2 \tag{13-17}$$

For an incompressible fluid the work of compression vanishes and the energy equation 13-15 simplifies to

$$\rho c_p \frac{DT}{Dt} = -\frac{\partial}{\partial x_i}(q_i^{\,c} + q_i^{\,r}) + S + \mu\Phi \tag{13-18}$$

Boundary Layer Simplifications

The foregoing continuity, momentum, and energy equations for a radiating fluid are similar to those for a nonradiating fluid except for the radiative heat flux term appearing in the energy equation. For a boundary layer type of flow these equations are simplified by utilizing an order-of-magnitude analysis described by Schlichting [28]; the resulting equations are immediately obtainable from those for a nonradiating fluid by proper cognizance of the radiative heat flux term in the energy equation.

We focus our attention on the steady, two-dimensional boundary layer flow of a radiating fluid over a body. The x and y coordinates are chosen, respectively, along the surface of the body in the direction of flow and perpendicular to the surface. Then the continuity, the momentum, and the energy equations with the boundary layer simplifications are given as

$$\frac{\partial}{\partial x}(\rho u) + \frac{\partial}{\partial y}(\rho v) = 0 \tag{13-19}$$

$$\rho\left(u\frac{\partial u}{\partial x} + v\frac{\partial u}{\partial y}\right) = -\frac{dp}{dx} + \frac{\partial}{\partial y}\left(\mu\frac{\partial u}{\partial y}\right) \tag{13-20}$$

$$\rho c_p\left(u\frac{\partial T}{\partial x} + v\frac{\partial T}{\partial y}\right) = \frac{\partial}{\partial x}\left(k\frac{\partial T}{\partial x}\right) + \frac{\partial}{\partial y}\left(k\frac{\partial T}{\partial y}\right)$$
$$- \frac{\partial q_x^{\,r}}{\partial x} - \frac{\partial q_y^{\,r}}{\partial y} + u\frac{dp}{dx} + \mu\left(\frac{\partial u}{\partial y}\right)^2 \tag{13-21}$$

where u and v are the x- and y-direction velocity components and $q_x^{\,r}$ and $q_y^{\,r}$ are the x- and y-direction radiative heat flux components, respectively. For generality, we have included in the energy equation the x-direction conductive and radiative heat transfer; in most applications, however, they are neglected. The equation of state for a perfect gas is given by

$$p = \rho RT \tag{13-22}$$

The thermal conductivity and the viscosity are prescribed as a function of the temperature

$$\mu = \mu(T) \qquad \text{and} \qquad k = k(T) \tag{13-23}$$

The y-direction momentum equation is not included in the above equations because, with the boundary layer simplifications, it reduces to $dp/dy = 0$ at any point x along the flow. This result implies that the pressure remains constant across the boundary layer in the y direction. By utilizing the constancy of the pressure across the boundary layer, Eq. 13-22 gives

$$\rho(x, y)T(x, y) = \rho_\infty(x)T_\infty(x) \tag{13-24}$$

where subscript ∞ refers to the external flow conditions. By evaluating the x-direction momentum equation 13-20 outside the boundary layer, the x-direction pressure gradient dp/dx is given as

$$\frac{dp}{dx} = -\rho_\infty(x)u_\infty(x)\frac{du_\infty(x)}{dx} \tag{13-25}$$

where the external flow velocity $u_\infty(x)$ is determined from the solution of the potential flow problem.

13-2 NONDIMENSIONAL RADIATION PARAMETERS

The nondimensional radiation parameters are useful to characterize the relative importance of energy transfer by radiation, as compared with energy transfer by convection and conduction. A discussion of these parameters is given by Viskanta [30a, pp. 213–219], Pai [31, Chapter 7], and Goulard [37]. In this section we present a brief description of the nondimensional radiation parameters.

Consider the steady, two-dimensional boundary layer energy equation, given in the form

$$\rho c_p\left(u\frac{\partial T}{\partial x} + v\frac{\partial T}{\partial y}\right) = \frac{\partial}{\partial y}\left(k\frac{\partial T}{\partial y}\right) + \frac{\partial}{\partial x}\left(k\frac{\partial T}{\partial x}\right) - \frac{\partial q_y^r}{\partial y} - \frac{\partial q_x^r}{\partial x} + u\frac{dp}{dx} + \mu\left(\frac{\partial u}{\partial y}\right)^2 \tag{13-26}$$

We define the following dimensionless quantities:

$$X = \frac{x}{L}, \qquad Y = \frac{y}{L}\,\mathrm{Re}^{1/2}, \qquad \mathrm{Re} = \frac{\rho_0 u_0 L}{\mu_0} \tag{13-27a}$$

$$U = \frac{u}{u_0}, \qquad V = \frac{v}{u_0}\,\mathrm{Re}^{1/2}, \qquad P = \frac{p}{\rho_0 {u_0}^2} \tag{13-27b}$$

$$\theta = \frac{T}{T_0}, \qquad Q_x^{\,r} = \frac{q_x^{\,r}}{4{n_0}^2\bar{\sigma}{T_0}^4}, \qquad Q_y^{\,r} = \frac{q_y^{\,r}}{4{n_0}^2\bar{\sigma}{T_0}^4} \tag{13-27c}$$

$$C = \frac{c_p}{c_{p0}}, \qquad K = \frac{k}{k_0}, \qquad \rho^* = \frac{\rho}{\rho_0}, \qquad \mu^* = \frac{\mu}{\mu_0} \tag{13-27d}$$

where L is a reference length, and subscript 0 refers to a reference condition.

Substitution of these dimensionless quantities into the energy equation 13-26 yields

$$\rho^* C\left(U\frac{\partial\theta}{\partial X} + V\frac{\partial\theta}{\partial Y}\right) = \frac{1}{\mathrm{Pr}}\frac{\partial}{\partial Y}\left(K\frac{\partial\theta}{\partial Y}\right) + \frac{1}{\mathrm{Pe}}\frac{\partial}{\partial X}\left(K\frac{\partial\theta}{\partial X}\right) - 4\frac{\mathrm{Re}^{1/2}}{\mathrm{Bo}}\frac{\partial Q_y^{\,r}}{\partial Y} - \frac{4}{\mathrm{Bo}}\frac{\partial Q_x^{\,r}}{\partial X} + \mathrm{E}U\frac{dP}{dX} + \mathrm{E}\mu^*\left(\frac{\partial U}{\partial Y}\right)^2 \tag{13-28}$$

where we have defined

$$\mathrm{Bo} = \frac{\rho_0 u_0 c_{p0}}{{n_0}^2\bar{\sigma}{T_0}^3} = \text{Boltzmann number} \tag{13-29a}$$

$$\mathrm{E} = \frac{{u_0}^2}{c_{p0}T_0} = \text{Eckert number} \tag{13-29b}$$

$$\mathrm{Pe} = \mathrm{Re}\,\mathrm{Pr} = \frac{\rho_0 u_0 c_{p0} L}{k_0} = \text{Peclet number} \tag{13-29c}$$

$$\mathrm{Pr} = \frac{c_{p0}\mu_0}{k_0} = \text{Prandtl number} \tag{13-29e}$$

$$\mathrm{Re} = \frac{\rho_0 u_0 L}{\mu_0} = \text{Reynolds number} \tag{13-29f}$$

The left-hand side of Eq. 13-28 represents the energy flow by convection; the first and the second terms on the right-hand side are the Y- and X-direction conductive energy transfer, the third and the fourth terms are the Y- and X-direction radiative energy transfer, the fifth term is the work of compression, and the last term is the viscous energy dissipation.

The dimensionless parameters E, Pe, Pr, and Re are well known. The Boltzmann number Bo characterizes the importance of energy transfer by axial convection in relation to that by radiation; the physical significance of the Boltzmann number is better envisioned if it is rearranged in the form

$$\text{Bo} = \frac{\rho_0 u_0 c_{p0} T_0}{n_0^2 \bar{\sigma} T_0^4} \tag{13-30a}$$

The role of the Boltzmann number in radiation is analogous to that of the Peclet number in conduction. When the Peclet number is large, the conductive energy transfer in the X direction is neglected; when the Boltzmann number is large, the radiative transfer in the X direction is neglected.

The Boltzmann number is related to the conduction-to-radiation parameter N, the Reynolds number Re, and the Prandtl number Pr in the form

$$\text{Bo} = 4\left(\frac{k_0 \beta_0}{4 n_0^2 \bar{\sigma} T_0^3}\right)\left(\frac{\rho_0 U_0 L}{\mu_0}\right)\left(\frac{C_{p0}\mu_0}{k_0}\right)\frac{1}{\beta_0 L} = 4N \text{ Re Pr} \frac{1}{\tau_0} \tag{13-30b}$$

where L is a characteristic length, and τ_0 is the corresponding optical thickness of the system. The physical significance of τ_0 is as follows:

$$\tau_0 = \frac{L}{1/\beta_0} = \frac{\text{characteristic length of the system}}{\text{mean free path (or mean penetration length) of radiation photon}} \tag{13-30c}$$

The Boltzmann number as defined by Eq. 13-30a is a ratio of the convective and radiative heat fluxes from the system. However, the characteristic optical thickness τ_0 plays a role in the importance of the radiative heat flux relative to the convective heat flux for the optically thick ($\tau_0 \gg 1$) and optically thin ($\tau_0 \ll 1$) cases. The radiative heat flux from the system for the optically thick and optically thin cases respectively, may be characterized, by $16 n_0^2 \bar{\sigma} T_0^4 / 3\tau_0$ and $2\tau_0 n_0^2 \bar{\sigma} T_0^4$; the convective heat flux from the system may be taken as $\rho_0 u_0 C_{p0} T_0$. With these considerations the ratio of the convective to the radiative heat flux for the optically thick and optically thin cases, respectively, may be characterized by

$$\left(\frac{\text{Convective flux}}{\text{Radiative flux}}\right)_{\substack{\text{opt.}\\\text{thick}}} = \frac{\rho_0 U_0 C_{p0} T_0}{16 n_0^2 \bar{\sigma} T_0^4 / 3\tau_0} = \tfrac{3}{16}\tau_0 \text{ Bo} \quad \text{for } \tau_0 \gg 1 \tag{13-31a}$$

$$\left(\frac{\text{Convective flux}}{\text{Radiative flux}}\right)_{\substack{\text{opt.}\\\text{thin}}} = \frac{\rho_0 U_0 C_{p0} T_0}{2\tau_0 n_0^2 \bar{\sigma} T_0^4} = \frac{\text{Bo}}{2\tau_0} \quad \text{for } \tau_0 \ll 1 \tag{13-31b}$$

Therefore the dimensionless parameters τ_0 Bo and Bo/$2\tau_0$ are useful to characterize the importance of energy transfer by convection in relation to that by radiation for the optically thick and optically thin cases, respectively.

We note from Eqs. 13-31a and 13-31b that the radiative heat flux, compared with the convective heat flux from the medium, tends to be negligible without requiring the Boltzmann number to be large, since $\tau_0 \gg 1$ for the optically thick case and $\tau_0 \ll 1$ for the optically thin case.

Optical Thickness of Thermal Boundary Layer

The presence of radiation increases the optical thickness of the thermal boundary layer. The dimensionless parameter characterizing the optical thickness of the thermal boundary layer is useful in heat transfer analysis. For the laminar boundary layer flow over a flat plate, for example, if the Prandtl number is assumed to be of the order of unity, the thickness δ_t of the thermal boundary layer is given by

$$\delta_t \cong 5\sqrt{\frac{\nu x}{u_\infty}} \tag{13-32}$$

which can be rearranged in the form

$$\tau_\delta \equiv \beta\delta_t = 5\beta\sqrt{\frac{\nu x}{u_\infty}} = 5\sqrt{\text{Pr}\, N\xi} \tag{13-33}$$

where β is the extinction coefficient, and various dimensionless quantities are defined as

$$N \equiv \frac{k\beta}{4n^2\bar{\sigma}T^3}, \qquad \text{Pr} \equiv \frac{c_p\mu}{k}, \qquad \text{and} \qquad \xi \equiv \frac{4n^2\bar{\sigma}T^3\beta x}{\rho c_p u_\infty} = \frac{4\beta x}{\text{Bo}} \tag{13-34}$$

Here the parameter $\sqrt{N\xi}$ characterizes the optical thickness of the thermal boundary layer.

The boundary layer is said to be optically thick when $\sqrt{N\xi} \gg 1$ and optically thin when $\sqrt{N\xi} \ll 1$. The magnitude of the parameters N and ξ is important in the estimation of the optical thickness of the boundary layer.

The values of the parameter N for gases such as carbon dioxide, water vapor, and ammonia at 1 atm pressure are given in Fig. 12-2 as a function of the temperature. To estimate the order of magnitude of the parameter ξ it is convenient to rearrange this term in the form

$$\xi = \frac{4\bar{\sigma}T^3\kappa x}{\rho c_p u_\infty} \equiv K\,\text{Re}_x \frac{1}{\text{E}} \tag{13-35}$$

where we have defined

$$K \equiv \frac{4\bar{\sigma}T^2\mu\kappa}{\rho^2 c_p^{\,2}} \tag{13-36a}$$

$$\text{Re} \equiv \frac{\rho u_\infty x}{\mu} \qquad \text{and} \qquad \text{E} \equiv \frac{u_\infty^{\,2}}{c_p T} \tag{13-36b}$$

replaced β by κ, since the scattering of thermal radiation by the gas molecules is negligible unless there are scattering particulates in the medium, and set the index of refraction n equal to unity. Figure 13-2 shows the value of the parameter K for gaseous CO_2 at 1 atm pressure as a function of temperature.

If a laminar boundary layer is assumed, the Reynolds number does not exceed about 5×10^5. For an external flow velocity corresponding to a Mach

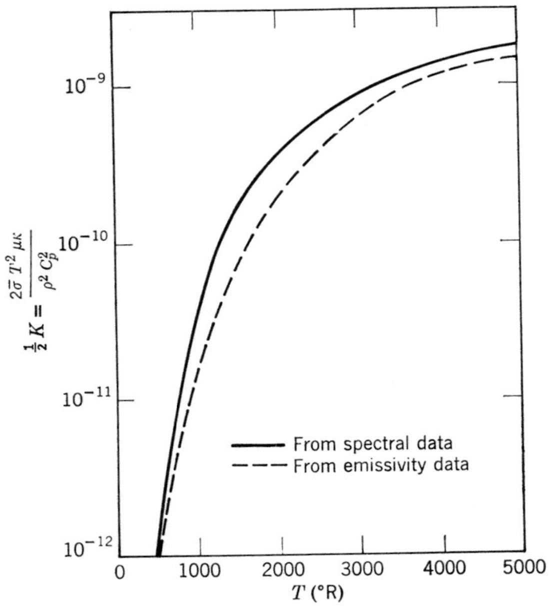

Fig. 13-2. Values of K for carbon dioxide at 1 atm. (From Y. Taitel and J. P. Hartnett [38].)

number no greater than 5, the Eckert number is about 7 or less. Then the thermal boundary layer for CO_2 at 1 atm pressure is optically thin, since the magnitude of ξ by Eq. 13-35 is of the order of 10^{-4} or less and N is much less than unity. However, the situation is different at low pressures; at reduced pressures the density decreases and K increases, which in turn thickens the boundary layer.

13-3 TRANSFORMATION OF THE BOUNDARY LAYER EQUATIONS

There are many situations in which the partial differential equations of laminar boundary layer flow can be transformed into a set of ordinary differential equations by means of a set of new variables called the *similarity variables*. Schlichting [28] gives a comprehensive treatment of the similarity transformation of boundary layer equations for nonradiating flow. Ames [39] describes the application of the one-parameter group theory developed by Morgan [40] to achieve a reduction in the number of independent variables in a system of partial differential equations. In this section we present the transformation of the steady, two-dimensional boundary layer equations for the laminar flow of a compressible, radiating fluid over a wedge. Then, the transformed equations for flow over a flat plate and for stagnation flow are immediately obtainable from the transformed equations for flow over a wedge.

Consider the continuity, momentum, and energy equations for laminar boundary layer flow, given in the form

$$\frac{\partial}{\partial x}(\rho u) + \frac{\partial}{\partial y}(\rho v) = 0 \tag{13-37}$$

$$\rho\left(u\frac{\partial u}{\partial x} + v\frac{\partial u}{\partial y}\right) = -\frac{dp(x)}{dx} + \frac{\partial}{\partial y}\left(\mu\frac{\partial u}{\partial y}\right) \tag{13-38}$$

$$\rho c_p\left(u\frac{\partial T}{\partial x} + v\frac{\partial T}{\partial y}\right) = \frac{\partial}{\partial y}\left(k\frac{\partial T}{\partial y}\right) - \frac{\partial q^r}{\partial y} + u\frac{dp(x)}{dx} + \mu\left(\frac{\partial u}{\partial y}\right)^2 \tag{13-39}$$

Assuming a perfect gas, we write the equation of state as

$$p(x) = \rho RT \tag{13-40}$$

The pressure gradient $dp(x)/dx$ is related to the external flow velocity $u_\infty(x)$ by

$$-\frac{dp(x)}{dx} = \rho_\infty(x)u_\infty(x)\frac{du_\infty(x)}{dx} \tag{13-41}$$

where the subscript ∞ refers to the condition in the external flow.

The thermal conductivity and viscosity of the fluid are assumed to be prescribed as a function of temperature.

Transformation of Equations

The foregoing continuity, momentum, and energy equations can be transformed so as to be more suitable for computational purposes by the application of the standard transformations used in the analysis of nonradiating boundary layer heat transfer.

We define a stream function $\psi(x, y)$ as

$$\frac{\partial \psi(x, y)}{\partial y} = \frac{\rho(x, y)}{\rho_0} u \tag{13-42}$$

$$\frac{\partial \psi(x, y)}{\partial x} = -\frac{\rho(x, y)}{\rho_0} v \tag{13-43}$$

and a new independent variable $Y(x, y)$ and a dimensionless temperature $\theta(x, y)$ as

$$Y(x, y) \equiv \int_0^y \frac{\rho(x, y')}{\rho_0} dy' \tag{13-44}$$

$$\theta(x, y) \equiv \frac{T(x, y)}{T_0} \tag{13-45}$$

where ρ_0 is a reference density evaluated at a reference temperature T_0.

Then the continuity equation 13-37 is identically satisfied. The momentum and the energy equations 13-38 and 13-39 are respectively transformed to

$$\frac{\partial \psi}{\partial Y}\frac{\partial^2 \psi}{\partial x\, \partial Y} - \frac{\partial \psi}{\partial x}\frac{\partial^2 \psi}{\partial Y^2} = \frac{\rho_\infty}{\rho} u_\infty \frac{du_\infty}{dx} + \frac{1}{\rho_0}\frac{\partial}{\partial Y}\left(\mu \frac{\rho}{\rho_0}\frac{\partial^2 \psi}{\partial Y^2}\right) \tag{13-46}$$

$$\frac{\partial \psi}{\partial Y}\frac{\partial \theta}{\partial x} - \frac{\partial \psi}{\partial x}\frac{\partial \theta}{\partial Y} = \frac{1}{\rho_0 c_p}\frac{\partial}{\partial Y}\left(k \frac{\rho}{\rho_0}\frac{\partial \theta}{\partial Y}\right) - \frac{1}{\rho_0 c_p T_0}\frac{\partial q^r}{\partial Y}$$
$$- \frac{\rho_\infty}{\rho c_p T_0} u_\infty \frac{du_\infty}{dx}\frac{\partial \psi}{\partial Y} + \frac{\rho \mu}{\rho_0^2 c_p T_0}\left(\frac{\partial^2 \psi}{\partial Y^2}\right)^2 \tag{13-47}$$

where

$$\rho_\infty \equiv \rho_\infty(x), \qquad u_\infty = u_\infty(x), \qquad \text{and} \qquad \rho \equiv \rho(x, Y)$$

Up to this point our analysis has been general and no restriction has been imposed on the external flow velocity $u_\infty(x)$. In the transformation of the boundary layer equations for a nonradiating flow it has been shown that the analysis is simplified significantly if the external flow velocity $u_\infty(x)$ is considered proportional to x^m or e^{mx}. Here we focus our attention on the external flow velocity $u_\infty(x)$, given in the form

$$u_\infty(x) = Cx^m \tag{13-48}$$

where the coefficient C and the exponent m are constants. An analysis of the external flow by the potential flow theory reveals that the velocity distribution given by Eq. 13-48 occurs in the neighborhood of the stagnation

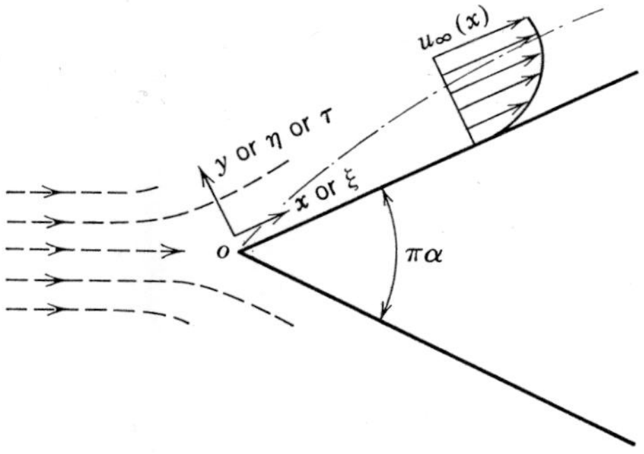

Fig. 13-3. Boundary layer flow over a wedge.

point of a wedge, as shown in Fig. 13-3. The exponent m is related to the wedge angle $\pi\alpha$ radians by [28, p. 140]

$$m = \frac{\alpha}{2-\alpha} \qquad \text{or} \qquad \alpha = \frac{2m}{m+1} \tag{13-49}$$

The special cases $\alpha = 0$ $(m = 0)$, $\alpha = \frac{1}{2}(m = \frac{1}{3})$, and $\alpha = 1 (m = 1)$ correspond, respectively, to flow over a flat plate, over a 90° wedge, and stagnation flow.

A new independent variable $\eta \equiv \eta(x, y)$ and a new dependent variable $f(\eta)$ are defined as

$$\eta \equiv Y\left[\frac{(m+1)u_\infty(x)}{j\nu_0 x}\right]^{1/2} = \left[\frac{(m+1)u_\infty(x)}{j\nu_0 x}\right]^{1/2}\int_0^y \frac{\rho(x, y')}{\rho_0}\,dy' \tag{13-50}$$

$$\psi(x, y) \equiv f(\eta)\left[\frac{j\nu_0 x u_\infty(x)}{m+1}\right]^{1/2} \tag{13-51}$$

Here the factor j, a nonvanishing positive constant, is included in these transformations for convenience later in the analysis. By assigning appropriate values to j, many cases considered in the literature are readily obtainable from the present results. For example, for flow over a wedge, by setting $j = m + 1$ we recover the transformation used by Kim, Özişik, and Mulligan [6], and by setting $j = 2$ we obtain the transformation used by Viskanta and Grosh [2]. In the case of flow over a flat plate (i.e., $m = 0$) by setting $j = 1$ we obtain the transformation used by Taitel and Hartnett [38].

By utilizing transformations 13-50 and 13-51 and the external flow velocity given by Eq. 13-48, the momentum and the energy equations 13-46 and 13-47 are respectively transformed to[5]

$$\frac{\partial}{\partial \eta}\left(\frac{\mu\rho}{\mu_0\rho_0}\frac{d^2 f}{d\eta^2}\right) + \frac{j}{2} f \frac{d^2 f}{d\eta^2} + \frac{jm}{1+m}\left[\frac{\rho_\infty(x)}{\rho} - \left(\frac{df}{d\eta}\right)^2\right] = 0 \tag{13-52}$$

$$\begin{aligned}\frac{1}{\mathrm{Pr}}\frac{\partial}{\partial \eta}\left(\frac{\kappa\rho}{k_0\rho_0}\frac{\partial\theta}{\partial\eta}\right) + \frac{j}{2} f \frac{\partial\theta}{\partial\eta} &= \frac{j}{1+m}\frac{df}{d\eta}\, x\frac{\partial\theta}{\partial x} + \left(\frac{j}{1+m}\right)^{1/2}\frac{4}{\mathrm{Bo}_x}\mathrm{Re}_x^{1/2}\frac{\partial Q^r}{\partial\eta}\\ &\quad + \frac{jm}{1+m}\frac{\rho_\infty(x)}{\rho}\mathrm{E}_\infty(x)\frac{df}{d\eta} - \frac{\mu\rho}{\mu_0\rho_0}\mathrm{E}_\infty(x)\left(\frac{d^2 f}{d\eta^2}\right)^2\end{aligned} \tag{13-53}$$

where various dimensionless quantities are defined as follows:

$$\mathrm{Bo}_x = \frac{c_p\rho_0 u_\infty(x)}{n_0^2\bar{\sigma}T_0^3} = \text{Boltzmann number (convection-to-radiation parameter)} \tag{13-54a}$$

$$\mathrm{E}_\infty(x) = \frac{u_\infty^2(x)}{c_p T_0} = \text{Eckert number} \tag{13-54b}$$

$$\mathrm{Pr} = \frac{c_p\mu_0}{k_0} = \text{Prandtl number} \tag{13-54c}$$

$$N = \frac{k_0\beta_0}{4n_0^2\bar{\sigma}T_0^3} = \text{conduction-to-radiation parameter} \tag{13-54d}$$

$$Q^r = \frac{q^r}{4n_0^2\bar{\sigma}T_0^4} = \text{dimensionless net radiative heat flux in the } y \text{ direction} \tag{13-54e}$$

$$\mathrm{Re}_x = \frac{x\rho_0 u_\infty(x)}{\mu_0} = \text{Reynolds number} \tag{13-54f}$$

Also, n_0 is the refractive index, and the subscript 0 refers to the conditions at a reference temperature T_0.

For a perfect gas the density ratio ρ/ρ_0 is related to the temperature by

$$\frac{\rho}{\rho_0} = \frac{T_0}{T} = \frac{1}{\theta} \tag{13-54g}$$

The velocity components u and v are given as[6]

$$u = u_\infty(x)\frac{df}{d\eta} \tag{13-55a}$$

$$v = -\frac{1}{2}\frac{\rho_0}{\rho}\sqrt{j\nu_0(1+m)\frac{1}{x}u_\infty(x)}\left(f + \frac{m-1}{m+1}\eta\frac{df}{d\eta}\right) \tag{13-55b}$$

Relation Between η and τ

In the above formulation the energy equation 13-53 involves the dimensionless net radiative heat flux term Q^r, defined as

$$Q^r = \frac{q^r}{4n_0^2\bar{\sigma}T_0^4} = \frac{2\pi\int_{-1}^{1} I(\tau, x, \mu)\mu\, d\mu}{4n_0^2\bar{\sigma}T_0^4} \tag{13-56}$$

The radiation intensity $I(\tau, x, \mu)$ is obtained from the solution of the equation of radiative transfer, which is generally given in the optical variable τ, defined as

$$\tau = \int_0^y \beta(x, y')\, dy' \tag{13-57a}$$

where $\beta(x, y)$ is the volumetric extinction coefficient. The energy equation 13-53, however, is given in the variable η, defined as

$$\eta = \left(\frac{(1 + m)u_\infty(x)}{j\nu_0 x}\right)^{1/2} \int_0^y \frac{\rho(x, y')}{\rho_0}\, dy' \tag{13-57b}$$

Therefore the relation between η and τ is needed. To obtain this relation we assume that the mass extinction coefficient β_m is constant. Then, by utilizing the relation between the mass and the volumetric extinction coefficients, we write[7]

$$\frac{\beta(x, y)}{\beta_0} = \frac{\rho(x, y)}{\rho_0} \tag{13-58}$$

Substitution of $\beta(x, y)$ from Eq. 13-58 into Eq. 13-57a yields

$$\tau = \beta_0 \int_0^y \frac{\rho(x, y')}{\rho_0}\, dy' \tag{13-59}$$

The relation between τ and η is obtained from Eqs. 13-57b and 13-59 and is given in various alternative forms as

$$\tau = \left(\frac{j\nu_0 x}{(1 + m)u_\infty(x)}\right)^{1/2} \beta_0 \eta \tag{13-60a}$$

$$= \left(\frac{j}{1 + m}\right)^{1/2} \mathrm{Re}_x^{-(1/2)} \beta_0 x \eta \tag{13-60b}$$

$$= \left(\frac{j}{1 + m}\right)^{1/2} \zeta \eta \tag{13-60c}$$

$$= \left(\frac{j}{1 + m}\, \xi N\, \mathrm{Pr}\right)^{1/2} \eta \tag{13-60d}$$

where we have defined the parameters ζ and ξ as

$$\zeta \equiv \beta_0 x \, \mathrm{Re}_x^{-(1/2)} \tag{13-61a}$$

$$\xi \equiv \frac{\zeta^2}{N \, \mathrm{Pr}} = \frac{4 n_0^2 \bar{\sigma} T_0^3 \beta_0 x}{\rho_0 c_p u_\infty(x)} \tag{13-61b}$$

The dimensionless group $4\mathrm{Re}_x^{1/2}/\mathrm{Bo}_x$ appearing in the energy equation 13-53 can be expressed in terms of the parameter ζ or ξ as

$$\frac{4 \, \mathrm{Re}_x^{1/2}}{\mathrm{Bo}_x} = \frac{\zeta}{N \, \mathrm{Pr}} = \left(\frac{\xi}{N \, \mathrm{Pr}}\right)^{1/2} \tag{13-61c}$$

Wall Heat Flux

The net heat flux at the wall is of interest in most engineering applications. For a wall that is impermeable to flow, the net heat flux at the wall q_w is composed of the conductive and radiative heat fluxes and given as

$$q_w = [q^c + q^r]_{y=0} = \left[-k \frac{\partial T}{\partial y} + q^r\right]_{y=0} \tag{13-62}$$

where k is the thermal conductivity of the fluid. The net heat flux at the wall is in the positive y direction when q_w is a positive quantity. Equation 13-62 is expressed in terms of the dimensionless quantities as

$$Q_w \equiv \frac{q_w}{4 n_0^2 \bar{\sigma} T_0^4} = \left[-\left(\frac{1+m}{j}\right)^{1/2} \frac{k\rho}{k_0 \rho_0} \frac{\mathrm{Bo}_x}{4 \, \mathrm{Re}_x^{1/2} \, \mathrm{Pr}} \frac{\partial \theta}{\partial \eta} + Q^r\right]_{\eta=0} \tag{13-63a}$$

$$= \left[-\left(\frac{1+m}{j}\right)^{1/2} \frac{k\rho}{k_0 \rho_0} \frac{N}{\zeta} \frac{\partial \theta}{\partial \eta} + Q^r\right]_{\eta=0} \tag{13-63b}$$

$$= \left[-\left(\frac{1+m}{j}\right)^{1/2} \frac{k\rho}{k_0 \rho_0} \left(\frac{N}{\xi \, \mathrm{Pr}}\right)^{1/2} \frac{\partial \theta}{\partial \eta} + Q^r\right]_{\eta=0} \tag{13-63c}$$

Velocity and Thermal Boundary Layers

In the boundary layer type of formulation considered above, two different boundary layers should be distinguished: the velocity boundary layer and the thermal boundary layer. The similarity transformation of the velocity boundary layer is determined only by the form of the external flow velocity $u_\infty(x)$. For the specific case considered here the choice of the external flow velocity profile of the form given by Eq. 13-48 has resulted in the momentum equation 13-52 with the function $f(\eta)$ depending only on the independent variable η. The thermal boundary layer with radiative heat transfer, however, is not in general similar; that is, in the energy equation 13-53 the temperature function $\theta(\eta, x)$ depends on two independent variables η and x. We shall see later in this chapter that only for optically thick radiation is complete similarity transformation of the thermal boundary layer possible, and the temperature becomes a function of η only.

Transformed Equations When μ and k Vary Linearly with Temperature

The transformed momentum and energy equations 13-52 and 13-53 involve the thermal conductivity and viscosity, which vary with temperature; the dependence of these physical properties on temperature is prescribed at the beginning for each problem studied. If it is assumed that μ and k vary linearly with temperature in the form[8]

$$\frac{\mu}{\mu_0} = \frac{k}{k_0} = \frac{T}{T_0} = \theta \tag{13-64a}$$

The momentum and the energy equations are simplified significantly, because for a perfect gas we have

$$\frac{\rho_0}{\rho} = \frac{T}{T_0} = \theta \tag{13-64b}$$

By combining Eqs. 13-64a and 13-64b we obtain

$$\mu\rho = \mu_0\rho_0 \qquad \text{and} \qquad k\rho = k_0\rho_0 \tag{13-64c}$$

In view of the relations in Eq. 14-64c the momentum equation 13-52 reduces to

$$\frac{d^3f}{d\eta^3} + \frac{j}{2}f\frac{d^2f}{d\eta^2} + \frac{jm}{1+m}\left[\frac{\rho_\infty}{\rho} - \left(\frac{df}{d\eta}\right)^2\right] = 0 \tag{13-65}$$

The energy equation 13-53 simplifies and can be written in alternative forms as

$$\frac{1}{\Pr}\frac{\partial^2\theta}{\partial\eta^2} + \frac{j}{2}f\frac{\partial\theta}{\partial\eta} = \frac{j}{1+m}\frac{df}{d\eta}x\frac{\partial\theta}{\partial x} + \left(\frac{j}{1+m}\right)^{1/2}\frac{4}{\mathrm{Bo}_x}\mathrm{Re}_x^{1/2}\frac{\partial Q^r}{\partial\eta}$$
$$+ \frac{jm}{1+m}\frac{\rho_\infty(x)}{\rho}\mathrm{E}_\infty(x)\frac{df}{d\eta} - \mathrm{E}_\infty(x)\left(\frac{d^2f}{d\eta^2}\right)^2 \tag{13-66a}$$

or

$$\frac{1}{\Pr}\frac{\partial^2\theta}{\partial\eta^2} + \frac{j}{2}f\frac{\partial\theta}{\partial\eta} = \frac{j(1-m)}{2(1+m)}\frac{df}{d\eta}\zeta\frac{\partial\theta}{\partial\zeta} + H$$
$$+ \frac{jm}{1+m}\frac{\rho_\infty(\zeta)}{\rho}\mathrm{E}_\infty(\zeta)\frac{df}{d\eta} - \mathrm{E}_\infty(\zeta)\left(\frac{d^2f}{d\eta^2}\right)^2 \tag{13-66b}$$

where

$$H \equiv \left(\frac{j}{1+m}\right)^{1/2}\frac{\zeta}{N\Pr}\frac{\partial Q^r}{\partial\eta} = \frac{j}{1+m}\frac{\zeta^2}{N\Pr}\frac{\partial Q^r}{\partial\tau}$$

or

$$\frac{1}{\mathrm{Pr}}\frac{\partial^2\theta}{\partial\eta^2}+\frac{j}{2}f\frac{\partial\theta}{\partial\eta}=\frac{j(1-m)}{1+m}\frac{df}{d\eta}\xi\frac{\partial\theta}{\partial\xi}+\frac{j}{1+m}\xi\frac{\partial Q^r}{\partial\tau}$$
$$+\frac{jm}{1+m}\frac{\rho_\infty(\xi)}{\rho}\mathrm{E}_\infty(\xi)\frac{df}{d\eta}-\mathrm{E}_\infty(\xi)\left(\frac{d^2f}{d\eta^2}\right)^2 \quad (13\text{-}66c)$$

To obtain these alternative forms of the energy equation the following relations are used:

$$x\frac{\partial\theta}{\partial x}=\frac{1-m}{2}\zeta\frac{\partial\theta}{\partial\zeta} \quad \text{by Eq. 13-61a} \quad (13\text{-}67a)$$

$$x\frac{\partial\theta}{\partial x}=(1-m)\zeta\frac{\partial\theta}{\partial\xi} \quad \text{by Eq. 13-61b} \quad (13\text{-}67b)$$

$$\frac{\partial Q^r}{\partial\eta}=\left(\frac{j}{1+m}\right)^{1/2}\zeta\frac{\partial Q^r}{\partial\tau} \quad \text{by Eq. 13-60c} \quad (13\text{-}67c)$$

$$\frac{4\,\mathrm{Re}_x^{1/2}}{\mathrm{Bo}_x}=\frac{\zeta}{N\,\mathrm{Pr}} \quad \text{and} \quad \xi=\frac{\zeta^2}{N\,\mathrm{Pr}} \quad \text{by Eq. 13-61c} \quad (13\text{-}67d)$$

The last term on the right-hand side of each of the above energy equations represents the viscous energy dissipation, and the term next to it represents the work of compression.

The relations for the net wall heat flux (Eqs. 13-63) simplify to

$$Q_w\equiv\frac{q_w}{4n_0{}^2\bar\sigma T_0{}^4}=\left[-\left(\frac{1+m}{j}\right)^{1/2}\frac{\mathrm{Bo}_x}{4\,\mathrm{Re}_x^{1/2}\,\mathrm{Pr}}\frac{\partial\theta}{\partial\eta}+Q^r\right]_{\eta=0} \quad (13\text{-}68a)$$

$$=\left[-\left(\frac{1+m}{j}\right)^{1/2}\frac{N}{\zeta}\frac{\partial\theta}{\partial\eta}+Q^r\right]_{\eta=0} \quad (13\text{-}68b)$$

$$=\left[-\left(\frac{1+m}{j}\frac{N}{\xi\,\mathrm{Pr}}\right)^{1/2}\frac{\partial\theta}{\partial\eta}+Q^r\right]_{\eta=0} \quad (13\text{-}68c)$$

The transformation variables η and ψ given by Eqs. 13-50 and 13-51, however, remain unaltered, that is,

$$\eta=\left(\frac{(1+m)u_\infty(x)}{j\nu_0 x}\right)^{1/2}\int_0^y\frac{\rho(x,y')}{\rho_0}\,dy' \quad (13\text{-}69a)$$

$$\psi(x,y)=f(\eta)\left(\frac{j\nu_0 x u_\infty(x)}{1+m}\right)^{1/2} \quad (13\text{-}69b)$$

where

$$u_\infty(x)=Cx^m \quad (13\text{-}69c)$$

If the reference temperature T_0 is chosen as the stagnation temperature of the external flow and a perfect gas is considered, the density ratio $\rho_\infty(x)/\rho(x, y)$ and the Eckert number $E_\infty(x)$ appearing in the above equations can be related to the Mach number $M_\infty(x)$ of the external flow by utilizing the relations given in the book by Shapiro [41].[9]

As special cases of the foregoing formulation we examine now the equations for flow over a flat plate and the stagnation flow.

(*a*) *Flow over a Flat Plate*

For flow over a flat plate we set $m = 0$ and the momentum equation 13-65 and the energy equation 13-66 are respectively simplified to

$$\frac{d^3 f}{d\eta^3} + \frac{j}{2} f \frac{d^2 f}{d\eta^2} = 0 \tag{13-70}$$

$$\frac{1}{\Pr}\frac{\partial^2 \theta}{\partial \eta^2} + \frac{j}{2} f \frac{\partial \theta}{\partial \eta} = j \frac{df}{d\eta} x \frac{\partial \theta}{\partial x} + \sqrt{j}\frac{4}{\mathrm{Bo}_x}\mathrm{Re}_x^{1/2}\frac{\partial Q^r}{\partial \eta} - \mathrm{E}_\infty(x)\left(\frac{d^2 f}{d\eta^2}\right)^2 \tag{13-71a}$$

or

$$\frac{1}{\Pr}\frac{\partial^2 \theta}{\partial \eta^2} + \frac{j}{2} f \frac{\partial \theta}{\partial \eta} = \frac{j}{2}\frac{df}{d\eta}\zeta\frac{\partial \theta}{\partial e} + H - \mathrm{E}_\infty\left(\frac{d^2 f}{d\eta^2}\right)^2 \tag{13-71b}$$

where

$$H \equiv \sqrt{j}\frac{\zeta}{N \Pr}\frac{\partial Q^r}{\partial \eta} = j\frac{\zeta^2}{N \Pr}\frac{\partial Q^r}{\partial \tau}$$

or

$$\frac{1}{\Pr}\frac{\partial^2 \theta}{\partial \eta^2} + \frac{j}{2} f \frac{\partial \theta}{\partial \eta} = j \frac{df}{d\eta}\xi\frac{\partial \theta}{\partial \xi} + j\xi\frac{\partial Q^r}{\partial \tau} - \mathrm{E}_\infty\left(\frac{d^2 f}{d\eta^2}\right)^2 \tag{13-71c}$$

We note that for flow over a flat plate the momentum equation 13-70 is uncoupled, and in the energy equation the term for the work of compression is dropped out.

The relations for the net wall heat flux given by Eqs. 13-68 simplify to

$$Q_w \equiv \frac{q_w}{4 n_0^{\,2}\bar{\sigma} T_0^{\,4}} = \left[-\left(\frac{1}{j}\right)^{1/2}\frac{\mathrm{Bo}_x}{4\,\mathrm{Re}_x^{1/2}\,\Pr}\frac{\partial \theta}{\partial \eta} + Q^r\right]_{\eta=0} \tag{13-72a}$$

$$= \left[-\left(\frac{1}{j}\right)^{1/2}\frac{N}{\zeta}\frac{\partial \theta}{\partial \eta} + Q^r\right]_{\eta=0} \tag{13-72b}$$

$$= \left[-\left(\frac{1}{j}\frac{N}{\xi \Pr}\right)^{1/2}\frac{\partial \theta}{\partial \eta} + Q^r\right]_{\eta=0} \tag{13-72c}$$

The transformation variables given by Eqs. 13-69 become

$$\eta = \left(\frac{u_\infty}{j\nu_0 x}\right)^{1/2} \int_0^y \frac{\rho(x, y')}{\rho_0}\, dy' \tag{13-73a}$$

$$\psi(x, y) = f(\eta)\sqrt{j\nu_0 x u_\infty} \tag{13-73b}$$

and

$$u_\infty = C = \text{constant} \tag{13-73c}$$

(*b*) *Stagnation Flow*

For the stagnation flow equations 13-65 through 13-69 are simplified by setting $m = 1$ in these equations.

Transformed Equations for Incompressible Fluid

For an incompressible fluid the density is constant. If it is further assumed that the viscosity and the thermal conductivity are independent of temperature, the momentum and energy equations 13-65 and 13-66 simplify to

$$\frac{d^3 f}{d\eta^3} + \frac{j}{2} f \frac{d^2 f}{d\eta^2} + \frac{jm}{1+m}\left[1 - \left(\frac{df}{d\eta}\right)^2\right] = 0 \tag{13-74}$$

$$\frac{1}{\Pr}\frac{\partial^2\theta}{\partial\eta^2} + \frac{j}{2} f \frac{\partial\theta}{\partial\eta} = \frac{j}{1+m}\frac{df}{d\eta}\, x \frac{\partial\theta}{\partial x} + \left(\frac{j}{1+m}\right)^{1/2} \frac{4}{\mathrm{Bo}_x} \mathrm{Re}_x^{1/2} \frac{\partial Q^r}{\partial\eta} - \mathrm{E}_\infty \left(\frac{d^2 f}{d\eta^2}\right)^2 \tag{13-75a}$$

or

$$\frac{1}{\Pr}\frac{\partial^2\theta}{\partial\eta^2} + \frac{j}{2} f \frac{\partial\theta}{\partial\eta} = \frac{j(1-m)}{2(1+m)}\frac{df}{d\eta}\, \zeta \frac{\partial\theta}{\partial\zeta} + H - \mathrm{E}_\infty \left(\frac{d^2 f}{d\eta^2}\right)^2 \tag{13-75b}$$

where

$$H \equiv \left(\frac{j}{1+m}\right)^{1/2} \frac{\zeta}{N \Pr}\frac{\partial Q^r}{\partial\eta} = \frac{j}{1+m}\frac{\zeta^2}{N \Pr}\frac{\partial Q^r}{\partial\tau}$$

or

$$\frac{1}{\Pr}\frac{\partial^2\theta}{\partial\eta^2} + \frac{j}{2} f \frac{\partial\theta}{\partial\eta} = \frac{j(1-m)}{1+m}\frac{df}{d\eta}\, \xi \frac{\partial\theta}{d\xi} + \frac{j}{1+m}\, \xi \frac{\partial Q^r}{\partial\tau} - \mathrm{E}_\infty \left(\frac{d^2 f}{d\eta^2}\right)^2 \tag{13-75c}$$

We note that the momentum equation is uncoupled and that the work of compression is dropped out in these equations. The flow velocities encountered in practice with incompressible flow are sufficiently small to assume $E_\infty \ll 1$; hence the viscous energy dissipation term in the energy equation can be neglected.

The relations for the net wall heat flux (Eqs. 13-68) remain unchanged, but the relations defining the transformation variables η and $\psi(x, y)$ (Eqs. 13-69) simplify to

$$\eta = \left[\frac{(1+m)u_\infty(x)}{j\nu_0 x}\right]^{1/2} y \tag{13-76a}$$

$$\psi(x, y) = f(\eta)\left[\frac{j\nu_0 x u_\infty(x)}{1+m}\right]^{1/2} \tag{13-76b}$$

where

$$u_\infty(x) = Cx^m \tag{13-76c}$$

13-4 OPTICALLY THICK, INCOMPRESSIBLE LAMINAR BOUNDARY LAYER FLOW OVER A WEDGE

Several investigators [2–6] used the optically thick limit approximation to study the interaction of radiation with boundary layer heat transfer. Although the range of validity of the optically thick approximation is small in boundary layer flow, it possesses the advantage of simplicity in the analysis because the governing energy equation can be transformed into an ordinary differential equation by the conventional similarity transformation. In this section we present the mathematical formulation of simultaneous convection and radiation for steady, laminar boundary layer flow over a wedge by using the optically thick limit approximation for the radiation part of the problem, and we discuss the method of solution and the results.

For simplicity in the analysis we consider an incompressible, constant-property, gray fluid flowing over a black wall that is impermeable to flow and is kept at a uniform temperature T_w. We note that the incompressible fluid approximation is good only for a small Eckert number. Therefore, by assuming $E_\infty \ll 1$, the viscous energy dissipation term in the energy equation becomes negligible. With these considerations the momentum and energy equations 13-74 and 13-75b are simplified; by setting the parameter $j = 1 + m$, we can write these equations as

$$\frac{d^3f}{d\eta^3} + \frac{1+m}{2} f \frac{d^2f}{d\eta^2} + m\left[1 - \left(\frac{df}{d\eta}\right)^2\right] = 0 \tag{13-77}$$

$$\frac{1}{\Pr}\frac{\partial^2\theta}{\partial\eta^2} + \frac{1+m}{2} f \frac{\partial\theta}{\partial\eta} = \frac{1-m}{2}\frac{df}{d\eta}\zeta\frac{\partial\theta}{\partial\zeta} + \frac{\zeta}{N\Pr}\frac{\partial Q^r}{\partial\eta} \tag{13-78}$$

and the transformation variables given by Eqs. 13-76 become

$$\eta = y\left(\frac{u_\infty(x)}{\nu x}\right)^{1/2} = \frac{y}{x}\,\mathrm{Re}_x^{1/2} = \frac{\tau}{\zeta} \tag{13-79a}$$

$$\psi(x, y) = f(\eta)\sqrt{\nu x u_\infty(x)} \tag{13-79b}$$

$$u_\infty(x) = Cx^m \tag{13-79c}$$

The dimensionless temperature $\theta(\zeta, \eta)$ and the dimensionless net radiative heat flux Q^r are defined as

$$\theta(\zeta, \eta) \equiv \frac{T}{T_\infty} \qquad \text{and} \qquad Q^r \equiv \frac{q^r}{4n^2\bar{\sigma}T_\infty{}^4} \tag{13-80}$$

where T_∞ is the temperature of the external flow and is constant.

Now a relation is needed for the net radiative heat flux q^r; for the optically thick limit approximation (i.e., the Rosseland approximation) the net radiative heat flux is given by Eq. 9-25 as

$$q^r = -\frac{4n^2\bar{\sigma}}{3\beta}\frac{\partial T^4}{\partial y} \tag{13-81}$$

Then Q^r is given by

$$Q^r = -\frac{1}{3\beta}\frac{\partial \theta^4}{\partial y} = -\frac{1}{3}\frac{\partial \theta^4}{\partial \tau} = -\frac{1}{3\zeta}\frac{\partial \theta^4}{\partial \eta} = -\frac{4}{3\zeta}\theta^3\frac{\partial \theta}{\partial \eta} \tag{13-82}$$

where we have utilized the relations given by Eq. 13-79a to transform the partial derivatives.

When the radiative heat flux Q^r is given as in Eq. 13-82, complete similarity transformation of the energy equation 13-78 into an ordinary differential equation in the variable η is possible;[10] then the energy equation 13-78 becomes

$$\frac{1}{\mathrm{Pr}}\frac{d^2\theta}{d\eta^2} + \frac{1+m}{2}f\frac{d\theta}{d\eta} = -\frac{4}{3N\,\mathrm{Pr}}\frac{d}{d\eta}\left(\theta^3\frac{d\theta}{d\eta}\right) \tag{13-83a}$$

or

$$\frac{1}{\mathrm{Pr}}\frac{d}{d\eta}\left[\left(1 + \frac{4}{3N}\theta^3\right)\frac{d\theta}{d\eta}\right] + \frac{1+m}{2}f\frac{d\theta}{d\eta} = 0 \tag{13-83b}$$

where

$$\theta \equiv \theta(\eta) \tag{13-83c}$$

The boundary conditions for the flow problem are taken as $u = v = 0$ at the wall and $u = u_\infty(x)$ outside the boundary layer; the transformed boundary conditions for the momentum equation 13-77 become

$$f = \frac{df}{d\eta} = 0 \quad \text{at } \eta = 0 \tag{13-84a}$$

$$\frac{df}{d\eta} = 1 \quad \text{at } \eta \to \infty \tag{13-84b}$$

The boundary conditions for the energy equation 13-83 are given as

$$\theta = \frac{T_w}{T_\infty} \equiv \theta_w \quad \text{at } \eta = 0 \tag{13-85a}$$

$$\theta = \frac{T_\infty}{T_\infty} = 1 \quad \text{at } \eta \to \infty \tag{13-85b}$$

Equations 13-77 are 13-83b are two ordinary differential equations for the functions $f(\eta)$ and $\theta(\eta)$. First, the momentum equation 13-77 is solved with the boundary conditions 13-84 and the function $f(\eta)$ is determined; then the energy equation 13-83b is solved subject to the boundary conditions 13-85 and the temperature distribution $\theta(\eta)$ is obtained.

Knowing the temperature distribution, we can evaluate the net radiative heat flux in the medium. However, the limitation to the accuracy of the optically thick limit approximation should be recognized. We saw in Chapter 9 that this approximation breaks down in the immediate vicinity of the boundaries because it does not take into account radiation from boundary surfaces. This is a serious restriction since heat transfer at the wall is important in boundary layer analysis. For the black wall condition considered here we used Deissler's extension (see Eq. 9-39) for the optically thick limit and replaced the factor $\frac{4}{3}$ by $\frac{2}{3}$ in Eq. 13-82; then the net wall heat flux q_w is given by

$$Q_w \equiv \frac{q_w}{4n^2\bar{\sigma}T_\infty^{\ 4}} = \left[-\frac{N}{\zeta}\frac{d\theta}{d\eta} + Q^r \right]_{\eta=0} \tag{13-86a}$$

$$= -\frac{N}{\zeta}\left(1 + \frac{2}{3N}\,\theta_w^{\ 3}\right)\frac{d\theta}{d\eta}\bigg|_{\eta=0} \tag{13-86b}$$

Linearized Equations

The energy equation 13-83b is nonlinear because it involves a higher power of temperature θ. It is instructive to linearize the energy equation in order to gain some insight into the first-order effects of radiation in boundary layer heat transfer. When the term T^4 is expanded in a Taylor series about T_∞

and the second- and higher-order terms are neglected, we obtain

$$T^4 \cong 4T_\infty^{\,3}T - 3T_\infty^{\,4} \tag{13-87}$$

or

$$\theta^4 \cong 4\theta - 3 \tag{13-88}$$

Substitution of Eq. 13-88 into the energy equation 13-83 yields the linearized equation in the form

$$\frac{1}{\mathrm{Pr}_m}\frac{d^2\theta}{d\eta^2} + \frac{1+m}{2} f \frac{d\theta}{d\eta} = 0 \tag{13-89}$$

where the modified Prandtl number Pr_m is defined as

$$\mathrm{Pr}_m = \frac{\mathrm{Pr}}{1 + (4/3N)} \tag{13-90}$$

We note that the linearized energy equation 13-89 is of exactly the same form as the energy equation for the nonradiating flow, with, however, the Prandtl number replaced by the modified Prandtl number.

Discussion of Results

In the problems of boundary layer heat transfer the temperature gradient at the wall is an important quantity because it affects the convective heat flux. Figure 13-4 shows the effect of radiation on the temperature gradient at the wall for flow over a flat plate with the cold wall condition. The case with $N = 10$ corresponds almost to nonradiating flow, and $N = 0.1$ to reasonably strong radiation effects. It appears from this figure that for

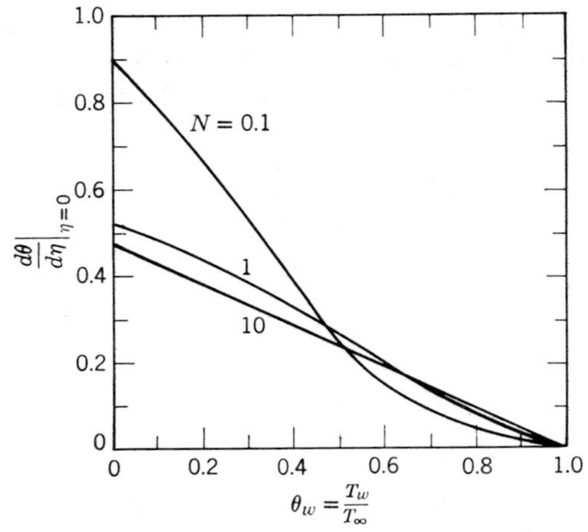

Fig. 13-4. Temperature gradients at the wall as a function of wall temperature for flow over a flat plate with cold wall condition. (From R. Viskanta and R. J. Grosh [2].)

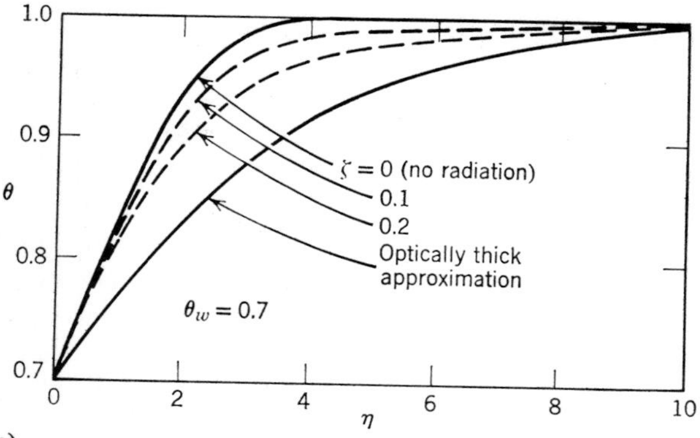

Fig. 13-5. (top)

approximate values of $\theta_w < 0.5$ radiation increases the temperature gradient at the wall, as compared with the nonradiating case, whereas for approximate values of $\theta_w > 0.5$ radiation decreases it. Similar calculations with hot wall conditions have shown that radiation always decreases the temperature gradient at the wall for all values of wall temperature [2].

Figure 13-5 shows the effect of radiation on the temperature distribution in the boundary layer for $N = 0.1$ and the cold wall condition. In order to

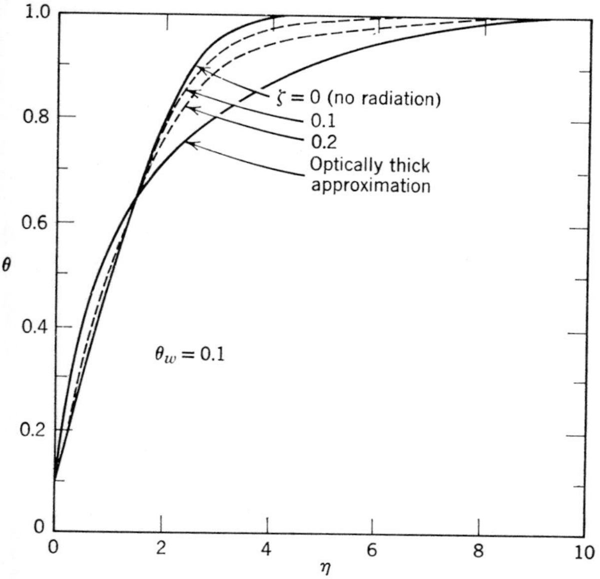

Fig. 13-5. (bottom) A comparison of temperature profiles obtained by the optically thick limit approximation and the exact analysis for flow over a 90° wedge for $N = 0.1$, $\Pr = 1$, absorbing and emitting fluid, and black wall condition. (From M. A. Boles and M. N. Özişik [15a].)

illustrate the limitations to the range of validity of the optically thick limit approximation we have included in this figure the temperature profiles at several different axial locations ζ, obtained from calculations in which the radiation part of the problem was treated exactly for $N = 0.1$, $\text{Pr} = 1$, a purely absorbing and emitting fluid (i.e., $\omega = 0$), and the black wall condition. In this figure the parameter $\zeta \equiv \kappa x/\text{Re}_x^{1/2}$, defined previously by Eq. 13-61a, characterizes the dimensionless distance from the leading edge, and the case $\zeta = 0$ corresponds to nonradiating flow. The exact temperature profiles at locations away from the wall appear to lie between those for the nonradiating case and the optically thick limit approximation, but the slopes of these curves at the wall differ greatly from those predicted by the optically thick limit profile.

In order to illustrate this point we present in Table 13-1 the temperature gradients, the net radiative heat flux, and the total heat flux at the wall for $\theta_w = 0.1$ and 0.7, for values of $N = 0.1$, $\text{Pr} = 1$, and for purely absorbing,

Table 13-1 Temperature Gradient, Net Radiative Heat Flux, and Total Heat Flux at the Wall for Laminar Flow over a 90° Wedge, Obtained from the Exact Solution and Optically Thick Limit Approximation for $N = 0.1$, Pr = 1.0, and Black Wall Conditions[a]

θ_w	$\zeta = \dfrac{\kappa x}{\text{Re}_x^{1/2}}$	$\left.\dfrac{\partial\theta}{\partial\eta}\right\rvert_{\eta=0}$	$-Q^r\rvert_{\eta=0}$	$-\dfrac{\zeta}{N} Q^t\rvert_{\eta=0}$
		Exact		
0.1	0.005	0.396	0.140	0.403
	0.05	0.401	0.193	0.498
	0.10	0.414	0.160	0.574
	0.20	0.446	0.107	0.660
		Optically Thick Limit Approximation		
		0.770		0.776[b]
				0.781[c]
		Exact		
0.7	0.005	0.132	0.126	0.138
	0.05	0.133	0.157	0.212
	0.10	0.137	0.136	0.273
	0.20	0.147	0.099	0.346
		Optically Thick Limit Approximation		
		0.084		0.275[b]
				0.467[c]

[a] From M. A. Boles and M. N. Özişik [42].
[b] Calculated using Deissler's extension.
[c] Calculated using the optically thick limit approximation only.

emitting flow ($\omega = 0$) and the black wall conditions. The exact results are given at several different axial locations ζ. It is apparent from this table that the gradient of temperature at the wall predicted by the optically thick limit approximation is subject to considerable error. This is to be expected since this approximation breaks down in the vicinity of boundaries. However, the optically thick limit approximation may be useful in the investigation of the general effects of radiation on the temperature profile in the boundary layer.

13-5 ABSORBING, EMITTING, COMPRESSIBLE LAMINAR BOUNDARY LAYER FLOW OVER A FLAT PLATE

In this section consideration is given to the interaction of radiation with heat transfer in the laminar boundary layer flow of an absorbing, emitting, compressible fluid over a flat plate. It is assumed that the fluid is a perfect gas and is gray, the viscosity varies linearly with temperature, the specific heat and the Prandtl number are constant, and the external flow temperature T_∞ is uniform. The surface of the wall is opaque and gray, is a diffuse emitter and diffuse reflector, is impervious to flow, and is subjected to an externally supplied constant heat flux q_w. Figure 13-6 shows the geometry and the coordinates.

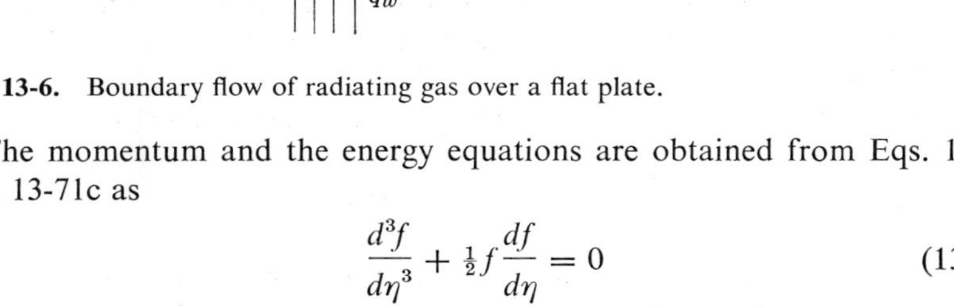

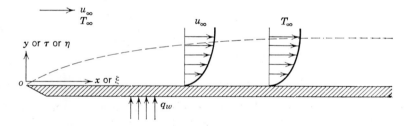

Fig. 13-6. Boundary flow of radiating gas over a flat plate.

The momentum and the energy equations are obtained from Eqs. 13-70 and 13-71c as

$$\frac{d^3f}{d\eta^3} + \tfrac{1}{2}f\frac{df}{d\eta} = 0 \tag{13-91}$$

$$\frac{1}{\mathrm{Pr}}\frac{\partial^2\theta}{\partial\eta^2} + \tfrac{1}{2}f\frac{\partial\theta}{\partial\eta} = \frac{df}{d\eta}\xi\frac{\partial\theta}{\partial\xi} + \xi\frac{\partial Q^r}{\partial\tau} - \mathrm{E}_\infty\left(\frac{d^2f}{d\eta^2}\right)^2 \tag{13-92}$$

where

$$\theta \equiv \theta(\xi, \eta) = \frac{T}{T_\infty} \tag{13-93}$$

In Eqs. 13-91 and 13-92 we set the parameter $j = 1$ in order to obtain the set of equations considered by Taitel and Hartnett [38]. If we had chosen the parameter $j = 2$, the resulting set of equations would have been similar to those considered by Oliver and McFadden [14].

The parameters N and ξ have already been defined by Eqs. 13-54d and 13-61b, respectively; for an absorbing, emitting, nonscattering fluid of the type considered here the extinction coefficient β should be replaced by the absorption coefficient κ in these expressions.

The boundary conditions for the momentum equation 13-91 are given as

$$f = \frac{df}{d\eta} = 0 \quad \text{at } \eta = 0 \tag{13-94a}$$

$$\frac{df}{d\eta} = 1 \qquad \text{at } \eta \to \infty \tag{13-94b}$$

which imply that the velocity components are zero at the wall and $u = u_\infty$ outside the boundary layer. The boundary conditions for the energy equation 13-92 are taken as follows: (a) the applied heat flux q_w at the wall surface is constant, (b) the temperature is equal to the external flow temperature T_∞ outside the thermal boundary layer, and (c) for $\xi = 0$ the solution of the energy equation is equal to the solution $\theta_0(\eta)$ of the same problem for the nonradiating flow. These boundary conditions are given as

$$\frac{q_w}{4n^2\bar{\sigma}T_\infty{}^4} = \left[-\sqrt{\frac{N}{\xi \Pr}}\frac{\partial\theta}{\partial\eta} + Q^r\right]_{\eta=0} \tag{13-95}$$

$$\theta = 1 \qquad \text{at } \eta \to \infty \tag{13-96a}$$

$$\theta = \theta_0(\eta) \quad \text{at } \xi = 0 \tag{13-96b}$$

where Eq. 13-95 is obtained from Eq. 13-72c by setting in the latter the parameter $j = 1$; $\theta_0(\eta)$ is the solution of the energy equation for the non-radiating case.

Equations 13-92 and 13-95 involve respectively the radiation terms

$$\frac{\partial Q^r}{\partial\tau} \qquad \text{and} \qquad Q^r|_{\eta=0}$$

which should be obtained from the solution of the equation of radiative transfer. In the following analysis we shall treat the radiation part of this problem (a) by the exact formulation and (b) by the application of the optically thin limit approximation, and then compare the results for the temperature distribution in the boundary layer obtained by these different methods, including the value obtained by the optically thick limit approximation.

(a) Exact Formulation

The fluid flowing over the flat plate is considered as an absorbing, emitting, gray, locally plane-parallel, semi-infinite ($0 \le \tau < \infty$) medium at a temperature $T(\tau, \xi)$. The boundary of $\tau = 0$ (i.e., $\eta = 0$) is a diffusely emitting, diffusely reflecting, opaque, gray surface at a temperature $T_w(\xi)$. The radiation part of this problem satisfies the equation of radiative transfer given in the form

$$\mu \frac{\partial I(\tau, \xi, \mu)}{\partial \tau} + I(\tau, \xi, \mu) = I_b[T(\tau, \xi)] \quad \text{in } 0 \le \tau < \infty,\ -1 \le \mu \le 1 \tag{13-97a}$$

with

$$I(0, \xi) = \varepsilon_w I_b[T_w(\xi)] + 2(1 - \varepsilon_w)\int_0^1 I(0, \xi, -\mu')\, d\mu', \quad \mu > 0 \tag{13-97b}$$

and intensity remaining finite as $\tau \to \infty$. Here ε_w is the wall emissivity, $T(\tau, \xi)$ is the fluid temperature, $T_w(\xi)$ is the wall temperature, and $1 - \varepsilon_w$ represents the wall reflectivity because Kirchhoff's law is assumed to be valid. We note that the intensity $I(\tau, \xi, \mu)$ depends on ξ because temperature $T(\tau, \xi)$ depends on ξ; therefore ξ appears in these equations merely as a parameter.

Once Eq. 13-97 is solved and the radiation intensity $I(\tau, \xi, \mu)$ is determined, the net radiative heat flux $q^r(\tau, \xi)$ is evaluated from

$$q^r(\tau, \xi) = 2\pi \int_{-1}^{1} I(\tau, \xi, \mu)\mu\, d\mu \tag{13-98}$$

and the desired expression for the radiation term appearing in the energy equation is readily obtained. However, we considered in Chapter 8 the formal solution of the equation of radiative transfer and presented expressions for the radiative heat flux term. Therefore we shall not repeat these calculations here, but obtain the formal solutions for $\partial q^r/\partial \tau$ and q^r directly from Eqs. 8-95 and 8-84, respectively, by setting in these equations $\tau_0 \to \infty$, $I_\nu^-(\tau_0) = 0$ and omitting the frequency dependence. We find

$$\frac{\partial q^r(\tau, \xi)}{\partial \tau} = 4\pi I_b[T(\tau, \xi)] - 2\pi E_2(\tau) I^+(0) - 2\pi \int_0^\infty I_b[T(\tau', \xi)] E_1(|\tau - \tau'|)\, d\tau' \tag{13-99}$$

$$q^r(0, \xi) = \pi I^+(0) - 2\pi \int_0^\infty I_b[T(\tau', \xi)] E_2(\tau')\, d\tau' \tag{13-100}$$

These equations involve the boundary surface intensity $I^+(0)$. The relation for $I^+(0)$ is obtained from Eq. 8-110a by setting in that equation $\tau_0 \to \infty$, $I^-(\tau_0) = 0$, $\rho = 1 - \varepsilon_w$ and omitting the frequency dependence:

$$I^+(0) = \varepsilon_w I_b[T_w(\xi)] + 2(1 - \varepsilon_w)\int_0^\infty I_b[T(\tau', \xi)]E_2(\tau')\,d\tau' \qquad (13\text{-}101)$$

Substitution of Eq. 13-101 into Eqs. 13-99 and 13-100, respectively, yields

$$\begin{aligned}\frac{\partial q^r(\tau, \xi)}{\partial \tau} &= 4n^2\bar{\sigma}T^4(\tau, \xi) - 2\varepsilon_w E_2(\tau)n^2\bar{\sigma}T_w^{\,4}(\xi)\\ &\quad - 4(1 - \varepsilon_w)E_2(\tau)\int_0^\infty n^2\bar{\sigma}T^4(\tau', \xi)E_2(\tau')\,d\tau'\\ &\quad - 2\int_0^\infty n^2\bar{\sigma}T^4(\tau', \xi)E_1(|\tau - \tau'|)\,d\tau' \qquad (13\text{-}102)\end{aligned}$$

$$q^r(0, \xi) = \varepsilon_w n^2\bar{\sigma}T_w^{\,4}(\xi) - 2\varepsilon_w\int_0^\infty n^2\bar{\sigma}T^4(\tau', \xi)E_2(\tau')\,d\tau' \qquad (13\text{-}103)$$

Assuming the refractive index is constant, we can write these equations in the dimensionless form as

$$\begin{aligned}\frac{\partial Q^r}{\partial \tau} \equiv \frac{\partial}{\partial \tau}\left[\frac{q^r(\tau, \xi)}{4n^2\bar{\sigma}T_\infty^{\,4}}\right] &= \theta^4(\tau, \xi) - \tfrac{1}{2}\varepsilon_w E_2(\tau)\theta_w^{\,2}(\xi)\\ &\quad - (1 - \varepsilon_w)E_2(\tau)\int_0^\infty \theta^4(\tau', \xi)E_2(\tau')\,d\tau'\\ &\quad - \frac{1}{2}\int_0^\infty \theta^4(\tau', \xi)E_1(|\tau - \tau'|)\,d\tau' \qquad (13\text{-}104)\end{aligned}$$

$$Q^r\big|_{\tau=0} \equiv \frac{q^r(0, \xi)}{4n^2\bar{\sigma}T_\infty^{\,4}} = \frac{\varepsilon_w}{4}\left[\theta_w^{\,4}(\xi) - 2\int_0^\infty \theta^4(\tau', \xi)E_2(\tau')\,d\tau'\right] \qquad (13\text{-}105a)$$

or

$$Q^r\big|_{\tau=0} = \frac{\varepsilon_w}{4}\left\{[\theta_w^{\,4}(\xi) - 1] - 2\int_0^\infty [\theta^4(\tau', \xi) - 1]E_2(\tau')\,d\tau'\right\} \qquad (13\text{-}105b)$$

These relations for the radiative heat flux are given in the optical variable τ, whereas the energy equation 13-92 and the boundary conditions 13-98 are expressed in the variable η. The relation between the independent variables τ and η is obtained from Eq. 13-60d by setting in that equation $j = 1$ and $m = 0$:

$$\tau = \eta\sqrt{\xi N\,\mathrm{Pr}} \qquad (13\text{-}106)$$

Once Eqs. 13-104 and 13-105 are introduced into the energy equation 13-92 and the boundary condition 13-95, respectively, the mathematical formulation of the problem is considered complete. The method of solution of the resulting system of equations can now be summarized. The momentum equation 13-94 is uncoupled; therefore it is solved subject to the boundary conditions 13-97 with a numerical scheme such as a Runge-Kutta method, and the functions $f(\eta)$, $df/d\eta$, and $d^2f/d\eta^2$ appearing in the energy equation are determined. The energy equation 13-92 is a nonlinear integro- partial differential equation because the radiative heat flux term involves the fourth power of temperature under the integral sign; hence its solution is not so straightforward. Taitel and Hartnett [38] used a finite difference scheme with respect to the ξ variable and solved the resulting equation numerically in the independent variable η at each axial station ξ_i. To perform these calculations, the η direction was divided into 0.4 intervals and the value of $\eta = 20$ was sufficiently large for the temperature profile to approach the temperature at infinity in an asymptotic manner. The step sizes in the ξ direction were chosen on an equal logarithmic basis. The numerical calculations were performed for both the constant applied wall heat flux and the adiabatic wall (i.e., the heat transfer from the fluid to the wall is balanced by radiation from the wall; hence the net wall heat flux is zero) conditions.

(b) Optically Thin Approximation

When $\xi \ll 1$ and N is of the order of unity or less, the thermal boundary layer is considered optically thin and the analysis is simplified by separating the temperature field $\theta(\eta, \xi)$ into two regions: (1) the *optically thin thermal boundary layer*, within which the optically thin limit approximation is applicable and the temperature gradients are steep; and (2) the *external radiation layer* whose optical thickness is large and within which the temperature gradients are small. This concept is analogous to separating the velocity field in the boundary layer analysis into a very thin boundary layer in which velocity gradients are steep, and an external potential flow field with negligible velocity gradients. We present now the governing equations that are applicable in these two regions, discuss their coupling at the edge of the thermal boundary layer, and present a method of solution of these equations.

Radiation Layer

In this region the energy equation 13-92 is simplified by neglecting the temperature gradients in the η direction and noting that

$$\frac{df}{d\eta} = 1 \qquad \text{and} \qquad \frac{d^2f}{d\eta^2} = 0$$

We find

$$\frac{\partial \theta_r}{\partial \xi} = -\frac{\partial Q^r}{\partial \tau} \tag{13-107}$$

where θ_r denotes the temperature in the radiation layer. The substitution of $\partial Q^r/\partial\tau$ from Eq. 13-104 into Eq. 13-107 yields the energy equation for the radiation layer as

$$\frac{\partial\theta_r(\tau,\xi)}{\partial\xi} = -\theta_r^4(\tau,\xi) + \tfrac{1}{2}\varepsilon_w E_2(\tau)\theta_w^4(\xi) + (1-\varepsilon_w)E_2(\tau) \times \int_0^\infty \theta_r^4(\tau',\xi)E_2(\tau')\,d\tau' + \frac{1}{2}\int_0^\infty \theta_r^4(\tau',\xi)E_1(|\tau-\tau'|)\,d\tau' \quad (13\text{-}108)$$

with the boundary condition

$$\theta_r(\tau,\xi) = 1 \quad \text{at } \xi = 0 \quad (13\text{-}109)$$

The solution of Eq. 13-108 with the boundary condition 13-109 gives the temperature distribution $\theta_r(\tau,\xi)$ in the external radiation layer. The solution obtained in this manner, however, is not applicable in the immediate vicinity of the wall because it does not satisfy the condition of temperature continuity at $\tau = 0$; but $\theta_r(\tau,\xi)$ evaluated for $\tau \to 0$ is used to characterize the temperature at the edge of the thermal boundary layer. This approach is analogous to the treatment of the potential flow field and its coupling to the velocity field in the boundary layer.

In the present analysis we are concerned with the solution of Eq. 13-108 for small values of ξ, since the optically thin boundary layer is valid for $\xi \ll 1$. For small values of ξ, Eq. 13-108 is simplified by setting on the right-hand side of this equation $\theta_r(\tau,\xi) = 1$ and $\theta_w(\xi) = \theta_{0w}$, where θ_{0w} is the wall temperature obtained from the solution of the energy equation for $\xi = 0$.

Then Eq. 13-108 simplifies to

$$\frac{d\theta_r(\tau,\xi)}{\partial\xi} = \tfrac{1}{2}\varepsilon_w E_2(\tau)(\theta_{0w}^4 - 1) \quad \text{for } \xi \ll 1 \quad (13\text{-}110)$$

with the boundary condition

$$\theta_r(\tau,\xi) = 1 \quad \text{at } \xi = 0 \quad (13\text{-}111)$$

The solution of Eq. 13-110 is given as

$$\theta_r(\tau,\xi) = 1 + \tfrac{1}{2}\varepsilon_w E_2(\tau)(\theta_{0w}^4 - 1)\xi \quad (13\text{-}112)$$

Optically Thin Thermal Boundary Layer

The energy equation 13-92 for an optically thin thermal boundary layer is simplified by introducing in that equation the relation for $\partial Q^r/\partial\tau$ as obtained by the optically thin limit approximation. If the boundary layer is considered a locally plane-parallel slab of optical thickness τ_0, the optically thin limit approximation for $\partial Q^r/\partial\tau$ is determined either directly from Eqs. 9-7 and 9-3a[11] or by simplifying Eq. 13-104 with the optically thin limit approximation. We obtain

$$\frac{\partial Q^r(\tau,\xi)}{\partial\tau} = [\theta^4(\tau,\xi) - 1] + \tfrac{1}{2}\varepsilon_w[1 - \theta_w^{\,4}(\xi)] \tag{13-113}$$

Substitution of Eq. 13-113 into Eq. 13-92 yields the energy equation for the optically thin boundary layer:

$$\frac{1}{\mathrm{Pr}}\frac{\partial^2\theta}{\partial\eta^2} + \tfrac{1}{2}f\frac{\partial\theta}{\partial\eta} + \mathrm{E}_\infty\left(\frac{d^2f}{d\eta^2}\right)^2 = \xi\left[\frac{df}{d\eta}\frac{\partial\theta}{d\xi} + (\theta^4 - 1) + \tfrac{1}{2}\varepsilon_w(1 - \theta_w^{\,4})\right] \tag{13-114}$$

The boundary conditions for Eq. 13-114 are taken as follows: (a) the net heat flux q_w at the wall $\eta = 0$ is constant, (b) the temperature at the outer edge of the boundary layer (i.e., $\eta \to \infty$) is equal to the temperature of the radiation layer at $\tau = 0$, and (c) for $\xi = 0$ the temperature is equal to the solution $\theta_0(\eta)$ of the energy equation for no radiation.

These three boundary conditions are written as

$$\frac{q_w}{4n^2\bar{\sigma}T_\infty^{\,4}} = -\sqrt{\frac{N}{\xi\,\mathrm{Pr}}}\frac{\partial\theta(\eta,\xi)}{\partial\eta} + Q^r(\eta,\xi) \quad \text{at } \eta = 0,\ \xi > 0 \tag{13-115}$$

$$\theta(\eta,\xi) = 1 + \tfrac{1}{2}\varepsilon_w(\theta_{0w}^{\,4} - 1)\xi \quad \text{at } \eta \to \infty \tag{13-116}$$

$$\theta(\eta,\xi) = \theta_0(\eta) \quad \text{at } \xi = 0 \tag{13-117}$$

where the boundary condition 13-116 is obtained from Eq. 13-112 by setting in that equation $\tau = 0$. The radiation flux term $Q^r(\eta,\xi)|_{\eta=0}$ appearing in Eq. 13-115 is obtained from Eq. 13-105b by simplifying the latter equation with the approximation applicable to $\xi \ll 1$. We find[12]

$$Q^r(\eta,\xi)\big|_{\eta=0} = \frac{\varepsilon_w}{4}\left\{[\theta_w^{\,4}(\xi) - 1] - 2\sqrt{\mathrm{Pr}\,N\xi}\int_0^\infty[\theta^4(\eta,\xi) - 1]\,d\eta - 4\varepsilon_w(\theta_{0w}^{\,4} - 1)\xi\int_0^\infty E_2^{\,2}(\tau)\,d\tau\right\} \tag{13-118}$$

Equations 13-114 through 13-118 constitute the complete formulation of the problem for the optically thin boundary layer. Solution of the energy equation 13-114 subject to the boundary conditions 13-115 to 13-117 yields the temperature distribution in the boundary layer. The radiative flux term $Q^r(\eta,\xi)$ appearing in the boundary condition is given by Eq. 13-118.

A power-series-expansion technique is used to solve Eq. 13-114 for small ξ. The temperature function $\theta(\eta, \xi)$ is expanded in powers of $\xi^{1/2}$ in the form

$$\theta(\eta, \xi) = \theta_0(\eta) + \xi^{1/2}\theta_1(\eta) + \xi\theta_2(\eta) + \xi^{3/2}\theta_3(\eta) + \cdots \qquad (13\text{-}119)$$

and $\theta^4(\eta, \xi)$ is approximated by

$$\theta^4(\eta, \xi) \cong \theta_0^{\,4}(\eta) + 4\theta_0^{\,3}(\eta)\theta_1(\eta)\xi^{1/2} + [4\theta_0^{\,3}(\eta)\theta_2(\eta) + 6\theta_0^{\,2}(\eta)\theta_1^{\,2}(\eta)]\xi \qquad (13\text{-}120)$$

where the functions $\theta_0(\eta)$, $\theta_1(\eta)$, $\theta_2(\eta)$, . . . , are to be determined. We note that the expansion given by Eq. 13-119 satisfies the boundary condition 13-117.

When Eqs. 13-119 and 13-120 are introduced into the energy equation 13-114 and the coefficients of like powers of ξ are collected, there results a set of ordinary differential equations for the functions $\theta_0(\eta)$, $\theta_1(\eta)$, $\theta_2(\eta)$, ... The first three of these equations are given as

$$\frac{1}{\mathrm{Pr}}\,\theta_0'' + \tfrac{1}{2}f\theta_0' + \mathrm{E}_\infty f'' = 0 \qquad (13\text{-}121)$$

$$\frac{1}{\mathrm{Pr}}\,\theta_1'' + \tfrac{1}{2}f\theta_1' - \tfrac{1}{2}f'\theta_1 = 0 \qquad (13\text{-}122)$$

$$\frac{1}{\mathrm{Pr}}\,\theta_2'' + \tfrac{1}{2}f\theta_2' - f'\theta_2 = (\theta_0^{\,4} - 1) + \frac{\varepsilon_w}{2}(1 - \theta_{0w}^{\,4}) \qquad (13\text{-}123)$$

where the primes denote differentiation with respect to η. The boundary conditions for these ordinary differential equations are obtained by introducing the expansion given in Eq. 13-119 into the boundary conditions 13-115 and 13-116 and collecting the coefficients of like powers of ξ.

We note that the first of these equations for the function $\theta_0(\eta)$ characterizes the nonradiative case. These ordinary differential equations are readily solved by a numerical scheme such as a Runge-Kutta method.

Discussion of Results

The combined convection and radiation problem described above was solved numerically by Taitel and Hartnett [38] for an absorbing and emitting fluid by using the exact formulation and the optically thin and the optically thick limit approximations. Recently Boles and Özişik [42] solved a similar problem for an absorbing, emitting, isotropically scattering fluid by exact formulation, using the normal-mode expansion technique. Figure 13-7 shows the temperature profile in the boundary layer for the adiabatic wall condition as a function of η at several different values of the parameter ξ for Pr = 1,

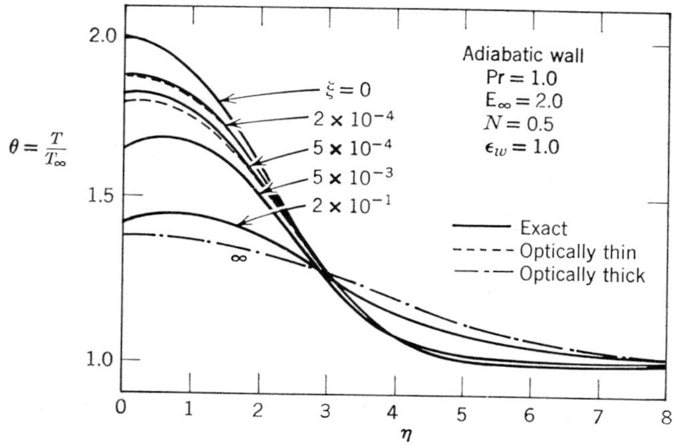

Fig. 13-7. Effects of radiation on temperature profile in boundary layer flow over an adiabatic flat plate for an absorbing, emitting fluid. (From Y. Taitel and J. P. Hartnett [38].)

$E_\infty = 2.0$, $\varepsilon_w = 1$, and $N = 0.5$. The temperature profile for $\xi = 0$ characterizes the problem without radiative heat flux. We note that the temperature in the boundary layer is a maximum for the nonradiative case. The radiative heat flux has the effect of reducing this maximum temperature resulting from viscous energy dissipation. As the value of the parameter ξ is increased, the temperature peak is reduced and the profile is flattened. For values of the parameter of the order of 10^{-4} or less, the boundary layer for the problem in question may be assumed to be optically thin; in this region the solution obtained with the optically thin limit approximation seems to agree reasonably well with the exact solution. However, care must be exercised in applying the optically thin limit approximation for combined convection and radiation in compressible boundary layer heat transfer. Pai and Tsao [12, Fig. 4] studied the flow of radiating, compressible gas over a flat plate and showed that the optically thin limit approximation may give entirely wrong results in comparison with the exact formulation.

We note from Fig. 13-7 that the exact solution approaches the result obtained by the optically thick limit approximation for values of ξ of the order of unity or larger. The temperature profile within the boundary layer is flattened, and a positive temperature gradient with respect to η is established at the wall with increasing ξ. The temperature gradient at the wall is controlled by a balance between the convective heat flux to the wall and the radiative heat flux away from the wall.

To illustrate the effects of scattering we present in Fig. 13-8 the exact solution of the same problem for an absorbing, emitting, isotropically scattering fluid, together with the results for no scattering. The curve for

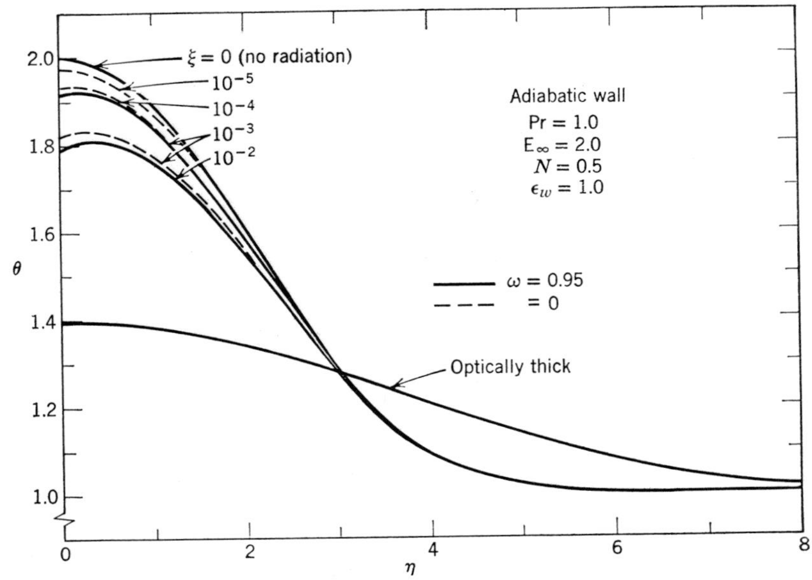

Fig. 13-8. Effect of scattering on temperature profile in boundary layer flow over an adiabatic flat plate. (From M. A. Boles and M. N. Özişik [42].)

$\xi = 0$ characterizes the nonradiating case. The presence of scattering reduces the effects of radiation; hence the temperature profiles with scattering are shifted toward the temperature profile for nonradiating flow.

13-6 LAMINAR FREE CONVECTION FROM A VERTICAL PLATE FOR AN ABSORBING, EMITTING, AND SCATTERING FLUID

The interaction of radiation with laminar free convection heat transfer from a vertical plate was investigated by Cess [25] for an absorbing, emitting fluid in the optically thick region, using the singular perturbation technique to solve the problem. Arpaci [26] considered a similar problem in both the optically thin and the optically thick regions and used the approximate integral technique and first-order profiles to solve the energy equation. Cheng and Özişik [27] considered a related problem for an absorbing, emitting, and isotropically scattering fluid, and treated the radiation part of the problem exactly with the normal-mode expansion technique. In this section we present the formulation of the problem of simultaneous radiation and free convection from a vertical plate, describe the methods of solution, and discuss some of the results.

Formulation

Consider a heated, vertical plate at a uniform temperature T_w submerged in an absorbing, emitting, isotropically scattering, incompressible, gray fluid of infinite extent and at a temperature T_∞. Figure 13-9 shows the geometry and coordinates for the case $T_w > T_\infty$ (i.e., heated plate). The continuity, momentum, and energy equations for two-dimensional, steady,

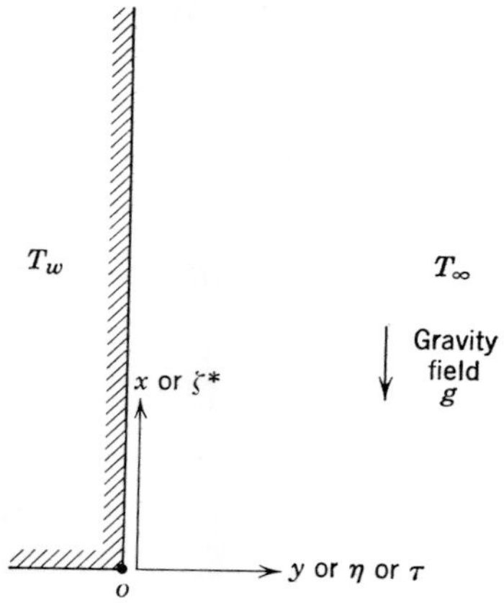

Fig. 13-9. Geometry and coordinate system for combined radiation and laminar free convection from a vertical hot plate ($T_w > T$).

laminar free convection with radiation are given as

$$\frac{\partial u}{\partial x} + \frac{\partial v}{\partial y} = 0 \tag{13-124}$$

$$u\frac{\partial u}{\partial x} + v\frac{\partial u}{\partial y} = \nu\frac{\partial^2 u}{\partial y^2} + g\lambda(T - T_\infty) \tag{13-125}$$

$$u\frac{\partial T}{\partial x} + v\frac{\partial T}{\partial y} = \frac{k}{\rho c_p}\frac{\partial^2 T}{\partial y^2} - \frac{1}{\rho c_p}\frac{\partial q^r}{\partial y} \tag{13-126}$$

where λ is the coefficient of thermal expansion, g is the acceleration of gravity, q^r is the net radiative heat flux in the y direction, and the term $g\lambda(T - T_\infty)$ in the momentum equation characterizes the buoyancy effects. We note that these equations are the familiar expressions for laminar free convection from a vertical plate except for the radiative heat flux term appearing in the energy equation.

The boundary conditions for the continuity and momentum equations are given as

$$u = v = 0 \quad \text{at } y = 0 \tag{13-127}$$

$$u = 0 \quad \text{at } y \to \infty \tag{13-128}$$

and for the energy equation as

$$T = T_w \quad \text{at } y = 0 \tag{13-129}$$

$$T = T_\infty \quad \text{at } y \to \infty \tag{13-130}$$

The radiative heat flux q^r in the energy equation is determined from the solution of the equation of radiative transfer, as described later in the analysis. At this point we focus our attention on the transformation of the above partial differential equations by the application of the standard techniques used in the treatment of similar problems for the nonradiative case.

A stream function $\psi(x, y)$ is defined as

$$u = \frac{\partial \psi(x, y)}{\partial y} \quad \text{and} \quad v = -\frac{\partial \psi(x, y)}{\partial x} \tag{13-131}$$

Then the continuity equation is identically satisfied, and the momentum and energy equations are transformed respectively to

$$\frac{\partial \psi}{\partial y}\frac{\partial^2 \psi}{\partial x\, \partial y} - \frac{\partial \psi}{\partial x}\frac{\partial^2 \psi}{\partial y^2} = \nu \frac{\partial^3 \psi}{\partial y^3} + g\lambda(T - T_\infty) \tag{13-132}$$

$$\frac{\partial \psi}{\partial y}\frac{\partial T}{\partial x} - \frac{\partial \psi}{\partial x}\frac{\partial T}{\partial y} = \alpha \frac{\partial^2 T}{\partial y^2} - \frac{1}{\rho c_p}\frac{\partial q^r}{\partial y} \tag{13-133}$$

where α is the thermal diffusivity of the fluid.

A new independent variable $\eta \equiv \eta(x, y)$ and a dependent variable $f(x, y)$ are introduced as

$$\eta \equiv \left(\frac{g\lambda}{4\nu^2}\right)^{1/4}(T_w - T_\infty)^{1/4}\frac{y}{x^{1/4}} = \left(\frac{\text{Gr}}{4}\right)^{1/4}\frac{y}{x} \tag{13-134a}$$

$$f(x, \eta) \equiv \frac{1}{4\nu}\left(\frac{\text{Gr}}{4}\right)^{-(1/4)}\psi(x, y) \tag{13-134b}$$

where Gr is the local Grashof number defined in the conventional manner, that is,

$$\text{Gr} \equiv \frac{g\lambda(T_w - T_\infty)x^3}{\nu^2} \tag{13-135}$$

When the above new variables are introduced into the momentum and energy equations 13-132 and 13-133, these equations are transformed respectively to[13]

$$\frac{\partial^3 f}{\partial \eta^3} + 3f\frac{\partial^2 f}{\partial \eta^2} - 2\left(\frac{\partial f}{\partial \eta}\right)^2 + \frac{\theta - \theta_\infty}{1 - \theta_\infty} = 4x\left(\frac{\partial^2 f}{\partial x\, \partial \eta}\frac{\partial f}{\partial \eta} - \frac{\partial f}{\partial x}\frac{\partial^2 f}{\partial \eta^2}\right) \tag{13-136}$$

$$\frac{1}{\text{Pr}}\frac{\partial^2 \theta}{\partial \eta^2} + 3f\frac{\partial \theta}{\partial \eta} = 4x\left(\frac{\partial f}{\partial \eta}\frac{\partial \theta}{\partial x} - \frac{\partial f}{\partial x}\frac{\partial \theta}{\partial \eta}\right) + \frac{\beta x}{N\,\text{Pr}}\left(\frac{4}{\text{Gr}}\right)^{1/4}\frac{\partial Q^r}{\partial \eta} \tag{13-137}$$

where we have defined

$$\theta \equiv \frac{T}{T_w}, \qquad \theta_\infty = \frac{T_\infty}{T_w} \tag{13-138a}$$

$$N \equiv \frac{k\beta}{4n^2\bar{\sigma}T_\infty{}^3}, \qquad Q^r = \frac{q^r}{4n^2\bar{\sigma}T_w{}^4} \tag{13-138b}$$

and $T_w > T_\infty$.

The velocity components given by Eqs. 13-131 becomes

$$u = \frac{4\nu}{x}\left(\frac{\mathrm{Gr}}{4}\right)^{1/2}\frac{\partial f}{\partial \eta} \tag{13-139a}$$

$$v = -\nu\left(\frac{\mathrm{Gr}}{4}\right)^{1/4}\left(-\frac{\eta}{x}\frac{\partial f}{\partial \eta} + 4\frac{\partial f}{\partial x} + 3\frac{f}{x}\right) \tag{13-139b}$$

Equations 13-136 and 13-137 are written in the alternative forms as

$$\frac{\partial^3 f}{\partial \eta^3} + 3f\frac{\partial^2 f}{\partial \eta^2} - 2\left(\frac{\partial f}{\partial \eta}\right)^2 + \frac{\theta - \theta_\infty}{1 - \theta_\infty} = \zeta^*\left(\frac{\partial^2 f}{\partial \zeta^* \partial \eta}\frac{\partial f}{\partial \eta} - \frac{\partial f}{\partial \zeta^*}\frac{\partial^2 f}{\partial \eta^2}\right) \tag{13-140}$$

$$\frac{1}{\mathrm{Pr}}\frac{\partial^2 \theta}{\partial \eta^2} + 3f\frac{\partial \theta}{\partial \eta} = \zeta^*\left(\frac{\partial f}{\partial \eta}\frac{\partial \theta}{\partial \zeta^*} - \frac{\partial f}{\partial \zeta^*}\frac{\partial \theta}{\partial \eta}\right) + \frac{\zeta^*}{N\mathrm{Pr}}\frac{\partial Q^r}{\partial \eta} \tag{13-141}$$

where we have defined the new independent variable ζ^* as

$$\zeta^* = \beta x\left(\frac{\mathrm{Gr}}{4}\right)^{-(1/4)} \tag{13-142a}$$

and utilized the relation

$$4x\frac{\partial}{\partial x} = \zeta^*\frac{\partial}{\partial \zeta^*} \tag{13-142b}$$

We note that the parameter ζ^* in free convection is analogous to the parameter ζ, defined by Eq. 13-61a, in forced convection.

The boundary conditions 13-127 and 13-128 for the momentum equation are transformed to

$$f = \frac{\partial f}{\partial \eta} = 0 \quad \text{at } \eta = 0 \tag{13-143a}$$

$$\frac{\partial f}{\partial \eta} = 0 \quad \text{at } \eta \to \infty \tag{13-143b}$$

The boundary conditions 13-129 and 13-130 for the energy equation become

$$\theta = 1 \quad \text{at } \eta = 0 \tag{13-143c}$$

$$\theta = \theta_\infty \quad \text{at } \eta \to \infty \tag{13-143d}$$

The transformed momentum and energy equations 13-140 and 13-141 involve partial derivatives with respect to the independent variable ζ^*; therefore additional boundary conditions are needed. These boundary conditions are chosen as

$$\theta = \theta_0 \quad \text{and} \quad f = f_0 \quad \text{for } \zeta^* = 0 \tag{13-144}$$

where f_0 and θ_0 are the solutions of the momentum and energy equations for the case of no radiation.

The net radiative heat flux term appearing in the energy equation is generally obtained from the solution of the equation of radiative transfer given in the optical variable τ, defined as

$$\tau = \beta y \tag{13-145}$$

whereas the energy equation 13-141 is given in the independent variable η, defined as

$$\eta = \left(\frac{\text{Gr}}{4}\right)^{1/4} \frac{y}{x} = \frac{y\beta}{\zeta^*} \tag{13-146}$$

The relation between τ and η is obtained from Eqs. 13-145 and 13-146 as

$$\tau = \zeta^* \eta \tag{13-147}$$

The mathematical formulation of the combined radiation and free convection problem here considered is now complete. The determination of temperature distribution in the fluid involves a simultaneous solution of the equations of momentum and energy and the equation of radiative transfer. We examine now the methods of solution of these equations by treating the radiation part of the problem (1) approximately with the optically thick limit approximation, and (2) exactly with the normal-mode expansion technique.

Optically Thick Limit Approximation

The dimensionless net radiative flux Q^r for the optically thick limit approximation is given by (see Eq. 13-82)

$$Q^r = -\tfrac{4}{3}\theta^3 \frac{\partial \theta}{\partial \tau} \tag{13-148}$$

Changing the variable from τ to η by the transformation given by Eq. 13-147, we obtain

$$Q^r = -\frac{4}{3\zeta^*}\,\theta^3 \frac{\partial \theta}{\partial \eta} \tag{13-149}$$

When this expression for the radiative heat flux is introduced into the energy equation, it can be shown that the analysis lends itself to complete similarity, and the momentum and the energy equations are transformed into ordinary differential equations in the η variable. We obtain

$$\frac{df}{d\eta^3} + 3f\frac{d^2f}{d\eta^2} - 2\left(\frac{df}{d\eta}\right)^2 + \frac{\theta - \theta_\infty}{1 - \theta_\infty} = 0 \tag{13-150}$$

$$\frac{1}{\Pr}\frac{d^2\theta}{d\eta^2} + 3f\frac{d\theta}{d\eta} = -\frac{4}{3N\Pr}\frac{d}{d\eta}\left(\theta^3\frac{d\theta}{d\eta}\right) \tag{13-151a}$$

or

$$\frac{1}{\Pr}\frac{d}{d\eta}\left[\left(1 + \frac{4}{3N}\theta^3\right)\frac{d\theta}{d\eta}\right] + 3f\frac{d\theta}{d\eta} = 0 \tag{13-151b}$$

where

$$\theta \equiv \theta(\eta) \qquad \text{and} \qquad f \equiv f(\eta) \tag{13-152}$$

The boundary conditions 13-143 are transformed to

$$f = \frac{df}{d\eta} = 0 \qquad \text{and} \qquad \theta = 1 \qquad \text{at } \eta = 0 \tag{13-153a}$$

$$\frac{df}{d\eta} = 0 \qquad \text{and} \qquad \theta = \theta_\infty \qquad \text{at } \eta \to \infty \tag{13-153b}$$

A simultaneous solution of Eqs. 13-150 and 13-151 subject to the boundary conditions 13-153 yields the temperature distribution in the boundary layer. A numerical scheme such as a Runge-Kutta method may be used to solve these equations.

Application of the Normal-Mode Expansion Technique

The normal-mode expansion technique can be applied to solve the radiation part of the problem and to determine an exact relation for the radiation term $\partial Q^r/\partial\eta$ (or $\partial Q^r/\partial\tau$) appearing in the energy equation.

We consider the fluid as an absorbing, emitting, isotropically scattering, gray, locally plane-parallel, semi-infinite ($0 \leq \tau < \infty$) medium at temperature $T(\tau, \zeta^*)$. The boundary at the wall is a diffusely emitting, specularly reflecting, opaque, gray surface at uniform temperature T_w. Then the equation of radiative transfer is given in the dimensionless form as

$$\mu\frac{\partial I(\tau, \zeta^*, \mu)}{\partial\tau} + I(\tau, \zeta^*, \mu) = (1 - \omega)\theta^4(\tau, \zeta^*) + \frac{\omega}{2}\int_{-1}^{1} I(\tau, \zeta^*, \mu')\, d\mu'$$

$$\text{in } 0 \leq \tau < \infty,\ -1 \leq \mu \leq 1 \tag{13-154}$$

with the boundary conditions

$$I(0, \zeta^*, \mu) = \varepsilon_w + (1 - \varepsilon_w)I(0, \zeta^*, -\mu), \quad \mu > 0 \tag{13-155a}$$

$$\lim_{\tau\to\infty} I(\tau, \zeta^*, \mu) \to I_p(\tau, \zeta^*, \mu) \tag{13-155b}$$

where $I_p(\tau, \zeta^*, \mu)$ is a particular solution of the equation of radiative transfer, ε_w is the wall emissivity, and $\theta(\tau, \zeta^*)$ is the dimensionless temperature as defined by Eq. 13-138a. The dimensionless net radiative heat flux Q^r is related to the dimensionless radiation intensity by

$$Q^r(\tau, \zeta^*) = \frac{1}{2}\int_{-1}^{1} I(\tau, \zeta^*, \mu)\mu \, d\mu \tag{13-156}$$

Therefore, if $I(\tau, \zeta^*, \mu)$ is known, the net radiative heat flux term appearing in the energy equation 13-141 can be evaluated.

The solution of the equation of radiative transfer (Eq. 13-154) can be written as a linear sum of the solutions to the homogeneous part and a particular solution $I_p(\tau, \zeta^*, \mu)$ in the form

$$I(\tau,\zeta^*,\mu) = A(\nu_0, \zeta^*)\phi(\nu_0, \mu)e^{-\tau/\nu_0} + \int_0^1 A(\nu,\zeta^*)\phi(\nu,\mu)e^{-\tau/\nu}\, d\nu + I_p(\tau,\zeta^*,\mu) \tag{13-157}$$

which satisfies the boundary condition given by Eq. 13-155b since we omitted in this solution the parts of the homogeneous solution that diverge at infinity. Here the discrete normal mode $\phi(\nu_0, \mu)$ and the continuum normal mode $\phi(\nu, \mu)$ were defined in Chapter 10 (see Eqs. 10-8 and 10-16), and the discrete eigenvalues ν_0 are the two roots of the dispersion relation given by Eq. 10-9. The two unknown expansion coefficients $A(\nu_0, \zeta^*)$ and $A(\nu, \zeta^*)$ can be determined by constraining the solution 13-157 to meet the boundary condition 13-155a and then by utilizing the orthogonality property of the normal modes and various normalization integrals as described in Chapters 10 and 11 or in reference 43.

However, the solution given by Eq. 13-157 involves a particular solution that cannot be determined until the temperature term $\theta^4(\tau, \zeta^*)$ appearing in the equation of radiative transfer is known. At this point we assume that an initial guess is available for the temperature distribution $\theta(\tau, \zeta^*)$ and that the function $\theta^4(\tau, \zeta^*)$ is represented in the form

$$\theta^4(\tau, \zeta^*) = \theta_\infty{}^4 + S(\tau, \zeta^*) \tag{13-158}$$

where the function $S(\tau, \zeta^*)$ is prescribed within the thermal boundary layer (i.e., $0 \le \tau \le \tau_0$ or $0 \le \eta \le \eta_0$) and vanishes outside the thermal boundary layer (i.e., $\tau > \tau_0$). We choose to represent function $S(\tau, \zeta^*)$ within the boundary layer by a truncated cosine series in the form

$$S(\tau, \zeta^*) = \sum_{m=0}^{M} B_m(\zeta^*) \cos\frac{m\pi\tau}{\tau_0} \quad \text{in } 0 \le \tau \le \tau_0 \tag{13-159}$$

$$= \sum_{m=0}^{M} B_m(\zeta^*) \cos\frac{m\pi\eta}{\tau_0} \quad \text{in } 0 \le \eta \le \eta_0 \tag{13-160}$$

and assume that the coefficients $B_m(\zeta^*)$ can be determined if $S(\tau, \zeta^*)$ is perscribed. Then the inhomogeneous term $(1-\omega)\theta^4(\tau, \zeta^*)$ is represented in the form

$$(1-\omega)\theta^4(\tau, \zeta^*) = (1-\omega)\left[\theta_\infty{}^4 + \sum_{m=0}^{M} B_m(\zeta^*) \cos \frac{m\pi\tau}{\tau_0}\right] \tag{13-161}$$

A particular solution $I_p(\tau, \zeta^*, \mu)$ of the equation of radiative transfer 13-155 for an inhomogeneous term represented as in Eq. 13-161 is readily obtainable from Table 10-6 as

$$I_p(\tau, \zeta^*, \mu) = \theta_\infty{}^4 + (1-\omega)\sum_{m=0}^{M} B_m(\zeta^*)$$

$$\times \frac{\zeta^*\eta_0/m\pi[(\zeta^*\eta_0/m\pi)\cos(m\pi\eta/\eta_0) + \mu \sin(m\pi\eta/\eta_0)]}{[1-(\omega\zeta^*\eta_0/m\pi)\tan^{-1}(m\pi/\zeta^*\eta_0)][(\zeta^*\eta_0/m\pi)^2 + \mu^2]} \tag{13-162}$$

Once a particular solution is available for a given guess as to the distribution of temperature, the corresponding expansion coefficients $A(\nu_0, \zeta^*)$ and $A(\nu, \zeta^*)$ are determined, the dimensionless radiation intensity $I^*(\tau, \zeta^*, \mu)$ is obtained from Eq. 13-157, and the dimensionless net radiative heat flux $Q^r(\tau, \zeta^*)$ is determined from Eq. 13-156 as

$$Q^r(\tau, \zeta^*) = \tfrac{1}{2}(1-\omega)\left[\nu_0 A(\nu_0, \zeta^*)e^{-\tau/\nu_0} + \int_0^1 \nu A(\nu, \zeta^*)e^{-\tau/\nu}\, d\nu\right.$$

$$\left. + \frac{1}{1-\omega}\int_{-1}^{1} I_p(\tau, \zeta^*, \mu')\mu'\, d\mu'\right] \tag{13-163}$$

By changing the independent variable from τ to η, Eq. 13-163 becomes

$$Q^r(\eta, \zeta^*) = \tfrac{1}{2}(1-\omega)\left[\nu_0 A(\nu_0, \zeta^*)e^{-\zeta^*\eta/\nu_0} + \int_0^1 \nu A(\nu, \zeta^*)e^{-\zeta^*\eta/\nu}\, d\nu\right.$$

$$\left. + \frac{1}{1-\omega}\int_{-1}^{1} I_p(\eta, \zeta^*, \mu')\mu'\, d\mu'\right] \tag{13-164}$$

The differentiation of Eq. 13-164 with respect to η yields

$$\frac{\partial Q^r(\eta, \zeta^*)}{\partial \eta} = -\tfrac{1}{2}\zeta^*(1-\omega)\left[A(\nu_0, \zeta^*)e^{-\zeta^*\eta/\nu_0} + \int_0^1 A(\nu, \zeta^*)e^{-\zeta^*\eta/\nu}\, d\nu\right.$$

$$\left. - \frac{1}{\zeta^*(1-\omega)}\int_{-1}^{1} \frac{\partial}{\partial \eta} I_p(\eta, \zeta^*, \mu')\mu'\, d\mu'\right] \tag{13-165}$$

where $I_p(\eta, \zeta^*, \mu)$ is given by Eq. 13-162. Equation 13-165 provides the desired expression for the radiative heat flux term $\partial Q^r/\partial \eta$ appearing in the energy equation 13-141.

The resulting momentum and energy equations 13-140 and 13-141 are given in the final form as

$$\frac{\partial^3 f}{\partial \eta^3} + 3f\frac{\partial^2 f}{\partial \eta^2} - 2\left(\frac{\partial f}{\partial \eta}\right)^2 + \frac{\theta - \theta_\infty}{1 - \theta_\infty} = \zeta^*\left(\frac{\partial^2 f}{\partial \zeta^* \partial \eta}\frac{\partial f}{\partial \eta} - \frac{\partial f}{\partial \zeta^*}\frac{\partial^2 f}{\partial \eta^2}\right) \quad (13\text{-}166)$$

$$\begin{aligned}\frac{1}{\mathrm{Pr}}\frac{\partial^2 \theta}{\partial \eta^2} + 3f\frac{\partial \theta}{\partial \eta} = \zeta^*\left(\frac{\partial f}{\partial \eta}\frac{\partial \theta}{\partial \zeta^*} - \frac{\partial f}{\partial \zeta^*}\frac{\partial \theta}{\partial \eta}\right) &- \tfrac{1}{2}(1-\omega)\frac{\zeta^{*2}}{N\,\mathrm{Pr}}\\ &\times\left[A(\nu_0, \zeta^*)e^{-\zeta^*\eta/\nu_0} + \int_0^1 A(\nu, \zeta^*)e^{-\zeta^*\eta/\nu}\,d\nu\right.\\ &\left.- \frac{1}{\zeta^*(1-\omega)}\int_{-1}^{1}\frac{\partial}{\partial \eta}I_p(\tau, \zeta^*, \mu')\mu'\,d\mu'\right] \quad (13\text{-}167)\end{aligned}$$

Equations 13-166 and 13-167 with the boundary conditions 13-143 and 13-144 can be solved numerically by an iterative scheme, and the temperature distribution in the fluid determined.

The total heat flux q_w at the wall is of interest in engineering applications. Once the temperature distribution is known, the total heat flux at the wall is evaluated from

$$Q_w \equiv \frac{q_w}{4n^2\bar{\sigma}T_w^{\,4}} = \left[-N\frac{\partial \theta}{\partial \tau} + Q^r\right]_{\tau=0} = \left[-\frac{N}{\zeta^*}\frac{\partial \theta}{\partial \eta} + Q^r\right]_{\eta=0} \quad (13\text{-}168)$$

where the dimensionless net radiative heat flux $Q^r|_{\eta=0}$ is given by (see Eq. 13-164)

$$\begin{aligned}Q^r|_{\eta=0} = \tfrac{1}{2}(1-\omega)\Bigg[\nu_0 A(\nu_0, \zeta^*) &+ \int_0^1 \nu A(\nu, \zeta^*)\,d\nu\\ &+ \frac{1}{1-\omega}\int_{-1}^{1} I_p(0, \zeta^*, \mu')\mu'\,d\mu'\Bigg] \quad (13\text{-}169)\end{aligned}$$

In heat transfer analysis a local Nusselt number Nu is defined as

$$\mathrm{Nu} \equiv \frac{hx}{k} = \frac{q_w x}{k(T_w - T_\infty)} \quad (13\text{-}170)$$

where h is the local heat transfer coefficient. When q_w from Eq. 13-168 is substituted into Eq. 13-170, the local Nusselt number for combined free convection and radiation becomes

$$\mathrm{Nu}\left(\frac{\mathrm{Gr}}{4}\right)^{-(1/4)} = \frac{1}{1-\theta_\infty}\left[-\frac{\partial \eta}{\partial \theta} + \frac{\zeta^*}{N}Q^r\right]_{\eta=0} \quad (13\text{-}171)$$

For very small values of ζ^*/N the radiation term in Eq. 13-171 is neglected, and the resulting equation agrees with the definition of local Nusselt number for free convection from a nonradiating fluid given by Ostrach [44].

Method of Solution and Discussion of Results

Cheng and Özişik [27] transformed the partial differential equations 13-166 and 13-167 into ordinary differential equations in η by a finite difference in the ζ^* variable and solved the resulting equations numerically with an iterative scheme subject to the boundary conditions 13-143 and 13-144. To start the calculations at each nodal point ζ_i^*, the temperature distribution $\theta(\tau, \zeta_{i-1}^*)$ at the previous station ζ_{i-1}^* is used as an initial guess, the fourth power of this temperature $\theta^4(\tau, \zeta_{i-1}^*)$ is represented in a 21-term cosine expansion as in Eq. 13-160, and the coefficients $B_m(\zeta^*)$ are determined. A particular solution $I_p(\tau, \zeta^*, \mu)$ of the equation of radiative transfer is obtained from Eq. 13-162, the expansion coefficients $A(\nu_0, \zeta_i^*)$ and $A(\nu, \zeta_i^*)$ are determined by an approach described in Chapters 10 and 11, and the radiative heat flux term $\partial Q^r/\partial \eta$ appearing in the energy equation is evaluated from Eq. 13-165. Knowing $\partial Q^r/\partial \eta$ at the station ζ_i^*, we can integrate the momentum and the energy equations numerically in the η direction by a Runge-Kutta method and obtain a first approximation for the temperature profile at ζ_i^*. This first approximation is utilized in a similar manner to obtain a second approximation, the second approximation to obtain a third approximation, and so forth. The calculations are repeated at the station ζ_i^* until the solution converges to a prescribed criterion.

Figure 13-10 shows the temperature profile at several different axial positions between $\zeta^* = 0$ and 0.9 for the case of a black wall with $\omega = 0.5$,

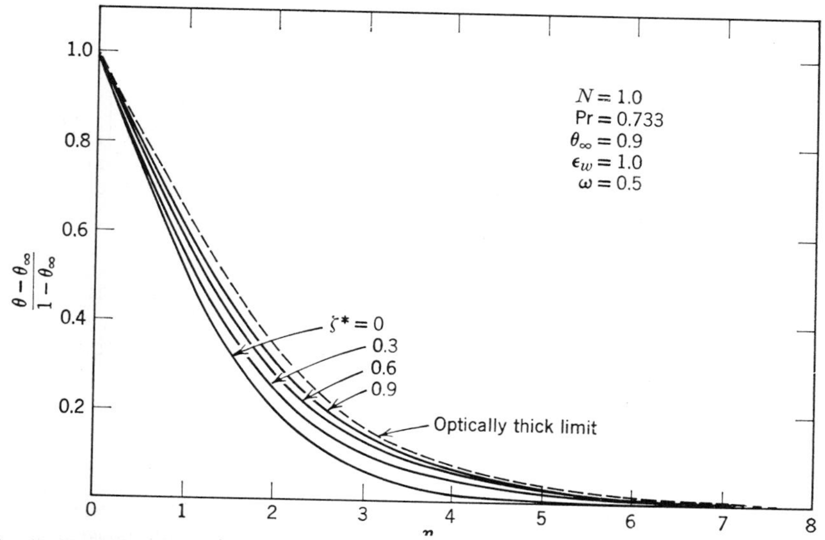

Fig. 13-10. Effects of ζ^* on temperature profile in free convection with radiation from a vertical hot plate. (From E. H. Cheng and M. N. Özişik [27].)

Table 13-2 Numerical Results for Local Nusselt Number[a]
($\theta_\infty = 0.9$, $\varepsilon_w = 1.0$, $T_w > T_\infty$)

Effects	ω	N	Pr	ζ^*	$\dfrac{Nu_x}{(Gr_x/4)^{1/4}}$
of N	0.0	0.1	0.733	0.2	1.2618
		1			0.5755
		10			0.5143
		∞			0.5083
of ω	0.0	0.1	0.733	0.1	0.7713
	0.5				0.6916
	0.9				0.6055
	1.0				0.5083
of ζ^*	0.0	0.1	0.733	0	0.5083
				0.025	0.5594
				0.05	0.6120
				0.10	0.7713
				0.15	0.9951
				0.20	1.2618
of Pr	0.0	0.1	0.733	0.1	0.7713
			10		1.3609
			100		2.3784

[a] From E. H. Cheng and M. N. Özişik [27].

$N = 1$, Pr $= 0.733$, and $\theta_\infty = 0.9$. Included in this figure is the temperature profile obtained with the optically thick limit approximation (i.e., the solution of Eqs. 13-150 and 13-151). The solution for $\zeta^* = 0$ characterizes the nonradiating case and is the same as that given by Ostrach [44]. As the value of ζ^* increases, the temperature profile approaches that obtained by the optically thick limit approximation.

Table 13-2 shows the effects of the parameters N, ω, ζ^*, and Pr on the local Nusselt number for a set of conditions specified in this table. The ratio $Nu_x/(Gr_x/4)^{1/4}$ increases with decreasing N, decreasing ω, increasing Pr, and increasing ζ^*. For large values of N or for $\omega = 1$ this ratio is the same as that for the nonradiative case.

REFERENCES

1. R. Goulard and M. Goulard, "Energy Transfer in the Couette Flow of a Radiant and Chemically Reacting Gas," *Proceedings of the Heat Transfer and Fluid Mechanics Institute*, pp. 126–139, Stanford University Press, Palo Alto, Calif., 1959.

2. R. Viskanta and R. J. Grosh, "Boundary Layer in Thermal Radiation Absorbing and Emitting Media," *Intern. J. Heat Mass Transfer*, **5**, 795–806, 1962.

3. J. L. Novotny and Kwang-Tzu Yang, "The Interaction of Thermal Radiation in Optically Thick Boundary Layers," *ASME Paper* No. 67-HT-9, 1967.

4. A. N. Rumynskii, "Boundary Layers in Radiating and Absorbing Media," *Am. Rocket Soc. J.*, **32**, 1135–1138, 1962.

5. R. Goulard, "The Transition from Black Body to Rosseland Formulations in Optically Thick Flows," *Intern. J. Heat Mass Transfer*, **7**, 1145–1146, 1964.

6. K. H. Kim, M. N. Özişik, and J. C. Mulligan, "Radiation Effects in an Optically Thick Non-Newtonian Boundary Layer with Injection and Suction," *Proceedings of the International Heat Transfer Conference*, Paris, Vol. 3, Section R2-4, pp. 1–10, 1970.

7. J. C. Y. Koh and C. N. de Silva, "Interaction Between Radiation and Convection in the Hypersonic Boundary Layer on a Flat Plate," *Am. Rocket Soc. J.*, **32**, 739–743, 1962.

8. R. D. Cess, "The Interaction of Thermal Radiation with Conduction and Convection Heat Transfer," in *Advances in Heat Transfer*, edited by T. F. Irvine and J. P. Hartnett, Vol. 1, pp. 1–50, Academic Press, New York, 1964.

9. R. J. Tabaczynski and L. A. Kennedy, "Thermal Radiation Effects in Laminar Boundary-Layer Flow," *AIAA J.*, **5**, 1893–1894, 1967.

10. R. D. Cess, "Radiation Effects upon Boundary Layer Flow of an Absorbing Gas," *J. Heat Transfer*, **86C**, 469–475, 1964.

11. A. M. Smith and H. A. Hassan, "Nongray Radiation Effects on the Boundary Layer at Low Eckert Numbers," *ASME Paper* No. 66-WA/HT-35, 1966.

12. S. I. Pai and C. K. Tsao, "A Uniform Flow of a Radiating Gas over a Flat Plate," *Proceedings of the Third International Heat Transfer Conference*, Chicago, Vol. 5, pp. 129–137, 1966.

13. S. I. Pai and A. P. Scaglione, "Unsteady Laminar Boundary Layers of an Infinite Plate in an Optically Thick Radiating Gas," *Appl. Sci. Res.*, **22**, 97–112, 1970.

14. C. C. Oliver and P. W. McFadden, "The Interaction of Radiation and Convection in the Laminar Boundary Layer," *J. Heat Transfer*, **88C**, 205–213, 1966.

15a. M. A. Boles and M. N. Özişik, (to be published).

15b. M. A. Boles, "Interactions of Radiation with Conductive and Convective Heat Transfer in Ablation and Boundary Layer Flow," Ph.D. Dissertation, Mechanical and Aerospace Engineering Department, North Carolina State University, Raleigh, N.C., 1972.

16. S. I. Pai and T. Hsieh, "A Perturbation Theory of Anisentropic Flow with Radiative Heat Transfer," *Z. Flugswiss.*, **18**, 44–50, 1970.

17. T. Hsieh and S. I. Pai, "A Perturbation Theory of Anisentropic Flow with Non-equilibrium Radiative Transfer," *Z. Flugswiss.*, **19**, 122–129, 1971.

18. J. D. Anderson, Jr., "Nongray Radiative Stagnation Point Heat Transfer," *AIAA J.*, **6**, 758–760, 1968.

19. W. B. Olstad, "Stagnation-Point Solutions for Inviscid Radiating Shock Layers," *NASA Tech. Note* TN D-5792, June 1970.

20. W. B. Olstad, "Nongray Radiating Flow about Smooth Symmetric Bodies," *AIAA J.*, **9**, 122–130, 1971.

21. J. D. Anderson, "Nongray Radiative Stagnation Point Heat Transfer," *AIAA J.*, **6,** 758–760, 1968.

22. R. B. Dirling, Jr., W. S. Rigdon and M. Thomas, "Stagnation Point Heating Including Spectral Radiative Transfer," *Proceedings of the Heat Transfer and Fluid Mechanics Institute*, pp. 141–162, Stanford University Press, Palo Alto, Calif., 1967.

23. K. H. Kim, M. N. Özişik and J. C. Mulligan, "A Non-Similar Solution of Heat Transfer in External Non-Newtonian Flow with Thermal Radiation," AIAA 77th *Aerospace Sciences Meeting*, Washington, D.C., Jan. 1973, (in press).

24. J. M. Elliott, R. I. Vachon, D. F. Dyer, and J. R. Dunn, "Application of the Patankar-Spalding Finite Difference Procedure to Turbulent Radiating Boundary Layer Flow," *Intern. J. Heat Mass Transfer*, **14,** 667–672, 1971.

25. R. D. Cess, "The Interaction of Thermal Radiation with Free Convection Heat Transfer," *Intern. J. Heat Mass Transfer*, **9,** 1269–1277, 1966.

26. Vedat S. Arpaci, "Effect of Thermal Radiation on the Laminar Free Convection from a Heated Vertical Plate," *Intern. J. Heat Mass Transfer*, **11,** 871–881, 1968.

27. E. H. Cheng and M. N. Özişik, "Radiation with Free Convection in an Absorbing, Emitting and Scattering Medium," *Intern. J. Heat Mass Transfer*, **15,** 1243–1252, 1972.

28. H. Schlichting, *Boundary Layer Theory*, 6th ed., McGraw-Hill Book Co., New York, 1968.

29. R. Goulard, "Fundamental Equations of Fluid Dynamics," in *High Temperature Aspects of Hypersonic Flow*, edited by W. C. Nelson, pp. 529–554, Macmillan Co., New York, 1964.

30a. R. Viskanta, "Radiation Transfer and Interaction of Convection with Radiation Heat Transfer, "in *Advances in Heat Transfer*, edited by T. F. Irvine and J. P. Hartnett, Vol. 3, pp. 175–251, Academic Press, New York, 1966.

30b. R. Viskanta, *Heat Transfer in Thermal Radiation Absorbing and Scattering Media* ANL-6170, Argonne National Laboratory, Argonne, Ill., 1960.

31. S. I. Pai, *Radiation Gas Dynamics*, Springer-Verlag, New York, 1966.

32. W. G. Vincenti and C. H. Kruger, Jr., *Introduction to Physical Gas Dynamics*, John Wiley and Sons, New York, 1965.

33. S. I. Pai, "Inviscid Flow of Radiation Gas Dynamics," *J. Math. Phys. Sci.*, **3,** 361–370, 1969.

34. E. A. Milne, "Thermodynamics of Stars," in *Handbuch der Astrophysik*, edited by G. Eberhard et al., Vol. 3, Part 1, pp. 65–255, Springer-Verlag, Berlin, 1930.

35. J. H. Jeans, "On the Radiation Viscosity and the Rotation of Astronomical Masses," *Monthly Notices Roy. Astron. Soc.*, **86,** 328–335, 1926.

36. J. H. Hazlehurst and W. L. W. Sargent, "Hydrodynamics in a Radiation Field—Covariant Treatment," *Astrophys. J.*, **130,** 276–285, 1959.

37. R. Goulard, "Similarity Parameters in Radiation Gas Dynamics," in *High Temperatures in Aeronautics*, edited by D. D. Tamburini, pp. 181–209, Pergamon Press, Oxford, 1963.

38. Y. Taitel and J. P. Hartnett, "Equilibrium Temperature in a Boundary Layer Flow over a Flat Plate of Absorbing-Emitting Gas," *ASME Paper* No. 66-WA/HT-48, 1966.

39. W. F. Ames, *Nonlinear Partial Differential Equations in Engineering*, Academic Press, New York, 1965.

40. A. J. A. Morgan, "The Reduction by One of the Number of Independent Variables in Some System of Partial Differential Equations," *Quart. J. Math. (Oxford)*, **3,** 250–259, 1951.

41. A. H. Shapiro, *Compressible Fluid Flow*, Vol. 1, Ronald Press Co., New York, 1953.

42. M. A. Boles and M. N. Özişik, (to be published).

43. M. N. Özişik and C. E. Siewert, "On the Normal-Mode Expansion Technique for Radiative Transfer in a Scattering, Absorbing and Emitting Slab with Specularly Reflecting Boundaries," *Intern. J. Heat Mass Transfer*, **12,** 611–620, 1969.

44. S. Ostrach, "An Analysis of Laminar Free-Convection Flow and Heat Transfer about a Flat Plate Parallel to the Direction of the Generating Body Force," *NACA Rept.* No. 1111, U.S. Government Printing Office, Washington, D.C., 1953.

NOTES

1

$$\begin{aligned}\sum_{i=1}^{3} n_i n_i = n_1^2 + n_2^2 + n_3^2 &= \cos^2\theta + \sin^2\theta\cos^2\phi + \sin^2\theta\sin^2\phi \\ &= \cos^2\theta + \sin^2\theta(\cos^2\phi + \sin^2\phi) \\ &= \cos^2\theta + \sin^2\theta = 1\end{aligned}$$

since $n_1 = \cos\theta$, $n_2 = \sin\theta\cos\phi$, and $n_3 = \sin\theta\sin\phi$.

2 The integral over the solid angle 4π in Eq. 13-5 is evaluated as

$$\int_{4\pi} n_i n_j \, d\Omega = \int_{\phi=0}^{2\pi}\int_{\theta=0}^{\pi/2} n_i n_j \sin\theta \, d\theta \, d\phi = \begin{cases} \dfrac{4\pi}{3} & \text{for } i = i \\ 0 & \text{for } i \neq j \end{cases}$$

since $d\Omega = \sin\theta \, d\theta \, d\phi$, $n_1 = \cos\theta$, $n_2 = \sin\theta\cos\phi$, and $n_3 = \sin\theta\sin\phi$.

3 If the effects of radiation stresses are included, the momentum equation is given in the form

$$\rho\left(\frac{\partial u_i}{\partial t} + u_j \frac{\partial u_i}{\partial j}\right) = F_i - \frac{\partial}{\partial x_i}(p + p^r) + \frac{\partial}{\partial x_j}(\tau_{ij} + \tau_{ij}^{\ r}) \tag{1}$$

For no radiation and, say, for $i = 1$ the momentum equation may be written explicitly in the form

$$\begin{aligned}&\rho\left(\frac{\partial u_1}{\partial t} + u_1\frac{\partial u_1}{\partial x_1} + u_2\frac{\partial u_1}{\partial x_2} + u_3\frac{\partial u_1}{\partial x_3}\right) \\ &\quad = F_1 - \frac{\partial p}{\partial x_1} + \frac{\partial}{\partial x_1}\left\{\mu\left[2\frac{\partial u_1}{\partial x_1} - \frac{2}{3}\left(\frac{\partial u_1}{\partial x_1} + \frac{\partial u_2}{\partial x_2} + \frac{\partial u_3}{\partial x_3}\right)\right]\right\} \\ &\qquad + \frac{\partial}{\partial x_2}\left[\mu\left(\frac{\partial u_1}{\partial x_2} + \frac{\partial u_2}{\partial x_1}\right)\right] + \frac{\partial}{\partial x_3}\left[\mu\left(\frac{\partial u_1}{\partial x_3} + \frac{\partial u_3}{\partial x_1}\right)\right]\end{aligned}$$

4 If the effects of the radiation energy density u^r, the radiation stress tensor $p_{ij}^{\ r} = -p^r\,\delta_{ij} + \tau_{ij}^{\ r}$, and the radiative heat flux vector $q_i^{\ r}$ are all included, the energy equation may be

written in the form

$$\frac{\partial}{\partial t}[\rho(\tfrac{1}{2}v^2 + e) + u^r] = -\frac{\partial}{\partial x_i}[\rho u_i(\tfrac{1}{2}v^2 + e) + u_i u^r] - \frac{\partial}{\partial x_i}(q_i{}^c + q_i{}^r) + S$$

$$- \frac{\partial}{\partial x_i}[u_i(p + p^r)] + \frac{\partial}{\partial x_i}[u_i(\tau_{ij} + \tau_{ij}{}^r)] + u_i F_i \quad (1)$$

where e is the internal energy per unit mass, u^r is the radiation energy density per unit volume, u_i is the velocity vector, v is the magnitude of the velocity, p and p^r are the hydrodynamic and radiation pressures, and τ_{ij} and $\tau_{ij}{}^r$ are the hydrodynamic and viscous stresses. The last term on the right-hand side represents the work done on the fluid per unit volume by the body force F_i.

It is convenient to rearrange Eq. 1 with the aid of the equations of continuity and momentum, that is,

$$\frac{\partial \rho}{\partial t} + \frac{\partial}{\partial x_i}(\rho u_i) = 0 \quad (2)$$

$$\rho\left(\frac{\partial u_i}{\partial t} + u_j \frac{\partial u_j}{\partial x_j}\right) = F_i - \frac{\partial}{\partial x_i}(p + p^r) + \frac{\partial}{\partial x_j}(\tau_{ij} + \tau_{ij}{}^r) \quad (3)$$

By carrying out the indicated differentiations in Eq. 1 and by utilizing the above continuity and momentum equations, the energy equation 1 is given in the alternative form as

$$\rho \frac{D}{Dt}\left(e + \frac{u^r}{\rho}\right) = -\frac{\partial}{\partial x_i}(q_i{}^c + q_i{}^r) + S - (p + p^r)\frac{\partial u_i}{\partial x_i} + (\tau_{ij} + \tau_{ij}{}^r)\frac{\partial u_i}{\partial x_i} \quad (4)$$

[5] The transformation of various partial derivatives from the x, Y coordinates to x, η coordinates is given as

$$\frac{dY}{d\eta} = \left[\frac{(1 + m)u_\infty(x)}{j\nu_0 x}\right]^{1/2}, \qquad \frac{d\eta}{dx} = \frac{m - 1}{2}\frac{\eta}{x} \quad (1)$$

$$\frac{\partial F(x, \eta)}{\partial Y} = \frac{\partial F}{\partial x}\cancel{\frac{dx}{dY}}^{0} + \frac{\partial F}{\partial \eta}\frac{d\eta}{dY} = \left[\frac{(1 + m)u_\infty(x)}{j\nu_0 x}\right]^{1/2}\frac{\partial F}{\partial \eta} \quad (2)$$

$$\frac{\partial F(x, \eta)}{\partial x} = \frac{\partial F}{\partial x}\frac{dx}{dx} + \frac{\partial F}{\partial \eta}\frac{d\eta}{dx} = \frac{\partial F}{\partial x} + \frac{m - 1}{2}\frac{\eta}{x}\frac{\partial F}{\partial \eta} \quad (3)$$

$$\frac{\partial \psi}{\partial Y} = \frac{\partial \psi}{\partial \eta}\frac{d\eta}{dY} = u_\infty(x)\frac{df}{d\eta} \quad (4)$$

$$\frac{\partial^2 \psi}{\partial Y^2} = \left[\frac{(1 + m)u_\infty(x)}{j\nu_0 x}\right]^{1/2} u_\infty(x)\frac{d^2 f}{d\eta^2} \quad (5)$$

$$\frac{\partial^2 \psi}{\partial x\, \partial Y} = \frac{\partial}{\partial x}\left(\frac{\partial \psi}{\partial Y}\right) = \left(\frac{\partial}{\partial x} + \frac{m - 1}{2}\frac{\eta}{x}\frac{\partial}{\partial \eta}\right)\left[u_\infty(x)\frac{df}{d\eta}\right] = \frac{u_\infty(x)}{x}\left(m\frac{df}{d\eta} + \frac{m - 1}{2}\eta\frac{d^2 f}{d\eta^2}\right) \quad (6)$$

$$\frac{\partial \psi}{\partial x} = \frac{\partial \psi}{\partial x} + \frac{m - 1}{2}\frac{\eta}{x}\frac{\partial \psi}{\partial \eta} = f\left(\frac{j\nu_0}{1 + m}\right)^{1/2}\frac{d}{dx}[x u_\infty(x)]^{1/2} + \frac{m - 1}{2}\frac{\eta}{x}\left(\frac{j\nu_0 x u_\infty(x)}{1 + m}\right)^{1/2}\frac{df}{d\eta}$$

$$= \frac{1}{2}\left[j\nu_0(1 + m)\frac{1}{x}u_\infty(x)\right]^{1/2}\left[f + \frac{m - 1}{m + 1}\eta\frac{df}{d\eta}\right] \quad (7)$$

[6] From Eqs. 4 and 7 of note 5 we have

$$\frac{\partial \psi}{\partial Y} = u_\infty(x)\frac{df}{d\eta} \tag{1}$$

$$\frac{\partial \psi}{\partial x} = \frac{1}{2}\sqrt{j\nu_0(1+m)\frac{1}{x}u_\infty(x)}\left[f + \frac{m-1}{m+1}\eta\frac{df}{d\eta}\right] \tag{2}$$

and from Eqs. 13-42 and 13-43 we have

$$u = \frac{\rho_0}{\rho}\frac{\partial \psi}{\partial y} = \frac{\partial \psi}{\partial Y} \tag{3}$$

$$v = -\frac{\rho_0}{\rho}\frac{\partial \psi}{\partial x} \tag{4}$$

Substituting Eqs. 1 and 2 into Eqs. 3 and 4, we obtain the results given by Eqs. 13-55.

[7] The mass extinction coefficient β_m and the volumetric extinction coefficient $\beta(x, y)$ are related by

$$\beta(x, y) = \beta_m \rho(x, y) \tag{1}$$

For constant β_m we obtain

$$\frac{\beta(x, y)}{\rho(x, y)} = \frac{\beta_0}{\rho_0} = \text{constant} \tag{2}$$

which is the relation given by Eq. 13-58.

[8] The following alternative assumptions also lead to a linear variation with temperature of μ and k as in Eq. 13-64a:

(a) The specific heat and the Prandtl number are constant, and μ varies linearly with temperature. In this case the requirement of constant c_p and Pr yields

$$\frac{\mu}{k} = \frac{\mu_0}{k_0} \quad \text{or} \quad \frac{k}{k_0} = \frac{\mu}{\mu_0} \tag{1}$$

If μ varies linearly with temperature, Eq. 1 gives

$$\frac{k}{k_0} = \frac{\mu}{\mu_0} = \frac{T}{T_0} = \theta \tag{2}$$

which is the same as Eq. 13-64a.

(b) The specific heat, the Prandtl number, and $\rho\mu$ are constant. In this case the assumption of constant c_p and Pr leads to the relation given by Eq. 1. For a perfect gas, the constancy of $\mu\rho$ yields

$$\frac{\mu}{\mu_0} = \frac{\rho_0}{\rho} = \frac{T}{T_0} = \theta \tag{3}$$

By combining Eqs. 1 and 3 we obtain

$$\frac{k}{k_0} = \frac{\mu}{\mu_0} = \frac{T}{T_0} = \theta \tag{4}$$

[9] From Shapiro [41, p. 80] we have

$$\frac{T_0}{T_\infty} = 1 + \frac{\gamma - 1}{2}M_\infty^{\ 2} \tag{1}$$

where T_0 is the stagnation temperature, T_∞ the temperature, M_∞ the Mach number for the external flow, and γ the specific heat ratio. Then the ratio of the densities becomes

$$\frac{\rho_\infty(x)}{\rho(x, y)} = \frac{T(x, y)}{T_\infty(x)} = \frac{T(x, y)}{T_0}\frac{T_0}{T_\infty(x)} = \theta(x, y)\left[1 + \frac{\gamma - 1}{2} M_\infty{}^2(x)\right] \tag{2}$$

The Eckert number $E_\infty(x)$ is related to the Mach number by

$$E_\infty(x) = \frac{u_\infty{}^2(x)}{c_p T_0} = \frac{v_{s,\infty}^2}{c_p T_\infty(x)}\frac{u_\infty{}^2(x)}{v_{s,\infty}^2}\frac{T_\infty(x)}{T_0} = (\gamma - 1)M_\infty{}^2(x)\left[1 + \frac{\gamma - 1}{2} M_\infty{}^2(x)\right]^{-1} \tag{3}$$

since T_0/T_∞ is given by Eq. 1, and the speed of sound $v_{s,\infty}$ in the external flow and c_p are given by [41, pp. 78–79]

$$v_{s,\infty}^2 = \gamma R T_\infty \qquad \text{and} \qquad c_p = \frac{\gamma R}{\gamma - 1} \tag{4}$$

hence

$$\frac{v_{s,\infty}^2}{c_p T_\infty} = \gamma - 1 \tag{5}$$

10 It is shown that the energy equation 13-78 is transformed into an ordinary differential equation in the variable η when Q^r is given by Eq. 13-82. We define a new dependent variable $g(\eta)$ as

$$\theta(\zeta, \eta) = \zeta^d g(\eta) \tag{1}$$

where d is an arbitrary constant. The application of this transformation to the energy equation 13-78 with q^r as given by Eq. 13-82 yields

$$\frac{1}{\mathrm{Pr}}\zeta^d\frac{d^2g}{d\eta^2} + \frac{1 + m}{2} f\zeta^d\frac{dg}{d\eta} = d\,\frac{1 - m}{2}\frac{df}{d\eta}\zeta^d g - \frac{4}{3N\,\mathrm{Pr}}\zeta^{4d}\frac{d}{d\eta}\left(g^3\frac{dg}{d\eta}\right) \tag{2}$$

If we set $d = 0$, Eq. 2 becomes

$$\frac{1}{\mathrm{Pr}}\frac{d^2g}{d\eta^2} + \frac{1 + m}{2} f\frac{dg}{d\eta} = -\frac{4}{3N\,\mathrm{Pr}}\frac{d}{d\eta}\left(g^3\frac{dg}{d\eta}\right) \tag{3}$$

where $g \equiv g(\eta)$ and Eq. 3 is an ordinary differential equation.

11 For an absorbing, emitting gray medium $\partial q^r/\partial\tau$ with the optically thin approximation is obtained from Eq. 9-7 as

$$\frac{\partial q^r}{\partial\tau} = 4\pi I_b(T) - 2\pi I^+(0) - 2\pi I^-(\tau_0) \tag{1}$$

and $I^+(0)$ is obtained from Eq. 9-3a as

$$I^+(0) = \varepsilon_w I_b(T_w) + (1 - \varepsilon_w)I^-(\tau_0) \tag{2}$$

Substitution of $I^+(0)$ from Eq. 2 into Eq. 1 yields

$$\frac{\partial q^r}{\partial\tau} = 4\pi[I_b(T) - I^-(\tau_0)] + 2\pi\varepsilon_w[I^-(\tau_0) - I_b(T_w)] \tag{3a}$$

or

$$\frac{\partial q^r}{\partial\tau} = 4(n^2\bar\sigma T^4 - n^2\bar\sigma T_\infty{}^4) + 2\varepsilon_w(n^2\bar\sigma T_\infty{}^4 - n^2\bar\sigma T_w{}^4) \tag{3b}$$

or

$$\frac{\partial Q^r}{\partial\tau} \equiv \frac{\partial}{\partial\tau}\left(\frac{q^r}{4n^2\bar\sigma T_\infty{}^4}\right) = (\theta^4 - 1) + \frac{\varepsilon_w}{2}(1 - \theta_w{}^4) \tag{4}$$

[12] The integral term in Eq. 13-105b can be rearranged as

$$\int_0^\infty (\theta^4 - 1)E_2(\tau)\,d\tau = \int_0^\infty (\theta^4 - \theta_r{}^4)E_2(\tau)\,d\tau + \int_0^\infty (\theta_r{}^4 - 1)E_2(\tau)\,d\tau \tag{1}$$

Since $\tau = \eta\sqrt{\Pr N\xi}$, the first integral on the right-hand side is transformed from τ to η, and Eq. 1 becomes

$$\int_0^\infty (\theta^4 - 1)E_2(\tau)\,d\tau = \sqrt{\Pr N\xi}\int_{\eta=0}^\infty (\theta^4 - \theta_{rw}{}^4)\,d\eta + \int_{\tau=0}^\infty (\theta_r{}^4 - 1)E_2(\tau)\,d\tau \tag{2}$$

From Eqs. 13-110 and 13-111, for $\xi \ll 1$, we obtain

$$\theta_r{}^4 \simeq 1 + 2\varepsilon_w E_2(\tau)(\theta_{0w}{}^4 - 1)\xi$$

or

$$\theta_r{}^4 - 1 = 2\varepsilon_w E_2(\tau)(\theta_{0w}{}^4 - 1)\xi \tag{3}$$

Substituting Eq. 3 into Eq. 2 and neglecting terms of the order of $\xi^{3/2}$, we obtain

$$\int_0^\infty (\theta^4 - 1)E_2(\tau)\,d\tau = \sqrt{\Pr N\xi}\int_{\eta=0}^\infty (\theta^4 - 1)\,d\eta + 2\varepsilon_w\xi(\theta_{0w}{}^4 - 1)\int_{\tau=0}^\infty E_2{}^2(\tau)\,d\tau \tag{4}$$

By replacing the integral term in Eq. 13-105b by Eq. 4 we obtain Eq. 13-118.

[13] The transformation of various partial derivatives from the x, y coordinates to the x, η coordinates is given as:

$$\eta = \left(\frac{\mathrm{Gr}}{4}\right)^{1/4}\frac{y}{x}; \qquad \frac{\partial\eta}{\partial x} = -\frac{1}{4}\frac{\eta}{x}, \qquad \frac{\partial\eta}{\partial y} = \frac{1}{x}\left(\frac{\mathrm{Gr}}{4}\right)^{1/4} \tag{1}$$

$$\frac{\partial F(x,\eta)}{\partial y} = \frac{\partial F}{\partial x}\frac{dx}{dy}\nearrow^{0} + \frac{\partial F}{\partial \eta}\frac{\partial \eta}{\partial y} = \frac{1}{x}\left(\frac{\mathrm{Gr}}{4}\right)^{1/4}\frac{\partial F}{\partial\eta} \tag{2}$$

$$\frac{\partial^2 F(x,\eta)}{\partial y^2} = \frac{1}{x^2}\left(\frac{\mathrm{Gr}}{4}\right)^{1/2}\frac{\partial^2 F}{\partial\eta^2} \tag{3}$$

$$\frac{\partial F(x,\eta)}{\partial x} = \frac{\partial F}{\partial x}\frac{dx}{dx} + \frac{\partial F}{\partial\eta}\frac{\partial\eta}{\partial x} = \frac{\partial F}{\partial x} - \frac{1}{4}\frac{\eta}{x}\frac{\partial F}{\partial\eta} \tag{4}$$

$$\psi(x,y) = 4\nu\left(\frac{\mathrm{Gr}}{4}\right)^{1/4} f(x,\eta); \qquad \frac{\partial\psi}{\partial y} = \frac{\partial\psi}{\partial\eta}\frac{\partial\eta}{\partial y} = \frac{4\nu}{x}\left(\frac{\mathrm{Gr}}{4}\right)^{1/2}\frac{\partial f}{\partial\eta} \tag{5}$$

$$\frac{\partial\psi}{\partial x} = \frac{\partial\psi}{\partial\eta}\frac{\partial\eta}{\partial x} + \frac{\partial\psi}{\partial x}\frac{dx}{dx} = \nu\left(\frac{\mathrm{Gr}}{4}\right)^{1/4}\left(-\frac{x}{\eta}\frac{\partial f}{\partial\eta} + 4\frac{\partial f}{\partial x} + 3\frac{f}{x}\right) \tag{6}$$

$$\frac{\partial^2\psi}{\partial x\,\partial y} = \frac{\nu}{x^2}\left(\frac{\mathrm{Gr}}{4}\right)^{1/2}\left(-\eta\frac{\partial^2 f}{\partial\eta^2} + 2\frac{\partial f}{\partial\eta} + 4x\frac{\partial^2 f}{\partial x\,\partial\eta}\right) \tag{7}$$

$$\frac{\partial^2\psi}{\partial y^2} = \frac{4\nu}{x^2}\left(\frac{\mathrm{Gr}}{4}\right)^{3/4}\frac{\partial^2 f}{\partial\eta^2} \tag{8}$$

$$\frac{\partial^3\psi}{\partial y^3} = \frac{4\nu}{x^3}\frac{\mathrm{Gr}}{4}\frac{\partial^3 f}{\partial\eta^3} \tag{9}$$

CHAPTER 14

Channel Flow with Radiation in a Participating Medium

Heat transfer in forced convection to a transparent fluid flowing inside channels has been well investigated. The basic theory and a survey of the pertinent literature can be found in the standard books on the subject, for example, those by Kays [1], Bird, Stewart, and Lightfoot [2], and Knudsen and Katz [3]. In the case of forced convection to a participating fluid flowing at temperatures encountered in engineering applications, the continuity and the momentum equations remain unchanged for the reasons stated in Chapter 13. The energy equation, however, contains an additional term: the divergence of the net radiative heat flux vector.

Goulard and Goulard [4] and Viskanta and Grosh [5] studied the interaction of radiation with heat transfer for an absorbing and emitting Couette flow. Viskanta [6, 7] considered an absorbing and emitting fluid in slug flow and laminar flow between two parallel, diffusely emitting, diffusely reflecting, isothermal infinite plates for a state of full thermal development. Einstein [8, 9] investigated the interaction of radiation for an absorbing, emitting, constant-property fluid with a parabolic velocity distribution in laminar flow between two parallel plates and inside a circular tube. Chen [10] considered slug flow between two parallel plates and applied the Schuster-Schwarzchild approximation for the radiation part of the problem. De Soto [11] and Pearce and Emery [12] examined heat transfer in the thermally developing region for an absorbing and emitting fluid flowing inside a circular tube with a parabolic velocity profile for both gray and nongray cases. Nichols [13] and Landram, Greif, and Habib [14] determined heat transfer to an absorbing, emitting, gray fluid in turbulent flow inside a circular tube under optically thin conditions, and Habib and Greif [15] considered heat transfer in the fully developed flow of a nongray radiating gas inside a circular tube and reported experimental and theoretical results. Thorsen [16, 17] used Green's function approach and gave a general formulation for determining temperature distribution in the thermally developing region for laminar flow inside a circular tube and presented numerical results for the case of optically thin radiation. Lii and Özişik [18] determined the effects of thermal radiation on the Nusselt number in the thermally developing region for an absorbing, emitting, scattering fluid in slug flow between two parallel plates and treated the radiation part of the problem exactly.

In this chapter we examine heat transfer to a participating fluid flowing in forced convection inside a channel, present the mathematical formulation of the problem, and discuss the results for a few representative cases.

14-1 ABSORBING, EMITTING COUETTE FLOW

Consider the steady flow of an absorbing and emitting fluid between two parallel infinite plates resulting from the motion of the upper plate at a constant velocity u, while the lower plate remains stationary. Figure 14-1 shows the geometry and the coordinate system. The lower and upper plates are at uniform temperatures T_1 and T_2, respectively, and are separated by a distance L. If an incompressible, constant-property fluid is assumed, the velocity and the temperature problems are uncoupled. The axial flow velocity $u(y)$ satisfies the momentum equation, given as

$$\frac{d}{dy}\left[\mu \frac{du(y)}{dy}\right] = 0 \tag{14-1}$$

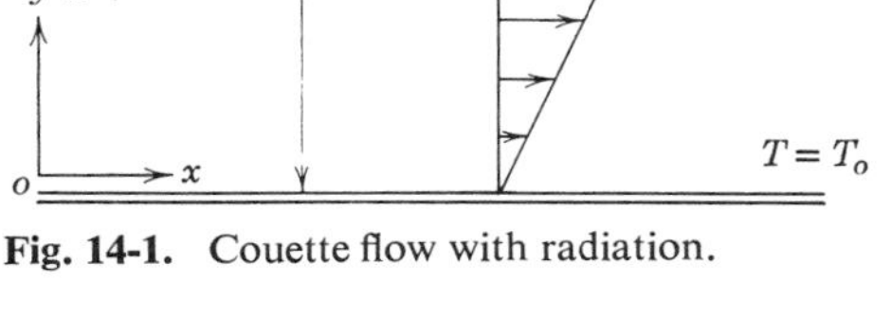

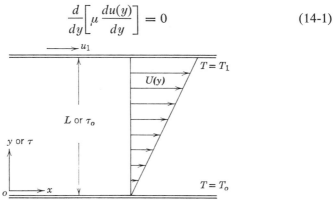

Fig. 14-1. Couette flow with radiation.

with the boundary conditions

$$u = 0 \quad \text{at} \quad y = 0 \tag{14-2a}$$

$$u = u_1 \quad \text{at} \quad y = L \tag{14-2b}$$

The solution of Eq. 14-1 for constant μ and subject to the boundary conditions 14-2 yields

$$u(y) = \frac{y}{L} u_1 \tag{14-3}$$

The temperature distribution in the medium satisfies the energy equation. Assuming that the conductive and radiative heat fluxes in the x direction are negligible and that there are no internal energy sources and no pressure gradients, we can write the energy equation as

$$\frac{d}{dy}\left(k\frac{dT}{dy} - q^r\right) + \mu\left(\frac{du}{dy}\right)^2 = 0 \tag{14-4}$$

Substituting $u(y)$ into Eq. 14-4 and arranging the resulting expression in the dimensionless form, we obtain

$$\frac{d}{d\tau}\left[N\frac{d\theta(\tau)}{d\tau} - Q^r(\tau)\right] + \left(\frac{u_1}{\tau_0}\right)^2 \frac{\mu\kappa}{4\bar{\sigma}T_r^4} = 0 \tag{14-5}$$

with the boundary conditions

$$\theta(\tau) = \frac{T_1}{T_r} \equiv \theta_1 \quad \text{at } \tau = 0 \tag{14-6a}$$

$$\theta(\tau) = \frac{T_2}{T_r} \equiv \theta_2 \quad \text{at } \tau = \tau_0 \tag{14-6b}$$

where we have defined

$$\left.\begin{aligned}
N &\equiv \frac{k\kappa}{4n^2\bar{\sigma}T_r^3} = \text{conduction-to-radiation parameter} \\
Q\,(\tau) &= \frac{q^r(\tau)}{4n^2\bar{\sigma}T_r^4} = \text{dimensionless net radiative heat flux} \\
\theta(\tau) &= \frac{T(\tau)}{T_r} = \text{dimensionless temperature} \\
\tau &= \kappa y = \text{optical variable} \\
\tau_0 &= \kappa L = \text{optical distance between plates}
\end{aligned}\right\} \tag{14-7}$$

Also, T_r is a reference temperature, and κ the absorption coefficient.

Assuming an absorbing, emitting, gray fluid and black walls, we readily obtain the net radiative heat flux $q^r(\tau)$ from Eq. 8-84. Then $Q^r(\tau)$ becomes

$$Q^r(\tau) = \tfrac{1}{2}[\theta_1{}^4E_3(\tau) - \theta_2{}^4E_3(\tau_0 - \tau)]$$
$$+ \tfrac{1}{2}\left[\int_0^\tau \theta^4(\tau')E_2(\tau - \tau')\,d\tau' - \int_\tau^{\tau_0} \theta^4(\tau')E_2(\tau' - \tau)\,d\tau'\right] \quad (14\text{-}8)$$

Equations 14-5, 14-6, and 14-8 provide the complete mathematical formulation of the problem under consideration. These equations are of a nonlinear integrodifferential type for which no analytical solution is available; however, they can be solved numerically. Goulard and Goulard [4] solved these equations numerically, using an iterative scheme for the optically thin limit and neglecting the viscous energy dissipation. Viskanta and Grosh [5] presented solutions obtained with the optically thick limit approximation and the exact formulation. We shall not discuss the results here, since we have already presented the temperature profile in a related problem involving combined conduction and radiation (see Fig. 12-3). Once the temperature distribution $\theta(\tau)$ is known, the heat transfer quantities defined above are readily calculated.

14-2 ABSORBING, EMITTING FLOW BETWEEN TWO PARALLEL PLATES

Consider an absorbing, emitting, gray, incompressible, constant-property fluid in fully developed laminar flow between two infinite parallel plates at a

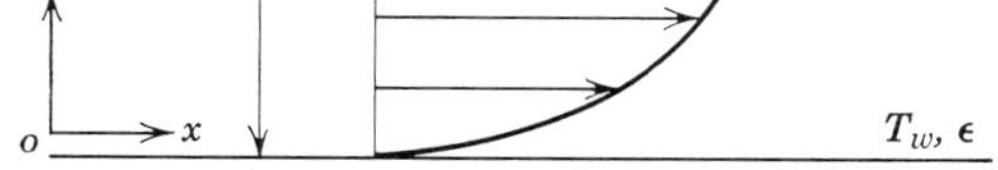

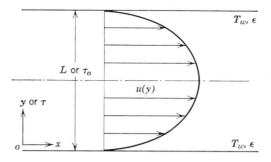

Fig. 14-2. Thermally developed laminar flow between parallel plates with radiation.

distance L apart. Figure 14-2 shows the geometry and the coordinate system. The velocity distribution for fully developed laminar flow between parallel plates is given as

$$\frac{u(y)}{u_m} = 6\left[\frac{y}{L} - \left(\frac{y}{L}\right)^2\right] \tag{14-9}$$

where u_m is the mean velocity.

The temperature distribution in the medium satisfies the energy equation

$$\rho c_p u_m \frac{\partial T}{\partial x} = k \frac{\partial^2 T}{\partial y^2} - \frac{\partial q^r}{\partial y} \quad \text{in } 0 \leq y \leq L \tag{14-10}$$

where we have assumed that the viscous energy dissipation and the x-direction conductive and radiative heat fluxes are negligible, and that the physical properties are constant.

If a state of full thermal development is assumed, analogously to the case with no radiation, the energy equation 14-10 can be simplified further; but care must be exercised in making such an assumption. The results by Pearce and Emery [12] and Lii and Özişik [18] for thermally developing flow with radiation show that when radiation is strong a state of full thermal development is not realized; hence the concept of thermally developed flow is unrealistic for such situations. However, with weak radiation it may be permissible to assume a thermally developed flow; the results from such an analysis may give some insight into the general effects of radiation on heat transfer, provided that one recognizes the limitations of the analysis. Viskanta [7] solved the above heat transfer problem for a thermally fully developed flow; here we present the formulation of the problem and discuss the results.

In a thermally fully developed region the axial temperature gradient $\partial T/\partial x$ may be represented, analogously to the case with no radiation, by [1, p. 107]

$$\frac{\partial T}{\partial x} = \frac{T_w - T}{T_w - T_m} \frac{dT_m}{dx} \tag{14-11}$$

where T_w is the wall temperature, and $T_m \equiv T_m(x)$ is the mixed mean temperature of the fluid. An overall energy balance over an elemental length dx about x at a distance away from the inlet gives

$$\rho c_p L u_m \, eT_m = 2 q_w \, dx \tag{14-12a}$$

or

$$\frac{dT_m}{dx} = \frac{2 q_w}{\rho c_p L u_m} \tag{14-12b}$$

where q_w is the net heat flux at the wall.

Substitution of Eqs. 14-9, 14-11, and 14-12b into the energy equation 14-10 yields

$$k \frac{d^2 T}{dy^2} = \frac{12}{L}\left[\frac{y}{L} - \left(\frac{y}{L}\right)^2\right] \frac{T_w - T}{T_w - T_m} q_w + \frac{dq^r}{dy} \quad \text{in } 0 \leq y \leq L \tag{14-13}$$

which is written in the dimensionless form as

$$N\frac{d^2\theta}{d\tau^2}=\frac{12}{\tau_0}\left[\frac{\tau}{\tau_0}-\left(\frac{\tau}{\tau_0}\right)^2\right]\left(\frac{\theta_w-\theta}{\theta_w-\theta_m}\right)Q_w+\frac{dQ^r}{d\tau},\quad 0\le\tau\le\tau_0 \tag{14-14}$$

with the boundary conditions

$$\theta=\theta_w \quad \text{at } \tau=0 \tag{14-15a}$$

$$\theta=\theta_w \quad \text{at } \tau=\tau_0 \tag{14-15b}$$

where

$$Q_w\equiv\frac{q_w}{4n^2\bar{\sigma}T_r^4}=\text{dimensionless net heat flux at the wall} \tag{14-16a}$$

$$\theta_w=\frac{T_w}{T_r},\qquad \theta_m=\frac{T_m}{T_r} \tag{14-16b}$$

and N, Q^r, τ, τ_0, and θ have already been defined (see Eqs. 14-7).

Equation 14-14 is an ordinary differential equation in the independent variable τ. The axial variable x enters this equation merely as a parameter through the mixed mean temperature θ_m, since the value of θ_m is different at different axial locations.

The dimensionless wall heat flux Q_w appearing in Eq. 14-14 is given as

$$Q_w=\left[-N\frac{d\theta}{d\tau}+Q^r\right]_{\tau=0} \tag{14-17}$$

The dimensionless net radiative heat flux Q^r and its derivative $dQ^r/d\tau$ are obtained from Eqs. 8-84 and 8-95, respectively, as

$$Q^r(\tau)=\tfrac{1}{2}[R(0)E_3(\tau)-R(\tau_0)E_3(\tau_0-\tau)]+\tfrac{1}{2}\left[\int_0^{\tau}\theta^4(\tau')E_2(\tau-\tau')\,d\tau'-\int_{\tau}^{\tau_0}\theta^4(\tau')E_2(\tau'-\tau)\,d\tau'\right] \tag{14-18}$$

$$\frac{dQ^r(\tau)}{d\tau}=\theta^4(\tau)-\frac{1}{2}\left[R(0)E_2(\tau)+R(\tau_0)E_2(\tau_0-\tau)+\int_0^{\tau_0}\theta^4(\tau')E(|\tau-\tau'|)\,d\tau'\right] \tag{14-19}$$

where $R(0)$ and $R(\tau_0)$ are the radiosities at the boundary surfaces at $\tau=0$ and $\tau=\tau_0$, respectively [i.e., $R(0)=\pi I^+(0)/n^2\bar{\sigma}T_r^4$ and $R(\tau_0)=\pi I^-(\tau_0)/n^2\bar{\sigma}T_r^4$]; the equations for $R(0)$ and $R(\tau_0)$ are obtained from Eqs. 8-110a and 8-110b as

$$R(0)=\varepsilon_1\theta_w^4+2\rho_1^d\left[R(\tau_0)E_3(\tau_0)+\int_0^{\tau_0}\theta^4(\tau')E_2(\tau')\,d\tau'\right] \tag{14-20}$$

$$R(\tau_0)=\varepsilon_2\theta_w^4+2\rho_2^d\left[R(0)E_3(\tau_0)+\int_0^{\tau_0}\theta^4(\tau')E_2(\tau_0-\tau')\,d\tau'\right] \tag{14-21}$$

where ε_1 and ε_2 are the emissivities, and ρ_1^d and ρ_2^d are the diffuse reflectivities of the boundary surfaces at $\tau = 0$ and $\tau = \tau_0$, respectively.

Equations 14-14 through 14-21 provide the complete mathematical formulation of the problem considered.

If it is further assumed that the boundary surfaces at $\tau = 0$ and $\tau = \tau_0$ are opaque and have the same emissivities $\varepsilon_1 = \varepsilon_2 = \varepsilon$ and the same reflectivities $\rho_1^d = \rho_2^d = \rho$, then the temperature profile $\theta(\tau)$ is symmetrical about the center line of the duct and the radiosities $R(0)$ and $R(\tau_0)$ are equal. In this case Eqs. 14-20 and 14-21 simplify to

$$R(0) = R(\tau_0) \equiv R = \frac{\varepsilon\theta_w^{\,4} + 2(1-\varepsilon)\displaystyle\int_0^{\tau_0} \theta^4(\tau')E_2(\tau')\,d\tau'}{1 - 2(1-\varepsilon)E_3(\tau_0)} \tag{14-22}$$

where the Kirchhoff law is assumed to be valid; thus ρ is replaced by $1 - \varepsilon$.

The Nusselt number is defined as

$$\text{Nu} \equiv \frac{hD_e}{k} = \frac{D_e q_w}{k(T_w - T_m)} \tag{14-23}$$

where D_e is the equivalent passage diameter (i.e., $D_e = 2L$). Equation 14-23 can be written alternatively in the form

$$\text{Nu} = \frac{2L}{k(T_w - T_m)}\left[-k\frac{dT}{dy} + q^r\right]_{y=0} = \frac{2\tau_0}{\theta_w - \theta_m}\left[-\frac{d\theta}{d\tau} + \frac{Q^r}{N}\right]_{\tau=0} \tag{14-24}$$

For large values of the conduction-to-radiation parameter (i.e., $N \to \infty$), Eq. 14-24 simplifies to the customary relation for the Nusselt number.

Once the temperature distribution $\theta(\tau)$ is determined, all of the heat transfer quantities defined above can be calculated.

Method of Solution

It is highly unlikely that an analytical solution can be found to the nonlinear integrodifferential equation 14-14 with $dQ^r/d\tau$ as given by Eq. 14-19; but the equation can be solved numerically by direct iteration after converting Eq. 14-14 into a nonlinear integral equation. However, the kernel $E_1(|\tau - \tau'|)$ being singular for $|\tau - \tau'| = 0$, the step size has to be kept very small in order to obtain accurate results; this requires long computing time. An approximate method was used by Viskanta [7] to solve these equations. He expanded $\theta^4(\tau')$ in a Taylor series in a neighborhood of τ in the form

$$\theta^4(\tau') = \theta^4(\tau) + \frac{(\tau' - \tau)}{1!} \frac{d\theta^4(\tau)}{d\tau}\bigg|_{\tau} + \frac{(\tau' - \tau)^2}{2!} \frac{d^2\theta^4(\tau)}{d\tau^2}\bigg|_{\tau} + \cdots \tag{14-25}$$

When this expansion is substituted for the integral term in Eq. 14-19, the integrations are performed, and the resulting expression for $dQ^r/d\tau$ is introduced into Eq. 14-14, one obtains a nonlinear differential equation for the temperature function $\theta(\tau)$. It is easier to solve this differential equation. However, the accuracy is affected by the number of terms contained in the truncated Taylor-series expansion of $\theta^4(\tau')$. The reader should refer to the original article for a discussion of the accuracy and the validity of truncating the series after the first three terms and for the evaluation of various integrals.

If the boundary surfaces at $\tau = 0$ and $\tau = \tau_0$ have the same emissivity ε and the same reflectivity ρ, the temperature $\theta(\tau)$ is symmetrical about the center line of the duct. In that case the differential equation is solved only for the region $0 \leq \tau \leq \tau_0/2$, and the boundary conditions 14-15 are replaced by

$$\theta = \theta_w \quad \text{at } \tau = 0 \tag{14-26a}$$

$$\frac{d\theta}{d\tau} = 0 \quad \text{at } \tau = \frac{\tau_0}{2} \tag{14-26b}$$

Results

When the boundary surfaces are opaque and have the same emissivity ε and the Kirchhoff law is valid, the above problem involves five independent variables: N, τ_0, ε, θ_w, and θ_m. Since the problem is nonlinear, one has to consider the heating of the fluid $\theta_w > \theta_m$ and the cooling of the fluid $\theta_w < \theta_m$ separately. Viskanta chose the reference temperature as $T_r = T_w$ for the heating of the fluid and replaced the mean temperature θ_m by the center line temperature θ_0 in presenting the results of the numerical calculations. Figure 14-3 shows the effects of the conduction-to-radiation parameter N on the temperature profile for the case of heating ($\theta_w = 1$) for $\tau_0 = 1$, $\varepsilon = 1$, and $\theta_c = 0.5$. The temperature profile for $N \to \infty$ characterizes the case with no radiation. As N decreases (i.e., radiation increases), the temperature profile deviates from the profile for the nonradiating case and the temperature gradient at the wall decreases.

The effects of optical thickness τ_0 and wall emissivity ε on the temperature profile were also investigated by Viskanta. For small optical thicknesses (i.e., $\tau_0 = 0.1$) the temperature profile with radiation has been found indistinguishable from that for pure conduction and convection. This is to be expected on physical grounds, since for small τ_0 the fluid becomes transparent to radiation and the effect of radiation is diminished. As the wall reflectivity

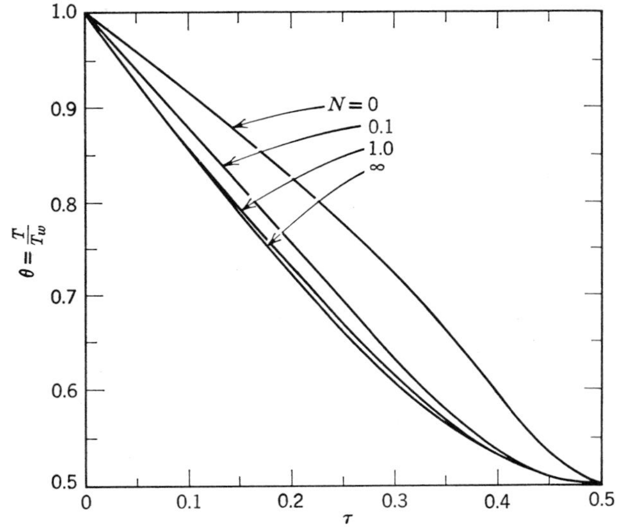

Fig. 14-3. Effect of N on temperature distribution for $\tau_0 = \varepsilon = 1.0$, $\theta_c = 0.5$, $\theta_w = 1.0$. (From R. Viskanta [7].)

is increased, the temperature profile with radiation has been found to approach that with no radiation.

14-3 ABSORBING, EMITTING, SCATTERING, THERMALLY DEVELOPING SLUG FLOW BETWEEN TWO PARALLEL PLATES

In this section consideration is given to the interaction of radiation in an absorbing, emitting, isotropically scattering, thermally developing slug flow between two infinite parallel plates at a distance $2L$ apart. The normal-mode expansion technique will be used in order to treat the radiation part of the problem exactly. The slug flow enters the heated section of the channel with a uniform temperature T_0 at the origin $x = 0$ of the axial coordinate, while the walls are kept at a uniform temperature T_w for $x > 0$. Figure 14-4 shows the geometry and the coordinate system. The plates are opaque, gray, diffuse emitters and specular reflectors, the emissivity ε is the same for both plates, and the Kirchhoff law is valid. This problem has been solved by Lii and Özişik [18]; here we present the formulation and discuss the method of solution and some of the results.

The energy equation is given in the form

$$\rho c_p u \frac{\partial T}{\partial x} = k \frac{\partial^2 T}{\partial y^2} - \frac{\partial q^r}{\partial y} \quad \text{in } 0 \leq y \leq L, \quad x \geq 0 \tag{14-27}$$

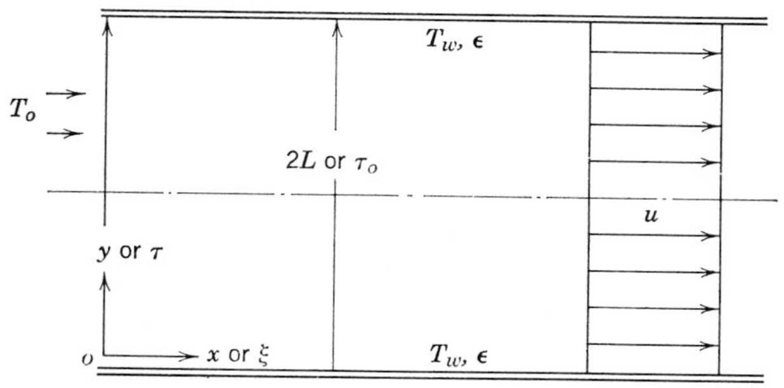

Fig. 14-4. Thermally developing slug flow between parallel plates with radiation.

where the flow velocity u is constant. This equation need be solved only for half of the region, $0 \leq y \leq L$, since the temperature is symmetrical about the center line of the duct. In the dimensionless form Eq. 14-27 is written as

$$\frac{\partial \theta(\tau, \xi)}{\partial \xi} = \frac{\partial^2 \theta(\tau, \xi)}{\partial \tau^2} - \frac{1}{N} \frac{\partial Q^r(\tau, \xi)}{\partial \tau} \quad \text{in } 0 \leq \tau \leq \tau_0, \quad \xi \geq 0 \tag{14-28}$$

with the boundary conditions

$$\theta(\tau, \xi) = 1 \quad \text{at } \tau = 0 \tag{14-29a}$$

$$\frac{\partial \theta(\tau, \xi)}{\partial \tau} = 0 \quad \text{at } \tau = \tau_0 \tag{14-29b}$$

$$\theta(\tau, \xi) = \theta_0 \quad \text{at } \xi = 0 \tag{14-29c}$$

where

$$D_e = 4L = \text{equivalent passage diameter} \tag{14-30a}$$

$$\text{Re} = \frac{\rho u D_e}{\mu}, \qquad \text{Pr} = \frac{c_p \mu}{k} \tag{14-30b}$$

$$\xi = \frac{k\beta^2 x}{\rho c_p u} = 16\tau_0^{\,2} \frac{x/D_e}{\text{Re Pr}} \tag{14-30c}$$

$$\theta(\tau, \xi) = \frac{T}{T_w}, \qquad \theta_0 = \frac{T_0}{T_w} \tag{14-30d}$$

and the dimensionless parameters N, Q^r, τ, and τ_0 are as defined by Eqs. 14-7, except that T_w is taken as the reference temperature and the extinction coefficient β has replaced κ.

The energy equation 14-28 with the boundary conditions 14-29 can be solved formally by the application of a finite Fourier transform described by

Özişik [19] if the radiation term $\partial Q^r/\partial\tau$ is treated as a prescribed function. We obtain

$$\theta(\tau, \xi) = 1 + \frac{2}{\tau_0}\sum_{m=1}^{\infty} \sin \nu_m\tau\left[\frac{\theta_0 - 1}{\nu_m} e^{-\nu_m^2\xi} - \frac{1}{N}\int_{\xi'=0}^{\xi} e^{-\nu_m^2(\xi-\xi')}\int_{\tau'=0}^{\tau_0} \sin \nu_m\tau' \frac{\partial Q^r(\tau', \xi')}{\partial\tau'}\, d\tau' d\xi'\right] \quad (14\text{-}31a)$$

where the eigenvalues ν_m are given as

$$\nu_m = \frac{(2m - 1)\pi}{2\tau_0}, \quad m = 1, 2, 3, \ldots \quad (14\text{-}31b)$$

Here we note that the solution given by Eq. 14-31 is not a solution in a real sense because the radiation term $\partial Q^r/\partial\tau$ is a function of the temperature $\theta(\tau, \xi)$. A relation will now be determined for $Q^r(\tau, \xi)$ and $\partial Q^r/\partial\tau$ from the solution of the equation of radiative transfer.

The dimensionless net radiative heat flux $Q^r(\tau, \xi)$ is related to the radiation intensity $I(\tau, \xi, \mu)$ by

$$Q^r(\tau, \xi) = \frac{2\pi\int_{-1}^{1} I(\tau, \xi, \mu)\mu\, d\mu}{4n^2\bar{\sigma}T_w^{\,4}} \equiv \frac{1}{2}\int_{-1}^{1}\psi(\tau, \xi, \mu)\mu\, d\mu \quad (14\text{-}32)$$

where the function $\psi(\tau, \xi, \mu)$ satisfies the equation of radiative transfer, given in the form

$$\mu\frac{\partial\psi(\tau, \xi, \mu)}{\partial\tau} + \psi(\tau, \xi, \mu) = (1 - \omega)\theta^4(\tau, \xi) + \frac{\omega}{2}\int_{-1}^{1}\psi(\tau, \xi, \mu')\, d\mu'$$

$$\text{in } 0 \leq \tau \leq \tau_0, \quad -1 \leq \mu \leq 1 \quad (14\text{-}33)$$

Assuming diffusely emitting, specularly reflecting boundaries and a symmetry condition along the channel axis at τ_0, we can write the boundary conditions as

$$\psi(0, \mu) = \varepsilon + (1 - \varepsilon)\psi(0, -\mu) \quad (14\text{-}34a)$$

$$\psi(\tau_0, -\mu) = \psi(\tau_0, \mu) \quad (14\text{-}34b)$$

where μ is the cosine of the angle between the direction of the radiation intensity and the positive τ axis, and ω is the single scattering albedo. In Eq. 14-33 the radiation intensity $\psi(\tau, \xi, \mu)$ depends on the independent variable ξ because the temperature $\theta(\tau, \xi)$ is a function of ξ; therefore ξ enters the radiation problem merely as a parameter. The solution of Eq. 14-33 can be written as a linear sum of the solutions to the homogeneous part of this equation and a particular solution $\psi_p(\tau, \xi, \mu)$ in the form (see Eq. 11-88)

$$\begin{aligned}\psi(\tau, \xi, \mu) = {} & A(\eta_0, \xi)\phi(\eta_0, \mu)e^{-\tau/\eta_0} + A(-\eta_0, \xi)\phi(-\eta_0, \mu)e^{\tau/\eta_0} \\ & + \int_0^1 A(\eta, \xi)\phi(\eta, \mu)e^{-\tau/\eta}\, d\eta \\ & + \int_0^1 A(-\eta, \xi)\phi(-\eta, \mu)e^{\tau/\eta}\, d\eta + \psi_p(\tau, \xi, \mu)\end{aligned} \tag{14-35}$$

where the discrete normal modes $\phi(\pm\eta_0, \mu)$ and the continuum normal modes $\phi(\pm\eta, \mu)$ have already been defined (see Eq. 10-18 or 11-89). A particular solution $\psi_p(\tau, \xi, \mu)$ of the equation of radiative transfer may be found if the function $\theta^4(\tau, \xi)$ appearing in the inhomogeneous term is prescribed; but the temperature distribution $\theta(\tau, \xi)$ in the medium is not yet known. Therefore, for the purpose of determining a particular solution, it is assumed that an initial guess is available for the temperature distribution $\theta^0(\tau, \xi)$, the fourth power of this temperature $[\theta^0(\tau, \xi)]^4$ is represented by a polynomial in τ in the form

$$(1 - \omega)[\theta^0(\tau, \xi)]^4 = (1 - \omega)\sum_{n=0}^{N} B_n(\xi)\tau^n \quad \text{in } 0 < \tau < \tau_0 \tag{14-36}$$

and the coefficients $B_n(\xi)$ can be determined by taking a finite number of terms in the expansion. A particular solution of the equation of radiative transfer for an inhomogeneous term of the form τ^n is given in Table 10-6 for $n = 0, 1, 2, 3, \ldots$. Then a particular solution $\psi_{p,n}(\tau, \xi, \mu)$ for an inhomogeneous term of the form $(1 - \omega)B_n(\xi)\tau^n$ is given as

$$\begin{aligned}\psi_{p,n}(\tau, \xi, \mu) = {} & B_n(\xi) \sum_{r=0,1,2,\ldots}^{n} (-1)^r \frac{n!}{(n-r)!}\tau^{n-r}\mu^r \\ & + \frac{\omega}{1-\omega} \sum_{s=1,2,\ldots}^{\substack{(n-1)/2(n \text{ odd}) \\ n/2(n \text{ even})}} \frac{n!}{(n-2s)!\,(2s+1)} \psi_{p,n-2s}(\tau, \xi, \mu)\end{aligned} \tag{14-37}$$

and a particular solution $\psi_p(\tau, \xi, \mu)$ of the equation of radiative transfer for an inhomogeneous term as represented by Eq. 14-36 can be determined from

$$\psi_p(\tau, \xi, \mu) = \sum_{n=0}^{N} \psi_{p,n}(\tau, \xi, \mu) \tag{14-38}$$

Knowing a particular solution $\psi_p(\tau, \xi, \mu)$, we can determine the expansion coefficients $A(\pm\eta_0, \xi)$ and $A(\pm\eta, \xi)$ by constraining the solution 14-35 to meet the boundary conditions 14-34 and by utilizing the orthogonality property of the normal modes and various normalization integrals as described in detail in Chapter 11.

Once a particular solution is found and the expansion coefficients $A(\pm\eta_0, \xi)$ and $A(\pm\eta, \xi)$ are determined, the corresponding dimensionless radiation intensity $\psi(\tau, \xi, \mu)$ is given by Eq. 14-35 and the dimensionless net radiative heat flux $Q^r(\tau, \xi)$ is obtained from Eqs. 14-32 and 14-35 as

$$Q^r(\tau, \xi) = \tfrac{1}{2}(1 - \omega)\Big[\eta_0 A(\eta_0, \xi)e^{-\tau/\eta_0} - \eta_0 A(\eta_0, \xi)e^{\tau/\eta_0} + \int_0^1 \eta A(\eta, \xi)e^{-\tau/\eta}\, d\eta - \int_0^1 \eta A(-\eta, \xi)e^{\tau/\eta}\, d\eta + \frac{1}{1-\omega}\int_{-1}^1 \psi_p(\tau, \xi, \mu)\mu\, d\mu\Big] \tag{14-39}$$

The differentiation of Eq. 14-39 with respect to τ yields

$$\frac{\partial Q^r(\tau, \xi)}{\partial \tau} = -\tfrac{1}{2}(1 - \omega)\Big[A(\eta_0, \xi)e^{-\tau/\eta_0} + A(-\eta_0, \xi)e^{\tau/\eta_0} + \int_0^1 A(\eta, \xi)e^{-\tau/\eta}\, d\eta + \int_0^1 A(-\eta, \xi)e^{\tau/\eta}\, d\eta - \frac{1}{1-\omega}\int_{-1}^1 \frac{\partial \psi_p(\tau, \xi, \mu)}{\partial \tau}\mu\, d\mu\Big] \tag{14-40}$$

The $\partial Q^r(\tau, \xi)/\partial\tau$ as given above is now considered a known function since the expansion coefficients and the particular solution are known for an initial guess as to distribution of temperature $\theta^0(\tau, \xi)$. Then Eq. 14-40 is introduced into Eq. 14-31a, the integrations are performed, and a first approximation is obtained for the temperature distribution at each axial location ξ_i. This first approximation is used to obtain a second approximation, the second approximation to obtain a third approximation, and so forth. The iterations are repeated until the solution converges to a prescribed criterion.

Once the temperature distribution $\theta(\tau, \xi)$ is determined at several difference locations ξ_i, various heat transfer quantities are readily calculated. For example, the local Nusselt number is defined as

$$\text{Nu} = \frac{D_e q_w(\xi)}{k[T_w - T_m(\xi)]} \tag{14-41}$$

which can be expressed in terms of the dimensionless quantities as

$$\text{Nu} = \frac{4\tau_0}{1 - \theta_m(\xi)}\Big[-\frac{\theta(\tau, \xi)}{\partial\tau} + \frac{Q^r(\tau, \xi)}{N}\Big]_{\tau=0} \tag{14-42}$$

The right-hand side of this expression differs by a factor of 2 from the definition of the Nusselt number given by Eq. 14-24, because of the fact that in the present analysis we have taken the channel width as $2L$. The value of the Nusselt number, however, remains invariant.

Discussion of Results

The combined convective and radiative heat transfer problem considered above for the heating of a fluid (i.e., $0 \leq \theta_0 < 1$ and $\theta_w = 1$) flowing inside

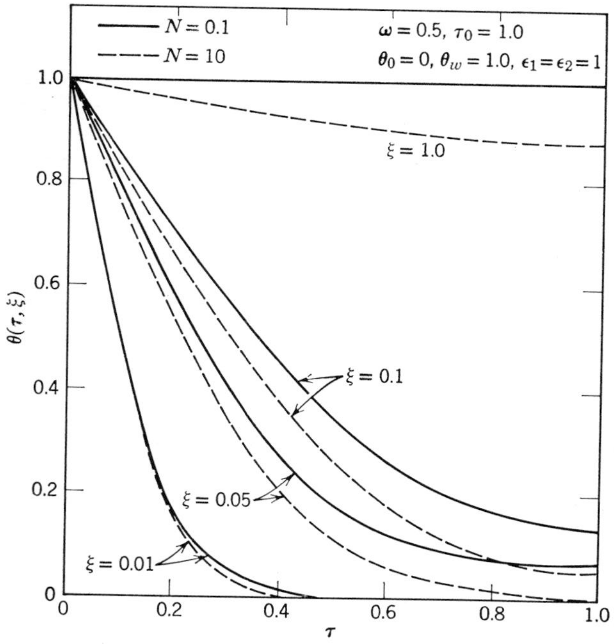

Fig. 14-5. Effect of ξ and N on temperature. (From C. C. Lii and M. N. Özişik [18].)

a parallel plate channel with black walls contains five independent variables: N, θ_0, τ_0, ω, and ε. Lii and Özişik [18] determined the temperature distribution and the local Nusselt number at several different axial locations along the channel.

Figure 14-5 shows the effects of the parameters ξ and N on the temperature profile for $\tau_0 = 1$, $\omega = 0.5$, $\theta_0 = 0$, $\theta_w = 1$, and $\varepsilon = 1$. The cases of $N = 10$ and 0.1 characterize, respectively, weak and reasonably strong radiation effects. Increased radiation flattens the temperature profile and decreases the temperature gradient at the wall. Radiation effects are less pronounced for small values of ξ.

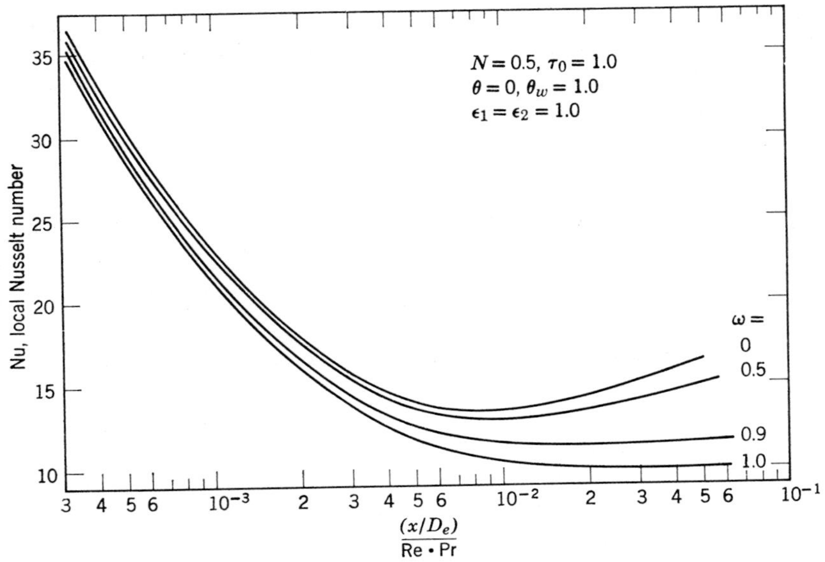

Fig. 14-6. Effects of ω on local Nusselt number. (From C. C. Lii and M. N. Özişik [18].)

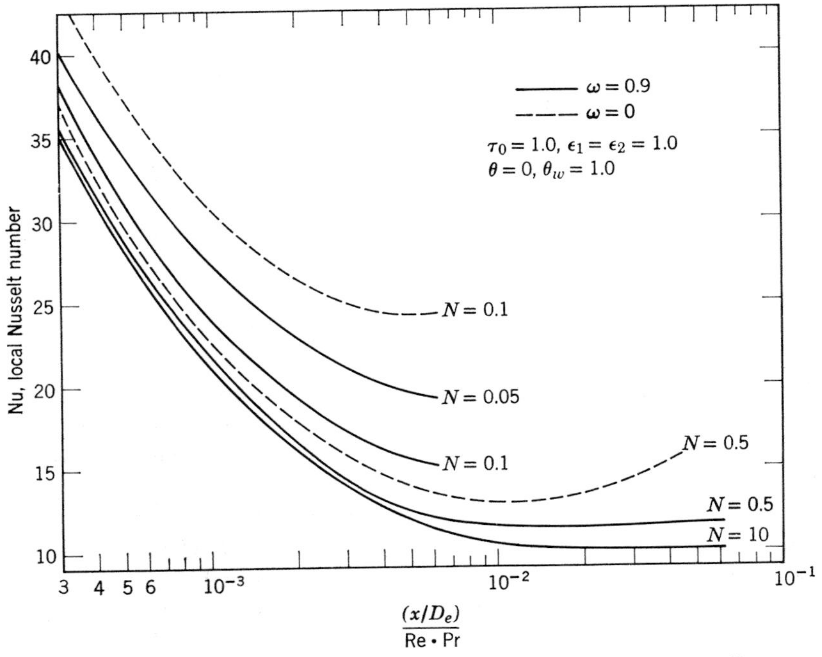

Fig. 14-7. Effects of N on local Nusselt number. (From C. C. Lii and M. N. Özişik [18].)

Figure 14-6 shows the effect of single scattering albedo ω on the local Nusselt number. The curve for $\omega = 1$ characterizes the case with no radiation since radiation does not interact with conduction and convection in a purely scattering medium; in this case the local Nusselt number decreases with the axial distance and reaches an asymptotic value π^2. For ω less than unity, the local Nusselt number increases as ω decreases and reaches a maximum when $\omega = 0$ (i.e., a purely absorbing and emitting medium).

Figure 14-7 shows the effect of the conduction-to-radiation parameter N on the local Nusselt number for the cases $\omega = 0.9$ and $\omega = 0$. The curve for $N = 10$ characterizes weak radiation interaction; in this case it is not possible to distinguish the curves for $\omega = 0.9$ and 0, and in fact the curve coincides with that for no radiation. Decreasing N, namely, increasing radiation, increases the Nusselt number at all axial positions.

Investigation with different values of τ_0 has shown that radiation has little effect on the local Nusselt number for $\tau_0 = 0.1$ or less, that is, the medium is optically thin.

REFERENCES

1. W. M. Kays, *Convective Heat and Mass Transfer*, McGraw-Hill Book Co., New York, 1966.
2. R. B. Bird, W. E. Stewart, and E. N. Lightfoot, *Transport Phenomena*, John Wiley and Sons, New York, 1960.
3. J. G. Knudsen and D. L. Katz, *Fluid Dynamics and Heat Transfer*, McGraw-Hill Book Co., New York, 1958.
4. R. Goulard and M. Goulard, "Energy Transfer in the Couette Flow of a Radiant and Chemically Reacting Gas," *Proceedings of the Heat Transfer and Fluid Mechanics Institute*, pp. 126–139, Stanford University Press, Palo Alto, Calif., 1959.
5. R. Viskanta and R. J. Grosh, "Temperature Distribution in Couette Flow with Radiation," *Am. Rocket Soc. J.*, **31**, 839–840, 1961.
6. R. Viskanta, "Heat Transfer in a Radiating Fluid with Slug Flow in a Parallel-Plate Channel," *Appl. Sci. Res.*, **13**, 291–311, 1964.
7. R. Viskanta, "Interaction of Heat Transfer by Conduction, Convection, and Radiation in a Radiating Fluid," *J. Heat Transfer*, **85C**, 318–328, 1963.
8. T. H. Einstein, "Radiant Heat Transfer to Absorbing Gases Enclosed Between Parallel Flat Plates with Flow and Conduction," *NASA Tech. Rept.* TR R-154, 1963.
9. T. H. Einstein, "Radiant Heat Transfer to Absorbing Gas Enclosed in a Circular Pipe with Conduction, Gas Flow, and Internal Heat Generation," *NASA Tech. Rept.* TR R-156, 1963.
10. J. C. Chen, "Simultaneous Radiative and Convective Heat Transfer in an Absorbing, Emitting and Scattering Medium in Slug Flow Between Parallel Plates," *A.I.Ch.E. J.*, **2**, 253–259, 1964.
11. Simon De Soto, "Coupled Radiation, Conduction and Convection in Entrance Region Flow," *Intern. J. Heat Mass Transfer*, **11**, 39–53, 1968.

12. B. E. Pearce and A. F. Emery, "Heat Transfer by Thermal Radiation and Laminar Forced Convection to an Absorbing Fluid in the Entry Region of a Pipe," *ASME Paper* No. 69-WA/HT-16, 1969.

13. L. D. Nichols, "Temperature Profiles in the Entrance Region of an Annular Passage Considering the Effects of Turbulent Convection and Radiation," *Intern. J. Heat Mass Transfer*, **8**, 589–607, 1965.

14. C. S. Landram, R. Greif, and I. S. Habib, "Heat Transfer in Turbulent Pipe Flow with Optically Thin Radiation," *J. Heat Transfer*, **91C**, 330–336, 1969.

15. I. S. Habib and R. Greif, "Heat Transfer to a Flowing Non-gray Radiating Gas: an Experimental and Theoretical Study," *Intern. J. Heat Mass Transfer*, **13**, 1571–1582, 1970.

16. R. S. Thorsen, "Combined Conduction, Convection and Radiation Effects in Internal Flows—Participating Gases," Mechanical Engineering Department, New York University, Report No. F-70-2, University Heights, New York, September 1970.

17. R. S. Thorsen, "Combined Conduction, Convection, and Radiation Effects in Optically Thin Tube Flow," *ASME Paper* No. 71-HT-17, 1971.

18. C. C. Lii and M. N. Özişik, "Heat Transfer in an Absorbing, Emitting and Scattering Slug Flow Between Parallel Plates," (to be published).

19. M. N. Özişik, *Boundary Value Problems of Heat Conduction*, International Textbook Co., Scranton, Penn., 1968.

APPENDIX

The Exponential Integrals $E_n(x)$

The nth exponential integral $E_n(x)$ of the argument x is defined by

$$E_n(x) = \int_1^{\infty} e^{-xt} t^{-n}\, dt = \int_0^1 e^{-(x/\mu)} \mu^{n-2}\, d\mu \tag{1}$$

Here we present a few useful relations for the functions $E_n(x)$ and a concise tabulation of the first four of these functions. We restrict ourselves to the positive integral values of n and real x. The reader should refer to the books by Chandrasekhar [1, pp. 373–378] and Kourganoff [2, pp. 253–271] for a detailed discussion of the properties of the functions $E_n(x)$, and to the publications by Case, de Hoffmann, and Placzek [3, pp. 153–162] and Placzek [4] and *Handbook of Mathematical Functions* [5, pp. 228–231] for the more complete tables.

From Eq. 1, for $x = 0$, we write

$$E_n(0) = \int_0^{\infty} t^{-n}\, dt = \begin{cases} +\infty & \text{for } n = 1 \quad (2a) \\ \dfrac{1}{n-1} & \text{for } n = 2, 3, 4, \ldots \quad (2b) \end{cases}$$

By direct differentiation of Eq. 1 we obtain

$$\frac{d}{dx} E_n(x) = \begin{cases} -\dfrac{1}{x} e^{-x} & \text{for } n = 1 \quad (3a) \\ -E_{n-1}(x) & \text{for } n = 2, 3, 4, \ldots \quad (3b) \end{cases}$$

Equation 1 can be written in the form

$$E_{n+1}(x) = -\frac{1}{n}\int_1^\infty e^{-xt}\frac{d}{dt}(t^{-n})\,dt \tag{4}$$

The integration of this equation by parts yields the following recurrence relation:

$$E_{n+1}(x) = \frac{1}{n}[e^{-x} - xE_n(x)], \quad n \geq 1 \tag{5}$$

It is apparent from Eq. 5 that all exponential integrals for $n > 1$ can be reduced to the first exponential integral, that is,

$$E_1(x) = \int_1^\infty e^{-xt}\frac{dt}{t} = \int_0^1 e^{-(x/\mu)}\frac{d\mu}{\mu} \tag{6}$$

A series expansion for $E_n(x)$ can be written in the form

$$E_n(x) = (-1)^n \frac{x^{n-1}}{(n-1)!}(\ln x + \psi_n) + \sum_{\substack{m=0 \\ m\neq n-1}}^{\infty} \frac{(-x)^m}{m!\,(n-1-m)} \tag{7}$$

where

$$\psi_n \equiv \begin{cases} \gamma & \text{for } n = 1 \\ \gamma - \sum\limits_{m=1}^{n-1}\dfrac{1}{m} & \text{for } n = 2, 3, \ldots \end{cases}$$

and $\gamma = 0.577216\cdots$ is Euler's constant.

Then the series expansions of $E_1(x)$, $E_2(x)$, and $E_3(x)$ become

$$E_1(x) = -(\gamma + \ln x) + \sum_{m=1}(-1)^{m-1}\frac{x^m}{m!\,m} \tag{8a}$$

$$= -(\gamma + \ln x) + x - \frac{x^2}{2!\,2} + \frac{x^3}{3!\,3} - \cdots \tag{8b}$$

$$E_2(x) = 1 + x(\gamma - 1 + \ln x) - \frac{x^2}{2!\,1} + \frac{x^3}{3!\,2} - \frac{x^4}{4!\,3} + \cdots \tag{9}$$

$$E_3(x) = \tfrac{1}{2} - x + \tfrac{1}{2}x^2(-\gamma + \tfrac{3}{2} - \ln x) + \frac{x^3}{3!\,1} - \frac{x^4}{4!\,2} + \cdots \tag{10}$$

For large values of x, the asymptotic expansion for $E_n(x)$ is given as

$$E_n(x) \cong \frac{e^{-x}}{x}\left[1 - \frac{n}{x} + \frac{n(n+1)}{x^2} - \frac{n(n+1)(n+2)}{x^3} + \cdots\right] \tag{11}$$

It is apparent from Eq. 11 that

$$E_n(x) \cong \frac{e^{-x}}{x} \quad \text{as } x \to \infty \tag{12}$$

Table 1 gives the first four of the functions $E_n(x)$ for values of x from 0 to 10.

Table 1 Functions[a] $E_n(x)$

(The figures in parantheses indicate the power of 10 by which the numbers to the left, and those below in the same column, are to be multiplied.)

x	E_1	E_2	E_3	E_4
0.00	∞	1.0000000	0.5000000	0.3333333
0.01	4.0379296	0.9496705	0.4902766	0.3283824
0.02	3.3547078	0.9131045	0.4809683	0.3235264
0.03	2.9591187	0.8816720	0.4719977	0.3187619
0.04	2.6812637	0.8535389	0.4633239	0.3140855
0.05	2.4678985	0.8278345	0.4549188	0.3094945
0.06	2.2953069	0.8040461	0.4467609	0.3049863
0.07	2.1508382	0.7818352	0.4388327	0.3005585
0.08	2.0269410	0.7609611	0.4311197	0.2962080
0.09	1.9187448	0.7412442	0.4236096	0.2919354
0.10	1.8229240	0.7225450	0.4162915	0.2877361
0.11	1.7371067	0.7047524	0.4091557	0.2836090
0.12	1.6595418	0.6877754	0.4021937	0.2795524
0.13	1.5888993	0.6715385	0.3953977	0.2755646
0.14	1.4241457	0.6559778	0.3887607	0.2716439
0.15	1.4644617	0.6410387	0.3822761	0.2677889
0.16	1.4091867	0.6266739	0.3759380	0.2639979
0.17	1.3577806	0.6128421	0.3697408	0.2602696
0.18	1.3097961	0.5995069	0.3636795	0.2566026
0.19	1.2648584	0.5866360	0.3577491	0.2529956
0.20	1.2226505	0.5742006	0.3519453	0.2494472
0.21	1.1829020	0.5621748	0.3462638	0.2459563
0.22	1.1453801	0.5505352	0.3407005	0.2425216
0.23	1.1098831	0.5392605	0.3352518	0.2391419
0.24	1.0762354	0.5283314	0.3299142	0.2358162
0.25	1.0442826	0.5177301	0.3246841	0.2325432
0.26	0.0138887	0.5074405	0.3195585	0.2293221
0.27	0.9849331	0.4974476	0.3145343	0.2261517
0.28	0.9573083	0.4877374	0.3096086	0.2230311
0.29	0.9309182	0.4782973	0.3047787	0.2199593
0.30	0.9056767	0.4691152	0.3000418	0.2169352
0.31	0.8815057	0.4601802	0.2953956	0.2139581
0.32	0.8583352	0.4514818	0.2908374	0.2110270
0.33	0.8361012	0.4430104	0.2863652	0.2081411

[a] This table is abbreviated from Case, de Hoffmann, and Placzek [3].

Table 1—(*Continued*)

x	E_1	E_2	E_3	E_4
0.34	0.8147456	0.4347568	0.2819765	0.2052994
0.35	0.7942154	0.4267127	0.2776693	0.2025013
0.36	0.7744622	0.4188699	0.2734416	0.1997458
0.37	0.7554414	0.4112210	0.2692913	0.1970322
0.38	0.7371121	0.4037588	0.2652165	0.1943597
0.39	0.7194367	0.3964766	0.2612155	0.1917276
0.40	0.7023801	0.3893680	0.2572864	0.1891352
0.41	0.6859103	0.3824270	0.2534276	0.1865816
0.42	0.6699973	0.3756479	0.2496373	0.1840664
0.43	0.6546134	0.3690253	0.2459141	0.1815887
0.44	0.6397328	0.3625540	0.2422563	0.1791479
0.45	0.6253313	0.3562291	0.2386625	0.1767433
0.46	0.6113865	0.3500458	0.2351313	0.1743744
0.47	0.5978774	0.3439999	0.2316612	0.1720405
0.48	0.5847843	0.3380869	0.2282508	0.1697410
0.49	0.5720888	0.3323029	0.2248990	0.1674753
0.50	0.5597736	0.3266439	0.2216044	0.1652428
0.55	0.5033641	0.3000996	0.2059475	0.1545596
0.60	0.4543795	0.2761839	0.1915506	0.1446271
0.65	0.4115170	0.2545597	0.1782910	0.1353855
0.70	0.3737688	0.2349471	0.1660612	0.1267808
0.75	0.3403408	0.2171109	0.1547667	0.1187638
0.80	0.3105966	0.2008517	0.1443238	0.1112900
0.85	0.2840193	0.1859986	0.1346581	0.1043185
0.90	0.2601839	0.1724041	0.1257030	0.0978123
0.95	0.2387375	0.1599404	0.1173988	0.0917374
1.00	0.2193839	0.1484955	0.1096920	0.0860625
1.05	0.2018728	0.1379713	0.1025339	0.0807590
1.10	0.1859909	0.1282811	0.0958809	0.0758007
1.15	0.1715554	0.1193481	0.0896932	0.0711632
1.20	0.1584084	0.1111041	0.0839347	0.0668242
1.25	0.1464134	0.1034881	0.0785723	0.0627631
1.30	0.1354510	0.0964455	0.0735763	0.0589609
1.35	0.1254168	0.0899275	0.0689191	0.0553998

Table 1—(*Continued*)

x	E_1	E_2	E_3	E_4
1.40	0.1162193	0.0838899	0.0645755	0.0520637
1.45	0.1077774	0.0782930	0.0605227	0.0489374
1.50	0.1000196	0.0731008	0.0567395	0.0460070
1.55	0.0928821	0.0682807	0.0532064	0.0432593
1.60	0.0863083	0.0638032	0.0499057	0.0406825
1.65	0.0802476	0.0596413	0.0468209	0.0382652
1.70	0.0746546	0.0557706	0.0439367	0.0359970
1.75	0.0694887	0.0521687	0.0412393	0.0338684
1.80	0.0647131	0.0488153	0.0387157	0.0318702
1.85	0.0602950	0.0456915	0.0363540	0.0299941
1.90	0.0562044	0.0427803	0.0341430	0.0282323
1.95	0.0524144	0.0400660	0.0320727	0.0265775
2.0	4.89005 (−2)	3.75343 (−2)	3.01334 (−2)	2.50228 (−2)
2.1	4.26143	3.29663	2.66136	2.21893
2.2	3.71911	2.89827	2.35207	1.96859
2.3	3.25023	2.55036	2.08002	1.74728
2.4	2.84403	2.24613	1.84054	1.55150
2.5	2.49149	1.97977	1.62954	1.37822
2.6	2.18502	1.74630	1.44349	1.22476
2.7	1.91819	1.54145	1.27932	1.08879
2.8	1.68553	1.36152	1.13437	0.96826
2 9	1.48240	1.20336	1.00629	0.86136
3.0	1.30484	1.06419	0.89306	0.76650
3.1	1.14944	0.94165	0.79290	0.68231
3.2	1.01330	0.83366	0.70425	0.60754
3.3	8.93904 (−3)	7.38433 (−3)	6.25744 (−3)	5.41120 (−3)
3.4	7.89097	6.54396	5.56190	4.82093
3.5	6.97014	5.80189	4.94538	4.29619
3.6	6.16041	5.14623	4.39865	3.82953
3.7	5.44782	4.56658	3.91360	3.41440
3.8	4.82025	4.05383	3.48310	3.04500
3.9	4.26715	3.60004	3.10087	2.71618
4.0	3.77935	3.19823	2.76136	2.42340
4.1	3.34888	2.84226	2.45969	2.16264
4.2	2.96876	2.52678	2.19156	1.93034

Table 1—(*Continued*)

x	E_1	E_2	E_3	E_4
4.3	2.63291	2.24704	1.95315	1.72334
4.4	2.33601	1.99890	1.74110	1.53883
4.5	2.07340 (−3)	1.77869 (−3)	1.55244 (−3)	1.37434 (−3)
4.6	1.84101	1.58321	1.38454	1.22765
4.7	1.63525	1.40960	1.23507	1.09682
4.8	1.45299	1.25538	1.10197	0.98010
4.9	1.29148	1.11831	0.98342	0.87594
5.0	1.14830	0.99647	0.87780	0.78298
5.1	1.02130	0.88812	0.78368	0.70000
5.2	9.0862 (−4)	7.9173 (−4)	6.9978 (−4)	6.2590 (−4)
5.3	8.0861	7.0597	6.2498	5.5974
5.4	7.1980	6.2964	5.5827	5.0064
5.5	6.4093	5.6168	4.9877	4.4784
5.6	5.7084	5.0116	4.4569	4.0067
5.7	5.0855	4.4725	3.9832	3.5852
5.8	4.5316	3.9922	3.5604	3.2084
5.9	4.0390	3.5641	3.1830	2.8716
6.0	3.6008	3.1826	2.8460	2.5704
6.1	3.2109	2.8424	2.5451	2.3012
6.2	2.8638	2.5390	2.2763	2.0603
6.3	2.5547	2.2683	2.0362	1.8449
6.4	2.2795	2.0269	1.8217	1.6522
6.5	2.0343	1.8115	1.6300	1.4798
6.6	1.8158	1.6192	1.4586	1.3256
6.7	1.6211	1.4475	1.3055	1.1875
6.8	1.4476	1.2942	1.1685	1.0639
6.9	1.2928	1.1573	1.0461	0 9533
7.0	1.1548	1.0351	0.9366	0.8543
7.1	1.0317	0.9259	0.8386	0.7656
7.2	9.2188 (−5)	8.2831 (−5)	7.5100 (−5)	6.8622 (−5)
7.3	8.2387	7.4112	6.7261	6.1511
7.4	7.3640	6.6319	6.0247	5.5142
7.5	6.5831	5.9353	5.3970	4.9437
7.6	5.8859	5.3125	4.8352	4.4326
7.7	5.2633	4.7556	4.3323	3.9747
7.8	4.7072	4.2576	3.8821	3.5644
7.9	4.2104	3.8122	3.4790	3.1967

Table 1—(*Continued*)

x	E_1	E_2	E_3	E_4
8.0	3.7666	4.4138	3.1181	2.8672
8.1	3.3700	3.0573	2.7949	2.5719
8.2	3.0155	2.7384	2.5054	2.3071
8.3	2.6986	2.4530	2.2461	2.0698
8.4	2.4154	2.1975	2.0138	1.8570
8.5	2.1621	1.9689	1.8057	1.6662
8.6	1.9356	1.7642	1.6192	1.4952
8.7	1.7331	1.5810	1.4521	1.3418
8 8	1.5519	1.4169	1.3024	1.2042
8.9	1.3898	1.2700	1.1682	1.0808
9.0	1.2447 −(5)	1.1384 (−5)	1.0479 (−5)	0.9701 (−5)
9.1	1.1150	1.0205	0.9400	0.8708
9.2	9.9881 −(6)	9.1492 (−6)	8.4335 (−6)	7.8169 (−6)
9.3	8.9485	8.2033	7.5668	7.0177
9.4	8.0179	7.3558	6.7896	6.3006
9.5	7.1848	6.5965	6.0927	5.6571
9.6	6.4388	5.9160	5.4677	5.0797
9.7	5.7709	5.3061	4.9071	4.5614
9.8	5.1727	4.7595	4.4044	4.0963
9.9	4.6369	4.2695	3.9533	3.6788
10.0	4.1570	3.8302	3.5488	3.3041

REFERENCES

1. S. Chandrasekhar, *Radiative Transfer*, Oxford University Press, London, 1950; also Dover Publications, New York, 1960.
2. V. Kourganoff, *Basic Methods in Transfer Problems*, Dover Publications, New York, 1963.
3. K. M. Case, F. de Hoffmann, and G. Placzek, *Introduction to the Theory of Neutron Diffusion*, Los Alamos Scientific Laboratory, Los Alamos, N.M., 1953.
4. G. Placzek, *The Functions $E_n(x)$*, Declassified Canadian Report No. MT-1 (NRC-1547).
5. *Handbook of Mathematical Functions*, edited by M. Abramowitz and I. A. Stegun, Dover Publications, New York, 1965.

Index